DISCRETE MATHEMATICS

Third Edition

Kenneth A. Ross
Charles R.B. Wright

*Department of Mathematics,
University of Oregon*

PRENTICE HALL   Englewood Cliffs, New Jersey 07632

Library of Congress Cataloging-in-Publication Data
Ross, Kenneth A.
    Discrete mathematics / Kenneth A. Ross, Charles R.B. Wright
3rd ed.
        p. cm.
    Includes index.
    ISBN 0-13-218157-6
    1. Computer science—Mathematics I. Wright, Charles R.B.
1937–   II. Title.
QA76.9.M35R67   1992
511—dc20.

                                        91-28006
                                        CIP

Editor-in-Chief: *Tim Bozik*
Acquisition Editor: *Steve Conmy*
Production: *Nicholas Romanelli*
Cover/Interior Design: *Amy Rosen/Lorraine Mullaney*
Cover Art: *Bo Richards/Superstock*
Prepress Buyer: *Paula Massenaro*
Manufacturing Buyer: *Lori Bulwin*

1992, 1988, 1985 by Prentice-Hall, Inc.
A Simon & Schuster Company
Englewood Cliffs, New Jersey, 07632

Printed in the United States of America
10 9 8 7 6

ISBN 0-13-218157-6

Prentice-Hall, International (UK) Limited, London
Prentice-Hall of Australia Pty. Limited, Sydney
Prentice-Hall Canada Inc., Toronto
Prentice-Hall Hispanoamericana, S.A. Mexico
Prentice-Hall of India Private Limited, New Delhi
Prentice-Hall of Japan, Inc., Tokyo
Prentice-Hall of Southeast Asia Pte. Ltd., Singapore
Prentice-Hall do Brasil, Ltda., Rio de Janeiro

To our wives, for their love and support

Ruth and Leslie

CONTENTS

13 PREDICATE CALCULUS AND INFINITE SETS   658

PREFACE TO THE THIRD EDITION

Our original plan for this edition was just to add a chapter on probability, to make some of the more difficult sections less abstract, and to revise the treatment of recursion and induction. When we got into the project, we saw that a massive reorganization could greatly increase the book's flexibility as a text as well as draw together topics that are closely related. The resulting major revision includes substantial new material as well as what we think are clearer accounts of topics in the second edition, all reordered to fit a wider range of courses than before.

One of our main goals is the development of mathematical maturity. We and our colleagues have used this material successfully for several years with average students at the level of beginning calculus, and we find that by the end of two terms they are ready for upperclass work. The presentation begins with an intuitive approach that becomes more and more rigorous as the students' appreciation for proofs and skill at building them increase. Our account is careful but informal. As we go along we illustrate the way mathematicians attack problems, and we show the power of an abstract approach. We have aimed to make the account simple enough that students can learn it, and complete enough that they won't have to learn it again.

The first two editions used the pedagogical strategy of developing topics at increasingly deeper levels, with references to earlier material wherever possible so that students could see the connections between various ideas. The drawback to that otherwise sound strategy was that it made constructing a course of core topics unnecessarily difficult. In the new arrangement, the first four chapters contain what we regard as the core material of any serious discrete mathematics course. Those topics can readily be covered in a quarter. A semester course can add combinatorics and some probability or pick up graphs, trees and recursive algorithms. The rooted tree on the back cover shows the dependencies among the various chapters. Instructors using the book for a year-long course will probably want to follow the sequence we use at Oregon and take the chapters in their order of appearance.

To include new material and still keep the size of the book reasonable, we have had to drop some topics, including a few we rather liked. We hope that if our deletions have made the book no longer suited to your course you will let us know.

The major casualty was Chapter 0, a friendly informal introduction to graph theory. The chapter was designed to show the reader that the subject is intrinsically interesting and to show that careful mathematical thinking is essential. Some of our readers and at least one of the authors very much liked this beginning, but the majority of those we heard from found the chapter distressing and felt that it set a misleading tone. In particular, many instructors found it difficult to cover the chapter without bogging down. In any case, Chapter 0 is gone. The essential introductory material that it contained is now sprinkled throughout Chapter 3. Here is a chapter-by-chapter report on the other changes from the second edition:

Chapter 1 now includes an introduction to functions and sequences. It also includes an entire section with a reasonably thorough introduction to big-oh notation. We hope that readers will find this section a useful reference throughout their study of the book.

Chapter 2 focuses on elementary logic as a working tool. We downplay formal proofs even more than in earlier editions, and we stress methods of proof and analysis of arguments. Induction is deferred to Chapter 4 for two reasons. To give meaningful applications we need to assume some common background, which we supply in Chapter 3, and it really makes sense to treat induction and recursion together in a coherent manner. Quantifiers appear informally in Chapter 2; it is good for students to get used to them, and they are handy abbreviations in the classroom. Chapter 13 contains more on quantifiers and a brief introduction to the predicate calculus. As the location suggests, this more formal account is not needed in the rest of the book.

Chapter 3, which we have simply called "Relations," brings together the basic tools of discrete mathematics: relations, graphs and digraphs, and matrices, giving several different views of the same set of ideas. Equivalence relations, which permeate all of abstract mathematics, get a careful treatment and are applied to modular arithmetic on $\mathbb{Z}$.

Most of the mathematical induction is in one chapter now, with a change in derivation. Chapter 4 begins with an account of loops and loop invariants, which our students find natural and relatively easy to grasp. We need this material later on in order to verify algorithm correctness, and we have discovered that students benefit in their computer science courses from having seen a clean, mathematical presentation of invariants. The ideas in a treatment of mathematical induction are similar to those associated with loop invariants, so we base our justification of induction on loops. The approach works at Oregon. We would welcome users' comments on experiences elsewhere. In any case, it is easy to give a more traditional presentation of induction in class.

Chapter 4 also includes an introduction to recursive definition of sequences and to recurrence relations. We have added a brief account of divide-and-conquer recurrences. The last section of the chapter, on variants of the Euclidean algorithm, is also new. The algorithms in it are good examples. The results of the section, however, are not referred to again until the end of Chapter 12.

Chapter 5 lost a topic and gained a topic. The section on infinite sets is fascinating material, but it is not needed in the rest of the text. It has been deferred to Chapter 13. We have added an elementary section on probability to introduce the language and to serve as preparation for Chapter 9.

The material on graphs and trees has been reorganized into three chapters. A reasonable introduction to the subject can be obtained from Chapter 6, after which one can study either Chapter 7 or Chapter 8. New material in Chapter 6 includes some fast graph-theoretic algorithms. Chapter 7 now emphasizes recursion, beginning with two new sections devoted to recursive definitions and general recursive algorithms. The remainder of the chapter is devoted to recursive algorithms that are related to trees, with some interesting applications. Chapter 8 is now exclusively on digraphs and digraph algorithms.

In the first two editions our development of Boolean algebras was based on special partially ordered systems. In response to our readers, we now give a purely algebraic account of Boolean algebras in Chapter 10, emphasizing Boolean functions. The applications to logic networks and Karnaugh maps have been retained.

The account of partial orderings in Chapter 11 has not changed much from the second edition. Since general relations are now discussed in some detail in Chapter 3, their treatment in Chapter 11 has been streamlined. The presentation of closure operators is now less abstract than in previous editions.

Chapter 12, on algebraic structures, has been completely rewritten. The first four sections are now devoted to permutations and to the relatively concrete context of a group acting on a set. We give some nontrivial applications to coloring problems. The remainder of the chapter covers traditional topics on groups and semigroups and ends with a very brief introduction to rings, including polynomial rings and fields.

Some topics have been de-emphasized or omitted. Wherever the associative law appears, a semigroup is lurking in the background, but we have avoided the semigroup terminology until Chapter 12. The free semigroup Σ^* appears throughout, without being called that, but the section on the semigroup $\mathscr{P}(\Sigma^*)$ has been dropped. The material on the lattice of partitions and its connection with relations is now gone. Much of the material on isomorphism invariants of digraphs has been omitted. And, as noted above, the theory of lattices has been dropped, though lattices are mentioned briefly in Chapter 11.

We have retained the general characteristics that seemed successful in the previous editions. There are still lots of examples for students to read for themselves, so instructors can spend class time going over selected topics and can assign the remaining material to be read outside class. With a few exceptions, each section can still be covered in a day. The instructor's manual, which is available from Prentice Hall, contains a number of suggestions on emphasis and teaching strategy, as well as answers to all the exercises.

We've received valuable comments and suggestions from a number of colleagues, including the following reviewers:

Richard Makohon, University of Portland
John Chollet, Towson State University
Bo Green, Abilene Christian University
Richard H. Austing, University of Maryland
Charles Searcy, New Mexico Highlands University
Hyeong-Ah Choi, George Washington University

It is also a great pleasure to thank the people at Prentice Hall who have worked to see this edition through. Priscilla McGeehon convinced us to take on the project, and Steve Conmy, our mathematics editor, has given us some fine ideas as well as overall support. We are delighted to have Nick Romanelli, our production editor on the first edition, back again. The professional appearance of the book is all his doing, with the able assistance of the compositors at Syntax International.

We believe the third edition is a big improvement over the first two editions. Let us know whether you agree or not. Write to us at the University of Oregon Department of Mathematics or send email messages to ross@math.uoregon.edu or wright@math.uoregon.edu.

To The Student Especially

We know that words like "obviously" and "clearly" can be very annoying; they have sometimes bothered us too. When you see them occasionally in this book they are intended as hints. If you don't find the passage obvious or clear, you are probably making the situation too complicated or reading something unintended into the text. Take a break; then back up and read the material again. Similarly, the examples are meant to be helpful. If you are pretty sure you know the ideas involved, but an example seems much too hard, skip over it on first reading and then come back later. If you aren't very sure of the ideas, though, take a more careful look at the example.

Exercises are an important part of the book. They give you a chance to check your understanding and to practice thinking and writing clearly and mathematically. As the book goes on, more and more exercises ask you for proofs. We use the words "show" and "prove" interchangeably, though "show" is more common when a calculation is enough of an answer and "prove" suggests some reasoning is called for. "Prove" means "give a convincing argument or discussion to show why the assertion is true." What you write should be convincing to an instructor, to a fellow student, and to yourself the next day. Proofs should include words and sentences, not just computations, so that the reader can follow your thought processes. Use the proofs in the book as models, especially at first. The discussion of logical proofs in Chapter 2 will also help. Perfecting the ability to write a "good" proof is like perfecting the ability to write a "good" essay or give a "good" oral presentation. They all take practice. Don't be discouraged when one of your proofs fails to convince an expert (say a teacher or a grader). Instead, try to see what failed to be convincing.

An article by Sandra Z. Keith in the May 1991 issue of *Ume Trends*, an undergraduate mathematics education journal, gives these general study suggestions:

> Keep a homework notebook, rework problems, try not to leaf forward and backward (to the answers), rewrite class notes, prepare a review sheet, exchange it with friends, and study for tests, even if you never did in high school!

To these excellent tips we would add another: read ahead. Look over the material before class, to get some idea of what's coming and to locate the hard spots. Then when the discussion gets to the tough points you can ask for clarification, confident that you're not wasting class time on things that would be obvious after you read the book. If you're prepared, you can take advantage of your instructor's help and save yourself a lot of struggling.

Each chapter ends with a list of the main points it covers and with some suggestions for how to use the list to review. One of the best ways to learn material which you plan to use again is to tie each new idea to as many familiar concepts and situations as you can, and to visualize settings in which the new fact would be helpful to you. We have included lots of examples in the text to make this process easier. The review lists can be used to go over the material in the same way by yourself or with fellow students.

Answers or hints to most odd-numbered exercises are given in the back of the book. Wise students will look at the answers only after trying seriously to do the problems. When a proof is called for we usually give a hint or an outline of a proof, which you should first understand and then expand upon. A symbols index appears on the inside of the front cover and the Greek alphabet on the inside of the back cover. At the back of the book there is an index of topics. After Chapter 13 there is a brief dictionary of terms that we use in the text without explanation, but which some readers may have forgotten or never encountered. Look at these items right now to see where they are and what they contain, and then join us for Chapter 1.

K.A. Ross / C.R.B. Wright

SETS, SEQUENCES AND FUNCTIONS

1

This chapter is introductory and contains a relatively large number of fundamental definitions and notations. Much of the material is probably familiar, although perhaps with different notation or at a different level of mathematical precision. Besides introducing concepts and methods, this chapter establishes the style of exposition that we will use for the remainder of the book.

§ 1.1   Some Special Sets

In the past few decades it has become traditional to use set theory as the underlying basis for mathematics. That is, the concepts of "set" and "membership" are taken as basic undefined terms and the rest of mathematics is defined or described in these terms. A set is a collection of objects; the definition of a set must be unambiguous in the sense that it must be possible to decide whether particular objects belong to the set. We will usually denote sets by capital letters, such as A, B, S or X. Objects are usually denoted by lowercase letters, such as a, b, s or x. An object a that belongs to a set S is called a **member** of S or **element** of S. If a is an object and A is a set, we write $a \in A$ to mean that a is a member of A and $a \notin A$ to mean that a is not a member of A. The symbol $\in$ can be read as a verb phrase "is an element of," "belongs to" or "is in" or as the preposition "in," depending on context.

Specific sets can be written in a variety of ways. A few especially common and important sets will be given their own names, i.e., their own symbols. We will reserve the symbol $\mathbb{N}$ for the set of **natural numbers**:

$$\mathbb{N} = \{0, 1, 2, 3, 4, 5, 6, \ldots\}.$$

Note that we include 0 among the natural numbers.

We write $\mathbb{P}$ for the set of **positive integers**:

$$\mathbb{P} = \{1, 2, 3, 4, 5, 6, 7, \ldots\}.$$

Many mathematics texts write this set as $\mathbb{N}$ instead. The set of all **integers**, positive, zero or negative, will be denoted by $\mathbb{Z}$ [for the German word "Zahl"]. Numbers of the form m/n where $m \in \mathbb{Z}$, $n \in \mathbb{Z}$ and $n \neq 0$ are called **rational numbers** [since they are ratios of integers]. The set of all rational numbers is denoted by $\mathbb{Q}$. The set of all **real numbers**, rational or not, is denoted by $\mathbb{R}$. Thus $\mathbb{R}$ contains all the numbers in $\mathbb{Q}$, and $\mathbb{R}$ also contains $\sqrt{2}, \sqrt{3}, \sqrt[3]{2}, -\pi, e$ and many, many other numbers. Small finite sets can be listed using braces $\{\ \}$ and commas. For example, $\{2, 4, 6, 8, 10\}$ is the set consisting of the five positive even integers less than 12, and $\{2, 3, 5, 7, 11, 13, 17, 19\}$ consists of the eight primes less than 20. Readers who need to be reminded what "even" or "prime" mean may consult the dictionary that appears after Chapter 13. Two sets are **equal** if they contain the same elements. Thus

$$\{2, 4, 6, 8, 10\} = \{10, 8, 6, 4, 2\} = \{2, 8, 2, 6, 2, 10, 4, 2\};$$

The same Elements. Not the Same # of Elements.

the order of the listing is irrelevant and there is no advantage [or harm] in listing elements more than once.

Large finite sets and even infinite sets can be listed with the aid of the mathematician's etcetera, namely three dots, $\ldots$, provided that the meaning of the three dots is clear. Thus $\{1, 2, 3, \ldots, 1000\}$ represents the set of positive integers less than or equal to 1000, and $\{3, 6, 9, 12, \ldots\}$ presumably represents the infinite set of positive integers that are divisible by 3. On the other hand, the meaning of $\{1, 2, 3, 5, 8, \ldots\}$ may be less than perfectly clear. The somewhat vague use of three dots is not always satisfactory, especially in computer science, and we will develop techniques for unambiguously describing such sets without using dots.

Sets are often described by properties of their elements using the notation

$$\{\quad : \quad \}.$$

A variable [n or x, for instance] is indicated before the colon, and the properties are given after the colon. For example,

$$\{n : n \in \mathbb{N} \text{ and } n \text{ is even}\}$$

represents the set of nonnegative even integers, i.e., the set $\{0, 2, 4, 6, 8, 10, \ldots\}$. The colon is always read "such that," so the above is read "the set of all n such that n is in $\mathbb{N}$ and n is even." Similarly,

$$\{x : x \in \mathbb{R} \text{ and } 1 \leq x < 3\}$$

represents the set of all real numbers that are greater than or equal to 1 and less than 3. The number 1 belongs to the set, but 3 does not. Just to streamline notation, the last two sets can be written as

$$\{n \in \mathbb{N} : n \text{ is even}\} \quad \text{and} \quad \{x \in \mathbb{R} : 1 \leq x < 3\}.$$

The first set is then read "the set of all n in $\mathbb{N}$ such that n is even."

Another way to list a set is to specify a rule for obtaining its elements using some other set of elements. For example, $\{n^2 : n \in \mathbb{N}\}$ represents the set of all integers that are the squares of integers in $\mathbb{N}$, i.e.,

$$\{n^2 : n \in \mathbb{N}\} = \{m \in \mathbb{N} : m = n^2 \text{ for some } n \in \mathbb{N}\}$$

$$= \{0, 1, 4, 9, 16, 25, 36, \ldots\}.$$

Note that this set equals $\{n^2 : n \in \mathbb{Z}\}$. Similarly, $\{(-1)^n : n \in \mathbb{N}\}$ represents the set obtained by evaluating $(-1)^n$ for all $n \in \mathbb{N}$, so that

$$\{(-1)^n : n \in \mathbb{N}\} = \{-1, 1\}.$$

This set has only two elements.

Now consider two sets S and T. We say that S is a **subset** of T provided that every element of S belongs to T. If S is a subset of T, we write $S \subseteq T$. The symbol $\subseteq$ can be read as "is a subset of." We also frequently say "S is contained in T" in case $S \subseteq T$, but note the potential for confusion if we also say "x is contained in T" when we mean "$x \in T$." Containment for a subset and containment for elements mean quite different things.

Two sets S and T are **equal** if they have exactly the same elements. Thus $S = T$ if and only if $S \subseteq T$ and $T \subseteq S$.

EXAMPLE 1   (a) We have $\mathbb{P} \subseteq \mathbb{N}$, $\mathbb{N} \subseteq \mathbb{Z}$, $\mathbb{Z} \subseteq \mathbb{Q}$, $\mathbb{Q} \subseteq \mathbb{R}$. As with the familiar inequality $\leq$, we can run these assertions together:

$$\mathbb{P} \subseteq \mathbb{N} \subseteq \mathbb{Z} \subseteq \mathbb{Q} \subseteq \mathbb{R}.$$

(b) Since 2 is the only even prime, we have

$$\{n \in \mathbb{P} : n \text{ is prime and } n \geq 3\} \subseteq \{n \in \mathbb{P} : n \text{ is odd}\}.$$

(c) Consider again any set S. Obviously $x \in S$ implies $x \in S$, and so $S \subseteq S$. That is, we regard a set as a subset of itself. This is why we use the notation $\subseteq$ rather than $\subset$. This usage is analogous to our usage of $\leq$ for real numbers. The inequality $x \leq 5$ is valid for many numbers, such as 3, 1 and -73. It is also valid for $x = 5$, i.e., $5 \leq 5$. This last inequality looks a bit peculiar because we actually know more, namely $5 = 5$. But $5 \leq 5$ says that "5 is less than 5 or else 5 is equal to 5," and this is a true statement. Similarly, $S \subseteq S$ is true even though we know more, namely $S = S$. Statements like "$5 = 5$," "$5 \leq 5$," "$S = S$" or "$S \subseteq S$" do no harm and are often useful to call attention to the fact that a particular case of a more general statement is valid. ■[1]

We will occasionally write $T \subset S$ to mean that $T \subseteq S$ and $T \neq S$, i.e., T is a subset of S different from S. This usage of $\subset$ is analogous to

---

[1] We will use ■ to signify the end of an example or proof.

our usage of $<$ for real numbers. If $T \subset S$, we say that T is a **proper subset** of S.

We next introduce notation for some special subsets of $\mathbb{R}$, called **intervals**. For $a, b \in \mathbb{R}$ with $a < b$, we define

$$[\boldsymbol{a}, \boldsymbol{b}] = \{x \in \mathbb{R} : a \leq x \leq b\}; \qquad (\boldsymbol{a}, \boldsymbol{b}) = \{x \in \mathbb{R} : a < x < b\};$$

$$[\boldsymbol{a}, \boldsymbol{b}) = \{x \in \mathbb{R} : a \leq x < b\}; \qquad (\boldsymbol{a}, \boldsymbol{b}] = \{x \in \mathbb{R} : a < x \leq b\}.$$

The general rule is that brackets $[\ ,\ ]$ signify that the endpoints are to be included and parentheses $(\ ,\ )$ signify that they are to be excluded. Intervals of the form $[a, b]$ are called **closed**; those of the form (a, b) are **open**. It is also convenient to use the term "interval" for some unbounded sets that we describe using the symbols ∞ and $-\infty$, which do not represent real numbers but are simply part of the notation for the sets. Thus

$$[\boldsymbol{a}, \infty) = \{x \in \mathbb{R} : a \leq x\}; \qquad (\boldsymbol{a}, \infty) = \{x \in \mathbb{R} : a < x\};$$

$$(-\infty, \boldsymbol{b}] = \{x \in \mathbb{R} : x \leq b\}; \qquad (-\infty, \boldsymbol{b}) = \{x \in \mathbb{R} : x < b\}.$$

Set and interval notation must be dealt with carefully. For example, $[0, 1]$, $(0, 1)$ and $\{0, 1\}$ all denote different sets. In fact, the intervals $[0, 1]$ and $(0, 1)$ are infinite sets, while $\{0, 1\}$ has only two elements.

Consider the following sets:

$$\{n \in \mathbb{N} : 2 < n < 3\}, \qquad \{x \in \mathbb{R} : x^2 < 0\},$$

$$\{r \in \mathbb{Q} : r^2 = 2\}, \qquad \{x \in \mathbb{R} : x^2 + 1 = 0\}.$$

These sets have one property in common: They contain no elements. From a strictly logical point of view, they all contain the same elements and so they are equal in spite of the different descriptions. This unique set having no elements at all is called the **empty set.** We will use two notations for it, the suggestive $\{\ \}$ and the standard $\varnothing$. The symbol $\varnothing$ is not a Greek phi ϕ; it is borrowed from the Norwegian alphabet and non-Norwegians should read it as "empty set." We regard $\varnothing$ as a subset of every set S because we regard the statement "$x \in \varnothing$ implies $x \in S$" as logically true in a vacuous sense. You can take this explanation on faith until you study §2.3.

Sets are objects, so they can be members of other sets. The set $\{\{1, 2\}, \{1, 3\}, \{2\}, \{3\}\}$ has four members, namely the sets $\{1, 2\}$, $\{1, 3\}$, $\{2\}$ and $\{3\}$. If we had a box containing two sacks full of marbles, we would consider it to be a box of sacks rather than a box of marbles, so it would contain two members. Similarly, if A is a set, then $\{A\}$ is a set with one member, namely A, no matter how many members A itself has. A box containing an empty sack contains something, namely a sack, so it is not an empty box. In the same way, $\{\varnothing\}$ is a set with one member, whereas $\varnothing$ is a set with no members, so $\{\varnothing\}$ and $\varnothing$ are different sets. We have $\varnothing \in \{\varnothing\}$ and even $\varnothing \subseteq \{\varnothing\}$, but $\varnothing \notin \varnothing$.

The set of all subsets of a set S is called the **power set** of S and will be denoted $\mathscr{P}(S)$. Clearly the empty set $\varnothing$ and the set S itself are elements of $\mathscr{P}(S)$, i.e., $\varnothing \in \mathscr{P}(S)$ and $S \in \mathscr{P}(S)$.

EXAMPLE 2   (a) We have $\mathscr{P}(\varnothing) = \{\varnothing\}$ since $\varnothing$ is the only subset of $\varnothing$.

(b) Consider a typical one-element set, say $S = \{a\}$. Then $\mathscr{P}(S) = \{\varnothing, \{a\}\}$ has two elements.

(c) If $S = \{a, b\}$ and $a \neq b$, then $\mathscr{P}(S) = \{\varnothing, \{a\}, \{b\}, \{a, b\}\}$ has four elements.

(d) If $S = \{a, b, c\}$ has three elements, then

$$\mathscr{P}(S) = \{\varnothing, \{a\}, \{b\}, \{c\}, \{a, b\}, \{a, c\}, \{b, c\}, \{a, b, c\}\}$$

has eight members.

(e) Let S be a finite set. Note that if S has n elements and if $n \leq 3$, then $\mathscr{P}(S)$ has 2^n elements, as shown in parts (a) to (d) above. This fact is not an accident, as we show in Example 4(b) of §4.2.

(f) If S is infinite, then $\mathscr{P}(S)$ is also infinite, of course. ■

We introduce one more special kind of set, denoted by Σ^*, that will recur throughout the book. Our goal is to allow a rather general, but precise mathematical treatment of languages. First we define an **alphabet** to be a finite nonempty set Σ [capital Greek sigma] whose members are symbols, often called **letters** of Σ, and which is subject to some minor restrictions that we will discuss at the end of this section. Given an alphabet Σ, a **word** is any finite string of letters from Σ. Finally, we denote the set of all words using letters from Σ by Σ^* [sigma-star]. Any subset of Σ^* is called a **language** over Σ.

EXAMPLE 3   (a) Let $\Sigma = \{a, b, c, d, \ldots, z\}$ consist of the twenty-six letters of the English alphabet. *Any* string of letters from Σ belongs to Σ^*. Thus Σ^* contains *math, is, fun, aint, lieblich, amour, zzyzzoomph, etcetera,* etc. Since Σ^* contains *a, aa, aaa, aaaa, aaaaa,* etc., Σ^* is clearly an infinite set. To be definite, we could define the **American language** L to be the subset of Σ^* consisting of words in the latest edition of *Webster's New World Dictionary of the American Language.* Thus

$$L = \{a, \text{aachen}, \text{aardvark}, \text{aardwolf}, \ldots, \text{zymurgy}\},$$

a large but finite set.

(b) To get simple examples and yet illustrate the ideas, we will frequently take Σ to be a 2-element set $\{a, b\}$. In this case Σ^* contains *a, b, ab, ba, bab, babbabb,* etc.; again Σ^* is infinite.

(c) If $\Sigma = \{0, 1\}$, then the set B of words in Σ^* that begin with 1 is exactly the set of binary notations for positive integers. That is,

$$B = \{1, 10, 11, 100, 101, 110, 111, 1000, 1001, \ldots\}. \quad \blacksquare$$

There is a special word in Σ^* somewhat analogous to the empty set, called the **empty word** or **null word**; it is the string with no letters at all and is denoted by λ [lowercase Greek lambda].

EXAMPLE 4

(a) If $\Sigma = \{a, b\}$, then

$$\Sigma^* = \{\lambda, a, b, aa, ab, ba, bb, aaa, aab, aba, abb, baa, bab, bba, \ldots\}.$$

(b) If $\Sigma = \{0, 1, 2\}$, then

$$\Sigma^* = \{\lambda, 0, 1, 2, 00, 01, 02, 10, 11, 12, 20, 21, 22, 000, 001, 002, \ldots\}.$$

(c) If $\Sigma = \{a\}$, then

$$\Sigma^* = \{\lambda, a, aa, aaa, aaaa, aaaaa, aaaaaa, \ldots\}.$$

This example doesn't contain any very useful languages, but it will serve to illustrate concepts later on.

(d) Various computer languages fit our definition of language. For example, the alphabet Σ for one version of ALGOL has 113 elements; Σ includes letters, the digits 0, 1, 2, . . . , 9 and a variety of operators, including sequential operators such as "go to" and "if." As usual, Σ^* contains all possible finite strings of letters from Σ, without regard to meaning. The subset of Σ^* consisting of those strings accepted for execution by an ALGOL compiler on a given computer is a well-defined and useful subset of Σ^*; we could call it the ALGOL language determined by the compiler. $\blacksquare$

As promised, we now discuss the restrictions needed for Σ. A problem can arise if the letters in Σ are themselves built out of other letters, either from Σ or from some other alphabet. For example, if Σ contains as letters the symbols a, b and ab, then the string aab could be taken to be a string of three letters a, a and b from Σ or as a string of two letters a and ab. There is no way to tell which it should be, and a machine reading in the letters a, a, b one at a time would find it impossible to assign a unique meaning to the input. To take another example, if Σ contained ab, aba and bab, the input string $ababab$ could be interpreted as either $(ab)(ab)(ab)$ or $(aba)(bab)$. To avoid these and related problems, we will not allow Σ to contain any letters that are themselves strings beginning with the letters in Σ. Thus $\Sigma = \{a, b, c\}$, $\Sigma = \{a, b, ca\}$ and $\Sigma = \{a, b, Ab\}$ are allowed, but $\Sigma = \{a, b, c, ac\}$ and even $\Sigma = \{a, b, ac\}$ are not.

With this agreement, we can unambiguously define **length**(w) for a word w in Σ^* to be the number of letters from Σ in w, counting each appearance of a letter. For example, if $\Sigma = \{a, b\}$, then length(aab) =

length(bab) = 3. If $\Sigma = \{a, b, Ab\}$, then length ($abbAb$) = 4. We also define length(λ) = 0. A more precise definition is given in §7.1.

One final note: We will use w, w_1, etc. as variable names for words. This practice should cause no confusion even though w also happens to be a letter of the English alphabet.

EXAMPLE 5   If $\Sigma = \{a, b\}$ and $A = \{w \in \Sigma^* : \text{length}(w) = 2\}$, then $A = \{aa, ab, ba, bb\}$. If

$$B = \{w \in \Sigma^* : \text{length}(w) \text{ is even}\},$$

then B is the infinite set $\{\lambda, aa, ab, ba, bb, aaaa, aaab, aaba, aabb, \ldots\}$. Note that A is a subset of B.   ■

EXERCISES 1.1

Terms such as "divisible," "prime" and "even" are defined in the dictionary that appears after Chapter 13.

1. List five elements in each of the following sets.
   (a) $\{n \in \mathbb{N} : n \text{ is divisible by 5}\}$ (b) $\{2n + 1 : n \in \mathbb{P}\}$
   (c) $\mathscr{P}(\{1, 2, 3, 4, 5\})$ (d) $\{2^n : n \in \mathbb{N}\}$
   (e) $\{1/n : n \in \mathbb{P}\}$ (f) $\{r \in \mathbb{Q} : 0 < r < 1\}$
   (g) $\{n \in \mathbb{N} : n + 1 \text{ is prime}\}$

2. List the elements in the following sets.
   (a) $\{1/n : n = 1, 2, 3, 4\}$ (b) $\{n^2 - n : n = 0, 1, 2, 3, 4\}$
   (c) $\{1/n^2 : n \in \mathbb{P}, n \text{ is even and } n < 11\}$ (d) $\{2 + (-1)^n : n \in \mathbb{N}\}$

3. List five elements in each of the following sets.
   (a) Σ^* where $\Sigma = \{a, b, c\}$
   (b) $\{w \in \Sigma^* : \text{length}(w) \leq 2\}$ where $\Sigma = \{a, b\}$
   (c) $\{w \in \Sigma^* : \text{length}(w) = 4\}$ where $\Sigma = \{a, b\}$
   Which sets above contain the empty word λ?

4. Determine the following sets; i.e., list their elements if they are nonempty, and write $\varnothing$ if they are empty.
   (a) $\{n \in \mathbb{N} : n^2 = 9\}$ (b) $\{n \in \mathbb{Z} : n^2 = 9\}$
   (c) $\{x \in \mathbb{R} : x^2 = 9\}$ (d) $\{n \in \mathbb{N} : 3 < n < 7\}$
   (e) $\{n \in \mathbb{Z} : 3 < |n| < 7\}$ (f) $\{x \in \mathbb{R} : x^2 < 0\}$
   (g) $\{n \in \mathbb{N} : n^2 = 3\}$ (h) $\{x \in \mathbb{Q} : x^2 = 3\}$
   (i) $\{x \in \mathbb{R} : x < 1 \text{ and } x \geq 2\}$ (j) $\{3n + 1 : n \in \mathbb{N} \text{ and } n \leq 6\}$
   (k) $\{n \in \mathbb{P} : n \text{ is prime and } n \leq 15\}$ [Recall that 1 isn't prime.]

5. How many elements are there in the following sets? Write ∞ if the set is infinite.

(a) $\{n \in \mathbb{N}: n^2 = 2\}$

(b) $\{n \in \mathbb{Z}: 0 \leq n \leq 73\}$

(c) $\{n \in \mathbb{Z}: 5 \leq |n| \leq 73\}$

(d) $\{n \in \mathbb{Z}: 5 < n < 73\}$

(e) $\{n \in \mathbb{Z}: n \text{ is even and } |n| \leq 73\}$

(f) $\{x \in \mathbb{Q}: 0 \leq x \leq 73\}$

(g) $\{x \in \mathbb{Q}: x^2 = 2\}$

(h) $\{x \in \mathbb{R}: x^2 = 2\}$

(i) $\{x \in \mathbb{R}: .99 < x < 1.00\}$

(j) $\mathscr{P}(\{0, 1, 2, 3\})$

(k) $\mathscr{P}(\mathbb{N})$

(l) $\{n \in \mathbb{N}: n \text{ is even}\}$

(m) $\{n \in \mathbb{N}: n \text{ is prime}\}$

(n) $\{n \in \mathbb{N}: n \text{ is even and prime}\}$

(o) $\{n \in \mathbb{N}: n \text{ is even or prime}\}$

6. How many elements are there in the following sets? Write ∞ if the set is infinite.

(a) $\{-1, 1\}$

(b) $[-1, 1]$

(c) $(-1, 1)$

(d) $\{n \in \mathbb{Z}: -1 \leq n \leq 1\}$

(e) Σ^* where $\Sigma = \{a, b, c\}$

(f) $\{w \in \Sigma^*: \text{length}(w) \leq 4\}$ where $\Sigma = \{a, b, c\}$

7. Consider the sets

$$A = \{n \in \mathbb{P}: n \text{ is odd}\}, \qquad B = \{n \in \mathbb{P}: n \text{ is prime}\},$$

$$C = \{4n + 3: n \in \mathbb{P}\}, \qquad D = \{x \in \mathbb{R}: x^2 - 8x + 15 = 0\}.$$

Which of these sets are subsets of which? Consider all sixteen possibilities.

8. Consider the sets $\{0, 1\}$, $(0, 1)$ and $[0, 1]$. True or False.

(a) $\{0, 1\} \subseteq (0, 1)$

(b) $\{0, 1\} \subseteq [0, 1]$

(c) $(0, 1) \subseteq [0, 1]$

(d) $\{0, 1\} \subseteq \mathbb{Z}$

(e) $[0, 1] \subseteq \mathbb{Z}$

(f) $[0, 1] \subseteq \mathbb{Q}$

(g) $1/2$ and $\pi/4$ are in $\{0, 1\}$

(h) $1/2$ and $\pi/4$ are in $(0, 1)$

(i) $1/2$ and $\pi/4$ are in $[0, 1]$

9. Consider the following three alphabets: $\Sigma_1 = \{a, b, c\}$, $\Sigma_2 = \{a, b, ca\}$ and $\Sigma_3 = \{a, b, Ab\}$. Determine to which of Σ_1^*, Σ_2^* and Σ_3^* each word below belongs, and give its length as a member of each set to which it belongs.

(a) *aba*

(b) *bAb*

(c) *cba*

(d) *cab*

(e) *caab*

(f) *baAb*

10. Here is a question to think about. Let $\Sigma = \{a, b\}$ and imagine, if you can, a dictionary for all the nonempty words of Σ^* with the words arranged in the usual alphabetical order. All the words $a, aa, aaa, aaaa$, etc. must appear before the word *ba*. How far into the dictionary will you have to dig to find

the word *ba*? How would the answer change if the dictionary contained only those words in Σ^* of length 5 or less?

11. Suppose that *w* is a nonempty word in Σ^*.

   (a) If the first [i.e., leftmost] letter of *w* is deleted, is the resulting string in Σ^*?

   (b) How about deleting letters from both ends of *w*? Are the resulting strings still in Σ^*?

   (c) If you had a device that could recognize letters in Σ and could delete letters from strings, how could you use it to determine if an arbitrary string of symbols is in Σ^*?

§ 1.2   Set Operations

In this section we introduce operations that allow us to create new sets from old sets. We define the **union** $A \cup B$ and **intersection** $A \cap B$ of sets *A* and *B* as follows:

$$A \cup B = \{x : x \in A \text{ or } x \in B \text{ or both}\};$$
$$A \cap B = \{x : x \in A \text{ and } x \in B\}.$$

We added "or both" to the definition of $A \cup B$ for emphasis and clarity. In ordinary English, the word "or" has two interpretations. Sometimes it is the **inclusive or** and means one or the other or both. This is the interpretation when a college catalog asserts: A student's program must include 2 years of science or 2 years of mathematics. At other times, "or" is the **exclusive or** and means one or the other but not both. This is the interpretation when a menu offers soup or salad. In mathematics we always interpret **or** as the "inclusive or" unless explicitly specified to the contrary. Sets *A* and *B* are said to be **disjoint** if they have no elements in common, i.e., if $A \cap B = \varnothing$.

Given sets *A* and *B*, the **relative complement** $A \backslash B$ is the set of objects that are in *A* and not in *B*:

$$A \backslash B = \{x : x \in A \text{ and } x \notin B\} = \{x \in A : x \notin B\}.$$

It is the set obtained by removing from *A* all the elements of *B* that happen to be in *A*.

The **symmetric difference** $A \oplus B$ of the sets *A* and *B* is the set

$$A \oplus B = \{x : x \in A \text{ or } x \in B \text{ but not both}\}.$$

Note the use of the "exclusive or" here. It follows from the definition that

$$A \oplus B = (A \cup B) \backslash (A \cap B) = (A \backslash B) \cup (B \backslash A).$$

It is sometimes convenient to illustrate relations between sets with pictures called **Venn diagrams**, in which sets correspond to subsets of the plane. See Figure 1, where the indicated sets have been shaded in.

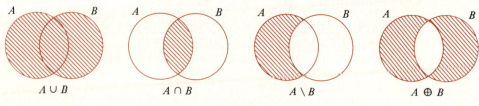

$$A \cup B \qquad\qquad A \cap B \qquad\qquad A \setminus B \qquad\qquad A \oplus B$$

FIGURE 1

EXAMPLE 1   (a) Let $A = \{n \in \mathbb{N} : n \le 11\}$, $B = \{n \in \mathbb{N} : n$ is even and $n \le 20\}$ and $E = \{n \in \mathbb{N} : n$ is even$\}$. Then we have

$$A \cup B = \{0, 1, 2, 3, 4, 5, 6, 7, 8, 9, 10, 11, 12, 14, 16, 18, 20\},$$

$$A \cap B = \{0, 2, 4, 6, 8, 10\},$$

$$A \backslash B = \{1, 3, 5, 7, 9, 11\},$$

$$B \backslash A = \{12, 14, 16, 18, 20\}.$$

$$A \oplus B = \{1, 3, 5, 7, 9, 11, 12, 14, 16, 18, 20\}.$$

We also have $E \cap B = B$, $B \backslash E = \{\ \ \}$,

$$E \backslash B = \{n \in \mathbb{N} : n \text{ is even and } n \ge 22\} = \{22, 24, 26, 28, \ldots\},$$

$$\mathbb{N} \backslash E = \{n \in \mathbb{N} : n \text{ is odd}\} = \{1, 3, 5, 7, 9, 11, \ldots\},$$

$$A \oplus E = \{1, 3, 5, 7, 9, 11\} \cup \{n \in \mathbb{N} : n \text{ is even and } n \ge 12\}$$

$$= \{1, 3, 5, 7, 9, 11, 12, 14, 16, 18, 20, 22, \ldots\}.$$

(b) Consider the intervals $[0, 2]$ and $(0, 1]$. Then $(0, 1] \subseteq [0, 2]$ and so

$$(0, 1] \cup [0, 2] = [0, 2] \quad \text{and} \quad (0, 1] \cap [0, 2] = (0, 1].$$

Moreover, we have

$$(0, 1] \backslash [0, 2] = \{\ \ \},$$

$$[0, 2] \backslash (0, 1] = \{0\} \cup (1, 2] \quad \text{and} \quad [0, 2] \backslash (0, 2) = \{0, 2\}.$$

(c) Let $\Sigma = \{a, b\}$, $A = \{\lambda, a, aa, aaa\}$, $B = \{\lambda, b, bb, bbb\}$ and $C = \{w \in \Sigma^* : \text{length}(w) \le 2\}$. Then we have

$$A \cup B = \{\lambda, a, b, aa, bb, aaa, bbb\}, \qquad A \cap B = \{\lambda\},$$

$$A \backslash B = \{a, aa, aaa\}, \qquad\qquad B \backslash A = \{b, bb, bbb\},$$

$$A \cap C = \{\lambda, a, aa\}, \qquad\qquad B \backslash C = \{bbb\},$$

$$C \backslash A = \{b, ab, ba, bb\}, \qquad\qquad A \backslash \Sigma = \{\lambda, aa, aaa\}. \quad\blacksquare$$

It is often convenient to fix some set U, such as $\mathbb{N}$, $\mathbb{R}$ or Σ^*, which we call the **universe** or **universal set**, and to consider only elements in U and only subsets of U. For $A \subseteq U$ the relative complement $U \backslash A$ is called in this setting the **absolute complement** or simply the **complement** of A and is denoted by A^c. Note that the relative complement $A \backslash B$ can be written in terms of the absolute complement: $A \backslash B = A \cap B^c$. In the Venn diagrams in Figure 2 we have drawn the universe U as a rectangle and shaded in the indicated sets.

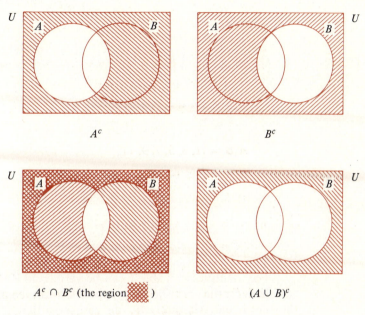

FIGURE 2

EXAMPLE 2   (a) If the universe is $\mathbb{N}$, and if A and E are as in Example 1(a), then

$$A^c = \{n \in \mathbb{N} : n \geq 12\} \quad \text{and} \quad E^c = \{n \in \mathbb{N} : n \text{ is odd}\}.$$

(b) If the universe is $\mathbb{R}$, then $[0, 1]^c = (-\infty, 0) \cup (1, \infty)$, $(0, 1)^c = (-\infty, 0] \cup [1, \infty)$ and $\{0, 1\}^c = (-\infty, 0) \cup (0, 1) \cup (1, \infty)$. For any $a \in \mathbb{R}$, $[a, \infty)^c = (-\infty, a)$ and $(a, \infty)^c = (-\infty, a]$. ∎

Note that the last two Venn diagrams in Figure 2 show that $A^c \cap B^c = (A \cup B)^c$. This set identity and many others are true in general. Table 1 lists some basic identities for sets and set operations. Don't be overwhelmed by them; look at them one at a time. As some of the names of the laws suggest, many of them are analogs of laws from algebra. The idempotent laws are new [certainly, $a + a = a$ fails for most numbers], and there is only one distributive law for numbers. Of course, the laws involving complementation are new. All sets in Table 1 are presumed to

TABLE 1. Laws of Algebra of Sets

1a. $A \cup B = B \cup A$
  b. $A \cap B = B \cap A$ $\Big\}$ commutative laws

2a. $(A \cup B) \cup C = A \cup (B \cup C)$
  b. $(A \cap B) \cap C = A \cap (B \cap C)$ $\Big\}$ associative laws

3a. $A \cup (B \cap C) = (A \cup B) \cap (A \cup C)$
  b. $A \cap (B \cup C) = (A \cap B) \cup (A \cap C)$ $\Big\}$ distributive laws

4a. $A \cup A = A$
  b. $A \cap A = A$ $\Big\}$ idempotent laws

5a. $A \cup \varnothing = A$
  b. $A \cup U = U$
  c. $A \cap \varnothing = \varnothing$
  d. $A \cap U = A$ $\Bigg\}$ identity laws

6. $(A^c)^c = A$ double complementation

7a. $A \cup A^c = U$
  b. $A \cap A^c = \varnothing$

8a. $U^c = \varnothing$
  b. $\varnothing^c = U$

9a. $(A \cup B)^c = A^c \cap B^c$
  b. $(A \cap B)^c = A^c \cup B^c$ $\Big\}$ DeMorgan laws

be subsets of some universal set U. Because of the associative laws, we can write the sets $A \cup B \cup C$ and $A \cap B \cap C$ without any parentheses and cause no confusion.

The identities in Table 1 can be verified in either of two ways. One can shade in the corresponding sets of a Venn diagram and observe that they are equal. Alternatively, one can show that sets S and T are equal by showing that $S \subseteq T$ and $T \subseteq S$; these inclusions can be verified by showing that $x \in S$ implies $x \in T$ and by showing that $x \in T$ implies $x \in S$. We give examples of both sorts of arguments, leaving most of the verifications to the interested reader.

EXAMPLE 3   The DeMorgan law 9a is illustrated by Venn diagrams in Figure 2. Here is a proof in which we show first that $(A \cup B)^c \subseteq A^c \cap B^c$ and then that $A^c \cap B^c \subseteq (A \cup B)^c$.

To show that $(A \cup B)^c \subseteq A^c \cap B^c$, we consider an element x in $(A \cup B)^c$. Then $x \notin A \cup B$. In particular, $x \notin A$, so we must have $x \in A^c$. Similarly, $x \notin B$, and so $x \in B^c$. Therefore, $x \in A^c \cap B^c$. We have shown that $x \in (A \cup B)^c$ implies $x \in A^c \cap B^c$; hence $(A \cup B)^c \subseteq A^c \cap B^c$.

To show the reverse inclusion, $A^c \cap B^c \subseteq (A \cup B)^c$, we consider x in $A^c \cap B^c$. Then $x \in A^c$, so $x \notin A$. Also, $x \in B^c$, and so $x \notin B$. Since $x \notin A$ and $x \notin B$, we conclude that $x \notin A \cup B$, i.e., $x \in (A \cup B)^c$. Hence $A^c \cap B^c \subseteq (A \cup B)^c$. ■

EXAMPLE 4 The Venn diagrams in Figure 3 show the distributive law 3b. The picture of the set $A \cap (B \cup C)$ is double-hatched in the diagram on the left; on the right, $(A \cap B) \cup (A \cap C)$ is represented by the set that is single- or double-hatched.

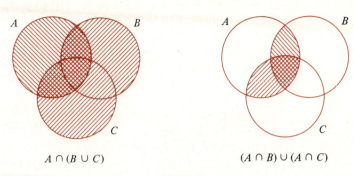

$A \cap (B \cup C)$ $(A \cap B) \cup (A \cap C)$

FIGURE 3

Here is a proof where we show that the sets are subsets of each other. First consider $x \in A \cap (B \cup C)$. Then x is in A for sure. Also, x is in $B \cup C$. So either $x \in B$, in which case $x \in A \cap B$, or else $x \in C$, in which case $x \in A \cap C$. In either case, we have $x \in (A \cap B) \cup (A \cap C)$. This shows that $A \cap (B \cup C) \subseteq (A \cap B) \cup (A \cap C)$.

Now consider $y \in (A \cap B) \cup (A \cap C)$. Either $y \in A \cap B$ or $y \in A \cap C$; we consider the two cases separately. If $y \in A \cap B$, then $y \in A$ and $y \in B$, so $y \in B \cup C$ and hence $y \in A \cap (B \cup C)$. Similarly, if $y \in A \cap C$, then $y \in A$ and $y \in C$, so $y \in B \cup C$ and thus $y \in A \cap (B \cup C)$. Since $y \in A \cap (B \cup C)$ in both cases, we've shown that $(A \cap B) \cup (A \cap C) \subseteq A \cap (B \cup C)$. We already proved the opposite inclusion, so the two sets are equal. ∎

The proofs using Venn diagrams seem much easier than the proofs where we analyze inclusions elementwise. Proofs by picture make many people nervous; on the other hand, the Venn diagram for A, B, C has eight regions [see Figure 4] and these comprise all the logical possibilities, so proofs using Venn diagrams are in fact valid. A much more serious objection to proofs via Venn diagrams is that they hide the thought process; the logic used to shade the diagrams is not specified. If we had written out the

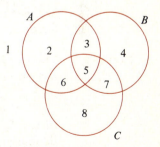

FIGURE 4

reasoning behind the diagrams in Figure 3, the proof would have been as long as the elementwise proof in Example 4, which relies only on logic. Another reason for avoiding Venn diagrams is that they are hard to draw whenever there are more than three sets. Still, nearly everyone who works with mathematics uses pictures, including Venn diagrams, to help understand mathematical situations.

Table 1 gives a few of the basic relationships in set theory. Many other relationships exist. They can be verified using one of three methods: (1) Venn diagrams, (2) elementwise arguments as in Examples 3 and 4, or (3) applying the laws in Table 1. Sometimes, proofs will combine methods (2) and (3).

EXAMPLE 5   We give three proofs for the relationship

$$(A \cup B) \cap A^c \subseteq B.$$

Proof 1. See Figure 5. The picture for $(A \cup B) \cap A^c$ is double-hatched and is clearly a subset of B.

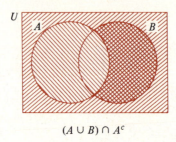
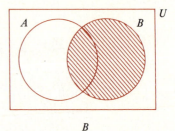

$(A \cup B) \cap A^c$ B

FIGURE 5

Proof 2. We show that $x \in (A \cup B) \cap A^c$ implies $x \in B$. Consider x in $(A \cup B) \cap A^c$. Then $x \in A^c$, and so $x \notin A$. Since x is also in $A \cup B$, it is in A or B, so it follows that x must be in B.

Proof 3. Using the laws of algebra in Table 1, we obtain

$$
\begin{aligned}
(A \cup B) \cap A^c &= A^c \cap (A \cup B) && \text{commutativity 1b} \\
&= (A^c \cap A) \cup (A^c \cap B) && \text{distributivity 3b} \\
&= (A \cap A^c) \cup (A^c \cap B) && \text{commutativity 1b} \\
&= \varnothing \cup (A^c \cap B) && \text{7b} \\
&= (A^c \cap B) \cup \varnothing && \text{commutativity 1a} \\
&= A^c \cap B. && \text{identity law 5a}
\end{aligned}
$$

This identity agrees, of course, with the picture on the left in Figure 5. Now it is clear that $A^c \cap B \subseteq B$, since if $x \in A^c \cap B$, then x must be in B.

The symmetric difference $\oplus$ is also an associative operation:

$$(A \oplus B) \oplus C = A \oplus (B \oplus C).$$

We can see this by looking at the Venn diagrams in Figure 6. On the left we have hatched $A \oplus B$ one way and C the other. Then $(A \oplus B) \oplus C$ is the set hatched one way or the other but not both. Doing the same sort of thing with A and $B \oplus C$ gives us the same set, so the sets $(A \oplus B) \oplus C$ and $A \oplus (B \oplus C)$ are equal.

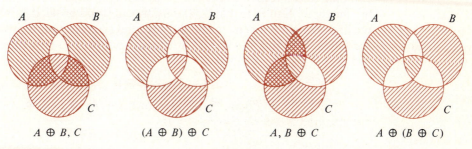

$A \oplus B, C$ $(A \oplus B) \oplus C$ $A, B \oplus C$ $A \oplus (B \oplus C)$

FIGURE 6

Of course, it is also possible to prove this fact without appealing to the pictures. You may want to construct such a proof yourself. Be warned, though, that a detailed argument will be fairly complicated.

Since $\oplus$ is associative, the expression $A \oplus B \oplus C$ is unambiguous. Note that an element belongs to this set provided that it belongs to exactly one or to all three of the sets A, B and C.

Consider two sets S and T. For each element s in S and each element t in T, we form an **ordered pair (s, t)**. Here s is the first element of the ordered pair, t is the second element, and the order is important. Thus $(s_1, t_1) = (s_2, t_2)$ if and only if $s_1 = s_2$ and $t_1 = t_2$. The set of all ordered pairs (s, t) is called the **product** of S and T and written $S \times T$:

$$S \times T = \{(s, t) : s \in S \text{ and } t \in T\}.$$

If $S = T$, we sometimes write S^2 for $S \times S$.

EXAMPLE 6   (a) Let $S = \{1, 2, 3, 4\}$ and $T = \{a, b, c\}$. Then $S \times T$ consists of the twelve ordered pairs listed on the left in Figure 7. We could also depict

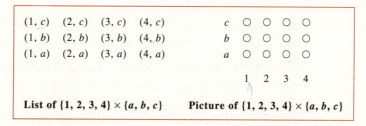

$(1, c)$	$(2, c)$	$(3, c)$	$(4, c)$	c	○	○	○	○
$(1, b)$	$(2, b)$	$(3, b)$	$(4, b)$	b	○	○	○	○
$(1, a)$	$(2, a)$	$(3, a)$	$(4, a)$	a	○	○	○	○

                                          1   2   3   4

List of $\{1, 2, 3, 4\} \times \{a, b, c\}$ **Picture of $\{1, 2, 3, 4\} \times \{a, b, c\}$**

FIGURE 7

these pairs as corresponding points in labeled rows and columns, in the manner shown on the right in the figure. The reader should list or draw $T \times S$ and note that $T \times S \neq S \times T$.

(b) If $S = \{1, 2, 3, 4\}$, then $S^2 = S \times S$ has sixteen ordered pairs; see Figure 8. Note that $(2, 4) \neq (4, 2)$; these ordered pairs involve the same two numbers, but in different orders. In contrast, the *sets* $\{2, 4\}$ and $\{4, 2\}$ are the same. Also note that $(2, 2)$ is a perfectly good ordered pair in which the first element happens to equal the second element. On the other hand, the set $\{2, 2\}$ is just the set $\{2\}$ in which 2 happens to be written twice.

$(1, 4)$	$(2, 4)$	$(3, 4)$	$(4, 4)$	4	O	O	O	O
$(1, 3)$	$(2, 3)$	$(3, 3)$	$(4, 3)$	3	O	O	O	O
$(1, 2)$	$(2, 2)$	$(3, 2)$	$(4, 2)$	2	O	O	O	O
$(1, 1)$	$(2, 1)$	$(3, 1)$	$(4, 1)$	1	O	O	O	O
					1	2	3	4

List of $\{1, 2, 3, 4\}^2$ Picture of $\{1, 2, 3, 4\}^2$

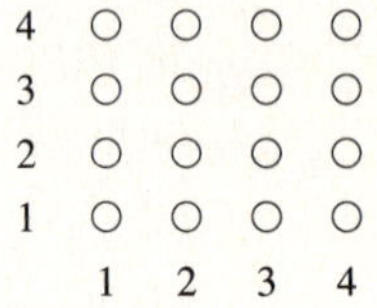

FIGURE 8

Our notation for ordered pairs, for instance $(2, 4)$, is in apparent conflict with our notation for intervals in §1.1 where $(2, 4)$ represented the set $\{x \in \mathbb{R} : 2 < x < 4\}$. Both uses of this notation are standard, however. Fortunately, the intended meaning is almost always clear from the context. ■

For any finite set S, we write $|S|$ for the number of elements in the set. Thus $|S| = |T|$ precisely when the finite sets S and T are of the same size. Observe that

$$|\varnothing| = 0 \text{ and } |\{1, 2, \ldots, n\}| = n \qquad \text{for } n \in \mathbb{P}.$$

Moreover, $|S \times T| = |S| \cdot |T|$. You can see where the notation $\times$ for the product of two sets came from. It turns out that $|\mathscr{P}(S)| = 2^{|S|}$, so some people also use the notation 2^S for $\mathscr{P}(S)$.

We can define the product of any finite collection $S_1, S_2, \ldots, S_n$ of sets. The **product set** $S_1 \times S_2 \times \cdots \times S_n$ consists of all **ordered n-tuples** $(s_1, s_2, \ldots, s_n)$ where $s_1 \in S_1$, $s_2 \in S_2$, etc. That is,

$$S_1 \times S_2 \times \cdots \times S_n = \{(s_1, s_2, \ldots, s_n) : s_k \in S_k \text{ for } k = 1, 2, \ldots, n\}.$$

Just as with ordered pairs, two ordered n-tuples $(s_1, s_2, \ldots, s_n)$ and $(t_1, t_2, \ldots, t_n)$ are regarded as equal if all the corresponding entries are equal: $s_k = t_k$ for $k = 1, 2, \ldots, n$. If the sets $S_1, S_2, \ldots, S_n$ are all equal, to S say, we may write S^n for the product $S_1 \times S_2 \times \cdots \times S_n$.

EXERCISES 1.2

1. Let $U = \{1, 2, 3, 4, 5, \ldots, 12\}$, $A = \{1, 3, 5, 7, 9, 11\}$, $B = \{2, 3, 5, 7, 11\}$, $C = \{2, 3, 6, 12\}$ and $D = \{2, 4, 8\}$. Determine the sets

(a) $A \cup B$ (b) $A \cap C$ (c) $(A \cup B) \cap C^c$

(d) $A \backslash B$ (e) $C \backslash D$ (f) $B \oplus D$

(g) How many subsets of C are there?

2. Let $A = \{1, 2, 3\}$, $B = \{n \in \mathbb{P} : n \text{ is even}\}$ and $C = \{n \in \mathbb{P} : n \text{ is odd}\}$.

(a) Determine $A \cap B$, $B \cap C$, $B \cup C$ and $B \oplus C$.

(b) List all subsets of A.

(c) Which of the following sets are infinite? $A \oplus B$, $A \oplus C$, $A \backslash C$, $C \backslash A$.

3. In this exercise the universe is $\mathbb{R}$. Determine the following sets.

(a) $[0, 3] \cap [2, 6]$ (b) $[0, 3] \cup [2, 6]$ (c) $[0, 3] \backslash [2, 6]$

(d) $[0, 3] \oplus [2, 6]$ (e) $[0, 3]^c$ (f) $[0, 3] \cap \varnothing$

4. Let $\Sigma = \{a, b\}$, $A = \{a, b, aa, bb, aaa, bbb\}$, $B = \{w \in \Sigma^* : \text{length}(w) \geq 2\}$ and $C = \{w \in \Sigma^* : \text{length}(w) \leq 2\}$.

(a) Determine $A \cap C$, $A \backslash C$, $C \backslash A$ and $A \oplus C$.

(b) Determine $A \cap B$, $B \cap C$, $B \cup C$ and $B \backslash A$.

(c) Determine $\Sigma^* \backslash B$, $\Sigma \backslash B$ and $\Sigma \backslash C$.

(d) List all subsets of Σ.

(e) How many sets are there in $\mathscr{P}(\Sigma)$?

5. In this exercise the universe is Σ^* where $\Sigma = \{a, b\}$. Let A, B and C be as in Exercise 4. Determine the following sets.

(a) $B^c \cap C^c$ (b) $(B \cap C)^c$ (c) $(B \cup C)^c$

(d) $B^c \cup C^c$ (e) $A^c \cap C$ (f) $A^c \cap B^c$

(g) Which of these sets are equal? Why?

6. The following statements involve subsets of some nonempty universal set U. Tell whether each is true or false. For each false one, give an example for which the statement is false.

(a) $A \cap (B \cup C) = (A \cap B) \cup C$ for all A, B, C.

(b) $A \cup B \subseteq A \cap B$ implies $A = B$.

(c) $(A \cap \varnothing) \cup B = B$ for all A, B.

(d) $A \cap (\varnothing \cup B) = A$ whenever $A \subseteq B$.

(e) $A \cap B = A^c \cup B^c$ for all A, B.

7. For any set A, what is $A \oplus A$? $A \oplus \varnothing$?

8. Use Venn diagrams to prove the following.

(a) $A \cap (B \oplus C) = (A \cap B) \oplus (A \cap C)$

(b) $A \oplus B \subseteq (A \oplus C) \cup (B \oplus C)$

9. Prove the generalized DeMorgan law $(A \cap B \cap C)^c = A^c \cup B^c \cup C^c$. *Hint:* First apply the DeMorgan law 9b to the sets A and $B \cap C$. The elementwise method can be avoided.

10. Prove the following without using Venn diagrams.

    (a) $A \cap B \subseteq A$ and $A \subseteq A \cup B$ for all sets A and B.

    (b) If $A \subseteq B$ and $A \subseteq C$, then $A \subseteq B \cap C$.

    (c) If $A \subseteq C$ and $B \subseteq C$, then $A \cup B \subseteq C$.

    (d) $A \subseteq B$ if and only if $B^c \subseteq A^c$.

11. Let $A = \{a, b, c\}$ and $B = \{a, b, d\}$.

    (a) List or draw the ordered pairs in $A \times A$.

    (b) List or draw the ordered pairs in $A \times B$.

    (c) List or draw the set $\{(x, y) \in A \times B : x = y\}$.

12. Let $S = \{0, 1, 2, 3, 4\}$ and $T = \{0, 2, 4\}$.

    (a) How many ordered pairs are in $S \times T$? $T \times S$?

    (b) List or draw the elements in $\{(m, n) \in S \times T : m < n\}$.

    (c) List or draw the elements in $\{(m, n) \in T \times S : m < n\}$.

    (d) List or draw the elements in $\{(m, n) \in S \times T : m + n \geq 3\}$.

    (e) List or draw the elements in $\{(m, n) \in T \times S : mn \geq 4\}$.

    (f) List or draw the elements in $\{(m, n) \in S \times S : m + n = 10\}$.

13. For each of the following sets, list all elements if the set has fewer than seven elements. Otherwise, list exactly seven elements of the set.

    (a) $\{(m, n) \in \mathbb{N}^2 : m = n\}$ (b) $\{(m, n) \in \mathbb{N}^2 : m + n \text{ is prime}\}$

    (c) $\{(m, n) \in \mathbb{P}^2 : m = 6\}$ (d) $\{(m, n) \in \mathbb{P}^2 : \min\{m, n\} = 3\}$

    (e) $\{(m, n) \in \mathbb{P}^2 : \max\{m, n\} = 3\}$ (f) $\{(m, n) \in \mathbb{N}^2 : m^2 = n\}$

14. Draw a Venn diagram for four sets A, B, C and D. Be sure to have a region for each of the sixteen possible sets such as $A \cap B^c \cap C^c \cap D$.

Note. In the remaining exercises, you may use any method of proof.

15. Prove or disprove. [A proof needs to be a general argument, but a single counterexample is sufficient for a disproof.]

    (a) $A \cap B = A \cap C$ implies $B = C$.

    (b) $A \cup B = A \cup C$ implies $B = C$.

    (c) $A \cap B = A \cap C$ and $A \cup B = A \cup C$ imply $B = C$.

    (d) $A \cup B \subseteq A \cap B$ implies $A = B$.

    (e) $A \oplus B = A \oplus C$ implies $B = C$.

16. (a) Prove that $A \subseteq B$ if and only if $A \cup B = B$. This is really two assertions:

    "$A \subseteq B$ implies $A \cup B = B$" and "$A \cup B = B$ implies $A \subseteq B$."

    (b) Prove that $A \subseteq B$ if and only if $A \cap B = A$.

17. (a) Show that relative complementation is not commutative; that is, $A \backslash B = B \backslash A$ can fail.

   (b) Show that relative complementation is not associative: $(A \backslash B) \backslash C = A \backslash (B \backslash C)$ can fail.

   (c) Show, however, that $(A \backslash B) \backslash C \subseteq A \backslash (B \backslash C)$ for all A, B and C.

§ 1.3 Functions

We begin with a descriptive working definition of "function." **A function** f assigns to each element x in some set S a unique element in a set T. We say such an f is **defined on** S with **values in** T. The set S is called the **domain** of f and is sometimes written **Dom(f)**. The element assigned to x is usually written **$f(x)$**. Care should be taken to avoid confusing a function f with its functional values $f(x)$, especially when people write, as we will later, "the function $f(x)$." A function f is completely specified by:

Dom(F) = sets

(a) the set on which f is defined, namely Dom(f);

(b) the assignment, rule or formula giving the value $f(x)$ for each $x \in$ Dom(f).

For x in Dom(f), $f(x)$ is called the **image** of x **under** f. The set of all images $f(x)$ is a subset of T called the **image** of f and written Im(f). Thus we have

$$\textbf{Im}(f) = \{f(x) : x \in \text{Dom}(f)\}.$$

It is often convenient to specify a set T of allowable images, i.e., a set T containing Im(f). Such a set is called a **codomain** of f. While a function f has exactly one domain Dom(f) and exactly one image Im(f), any set containing Im(f) can serve as a codomain. Of course, when we specify a codomain we will try to choose one that is useful or informative in context. The notation $f : S \to T$ is shorthand for: "f is a function with domain S and codomain T." We sometimes refer to a function as a **map** or **mapping** and say that f **maps** S into T. When we feel the need of a picture, we sometimes draw sketches such as those in Figure 1.

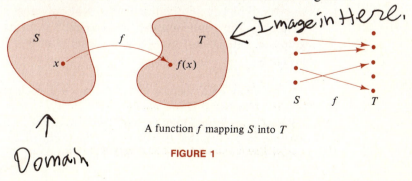

← Image in Here.

↑ Domain

A function f mapping S into T

FIGURE 1

EXAMPLE 1

(a) Consider a function $f: \mathbb{R} \to \mathbb{R}$. This means that $\text{Dom}(f) = \mathbb{R}$, and for each $x \in \mathbb{R}$, $f(x)$ represents a unique number in $\mathbb{R}$. Thus $\mathbb{R}$ is a codomain for f but the image $\text{Im}(f)$ may be a much smaller set. For example, if $f_1(x) = x^2$ for all $x \in \mathbb{R}$, then $\text{Im}(f_1) = [0, \infty)$ and we could write $f_1: \mathbb{R} \to [0, \infty)$. If f_2 is defined by

$$f_2(x) = \begin{cases} 1 & \text{if } x \geq 0, \\ 0 & \text{if } x < 0, \end{cases}$$

then $\text{Im}(f_2) = \{0, 1\}$ and we could write $f_2: \mathbb{R} \to [0, \infty)$ or $f_2: \mathbb{R} \to \mathbb{N}$ or $f_2: \mathbb{R} \to \{0, 1\}$ among other choices.

(b) Recall that the **absolute value** $|x|$ of x in $\mathbb{R}$ is defined by the rule

$$|x| = \begin{cases} x & \text{if } x \geq 0, \\ -x & \text{if } x < 0. \end{cases}$$

The function f given by $f(x) = |x|$ is a function with domain $\mathbb{R}$ and image $[0, \infty)$; note that $|x| \geq 0$ for all $x \in \mathbb{R}$. Absolute value has two important properties that we will use in §1.6 and return to in the next chapter: $|x \cdot y| = |x| \cdot |y|$ and $|x + y| \leq |x| + |y|$ for all $x, y \in \mathbb{R}$.

(c) Consider the function $g: \mathbb{N} \to \mathbb{N}$ defined by $g(n) = n^2 - n$. Here it is useful to specify $\mathbb{N}$ as a codomain, since we might not be interested in the exact set $\text{Im}(g)$. ∎

We will avoid the terminology "range of a function f" because many authors use "range" for what we call the image of f and many others use "range" for what we call a codomain.

Consider a function $f: S \to T$. The **graph of** f is the following subset of $S \times T$:

$$\text{Graph}(f) = \{(x, y) \in S \times T : y = f(x)\}.$$

This definition is compatible with the use of the term in algebra and calculus. The graphs of the functions in Example 1(a) are sketched in Figure 2 on the next page.

Our working definition of "function" is incomplete; in particular, the term "assigns" is undefined. A very precise set-theoretical definition can be given. The key observation is this: Not only does a function determine its graph, but a function can be recovered from its graph. In fact, the graph of a function $f: S \to T$ is a subset G of $S \times T$ with the following property:

for each $x \in S$ there is exactly one $y \in T$ such that $(x, y) \in G$.

Given G, we have $\text{Dom}(f) = S$, and, for each $x \in S$, $f(x)$ is the unique element in T such that $(x, f(x)) \in G$. The point to observe is that nothing is lost if we regard functions and their graphs as the same, and we gain

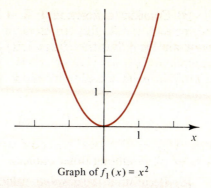

Graph of $f_1(x) = x^2$

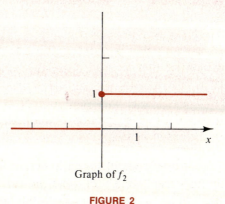

Graph of f_2

FIGURE 2

some precision in the process. A **function** with domain S and codomain T is a subset G of $S \times T$ satisfying:

for each $x \in S$ there is exactly one $y \in T$ such that $(x, y) \in G$.

If S and T are subsets of $\mathbb{R}$ and if $S \times T$ is drawn so that S is part of the horizontal axis and T is part of the vertical axis, then a subset G of $S \times T$ is a function [or the graph of a function] if every vertical line through a point in S intersects G in exactly one point.

A function $f: S \to T$ is called **one-to-one** in case distinct elements in S have distinct images in T under f:

if $x_1, x_2 \in S$ and $x_1 \neq x_2$ then $f(x_1) \neq f(x_2)$.

This condition is logically equivalent to:

$$\text{if} \quad x_1, x_2 \in S \quad \text{and} \quad f(x_1) = f(x_2) \quad \text{then} \quad x_1 = x_2,$$

a form that is often useful. In terms of the graph G of f, f is one-to-one if and only if:

for each $y \in T$ there is at most one $x \in S$ such that $(x, y) \in G$.

If S and T are subsets of $\mathbb{R}$ and f is graphed as above, this condition states that horizontal lines intersect G at most once.

Given $f: S \to T$, we say that f maps **onto** a subset B of T provided that $B = \text{Im}(f)$. In particular, we say f maps **onto** T provided that $\text{Im}(f) = T$. In terms of the graph G of f, f maps S onto T if and only if:

for each $y \in T$ there is at least one $x \in S$ such that $(x, y) \in G$.

A function $f: S \to T$ that is one-to-one and maps onto T is called a **one-to-one correspondence** between S and T. Thus f is a one-to-one correspondence if and only if:

for each $y \in T$ there is exactly one $x \in S$ such that $(x, y) \in G$.

These three kinds of special functions are illustrated in Figure 3.

This may sound confusing, & it is but the pictures make it clear

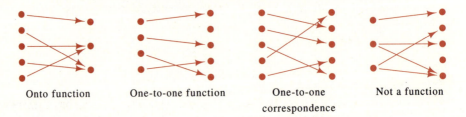

Onto function   One-to-one function   One-to-one correspondence   Not a function

FIGURE 3

Before we turn to mathematical examples, we illustrate the ideas in a nonmathematical setting.

EXAMPLE 2   Suppose that each student in a class S is assigned a seat number from the set $T = \{1, 2, \ldots, 75\}$. This assignment provides a function $f: S \to T$; thus for each student s, $f(s)$ represents his or her seat number. The function will be one-to-one provided that no two students are assigned the same seat number. In this case, the class cannot have more than 75 students. The function will map S onto T provided that every number in T has been assigned to at least one student. Note that for this to happen the class must have at least 75 students. The only way that f can possibly be a one-to-one correspondence of S onto T is if the class has exactly 75 students.

If we view the function f as a set of ordered pairs, it will consist of pairs in $S \times T$ like (Ann Nelson, 73). ∎

EXAMPLE 3   (a) We define $f: \mathbb{N} \to \mathbb{N}$ by the rule $f(n) = 2n$. Then f is one-to-one since

$$f(n_1) = f(n_2) \quad \text{implies} \quad 2n_1 = 2n_2 \quad \text{implies} \quad n_1 = n_2.$$

However, f does not map $\mathbb{N}$ onto $\mathbb{N}$ since $\text{Im}(f)$ consists only of the even natural numbers.

(b) Let Σ be an alphabet. Then length$(w) \in \mathbb{N}$ for each word w in Σ^*; see §1.1. Thus "length" is a function from Σ^* onto $\mathbb{N}$. [Note that functions can have fancier names than "f."] To see this, recall that Σ is nonempty, so Σ contains some letter, say a. Now $0 = $ length(λ), $1 = $ length(a), $2 = $ length(aa), etc. The function length is not one-to-one unless Σ has only one element. ∎

EXAMPLE 4   We prove that $f: \mathbb{R} \to \mathbb{R}$ defined by $f(x) = 3x - 5$ is a one-to-one correspondence of $\mathbb{R}$ onto $\mathbb{R}$. To check that f is one-to-one we need to show that

$$\text{if} \quad f(x) = f(x') \quad \text{then} \quad x = x',$$

i.e.,

$$\text{if} \quad 3x - 5 = 3x' - 5 \quad \text{then} \quad x = x'.$$

Could be an important Proof.

But if $3x - 5 = 3x' - 5$, then $3x = 3x'$ [add 5 to both sides] and this implies that $x = x'$ [divide both sides by 3].

To show that f maps $\mathbb{R}$ onto $\mathbb{R}$ we consider an element y in $\mathbb{R}$. We need to find an x in $\mathbb{R}$ such that $f(x) = y$, i.e., $3x - 5 = y$. So we solve for x and obtain $x = (y + 5)/3$. Thus, given y in $\mathbb{R}$, $(y + 5)/3$ belongs to $\mathbb{R}$ and $f((y + 5)/3) = 3((y + 5)/3) - 5 = y$. This shows that every y in $\mathbb{R}$ belongs to $\text{Im}(f)$ so that f maps $\mathbb{R}$ onto $\mathbb{R}$. ∎

Some special functions occur so often that they have special names. Let S be a nonempty set. The **identity function** 1_S on S is the function that maps each element of S to itself:

$$1_S(x) = x \qquad \text{for all} \quad x \in S.$$

Thus the identity function is a one-to-one correspondence of S onto S.

A function $f: S \to T$ is called a **constant function** if there is some $y_0 \in T$ so that $f(x) = y_0$ for all $x \in S$. The value a constant function takes does not change or vary as x varies over S.

Consider a set S and a subset A of S. The function on S that takes the value 1 at members of A and the value 0 at the other members of S is called the **characteristic function** of A and is denoted χ_A [lowercase Greek chi, sub A]. Thus

with this function, if x is in subset a, the value of F(x) = 1. If it is not in A but is in the set S F(x) = ∅

$$\chi_A(x) = \begin{cases} 1 & \text{for} \quad x \in A, \\ 0 & \text{for} \quad x \in S \backslash A. \end{cases}$$

Note that $\chi_A: S \to \{0, 1\}$ is rarely one-to-one and is usually an onto map. In fact, χ_A maps S onto $\{0, 1\}$ unless $A = S$ or $A = \varnothing$. If either A or $S \backslash A$ has at least two members, then χ_A is not one-to-one.

Now consider functions $f: S \to T$ and $g: T \to U$; see Figure 4. We define the **composition** $g \circ f: S \to U$ by the rule

$$g \circ f(x) = g(f(x)) \qquad \text{for all} \quad x \in S.$$

One might read the left side "g circle f of x" or "g of f of x." Complicated operations that are performed in calculus or on a calculator can be viewed as the composition of simpler functions.

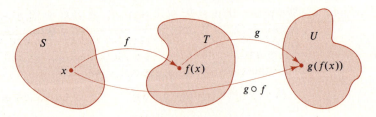

Composing functions

FIGURE 4

EXAMPLE 5   (a) Consider the function $h: \mathbb{R} \to \mathbb{R}$ given by

$$h(x) = (x^3 + 2x)^7.$$

The value $h(x)$ is obtained by first calculating $x^3 + 2x$ and then taking its seventh power. We write f for the first or inside function: $f(x) = x^3 + 2x$. We write g for the second or outside function: $g(x) = x^7$. The name of the variable x is irrelevant; we could just as well have written $g(y) = y^7$ for $y \in \mathbb{R}$. Either way, we see that

$$g(f(x)) = g(x^3 + 2x) = (x^3 + 2x)^7 = h(x) \qquad \text{for} \quad x \in \mathbb{R}.$$

Thus $h = g \circ f$. The ability to view complicated functions as the composition of simpler functions is a critical skill in calculus. Note that the order of f and g is important. In fact,

$$f \circ g(x) = f(x^7) = (x^7)^3 + 2(x^7) = x^{21} + 2x^7 \qquad \text{for} \quad x \in \mathbb{R}.$$

(b) Suppose that one wishes to calculate $h(x) = \sqrt{\log x}$ for certain positive values of x on a hand-held calculator. The calculator has the functions $\sqrt{x}$ and $\log x$, which stands for $\log_{10} x$. One works from the inside out. For example, if $x = 73$, one keys in this value, performs $\log x$ to obtain 1.8633, and then performs $\sqrt{x}$ to obtain 1.3650. Note that $h = g \circ f$ where $f(x) = \log x$ and $g(x) = \sqrt{x}$. As in part (a), order is important: $h \neq f \circ g$, i.e., $\sqrt{\log x}$ is not generally equal to $\log \sqrt{x}$. For example, if $x = 73$, then $\sqrt{x}$ is approximately 8.5440 and $\log \sqrt{x}$ is approximately .9317.

(c) Of course, some functions f and g do **commute** under composition, i.e., satisfy $f \circ g = g \circ f$. For example, if $f(x) = \sqrt{x}$ and $g(x) = 1/x$ for $x \in (0, \infty)$, then $f \circ g = g \circ f$ because

$$\sqrt{\frac{1}{x}} = \frac{1}{\sqrt{x}} \qquad \text{for} \quad x \in (0, \infty)$$

For example, for $x = 9$ we have $\sqrt{1/9} = 1/3 = 1/\sqrt{9}$. ∎

We can compose more than two functions if we wish.

EXAMPLE 6 Define the functions f, g and h that map $\mathbb{R}$ into $\mathbb{R}$ by

$$f(x) = x^4, \qquad g(y) = \sqrt{y^2 + 1}, \qquad h(z) = z^2 + 72.$$

We've used the different variable names x, y and z to help clarify our computations below. Let's calculate $h \circ (g \circ f)$ and $(h \circ g) \circ f$ and compare the answers. First, for $x \in \mathbb{R}$ we have

$$
\begin{aligned}
(h \circ (g \circ f))(x) &= h(g \circ f(x)) && \text{by definition of } h \circ (g \circ f) \\
&= h(g(f(x))) && \text{by definition of } g \circ f \\
&= h(g(x^4)) && \text{since } f(x) = x^4 \\
&= h(\sqrt{x^8 + 1}) && y = x^4 \text{ in definition of } g \\
&= (\sqrt{x^8 + 1})^2 + 72 && z = \sqrt{x^8 + 1} \text{ in definition of } h \\
&= x^8 + 73 && \text{algebra.}
\end{aligned}
$$

On the other hand,

$$
\begin{aligned}
((h \circ g) \circ f)(x) &= (h \circ g)(f(x)) && \text{by definition of } (h \circ g) \circ f \\
&= h(g(f(x))) && \text{by definition of } h \circ g \\
&= x^8 + 73 && \text{exactly as above.}
\end{aligned}
$$

We conclude that

$$(h \circ (g \circ f))(x) = ((h \circ g) \circ f)(x) = x^8 + 73 \qquad \text{for all} \quad x \in \mathbb{R},$$

so the functions $h \circ (g \circ f)$ and $(h \circ g) \circ f$ are exactly the same function. This is no accident, as we observe in the next general theorem. ∎

Consider functions $f: S \to T$, $g: T \to U$ and $h: U \to V$. Then $h \circ (g \circ f) = (h \circ g) \circ f$.

The proof of this basic result amounts to checking that the functions $h \circ (g \circ f)$ and $(h \circ g) \circ f$ both map S into V and that, just as in Example 6, for each $x \in S$ the values $(h \circ (g \circ f))(x)$ and $((h \circ g) \circ f)(x)$ are both equal to $h(g(f(x)))$.

Since composition is associative, we can write $h \circ g \circ f$ unambiguously without any parentheses. We can also compose any finite number of functions without using parentheses.

EXAMPLE 7 (a) If $f(x) = x^4$ for $x \in [0, \infty)$, $g(x) = \sqrt{x + 2}$ for $x \in [0, \infty)$ and $h(x) = x^2 + 1$ for $x \in \mathbb{R}$, then

$$h \circ g \circ f(x) = h(g(x^4)) = h(\sqrt{x^4 + 2}) = (x^4 + 2) + 1$$
$$= x^4 + 3 \qquad \text{for} \quad x \in [0, \infty),$$
$$f \circ g \circ h(x) = f(g(x^2 + 1)) = f(\sqrt{x^2 + 1 + 2})$$
$$= (x^2 + 3)^2 \qquad \text{for} \quad x \in \mathbb{R},$$
$$f \circ h \circ g(x) = f(h(\sqrt{x + 2})) = f(x + 2 + 1)$$
$$= (x + 3)^4 \qquad \text{for} \quad x \in [0, \infty).$$

(b) The function F given by

$$F(x) = (\sqrt{x^2 + 1} + 3)^5 \qquad \text{for} \quad x \in \mathbb{R}$$

can be written as $k \circ h \circ g \circ f$ where

$$f(x) = x^2 + 1 \qquad \text{for} \quad x \in \mathbb{R},$$
$$g(x) = \sqrt{x} \qquad \text{for} \quad x \in [0, \infty),$$
$$h(x) = x + 3 \qquad \text{for} \quad x \in \mathbb{R},$$
$$k(x) = x^5 \qquad \text{for} \quad x \in \mathbb{R}. \quad \blacksquare$$

EXERCISES 1.3

1. We define $f: \mathbb{R} \to \mathbb{R}$ as follows:

$$f(x) = \begin{cases} x^3 & \text{if } x \geq 1, \\ x & \text{if } 0 \leq x < 1, \\ -x^3 & \text{if } x < 0. \end{cases}$$

(a) Calculate $f(3)$, $f(1/3)$, $f(-1/3)$ and $f(-3)$.

(b) Sketch a graph of f.

(c) Find Im(f).

2. The functions sketched in Figure 5 have domain and codomain both equal to $[0, 1]$.

(a) Which of these functions are one-to-one?

(b) Which of these functions map $[0, 1]$ onto $[0, 1]$?

(c) Which of these functions are one-to-one correspondences?

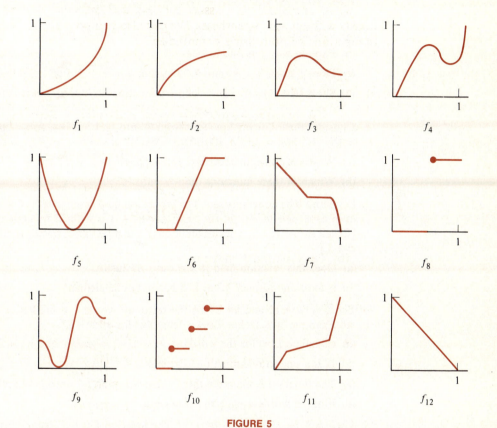

FIGURE 5

3. Let $S = \{1, 2, 3, 4, 5\}$ and $T = \{a, b, c, d\}$. For each question below: if the answer is YES, give an example; if the answer if NO, explain briefly.

(a) Are there any one-to-one functions from S into T?

(b) Are there any one-to-one functions from T into S?

(c) Are there any functions mapping S onto T?

(d) Are there any functions mapping T onto S?

(e) Are there any one-to-one correspondences between S and T?

4. Let $S = \{1, 2, 3, 4, 5\}$ and consider the following functions from S into S: $1_S(n) = n$, $f(n) = 6 - n$, $g(n) = \max\{3, n\}$, $h(n) = \max\{1, n - 1\}$.

   (a) Write each of these functions as a set of ordered pairs; i.e., list the elements in their graphs.

   (b) Sketch a graph of each of these functions.

   (c) Which of these functions are both one-to-one and onto?

5. The rule $f((m, n)) = 2^m 3^n$ defines a one-to-one function from $\mathbb{N} \times \mathbb{N}$ into $\mathbb{N}$. *Note.* When functions are defined on ordered pairs, it is customary to omit one set of parentheses. Thus we will write $f(m, n) = 2^m 3^n$.

   (a) Calculate $f(m, n)$ for five different elements (m, n) in $\mathbb{N} \times \mathbb{N}$.

   (b) Explain why f is one-to-one.

   (c) Does f map $\mathbb{N} \times \mathbb{N}$ onto $\mathbb{N}$? Explain.

   (d) Show that $g(m, n) = 2^m 4^n$ defines a function on $\mathbb{N} \times \mathbb{N}$ that is not one-to-one.

6. Consider the following functions from $\mathbb{N}$ into $\mathbb{N}$: $1_{\mathbb{N}}(n) = n$, $f(n) = 3n$, $g(n) = n + (-1)^n$, $h(n) = \min\{n, 100\}$, $k(n) = \max\{0, n - 5\}$.

   (a) Which of these functions are one-to-one?

   (b) Which of these functions map $\mathbb{N}$ onto $\mathbb{N}$?

7. Let A and B be nonempty sets. The projection map PROJ picks the first element from each pair in $A \times B$, i.e., PROJ: $A \times B \to A$ is defined by PROJ$(a, b) = a$. [Recall the convention about functions on ordered pairs mentioned in Exercise 5.]

   (a) Does this function map $A \times B$ onto A? Justify.

   (b) Is PROJ one-to-one? What if B has only one element?

8. Let $\Sigma = \{a, b, c\}$ and let Σ^* be the set of all words w using letters from Σ; see Example 3(b). Define $L(w) = \text{length}(w)$ for all $w \in \Sigma^*$.

   (a) Calculate $L(w)$ for the words $w_1 = cab$, $w_2 = ababac$ and $w_3 = \lambda$.

   (b) Is L a one-to-one function? Explain.

   (c) The function L maps Σ^* into $\mathbb{N}$. Does L map Σ^* onto $\mathbb{N}$? Explain.

   (d) Find all words w such that $L(w) = 2$.

9. For $n \in \mathbb{Z}$, let $f(n) = \frac{1}{2}[(-1)^n + 1]$. The function f is the characteristic function for some subset of $\mathbb{Z}$. Which subset?

10. In Example 5(b), we compared the functions $\sqrt{\log x}$ and $\log \sqrt{x}$. Show that these functions take the same value for $x = 10,000$.

11. We define three functions mapping $\mathbb{R}$ into $\mathbb{R}$ as follows: $f(x) = x^3 - 4x$, $g(x) = 1/(x^2 + 1)$, $h(x) = x^4$. Find

   (a) $f \circ g \circ h$ (b) $f \circ h \circ g$ (c) $h \circ g \circ f$

   (d) $f \circ f$ (e) $g \circ g$ (f) $h \circ g$

   (g) $g \circ h$

12. Show that if $f: S \to T$ and $g: T \to U$ are one-to-one, then $g \circ f$ is one-to-one.

13. Prove that the composition of functions is associative.

14. Several important functions can be found on hand-held calculators. Why isn't the identity function, i.e., the function $1_{\mathbb{R}}$ where $1_{\mathbb{R}}(x) = x$ for all $x \in \mathbb{R}$, among them?

15. Consider the functions f and g mapping $\mathbb{Z}$ into $\mathbb{Z}$, where $f(n) = n - 1$ for $n \in \mathbb{Z}$ and g is the characteristic function χ_E of $E = \{n \in \mathbb{Z}: n \text{ is even}\}$.

   (a) Calculate $(g \circ f)(5)$, $(g \circ f)(4)$, $(f \circ g)(7)$ and $(f \circ g)(8)$.

   (b) Calculate $(f \circ f)(11)$, $(f \circ f)(12)$, $(g \circ g)(11)$ and $(g \circ g)(12)$.

   (c) Determine the functions $g \circ f$ and $f \circ f$.

   (d) Show that $g \circ g = g \circ f$ and that $f \circ g$ is the negative of $g \circ f$.

§ 1.4 Inverses of Functions

Roughly speaking, an inverse for the function f is a function that undoes the action of f. Applying f first and then the inverse restores every member of the domain of f to where it started. See Figure 1.

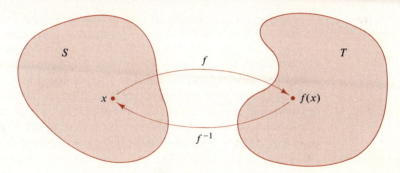

A function and its inverse

FIGURE 1

EXAMPLE 1　(a) The functions x^2 and $\sqrt{x}$ with domains $[0, \infty)$ are "inverses" to each other. If you apply these operations in either order to some value, the original value is obtained. Try it on a calculator! In symbols,

$$\sqrt{x^2} = x \quad \text{and} \quad (\sqrt{x})^2 = x \qquad \text{for} \quad x \in [0, \infty).$$

(b) The function $1/x$ is its own "inverse." If you apply the operation twice to some value, you get the original value. That is,

$$\frac{1}{1/x} = x \qquad \text{for all nonzero } x \text{ in } \mathbb{R}. \quad \blacksquare$$

Here is the precise definition. An **inverse** of a function $f: S \to T$ is a function $\boldsymbol{f^{-1}}: T \to S$ such that $f^{-1} \circ f = 1_S$ and $f \circ f^{-1} = 1_T$, i.e., such that

$$f^{-1}(f(x)) = x \qquad \text{for all} \quad x \in S$$

and

$$f(f^{-1}(y)) = y \qquad \text{for all} \quad y \in T.$$

Not all functions have inverses; those that do are called **invertible** functions. We will see in the proof of the theorem that the defining conditions completely determine f^{-1} if it exists, so an invertible function can't have two different inverses.

EXAMPLE 2   Consider a positive real number b with $b \neq 1$. The important examples of b are 2, 10 and the number e that appears in calculus and is approximately 2.718. The function f_b given by $f_b(x) = b^x$ for $x \in \mathbb{R}$ has an inverse f_b^{-1} with domain $(0, \infty)$, which is called a **logarithm function**. We write $f_b^{-1}(y) = \log_b y$; by the definition of an inverse we have

$$\log_b b^x = x \qquad \text{for every} \quad x \in \mathbb{R}$$

and

$$b^{\log_b y} = y \qquad \text{for every} \quad y \in (0, \infty).$$

In particular, e^x and $\log_e x$ are inverse functions. The function $\log_e x$ is called the **natural logarithm** and is often denoted $\ln x$. The functions 10^x and $\log_{10} x$ are inverses, and so are 2^x and $\log_2 x$. The functions $\log_{10} x = \log$ and $\log_e x = \ln$ appear on many calculators; such calculators also allow one to compute the inverses 10^x and e^x of these functions. To compute $\log_2 x$ on a calculator, use one of the formulas

$$\log_2 x = \frac{\log_{10} x}{\log_{10} 2} \approx 3.321928 \cdot \log x$$

or

$$\log_2 x = \frac{\log_e x}{\log_e 2} = \frac{\ln x}{\ln 2} \approx 1.442695 \cdot \ln x. \quad \blacksquare$$

The next theorem tells us which functions are invertible.

Theorem | The function $f: S \to T$ is invertible if and only if f is one-to-one and maps S onto T.

Proof. Suppose that f has an inverse, f^{-1}. If $x_1, x_2 \in S$ with $f(x_1) = f(x_2)$, then

$$x_1 = f^{-1}(f(x_1)) = f^{-1}(f(x_2)) = x_2.$$

Thus f is one-to-one. Moreover, if $y \in T$, then $f^{-1}(y)$ belongs to S and $f(f^{-1}(y)) = y$; so $y \in \text{Im}(f)$. Hence $T = \text{Im}(f)$ and f maps S onto T.

Conversely, if f maps S onto T, then for each y in T there is some $x \in S$ with $f(x) = y$. If f is also one-to-one, then there is exactly one such x, and we get a formula for f^{-1}, namely:

(∗) $\qquad\qquad f^{-1}(y) = $ that unique $x \in S$ such that $f(x) = y$.

This definition immediately gives $f(f^{-1}(y)) = y$, and $f^{-1}(f(x))$ is the unique member of S that f maps to $f(x)$, namely x itself. Thus f^{-1}, as defined by (∗), meets the conditions to be the inverse of f. ∎

This proof also shows how to get $f^{-1}(y)$ if f is invertible. Simply solve for x in terms of y.

EXAMPLE 3 Consider the function $f: \mathbb{R} \to \mathbb{R}$ given by $f(x) = x^3 + 1$. To see that f is one-to-one, we note that

$$f(x_1) = f(x_2) \quad \text{implies} \quad x_1^3 + 1 = x_2^3 + 1$$
$$\text{implies} \quad x_1^3 = x_2^3 \quad \text{implies} \quad x_1 = x_2;$$

the last implication holds because each real number has a unique cube root.

To check that f maps $\mathbb{R}$ onto $\mathbb{R}$, consider y in $\mathbb{R}$. We need to find $x \in \mathbb{R}$ so that $f(x) = y$; i.e., we need to solve $x^3 + 1 = y$ for x. When we do, we get $x = \sqrt[3]{y - 1}$, which belongs to $\mathbb{R}$. Hence f maps $\mathbb{R}$ onto $\mathbb{R}$.

Since f is one-to-one and maps $\mathbb{R}$ onto $\mathbb{R}$, f is invertible, by the theorem. We found f^{-1} in the last paragraph when we solved for x. Thus $f^{-1}(y) = \sqrt[3]{y - 1}$. This formula makes sense for each y in $\mathbb{R}$, and so f^{-1} is completely determined. ∎

EXAMPLE 4 Consider the function $g: \mathbb{Z} \times \mathbb{Z} \to \mathbb{Z} \times \mathbb{Z}$ given by $g(m, n) = (-n, -m)$. We will check that g is one-to-one and onto, and then we'll find its inverse. To show that g is one-to-one we need to show that

$$g(m, n) = g(m', n') \quad \text{implies} \quad (m, n) = (m', n').$$

If $g(m, n) = g(m', n')$ then $(-n, -m) = (-n', -m')$. Since these ordered pairs are equal we must have $-n = -n'$ and $-m = -m'$. Hence $m = m'$ and $n = n'$ so that $(m, n) = (m', n')$, as desired.

To show that g maps onto $\mathbb{Z} \times \mathbb{Z}$, consider (p, q) in $\mathbb{Z} \times \mathbb{Z}$. We need to find (m, n) in $\mathbb{Z} \times \mathbb{Z}$ so that $g(m, n) = (p, q)$. Thus we need $(-n, -m) = (p, q)$, and this tells us that n should be $-p$ and m should be $-q$. In other words, given (p, q) in $\mathbb{Z} \times \mathbb{Z}$ we see that $(-q, -p)$ is an element in $\mathbb{Z} \times \mathbb{Z}$ such that $g(-q, -p) = (p, q)$. Thus g maps $\mathbb{Z} \times \mathbb{Z}$ onto $\mathbb{Z} \times \mathbb{Z}$.

To find the inverse of g we need to take (p, q) in $\mathbb{Z} \times \mathbb{Z}$ and find $g^{-1}(p, q)$. We just did this in the last paragraph; g maps $(-q, -p)$ onto (p, q), and hence $g^{-1}(p, q) = (-q, -p)$ for all (p, q) in $\mathbb{Z} \times \mathbb{Z}$.

It is interesting to note that $g = g^{-1}$ in this case.   ■

Inverses of functions are so useful that we sometimes restrict functions that are not one-to-one to smaller domains on which they are one-to-one. If we then arrange for the codomain to equal the image of the function, we obtain an invertible function.

EXAMPLE 5 (a) Consider $f \colon \mathbb{R} \to \mathbb{R}$ defined by $f(x) = x^2$. Then f is not one-to-one, but it is one-to-one if we restrict the domain to $[0, \infty)$. Thus we define a new function F by the same rule, $F(x) = x^2$, but with $\mathrm{Dom}(F) = [0, \infty)$. Then F is one-to-one. In fact, $F \colon [0, \infty) \to [0, \infty)$ is one-to-one and onto. It is this function that has $F^{-1}(x) = \sqrt{x}$ as its inverse; see Example 1(a).

The function F is called the **restriction** of f to $[0, \infty)$. This sort of restriction is clearly possible and desirable in many settings of interest.

(b) You should be able to follow this example even if you know no trigonometry. It turns out that none of the trigonometric functions are one-to-one. For example, consider the graph of $\sin x$ in Figure 2. Nevertheless, $\sin x$ is one-to-one if its domain is restricted to, say, $[-\pi/2, \pi/2]$.

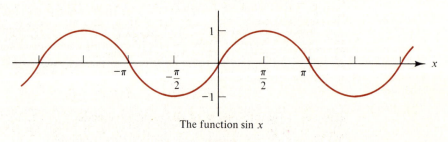

The function $\sin x$

FIGURE 2

See Figure 3(a), where we have denoted the restriction by $\mathrm{Sin}\, x$. With codomain $[-1, 1]$, we obtain an invertible function; the inverse is given in Figure 3(b). This is the inverse sine or arc sin encountered in trigonometry, calculus and many hand-held calculators.   ■

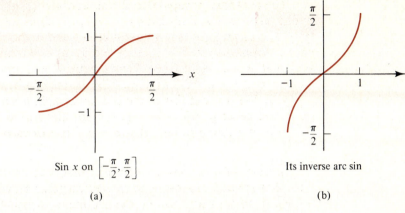

Sin x on $\left[-\dfrac{\pi}{2}, \dfrac{\pi}{2}\right]$

(a)

Its inverse arc sin

(b)

FIGURE 3

Consider a function $f: S \to T$. If A is a subset of S, we define

$$f(A) = \{f(x) : x \in A\}.$$

Thus $f(A)$ is the set of images $f(x)$ as x varies over A. We call $f(A)$ the **image of the set** A **under** f. We are also interested in the **inverse image of a set** B in T:

$$f^{\leftarrow}(B) = \{x \in S : f(x) \in B\}.$$

The set $f^{\leftarrow}(B)$ is called the **pre-image of the set** B **under** f.
   If f is invertible, then [Exercise 16] the pre-image of the subset B of T under f equals the image of B under f^{-1}, i.e., in this case

$$f^{\leftarrow}(B) = \{f^{-1}(y) : y \in B\} = f^{-1}(B).$$

If f is not invertible, it makes no sense to write $f^{-1}(y)$ or $f^{-1}(B)$, of course. Because $f^{-1}(B)$ can't have any meaning *unless* it means $f^{\leftarrow}(B)$, some people extend the notation and write $f^{-1}(B)$ for what we denote by $f^{\leftarrow}(B)$, even if f is not invertible. Beware!
   For $y \in T$ we write $f^{\leftarrow}(y)$ for the set $f^{\leftarrow}(\{y\})$. That is,

$$f^{\leftarrow}(y) = \{x \in S : f(x) = y\}.$$

This set is the **pre-image of the element** y **under** f. Note that solving the equation $f(x) = y$ for x is equivalent to finding the set $f^{\leftarrow}(y)$. That is, $f^{\leftarrow}(y)$ is the **solution set** for the equation $f(x) = y$. As with equations in algebra, the set $f^{\leftarrow}(y)$ might have one element, several elements, or no elements at all.

EXAMPLE 6   (a) Consider $f: \mathbb{R} \to \mathbb{R}$ given by $f(x) = x^2$. Then

$$f^{\leftarrow}(4) = \{x \in \mathbb{R} : x^2 = 4\} = \{-2, 2\},$$

which is the solution set of the equation $x^2 = 4$. The pre-image of the set $[1, 9]$ is

$$f^{\leftarrow}([1, 9]) = \{x \in \mathbb{R} : x^2 \in [1, 9]\} = \{x \in \mathbb{R} : 1 \leq x^2 \leq 9\}$$
$$= [-3, -1] \cup [1, 3].$$

Also we have $f^{\leftarrow}([-1, 0]) = \{0\}$ and $f^{\leftarrow}([-1, 1]) = [-1, 1]$.

(b) Consider the function $g : \mathbb{N} \times \mathbb{N} \to \mathbb{N}$ defined by $g(m, n) = m^2 + n^2$. Then $g^{\leftarrow}(0) = \{(0, 0)\}$, $g^{\leftarrow}(1) = \{(0, 1), (1, 0)\}$, $g^{\leftarrow}(2) = \{(1, 1)\}$, $g^{\leftarrow}(3) = \varnothing$, $g^{\leftarrow}(4) = \{(0, 2), (2, 0)\}$, etc. For instance, we have $g^{\leftarrow}(25) = \{(0, 5), (3, 4), (4, 3), (5, 0)\}$. ■

EXAMPLE 7   (a) Let Σ be an alphabet and let L be the length function on Σ^*; $L(w) = \text{length}(w)$ for $w \in \Sigma^*$. As we noted in Example 3(b) of § 1.3, L maps Σ^* onto $\mathbb{N}$. For each $k \in \mathbb{N}$,

$$L^{\leftarrow}(k) = \{w \in \Sigma^* : L(w) = k\} = \{w \in \Sigma^* : \text{length}(w) = k\}.$$

Note that the various sets $L^{\leftarrow}(k)$ are disjoint and that their union is Σ^*:

$$L^{\leftarrow}(0) \cup L^{\leftarrow}(1) \cup L^{\leftarrow}(2) \cup \cdots = \Sigma^*.$$

(b) Consider $h : \mathbb{Z} \to \{-1, 1\}$ where $h(n) = (-1)^n$. Then

$$h^{\leftarrow}(1) = \{n \in \mathbb{Z} : n \text{ is even}\} \quad \text{and} \quad h^{\leftarrow}(-1) = \{n \in \mathbb{Z} : n \text{ is odd}\}.$$

These two sets are disjoint and their union is all of $\mathbb{Z}$:

$$h^{\leftarrow}(1) \cup h^{\leftarrow}(-1) = \mathbb{Z}. \quad ■$$

It is not a fluke that the pre-images of elements cut the domains into slices in these last two examples. We will see in § 3.5 that something like this always happens, and we will learn how to exploit the resulting partition into disjoint pieces.

EXERCISES 1.4

1. Find the inverses of the following functions mapping $\mathbb{R}$ into $\mathbb{R}$.

   (a) $f(x) = 2x + 3$ (b) $g(x) = x^3 - 2$

   (c) $h(x) = (x - 2)^3$ (d) $k(x) = \sqrt[3]{x} + 7$

2. Many hand-held calculators have the functions $\log x$, x^2, $\sqrt{x}$ and $1/x$.

   (a) Specify the domains of these functions.

   (b) Which of these functions are inverses of each other?

   (c) Which pairs of these functions commute with respect to composition?

(d) Some hand-held calculators also have the functions sin x, cos x and tan x. If you know a little trigonometry, repeat parts (a), (b) and (c) for these functions.

3. Here are some functions from $\mathbb{N} \times \mathbb{N}$ to $\mathbb{N}$: SUM$(m, n) = m + n$, PROD$(m, n) = m * n$, MAX$(m, n) = \max\{m, n\}$, MIN$(m, n) = \min\{m, n\}$; here $*$ denotes multiplication of integers.

(a) Which of these functions map $\mathbb{N} \times \mathbb{N}$ onto $\mathbb{N}$?

(b) Show that none of these functions are one-to-one.

(c) For each of these functions F, how big is the set $F^{\leftarrow}(4)$?

4. Here are some functions mapping $\mathscr{P}(\mathbb{N}) \times \mathscr{P}(\mathbb{N})$ into $\mathscr{P}(\mathbb{N})$: UNION$(A, B) = A \cup B$, INTER$(A, B) = A \cap B$ and SYM$(A, B) = A \oplus B$.

(a) Show that each of these functions maps $\mathscr{P}(\mathbb{N}) \times \mathscr{P}(\mathbb{N})$ onto $\mathscr{P}(\mathbb{N})$.

(b) Show that none of these functions are one-to-one.

(c) For each of these functions F, how big is the set $F^{\leftarrow}(\varnothing)$? the set $F^{\leftarrow}(\{0\})$?

5. Here are two "shift functions" mapping $\mathbb{N}$ into $\mathbb{N}$: $f(n) = n + 1$ and $g(n) = \max\{0, n - 1\}$ for $n \in \mathbb{N}$.

(a) Calculate $f(n)$ for $n = 0, 1, 2, 3, 4, 73$.

(b) Calculate $g(n)$ for $n = 0, 1, 2, 3, 4, 73$.

(c) Show that f is one-to-one but does not map $\mathbb{N}$ onto $\mathbb{N}$.

(d) Show that g maps $\mathbb{N}$ onto $\mathbb{N}$ but is not one-to-one.

(e) Show that $g \circ f = 1_{\mathbb{N}}$ but that $f \circ g \neq 1_{\mathbb{N}}$.

6. We define $f: \mathbb{N} \rightarrow \mathbb{N}$ and $g: \mathbb{N} \rightarrow \mathbb{N}$ as follows: $f(n) = 2n$ for all $n \in \mathbb{N}$, $g(n) = n/2$ if n is even and $g(n) = (n - 1)/2$ if n is odd.

(a) Calculate $g(n)$ for $n = 0, 1, 2, 3, 4, 73$.

(b) Show that $g \circ f = 1_{\mathbb{N}}$ but that $f \circ g \neq 1_{\mathbb{N}}$.

7. If $f: S \rightarrow S$ and $f \circ f = 1_S$, then f is its own inverse. Show that the following functions are their own inverses.

(a) The function $f: (0, \infty) \rightarrow (0, \infty)$ given by $f(x) = 1/x$.

(b) The function $\phi: \mathscr{P}(S) \rightarrow \mathscr{P}(S)$ defined by $\phi(A) = A^c$.

(c) The function $g: \mathbb{R} \rightarrow \mathbb{R}$ given by $g(x) = 1 - x$.

(d) The function REV: $C \times C \rightarrow C \times C$ [for C a set] defined by REV$(x, y) = (y, x)$.

8. Let A be a subset of some set S and consider the characteristic function χ_A of A. Find $\chi_A^{\leftarrow}(1)$ and $\chi_A^{\leftarrow}(0)$.

9. Let $f: S \rightarrow T$ and $g: T \rightarrow U$ be invertible functions. Show that $g \circ f$ is invertible and that $(g \circ f)^{-1} = f^{-1} \circ g^{-1}$.

10. Let $f: S \rightarrow T$ be an invertible function. Show that f^{-1} is invertible and that $(f^{-1})^{-1} = f$.

11. Consider functions $f: S \to T$ and $g: T \to S$ such that $g \circ f = 1_S$. Nontrivial examples of such pairs of functions appear in Exercises 5 and 6.

(a) Prove that f is one-to-one.   (b) Prove that g maps T onto S.

12. Consider the function $f: \mathbb{R} \times \mathbb{R} \to \mathbb{R} \times \mathbb{R}$ defined by

$$f(x, y) = (x + y, x - y).$$

(a) Prove that f is one-to-one on $\mathbb{R} \times \mathbb{R}$.

(b) Prove that f maps $\mathbb{R} \times \mathbb{R}$ onto $\mathbb{R} \times \mathbb{R}$.

(c) Find the inverse function f^{-1}.

(d) Find the composite functions $f \circ f^{-1}$ and $f \circ f$.

13. Let $f: S \to T$.

(a) Show that $f(f^{\leftarrow}(B)) \subseteq B$ for subsets B of T.

(b) Show that $A \subseteq f^{\leftarrow}(f(A))$ for subsets A of S.

(c) Show that $f^{\leftarrow}(B_1 \cap B_2) = f^{\leftarrow}(B_1) \cap f^{\leftarrow}(B_2)$ for subsets B_1 and B_2 of T.

(d) Under what conditions on B does equality hold in part (a)?

14. Let $f: S \to T$. Prove or disprove. If false, a single example will suffice.

(a) $f(A_1 \cap A_2) = f(A_1) \cap f(A_2)$ for subsets A_1 and A_2 of S.

(b) $f(A_1 \backslash A_2) = f(A_1) \backslash f(A_2)$ for subsets A_1 and A_2 of S.

(c) $f(A_1) = f(A_2)$ implies $A_1 = A_2$.

15. (a) One can show that if $f: T \to U$ is one-to-one and if $g: S \to T$ and $h: S \to T$ satisfy $f \circ g = f \circ h$, then $g = h$. Give examples of functions f, g and h for which $f \circ g = f \circ h$ but $g \neq h$.

(b) Give examples of functions f, g and h for which $g \circ f = h \circ f$ but $g \neq h$.

(c) Give a condition on f so that $g \circ f = h \circ f$ implies $g = h$.

16. Suppose that the function $f: S \to T$ is invertible and $B \subseteq T$.

(a) Show that if $y \in B$, then $f^{-1}(y) \in f^{\leftarrow}(B)$.

(b) Show that if $x \in f^{\leftarrow}(B)$, then $x \in f^{-1}(B)$.

(c) Show that $f^{\leftarrow}(B) = f^{-1}(B)$.

§ 1.5   Sequences

This section is concerned with lists of things. Subscript notation comes in handy in this context and whenever we deal with large collections of objects; here "large" often means "more than 3 or 4." For example, letters x, y and z are adequate when dealing with equations involving three or fewer unknowns. But if there are ten unknowns, or if we wish to discuss the general situation of n unknowns for some unspecified integer n in $\mathbb{P}$, then $x_1, x_2, \ldots, x_n$ would be a good choice for the names of the unknowns. Here we distinguish the unknowns by the little numbers $1, 2, \ldots, n$ written below the x's, which are called **subscripts**. As another example, a general nonzero polynomial has the form

$$a_n x^n + a_{n-1} x^{n-1} + \cdots + a_2 x^2 + a_1 x + a_0,$$

where $a_n \neq 0$. Here n is the degree of the polynomial and the $n+1$ possible coefficients are labeled $a_0, a_1, \ldots, a_n$ using subscripts. For example, the polynomial $x^3 + 4x^2 - 73$ fits this general scheme, with $n = 3$, $a_3 = 1$, $a_2 = 4$, $a_1 = 0$ and $a_0 = -73$.

We have used the symbol Σ as a name for an alphabet. In mathematics the big Greek sigma $\sum$ has a standard use as a general summation sign. The terms following it are to be summed according to how $\sum$ is decorated. For example, consider the expression

$$\sum_{k=1}^{10} k^2.$$

The decorations "$k = 1$" and "10" tell us to add up the numbers k^2 obtained by successively setting $k = 1$, then $k = 2$, then $k = 3$, etc. on up to $k = 10$. That is,

$$\sum_{k=1}^{10} k^2 = 1 + 4 + 9 + 16 + 25 + 36 + 49 + 64 + 81 + 100 = 385.$$

The letter k is a variable [it varies from 1 to 10] that could be replaced here by any other variable. Thus

$$\sum_{k=1}^{10} k^2 = \sum_{j=1}^{10} j^2 = \sum_{r=1}^{10} r^2.$$

We can also consider more general sums like

$$\sum_{k=1}^{n} k^2$$

in which the stopping point n can take on different values. Each value of n gives a particular value of the sum; for each choice of n the variable k takes on the values from 1 to n. Here are some of the sums represented by $\sum_{k=1}^{n} k^2$.

Value of n	The sum
$n = 1$	$1^2 = 1$
$n = 2$	$1^2 + 2^2 = 1 + 4 = 5$
$n = 3$	$1^2 + 2^2 + 3^2 = 14$
$n = 4$	$1^2 + 2^2 + 3^2 + 4^2 = 30$
$n = 10$	$1^2 + 2^2 + 3^2 + 4^2 + 5^2 + 6^2 + 7^2 + 8^2 + 9^2 + 10^2 = 385$
$n = 73$	$1^2 + 2^2 + 3^2 + 4^2 + \cdots + 73^2 = 132{,}349$

We can also discuss even more general sums such as

$$\sum_{k=1}^{n} x_k \quad \text{and} \quad \sum_{j=m}^{n} a_j.$$

Here it is understood that $\{x_k : 1 \le k \le n\}$ and $\{a_j : m \le j \le n\}$ represent collections of numbers. Presumably $m \le n$, since otherwise there would be nothing to sum.

In analogy with $\sum$, the big Greek pi $\prod$ is a general product sign. For $n \in \mathbb{P}$ the product of the first n integers is called **n factorial** and is written $n!$. Thus

$$n! = 1 \cdot 2 \cdot 3 \cdots n = \prod_{k=1}^{n} k.$$

The expression $1 \cdot 2 \cdot 3 \cdots n$ is somewhat confusing for small values of n like 1 and 2; it really means "multiply consecutive integers until you reach n." The expression $\prod_{k=1}^{n} k$ is less ambiguous. Here are the first few values of $n!$: $1! = 1$, $2! = 1 \cdot 2 = 2$, $3! = 1 \cdot 2 \cdot 3 = 6$, $4! = 1 \cdot 2 \cdot 3 \cdot 4 = 24$. More values of $n!$ appear in Figures 1 and 2. For technical reasons $n!$ is also defined for $n = 0$; $0!$ is defined to be 1. The definition of $n!$ will be reexamined in §4.1.

An infinite string of objects can be listed by using subscripts from the set $\mathbb{N} = \{0, 1, 2, \ldots\}$ of natural numbers [or $\{m, m + 1, m + 2, \ldots\}$ for some integer m]. Such strings are called **sequences**. Thus a sequence on $\mathbb{N}$ is a list $s_0, s_1, \ldots, s_n, \ldots$ that has a specified value s_n for each integer $n \in \mathbb{N}$. We frequently call s_n the nth **term** of the sequence. It is often convenient to denote the sequence itself by (s_n) or $(s_n)_{n \in \mathbb{N}}$ or $(s_0, s_1, s_2, \ldots)$. Sometimes we will write $s(n)$ instead of s_n. Computer scientists commonly use the notation $s[n]$, in part because it is easy to type on a terminal.

The notation $s(n)$ looks like our notation for functions. In fact, a sequence *is* a function whose domain is the set $\mathbb{N} = \{0, 1, 2, \ldots\}$ of natural numbers or is $\{m, m + 1, m + 2, \ldots\}$ for some integer m. Each integer n in the domain of the sequence determines the value $s(n)$ of the nth term.

Our first examples will be sequences of real numbers.

EXAMPLE 1   (a) The sequence $(s_n)_{n \in \mathbb{N}}$ where $s_n = n!$ is just the sequence $(1, 1, 2, 6, 24, \ldots)$ of factorials. The *set* of values is $\{1, 2, 6, 24, \ldots\} = \{n! : n \in \mathbb{P}\}$.

(b) The sequence $(a_n)_{n \in \mathbb{N}}$ given by $a_n = (-1)^n$ for $n \in \mathbb{N}$ is the sequence $(1, -1, 1, -1, 1, -1, 1, \ldots)$ whose *set* of values is $\{-1, 1\}$. ∎

As the last example suggests, it is important to distinguish between a sequence and its set of values. We always use braces $\{\ \ \}$ to list or describe a set and never use them to describe a sequence. The sequence

$(a_n)_{n \in \mathbb{N}}$ given by $a_n = (-1)^n$ in Example 1(b) has an infinite number of terms, even though their values are repeated over and over. On the other hand, the set of values $\{(-1)^n : n \in \mathbb{N}\}$ is exactly the set $\{-1, 1\}$ consisting of two numbers.

Sequences are frequently given suggestive abbreviated names, such as SEQ, FACT, SUM, and the like.

EXAMPLE 2   (a) Let FACT$(n) = n!$ for $n \in \mathbb{N}$. This is exactly the same sequence as in Example 1(a); only its name [FACT, instead of s] has been changed. Note that FACT$(n + 1) = (n + 1) * \text{FACT}(n)$ for $n \in \mathbb{N}$, where $*$ denotes multiplication of integers.

(b) For $n \in \mathbb{N}$, let TWO$(n) = 2^n$. Then TWO is a sequence. Note that TWO$(n + 1) = 2 * \text{TWO}(n)$ for $n \in \mathbb{N}$. ■

Our definition of sequence allows the domain to be any set of the form $\{m, m + 1, m + 2, \ldots\}$ where m is an integer.

EXAMPLE 3   (a) The sequence (b_n) given by $b_n = 1/n^2$ for $n \geq 1$ clearly needs to have its domain avoid the value $n = 0$. The first few values of the sequence are $1, \frac{1}{4}, \frac{1}{9}, \frac{1}{16}, \frac{1}{25}$.

(b) Consider the sequence whose nth term is $\log_2 n$. Note that $\log_2 0$ makes no sense, so this sequence must begin with $n = 1$. We have $\log_2 1 = 0$ since $2^0 = 1$, $\log_2 2 = 1$ since $2^1 = 2$, $\log_2 4 = 2$ since $2^2 = 4$, $\log_2 8 = 3$ since $2^3 = 8$, etc. The intermediate values of $\log_2 n$ can only be approximated. See Figure 1. For example, $\log_2 5 \approx 2.3219$ is only an approximation since $2^{2.3219} \approx 4.9999026$. ■

$\log_2 n$	$\sqrt{n}$	n	n^2	2^n	$n!$	n^n
0	1.0000	**1**	1	2	1	1
1.0000	1.4142	**2**	4	4	2	4
1.5850	1.7321	**3**	9	8	6	27
2.0000	2.0000	**4**	16	16	24	256
2.3219	2.2361	**5**	25	32	120	3125
2.5850	2.4495	**6**	36	64	720	46,656
2.8074	2.6458	**7**	49	128	5040	823,543
3.0000	2.8284	**8**	64	256	40,320	$1.67 \cdot 10^7$
3.1699	3.0000	**9**	81	512	362,880	$3.87 \cdot 10^8$
3.3219	3.1623	**10**	100	1,024	3,628,800	10^{10}

FIGURE 1

EXAMPLE 4   (a) We will be interested in comparing the growth rates of familiar sequences like $\log_2 n, \sqrt{n}, n^2, 2^n, n!$ and n^n. Even for relatively small values of n it seems clear from Figure 1 that n^n grows a lot faster than $n!$, which grows a lot faster than 2^n, etc., although $\log_2 n$ and $\sqrt{n}$ seem to be running close to each other. In § 1.6 we will make these ideas more precise and give arguments that don't rely on appearances based on a few calculations.

(b) We are really interested in comparing the sequences in part (a) for large values of n. See Figure 2.[1] It now appears that $\log_2 n$ does grow [a lot] slower than $\sqrt{n}$, a fact that we will verify in the next section. The growth is slower because 2^n grows [a lot] faster than n^2, and $\log_2 x$ and $\sqrt{x}$ are the inverse functions of 2^x and x^2, respectively. ■

$\log_2 n$	$\sqrt{n}$	n	n^2	2^n	$n!$	n^n
3.32	3.16	**10**	100	1,024	$3.63 \cdot 10^6$	10^{10}
6.64	10	**100**	10,000	$1.27 \cdot 10^{30}$	$9.33 \cdot 10^{157}$	10^{200}
9.97	31.62	**1,000**	10^6	$1.07 \cdot 10^{301}$	$4.02 \cdot 10^{2567}$	10^{3000}
13.29	100	**10,000**	10^8	$2.00 \cdot 10^{3010}$	$2.85 \cdot 10^{35659}$	10^{40000}
16.61	316.2	**100,000**	10^{10}	$1.00 \cdot 10^{30103}$	$2.82 \cdot 10^{456573}$	10^{500000}
19.93	1000	**10^6**	10^{12}	$9.90 \cdot 10^{301029}$	$8.26 \cdot 10^{5565708}$	$10^{6000000}$
39.86	10^6	**10^{12}**	10^{24}	big	bigger	biggest

FIGURE 2

So far, all of our sequences have had real numbers as values. However, there is no such restriction in the definition and, in fact, we will be interested in sequences with values of other sorts.

EXAMPLE 5   The following sequences have values that are sets.

(a) A sequence $(D_n)_{n \in \mathbb{N}}$ of subsets of $\mathbb{Z}$ is defined by

$$D_n = \{m \in \mathbb{Z} : m \text{ is a multiple of } n\}$$

$$= \{0, \pm n, \pm 2n, \pm 3n, \ldots\}.$$

(b) Let Σ be an alphabet. For each $k \in \mathbb{N}$, Σ^k is defined to be the set of all words in Σ^* having length k. In symbols,

$$\Sigma^k = \{w \in \Sigma^* : \text{length}(w) = k\}.$$

The sequence $(\Sigma^k)_{k \in \mathbb{N}}$ is a sequence of subsets of Σ^* whose union, $\bigcup_{k \in \mathbb{N}} \Sigma^k$, is Σ^*. Note that the sets Σ^k are disjoint, that $\Sigma^0 = \{\lambda\}$, and that $\Sigma^1 = \Sigma$. In case $\Sigma = \{a, b\}$, we have $\Sigma^0 = \{\lambda\}$, $\Sigma^1 = \Sigma = \{a, b\}$, $\Sigma^2 = \{aa, ab, ba, bb\}$, etc.

---

[1] We thank our colleague Richard M. Koch for supplying the larger values in this table. He used *Mathematica*.

The sets Σ^k appeared briefly in Example 7 of §1.4 with the names $L^{\leftarrow}(k)$ where L was the length function on Σ^*. Henceforth we will use the notation Σ^k for these sets. ■

In the last example we wrote $\bigcup_{k \in \mathbb{N}} \Sigma^k$ for the union of an infinite sequence of sets. A word is a member of this set if it belongs to one of the sets Σ^k. The sets Σ^k are disjoint, but in general we consider unions of sets that may overlap. To be specific, if $(A_k)_{k \in \mathbb{N}}$ is a sequence of sets, then we define

$$\bigcup_{k \in \mathbb{N}} A_k = \{x : x \in A_k \text{ for at least one } k \text{ in } \mathbb{N}\}.$$

This definition makes sense, of course, if the sets are defined for k in $\mathbb{P}$ or in some other set. Similarly, we define

$$\bigcap_{k \in \mathbb{N}} A_k = \{x : x \in A_k \text{ for all } k \in \mathbb{N}\}.$$

The notation $\bigcup_{k=0}^{\infty} A_k$ has a similar interpretation except that ∞ plays a special role. The notation $\bigcup_{k=0}^{\infty}$ signifies that k takes the values $0, 1, 2, \ldots$ without stopping; but k does *not* take the value ∞. Thus

$$\bigcup_{k=0}^{\infty} A_k = \{x : x \in A_k \text{ for at least one integer } k \geq 0\}$$

whereas

$$\bigcap_{k=1}^{\infty} A_k = \{x : x \in A_k \text{ for all integers } k \geq 1\}.$$

In Example 5 we could just as well have written $\Sigma^* = \bigcup_{k=0}^{\infty} \Sigma^k$.

Some lists are not infinite. A **finite sequence** is a string of objects that are listed using subscripts from a finite subset of $\mathbb{Z}$ of the form $\{m, m+1, \ldots, n\}$. Frequently, m will be 0 or 1. Such a sequence $(a_m, a_{m+1}, \ldots, a_n)$ is a function with domain $\{m, m+1, \ldots, n\}$, just as an infinite sequence $(a_m, a_{m+1}, \ldots)$ has domain $\{m, m+1, \ldots\}$.

EXAMPLE 6   (a) At the beginning of this section we mentioned general sums such as $\sum_{j=m}^{n} a_j$. The values to be summed are from the finite sequence $(a_m, a_{m+1}, \ldots, a_n)$.

(b) The digits in the base-10 representation of an integer form a finite sequence. The digit sequence of 8832 is $(8, 8, 3, 2)$ if we take the most significant digits first, but is $(2, 3, 8, 8)$ if we start at the least significant end. ■

EXERCISES 1.5

1. Calculate

   (a) $\dfrac{7!}{5!}$ (b) $\dfrac{10!}{6!4!}$ (c) $\dfrac{9!}{0!9!}$

   (d) $\dfrac{8!}{4!}$ (e) $\displaystyle\sum_{k=0}^{5} k!$ (f) $\displaystyle\prod_{j=3}^{6} j$

2. Simplify

   (a) $\dfrac{n!}{(n-1)!}$ (b) $\dfrac{(n!)^2}{(n+1)!\,(n-1)!}$

3. Calculate

   (a) $\displaystyle\sum_{k=1}^{n} 3^k$ for $n = 1, 2, 3$ and 4 (b) $\displaystyle\sum_{k=3}^{n} k^3$ for $n = 3, 4$ and 5

   (c) $\displaystyle\sum_{j=n}^{2n} j$ for $n = 1, 2$ and 5

4. Calculate

   (a) $\displaystyle\sum_{i=1}^{10} (-1)^i$ (b) $\displaystyle\sum_{k=0}^{3} (k^2 + 1)$ (c) $\left(\displaystyle\sum_{k=0}^{3} k^2\right) + 1$

   (d) $\displaystyle\prod_{n=1}^{5} (2n+1)$ (e) $\displaystyle\prod_{j=4}^{8} (j-1)$

5. (a) Calculate $\displaystyle\prod_{r=1}^{n} (r-3)$ for $n = 1, 2, 3, 4$ and 73.

   (b) Calculate $\displaystyle\prod_{k=1}^{m} \dfrac{k+1}{k}$ for $m = 1, 2$ and 3. Give a formula for this product
      for all $m \in \mathbb{P}$.

6. (a) Calculate $\displaystyle\sum_{k=0}^{n} 2^k$ for $n = 1, 2, 3, 4$ and 5.

   (b) Use your answers to part (a) to guess a general formula for this sum.

7. Consider the sequence given by $a_n = \dfrac{n-1}{n+1}$ for $n \in \mathbb{P}$.

   (a) List the first six terms of this sequence.

   (b) Calculate $a_{n+1} - a_n$ for $n = 1, 2, 3$.

   (c) Show that $a_{n+1} - a_n = \dfrac{2}{(n+1)(n+2)}$ for $n \in \mathbb{P}$.

8. Consider the sequence given by $b_n = \frac{1}{2}[1 + (-1)^n]$ for $n \in \mathbb{N}$.

   (a) List the first seven terms of this sequence.

   (b) What is its set of values?

9. For $n \in \mathbb{N}$, let $\text{SEQ}(n) = n^2 - n$.

   (a) Calculate $\text{SEQ}(n)$ for $n \le 6$.

(b) Show that $\text{SEQ}(n + 1) = \text{SEQ}(n) + 2n$ for all $n \in \mathbb{N}$.

(c) Show that $\text{SEQ}(n + 1) = \dfrac{n + 1}{n - 1} * \text{SEQ}(n)$ for $n \geq 2$.

10. For $n = 1, 2, 3, \ldots$, let $\text{SSQ}(n) = \displaystyle\sum_{i=1}^{n} i^2$.

(a) Calculate $\text{SSQ}(n)$ for $n = 1, 2, 3$ and 5.

(b) Observe that $\text{SSQ}(n + 1) = \text{SSQ}(n) + (n + 1)^2$ for $n \geq 1$.

(c) It turns out that $\text{SSQ}(73) = 132{,}349$. Use this to calculate $\text{SSQ}(74)$ and $\text{SSQ}(72)$.

11. For the following sequences, write the first several terms until the behavior of the sequence is clear.

(a) $a_n = [2n - 1 + (-1)^n]/4$ for $n \in \mathbb{N}$.

(b) (b_n) where $b_n = a_{n+1}$ for $n \in \mathbb{N}$ and a_n is as in part (a).

(c) $\text{VEC}(n) = (a_n, b_n)$ for $n \in \mathbb{N}$.

12. Find the values of the sequences $\log_2 n$ and $\sqrt{n}$ for $n = 16, 64, 256$, and 4096, and compare.

13. (a) Using a calculator or other device, complete the table in Figure 3. [Write E if the calculation is beyond the capability of your calculator.]

(b) Discuss the apparent relative growth behaviors of n^4, 4^n, n^{20}, 20^n and $n!$.

n	n^4	4^n	n^{20}	20^n	$n!$
5			$9.54 \cdot 10^{13}$	$3.2 \cdot 10^6$	
10				$1.02 \cdot 10^{13}$	$3.63 \cdot 10^6$
25	$3.91 \cdot 10^5$				
50		$1.27 \cdot 10^{30}$			

FIGURE 3

14. Repeat Exercise 13 for the table in Figure 4.

n	$\log_{10} n$	$\sqrt{n}$	$20 \cdot \sqrt[4]{n}$	$\sqrt[4]{n} \cdot \log_{10} n$
50	1.70	7.07	53.18	4.52
100				
10^4				
10^6				

FIGURE 4

15. Note the exponents that appear in the 2^n column of Figure 2. Note also that $\log_{10} 2 \approx .30103$. What's going on here?

§ 1.6 Big-Oh Notation

One way that sequences arise naturally in computer science is as lists of successive computed values; the sequences FACT and TWO in Example 2 of § 1.5 are of this sort. Another important application of sequences, especially later when we analyze algorithms, is to the problem of estimating how long a computation will take for a given input.

For example, think of sorting a list of n given integers into increasing order. There are lots of algorithms available for doing this job; you can probably think of several different methods yourself. Some algorithms are faster than others, and all of them take more time as n gets larger. If n is small, it probably doesn't make much difference which method we choose, but for large values of n a good choice may lead to a substantial saving in time. We need a way to describe the time behavior of our algorithms.

In our sorting example, say the sequence t measures the time a particular algorithm takes, so $t(n)$ is the time to sort a list of length n. On a faster computer we might cut all the values of $t(n)$ by a factor of 2 or 100 or even 1000. But then *all* of our algorithms would get faster, too. For choosing between methods, what really matters is some measure, not of the absolute size of $t(n)$, but of the rate at which $t(n)$ grows as n gets large. Does it grow like 2^n, n^2, n or $\log_2 n$, or like some other function of n that we looked at in § 1.5?

The main point of this section is to develop notation to describe rates of growth. Before we do so, we study more closely the relationships among familiar sequences such as $\log_2 n$, $\sqrt{n}$, n, n^2 and 2^n.

EXAMPLE 1   (a) For all positive integers n we have

$$\cdots \leq \sqrt[4]{n} \leq \sqrt[3]{n} \leq \sqrt{n} \leq n \leq n^2 \leq n^3 \leq n^4 \leq \cdots .$$

Of course, other exponents of n can be inserted into this string of inequalities. For example,

$$n \leq n\sqrt{n} \leq n^2 \qquad \text{for all } n;$$

recall that $n\sqrt{n} = n^{3/2}$.

(b) We have $n < 2^n$ for all $n \in \mathbb{N}$. Actually, we have $n \leq 2^{n-1}$ for all n; this is clear for small values of n like 1, 2 and 3. In general,

$$n = 2 \cdot \frac{3}{2} \cdot \frac{4}{3} \cdot \frac{5}{4} \cdots \frac{n-1}{n-2} \cdot \frac{n}{n-1}.$$

There are $n - 1$ factors on the right and each of them is bounded by 2, so $n \leq 2^{n-1}$.

(c) $n^2 \leq \frac{9}{8} \cdot 2^n$ for $n \geq 1$. This is easily checked for $n = 1, 2, 3$ and 4. Note that we get equality with $n = 3$. For $n > 4$ observe that

$$n^2 = 4^2 \cdot \left(\frac{5}{4}\right)^2 \cdots \left(\frac{n-1}{n-2}\right)^2 \cdots \left(\frac{n}{n-1}\right)^2.$$

Each factor on the right except the first one is at most $(\frac{5}{4})^2$, and there are $n - 4$ such factors. Since $(\frac{5}{4})^2 = 1.5625 < 2$,

$$n^2 < 4^2 \cdot 2^{n-4} = 2^4 \cdot 2^{n-4} = 2^n \qquad \text{if} \quad n > 4. \quad \blacksquare$$

EXAMPLE 2   (a) Since $n \leq 2^{n-1}$ for all n in $\mathbb{N}$, we have

$$\log_2 n \leq \log_2 2^{n-1} = n - 1 \qquad \text{for} \quad n \geq 1.$$

We can use this fact to show that

$$\log_2 x < x \qquad \text{for all real numbers} \quad x > 0;$$

see Figure 1. Indeed, if n is the least integer bigger than x, then we have $n - 1 \leq x < n$, so $\log_2 x < \log_2 n \leq n - 1 \leq x$.

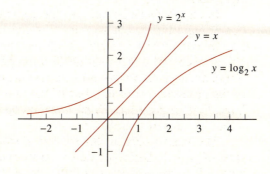

FIGURE 1

(b) In Example 1(c) we saw that $n^2 < 2^n$ for $n > 4$. In fact, given any positive constant m, we have

$$n^m < 2^n \qquad \text{for all sufficiently large } n.$$

To see this, first note that $\frac{1}{2} \log_2 n = \log_2 n^{1/2} = \log_2 \sqrt{n} < \sqrt{n}$ by (a), and so

$$\log_2 n^m = m \cdot \log_2 n < 2m \cdot \sqrt{n} = \frac{2m}{\sqrt{n}} \cdot n.$$

Now $2m/\sqrt{n} \leq 1$ for $n \geq 4m^2$, and so

$$\log_2 n^m < n \qquad \text{for} \quad n \geq 4m^2.$$

Hence we have

$$n^m < 2^n \qquad \text{for} \quad n \geq 4m^2.$$

(c) Let m be a positive integer. From part (b) we have

$$\log_2 n^m < n \qquad \text{for} \quad n \geq 4m^2.$$

This inequality holds even if n is not an integer, so we can replace n by $\sqrt[m]{n}$ to obtain $\log_2 (\sqrt[m]{n})^m < \sqrt[m]{n}$ for $\sqrt[m]{n} \geq 4m^2$, i.e.,

$$\log_2 n < \sqrt[m]{n} \qquad \text{for} \quad n \geq (4m^2)^m.$$

Thus $\log_2 n < \sqrt[m]{n}$ for sufficiently large n. ■

Roughly speaking, Example 2(b) shows that 2^n grows faster than any power of n, and Example 2(c) tells us that $\log_2 n$ grows more slowly than any root of n. Before making these ideas precise, we observe some more inequalities.

EXAMPLE 3   (a) We have

$$2^n < n! < n^n \qquad \text{for} \quad n \geq 4.$$

For $n = 4$ these inequalities are evident: $16 < 24 < 256$. For $n > 4$ we have $n! = (4!) \cdot 5 \cdot 6 \cdots (n-1) \cdot n$. The first factor $4!$ exceeds 2^4 and the remaining $n - 4$ factors each exceed 2. So $n! > 2^4 \cdot 2^{n-4} = 2^n$.

The inequality $n! < n^n$ is clear since $n!$ is a product of integers all but one of which are less than n.

(b) Let's be greedy and claim that $40^n < n!$ for sufficiently large n. This will be a little trickier than $2^n < n!$ to verify. Observe that for $n > 80$ we can write

$$n! > n(n-1) \cdots 81 \qquad [n - 80 \text{ factors}]$$

$$> 80 \cdot 80 \cdots 80 \qquad [n - 80 \text{ factors}]$$

$$= 80^{n-80} = 40^n \left\{ 2^n \cdot \frac{1}{80^{80}} \right\} > 40^n$$

provided that $2^n > 80^{80}$ or $n > \log_2 (80^{80}) = 80 \cdot \log_2 80 \approx 505.8$. This was a "crude" argument in the sense that we lost a lot [$80!$ in fact] when we wrote $n! > n(n-1) \cdots 81$. If we had wanted better information about where $n!$ actually overtakes 40^n, we could have made more careful estimates. Our work was enough, though, to show that $40^n < n!$ for all sufficiently large n. ■

To be more precise about what we mean when we say ". . . grows like . . . for large n," we need to develop some new notation, called the **big-oh notation**. Our main use for the notation will be to describe algorithm running times.

Given two sequences, say s and t, with nonnegative real values, the statement "$s(n) = O(t(n))$" [read "$s(n)$ is big oh of $t(n)$"] is intended to mean that as n gets large the values of s are no larger than the values of some constant multiple of t.

EXAMPLE 4   (a) Example 1(a) tells us that $\sqrt{n} = O(n)$, $n = O(n^2)$, etc. Example 1(b) tells us that $n = O(2^n)$, and Example 1(c) tells us that $n^2 = O(2^n)$. In fact, $n^m = O(2^n)$ for each m by Example 2(b). Example 3(a) shows that $2^n = O(n!)$ and that $n! = O(n^n)$.

(b) The dominant term of $6n^4 + 20n^2 + 2000$ is $6n^4$, since for large n the value of n^4 is much greater than n^2 or the constant 2000. We will write this observation as

$$6n^4 + 20n^2 + 2000 = O(n^4),$$

to indicate that the expression on the left "grows no worse than a multiple of n^4." It actually grows a little bit *faster* than $6n^4$. What matters, though, is that for large enough n [in this case, $n \geq 8$ is big enough] $20n^2 + 2000 < n^4$, so $6n^4 + 20n^2 + 2000 < 7n^4$; i.e., the total quantity is no larger than some fixed multiple of n^4. We could also say, of course, that it grows no faster than a multiple of n^5, but that's not as useful a piece of information as the one we've given. ∎

Here is the precise definition. Let f and g be sequences of real numbers. We write

$$f(n) = O(g(n))$$

in case there is some positive constant C such that

$$|f(n)| \leq C \cdot |g(n)| \qquad \text{for all sufficiently large values of } n.$$

Here we are willing to have the inequality fail for a few small values of n, perhaps because $f(n)$ or $g(n)$ fails to be defined for those values. All we really care about are the large values of n. In practice, $f(n)$ will represent some sequence of current interest [such as an upper bound on the time some algorithm will run], while $g(n)$ will be some simple sequence, like n, $\log_2 n$, n^3, etc., whose growth we understand. The next theorem lists some of what we have learned in Examples 1 to 3.

Theorem 1

> Here is the hierarchy of several familiar sequences in the sense that each sequence is big-oh of any sequence to its right:
>
> $1, \log_2 n, \ldots, \sqrt[4]{n}, \sqrt[3]{n}, \sqrt{n}, n, n \cdot \log_2 n, n\sqrt{n}, n^2, n^3, n^4, \ldots, 2^n, n!, n^n.$

The constant sequence 1 in the theorem is defined by $1(n) = 1$ for all n. It doesn't grow at all.

EXAMPLE 5   (a) Suppose that $g(n) = n$ for all $n \in \mathbb{N}$. The statement $f(n) = O(n)$ means that $|f(n)|$ is bounded by a constant multiple of n, i.e., that there is some $C > 0$ so that $|f(n)| \leq Cn$ for all large enough $n \in \mathbb{N}$.

(b) Suppose that $g(n) = 1$ for all $n \in \mathbb{N}$. We say that $f(n)$ is $O(1)$ if there is a constant C such that $|f(n)| \leq C$ for all large n, that is, in case the values of $|f|$ are bounded above by some constant.

(c) The sequence s defined by $s_n = 3n^2 + 15n$ satisfies $s_n = O(n^2)$, because $n \leq n^2$ for $n \geq 1$, and thus $|s_n| \leq 3n^2 + 15n^2 = 18n^2$ for all large enough n.

(d) The sequence t given by $t_n = 3n^2 + (-1)^n 15n$ also satisfies $t_n = O(n^2)$. As in part (c), we have $|t_n| \leq 3n^2 + 15n^2 \leq 18n^2$ for $n \geq 1$.

(e) We can generalize the examples in parts (c) and (d). If $f(n) = a_m n^m + a_{m-1} n^{m-1} + \cdots + a_0$ is a polynomial in n of degree m with $a_m \neq 0$, then $|a_k n^k| \leq |a_k| \cdot n^m$ for $k = 0, 1, \ldots, m-1$, so

$$|f(n)| \leq |a_m n^m| + |a_{m-1} n^{m-1}| + \cdots + |a_0|$$

$$\leq (|a_m| + |a_{m-1}| + \cdots + |a_0|)n^m,$$

and hence $f(n) = O(n^m)$. The first inequality holds because

$$|x_1 + x_2 + \cdots + x_i| \leq |x_1| + |x_2| + \cdots + |x_i|$$

for any finite sequence $x_1, x_2, \ldots, x_i$ in $\mathbb{R}$. ∎

Because of Example 5(e) a sequence $f(n)$ is said to have **polynomial growth** if $f(n) = O(n^m)$ for some positive integer m. From a theoretical point of view, algorithms whose time behaviors have polynomial growth are regarded as manageable. They certainly are, compared to those with time behavior $O(2^n)$, say. In practice, efficient time behavior like $O(n)$ or $O(n \cdot \log_2 n)$ is most desirable. The next example concerns two sequences that arise in estimating the time behavior of algorithms.

EXAMPLE 6

(a) Let $s_n = 1 + \dfrac{1}{2} + \cdots + \dfrac{1}{n}$ for $n \geq 1$. We claim that

$$s_n = O(\log_2 n).$$

Observe that

$$s_2 = 1 + \tfrac{1}{2} < 2$$

$$s_4 = s_2 + (\tfrac{1}{3} + \tfrac{1}{4}) < 2 + (\tfrac{1}{2} + \tfrac{1}{2}) = 3$$

$$s_8 = s_4 + (\tfrac{1}{5} + \tfrac{1}{6} + \tfrac{1}{7} + \tfrac{1}{8}) < 3 + (\tfrac{1}{4} + \tfrac{1}{4} + \tfrac{1}{4} + \tfrac{1}{4}) = 4,$$

etc. In general $s_{2^k} < k + 1$. [An induction argument from §4.2 makes this completely persuasive.] Now consider any integer $n > 2$. Trap n between powers of 2, say $2^k < n \leq 2^{k+1}$. Since $k < \log_2 n$, we have

$$s_n \leq s_{2^{k+1}} < (k+1) + 1 < \log_2 n + 2.$$

If $n \geq 4$, then $\log_2 n \geq 2$, and so

$$s_n < 2 \log_2 n \qquad \text{for} \quad n \geq 4.$$

Hence $s_n = O(\log_2 n)$.

(b) If $t_n = n + \dfrac{n}{2} + \dfrac{n}{3} + \cdots + \dfrac{n}{n}$ for $n \geq 1$, then we have $t_n = n \cdot s_n$, so $t_n < 2n \cdot \log_2 n$ for $n \geq 4$, by (a). Therefore, we have $t_n = O(n \cdot \log_2 n)$. ∎

The next theorem lists some general facts about big-oh notation. Remember that writing $f(n) = O(g(n))$ just signifies that $f(n)$ is some sequence that is $O(g(n))$.

Theorem 2

(a) If $f(n) = O(g(n))$ and if c is a constant, then $c \cdot f(n) = O(g(n))$.

(b) If $f(n) = O(g(n))$ and $h(n) = O(g(n))$, then $f(n) + h(n) = O(g(n))$.

(c) If $f(n) = O(a(n))$ and $g(n) = O(b(n))$, then we have $f(n) \cdot g(n) = O(a(n) \cdot b(n))$.

(d) If $a(n) = O(b(n))$ and $b(n) = O(c(n))$, then $a(n) = O(c(n))$.

Proof. Parts (a) and (c) are left to Exercise 13.

(b) If $f(n) = O(g(n))$ and $h(n) = O(g(n))$, there exist positive constants C and D such that

$$|f(n)| \leq C \cdot |g(n)| \qquad \text{for all sufficiently large } n,$$

and

$$|h(n)| \leq D \cdot |g(n)| \qquad \text{for all sufficiently large } n.$$

Since $|x + y| \leq |x| + |y|$ for $x, y \in \mathbb{R}$ we conclude that

$$|f(n) + h(n)| \leq |f(n)| + |h(n)| \leq (C + D) \cdot |g(n)|$$

for all sufficiently large n. Consequently, we have $f(n) + h(n) = O(g(n))$.

(d) If $a(n) = O(b(n))$ and $b(n) = O(c(n))$, there exist positive constants C and D such that

$$|a(n)| \leq C \cdot |b(n)| \quad \text{and} \quad |b(n)| \leq D \cdot |c(n)|$$

for all sufficiently large n. Thus

$$|a(n)| \leq C \cdot |b(n)| \leq C \cdot D \cdot |c(n)| \qquad \text{for all sufficiently large } n,$$

and hence $a(n) = O(c(n))$. ∎

The general principles in Theorem 2 would have shortened some arguments in Examples 5 and 6. For instance, we know that $n^k = O(n^m)$ if $k \le m$, so Theorem 2 gives

$$a_m n^m + a_{m-1} n^{m-1} + \cdots + a_1 n + a_0$$

$$= O(n^m) + O(n^{m-1}) + \cdots + O(n) + O(1) \qquad \text{[by (a)]}$$

$$= O(n^m) + O(n^m) + \cdots + O(n^m) + O(n^m) \qquad \text{[by (d)]}$$

$$= O(n^m) \qquad\qquad\qquad\qquad\qquad\qquad \text{[by (b)]}.$$

We are able to use (b) here because the number of summands, $m + 1$, does not depend on n.

In Example 6(a) we were really done when we got $s_n < \log_2 n + 2$, since

$$\log_2 n + 2 = O(\log_2 n) + O(1) = O(\log_2 n).$$

Example 6(b) is immediate, because

$$t_n = n \cdot s_n = O(n \cdot \log_2 n)$$

by Theorem 2(c).

Looking at Example 4(b) and Example 5, we see that it might be useful to describe a sequence by giving its dominant term plus terms that contribute less for large n. Thus we might write

$$6n^4 + 20n^2 + 1000 = 6n^4 + O(n^2),$$

where the big-oh term here stands for the expression $20n^2 + 1000$, which we know satisfies the condition $20n^2 + 1000 = O(n^2)$. This usage of the big-oh notation in expressions such as $f(n) + O(g(n))$ is slightly different in spirit from the usage in equations like $f(n) = O(g(n))$, but both are consistent with an interpretation that $O(g(n))$ represents "a sequence whose values are bounded by some constant multiple of $g(n)$ for all large enough n." It is this meaning that we shall use from now on.

Just as we made sense out of $a(n) + O(b(n))$, we can write $a(n) \cdot O(b(n))$ to mean "$a(n) \cdot f(n)$ where $f(n) = O(b(n))$." With this interpretation we can write "equations" like those in the next theorem.

Theorem 3

For any sequences $a(n)$ and $b(n)$, we have

(a) $O(a(n)) + O(b(n)) = O(\max\{|a(n)|, |b(n)|\})$.

(b) $O(a(n)) \cdot O(b(n)) = O(a(n) \cdot b(n))$.

These statements simply say: If $f(n) = O(a(n))$ and $g(n) = O(b(n))$, then $f(n) + g(n) = O(\max\{|a(n)|, |b(n)|\})$ and $f(n) \cdot g(n) = O(a(n) \cdot b(n))$. For instance, we have $O(n^3) + O(n^4) = O(n^4)$, since $\max\{n^3, n^4\} = n^4$, and $O(n^3) \cdot O(n^4) = O(n^7)$. Unlike true equations, statements like these really don't mean much if we read them from right to left.

Proof of Theorem 3. (a) Let $f(n) = O(a(n))$ and $g(n) = O(b(n))$. Then there exist positive constants C and D such that

$$|f(n)| \le C \cdot |a(n)| \quad \text{and} \quad |g(n)| \le D \cdot |b(n)| \qquad \text{for sufficiently large } n.$$

Then we have

$$|f(n) + g(n)| \le |f(n)| + |g(n)| \le C \cdot |a(n)| + D \cdot |b(n)|$$
$$\le C \cdot \max\{|a(n)|, |b(n)|\} + D \cdot \max\{|a(n)|, |b(n)|\}$$
$$= (C + D) \cdot \max\{|a(n)|, |b(n)|\}$$

for sufficiently large n. Therefore, $f(n) + g(n) = O(\max\{|a(n)|, |b(n)|\})$.

Part (b) is immediate from Theorem 2(c). ∎

EXAMPLE 7 (a) Since $n^2 + 13n = O(n^2)$ and $(n + 1)^3 = O(n^3)$ [think about this], $(n^2 + 13n) + (n + 1)^3 = O(n^3)$ by Theorem 3(a), and $(n^2 + 13n)(n + 1)^3 = O(n^5)$ by Theorem 3(b).

(b) If $a(n) = O(n^4)$ and $b(n) = O(\log_2 n)$, then we have $a(n) \cdot b(n) = O(n^4 \cdot \log_2 n)$, $a(n)^2 = O(n^8)$ and $b(n)^2 = O(\log_2^2 n)$. Note that $\log_2^2 n$ denotes $(\log_2 n)^2$.

(c) As a consequence of Example 5(e), we can also write

$$\sum_{k=0}^{m} a_k n^k = a_m n^m + O(n^{m-1}) \qquad \text{if} \quad a_m \ne 0. \quad ∎$$

EXERCISES 1.6

1. For each sequence below find the smallest number k such that $f(n) = O(n^k)$. You need not prove that your k is best possible.

   (a) $f(n) = 13n^2 + 4n - 73$ (b) $f(n) = (n^2 + 1)(2n^4 + 3n - 8)$

   (c) $f(n) = (n^3 + 3n - 1)^4$ (d) $\sqrt{n + 1}$

2. Repeat Exercise 1 for

   (a) $f(n) = (n^2 - 1)^7$ (b) $f(n) = \sqrt{n^2 - 1}$

   (c) $f(n) = \sqrt{n^2 + n}$ (d) $f(n) = (n^2 + n + 1)^2 \cdot (n^3 + 5)$

3. For each sequence below give the sequence $a(n)$ in the hierarchy of Theorem 1 such that $f(n) = O(a(n))$ and such that $a(n)$ is as far to the left as possible in the hierarchy.

   (a) $f(n) = 3^n$ (b) $f(n) = n^3 \cdot \log_2 n$ (c) $f(n) = \sqrt{\log_2 n}$

4. Repeat Exercise 3 for

   (a) $f(n) = n + 3 \cdot \log_2 n$ (b) $f(n) = (n \cdot \log_2 n + 1)^2$

   (c) $f(n) = (n + 1)!$

5. State whether each of the following is true or false. In each case give a reason for your answer.

   (a) $2^{n+1} = O(2^n)$

   (b) $(n+1)^2 = O(n^2)$

   (c) $2^{2n} = O(2^n)$

   (d) $(200n)^2 = O(n^2)$

6. Repeat Exercise 5 for

   (a) $\log_2^{73} n = O(\sqrt{n})$

   (b) $\log_2(n^{73}) = O(\log_2 n)$

   (c) $\log_2 n^n = O(\log_2 n)$

   (d) $(\sqrt{n} + 1)^4 = O(n^2)$

7. True or false. In each case give a reason.

   (a) $40^n = O(2^n)$

   (b) $(40n)^2 = O(n^2)$

   (c) $(2n)! = O(n!)$

   (d) $(n+1)^{40} = O(n^{40})$

8. Let A be a positive constant. Show that $A^n < n!$ for sufficiently large n. *Hint:* Analyze Example 3(b).

9. (a) For $n \in \mathbb{P}$ let $s_n = \sum_{k=1}^{n} \frac{1}{k^2}$, so that $s_n = 1 + \frac{1}{4} + \frac{1}{9} + \cdots + \frac{1}{n^2}$ for $n \geq 4$.

   Show that $s_n = O(1)$. *Hint:* Show that $\frac{1}{k^2} \leq \left(\frac{1}{k-1} - \frac{1}{k} \right)$ for $k \geq 2$, so that

   $$s_n \leq 1 + \sum_{k=2}^{n} \left(\frac{1}{k-1} - \frac{1}{k} \right) = 2 - \frac{1}{n}.$$

   (b) Show that $t_n = O(n^2)$ where $t_n = \sum_{k=1}^{n} k = 1 + 2 + \cdots + n$.

10. (a) Show that if $s_n = \sum_{k=1}^{n} k^2$, then $s_n = O(n^3)$.

    (b) Fix m in $\mathbb{P}$ and define $t_n = \sum_{k=1}^{n} k^m$. Show that $t_n = O(n^{m+1})$.

11. Show that if $f(n) = 3n^4 + O(n)$ and $g(n) = 2n^3 + O(n)$, then

    (a) $f(n) + g(n) = 3n^4 + O(n^3)$

    (b) $f(n) \cdot g(n) = 6n^7 + O(n^5)$

12. Show that

    (a) $(5n^3 + O(n^2)) \cdot (3n^4 + O(n^3)) = 15n^7 + O(n^6)$

    (b) $(5n^3 + O(n)) \cdot (3n^4 + O(n^2)) = 15n^7 + O(n^5)$

13. Explain why parts (a) and (c) of Theorem 2 are true.

14. (a) Prove directly from Theorem 2 that if $a(n) = O(c(n))$ and $b(n) = O(c(n))$, then $O(a(n)) + O(b(n)) = O(c(n))$.

    (b) Observe that this gives another proof of Theorem 3(a).

15. This exercise shows that one must be careful using division in big-oh calculations.

    (a) Let $a(n) = n^5$ and $b(n) = n$. Observe that $a(n) = O(n^5)$ and $b(n) = O(n^2)$ but that $a(n)/b(n)$ is not $O(n^3)$.

    (b) Give examples of sequences $a(n)$ and $b(n)$ such that $a(n) = O(n^6)$ and $b(n) = O(n^2)$ but $a(n)/b(n)$ is not $O(n^4)$.

16. Show that $\log_{10} n = O(\log_2 n)$.

17. For each $n \in \mathbb{P}$ let $\text{DIGIT}(n)$ be the number of digits in the decimal expansion of n.

    (a) Show that $10^{\text{DIGIT}(n)-1} \le n < 10^{\text{DIGIT}(n)}$.

    (b) Show that $\log_{10} n$ is $O(\text{DIGIT}(n))$.

    (c) Show that $\text{DIGIT}(n)$ is $O(\log_{10} n)$.

    (d) Let $\text{DIGIT2}(n)$ be the number of digits in the binary expansion of n. How are $O(\text{DIGIT}(n))$ and $O(\text{DIGIT2}(n))$ related?

CHAPTER HIGHLIGHTS

To check your understanding of the material in this chapter, we recommend that you consider each item listed below and:

   (a) Satisfy yourself that you can define each concept and notation and can describe each method.

   (b) Give at least one reason why the item was included in this chapter.

   (c) Think of at least one example of each concept and at least one situation in which each fact or method would be useful.

This chapter is introductory and contains a relatively large number of fundamental definitions and notations. It is important to be comfortable with this material now, since the rest of the book is built on it.

CONCEPTS

set [undefined]
   member = element, subset
   equal, disjoint
   set operations
   universe, complement
   Venn diagram
ordered pair, product of sets
alphabet, language, word, length of word
function = map = mapping
   domain, codomain
   image of x, image of $f = \text{Im}(f)$
   graph of a function
   one-to-one, onto, one-to-one correspondence
   composition of functions
   inverse function

restriction, $f(A)$
pre-image, $f^{\leftarrow}(B)$
sequence, finite sequence
"big-oh" notation

EXAMPLES AND NOTATION

$\mathbb{N}, \mathbb{P}, \mathbb{Z}, \mathbb{Q}, \mathbb{R}$

$\in, \notin, \{\ :\ \}, \subseteq, \subset$

$\varnothing = \{\quad\} = $ empty set

$[a, b], (a, b), [a, b), (a, b]$ notation for intervals

$\mathcal{P}(S), \cup, \cap, A \backslash B, A \oplus B, A^c$

(s, t) notation for ordered pairs, $S \times T$, S^n

$|S| = $ number of elements in the set S

$\sum$ notation for sums, $\prod$ notation for products

$n!$ for n factorial

$\Sigma^*, \lambda = $ empty word

special functions $1_S, \chi_A, \log_b$

$f(n) = O(g(n)), f(n) = g(n) + O(h(n))$

FACTS

Basic laws of set algebra [Table 1 of §1.2].

Composition of functions is associative.

A function is invertible if and only if it is a one-to-one correspondence.

Comparative growth rates of common sequences [Figure 2 of §1.5, Theorem 1 of §1.6].

METHODS

Use of Venn diagrams.

Reasoning from definitions and previously established facts.

ELEMENTARY LOGIC

2

This chapter contains an informal introduction to logic, both as a set of tools for building arguments and as an object of study in its own right. Everyone who depends on making inferences needs to be able to recognize and distinguish between valid and invalid arguments. Mathematicians place a great deal of emphasis on creating proofs that are logically watertight. Computer scientists need to be able to reason logically, of course, but in addition they need to know the formal rules of logic that their machines follow. Our emphasis will be on logic as a working tool. We will try to be informal, but we will also hint at what would be involved in a careful formal development of the subject. We will also develop some of the symbolic techniques required for computer logic. The connection with hardware and logical circuits will be made more explicit in Chapter 10.

Section 2.1 introduces some terminology and common notation, including the useful quantifiers ∀ and ∃. In §2.2 we give the basic framework of the propositional calculus. The general concepts are important and should be mastered. Some rather intimidating tables, which need not be memorized, are provided for easy reference. Ultimately, the purpose of proofs is to communicate by providing convincing arguments. We discuss proofs as encountered in the "real" world in §2.3, formalize the ideas in §2.4, and then return to analyze informal arguments in §2.5.

§ 2.1   Informal Introduction

Propositional calculus is the study of the logical relationships between propositions, which are usually interpretable as meaningful assertions in real-life contexts. For us, a **proposition** will be any sentence that is either true or false, but not both. That is, it is a sentence that can be assigned the truth value **true** or the truth value **false**, and not both. We do not need to know what its truth value is in order to consider a proposition.

EXAMPLE 1   The following are propositions:

(a) Julius Caesar was president of the United States.
(b) Russia is the world's largest country in area.
(c) $2 + 2 = 4$.
(d) $2 + 3 = 7$.
(e) The number 4 is positive and the number 3 is negative.
(f) If a set has n elements, then it has 2^n subsets.
(g) $2^n + n$ is a prime number for infinitely many n.
(h) Every even integer greater than 4 is the sum of two prime numbers.

Note that propositions (c) and (d) are mathematical sentences, where "$=$" serves as the verb "equals" or "is equal to." Proposition (e) is false, since 3 is not negative. If this is not clear now, it will become clear soon, since (e) is the compound proposition: "4 is positive *and* 3 is negative." We have no idea whether proposition (g) is true or false, although some mathematicians may know the answer. On the other hand, as of the writing of this book *no one knows* whether proposition (h) is true; its truth is known as "Goldbach's conjecture." ■

EXAMPLE 2   Here are some more propositions:

(a) $x + y = y + x$ for all $x, y \in \mathbb{R}$.
(b) $2^n = n^2$ for some $n \in \mathbb{N}$.
(c) It is not true that 3 is an even integer or 7 is a prime.
(d) If the world is flat, then $2 + 2 = 4$.

Proposition (a) is really an infinite set of propositions covered by the phrase "Every" or "for all." Proposition (b) is a special sort of proposition because of the phrase "for some." Propositions of these types will be discussed later in this section and studied systematically in Chapter 13. Proposition (c) is a somewhat confusing compound proposition whose truth value will be easy to analyze after the study of this chapter. Our propositional calculus will allow us to construct propositions like that in (d), even when they may appear silly or even paradoxical. ■

EXAMPLE 3   The following sentences are not propositions:

(a) Your place or mine?
(b) Why is induction important?
(c) Go directly to jail.
(d) Either that dog goes or I do.
(e) $x - y = y - x$.

The trouble with (d) is that only the person making the statement can know whether or not it is true. If the truth value is known, this is a perfectly good proposition. The reason that sentence (e) is not a proposition is that the symbols are not specified. If the intention is

(e') $x - y = y - x$ for all $x, y \in \mathbb{R}$,

then this is a false proposition. If the intention is

(e'') $x - y = y - x$ for some $x, y \in \mathbb{R}$,

or

(e''') $x - y = y - x$ for all x, y in $\{0\}$,

then this is a true proposition. The problem of unspecified symbols will be dealt with at the end of this section. ■

EXAMPLE 4   Of course, in the real world there are ambiguous propositions:

(a) Teachers are overpaid.
(b) Doctors are rich.
(c) It was cold in Minneapolis in January 1924.
(d) Math is fun.
(e) $A^2 = 0$ implies $A = 0$ for all A.

The difficulty with sentence (e) is that the set of allowable A's is not specified. The statement in (e) is true for all $A \in \mathbb{R}$. It turns out that (e) is meaningful, but false, for the set of all 2×2 matrices A. Ambiguous propositions should either be made unambiguous or abandoned. We will not concern ourselves with this process, but assume that our propositions are unambiguous. [See Exercises 17 and 18, though, for more potentially ambiguous examples.] ■

In the propositional calculus, we will generally use lowercase letters such as $p, q, r, \ldots$ to stand for propositions, and we will combine propositions to obtain compound propositions using standard connective symbols:

$\neg$ for "not" or negation;
$\wedge$ for "and";
$\vee$ for "or" [inclusive];
$\rightarrow$ for "implies" or the conditional implication;
$\leftrightarrow$ for "if and only if" or the biconditional.

Other connectives, such as $\oplus$, appear in the exercises of §§ 2.2 and 2.4. Sometimes we will be thinking of building compound propositions from simpler ones, and sometimes we will be trying to analyze a complicated proposition in terms of its constituent parts.

In the next section we will carefully discuss each of the connective symbols and explain how each affects the truth values of compound propositions. At this point, though, we will just treat the symbols informally, as a kind of abbreviation for English words or phrases, and we will try to get a little experience using them to modify propositions or to link propositions together. Our first illustration shows how some of the propositions in Examples 1 and 2 can be viewed as compound propositions.

EXAMPLE 5   (a) Recall proposition (e) of Example 1: "The number 4 is positive and the number 3 is negative." This can be viewed as the compound proposition $p \wedge q$ [read "p and q"] where $p =$ "4 is positive" and $q =$ "3 is negative."

(b) Proposition (d) of Example 2, "If the world is flat, then $2 + 2 = 4$," can be viewed as the compound proposition $r \rightarrow s$ [read "r implies s"] where $r =$ "the world is flat" and $s =$ "$2 + 2 = 4$."

(c) Proposition (c) of Example 2 says "It is not true that 3 is an even integer or 7 is a prime." This is $\neg(p \vee q)$ where $p =$ "3 is even" and $q =$ "7 is a prime." Actually, this proposition is poorly written and can also be interpreted to mean $(\neg p) \vee q$. When we read (c) aloud as "not p or q" we need to make it clear somehow where the parentheses go. Is it not p, or is it not both?

(d) The English word "unless" can be interpreted symbolically in several ways. For instance, "We eat at six unless the train is late" means "We eat at six or the train is late," or "If the train is not late, then we eat at six," or even "If we do not eat at six, then the train is late." ∎

The compound proposition $p \rightarrow q$ is read "p implies q," but has several other English language equivalents, such as "if p, then q." In fact, in Example 5(b) the compound proposition $r \rightarrow s$ was a translation of "if r, then s." So that proposition could have been written: "The world is flat implies that $2 + 2 = 4$." Other English language equivalents for $p \rightarrow q$ are: "p only if q," "q if p." We will usually avoid these; but see Exercises 15 to 18.

Compound propositions of the form $p \rightarrow q$, $q \rightarrow p$, $\neg p \rightarrow \neg q$, etc., appear to be related and are sometimes confused with each other. It is important to keep them straight. The proposition $q \rightarrow p$ is called the **converse** of the proposition $p \rightarrow q$. As we will see, it has a different meaning from $p \rightarrow q$. It turns out that $p \rightarrow q$ is equivalent to $\neg q \rightarrow \neg p$, which is called the **contrapositive** of $p \rightarrow q$.

EXAMPLE 6 Consider the sentence: "If it is raining, then there are clouds in the sky." This is the compound proposition $p \to q$ where $p =$ "it is raining" and $q =$ "there are clouds in the sky." This is a true proposition. Its converse $q \to p$ reads: "If there are clouds in the sky, then it is raining." Fortunately, this is a false proposition. The contrapositive $\neg q \to \neg p$ says: "If there are no clouds in the sky, then it is not raining." Not only is this a true proposition, but most people would agree that this follows "logically" from $p \to q$, without having to think again about the physical connection between rain and clouds. It does follow, and this logical connection will be made more precise in §2.2 [Table 1, item 9]. ∎

In logic we are concerned with determining the truth values of propositions from related propositions. Example 6 illustrates that the truth of $p \to q$ does not imply that $q \to p$ is true, but it suggests that the truth of the contrapositive $\neg q \to \neg p$ follows from that of $p \to q$. Here is another illustration of why one must be careful in manipulating logical expressions.

EXAMPLE 7 Consider the argument "If Tom doesn't go to work tomorrow, he will not keep his job. He will go to work tomorrow, so he will keep his job." We are not concerned with whether Tom actually keeps his job [that's Tom's problem], but with whether Tom's keeping his job follows *logically* from the previous two assertions: "If Tom doesn't go to work tomorrow, he will not keep his job," and "He will go to work tomorrow." It turns out that this reasoning is not valid. The first sentence only tells us that Tom is in trouble if he doesn't go to work tomorrow; it tells us nothing otherwise. Perhaps Tom will lose his job for incompetence. If the reasoning above were valid, the following would also be: "If Carol doesn't buy a lottery ticket, then she will not win \$1,000,000. Carol does buy a lottery ticket, so she will win \$1,000,000."

Symbolically, these invalid arguments both take the form: If $\neg p \to \neg q$ and p are true, then q is true. The propositional calculus we develop in §§2.2 and 2.4 will provide a formal framework with which to analyze the validity of arguments such as these. ∎

The truth of $p \to q$ is sometimes described by saying that p is a **sufficient condition** for q or that q is a **necessary condition** for p. Saying that p is a necessary and sufficient condition for q is another way of saying that $p \leftrightarrow q$ is true.

EXAMPLE 8 (a) To pass this course it is necessary to work hard. That is, pass $\to$ work hard is true. Hard work is not sufficient, though. We know of examples to show that work hard $\to$ pass is not always true.

(b) To kill a fly it is sufficient to hit it with a cannonball, but it is not necessary. Thus cannonball → wipeout is true, but wipeout → cannonball is not.

(c) A necessary and sufficient condition for a prime number p to be even is that $p = 2$. ∎

We return now to compound propositions that include the phrase "for all" or "for every" and involve several propositions, possibly an infinite number.

EXAMPLE 9   Consider again Goldbach's conjecture from Example 1: "Every even integer greater than 4 is the sum of two prime numbers." This proposition turns out to be decomposable as

$$\text{"}p(6) \wedge p(8) \wedge p(10) \wedge \cdots\text{"}$$

or

$$\text{"}p(n) \text{ for every even } n \text{ in } \mathbb{N} \text{ greater than 4"}$$

where $p(n)$ is the simple proposition "n is the sum of two prime numbers." However, the rules for connectives in the propositional calculus do not allow constructions, such as these, which involve more than a finite number of propositions or phrases like "for every" or "for some." ∎

There are two useful connectives from what is called the "predicate calculus" that will allow us to symbolize propositions like those in Example 9. We will discuss the logic of these connectives, called **quantifiers**, in more detail in Chapter 13; for now we just treat them as informal abbreviations.

Suppose that $\{p(x) : x \in U\}$ is a family of propositions, where U is a set that may be infinite. In other words, p is a proposition-valued function defined on the set U. The **universal quantifier ∀**, an upside-down A as in "for All," is used to build compound propositions of the form

$$\forall x \, p(x),$$

which we read as "for all x, $p(x)$." Other translations of ∀ are "for each," "for every," "for any." Beware of "for any," though, as it might be misinterpreted as meaning "for some." The phrase "if you have income for any month" is not the same as "if you have income for every month." The compound proposition $\forall x \, p(x)$ is assigned truth values as follows:

$\forall x \, p(x)$ is true if $p(x)$ is true for every x in U; otherwise, $\forall x \, p(x)$ is false.

The **existential quantifier** ∃, a backward E as in "there Exists," is used to form propositions like

$$\exists x\, p(x),$$

which we read as "there exists an x such that $p(x)$," "there is an x such that $p(x)$," or "for some x, $p(x)$." The compound proposition $\exists x\, p(x)$ has these truth values:

> $\exists x\, p(x)$ is true if $p(x)$ is true for at least one x in U;
> $\exists x\, p(x)$ is false if $p(x)$ is false for every x in U.

EXAMPLE 10   (a) For each n in $\mathbb{N}$ let $p(n)$ be the proposition "$n^2 = n$." Then $\forall n\, p(n)$ is false because, for example, $p(3)$, i.e., $3^2 = 3$, is false. On the other hand, $\exists n\, p(n)$ is true because at least one proposition $p(n)$ is true; in fact, exactly two of them are true, namely $p(0)$ and $p(1)$.

For $n \in \mathbb{N}$, let $q(n)$ be the proposition "$(n + 1)^2 = n^2 + 2n + 1$." We can use $p(n)$, $q(n)$ and connectives from the propositional calculus to obtain other quantified propositions. For example, $\exists n[\neg q(n)]$ is a proposition which is false; every proposition $q(n)$ is true, so every proposition $\neg q(n)$ is false. The proposition $\forall n[p(n) \vee q(n)]$ is true because each proposition $p(n) \vee q(n)$ is true. [Remember, $\vee$ means "or."] Of course, the weaker proposition $\exists n[p(n) \vee q(n)]$ is also true.

(b) Let $p(x)$ be "$x \leq 2x$" and $q(x)$ be "$x^2 \geq 0$" for $x \in \mathbb{R}$. Since $\mathbb{R}$ cannot be listed as a sequence, it would really be impossible to symbolize $\exists x\, p(x)$ or $\forall x\, q(x)$ in the propositional calculus. Clearly, $\exists x\, p(x)$ is true; $\forall x\, p(x)$ is false because $p(x)$ is false for negative x. Both $\forall x\, q(x)$ and $\exists x\, q(x)$ are true and $\exists x[\neg q(x)]$ is false.

(c) Quantifiers are also useful when one is dealing with a finite, but large, set of propositions. Suppose, for example, that we have propositions $p(n)$ for n in the set $\{n \in \mathbb{N} : 0 \leq n \leq 65{,}535\}$. The notation $\forall n\, p(n)$ is clearly preferable to

$$p(0) \wedge p(1) \wedge p(2) \wedge p(3) \wedge \cdots \wedge p(65{,}535)$$

although we might invent the acceptable $\displaystyle\bigwedge_{n=0}^{65535} p(n)$.

(d) The Goldbach conjecture can now be written as $\forall n\, p(n)$ is true, where $p(n) =$ "if n in $\mathbb{N}$ is even and greater than 4, then n is the sum of two prime numbers." ■

EXAMPLE 11   (a) The quantifiers $\forall$ and $\exists$ are often used informally as abbreviations. The first two propositions in Example 2 might be written as "$x + y = y + x \; \forall x, y \in \mathbb{R}$" and "$\exists n \in \mathbb{N}$ so that $2^n = n^2$."

(b) In practice, it is common to omit understood quantifiers. The associative and cancellation laws for $\mathbb{R}$ are often written

(A) $$(x + y) + z = x + (y + z),$$

(C) $$xz = yz \quad \text{and} \quad z \neq 0 \quad \text{imply} \quad x = y.$$

The intended meanings are

(A) $$\forall x \, \forall y \, \forall z [(x + y) + z = x + (y + z)],$$

(C) $$\forall x \, \forall y \, \forall z [(xz = yz \land z \neq 0) \rightarrow x = y],$$

where x, y and z are in $\mathbb{R}$. In everyday usage (A) might also be written as

$$(x + y) + z = x + (y + z) \quad \forall x \, \forall y \, \forall z,$$

or

$$(x + y) + z = x + (y + z) \quad \forall x, y, z \in \mathbb{R},$$

or

$$(x + y) + z = x + (y + z) \quad \text{for all} \quad x, y, z \in \mathbb{R}. \quad \blacksquare$$

We will often be able to prove propositions $\forall n \, p(n)$, where $n \in \mathbb{N}$, by using an important technique, mathematical induction, which we describe in Chapter 4.

A compound proposition of the form $\forall x \, p(x)$ will be false if any one [or more] of its propositions $p(x)$ is false. So to **disprove** such a compound proposition it is enough to show that one of its propositions is false. In other words, it is enough to supply an example that is counter to, or contrary to, the general proposition, i.e., a **counterexample**.

Goldbach's conjecture is still unsettled because no one has been able to prove that *every* even integer greater than 4 is the sum of two primes, and no one has found a counterexample. The conjecture has been verified for a great many even integers.

EXAMPLE 12   (a) The number 2 provides a counterexample to the assertion "All prime numbers are odd numbers."

(b) The number 7 provides a counterexample to the statement "Every positive integer is the sum of three squares of integers." Incidentally, it can be proved that every positive integer is the sum of four squares of integers, e.g., $1 = 1^2 + 0^2 + 0^2 + 0^2$, $7 = 2^2 + 1^2 + 1^2 + 1^2$, $73 = 8^2 + 3^2 + 0^2 + 0^2$.

(c) The value $n = 3$ provides a counterexample to the statement: "$n^2 \leq 2^n$ for all $n \in \mathbb{N}$," which we may write as "$n^2 \leq 2^n \, \forall n \in \mathbb{N}$." There are no other counterexamples, as we showed in Example 1(c) of § 1.6.

(d) Gerald Ford is a counterexample to the assertion: "All American presidents have been right-handed." There are three other counterexamples. $\blacksquare$

Given a general assertion whose truth value is unknown, often the only strategy is to make a guess and go with it. If you guess that the assertion is true, then analyze the situation to see why it always seems to be true. This analysis might lead you to a proof. If you fail to find a proof and you can see why you have failed, you might discover a counterexample. Then again, if you can't find a counterexample, you might begin to suspect once more that the result is true and formulate reasons why it must be. It is very common, especially on difficult problems, to spend considerable efforts trying to establish *each* of the two possibilities, until one wins out. One of the authors spent a good deal of energy searching for a counterexample to a result that he felt was false, only to have a young Englishman later provide a proof that it was true.

EXERCISES 2.1

1. Let p, q, r be the following propositions:

$$p = \text{``it is raining,''}$$

$$q = \text{``the sun is shining,''}$$

$$r = \text{``there are clouds in the sky.''}$$

Translate the following into logical notation, using p, q, r and logical connectives.

(a) It is raining and the sun is shining.

(b) If it is raining, then there are clouds in the sky.

(c) If it is not raining, then the sun is not shining and there are clouds in the sky.

(d) The sun is shining if and only if it is not raining.

(e) If there are no clouds in the sky, then the sun is shining.

2. Let p, q, r be as in Exercise 1. Translate the following into English sentences.

(a) $(p \wedge q) \to r$ (b) $(p \to r) \to q$

(c) $\neg p \leftrightarrow (q \vee r)$ (d) $\neg(p \leftrightarrow (q \vee r))$

(e) $\neg(p \vee q) \wedge r$

3. (a) Give the truth values of the propositions in parts (a) to (e) of Example 1.

(b) Do the same for parts (a) and (b) of Example 2.

4. Which of the following are propositions? Give the truth values of the propositions.

(a) $x^2 = x \, \forall x \in \mathbb{R}$.   (b) $x^2 = x$ for some $x \in \mathbb{R}$.

(c) $x^2 = x$.   (d) $x^2 = x$ for exactly one $x \in \mathbb{R}$.

(e) $xy = xz$ implies $y = z$.

(f) $xy = xz$ implies $y = z$ $\forall x, y, z \in \mathbb{R}$.

(g) $w_1 w_2 = w_1 w_3$ implies $w_2 = w_3$ for all words $w_1, w_2, w_3 \in \Sigma^*$.

5. Consider the ambiguous sentence "$x^2 = y^2$ implies $x = y$ $\forall x, y$."

(a) Make the sentence into an unambiguous proposition whose truth value is true.

(b) Make the sentence into an unambiguous proposition whose truth value is false.

6. Give the converses of the following propositions.

(a) $q \rightarrow r$.   (b) If I am smart, then I am rich.

(c) If $x^2 = x$, then $x = 0$ or $x = 1$.   (d) If $2 + 2 = 4$, then $2 + 4 = 8$.

7. Give the contrapositives of the propositions in Exercise 6.

8. (a) Verify that Goldbach's conjecture is true for some small values such as 6, 8 and 10.

(b) Do the same for 98.

9. (a) Show that the value $n = 3$ provides a counterexample to the assertion "$n^3 < 3^n$ $\forall n \in \mathbb{N}$."

(b) Can you find any other counterexamples?

10. (a) Show that $(m, n) = (4, -4)$ gives a counterexample to the assertion: "If m, n are nonzero integers that divide each other, then $m = n$."

(b) Give another counterexample.

11. (a) Show that $x = -1$ is a counterexample to "$(x + 1)^2 \geq x^2$ $\forall x \in \mathbb{R}$."

(b) Find another counterexample.

(c) Can a nonnegative number serve as a counterexample? Explain.

12. Find counterexamples to the following assertions.

(a) $2^n - 1$ is prime for every $n \geq 2$.

(b) $2^n + 3^n$ is prime $\forall n \in \mathbb{N}$.

(c) $2^n + n$ is prime for every positive odd integer n.

13. (a) Give a counterexample to: "$x > y$ implies $x^2 > y^2$ $\forall x, y \in \mathbb{R}$." Your answer should be an ordered pair (x, y).

(b) How might you restrict x and y so that the proposition in part (a) is true?

14. Let S be a nonempty set. Determine which of the following assertions are true. For the true ones, give a reason. For the false ones, provide a counterexample.

(a) $A \cup B = B \cup A$ $\forall A, B \in \mathscr{P}(S)$.

(b) $(A \backslash B) \cup B = A$ $\forall A, B \in \mathscr{P}(S)$.

(c) $(A \cup B) \backslash A = B$ $\forall A, B \in \mathscr{P}(S)$.

(d) $(A \cap B) \cap C = A \cap (B \cap C)$ $\forall A, B, C \in \mathscr{P}(S)$.

15. Even though we will normally use "implies" and "if . . . , then" to describe implication, other word orders and phrases often arise in practice, as in the examples below. Let p, q and r be the propositions:

$$p = \text{"the flag is set,"}$$

$$q = \text{"}I = 0\text{,"}$$

$$r = \text{"subroutine } S \text{ is completed."}$$

Translate each of the following propositions into symbols, using the letters p, q, r and the logical connectives.

(a) If the flag is set, then $I = 0$.

(b) Subroutine S is completed if the flag is set.

(c) The flag is set if subroutine S is not completed.

(d) Whenever $I = 0$ the flag is set.

(e) Subroutine S is completed only if $I = 0$.

(f) Subroutine S is completed only if $I = 0$ or the flag is set. Note the ambiguity; there are two different answers, each with its own claim to validity. Would punctuation help?

16. Consider the following propositions:

$$r = \text{"ODD}(N) = T\text{,"}$$

$$m = \text{"the output goes to the monitor,"}$$

$$p = \text{"the output goes to the printer."}$$

Translate the following, as in Exercise 15.

(a) The output goes to the monitor if $\text{ODD}(N) = T$.

(b) The output goes to the printer whenever $\text{ODD}(N) = T$ is not true.

(c) $\text{ODD}(N) = T$ only if the output goes to the monitor.

(d) The output goes to the monitor if the output goes to the printer.

(e) $\text{ODD}(N) = T$ or the output goes to the monitor if the output goes to the printer.

17. Each of the following sentences expresses an implication. Rewrite each one in the form "If p, then q."

(a) Touch those cookies if you want a spanking.

(b) Touch those cookies and you'll be sorry.

(c) You leave or I'll set the dog on you.

(d) I will if you will.

(e) I will go unless you stop that.

18. Express the contrapositive of each sentence in Exercise 17 in the form "If p, then q."

§ 2.2   **Propositional Calculus**

We have two goals in this section. We want to develop a set of formal rules for analyzing and manipulating propositions, a sort of algebra of propositions similar in some ways to the algebra of numbers, and we also want a mechanical way to calculate truth values of complicated propositions. It is the calculation aspect that has given the subject the name "calculus."

If a proposition is constructed from other propositions by using logical connectives, then its truth or falsity is completely determined by the truth values of the simpler propositions, together with the way the compound proposition is built up from them. Given propositions p and q, the truth values of the compound propositions $\neg p$, $p \wedge q$, $p \vee q$, $p \rightarrow q$ and $p \leftrightarrow q$ will be determined by the truth values of p and q. Since there are only four different combinations of truth values for p and q, we can simply give tables to describe the truth values of these compound propositions for all of the possible combinations.

One way to indicate truth values in a table would be to use the letters T and F. We have chosen instead to be consistent with the usage for Boolean variables in most computer languages and to use 1 for true and 0 for false.

The proposition $\neg p$ should be false when p is true and true when p is false. Thus our table for the connective $\neg$ is

p	$\neg p$
0	1
1	0

The column to the left of the vertical line lists the possible truth values of p. To the right of the line are the corresponding truth values for $\neg p$.

The truth table for $\wedge$ is

p	q	$p \wedge q$
0	0	0
0	1	0
1	0	0
1	1	1

Here the four possible combinations of truth values for p and q are listed to the left of the line, and the corresponding truth values of $p \wedge q$ are shown to the right. Note that $p \wedge q$ has truth value true exactly when both p *and* q are true.

As we explained at the beginning of §1.2, the use of "or" in the English language is somewhat ambiguous, but our use of $\vee$ will not be ambiguous. We define ∨ as follows:

p	q	$p \lor q$
0	0	0
0	1	1
1	0	1
1	1	1

Most people would agree with the truth value assignments for the first three lines. The fourth line states that we regard $p \lor q$ to be true if both p and q are true. This is the "inclusive or," sometimes written "and/or." Thus $p \lor q$ is true if p is true or q is true *or both*. The "exclusive or," symbolized ⊕, means that one or the other is true but not both; see Exercise 13.

   The **conditional implication** $p \rightarrow q$ means that the truth of p implies the truth of q. In other words, if p is true, then q must be true. The only way that this implication can fail is if p is true while q is false.

p	q	$p \rightarrow q$
0	0	1
0	1	1
1	0	0
1	1	1

   The first two lines of the truth table for $p \rightarrow q$ may bother some people because it looks as if false propositions imply anything. In fact, we are simply defining the *compound proposition* $p \rightarrow q$ to be true if p is false. This usage of implication appears in ordinary English. Suppose that a politician promises "If I am elected, then taxes will be lower next year." If the politician is not elected, we would surely not regard him or her as a liar, no matter how the tax rates changed.

   We will discuss the biconditional $p \leftrightarrow q$ after we introduce general truth tables.

   A **truth table** for a compound proposition built up from propositions $p, q, r, \ldots$ is a table giving the truth values of the compound proposition in terms of the truth values of $p, q, r, \ldots$. We call $p, q, r, \ldots$ the **variables** of the table and of the compound proposition. One can determine the truth values of the compound proposition by determining the truth values of subpropositions working from the inside out, as we now illustrate.

EXAMPLE 1   Here is a truth table for the compound proposition $(p \land q) \lor \neg(p \rightarrow q)$. Note that there are still only four rows, because there are still only four distinct combinations of truth values for p and q.

column	1	2	3	4	5	6
	p	q	$p \wedge q$	$p \to q$	$\neg(p \to q)$	$(p \wedge q) \vee \neg(p \to q)$
	0	0	0	1	0	0
	0	1	0	1	0	0
	1	0	0	0	1	1
	1	1	1	1	0	1

The values in columns 3 and 4 are determined by the values in columns 1 and 2. The values in column 5 are determined by the values in column 4. The values in column 6 are determined by the values in columns 3 and 5. The sixth column gives the truth values of the complete compound proposition.

One can use a simpler truth table, with the same thought processes, by writing the truth values under the connectives, as follows:

p	q	$(p \wedge q)$	$\vee$	$\neg(p \to q)$	
0	0	0	0	0	1
0	1	0	0	0	1
1	0	0	1	1	0
1	1	1	1	0	1
step 1	1	2	4	3	2

The truth values at each step are determined by the values at earlier steps. For example, the values at the third step were determined by the values in the last column. The values at the fourth step were determined by the values in the third and fifth columns. The column created at the last step gives the truth values of the compound proposition. ■

The simpler truth tables become more advantageous as the compound propositions get more complicated.

EXAMPLE 2 Here is the truth table for

$$(p \to q) \wedge [(q \wedge \neg r) \to (p \vee r)].$$

p	q	r	$(p \to q)$	$\wedge$	$[(q \wedge \neg r) \to (p \vee r)]$			
0	0	0	1	1	0	1	1	0
0	0	1	1	1	0	0	1	1
0	1	0	1	0	1	1	0	0
0	1	1	1	1	0	0	1	1
1	0	0	0	0	0	1	1	1
1	0	1	0	0	0	0	1	1
1	1	0	1	1	1	1	1	1
1	1	1	1	1	0	0	1	1
step 1	1	1	2	5	3	2	4	2

Notice that the rows of a truth table could be given in any order. We've chosen a systematic order for the truth combinations of p, q, r partly to be sure that we have listed them all. ■

The **biconditional** $p \leftrightarrow q$ is defined by the truth table for $(p \to q) \wedge (q \to p)$:

p	q	$(p \to q)$	$\wedge$	$(q \to p)$
0	0	1	1	1
0	1	1	0	0
1	0	0	0	1
1	1	1	1	1
step 1	1	2	3	2

That is,

p	q	$p \leftrightarrow q$
0	0	1
0	1	0
1	0	0
1	1	1

Thus $p \leftrightarrow q$ is true if both p and q are true or if both p and q are false. The following are English language equivalents to $p \leftrightarrow q$: "p if and only if q," "p is a necessary and sufficient condition for q" and "p precisely if q."

It is worth emphasizing that the compound proposition $p \to q$ and its converse $q \to p$ are quite different; they have different truth tables.

An important class of compound propositions consists of those that are always true no matter what the truth values of the variables p, q, etc., are. Such a compound proposition is called a **tautology**. Why would we ever be interested in a proposition that is always true, and hence is pretty boring? The answer is that we are going to be dealing with some rather complicated-looking propositions that we hope to show are true, and the way in which we will show their truth will be by using other propositions that are known to be true always. Just wait. We begin with a very simple tautology.

EXAMPLE 3 (a) The classical tautology is the compound proposition $p \to p$:

p	$p \to p$
0	1
1	1

(b) The compound proposition $[p \wedge (p \rightarrow q)] \rightarrow q$ is a tautology:

p	q	$[p \wedge (p \rightarrow q)]$		$\rightarrow$	q
0	0	0	1	1	
0	1	0	1	1	
1	0	0	0	1	
1	1	1	1	1	
step	1 1	3	2	4	

(c) $\neg(p \vee q) \leftrightarrow (\neg p \wedge \neg q)$ is a tautology:

p	q	$\neg(p \vee q)$		$\leftrightarrow$	$(\neg p \wedge \neg q)$		
0	0	1	0	1	1	1	1
0	1	0	1	1	1	0	0
1	0	0	1	1	0	0	1
1	1	0	1	1	0	0	0
step	1 1	3	2	4	2	3	2

A compound proposition that is always false is called a **contradiction**. Clearly, a compound proposition P is a contradiction if and only if $\neg P$ is a tautology.

EXAMPLE 4   The classical contradiction is the compound proposition $p \wedge \neg p$:

p	p	$\wedge$	$\neg p$
0	0	0	1
1	1	0	0

Two compound propositions P and Q are regarded as **logically equivalent** if they have the same truth values for all choices of truth values of the variables p, q, etc. In other words, the final columns of their truth tables are the same. When this occurs, we write $P \Leftrightarrow Q$. Since the table for $P \leftrightarrow Q$ has truth values true precisely where the truth values of P and Q agree, we see that

$P \Leftrightarrow Q$ *if and only if* $P \leftrightarrow Q$ *is a tautology.*

Observing that $P \Leftrightarrow Q$ will be especially useful in cases where P and Q look quite different from each other. See, for instance, the formulas in Table 1.

TABLE 1. Logical Equivalences[a]

1. $\neg\neg p \Leftrightarrow p$	double negation
2a. $(p \vee q) \Leftrightarrow (q \vee p)$ b. $(p \wedge q) \Leftrightarrow (q \wedge p)$ c. $(p \leftrightarrow q) \Leftrightarrow (q \leftrightarrow p)$	commutative laws
3a. $[(p \vee q) \vee r] \Leftrightarrow [p \vee (q \vee r)]$ b. $[(p \wedge q) \wedge r] \Leftrightarrow [p \wedge (q \wedge r)]$	associative laws
4a. $[p \vee (q \wedge r)] \Leftrightarrow [(p \vee q) \wedge (p \vee r)]$ b. $[p \wedge (q \vee r)] \Leftrightarrow [(p \wedge q) \vee (p \wedge r)]$	distributive laws
5a. $(p \vee p) \Leftrightarrow p$ b. $(p \wedge p) \Leftrightarrow p$	idempotent laws
6a. $(p \vee c) \Leftrightarrow p$ b. $(p \vee t) \Leftrightarrow t$ c. $(p \wedge c) \Leftrightarrow c$ d. $(p \wedge t) \Leftrightarrow p$	identity laws
7a. $(p \vee \neg p) \Leftrightarrow t$ b. $(p \wedge \neg p) \Leftrightarrow c$	
8a. $\neg(p \vee q) \Leftrightarrow (\neg p \wedge \neg q)$ b. $\neg(p \wedge q) \Leftrightarrow (\neg p \vee \neg q)$ c. $(p \vee q) \Leftrightarrow \neg(\neg p \wedge \neg q)$ d. $(p \wedge q) \Leftrightarrow \neg(\neg p \vee \neg q)$	DeMorgan laws
9. $(p \rightarrow q) \Leftrightarrow (\neg q \rightarrow \neg p)$	contrapositive
10a. $(p \rightarrow q) \Leftrightarrow (\neg p \vee q)$ b. $(p \rightarrow q) \Leftrightarrow \neg(p \wedge \neg q)$	implication
11a. $(p \vee q) \Leftrightarrow (\neg p \rightarrow q)$ b. $(p \wedge q) \Leftrightarrow \neg(p \rightarrow \neg q)$	
12a. $[(p \rightarrow r) \wedge (q \rightarrow r)] \Leftrightarrow [(p \vee q) \rightarrow r]$ b. $[(p \rightarrow q) \wedge (p \rightarrow r)] \Leftrightarrow [p \rightarrow (q \wedge r)]$	
13. $(p \leftrightarrow q) \Leftrightarrow [(p \rightarrow q) \wedge (q \rightarrow p)]$	equivalence
14. $[(p \wedge q) \rightarrow r] \Leftrightarrow [p \rightarrow (q \rightarrow r)]$	exportation law
15. $(p \rightarrow q) \Leftrightarrow [(p \wedge \neg q) \rightarrow c]$	reductio ad absurdum

In this table, t represents any tautology and c represents any contradiction.

EXAMPLE 5   (a) In view of Example 3(c), the compound propositions $\neg(p \vee q)$ and $\neg p \wedge \neg q$ are logically equivalent. That is, $\neg(p \vee q) \Leftrightarrow (\neg p \wedge \neg q)$. To say that I will not go for a walk *or* watch television is the same as saying that I will not go for a walk *and* I will not watch television.

(b) The very nature of the connectives $\vee$ and $\wedge$ suggests that $p \vee q \Leftrightarrow q \vee p$ and $p \wedge q \Leftrightarrow q \wedge p$. Of course, one can verify these assertions by showing that $(p \vee q) \leftrightarrow (q \vee p)$ and $(p \wedge q) \leftrightarrow (q \wedge p)$ are tautologies. ∎

It is worth stressing the difference between ↔ and ⇔. The expression "$P \leftrightarrow Q$" simply represents some compound proposition that might or might not be a tautology. The expression "$P \Leftrightarrow Q$" is an *assertion about propositions*, namely that P and Q are logically equivalent, i.e., that $P \leftrightarrow Q$ *is* a tautology.

Table 1 lists a number of logical equivalences selected for their usefulness. To obtain a table of tautologies, simply replace each ⇔ by ↔. These tautologies can all be verified by truth tables. However, most of them should be intuitively reasonable.

Many of the entries in the table have names, which we have given, but there is no need to memorize most of them. The logical equivalences 2, 3 and 4 have familiar names. Equivalences 8, the DeMorgan laws, and 9, the **contrapositive** rule, come up often enough to make their names worth remembering.

Notice that in the DeMorgan laws the proposition on one side of the ⇔ has an ∧, while the one on the other side has an ∨. We will see in §2.4 that we can use the DeMorgan laws to replace a given proposition by a logically equivalent one in which some or all of the ∧'s have been converted into ∨'s, or vice versa.

Given two compound propositions P and Q, we say that P **logically implies** Q in case Q has truth value true whenever P has truth value true. We write $P \Rightarrow Q$ when this occurs. Note that

$P \Rightarrow Q$ *if and only if the compound proposition* $P \to Q$ *is a tautology.*

Equivalently, $P \Rightarrow Q$ means that P and Q never simultaneously have the truth values 1 and 0, respectively; when P is true, Q is true, and when Q is false, P is false.

EXAMPLE 6   (a) We have $[p \land (p \to q)] \Rightarrow q$ since $[p \land (p \to q)] \to q$ is a tautology by Example 3(b).

(b) The statement $(A \land B) \Rightarrow C$ means that $(A \land B) \to C$ is a tautology. Since $(A \land B) \to C \Leftrightarrow A \to (B \to C)$ by Rule 14 [exportation], $(A \land B) \to C$ is a tautology if and only if $A \to (B \to C)$ is a tautology, i.e., if and only if $A \Rightarrow (B \to C)$. Thus the statements $(A \land B) \Rightarrow C$ and $A \Rightarrow (B \to C)$ mean the same thing. ∎

In Table 2 we list some useful logical implications. Each entry becomes a tautology if ⇒ is replaced by →. As with Table 1, many of the implications have names that need not be memorized.

To check the logical implication $P \Rightarrow Q$ it is only necessary to look for rows of the truth table where P is true and Q is false. If there are any, then $P \Rightarrow Q$ is not true. Otherwise, $P \Rightarrow Q$ is true. Thus we can ignore the rows in which P is false, as well as the rows in which Q is true.

TABLE 2. Logical Implications

16.	$p \Rightarrow (p \vee q)$	addition
17.	$(p \wedge q) \Rightarrow p$	simplification
18.	$(p \rightarrow c) \Rightarrow \neg p$ [c any contradiction]	absurdity
19.	$[p \wedge (p \rightarrow q)] \Rightarrow q$	modus ponens
20.	$[(p \rightarrow q) \wedge \neg q] \Rightarrow \neg p$	modus tollens
21.	$[(p \vee q) \wedge \neg p] \Rightarrow q$	disjunctive syllogism
22.	$p \Rightarrow [q \rightarrow (p \wedge q)]$	
23.	$[(p \leftrightarrow q) \wedge (q \leftrightarrow r)] \Rightarrow (p \leftrightarrow r)$	transitivity of $\leftrightarrow$
24.	$[(p \rightarrow q) \wedge (q \rightarrow r)] \Rightarrow (p \rightarrow r)$	transitivity of $\rightarrow$ *or* hypothetical syllogism
25a.	$(p \rightarrow q) \Rightarrow [(p \vee r) \rightarrow (q \vee r)]$	
b.	$(p \rightarrow q) \Rightarrow [(p \wedge r) \rightarrow (q \wedge r)]$	
c.	$(p \rightarrow q) \Rightarrow [(q \rightarrow r) \rightarrow (p \rightarrow r)]$	
26a.	$[(p \rightarrow q) \wedge (r \rightarrow s)] \Rightarrow [(p \vee r) \rightarrow (q \vee s)]$ ⎫	constructive dilemmas
b.	$[(p \rightarrow q) \wedge (r \rightarrow s)] \Rightarrow [(p \wedge r) \rightarrow (q \wedge s)]$ ⎬	
27a.	$[(p \rightarrow q) \wedge (r \rightarrow s)] \Rightarrow [(\neg q \vee \neg s) \rightarrow (\neg p \vee \neg r)]$ ⎫	destructive dilemmas
b.	$[(p \rightarrow q) \wedge (r \rightarrow s)] \Rightarrow [(\neg q \wedge \neg s) \rightarrow (\neg p \wedge \neg r)]$ ⎬	

EXAMPLE 7   (a) We verify the implication $(p \wedge q) \Rightarrow p$. We need only consider the case where $p \wedge q$ is true, i.e., both p and q are true. Thus we consider the truncated table

p	q	$(p \wedge q)$	$\rightarrow$	p
1	1	1		1

(b) To verify $\neg p \Rightarrow (p \rightarrow q)$ we need only look at the cases in which $\neg p$ is true, i.e., in which p is false. The truncated table is:

p	q	$\neg p$	$\rightarrow$	$(p \rightarrow q)$
0	0	1	1	1
0	1	1	1	1

Even quicker, we could just consider the case in which $p \rightarrow q$ is false, i.e., the case p true and q false. The only row is:

p	q	$\neg p$	$\rightarrow$	$(p \rightarrow q)$
1	0	0	1	0

(c) We verify the implication 26a. The full truth table would require 16 rows. However, we need only consider the cases where the implication $(p \vee r) \rightarrow (q \vee s)$ might be false. Thus it is enough to look at the cases for which $q \vee s$ is false, that is, with both q and s false.

p q r s	[(p → q) ∧ (r → s)]	→	[(p ∨ r) → (q ∨ s)]
0 0 0 0	1 1 1	1	0 1 0
0 0 1 0	1 0 0	1	1 0 0
1 0 0 0	0 0 1	1	1 0 0
1 0 1 0	0 0 0	1	1 0 0
step 1 1 1 1	2 3 2	4	2 3 2	∎

Just from the definition of $\Leftrightarrow$ we can see that if $P \Leftrightarrow Q$ and $Q \Leftrightarrow R$, then $P \Leftrightarrow R$ and all three propositions are logically equivalent to each other. If we have a chain $P_1 \Leftrightarrow P_2 \Leftrightarrow \cdots \Leftrightarrow P_n$, then all the propositions P_i are equivalent. Similarly [Exercise 23(a)], if $P \Rightarrow Q$ and $Q \Rightarrow R$, then $P \Rightarrow R$. The symbols $\Leftrightarrow$ and $\Rightarrow$ behave somewhat analogously to the symbols $=$ and $\geq$ in algebra. We sometimes say that P is **stronger than** Q or Q is **weaker than** P in case $P \Rightarrow Q$.

EXERCISES 2.2

1. Give the converse and contrapositive for each of the following propositions.
   (a) $p \to (q \wedge r)$.
   (b) If $x + y = 1$, then $x^2 + y^2 \geq 1$.
   (c) If $2 + 2 = 4$, then $3 + 3 = 8$.

2. Consider the proposition "if $x > 0$, then $x^2 > 0$." Here $x \in \mathbb{R}$.
   (a) Give the converse and contrapositive of the proposition.
   (b) Which of the following are true propositions: The original proposition, its converse, its contrapositive?

3. Consider the following propositions:

$$p \to q, \qquad \neg p \to \neg q, \qquad q \to p, \qquad \neg q \to \neg p,$$

$$q \wedge \neg p, \qquad \neg p \vee q, \qquad \neg q \vee p, \qquad p \wedge \neg q.$$

   (a) Which proposition is the converse of $p \to q$?
   (b) Which proposition is the contrapositive of $p \to q$?
   (c) Which propositions are logically equivalent to $p \to q$?

4. Determine the truth values of the following compound propositions.
   (a) If $2 + 2 = 4$, then $2 + 4 = 8$.
   (b) If $2 + 2 = 5$, then $2 + 4 = 8$.
   (c) If $2 + 2 = 4$, then $2 + 4 = 6$.
   (d) If $2 + 2 = 5$, then $2 + 4 = 6$.
   (e) If the earth is flat, then Julius Caesar was the first president of the United States.

(f) If the earth is flat, then George Washington was the first president of the United States.

(g) If George Washington was the first president of the United States, then the earth is flat.

(h) If George Washington was the first president of the United States, then $2 + 2 = 4$.

5. Suppose that $p \to q$ is known to be false. Give the truth values for

   (a) $p \wedge q$ (b) $p \vee q$ (c) $q \to p$

6. Construct truth tables for

   (a) $p \wedge \neg p$ (b) $p \vee \neg p$

   (c) $p \leftrightarrow \neg p$ (d) $\neg\neg p$

7. Construct truth tables for

   (a) $\neg(p \wedge q)$ (b) $\neg(p \vee q)$

   (c) $\neg p \wedge \neg q$ (d) $\neg p \vee \neg q$

8. Construct the truth table for $(p \to q) \to [(p \vee \neg q) \to (p \wedge q)]$.

9. Construct the truth table for $[(p \vee q) \wedge r] \to (p \wedge \neg q)$.

10. Construct the truth table for $[(p \leftrightarrow q) \vee (p \to r)] \to (\neg q \wedge p)$.

11. Construct truth tables for

    (a) $\neg(p \vee q) \to r$ (b) $\neg((p \vee q) \to r)$

    This exercise shows that one must be careful with parentheses. We will discuss this issue further in §7.2, particularly in Exercise 15 of that section.

12. In which of the following statements is the "or" an "inclusive or"?

    (a) Choice of soup or salad.

    (b) To enter the university, a student must have taken a year of chemistry or physics in high school.

    (c) Publish or perish.

    (d) Experience with FORTRAN or PASCAL is desirable.

    (e) The task will be completed on Thursday or Friday.

    (f) Discounts are available to persons under 20 or over 60.

    (g) No fishing or hunting allowed.

    (h) The school will not be open in July or August.

13. The **exclusive or** connective $\oplus$ [the computer scientists' **XOR**] is defined by the truth table

p	q	$p \oplus q$
0	0	0
0	1	1
1	0	1
1	1	0

(a) Show that $p \oplus q$ has the same truth tables as $\neg(p \leftrightarrow q)$.

(b) Construct a truth table for $p \oplus p$.

(c) Construct a truth table for $(p \oplus q) \oplus r$.

(d) Construct a truth table for $(p \oplus p) \oplus p$.

14. (a) Write a compound proposition that is true when exactly one of the three propositions p, q and r is true.

(b) Write a compound proposition that is true when exactly two of the three propositions p, q, and r are true.

15. (a) Rewrite (g) and (h) of Exercise 12 using DeMorgan's law.

(b) Does DeMorgan's law hold using the exclusive or $\oplus$ instead of $\vee$? Discuss.

16. Prove or disprove.

(a) $[p \rightarrow (q \rightarrow r)] \Leftrightarrow [(p \rightarrow q) \rightarrow (p \rightarrow r)]$

(b) $[p \oplus (q \rightarrow r)] \Leftrightarrow [(p \oplus q) \rightarrow (p \oplus r)]$

(c) $[(p \rightarrow q) \rightarrow r] \Leftrightarrow [p \rightarrow (q \rightarrow r)]$

(d) $[(p \leftrightarrow q) \leftrightarrow r] \Leftrightarrow [p \leftrightarrow (q \leftrightarrow r)]$

(e) $(p \oplus q) \oplus r \Leftrightarrow p \oplus (q \oplus r)$

17. Verify the following logical equivalences using truth tables.

(a) rule 12a

(b) the exportation law, rule 14

(c) rule 15

18. Verify the following logical implications using truth tables.

(a) modus tollens, rule 20

(b) disjunctive syllogism, rule 21

19. Verify the following logical implications using truth tables and shortcuts as in Example 7.

(a) rule 25b   (b) rule 25c   (c) rule 26b

20. Prove or disprove the following. Note that only *one* line of the truth table is needed to show that a proposition is *not* a tautology.

(a) $(q \rightarrow p) \Leftrightarrow (p \wedge q)$ (b) $(p \wedge \neg q) \Rightarrow (p \rightarrow q)$

(c) $(p \wedge q) \Rightarrow (p \vee q)$

21. A logician told her son "If you don't finish your dinner, you will not get to stay up and watch TV." He finished his dinner and then was sent straight to bed. Discuss.

22. Consider the statement "Concrete does not grow if you do not water it."

(a) Give the contrapositive.

(b) Give the converse.

(c) Give the converse of the contrapositive.

(d) Which among the original statement and the ones in parts (a), (b) and (c) are true?

23. (a) Show that if $A \Rightarrow B$ and $B \Rightarrow C$, then $A \Rightarrow C$.

(b) Show that if $P \Leftrightarrow Q$, $Q \Rightarrow R$ and $R \Leftrightarrow S$, then $P \Rightarrow S$.

(c) Show that if $P \Rightarrow Q$, $Q \Rightarrow R$ and $R \Rightarrow P$, then $P \Leftrightarrow Q$.

§ 2.3   Methods of Proof

The constant emphasis on logic and proofs is what sets mathematics apart from other pursuits. The proofs used in everyday working mathematics are based on the logical framework that we have introduced using the propositional calculus. In §2.4 we will develop a concept of formal proof, using some of the symbolism of the propositional calculus, and in §2.5 we will apply the formal development to analyze typical real-life arguments.

This section contains a preliminary, informal account of some common methods of proof and the standard terminology that goes with them. We are typically faced with a set of hypotheses $H_1, \ldots, H_n$ from which we want to infer the conclusion C. One of the most natural sorts of proof is the **direct proof** in which we show

(1) $$H_1 \wedge H_2 \wedge \cdots \wedge H_n \Rightarrow C.$$

The proofs that we gave in Chapter 1 were mainly of this sort.

One type of **indirect proof** is a proof of the **contrapositive**

(2) $$\neg C \Rightarrow \neg(H_1 \wedge H_2 \wedge \cdots \wedge H_n).$$

According to rule 9 in Table 1 of §2.2, the implication (2) is true if and only if the implication (1) is true, so a proof of (2) will let us deduce C from $H_1, \ldots, H_n$.

EXAMPLE 1   Let $m, n \in \mathbb{N}$. We wish to prove that if $m + n \geq 73$, then $m \geq 37$ or $n \geq 37$. To do this, we prove the contrapositive: not "$m \geq 37$ or $n \geq 37$" implies not "$m + n \geq 73$." By DeMorgan's law, the negation of "$m \geq 37$ or $n \geq 37$" is "not $m \geq 37$ and not $n \geq 37$," i.e., "$m \leq 36$ and $n \leq 36$." So the contrapositive proposition is: If $m \leq 36$ and $n \leq 36$, then $m + n \leq 72$. This proposition follows immediately from a general property about inequalities: $a \leq c$ and $b \leq d$ imply that $a + b \leq c + d$ for real numbers a, b, c, d. ■

Another type of **indirect proof** is a **proof by contradiction**:

(3) $$H_1 \wedge H_2 \wedge \cdots \wedge H_n \wedge \neg C \Rightarrow \text{a contradiction.}$$

Rule 15 in Table 1 of §2.2 tells us that (3) is true if and only if the implication (1) is true.

EXAMPLE 2   We wish to prove that $\sqrt{2}$ is irrational. That is, if x is in $\mathbb{R}$ and $x^2 = 2$, then x is not a rational number. The property of irrationality is a negative sort of property and not easily verified directly. We can show, however, that the hypotheses $x^2 = 2$ and x rational [i.e., not irrational] together lead to a contradiction.

Assume that $x \in \mathbb{R}$, that $x^2 = 2$, and that x is rational. Then by the definition of rational number, we have $x = p/q$ with $p, q \in \mathbb{Z}$ and $q \neq 0$. By reducing the fraction if necessary, we may assume that p and q have no common factors. In particular, p and q are not both even. Since $2 = x^2 = p^2/q^2$ we have $p^2 = 2q^2$, and so p^2 is even. This implies that p is even, as we will show in Example 4. Hence $p = 2k$ for some $k \in \mathbb{Z}$. Then $(2k)^2 = 2q^2$ and therefore $2k^2 = q^2$. Thus q^2 and q are also even. But then p and q are both even, contradicting our earlier statement. Hence $\sqrt{2}$ is irrational. ∎

EXAMPLE 3   We prove by contradiction that there are infinitely many primes. Thus assume that there are finitely many primes, say k of them. We write them as $p_1, p_2, \ldots, p_k$ so that $p_1 = 2$, $p_2 = 3$, etc. Consider $n = 1 + p_1 p_2 \cdots p_k$. Since $n > p_j$ for all $j = 1, 2, \ldots, k$, the integer n itself is not prime. However, n is a product of primes [this believable fact is not trivial to prove and, in fact, we provide a proof later in Example 1 of §4.5]. Therefore, at least one of the p_j's must divide n. Since each p_j divides $n - 1$, at least one p_j divides both $n - 1$ and n, but this is impossible. Indeed, if p_j divides both $n - 1$ and n, it divides their difference, 1, which is absurd. ∎

One should avoid artificial proofs by contradiction such as in the next example.

EXAMPLE 4   We prove by contradiction that the product of two odd integers is an odd integer. Assume that $m, n \in \mathbb{N}$ are odd integers but that mn is even. There exist k, l in $\mathbb{N}$ so that $m = 2k + 1$ and $n = 2l + 1$. Then

$$mn = 4kl + 2k + 2l + 1 = 2(2kl + k + l) + 1,$$

an odd number, contradicting the assumption that mn is even.

This proof by contradiction is artificial because we did not use the assumption "mn is even" until after we established that "mn is odd." The following direct proof is far preferable.

Consider odd integers m, n, in $\mathbb{N}$. There exist $k, l \in \mathbb{N}$ so that $m = 2k + 1$ and $n = 2l + 1$. Then

$$mn = 4kl + 2k + 2l + 1 = 2(2kl + k + l) + 1,$$

which is odd. ∎

An implication of the form

$$H_1 \vee H_2 \vee \cdots \vee H_n \Rightarrow C$$

is equivalent to

$$(H_1 \Rightarrow C) \quad \text{and} \quad (H_2 \Rightarrow C) \quad \text{and} \quad \cdots \quad \text{and} \quad (H_n \Rightarrow C)$$

[for $n = 2$, compare rule 12a, Table 1 of §2.2] and hence can be proved by **cases**, i.e., by proving each of $H_1 \Rightarrow C, \dots, H_n \Rightarrow C$ separately. The next example illustrates how boring and repetitive a proof by cases can be.

EXAMPLE 5   Recall that the **absolute value** $|x|$ of x in $\mathbb{R}$ is defined by the rule:

$$|x| = \begin{cases} x & \text{if } x \geq 0 \\ -x & \text{if } x < 0. \end{cases}$$

Assuming the familiar order properties of $\leq$ on $\mathbb{R}$, we prove that

$$|x + y| \leq |x| + |y| \qquad \text{for} \quad x, y \in \mathbb{R}.$$

We consider four cases: (i) $x \geq 0$ and $y \geq 0$; (ii) $x \geq 0$ and $y < 0$; (iii) $x < 0$ and $y \geq 0$; (iv) $x < 0$ and $y < 0$.

Case (i). If $x \geq 0$ and $y \geq 0$, then $x + y \geq 0$, so $|x + y| = x + y = |x| + |y|$.

Case (ii). If $x \geq 0$ and $y < 0$, then

$$x + y < x + 0 = |x| \leq |x| + |y|$$

and

$$-(x + y) = -x + (-y) \leq 0 + (-y) = |y| \leq |x| + |y|.$$

Either $|x + y| = x + y$ or $|x + y| = -(x + y)$; either way we conclude that $|x + y| \leq |x| + |y|$ by the inequalities above.

Case (iii). The case $x < 0$ and $y \geq 0$ is similar to Case (ii).

Case (iv). If $x < 0$ and $y < 0$, then $x + y < 0$ and $|x + y| = -(x + y) = -x + (-y) = |x| + |y|$.
So in all four cases, $|x + y| \leq |x| + |y|$. ∎

EXAMPLE 6   For every $n \in \mathbb{N}$, $n^3 + n$ is even. We can prove this fact by cases.

Case (i). Suppose that n is even. Then $n = 2k$ for some $k \in \mathbb{N}$, so

$$n^3 + n = 8k^3 + 2k = 2(4k^3 + k),$$

which is even.

Case (ii). Suppose that n is odd; then $n = 2k + 1$ for some $k \in \mathbb{N}$, so

$$n^3 + n = (8k^3 + 12k^2 + 6k + 1) + (2k + 1) = 2(4k^3 + 6k^2 + 4k + 1),$$

which is even.

Here is a more elegant proof by cases. Given n in $\mathbb{N}$, we have $n^3 + n = n(n^2 + 1)$. If n is even, so is $n(n^2 + 1)$. If n is odd, then n^2 is odd, hence $n^2 + 1$ is even, and so $n(n^2 + 1)$ is even. ■

An implication $P \Rightarrow Q$ is said to be **vacuously true** or **true by default** if P is false. Example 7(b) of §2.2 showed that $\neg p \Rightarrow (p \to q)$, so if $\neg P$ is true, then $P \to Q$ is also true. A **vacuous proof** is a proof of an implication $P \Rightarrow Q$ by showing that P is false. Such implications rarely have intrinsic interest. They typically arise in proofs of general assertions by looking at cases. A case handled by a vacuous proof is usually one in which P has been ruled out; in a sense, there is no hypothesis to check. Although $P \Rightarrow Q$ is true in such a case, we learn nothing about Q.

EXAMPLE 7   (a) Consider finite sets A and B. The assertion

If A has fewer elements than B, then there is a one-to-one mapping of A onto a proper subset of B

is vacuously true if B is the empty set, because the hypothesis must be false. A vacuous proof here consists of simply observing that the hypothesis is impossible.

(b) The assertion $n \geq 4m^2$ implies $n^m < 2^n$ that we saw in Example 2(b) of §1.6 is vacuously true for $n = 0, 1, 2, 3, \ldots, 4m^2 - 1$. For these values of n, its truth does not depend on—or tell us—whether $n^m < 2^n$ is true. ■

An implication $P \Rightarrow Q$ is sometimes said to be **trivially true** if Q is true. In this case, the truth value of P is irrelevant. A **trivial proof** of $P \Rightarrow Q$ is one in which Q is shown to be true without any reference to P.

EXAMPLE 8   If x and y are real numbers such that $xy = 0$, then $(x + y)^n = x^n + y^n$ for $n \in \mathbb{P}$. This proposition is trivially true for $n = 1$; $(x + y)^1 = x^1 + y^1$ is obviously true, and this fact does not depend on the hypothesis $xy = 0$. For $n \geq 2$, the hypothesis is needed. ■

One sometimes encounters references to **constructive proofs** and **nonconstructive proofs** for the existence of mathematical objects satisfying certain properties. A constructive proof either specifies the object [a number or a matrix, say] or indicates how it can be determined by some explicit procedure or algorithm. A nonconstructive proof establishes the existence of objects by some indirect means such as a proof by contradiction, without giving directions for how to find them.

EXAMPLE 9   In Example 3 we proved by contradiction that there are infinitely many primes. We did not construct an infinite list of primes. Our proof can be revised to give a constructive procedure for building an arbitrarily long list of distinct primes, provided that we have some way of factoring integers. [This is Exercise 14.] ■

EXAMPLE 10   If the infinite sequence $a_1, a_2, a_3, \ldots$ has all of its values in the finite set S, then some two members of the sequence must have the same value. In fact, there must be some s in S such that $a_i = s$ for infinitely many values of i. We can only say that such an s exists; we can't say which element it is, nor which of the a_i's equal s, without knowing more about the sequence. Still, the existence information alone may be useful.

For instance, we claim that there are two members of the sequence $1, 2, 2^2, 2^3, 2^4, \ldots$ that differ by a multiple of the prime 7. That's pretty easy: $2^3 - 1 = 7$ by inspection, and in fact $2^4 - 2 = 16 - 2 = 2 \cdot 7$, $2^5 - 2^2 = 4 \cdot 7$, and $2^6 - 1 = 9 \cdot 7$ as well. What if we take some large prime p, though, say 8191? Must some two members of the sequence differ by a multiple of p? Yes. Think of taking each power 2^k, dividing it by p and putting it in a box according to what the remainder is. The ones with remainder 1 go in the 1's box, the ones with remainder 2 in the 2's box, and so on. There are only p boxes, corresponding to the possible remainders $0, 1, 2, \ldots, p - 1$. At least one box has at least two members in it. [This is our existential statement.] Say 2^k and 2^l are both in the m's box, so $2^k = s \cdot p + m$ and $2^l = t \cdot p + m$ for some integers s and t. Then $2^k - 2^l = s \cdot p + m - t \cdot p - m = (s - t) \cdot p$, a multiple of p.

This nonconstructive argument doesn't tell us how to find k and l, but it does convince us that they must exist. Given the existence information, we can say somewhat more. If $2^l < 2^k$, say, then $2^k - 2^l = 2^l \cdot (2^{k-l} - 1)$. Since p is an odd prime that divides this difference, $2^{k-l} - 1$ must be a multiple of p, so there must be at least two numbers in the 1's box, namely 1 and 2^{k-l}. ■

EXAMPLE 11   Every positive integer n has the form $2^k m$ where $k \in \mathbb{N}$ and m is odd. This fact can be proved in several ways that suggest the following constructive procedure. If n is odd, let $k = 0$ and $m = n$. Otherwise, divide n by 2 and apply the procedure to $n/2$. Continue until an odd number is reached. Then k will equal the number of times division by 2 was necessary. Exercise 15 asks you to check out this procedure. ■

We began this chapter with a discussion of a very limited system of logic, namely the propositional calculus. In this section we relaxed the formality in order to discuss several methods of proof encountered in this book and elsewhere. We hope that you now have a better idea of what a mathematical proof is. In §§2.4 and 2.5 we return to the propositional calculus and its applications. Chapter 13 deals with some more sophisticated aspects of logic.

Outside the realm of logic, and in particular in most of this book, proofs are communications intended to convince the reader of the truths of assertions. Logic will serve as the foundation of the process but will recede into the background. That is, it should usually not be necessary to think consciously of the logic presented in this chapter, but if a particular proof in this book or elsewhere is puzzling, you may wish to analyze it more closely. What are the exact hypotheses? Is the author using hidden assumptions? Is the author giving an indirect proof?

Finally, there is always the possibility that the author has made an error or has not stated what was intended. Maybe you can show that the assertion is false—or at least show that the reasoning is fallacious. Even some good mathematicians have made the mistake of trying to prove $P \Leftrightarrow Q$ by showing both $P \Rightarrow Q$ and $\neg Q \Rightarrow \neg P$, probably in some disguise. We will come back to these themes after the next section.

EXERCISES 2.3

In all exercises with proofs, indicate the methods of proof used.

1. Prove that the product of two even integers is a multiple of 4.

2. Prove that the product of an even integer and an odd integer is even.

3. Prove that $|xy| = |x| \cdot |y|$ for $x, y \in \mathbb{R}$.

4. Prove that $n^4 - n^2$ is divisible by 3 for all $n \in \mathbb{N}$.

5. Prove that $n^2 - 2$ is never divisible by 3 for $n \in \mathbb{N}$.

6. (a) Prove that $\sqrt{3}$ is irrational.

   (b) Prove that $\sqrt[3]{2}$ is irrational.

7. Prove or disprove:

   (a) The sum of two even integers is an even integer.

   (b) The sum of two odd integers is an odd integer.

   (c) The sum of two primes is never a prime.

   (d) The sum of three consecutive integers is divisible by 3.

   (e) The sum of four consecutive integers is divisible by 4.

   (f) The sum of five consecutive integers is divisible by 5.

8. (a) It is not known whether there are infinitely many **prime pairs**, i.e., odd primes whose difference is 2. Examples of prime pairs are (3, 5), (5, 7), (11, 13) and (71, 73). Give three more examples of prime pairs.

   (b) Prove that (3, 5, 7) is the only "prime triple." *Hint:* Given $2k + 1, 2k + 3, 2k + 5$ where $k \in \mathbb{N}$, show that one of these must be divisible by 3.

9. Prove the following assertions for a real number x and $n = 1$.

   (a) If $x \geq 0$, then $(1 + x)^n \geq 1 + nx$.

   (b) If $x^n = 0$, then $x = 0$.

   (c) If n is even, then $x^n \geq 0$.

10. Prove the result in Example 8. Use the fact that if $xy = 0$, then $x = 0$ or $y = 0$.

11. Prove that there are two different primes p and q whose last six decimal digits are the same. Is your proof constructive? If so, produce the primes.

12. Show that the sequence $a_1, a_2, \ldots$ defined by $a_n = 40^n - n!$ has a largest value. Is your proof constructive? *Hint:* Example 3(b) of § 1.6 shows that if $n > 80 \cdot \log_2 80$, then $n! > 40^n$.

13. (a) Prove that given n in $\mathbb{N}$, there exist n consecutive positive integers that are not prime; i.e., the set of prime integers has arbitrarily large gaps. *Hint:* Start with $(n + 1)! + 2$.

    (b) Is the proof constructive? If so, use it to give six consecutive nonprimes.

    (c) Give seven consecutive nonprimes.

14. Suppose that $p_1, p_2, \ldots, p_k$ is a given list of distinct primes. Explain how one could use an algorithm that factors integers into prime factors to construct a prime that is not in the list. *Suggestion:* Factor $1 + p_1 p_2 \cdots p_k$.

15. Use the procedure in Example 11 to write the following positive integers in the form $2^k m$ where $k \in \mathbb{N}$ and m is odd.

    (a) 14   (b) 73   (c) 96   (d) 1168

16. (a) The argument in Example 10 applies to $p = 5$. Find two powers of 2 that differ by a multiple of 5.

    (b) Do the same for $p = 11$.

§ 2.4   More Propositional Calculus

In this section we first discuss how to get new logical equivalences and implications from old ones without using truth tables. After that we formalize the ideas of deduction and valid reasoning.

We begin with two useful substitution rules that are natural but that must be handled with some care.

Substitution Rule (a)   If a compound proposition P is a tautology and if all occurrences of some variable of P, say p, are replaced by the same proposition E, then the resulting compound proposition P^* is also a tautology.

EXAMPLE 1   (a) According to the modus ponens rule 19 of Table 2 in § 2.2,

$$P = \text{``}[p \wedge (p \to q)] \to q\text{''}$$

is a tautology. Thus the truth table entries for P are all 1 regardless of the truth values for p and q. [See the truth table in Example 3(b) of § 2.2.] Suppose that we replace each occurrence of p by the proposition $E = "q \to r"$ to obtain the new proposition

$$P^* = "[(q \to r) \wedge ((q \to r) \to q)] \to q."$$

According to Substitution Rule (a), P^* is a tautology. Let's see why. The truth table for P^* is

q	r	$[(q \to r) \wedge ((q \to r) \to q)]$				$\to$	q
0	0	1	0	1	0		1
0	1	1	0	1	0		1
1	0	0	0	0	1		1
1	1	1	1	1	1		1

In the first two rows, E has truth value 1 [in red] and q has truth value 0. Since we replaced every p in P by E, the rest of the truth values in these rows will be exactly the same as in the $p = 1$, $q = 0$ row for the *original* proposition P. Since P is a tautology, the final truth values for P, and hence for P^*, must be 1. In the third row, E has truth value 0 and q has truth value 1, so the rest of the truth values will be exactly the same as in the $p = 0$, $q = 1$ row for P. Again the final truth value will be 1. Similarly, the fourth row corresponds to the $p = 1$, $q = 1$ row for P, so the final truth value will be 1.

Observe that as soon as we determined the truth value of the proposition E we knew which row in the truth table for P to look at. But it didn't matter. *All* the rows for the tautology P gave the final truth value 1 for P anyway, so P^* also had truth value 1. Thus P^* was also a tautology. This argument can be generalized to show that Substitution Rule (a) is always valid.

(b) If instead we replace each occurrence of q by $E = "q \to r,"$ we obtain the tautology

$$[p \wedge (p \to (q \to r))] \to (q \to r).$$

Note that the truth table for this tautology will have 8 rows. ■

Substitution Rule (a) can be used to produce new logical equivalences from old. Suppose that $A \Leftrightarrow B$ is known, for example from Table 1 of § 2.2. Then $A \leftrightarrow B$ is a tautology. If every occurrence of p in both A and B is replaced by E to yield A^* and B^*, respectively, then Rule (a) says that $A^* \leftrightarrow B^*$ is also a tautology. Hence $A^* \Leftrightarrow B^*$ is true. That is, if A and B are logically equivalent, then so are A^* and B^*. Similarly, if $A \Rightarrow B$ is true, then so is $A^* \Rightarrow B^*$.

EXAMPLE 2   (a) Here are some simple illustrations. Associative law 3a of Table 1 in §2.2 is

$$[(p \lor q) \lor r] \Leftrightarrow [p \lor (q \lor r)].$$

We may replace p by $p \land q$ everywhere to get

$$[((p \land q) \lor q) \lor r] \Leftrightarrow [(p \land q) \lor (q \lor r)]$$

and then replace all q's by p's to get

$$[((p \land p) \lor p) \lor r] \Leftrightarrow [(p \land p) \lor (p \lor r)].$$

Making the replacements in the other order would have yielded first

$$[(p \lor p) \lor r] \Leftrightarrow [p \lor (p \lor r)]$$

and finally,

$$[((p \land q) \lor (p \land q)) \lor r] \Leftrightarrow [(p \land q) \lor ((p \land q) \lor r)].$$

We may also simultaneously replace each p by $p \land q$ and each q by p to get the equivalence

$$[((p \land q) \lor p) \lor r] \Leftrightarrow [(p \land q) \lor (p \lor r)].$$

This is different from either of the equivalences obtained in the last paragraph. Simultaneous substitution may look illegal, since it's not mentioned in Substitution Rule (a), but it can be justified by turning it into a sequence of single substitutions, using temporary names. In our example we could replace each q by the temporary letter s, then each p by $p \land q$ and, finally, each s by p. Thus

$$[(p \lor q) \lor r] \Leftrightarrow [p \lor (q \lor r)]$$

becomes

$$[(p \lor s) \lor r] \Leftrightarrow [p \lor (s \lor r)],$$

which becomes

$$[((p \land q) \lor s) \lor r] \Leftrightarrow [(p \land q) \lor (s \lor r)]$$

and finally,

$$[((p \land q) \lor p) \lor r] \Leftrightarrow [(p \land q) \lor (p \lor r)].$$

(b) Implication law 10a of Table 1 in §2.2 is

$$(p \to q) \Leftrightarrow (\neg p \lor q),$$

corresponding to the tautology $(p \to q) \leftrightarrow (\neg p \lor q)$. Replacing each p by $\neg p$ and each q by $p \to q$ in this tautology gives

$$[\neg p \to (p \to q)] \leftrightarrow [\neg \neg p \lor (p \to q)],$$

which Rule (a) says is also a tautology. Hence $\neg p \to (p \to q)$ is logically equivalent to $\neg\neg p \vee (p \to q)$.

(c) Law 18, absurdity, is $(p \to c) \Rightarrow \neg p$. It corresponds to the tautology $(p \to c) \to \neg p$. Replacing p by $q \vee r$ with Rule (a) gives the tautology $((q \vee r) \to c) \to \neg(q \vee r)$, which corresponds to the logical implication $[(q \vee r) \to c] \Rightarrow \neg(q \vee r)$.

Replacing c by p in $(p \to c) \to \neg p$ gives $(p \to p) \to \neg p$, which is *not* a tautology. The trouble is that c is not a variable, so Rule (a) does not apply. ∎

Our second substitution rule is somewhat like the algebra rule which says that we can always replace quantities by others equal to them. In the present case, our replacement proposition need only be equivalent to the one it substitutes for, and the result will just be equivalent to what we started with.

<table>
<tr><td>*Substitution*
Rule (b)</td><td>If a compound proposition P contains a proposition Q, and if Q is replaced by a logically equivalent proposition Q^*, then the resulting compound proposition P^* is logically equivalent to P.</td></tr>
</table>

EXAMPLE 3   (a) Consider the proposition

$$P = \text{``}\neg[(p \to q) \wedge (p \to r)] \to [q \to (p \to r)]\text{''}$$

which is not a tautology. By Substitution Rule (b) we obtain a logically equivalent proposition P^* if we replace $Q = (p \to q)$ by the logically equivalent $Q^* = (\neg p \vee q)$. Similarly, we could replace one or both occurrences of $(p \to r)$ by $(\neg p \vee r)$; see rule 10a. We could also replace $[(p \to q) \wedge (p \to r)]$ by $[p \to (q \wedge r)]$ thanks to rule 12b. Thus P is logically equivalent to each of the following propositions, among many others:

$$\neg[(\neg p \vee q) \wedge (p \to r)] \to [q \to (p \to r)],$$

$$\neg[(p \to q) \wedge (\neg p \vee r)] \to [q \to (p \to r)],$$

$$\neg[p \to (q \wedge r)] \to [q \to (\neg p \vee r)].$$

(b) Let's see why $P = \text{``}\neg[(p \to q) \wedge (\boldsymbol{p \to r})] \to [q \to (p \to r)]\text{''}$ is equivalent to $P^* = \text{``}\neg[(p \to q) \wedge (\boldsymbol{\neg p \vee r})] \to [q \to (p \to r)]\text{.''}$ We chose this substitution to analyze in order to show that the appearance of another $(p \to r)$, which was not changed, does not matter. Think about the columns of the truth tables for P and P^* corresponding to the boldface $\to$ and $\vee$. Since $\boldsymbol{p \to r}$ and $\boldsymbol{\neg p \vee r}$ are equivalent, the truth values in these two columns are the same. The rest of the truth values in the tables for the propositions P and P^* will be identical, so the final values will agree.

This sort of reasoning generalizes to explain why Substitution Rule (b) is valid. ∎

It is worth emphasizing that, unlike Substitution Rule (b), Substitution Rule (a) requires that *all* occurrences of some variable be replaced by the same proposition.

EXAMPLE 4   (a) We can use Rules (a) and (b) to build equivalence chains. For example, we show $(p \vee q) \vee (p \vee r) \Leftrightarrow (p \vee q) \vee r$ by listing a sequence of equivalent propositions, starting with $(p \vee q) \vee (p \vee r)$ and ending with $(p \vee q) \vee r$. Exercise 23 of §2.2 shows that all of the propositions in such a chain must be equivalent to each other.

Equivalent propositions	*Explanations*
$(p \vee q) \vee (p \vee r)$	Given
$[(p \vee q) \vee p] \vee r$	Law 3a with $p \vee q$ for p and p for q, using Substitution Rule (a); compare Example 2(a)
$[p \vee (q \vee p)] \vee r$	Law 3a with p for r, using Rule (a) and Rule (b)
$[p \vee (p \vee q)] \vee r$	Law 2a and Rule (b)
$[(p \vee p) \vee q] \vee r$	Law 3a and Rule (a) with p for q and q for r and Rule (b)
$[p \vee q] \vee r$	Law 5a and Rule (b)

Here we have carefully mentioned each application of Substitution Rules (a) and (b), but one would normally explain the substitution involved at a given step only if it appeared that the reader might not see it without help.

(b) We derive the tautology

$$[(p \rightarrow q) \vee (p \rightarrow r)] \rightarrow [p \rightarrow (q \vee r)].$$

The associative law 3a gives the tautology

$$[(p \vee q) \vee r] \rightarrow [p \vee (q \vee r)].$$

Part (a) tells us that $(p \vee q) \vee (p \vee r)$ is logically equivalent to $(p \vee q) \vee r$, so Rule (b) says that

$$[(p \vee q) \vee (p \vee r)] \rightarrow [p \vee (q \vee r)]$$

is a tautology. Replacing each occurrence of p by $\neg p$ using Rule (a), we get the tautology

$$(*) \qquad [(\neg p \vee q) \vee (\neg p \vee r)] \rightarrow [\neg p \vee (q \vee r)].$$

Rule 10a says that $\neg p \vee q \Leftrightarrow p \to q$; Rule (a) then also yields

$$\neg p \vee r \Leftrightarrow p \to r \quad \text{and} \quad \neg p \vee (q \vee r) \Leftrightarrow p \to (q \vee r).$$

Three applications of Rule (b) to (∗) show that

$$[(p \to q) \vee (p \to r)] \to [p \to (q \vee r)]$$

is a tautology.   ■

One could use the substitution rules to deal with incredibly complicated propositions. One could, but we won't. Our point here is simply to show that there are techniques, analogous to those we are familiar with from algebra, that let us manipulate and rewrite logical expressions to get them into more convenient forms. What is "convenient" depends, of course, on what we intend to use them for.

EXAMPLE 5   We use the DeMorgan law 8d and substitution to find a proposition that is logically equivalent to $(p \wedge q) \to (\neg p \wedge q)$ but that does not use the connective $\wedge$. Since $p \wedge q$ is equivalent to $\neg(\neg p \vee \neg q)$, and $\neg p \wedge q$ is equivalent to $\neg(\neg\neg p \vee \neg q)$, the given proposition is equivalent by Rule (b) to

$$\neg(\neg p \vee \neg q) \to \neg(\neg\neg p \vee \neg q)$$

and so, again by Rule (b), to

$$\neg(\neg p \vee \neg q) \to \neg(p \vee \neg q).$$

If desired, we could apply rule 10a to obtain the equivalent

$$\neg(p \to \neg q) \to \neg(q \to p),$$

which uses neither $\wedge$ nor $\vee$. On the other hand, we could avoid the use of the connective $\to$ by applying rule 10a.

The sort of rewriting we have done here to eliminate one or more symbols has important applications in logical circuit design, as we will see in Chapter 10.

Notice that we are *not* claiming here that $(p \wedge q) \to (\neg p \wedge q)$ or $\neg(\neg p \vee \neg q) \to \neg(\neg\neg p \vee \neg q)$ or $\neg(\neg p \vee \neg q) \to \neg(p \vee \neg q)$ are tautologies. We have simply shown that these three propositions are logically equivalent to each other.   ■

Tautologies give us one way of looking at the logical dependence between two different propositions: Check the possibilities for the variables and see how the truth values for the propositions compare. Another approach is to see whether there is a logical way to deduce one proposition from the other, perhaps by a sequence of small steps that everyone agrees are legal. Our next goal is to formalize this deductive idea. Later we will discuss the connection between truth and deduction as two ways to describe logical reasoning.

The key to the deductive approach is to formalize the concept of "proof." Suppose that we are given some set of propositions, our **hypotheses**, and some **conclusion** C. A **formal proof** of C from the hypotheses consists of a chain $P_1, P_2, \ldots, P_n, C$ of propositions, ending with C, in which each P_i is either

(i) a hypothesis, or

(ii) a tautology, or

(iii) a consequence of previous members of the chain by using an allowable rule of inference.

A **theorem** is a statement of the form "If H, then C," where H is a set of hypotheses and C is a conclusion. A formal proof of C from H is called a formal proof of the theorem.

The **rules of inference**, or logical rules, that we allow in formal proofs are Substitution Rules (a) and (b) and rules based on logical implications of the form $H_1 \wedge H_2 \wedge \cdots \wedge H_m \Rightarrow Q$. If $H_1, H_2, \ldots, H_m$ have already appeared in the chain of a proof, and if $H_1 \wedge H_2 \wedge \cdots \wedge H_m \Rightarrow Q$ is true, then we are allowed to add Q to the chain.

One way to list the members of the chain for a formal proof is just to write them in sequence, like ordinary prose. Another way is to list them vertically, which gives us room to keep a record of the reason for allowing each member into the chain.

EXAMPLE 6   Here is a very short formal proof written out vertically. The hypotheses are $B \wedge S$ and $B \vee S \to P$, and the conclusion is P. The chain gives a formal proof of P from $\{B \wedge S, B \vee S \to P\}$. To give some meaning to the symbolism, you might think of B as "I am wearing a belt," S as "I am wearing suspenders" and P as "my pants stay up."

Proof	*Reason*
1. $B \wedge S$	hypothesis
2. B	1; simplification rule 17
3. $B \vee S$	2; addition rule 16
4. $B \vee S \to P$	hypothesis
5. P	3, 4; modus ponens rule 19

This proof could also have been arranged in different orders. Figures 1(a) to (d) show four other formal proofs with the same hypotheses and conclusion. Figure 1(e) displays the logical dependence of the propositions.

$B \lor S \to P$	$B \land S$	$B \land S$	$B \land S$	$B \land S$
$B \land S$	$B \lor S \to P$	B	$S \land B$	$\mid$
B	B	$B \lor S \to P$	S	S
$B \lor S$	$B \lor S$	$B \lor S$	$S \lor B$	$\mid$
P	P	P	$B \lor S \to P$	$B \lor S \qquad B \lor S \to P$
			$S \lor B \to P$	$\diagdown \quad \diagup$
			P	P
(a)	(b)	(c)	(d)	(e)

FIGURE 1   ■

The proof in Example 6 used three rules of inference, based on the implications $(p \land q) \Rightarrow p$, $p \Rightarrow (p \lor q)$ and $[p \land (p \to q)] \Rightarrow q$. Table 1 lists these rules and four others that are frequently useful. Each has a corresponding logical implication of the form $H_1 \land \cdots \land H_m \Rightarrow C$ in Table 2 of §2.2, except Rule 34, which corresponds to $p \land q \Rightarrow p \land q$. We have used the format

$$H_1$$
$$H_2$$
$$\vdots$$
$$\underline{H_m}$$
$$\therefore C \qquad \text{[Read } \therefore \text{ as "therefore" or "hence."]}$$

to describe the inference of C from $H_1 \land \cdots \land H_m$. When we write out the reasons for formal proofs we can use either the rules from §2.2 or their counterparts from Table 1.

TABLE 1. Rules of Inference

28.	P		29.	$P \land Q$	
	$\therefore \overline{P \lor Q}$	addition		$\therefore \overline{P}$	simplification
30.	P		31.	$P \to Q$	
	$P \to Q$			$\neg Q$	
	$\therefore \overline{Q}$	modus ponens		$\therefore \overline{\neg P}$	modus tollens
32.	$P \lor Q$		33.	$P \to Q$	
	$\neg P$			$Q \to R$	
	$\therefore \overline{Q}$	disjunctive syllogism		$\therefore \overline{P \to R}$	hypothetical syllogism
34.	P				
	Q				
	$\therefore \overline{P \land Q}$	conjunction			

Observe that logical equivalences of the form $H_1 \wedge \cdots \wedge H_m \Leftrightarrow C$, such as $[(p \to r) \wedge (q \to r)] \Leftrightarrow [(p \vee q) \to r]$ from Table 1 of §2.2, yield logical implications $H_1 \wedge \cdots \wedge H_m \Rightarrow C$. So they also give rules of inference.

Here are two additional formal proofs to illustrate how steps are justified.

EXAMPLE 7   (a) We conclude $s \to r$ from $p \to (q \to r)$, $p \vee \neg s$ and q.

1.	$p \to (q \to r)$	hypothesis
2.	$p \vee \neg s$	hypothesis
3.	q	hypothesis
4.	$\neg s \vee p$	3; commutative law 2a
5.	$s \to p$	4; implication rule 10a
6.	$s \to (q \to r)$	1, 5; hypothetical syllogism rule 33
7.	$(s \wedge q) \to r$	6; exportation rule 14
8.	$q \to [s \to (q \wedge s)]$	rule 22
9.	$s \to (q \wedge s)$	3, 8; modus ponens rule 30
10.	$s \to (s \wedge q)$	9; commutative law 2b
11.	$s \to r$	7, 10; hypothetical syllogism rule 33

(b) Here we conclude $\neg p$ by contradiction from $p \to (q \wedge r)$, $r \to s$ and $\neg(q \wedge s)$.

1.	$p \to (q \wedge r)$	hypothesis
2.	$r \to s$	hypothesis
3.	$\neg(q \wedge s)$	hypothesis
4.	$\neg(\neg p)$	negation of the conclusion
5.	p	4; double negation rule 1
6.	$q \wedge r$	1, 5; modus ponens rule 30
7.	q	6; simplification rule 29
8.	$r \wedge q$	6; commutative law 2b
9.	r	8; simplification rule 29
10.	s	2, 9; modus ponens rule 30
11.	$q \wedge s$	7, 10; conjunction rule 34
12.	$(q \wedge s) \wedge \neg(q \wedge s)$	3, 11; conjunction rule 34
13.	contradiction	12; rule 7b

■

You might wonder why anyone would want to be so formal. We are trying to build a symbolic model that shows how to construct logical proofs and how to tell if a chain of inferences is valid. No one, not even the formal logician or the developer of a computerized proof checker, is enchanted about writing out formal proofs. The idea is to learn what kinds of proofs are possible and to see ways the rules of inference apply, since we use these rules implicitly even when we write proofs informally.

Now that we have found a way to describe logical arguments, what does all of this have to do with *truth*? Truth tables are built into our rules of inference, since each logical implication $P \Rightarrow Q$ is really a statement about truth tables. A long, tedious argument shows that if there is a formal proof of C from H, and if H is true, then so is C. We aren't allowed to prove false propositions from true ones. One can also show that if H logically implies C, then a formal proof of C from H does exist. That is, there is a formal proof of C from H if and only if $H \Rightarrow C$ is true.

EXAMPLE 8   Recall from Example 6(b) of §2.2 that $A \wedge B \Rightarrow C$ is true if and only if $A \Rightarrow (B \rightarrow C)$ is true. Hence there is a formal proof of C from $A \wedge B$, i.e., from A and B, if and only if there is a proof of $B \rightarrow C$ from A.   ■

EXERCISES 2.4

1. Use Substitution Rule (a) with $p \rightarrow q$ replacing q to make a tautology out of each of the following.

   (a) $\neg q \rightarrow (q \rightarrow p)$

   (b) $[p \wedge (p \rightarrow q)] \rightarrow q$

   (c) $p \vee \neg p$

   (d) $(p \vee q) \leftrightarrow ((\neg q) \rightarrow p)$

2. Give the rules of inference corresponding to the logical implications 23, 26a and 27b.

3. Find a logical equivalence in Table 1 of §2.2 from which the use of Substitution Rule (a) yields the indicated equivalence.

   (a) $[(p \wedge (q \wedge r)) \rightarrow r] \Leftrightarrow [p \rightarrow ((q \wedge r) \rightarrow r)]$

   (b) $[p \vee (q \wedge (r \wedge s))] \Leftrightarrow [(p \vee q) \wedge (p \vee (r \wedge s))]$

   (c) $\neg[(\neg p \wedge r) \vee (q \rightarrow r)] \Leftrightarrow [\neg(\neg p \wedge r) \wedge \neg(q \rightarrow r)]$

4. Find a logical implication in Table 2 of §2.2 from which Substitution Rule (a) yields the indicated implication.

   (a) $[\neg p \vee q] \Rightarrow [q \rightarrow ((\neg p \vee q) \wedge q)]$

   (b) $[(p \rightarrow q) \wedge (r \rightarrow q)] \Rightarrow [(\neg q \vee \neg q) \rightarrow (\neg p \vee \neg r)]$

   (c) $[((p \rightarrow s) \rightarrow (q \wedge s)) \wedge \neg(q \wedge s)] \Rightarrow \neg(p \rightarrow s)$

5. Give reasons for each of the following. Try not to use truth tables.

   (a) $(p \vee q) \wedge s \Leftrightarrow (q \vee p) \wedge s$

   (b) $s \rightarrow (\neg(p \vee q)) \Leftrightarrow s \rightarrow [(\neg p) \wedge (\neg q)]$

   (c) $(p \rightarrow q) \wedge (p \vee q) \Leftrightarrow (\neg p \vee q) \wedge (\neg \neg p \vee q)$

   (d) $t \wedge (s \vee p) \Leftrightarrow t \wedge (p \vee s)$

6. Repeat Exercise 5 for the following.

   (a) $s \wedge p \Leftrightarrow s \wedge (p \wedge p)$

(b) $[(a \vee b) \leftrightarrow (p \rightarrow q)] \Leftrightarrow [(a \vee b) \leftrightarrow (\neg p \vee q)]$

(c) $[(a \vee b) \leftrightarrow \neg (p \wedge q)] \Leftrightarrow [(b \vee a) \leftrightarrow (\neg p \vee \neg q)]$

7. Give a reason for each equivalence in the following chain.

$$(p \rightarrow s) \vee (\neg s \rightarrow t)$$

(a) $\Leftrightarrow (\neg p \vee s) \vee (s \vee t)$

(b) $\Leftrightarrow [(\neg p \vee s) \vee s] \vee t$

(c) $\Leftrightarrow [\neg p \vee (s \vee s)] \vee t$

(d) $\Leftrightarrow (\neg p \vee s) \vee t$

(e) $\Leftrightarrow \neg p \vee (s \vee t)$

(f) $\Leftrightarrow p \rightarrow (s \vee t)$

8. Repeat Exercise 7 with the following.

$$[(a \wedge p) \vee p] \rightarrow p$$

(a) $\Leftrightarrow \neg [(a \wedge p) \vee p] \vee p$

(b) $\Leftrightarrow [\neg (a \wedge p) \wedge \neg p] \vee p$

(c) $\Leftrightarrow [(\neg a \vee \neg p) \wedge \neg p] \vee p$

(d) $\Leftrightarrow p \vee [(\neg a \vee \neg p) \wedge \neg p]$

(e) $\Leftrightarrow [p \vee (\neg a \vee \neg p)] \wedge (p \vee \neg p)$

(f) $\Leftrightarrow [(\neg a \vee \neg p) \vee p] \wedge t$ [t for tautology here]

(g) $\Leftrightarrow (\neg a \vee \neg p) \vee p$

(h) $\Leftrightarrow \neg a \vee (\neg p \vee p)$

(i) $\Leftrightarrow \neg a \vee t$

(j) $\Leftrightarrow t$

9. Show by the methods of Example 2. Give reasons for your statements.

(a) $[(p \vee r) \wedge (q \rightarrow r)] \Leftrightarrow [(p \wedge \neg q) \vee r]$

(b) $[(p \wedge \neg q) \vee r] \Leftrightarrow [(p \rightarrow q) \rightarrow r]$

(c) $p \vee \neg q \Leftrightarrow p \vee (\neg p \wedge \neg q)$

10. Repeat Exercise 9 for the following.

(a) $\neg q \Rightarrow \neg q \vee p$ (b) $(p \vee q) \rightarrow p \Leftrightarrow p \vee (\neg p \wedge \neg q)$

(c) $p \vee (\neg p \wedge \neg q) \Leftrightarrow \neg q \vee p$ (d) $\neg q \Rightarrow [(p \vee q) \rightarrow p]$

11. Let P be the proposition $[p \wedge (q \vee r)] \vee \neg [p \vee (q \vee r)]$. Replacing all occurrences of $q \vee r$ by $q \wedge r$ yields

$$P^* = \text{``}[p \wedge (q \wedge r)] \vee \neg [p \vee (q \wedge r)].\text{''}$$

Since $q \wedge r \Rightarrow q \vee r$, one might suppose that $P \Rightarrow P^*$ or that $P^* \Rightarrow P$. Show that neither of these is the case.

12. Show that if the first p in the tautology $p \rightarrow [q \rightarrow (p \wedge q)]$ is replaced by the proposition $p \vee q$, then the new proposition is not a tautology. This shows that Substitution Rule (a) must be applied with care.

13. Give an explanation for each step in the following formal proof of $\neg s$ from $\{(s \vee g) \to p, \neg a, p \to a\}$.

    1. $(s \vee g) \to p$
    2. $\neg a$
    3. $p \to a$
    4. $s \to (s \vee g)$
    5. $s \to p$
    6. $s \to a$
    7. $\neg s$

14. Rearrange the steps in Exercise 13 in three different ways, all of which prove $\neg s$ from $\{(s \vee g) \to p, \neg a, p \to a\}$.

15. (a) Show that if A is a tautology and if there is a formal proof of C from A, then there is a proof of C with no hypotheses at all.

    (b) Show that if there is a formal proof of C from B, then there is a proof of $B \to C$ with no hypotheses at all.

16. Every compound proposition is equivalent to one that uses only the connectives $\neg$ and $\vee$. This fact follows from the equivalences $(p \to q) \Leftrightarrow (\neg p \vee q)$, $(p \wedge q) \Leftrightarrow \neg(\neg p \vee \neg q)$ and $(p \leftrightarrow q) \Leftrightarrow [(p \to q) \wedge (q \to p)]$. Find propositions logically equivalent to the following but using only the connectives $\neg$ and $\vee$.

    (a) $p \leftrightarrow q$ (b) $(p \wedge q) \to (\neg q \wedge r)$

    (c) $(p \to q) \wedge (q \vee r)$ (d) $p \oplus q$

17. (a) Show that $p \vee q$ and $p \wedge q$ are logically equivalent to propositions using only the connectives $\neg$ and $\to$.

    (b) Show that $p \vee q$ and $p \to q$ are logically equivalent to propositions using only the connectives $\neg$ and $\wedge$.

    (c) Is $p \to q$ logically equivalent to a proposition using only the connectives $\wedge$ and $\vee$? Explain.

18. The **Sheffer stroke** is the connective $|$ defined by the truth table:

p	q	$p \vert q$
0	0	1
0	1	1
1	0	1
1	1	0

Thus $p|q \Leftrightarrow \neg(p \wedge q)$ [hence the computer scientists' name **NAND** for this connective]. All compound propositions are equivalent to ones that use only $|$, a useful fact that follows from the remarks in Exercise 16 and parts (a) and (b) below.

    (a) Show that $\neg p \Leftrightarrow p|p$.

    (b) Show that $p \vee q \Leftrightarrow (p|p)|(q|q)$.

(c) Find a proposition equivalent to $p \wedge q$ using only the Sheffer stroke.

(d) Do the same for $p \rightarrow q$.

(e) Do the same for $p \oplus q$.

19. Verify the following **absorption laws**.

(a) $[p \vee (p \wedge q)] \Leftrightarrow p$ [Compare with Exercise 8.]

(b) $[p \wedge (p \vee q)] \Leftrightarrow p$

§ 2.5   Analysis of Arguments

Section 2.4 described a formalization of the notion of "proof." Our aim in this section is to use that abstract version as a guide to understanding more about the proofs and fallacies that we encounter in everyday situations. We will look at examples of how to construct arguments, and we will see some errors to watch out for. The same kind of analysis that lets us judge validity of proofs will also help us to untangle complicated logical descriptions.

Our formal rules of inference are sometimes called **valid inferences**, and formal proofs are called **valid proofs**. We will extend the use of the word "valid" to describe informal arguments and inferences that correspond to formal proofs and their rules. A sequence of propositions that fails to meet the requirements for being a formal proof is called a **fallacy**. We will use this term also to mean an argument whose formal counterpart is not a valid proof.

We will see a variety of examples of fallacies. First, though, we construct some valid arguments.

EXAMPLE 1   We analyze the following. "If I study or if I am a genius, then I will pass the course. If I pass the course, then I will be allowed to take the next course. Therefore, if I am not allowed to take the next course, then I am not a genius." Let

$$s = \text{"I study,"}$$

$$g = \text{"I am a genius,"}$$

$$p = \text{"I will pass the course,"}$$

$$n = \text{"I will be allowed to take the next course."}$$

We are given the hypotheses $s \vee g \rightarrow p$ and $p \rightarrow n$, and want a proof of $\neg n \rightarrow \neg g$. Here is our scratchwork. Since $\neg n \rightarrow \neg g$ is the contrapositive of $g \rightarrow n$, it will be enough to get a proof of $g \rightarrow n$ and then quote the contrapositive law, rule 9. From $s \vee g \rightarrow p$ we can surely infer $g \rightarrow p$ [details in the next sentence] and then combine this with $p \rightarrow n$ to get $g \rightarrow n$ using rule 33. To get $g \rightarrow p$ from $s \vee g \rightarrow p$, rule 33 shows that it is enough to observe that $g \to s \vee g$ by the addition rule. Well, almost: The addition rule gives $g \to g \vee s$, so we need to invoke commutativity, too. Here is the corresponding formal proof.

Proof	*Explanations*
1. $s \vee g \to p$	hypothesis
2. $p \to n$	hypothesis
3. $g \to g \vee s$	addition (rule 16)
4. $g \to s \vee g$	3; commutative law 2a
5. $g \to p$	4, 1; hypothetical syllogism (rule 33)
6. $g \to n$	5, 2; hypothetical syllogism (rule 33)
7. $\neg n \to \neg g$	6; contrapositive (rule 9)

So there is a valid proof of the conclusion from the hypotheses. Isn't that reassuring? Of course we are accustomed to handling this sort of simple reasoning in our heads every day, and the baggage of formal argument here just seems to complicate an easy problem. ∎

Still, we continue.

EXAMPLE 2 "If I study or if I am a genius, then I will pass the course. I will not be allowed to take the next course. If I pass the course, then I will be allowed to take the next course. Therefore, I did not study." With the notation of Example 1, we want a proof of: If $s \vee g \to p$, $\neg n$ and $p \to n$, then $\neg s$. The only rule of inference in Table 1 of §2.4 that might help is rule 31 [modus tollens]. Given the hypothesis $\neg n$, this rule would allow us to infer $\neg s$ if we could just get $s \to n$, which we can deduce from the first and third hypotheses, as in Example 1. Here is a formal version.

Proof	*Explanations*
1. $s \vee g \to p$	hypothesis
2. $s \to s \vee g$	addition (rule 16)
3. $s \to p$	2, 1; hypothetical syllogism (rule 33)
4. $p \to n$	hypothesis
5. $s \to n$	3, 4; hypothetical syllogism (rule 33)
6. $\neg n$	hypothesis
7. $\neg s$	5, 6; modus tollens (rule 31)

In words, the argument would go:

If I study, then since either studying or being a genius will get me through, I will pass. But then, because passing will let me in the next course, I can take the next one. So if I can't take the next one, something's gone wrong—I must not have studied.

Rule 16 is only used implicitly, and the three inferences made are clearly "logical." ■

This example illustrates a general strategy that can be helpful in constructing proofs. Look at the conclusion C. What inference would give it? For example, do we know of a proof of something like $B \to C$? How hard would it be to prove B? And so on, working backward from C to B to A to ... Alternatively, look at the hypotheses. What can we quickly deduce from them? Do any of the new deductions seem related to C? If you have ever had to prove trigonometric identities, you will recognize the strategy: You worked with the right-hand side, then the left-hand side, then the right-hand, etc., trying to bring the two sides of the identity together.

EXAMPLE 3   We can also obtain the conclusion in Example 2 by contradiction, i.e., by assuming the negation of the conclusion, that I *did* study, and reaching an absurdity. The formal proof looks a little more complicated. Given hypotheses $s \vee g \to p$, $\neg n$, $p \to n$ and now $\neg(\neg s)$, we want to get a contradiction like $s \wedge (\neg s)$, $n \wedge (\neg n)$, $g \wedge (\neg g)$ or $p \wedge (\neg p)$. Since we already have $\neg n$ and s [from $\neg(\neg s)$], the first two contradictions look easiest to reach. We aim for $n \wedge (\neg n)$, because we already have $p \to n$. We have s, and we can get $s \to p$ from $s \vee g \to p$, so we can get p. Here is one formal proof that develops; see Exercise 11 for a proof with modus ponens instead of hypothetical syllogism.

Proof	*Explanations*
1. $s \to s \vee g$	addition (rule 16)
2. $s \vee g \to p$	hypothesis
3. $s \to p$	1, 2; hypothetical syllogism (rule 33)
4. $p \to n$	hypothesis
5. $s \to n$	3, 4; hypothetical syllogism (rule 33)
6. $\neg(\neg s)$	negation of the conclusion
7. s	6; double negation (rule 1)
8. n	7, 5; modus ponens (rule 30)
9. $\neg n$	hypothesis
10. $n \wedge (\neg n)$	8, 9; conjunction (rule 34)
11. contradiction	10; rule 7b

The English version goes like this:

As in Example 2, if I study then I can take the next course. Just suppose I do study [implicit use of double negation here]. Then I can take the next course. But I can't take it, by hypothesis. This contradiction means that my supposition was wrong.

In general, a proof by contradiction begins with a supposition and runs for a while until it reaches a contradiction. At that point we look back to the last "suppose" and say, "Well, that supposition must have been wrong." We could also conclude simply that the hypotheses and $\neg C$ are mutually inconsistent, and that if we want to keep $\neg C$, then we must negate one of the hypotheses.

The formal proof by contradiction in this example is longer than the proof in Example 2, but the English version is about the same length, and it may be conceptually more straightforward. In the real world, one resorts to proofs by contradiction when it is easier to use $\neg C$ *together with* the hypotheses than it is to derive *C from* the hypotheses. We illustrated this point in §2.3. ∎

If it were easy to construct proofs, everybody would do it, and the mathematicians would be out of work. Still, with practice and by looking closely at the proofs you read, you can get quite good at building your own proofs and at spotting holes in other people's.

This book is full of examples of correct proofs. Let's look now at a few fallacies.

EXAMPLE 4   We are given that if a program does not fail, then it begins and terminates, and we know that our program began and failed. We conclude that it did not terminate. Is this reasoning valid? Let

$$B = \text{"the program begins,"}$$

$$T = \text{"the program terminates,"}$$

$$F = \text{"the program fails."}$$

Our hypotheses are $\neg F \to (B \wedge T)$ and $B \wedge F$, and the conclusion is $\neg T$. Here is an attempt at a formal proof.

Proof?	*Explanations*
1. $\neg F \to (B \wedge T)$	hypothesis
2. $B \wedge F$	hypothesis
3. $(B \wedge T) \to T$	simplification (rule 17)
4. $\neg F \to T$	1, 3; hypothetical syllogism (rule 33)
5. F	2; simplification (rule 29)
6. $\neg T$	4, 5; **??**

How can we infer proposition 6 from propositions 4 and 5? It appears that the best hope is modus tollens, but a closer look shows that modus tollens does not apply. What is needed is $P \to Q, \neg P, \therefore \neg Q$ and this is *not* a valid rule of inference. The alleged proof above is not valid; it is a fallacy. Did we perhaps simply go about the proof in the wrong way? No. In fact, no correct proof exists in this case, because the conclusion does not follow from the hypotheses. That is,

$$\{[\neg F \rightarrow (B \wedge T)] \wedge (B \wedge F)\} \rightarrow \neg T$$

is not a tautology. To see this, consider the row in its truth table where B, F and T are all true. In terms of the original hypotheses, the program might begin and terminate, yet fail for some other reason. ■

EXAMPLE 5   (a) I am a famous basketball player. Famous basketball players make lots of money. If I make lots of money, then you should do what I say. I say you should buy Pearly Maid shoes. Therefore, you should buy Pearly Maid shoes.

We set

$$B = \text{``basketball player,''}$$

$$M = \text{``makes lots of money,''}$$

$$D = \text{``do what I say,'' and}$$

$$S = \text{``buy those shoes!''}$$

The hypotheses are B, $B \rightarrow M$, $M \rightarrow D$ and $D \rightarrow S$, and the conclusion S surely follows by three applications of modus ponens.

(b) Suppose that we leave out $M \rightarrow D$ in (a). Then there is no valid way to conclude S from B, $B \rightarrow M$ and $D \rightarrow S$. [Look at the case S and D false, B and M true, for example.] The argument would require filling in the hidden assumption that we should do what people with lots of money tell us to. [As usually presented, the argument leaves both $B \rightarrow M$ and $M \rightarrow D$ unstated.]

(c) Suppose that instead of $M \rightarrow D$ in the original version we have the appeal to sympathy, "I will make money only if you buy our shoes." The hypotheses now are B, $B \rightarrow M$, $M \rightarrow S$ and $D \rightarrow S$. Again, S is a valid conclusion. The hypothesis $M \rightarrow S$ seems suspect, but given the truth of that hypothesis, the argument is flawless.

(d) We could replace $M \rightarrow D$ in the original by the more believable statement $B \vee S \rightarrow M$, "I will make money if I am a famous player *or* if you buy our shoes." We already have the stronger hypothesis $B \rightarrow M$, though, so the analysis in (b) shows that we can't deduce S in this case either. ■

Sometimes a situation is so complicated that it is not at all clear at first whether one should be trying to show that an alleged conclusion does follow or that it doesn't. In such a case it is often a good plan to see what you can deduce from the hypotheses, in the hope that you'll either learn a great deal about the possibilities or arrive at a contradiction. Either way, you win. If nothing else, this approach forces you to look at the problem slowly and methodically, which by itself is often enough to unravel it.

EXAMPLE 6   (a) Let

$$A = \text{"I am an adult,"}$$

$$B = \text{"I am big and brave,"}$$

$$Y = \text{"a year goes by,"}$$

$$L = \text{"life is tough," and}$$

$$N = \text{"nobody loves me."}$$

Here is part of an argument. If I am an adult, then I am big and brave. If I am big and brave or a year goes by, then life is tough and nobody loves me. If I am big and brave and nobody loves me, then I am an adult. If I am big and brave or a year does not go by, then I am an adult. If I am big and brave or a year goes by, then nobody loves me. Either a year goes by, or it doesn't. If I am big and brave and somebody loves me, then I am an adult. Therefore, I am an adult [and this is an example of adult reasoning].

What a mess! Does the conclusion follow? Let's see. The hypotheses are $A \to B, (B \vee Y) \to (L \wedge N), (B \wedge N) \to A, (B \vee \neg Y) \to A, (B \vee Y) \to N$ [a consequence of $(B \vee Y) \to (L \wedge N)$], $Y \vee \neg Y$ [a tautology] and $(B \wedge \neg N) \to A$. For all of these to be true and yet A to be false, we must have p false for every hypothesis $p \to A$. Thus if A is false, then $B \wedge N$, $B \vee \neg Y$ and $B \wedge \neg N$ must be false. If $B \vee \neg Y$ is false, then B must be false and Y true. Then since $(B \vee Y) \to (L \wedge N)$ is true, so is $Y \to (L \wedge N)$, and thus L and N are also true. At this point we know that if A is false, then so is B, but Y, L and N must be true. We check that these truth values satisfy all the hypotheses, so A does not follow from the hypotheses by any valid argument.

We did not need to go through all the analysis of the last paragraph. All we *really* needed was a set of truth values that satisfied the hypotheses with A false. But how were we to find such values? The way we went about the analysis meant that we might be led to such truth values or we might be led to a proof that no such values are possible. In either case we would get a useful answer.

The original argument here was really just a lot of hypotheses and a conclusion, with no steps shown in between. It is possible for an argument to have lots of steps filled in correctly but still to be a fallacy if there is a gap somewhere in the middle.

(b) Suppose that we want to show that life is tough. A standard sort of argument goes like this.

Assume that life is tough.

Blah, blah, blah.

Therefore, life is tough.

Assuming the conclusion in this way seems too obviously wrong to be worth mentioning, but experience teaches us that this fallacy, perhaps disguised, is a common one.

(c) Suppose that we want to show that "If I am an adult, then life is tough." Try this.

Assume L. Since $L \to (A \to L)$ is a tautology, $A \to L$ must be true.

Here again we start out assuming an additional hypothesis, conveniently strong enough to get the conclusion. ■

EXAMPLE 7 (a) We have already mentioned in §2.3 the fallacy of arguing $A \Leftrightarrow B$ by showing $A \Rightarrow B$ and $\neg B \Rightarrow \neg A$.

(b) A related fallacy is proving the converse of what's called for, i.e., trying to show $A \Rightarrow B$ by showing $B \Rightarrow A$. The false inference $A, B \to A$, $\therefore B$ is a similar fallacy. It is amazing how many people fall into these traps. Sometimes, perhaps, confusing wording trips them up. Thus suppose that I am rich and that people are rich whenever they live in big houses. It does not follow that I live in a big house. I might live on board my yacht. Here we have R and $H \to R$ and want to deduce H. To do so, though, we would want $R \to H$; that is, all rich people live in big houses. ■

EXAMPLE 8 We will argue the case for "If I get a grade of F, then I will pass the course," using proof by contradiction. Assume, then, that I will not pass. That would mean that I *would get* an F [the converse of what we want to show, but certainly true]. But the contrapositive of "if I get an F, then I will pass" is "if I do not pass, then I *will not get* an F." I can't both get an F and not get an F, so there's contradiction. The only assumption we have made—that I will not pass—must be wrong. Hooray! I will pass.

What happened here is that we thrashed around for a while, getting more and more confused, until eventually we stated some proposition and also its negative. Then we seized upon our statements as a contradiction.

Proofs by contradiction are especially tricky. If they are valid, then the propositions describe a situation that cannot in fact be true, and that should conflict with our intuition. The example above was deliberately clumsy, of course. Real-life proofs by contradiction are often exceedingly complicated and consist of building up more and more evidence until eventually the situation described is seen to be impossible. One small mistake can produce a contradiction where none in fact exists, and can make a false argument seem valid. ■

A few key phrases warn you to be especially careful. Whenever someone says "it stands to reason," you can be pretty certain that there is a large gap in the argument. The person sees no way to bridge the gap legally and is trying to manufacture the needed tautology or rule of inference on the spot without specifying what it is. "Proof by intimidation" and "proof by authority" are more powerful versions of this fallacy.

"Clearly" and "obviously" are also clues to look for errors, even in printed material. Obviously, there are no mistakes in *this* book, but previous editions did have one or two logical errors.

When you are constructing proofs to answer exercises, you should check for unused hypotheses. Although real-life problems often have more than enough information, textbook exercises *usually* have none to spare. If you haven't used some of what was given, perhaps your argument has a flaw, for instance an overlooked case.

The thought patterns that help us verify proofs can also be useful for untangling complicated logical formulations.

EXAMPLE 9   We look at a fairly typical excerpt from the manual of a powerful computer operating system. Let

A = "a signal condition arises for a process,"

P = "the signal is added to a set of signals pending for the process,"

B = "the signal is currently blocked by the process,"

D = "the signal is delivered to the process,"

S = "the current state of the process is saved,"

M = "a new signal mask is calculated,"

H = "the signal handler is invoked,"

N = "the handling routine is invoked normally,"

R = "the process will resume execution in the previous context,"

I = "the process must arrange to restore the previous context itself."

The manual says "$A \rightarrow P, (P \wedge \neg B) \rightarrow D, D \rightarrow (S \wedge M \wedge H), (H \wedge N) \rightarrow R,$ $(H \wedge \neg R) \rightarrow I$." It really does. We have just translated the manual from English into letters and symbols and left out a few words.

What can we conclude from these hypotheses? In particular, what will happen if $A \wedge \neg B \wedge \neg R$ is true, i.e., if a signal condition arises, the signal is not blocked by a process, but the process does not wish to resume execution in its previous context?

We can deduce $P \wedge \neg B$ from $A \rightarrow P$ and $A \wedge \neg B$ [Exercise 13(a)]. Hence, using $(P \wedge \neg B) \rightarrow D$ and $D \rightarrow (S \wedge M \wedge H)$, we have $S \wedge M \wedge H$. In particular, H is true. Another short proof [Exercise 13(b)] deduces $\neg N$ from $H \wedge \neg R$ and $(H \wedge N) \rightarrow R$. Thus since $(H \wedge \neg R) \rightarrow I$, $H \wedge \neg R$ yields $I \wedge \neg N$. We have been able to show that if A is true and B and R are false, then I is true and N is false; i.e., signal handling is not invoked normally, and the process must restore its own previous context. Along the way we also showed that P, D, S, M and H are true.

Of course, one would not usually write this analysis out as a formal proof. It is helpful, though, to dissect the verbal description into separate statements, to give the statements labels, and to write out symbolically what the hypotheses are, as well as any conclusions that follow from them. ■

EXERCISES 2.5

1. We observe C and observe that A implies C. We reason that this means that if A were false, then C would be false too, which it isn't. So A must be true. Is this argument valid? Explain.

2. Give two examples of fallacies drawn from everyday life. Explain why the arguments are fallacies. *Suggestion:* Advertising and letters to the editor are good sources.

3. (a) Show that there is no valid proof of C from the hypothesis $A \rightarrow C$.

   (b) Is there a valid proof of C from $A \rightarrow C$ and $B \rightarrow C$? Explain.

   (c) How strong is the argument for C if $A_1 \rightarrow C$, $A_2 \rightarrow C, \ldots, A_{1000000} \rightarrow C$ are true? Explain.

4. (a) If we leave out the hypothesis $B \rightarrow M$ in Example 5(a), do the remaining hypotheses imply S? Explain.

   (b) Would the argument for S be stronger in part (a) with the added hypothesis $M \rightarrow S$, but still without $B \rightarrow M$? Explain.

5. For the following sets of hypotheses, state a conclusion that can be inferred and specify the rules of inference used.

   (a) If the TV set is not broken, then I will not study. If I study, then I will pass the course. I will not pass the course.

   (b) If I passed the midterm and the final, then I passed the course. If I passed the course, then I passed the final. I failed the course.

   (c) If I pass the midterm or the final, then I will pass this course. I will take the next course only if I pass this course. I will not take the next course.

6. Consider the following hypotheses. If I take the bus or subway, then I will be late for my appointment. If I take a cab, then I will not be late for my appointment and I will be broke. I will be on time for my appointment.

   Which of the following conclusions *must* follow, i.e., can be inferred from the hypotheses? Justify your answers.

   (a) I will take a cab.

   (b) I will be broke.

   (c) I will not take the subway.

   (d) If I become broke, then I took a cab.

   (e) If I take the bus, then I won't be broke.

7. Assume the hypotheses of Example 6(a). Which of the following are valid conclusions? Justify your answer in each case.

   (a) I am an adult if and only if I am big and brave.

   (b) If I am big and brave, then somebody loves me.

   (c) Either a year goes by or I am an adult.

   (d) Life is tough and nobody loves me.

8. Either Pat did it or Quincy did. Quincy could not have been reading and done it. Quincy was reading. Who did it? Explain, using an appropriate formal proof with the variables P, Q, R.

9. Convert each of the following arguments into logical notation using the suggested variables. Then provide a formal proof.

   (a) "If my computations are correct and I pay the electric bill, then I will run out of money. If I don't pay the electric bill, the power will be turned off. Therefore, if I don't run out of money and the power is still on, then my computations are incorrect." (c, b, r, p)

   (b) "If the weather bureau predicts dry weather, then I will take a hike or go swimming. I will go swimming if and only if the weather bureau predicts warm weather. Therefore, if I don't go on a hike, then the weather bureau predicts wet or warm weather." (d, h, s, w)

   (c) "If I get the job and work hard, then I will get promoted. If I get promoted, then I will be happy. I will not be happy. Therefore, either I will not get the job or I will not work hard." (j, w, p, h)

   (d) "If I study law, then I will make a lot of money. If I study archaeology, then I will travel a lot. If I make a lot of money or travel a lot, then I will not be disappointed. Therefore, if I am disappointed, then I did not study law and I did not study archaeology." (l, m, a, t, d)

10. (a) Revise the proof in Example 3 to reach the contradiction $s \wedge (\neg s)$. *Hint:* Use lines 5 and 9 and rearrange steps.

    (b) Revise the proof in Example 3 to reach the contradiction $p \wedge (\neg p)$. *Hint:* Use lines 3 and 7, then lines 4 and 9, and rearrange.

11. Give an alternative formal proof for Example 3 that does not use rule 33.

12. For each of the following, give a formal proof of the theorem or show that it is false by exhibiting a suitable row of a truth table.

    (a) If $(q \wedge r) \to p$ and $q \to \neg r$, then p.

    (b) If $q \vee \neg r$ and $\neg(r \to q) \to \neg p$, then p.

    (c) If $p \to (q \vee r)$, $q \to s$ and $r \to \neg p$, then $p \to s$.

13. Give formal proofs of the following.

    (a) $P \wedge \neg B$ from $A \to P$ and $A \wedge \neg B$.

    (b) $\neg N$ from $H \wedge \neg R$ and $(H \wedge N) \to R$. *Suggestion:* Proof by contradiction.

14. If you are here, then today must be Friday. If I am not confused, then today is Saturday. Either today is not yesterday, or today is Friday. I must not be

confused if you are here. Friday is not Saturday. If today is Saturday, then yesterday was Friday.

(a) Are you here?

(b) Suppose that I am not confused. Can you conclude that yesterday is not today?

(c) If I am not confused, what day was yesterday? Explain.

CHAPTER HIGHLIGHTS

To check your understanding of the material in this chapter, we recommend that you consider the items listed below and:

(a) Satisfy yourself that you can define each concept and notation and can describe each method.

(b) Give a least one reason why the item was included in this chapter.

(c) Think of at least one example of each concept and at least one situation in which each fact or method would be useful.

CONCEPTS AND NOTATION

propositional calculus
  proposition
  logical connectives $\neg$, $\vee$, $\wedge$, $\rightarrow$, $\leftrightarrow$
  $\forall$, $\exists$ notation
  converse, contrapositive, counterexample
  necessary condition, sufficient condition
  compound proposition
    truth table
    tautology, contradiction
    logical equivalence $\Leftrightarrow$, logical implication $\Rightarrow$
methods of proof
  direct, indirect, by contradiction, by cases
  vacuous, trivial
  constructive, nonconstructive
formal proof
  hypothesis, conclusion, theorem
  rule of inference
  vertical listing of proof
analysis of arguments
  valid inference, valid proof, fallacy

Basic logical equivalences [Table 1 of §2.2].

Basic logical implications [Table 2 of §2.2].

Substitution Rules (a) and (b) in §2.4.

Basic rules of inference [Table 1 of §2.4].

There is a formal proof of C from H if and only if $H \Rightarrow C$ is true.

METHODS

Use of:

Truth tables, especially to verify logical equivalences and implications.

DeMorgan laws to eliminate $\vee$ or $\wedge$.

Rules of inference to construct formal proofs.

Formal symbolism to analyze informal arguments.

RELATIONS

3

One frequently wants to compare or contrast various members of a set, perhaps to arrange them in some appropriate order or to group together those with similar properties. The mathematical framework to describe this kind of organization of sets is the theory of relations. In this chapter we introduce the idea of relations and their connections with digraphs and matrices.

§ 3.1 Relations

Given sets S and T, a **binary relation** from S to T is any subset R of $S \times T$. This is a general, all-purpose definition. Throughout the book we will see a variety of special kinds of interesting relations.

EXAMPLE 1 A mail-order record company has a list L of customers. Each customer indicates interest in certain categories of recordings: classical, easy-listening, Latin, religious, popular, rock, etc. Let C be the set of possible categories. The set of all ordered pairs (name, category selected) is a relation R from L to C. This relation might contain such pairs as (K. A. Ross, classical), (C. R. B. Wright, classical) and (C. R. B. Wright, thrash metal). ∎

EXAMPLE 2 A university would be interested in the relation R_1 consisting of all ordered pairs whose first entries are students and whose second entries are the courses the students are currently enrolled in. This relation is from the set S of university students to the set C of courses offered. Note that if student s in S is fixed, then $\{c \in C : (s, c) \in R_1\}$ is the set of courses taken by s. On the other hand, if course c in C is fixed, then $\{s \in S : (s, c) \in R_1\}$ is the class list for the course.

Another relation R_2 consists of all ordered pairs whose first entries are courses and whose second entries are the departments for which the

$(\text{student}, \text{courses})$

$\downarrow \qquad\qquad \downarrow$

$\text{set } S \qquad \text{set } C$

(courses, requirement of)

course is a major requirement. Thus R_2 is a relation from C to the set D of departments in the university. For fixed $c \in C$, $\{d \in D : (c, d) \in R_2\}$ is the set of departments for which c is a major requirement. For fixed $d \in D$, $\{c \in C : (c, d) \in R_2\}$ is the list of courses required for that department's majors. A computerized degree-checking program would need to use a data structure that contained enough information to determine relations R_1 and R_2. ∎

EXAMPLE 3   (a) Consider a set P of programs written to be carried out on a computer, and a catalog C of canned programs available for use. We get a relation from C to P if we say that a canned program c is related to a program p in P provided that p calls c as a subroutine. A frequently used c might be related to a number of p's, while a c that is never called is related to no p.

(b) A translator from decimal representations to binary representations can be viewed as the relation consisting of all ordered pairs whose first entries are allowable decimal representations and whose second entries are the corresponding binary representations. Actually, this relation is a function. ∎

Recall that in § 1.3 we indicated how functions can be identified with their graphs and, regarded that way, as sets of ordered pairs. In fact, if $f : S \to T$ we identified f with the set

$$R_f = \{(x, y) \in S \times T : y = f(x)\},$$

which is a relation from S to T. Of course, not all relations are functions. From this point of view a **function from S to T** is a special kind of relation R from S to T, one such that

for each $x \in S$ there is exactly one $y \in T$ with $(x, y) \in R$.

This characterization is simply a restatement of the definition in § 1.3. Thus functions are the relations for which functional notation f makes sense: $f(x)$ is that unique element in T such that $(x, f(x))$ belongs to R_f.

In case $S = T$, we will say that a subset R of $S \times S$ is a **relation on S**.

EXAMPLE 4   (a) Any set S has the very basic "equality relation": $E = \{(x, x) : x \in S\}$. Thus two elements in S satisfy this relation if and only if they are identical. We normally write $=$ for this relation. Thus $(x, y) \in E$ if and only if $x = y$.

(b) For the set $\mathbb{R}$ the familiar inequality relation $\leq$ can be viewed as a subset R of $\mathbb{R} \times \mathbb{R}$, namely the set $R = \{(x, y) : x \leq y\}$. Since $(x, y) \in R$ if and only if $x \leq y$, we normally write the relation as $\leq$. Note that the familiar properties

(R) $x \leq x$ for all $x \in \mathbb{R}$,
(AS) $x \leq y$ and $y \leq x$ imply that $x = y$,
(T) $x \leq y$ and $y \leq z$ imply that $x \leq z$,

can be written as

(R) $(x, x) \in R$ for all $x \in \mathbb{R}$.
(AS) $(x, y) \in R$ and $(y, x) \in R$ imply that $x = y$,
(T) $(x, y) \in R$ and $(y, z) \in R$ imply that $(x, z) \in R$.

Here the labels (R), (AS) and (T) refer to "reflexive," "antisymmetric" and "transitive," terms that we will be using in connection with arbitrary relations on a set S.

(c) The strict inequality relation $<$ on $\mathbb{R}$ is also a relation, of course, and corresponds to the set $R = \{(x, y) : x < y\}$. This relation satisfies the following properties:

(AR) $x < x$ *never* holds,
(T) $x < y$ and $y < z$ imply that $x < z$.

These can be rewritten as

(AR) $(x, x) \notin R$ for all $x \in \mathbb{R}$,
(T) $(x, y) \in R$ and $(y, z) \in R$ imply that $(x, z) \in R$.

Here (AR) refers to "antireflexive" and (T) again refers to "transitive." ■

Let p be a fixed integer greater than 1. Consider integers m and n. We say that m **is congruent to** n **modulo** p and we write $m \equiv n \ (\textbf{mod } p)$ provided that $m - n$ is a multiple of p. This defines what is called a **congruence relation** on the set $\mathbb{Z}$ of integers. We will return to this in §3.6. In that section we will see that

(R) $m \equiv m \ (\text{mod } p)$ for all $m \in \mathbb{Z}$,
(S) $m \equiv n \ (\text{mod } p)$ implies $n \equiv m \ (\text{mod } p)$,
(T) $m \equiv n \ (\text{mod } p)$ and $n \equiv r \ (\text{mod } p)$ imply $m \equiv r \ (\text{mod } p)$.

For the corresponding formal relation

$$R = \{(m, n) \in \mathbb{Z} \times \mathbb{Z} : m \equiv n \ (\text{mod } p)\},$$

these properties become

(R) $(m, m) \in R$ for all $m \in \mathbb{Z}$,
(S) $(m, n) \in R$ implies $(n, m) \in R$,
(T) $(m, n) \in R$ and $(n, r) \in R$ imply $(m, r) \in R$.

Here the labels (R), (S) and (T) refer to "reflexive," "symmetric" and "transitive." Note that this usage of reflexive and transitive is consistent with that in Example 4.

In general, we define a relation R on a set S to be **reflexive, antireflexive, symmetric, antisymmetric** or **transitive** if it satisfies the corresponding condition:

(R) $(x, x) \in R$ for all $x \in S$,
(AR) $(x, x) \notin R$ for all $x \in S$,
(S) $(x, y) \in R$ implies $(y, x) \in R$ for all $x, y \in S$,
(AS) $(x, y) \in R$ and $(y, x) \in R$ imply $x = y$,
(T) $(x, y) \in R$ and $(y, z) \in R$ imply $(x, z) \in R$.

Consider again an arbitrary relation R from a set S to a set T. That is, $R \subseteq S \times T$. The **converse relation** $R^{\leftarrow}$ is the relation from T to S defined by

$$R^{\leftarrow} = \{(y, x) \in T \times S : (x, y) \in R\}.$$

Since every function $f: S \to T$ is a relation, its converse $f^{\leftarrow}$ always exists:

As a relation $f^{\leftarrow} = \{(y, x) \in T \times S : y = f(x)\}.$

This relation is a function precisely when f is an invertible function as defined in §1.4 and, in this case, we have $f^{\leftarrow} = f^{-1}$.

EXAMPLE 5   (a) Recall that if $f: S \to T$ is a function and $A \subseteq S$, then the image of A under f is

$$f(A) = \{f(x) : x \in A\} = \{y \in T : y = f(x) \text{ for some } x \in A\}.$$

If we view f as the relation R_f, then this set equals

$$\{y \in T : (x, y) \in R_f \text{ for some } x \in A\}.$$

Similarly, for any relation R from S to T we can define

$$R(A) = \{y \in T : (x, y) \in R \text{ for some } x \in A\}.$$

Since $R^{\leftarrow}$ is a relation from T to S, for $B \subseteq T$ we also have

$$R^{\leftarrow}(B) = \{x \in S : (y, x) \in R^{\leftarrow} \text{ for some } y \in B\}$$

$$= \{x \in S : (x, y) \in R \text{ for some } y \in B\}.$$

If R is actually R_f for a function f from S to T, this gives

$$R_f^{\leftarrow}(B) = \{x \in S : y = f(x) \text{ for some } y \in B\} = \{x \in S : f(x) \in B\},$$

which is exactly the definition we gave for $f^{\leftarrow}(B)$ in §1.4.

(b) For a concrete example of part (a), let S be a set of suppliers and T a set of products, and define $(x, y) \in R$ if supplier x sells product y. For a given set A of suppliers, the set $R(A)$ is the set of products sold by at least one member of A. For a given set B of products, $R^{\leftarrow}(B)$ is the set of suppliers who sell at least one product in B. The relation R is R_f for a function f from S to T if and only if each supplier sells exactly one product. ■

It is sometimes handy to draw pictures of relations on small sets.

EXAMPLE 6 (a) Consider the relation R_1 on the set $\{0, 1, 2, 3\}$ defined by $\leq$; thus $(m, n) \in R_1$ if and only if $m \leq n$. A picture of R_1 is given in Figure 1(a). Observe that we have drawn an arrow from m to n whenever $(m, n) \in R_1$, although we left off the arrowheads on the "loops" $0 \to 0$, $1 \to 1$, etc.

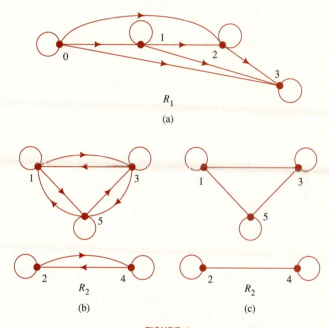

FIGURE 1

(b) Let R_2 be the relation on $\{1, 2, 3, 4, 5\}$ defined by $(m, n) \in R_2$ if and only if $m - n$ is even. A picture is given in Figure 1(b).

(c) The picture of the converse $R_1^{\leftarrow}$ relation is obtained by reversing all the arrows in Figure 1(a). The loops are unchanged.

(d) The picture of the converse $R_2^{\leftarrow}$ is also obtained by reversing all the arrows [in Figure 1(b)], but this time we obtain the same picture. This is because R_2 is symmetric and so $R_2^{\leftarrow} = R_2$; see Exercise 11(a). ■

When a relation is symmetric, such as R_2 in Figure 1(b), for every arrow drawn there is a reverse arrow. So it is redundant to draw each such pair of arrows. An equally informative picture is drawn in Figure 1(c). Just as with city maps, arrows signify one-way streets and plain lines signify two-way streets.

EXERCISES 3.1

1. For the following relations on $S = \{0, 1, 2, 3\}$, specify which of the properties (R), (AR), (S), (AS) and (T) the relations satisfy.

   (a) $(m, n) \in R_1$ if $m + n = 3$ (b) $(m, n) \in R_2$ if $m - n$ is even

   (c) $(m, n) \in R_3$ if $m \leq n$ (d) $(m, n) \in R_4$ if $m + n \leq 4$

   (e) $(m, n) \in R_5$ if $\max\{m, n\} = 3$

2. Let $A = \{0, 1, 2\}$. Each of the statements below defines a relation R on A by $(m, n) \in R$ if the statement is true for m and n. Write each of the relations as a set of ordered pairs.

   (a) $m \leq n$ (b) $m < n$ (c) $m = n$

   (d) $mn = 0$ (e) $mn = m$ (f) $m + n \in A$

   (g) $m^2 + n^2 = 2$ (h) $m^2 + n^2 = 3$ (i) $m = \max\{n, 1\}$

3. Which of the relations in Exercise 2 are reflexive? symmetric?

4. The following binary relations are defined on $\mathbb{N}$.

   (a) Write the binary relation R_1 defined by $m + n = 5$ as a set of ordered pairs.

   (b) Do the same for R_2 defined by $\max\{m, n\} = 2$.

   (c) The binary relation R_3 defined by $\min\{m, n\} = 2$ consists of infinitely many ordered pairs. List five of them.

5. For each of the relations in Exercise 4, specify which of the properties (R), (AR), (S), (AS) and (T) it satisfies.

6. Consider the relation R on $\mathbb{Z}$ defined by $(m, n) \in R$ if and only if $m^3 - n^3 \equiv 0 \pmod 5$. Which of the properties (R), (AR), (S), (AS) and (T) are satisfied by R?

7. (a) Consider the **empty relation** $\varnothing$ on a nonempty set S. Which of the properties (R), (AR), (S), (AS) and (T) does $\varnothing$ possess?

   (b) Repeat part (a) for the **universal relation** $U = S \times S$ on S.

8. Give an example of a relation that is:

   (a) antisymmetric and transitive but not reflexive,

   (b) symmetric but not reflexive or transitive.

9. Let R_1 and R_2 be binary relations on a set S.

   (a) Show that $R_1 \cap R_2$ is reflexive if R_1 and R_2 are.

   (b) Show that $R_1 \cap R_2$ is symmetric if R_1 and R_2 are.

   (c) Show that $R_1 \cap R_2$ is transitive if R_1 and R_2 are.

10. Let R_1 and R_2 be binary relations on a set S.

    (a) Must $R_1 \cup R_2$ be reflexive if R_1 and R_2 are?

    (b) Must $R_1 \cup R_2$ be symmetric if R_1 and R_2 are?

    (c) Must $R_1 \cup R_2$ be transitive if R_1 and R_2 are?

11. Let R be a binary relation on a set S.

   (a) Prove that R is symmetric if and only if $R = R^\leftarrow$.

   (b) Prove that R is antisymmetric if and only if $R \cap R^\leftarrow \subseteq E$, where $E = \{(x, x) : x \in S\}$.

12. Let R_1 and R_2 be binary relations from a set S to a set T.

   (a) Show that $(R_1 \cup R_2)^\leftarrow = R_1^\leftarrow \cup R_2^\leftarrow$.

   (b) Show that $(R_1 \cap R_2)^\leftarrow = R_1^\leftarrow \cap R_2^\leftarrow$.

   (c) Show that if $R_1 \subseteq R_2$, then $R_1^\leftarrow \subseteq R_2^\leftarrow$.

13. Draw pictures of each of the relations in Exercise 1. Don't use arrows if the relation is symmetric.

14. Draw pictures of each of the relations in Exercise 2. Don't use arrows if the relation is symmetric.

§ 3.2   Digraphs and Graphs

You are already undoubtedly familiar with the idea of a graph as a picture of a function. The word "graph" is also used to described a different kind of structure that arises in a variety of natural settings. In a loose sense these new graphs are diagrams which, properly interpreted, contain information. The graphs we are concerned with are like road maps, circuit diagrams or flowcharts in the sense that they depict connections or relationships between various parts of the diagram.

The diagrams in Figure 1 on the next page are from a variety of settings. Figure 1(a) shows a simple flowchart. Figure 1(b) might represent five warehouses of a trucking firm and truck routes between them, labeled with their distances. Figure 1(c) could be telling us about the probability that a rat located in one of four cages will move to one of the other three or stay in its own cage. Figure 1(d) might depict possible outcomes of a repeated experiment such as coin tossing [Heads or Tails]. The diagrams in Figure 1 of §3.1 tell us which pairs of vertices belong to the relations R_1 and R_2. What do all of these diagrams have in common? Each consists of a collection of objects—boxes, circles or dots—and some lines, possibly curved, that connect objects. Sometimes the lines are directed; that is, they are arrows.

The essential features of a directed graph [digraph for short] are its objects and directed lines. Specifically, a **digraph** G consists of two sets, the nonempty set $V(G)$ of **vertices of** G and the set $E(G)$ of **edges** of G, together with a function γ [lowercase Greek gamma] from $E(G)$ to $V(G) \times V(G)$. If e is an edge of G and $\gamma(e) = (p, q)$, then p is called the **initial vertex** of e and q the **terminal vertex** of e, and we say that e **goes from p to q**. This definition makes sense if $V(G)$ or $E(G)$ is infinite, but because our applications are to finite sets we will assume in this section that $V(G)$ and $E(G)$ are finite.

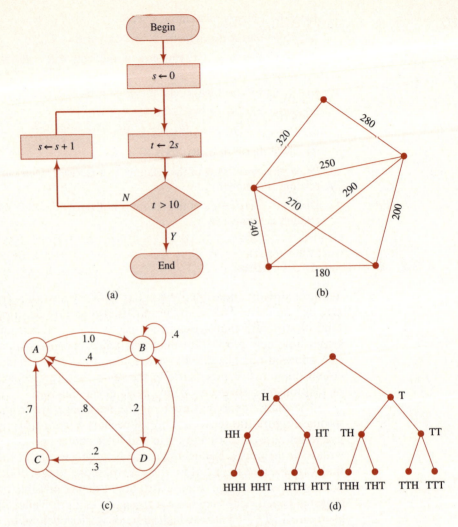

FIGURE 1

A **picture** of the digraph G is a diagram consisting of points, corresponding to the members of $V(G)$, and arrows, corresponding to the members of $E(G)$, such that if $\gamma(e) = (p, q)$, then the arrow corresponding to e goes from the point labeled p to the point labeled q.

EXAMPLE 1   Consider the digraph G with $V(G) = \{w, x, y, z\}$, $E(G) = \{a, b, c, d, e, f, g, h\}$ and γ given by the table in Figure 2(a). The diagrams in Figures 2(b) and 2(c) are both pictures of G. In Figure 2(b) we labeled the arrows to make the correspondence to $E(G)$ plain. In Figure 2(c) we simply labeled the points and let the arrows take care of themselves. This causes no confusion because, in this case, there are no **parallel edges**; i.e., there is at

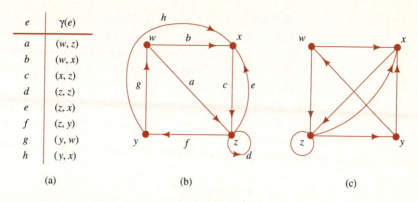

e	$\gamma(e)$
a	(w, z)
b	(w, x)
c	(x, z)
d	(z, z)
e	(z, x)
f	(z, y)
g	(y, w)
h	(y, x)

(a)   (b)   (c)

FIGURE 2

most one edge with a given initial vertex and terminal vertex. In other words, the function γ is one-to-one. Note also that we omitted the arrow-head on edge d since z is clearly both the initial and terminal vertex. ∎

If $\gamma: E(G) \to V(G) \times V(G)$ is one-to-one, then we can identify the edges e with their images $\gamma(e)$ in $V(G) \times V(G)$, and consider $E(G)$ to be a subset of $V(G) \times V(G)$. In fact, some people define digraphs to have $E(G) \subseteq V(G) \times V(G)$, and call the more general digraphs we are considering "directed multigraphs."

Given a picture of G we can reconstruct G itself, since the arrows tell us all about γ. We will commonly describe digraphs by giving pictures of them, rather than tables of γ, but the pictorial description is chosen just for human convenience. A computer stores a digraph by storing the function γ in one way or another.

Many of the important questions connected with digraphs can be stated in terms of sequences of edges leading from one vertex to another. A **path** in a digraph G is a sequence of edges such that the terminal vertex of one edge is the initial vertex of the next. Thus if $e_1, \ldots, e_n$ are in $E(G)$, then $e_1 e_2 \cdots e_n$ is a path provided that there are vertices $x_1, x_2, \ldots, x_n, x_{n+1}$ so that $\gamma(e_i) = (x_i, x_{i+1})$ for $i = 1, 2, \ldots, n$. We say that $e_1 e_2 \cdots e_n$ is a path of **length** n **from** x_1 to x_{n+1}. The path is **closed** if $x_1 = x_{n+1}$.

EXAMPLE 2   In the digraph G in Figure 2 the sequence $fgae$ is a path of length 4 from z to x. The sequences $cecec$ and $fgafhc$ are also paths, but fa is not a path since $\gamma(f) = (z, y)$, $\gamma(a) = (w, z)$ and $y \neq w$. The paths $fgafhc$, $cece$ and d are closed; $fhce$ and df are not. ∎

A path $e_1 \cdots e_n$ with $\gamma(e_i) = (x_i, x_{i+1})$ has an associated sequence of vertices $x_1 x_2 \cdots x_n x_{n+1}$. If each e_i is the only edge from x_i to x_{i+1}, then this sequence of vertices uniquely determines the path, and we can describe the path simply by listing the vertices in succession.

EXAMPLE 3   (a) In Figure 2 the path $f\,g\,a\,e$ has vertex sequence $z\,y\,w\,z\,x$. Observe that this vertex sequence alone determines the path. The path can be recovered from $z\,y\,w\,z\,x$ by looking at Figure 2(b) or 2(c) or using the table of γ in Figure 2(a). Since the digraph has no parallel edges, all its paths are determined by their vertex sequences.

(b) For the digraph pictured in Figure 3 the vertex sequence $y\,z\,z\,z$ corresponds only to the path fgg, but the sequence $y\,v\,w\,z$ belongs to both $c\,a\,e$ and $c\,b\,e$. ■

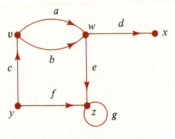

FIGURE 3

A closed path of length at least 1 with vertex sequence $x_1\,x_2\cdots x_n\,x_1$ is called a **cycle** if $x_1,\ldots,x_n$ are all different. The language of graph theory has not been standardized: various authors use "circuit" and "loop" for what we call a cycle, and "cycle" is sometimes used as a name for a closed path. A digraph with no cycles is called **acyclic**. Some people call an acyclic digraph a **DAG**, short for "directed acyclic graph." A path is **acyclic** if the digraph consisting of the vertices and edges of the path is acyclic.

EXAMPLE 4   In Figure 2 the path $a\,f\,g$ is a cycle since its vertex sequence is $w\,z\,y\,w$. Similarly, the paths $c\,f\,h$ and $c\,f\,g\,b$, with vertex sequences $x\,z\,y\,x$ and $x\,z\,y\,w\,x$, are cycles. The short path $c\,e$ and the loop d are also cycles, since their vertex sequences are $x\,z\,x$ and $z\,z$, respectively. The path $c\,f\,g\,a\,e$ is not a cycle, since its vertex sequence is $x\,z\,y\,w\,z\,x$ and the vertex z is repeated. ■

As we saw in Example 6 of §3.1, a relation R on a set S determines a digraph G in a natural way: let $V(G) = S$ and put an edge from v to w whenever $(v, w) \in R$. We can reverse this procedure. Given a digraph G and vertices v and w in $V(G)$, call v **adjacent** to w if there is an edge in $E(G)$ from v to w. The **adjacency relation** A on the set $V(G)$ of vertices is defined by $(v, w) \in A$ if and only if v is adjacent to w. The relation A can be perfectly general and does not need to have any special properties. It will be reflexive only if G has a loop at each vertex, and symmetric only if there is an edge from v to w whenever there is one from w to v.

EXAMPLE 5   (a) Consider the digraph in Figure 2. The adjacency relation consists of the ordered pairs (w, z), (w, x), (x, z), (z, z), (z, x), (z, y), (y, w) and (y, x). In other words, A consists of the images of γ listed in Figure 2(a). In general, $A = \gamma(E(G)) \subseteq V(G) \times V(G)$.

(b) The adjacency relation A for the digraph in Figure 3 consists of the ordered pairs (v, w), (w, x), (w, z), (z, z), (y, z) and (y, v). We cannot recover the digraph from A because A does not convey information about multiple edges. Since (v, w) belongs to A we know that the digraph has *at least one* edge from v to w, but we cannot tell that it has exactly two edges.

(c) The digraph obtained from Figure 3 by removing the edge a has the same adjacency relation as the original digraph. ∎

The last example shows that different digraphs may have the same adjacency relation. However, if we restrict our attention to digraphs without multiple edges, there is a one-to-one correspondence between such digraphs and relations.

If we ignore the arrows on our edges, i.e., their directions, we obtain what are called graphs. Figures 1(b) and 1(d) of this section and Figure 1(c) in §3.1 are all pictures of graphs. Graphs can have multiple edges; to see an example, just omit the arrowheads in Figure 3. The ideas and terminology for studying graphs are similar to those for digraphs.

Instead of being associated with an ordered pair of vertices, as edges in digraphs are, an undirected edge has an unordered set of vertices. Following the pattern we used for digraphs, we define an [undirected] **graph** to consist of two sets, the set $V(G)$ of **vertices** of G and set $E(G)$ of **edges** of G, together with a function γ from $E(G)$ to the set $\{\{u, v\} : u, v \in V(G)\}$ of all subsets of $V(G)$ with one or two members. For an edge e in $E(G)$ the members of $\gamma(e)$ are called the **vertices** of e or the **endpoints** of e; we say that e **joins** its endpoints. A **loop** is an edge with only one endpoint. Distinct edges e and f with $\gamma(e) = \gamma(f)$ are called **parallel** or **multiple** edges.

These precise definitions make it clear that a computer might view a graph as two sets together with a function γ that specifies the endpoints of the edges.

What we have just described is called a **multigraph** by some authors, who reserve the term "graph" for those graphs with no loops or parallel edges. If there are no parallel edges, then γ is one-to-one and the sets $\gamma(e)$ uniquely determine the edges e. That is, there is only one edge for each set $\gamma(e)$. In this case, we often dispense with the set $E(G)$ and the function γ and simply write the edges as sets, like $\{u, v\}$ or $\{u\}$, or as vertex sequences, like uv, vu or uu. Thus we will commonly write $e = \{u, v\}$; we will also sometimes write $e = \{u, u\}$ instead of $e = \{u\}$ if e is a loop with vertex u.

A **picture** of a graph G is a diagram consisting of points corresponding to the vertices of G and arcs or lines corresponding to edges, such that if $\gamma(e) = \{u, v\}$, then the arc for the edge e joins the points labeled u and v.

EXAMPLE 6   (a) The graphs in Figures 4(a) and 4(b) have 5 vertices and 7 edges. The crossing of the two lines in Figure 4(b) is irrelevant and is just a peculiarity of our drawing. The graph in Figure 4(c) has 4 vertices and 6 edges. It has multiple edges joining v and w, and a loop at w.

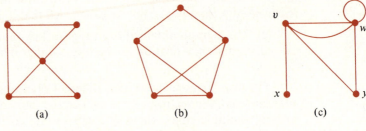

(a)   (b)   (c)

FIGURE 4

(b) If we omit the direction arrows from the edges in Figure 2 we get the picture in Figure 5(a). A table of γ for this graph is given in Figure 5(b). This graph has parallel edges c and e joining x and z, and has a loop d with vertex z. The graph pictured in Figure 5(c) has no parallel edges, so for that graph a description such as "the edge $\{x, z\}$" is unambiguous. The same phrase could not be applied to the graph in Figure 5(a). ■

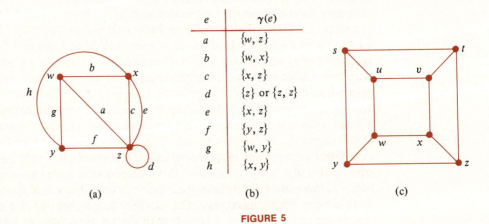

e	$\gamma(e)$
a	$\{w, z\}$
b	$\{w, x\}$
c	$\{x, z\}$
d	$\{z\}$ or $\{z, z\}$
e	$\{x, z\}$
f	$\{y, z\}$
g	$\{w, y\}$
h	$\{x, y\}$

(a)   (b)   (c)

FIGURE 5

A sequence of edges that link up with each other is called a **path**. To illustrate the idea, we redraw Figure 1(c) in Figure 6(a) and label the edges as well as the vertices. Examples of paths are $d b f e$ [Figure 6(b)] and $cfeba$ [Figure 6(c)]. Note that the drawing alone does not tell us which path we have in mind: Figure 6(c) is also a picture of the paths $bafec$ and $cafeb$. Paths can repeat edges: $babefaab$ is a path. The **length** of a path is the number of edges in the path. Thus $babefaab$ has length 8.

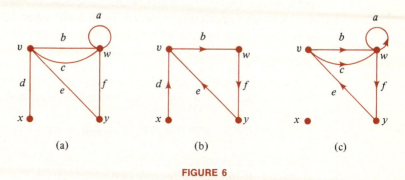

FIGURE 6

Consecutive edges in a path must have a vertex in common. So a path determines a sequence of vertices. The vertex sequences for the paths discussed above are shown in Figure 7.

Note that the number of vertices in a vertex sequence is one larger than the number of edges in the path. When a loop appears in a path, its vertex is repeated in the vertex sequence. Vertex sequences treat parallel edges the same, so different paths, such as $bafec$ and $cafeb$, can have the same vertex sequence. If a graph has no parallel edges or multiple loops,

The path	Its vertex sequence
$dbfe$	$x\,v\,w\,y\,v$
$cfeba$	$v\,w\,y\,v\,w\,w$
$bafec$	$v\,w\,w\,y\,v\,w$
$cafeb$	$v\,w\,w\,y\,v\,w$
$babefaab$	$v\,w\,w\,v\,y\,w\,w\,w\,v$

FIGURE 7

then vertex sequences do uniquely determine paths. In this case the edges can be described by just listing the two vertices that they connect, and we may describe a path by its vertex sequence.

In general, a **path** of **length** n from the vertex u to the vertex v is a sequence $e_1 \cdots e_n$ of edges together with a sequence $x_1 \cdots x_{n+1}$ of vertices with $\gamma(e_i) = \{x_i, x_{i+1}\}$ for $i = 1, \ldots, n$ and $x_1 = u$, $x_{n+1} = v$. We will not have occasion to use paths of length 0. If $e_1 e_2 \cdots e_n$ is a path from u to v with vertex sequence $x_1 x_2 \cdots x_{n+1}$, then $e_n \cdots e_2 e_1$ with vertex sequence $x_{n+1} x_n \cdots x_1$ is a path from v to u. We may speak of either of these paths as a **path between** u and v. If $u = v$, the path is **closed**.

The edge sequence of a path usually determines the vertex sequence, and we will sometimes use phrases such as "the path $e_1 e_2 \cdots e_n$" without mentioning vertices. There is a slight fuzziness here, since ee, eee, etc., don't specify which end of e to start at. A similar problem arises with ef in case e and f are parallel. If a graph has no parallel edges, then the vertex sequence completely determines the edge sequence. In that setting, or if the actual choice of edges is unimportant, we will commonly use vertex sequences as descriptions for paths.

Just as with digraphs, we can define an **adjacency relation** A for a graph: $(u, v) \in A$ provided that $\{u, v\}$ is an edge of the graph. For the remainder of this section, A will denote an adjacency relation either for a digraph or for a graph. To get a transitive relation from A we must consider chains of edges, from u_1 to u_2, u_2 to u_3, etc. The appropriate notion is reachability. Define the **reachable relation** R on $V(G)$ by

$(v, w) \in R$ if there is a path of length at least 1 in G from v to w.

Then R is a transitive relation. Since we require all paths to have length at least 1, R might not be reflexive.

EXAMPLE 7   (a) All the vertices in the digraph of Figure 2 are reachable from all other vertices. Hence the reachable relation R consists of all possible ordered pairs; this is the so-called universal relation.

(b) The reachable relation for the digraph in Figure 3 consists of (v, w), (v, x), (v, z), (w, x), (w, z), (y, v), (y, w), (y, x), (y, z) and (z, z). Note that every vertex can be reached from y except y itself. Also, z is the only vertex that can be reached from itself.

(c) All the graphs in Figures 4 and 5 are connected, in the sense that every vertex can be reached from every other vertex. So in each case the reachable relation is the universal relation. ∎

EXERCISES 3.2

1. Give a table of the function γ for each of the digraphs pictured in Figure 8.
2. Draw a picture of the digraph G with $V(G) = \{w, x, y, z\}$, $E(G) = \{a, b, c, d, e, f, g\}$ and γ given by the following table.

e	a	b	c	d	e	f	g
$\gamma(e)$	(x, w)	(w, x)	(x, x)	(w, z)	(w, y)	(w, z)	(z, y)

3. Which of the following vertex sequences describe paths in the digraph pictured in Figure 9(a)?

   (a) $zyvwt$

   (b) $xzwt$

   (c) $vstx$

   (d) $zysu$

   (e) $xzyvs$

   (f) $suxt$

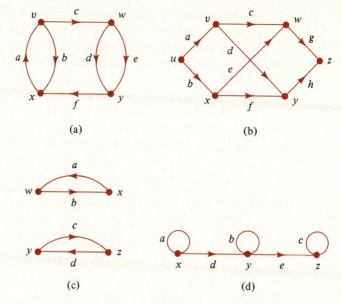

(a)   (b)

(c)   (d)

FIGURE 8

4. Find the length of a shortest path from x to w in the digraph shown in Figure 9(a).

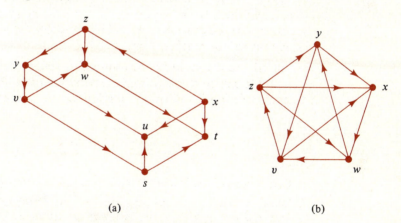

(a)   (b)

FIGURE 9

5. Consider the digraph pictured in Figure 9(b). Describe an acyclic path

   (a) from x to y (b) from y to z (c) from v to w

   (d) from x to z (e) from z to v

6. There are four basic blood types: A, B, AB and O. Type O can donate to any of the four types, A and B can donate to AB as well as to their own types, but type AB can only donate to AB. Draw a digraph that presents this information. Is the digraph acyclic?

7. Give an example of a digraph with vertices x, y and z in which there is a cycle with x and y as vertices and another cycle with y and z, but there is no cycle with x and z as vertices.

8. Determine the reachable relation for the digraphs in Figure 8, parts (a), (c) and (d).

9. (a) Which of the following belong to the reachable relation for the digraph in Figure 8(b)? (v, u), (v, v), (v, w), (v, x), (v, y), (v, z)

   (b) Which of the following belong to the reachable relation for the digraph in Figure 9(a)? (v, s), (v, t), (v, u), (v, v), (v, w), (v, x), (v, y), (v, z)

   (c) Which of the following belong to the reachable relation for the digraph in Figure 9(b)? (v, v), (v, w), (v, x), (v, y), (v, z)

10. Which of the following vertex sequences correspond to paths in the graph of Figure 10(a)?

   (a) $z\,x\,w$ (b) $w\,x\,z\,x\,w\,y\,w\,w$ (c) $w\,w\,x\,z$

   (d) $w\,x\,z\,z$ (e) $z\,x\,w\,y\,y\,w\,z$ (f) $w\,x\,w$

   (g) $y\,y\,w\,w$

11. Give the length of each path in Exercise 10.

12. Which paths in Exercise 10 are closed paths?

13. List the parallel edges in the graphs of Figure 10.

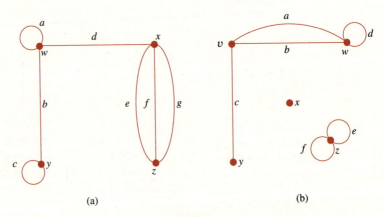

(a)   (b)

FIGURE 10

14. How many loops are there in each of the graphs of Figure 10?

15. Give the adjacency relation A and the reachable relation R for each of the graphs of Figure 10.

16. For the graph in Figure 10(a), give an example of each of the following. Be sure to specify the edge sequence and the vertex sequence.

   (a) a path of length 2 from w to z.

   (b) a path of length 4 from z to itself.

(c) a path of length 5 from z to itself.

(d) a path of length 3 from w to x.

17. For the graph in Figure 10(b), given an example of each of the following. Be sure to specify the edge sequence and the vertex sequence.

    (a) a path of length 3 from y to w.

    (b) a path of length 3 from v to y.

    (c) a path of length 4 from v to y.

    (d) a path of length 3 from z to itself.

§ 3.3   Matrices

We saw the close connection between digraphs and relations in §3.2. Matrices, the subject of this section, are an important tool for describing both digraphs and relations, as well as for many other purposes. We will see that the topics of the first three sections of this chapter give us three different ways of viewing equivalent structures: relations, digraphs without parallel edges, and matrices whose entries are 0's and 1's.

In general a **matrix** is a rectangular array. It is traditional to use capital letters, such as **A**, for matrices. If we denote the entry in the ith row and jth column by a_{ij}, then we can write

$$\mathbf{A} = \begin{bmatrix} a_{11} & a_{12} & \cdots & a_{1n} \\ a_{21} & a_{22} & \cdots & a_{2n} \\ a_{31} & a_{32} & \cdots & a_{3n} \\ \vdots & \vdots & & \vdots \\ a_{m1} & a_{m2} & \cdots & a_{mn} \end{bmatrix},$$

or simply $\mathbf{A} = [a_{ij}]$. This matrix has m horizontal rows and n vertical columns and is called an $m \times n$ **matrix**. Whenever double indexes are used in matrix theory, rows precede columns! Sometimes we denote the entry in the ith row and jth column by $\mathbf{A}[i, j]$; this notation is preferable in computer science since it avoids subscripts. In this book a **matrix** has real entries unless otherwise specified.

EXAMPLE 1   (a) The matrix

$$\mathbf{A} = \begin{bmatrix} 2 & -1 & 0 & 3 & 2 \\ 1 & -2 & 1 & -1 & 3 \\ 3 & 0 & 1 & 2 & -3 \end{bmatrix}$$

is a 3×5 matrix. If we write $\mathbf{A} = [a_{ij}]$, then $a_{11} = 2$, $a_{31} = 3$, $a_{13} = 0$, $a_{35} = -3$, etc. If we use the notation $\mathbf{A}[i, j]$, then $\mathbf{A}[1, 2] = -1$, $\mathbf{A}[2, 1] = 1$, $\mathbf{A}[2, 2] = -2$, etc.

(b) If **B** is the 3×4 matrix defined by $\mathbf{B}[i, j] = i - j$, then $\mathbf{B}[1, 1] = 1 - 1 = 0$, $\mathbf{B}[1, 2] = 1 - 2 = -1$, etc., and so

$$\mathbf{B} = \begin{bmatrix} 0 & -1 & -2 & -3 \\ 1 & 0 & -1 & -2 \\ 2 & 1 & 0 & -1 \end{bmatrix}. \quad \blacksquare$$

Matrices are used in all the mathematical sciences. Because they are intimately connected with relations, they provide a convenient way to store information that connects two sets of data, so they have important uses in business, economics and computer science. They also arise in solving systems of linear equations. Many physical phenomena in nature can be described, at least approximately, by means of matrices. In addition, the set of $n \times n$ matrices has a very rich algebraic structure, which is of interest in itself and is also a source of inspiration in the study of more abstract algebraic structures. We will introduce various algebraic operations on matrices in this section and the next.

For positive integers m and n, we write $\mathfrak{M}_{m,n}$ for the set of all $m \times n$ matrices. Two matrices **A** and **B** in $\mathfrak{M}_{m,n}$ are **equal** provided that all their corresponding entries are equal; i.e., $\mathbf{A} = \mathbf{B}$ provided that $a_{ij} = b_{ij}$ for all i and j with $1 \le i \le m$ and $1 \le j \le n$. Matrices that have the same number of rows as columns are called **square matrices**. Thus **A** is a square matrix if **A** belongs to $\mathfrak{M}_{n,n}$ for some $n \in \mathbb{P}$. The **transpose** $\mathbf{A}^T$ of a matrix $\mathbf{A} = [a_{ij}]$ in $\mathfrak{M}_{m,n}$ is the matrix in $\mathfrak{M}_{n,m}$ whose entry in the ith row and jth column is a_{ji}. That is, $\mathbf{A}^T[i, j] = \mathbf{A}[j, i]$. For example, if

$$\mathbf{A} = \begin{bmatrix} 2 & -1 & 0 & 4 \\ 3 & 2 & -1 & 2 \\ 4 & 0 & 1 & 3 \end{bmatrix}, \quad \text{then} \quad \mathbf{A}^T = \begin{bmatrix} 2 & 3 & 4 \\ -1 & 2 & 0 \\ 0 & -1 & 1 \\ 4 & 2 & 3 \end{bmatrix}.$$

The first row in **A** becomes the first column in $\mathbf{A}^T$, etc.

Matrices that have only one row, i.e., $1 \times n$ matrices, are often called **row vectors**, while matrices that have only one column, i.e., $m \times 1$ matrices, are called **column vectors**. The transpose of a row vector is a column vector and the transpose of a column vector is a row vector. Thus $[2 \quad 4 \quad -3 \quad -1]$ is a row vector and its transpose

$$\begin{bmatrix} 2 \\ 4 \\ -3 \\ -1 \end{bmatrix}$$

is a column vector. We sometimes view an $m \times n$ matrix as composed of m row vectors or of n column vectors. Two matrices **A** and **B** can be added if they are the same size, that is, if they belong to the same set

$\mathfrak{M}_{m,n}$. In this case, the **matrix sum** is obtained by adding corresponding entries. More explicitly, if both $A = [a_{ij}]$ and $B = [b_{ij}]$ are in $\mathfrak{M}_{m,n}$, then $A + B$ is the matrix $C = [c_{ij}]$ in $\mathfrak{M}_{m,n}$ defined by

$$c_{ij} = a_{ij} + b_{ij} \quad \text{for} \quad 1 \le i \le m \quad \text{and} \quad 1 \le j \le n.$$

Equivalently, we define

$$(A + B)[i, j] = A[i, j] + B[i, j] \quad \text{for} \quad 1 \le i \le m \quad \text{and} \quad 1 \le j \le n.$$

Since m or n can be 1, this definition applies in particular to row vectors and to column vectors.

EXAMPLE 2   (a) Consider

$$A = \begin{bmatrix} 2 & 4 & 0 \\ -1 & 3 & 2 \\ -3 & 1 & 2 \end{bmatrix}, \quad B = \begin{bmatrix} 1 & 0 & 5 & 3 \\ 2 & 3 & -2 & 1 \\ 4 & -2 & 0 & 2 \end{bmatrix},$$

$$C = \begin{bmatrix} 3 & 1 & -2 \\ -5 & 0 & 2 \\ -2 & 4 & 1 \end{bmatrix}.$$

Then we have

$$A + C = \begin{bmatrix} 5 & 5 & -2 \\ -6 & 3 & 4 \\ -5 & 5 & 3 \end{bmatrix},$$

but $A + B$ and $B + C$ are not defined. Of course, the sums $A + A$, $B + B$ and $C + C$ are also defined; for example,

$$B + B = \begin{bmatrix} 2 & 0 & 10 & 6 \\ 4 & 6 & -4 & 2 \\ 8 & -4 & 0 & 4 \end{bmatrix}.$$

(b) Consider the row vectors

$$v_1 = [-2 \ \ 1 \ \ 2 \ \ 3], \quad v_2 = [4 \ \ 0 \ \ 3 \ \ -2], \quad v_3 = [1 \ \ 3 \ \ 5]$$

and the column vectors

$$v_4 = \begin{bmatrix} 1 \\ 2 \\ -3 \\ 2 \end{bmatrix}, \quad v_5 = \begin{bmatrix} 0 \\ 3 \\ -2 \end{bmatrix}, \quad v_6 = \begin{bmatrix} 4 \\ 1 \\ 5 \end{bmatrix}.$$

The only sums of distinct vectors here that are defined are

$$v_1 + v_2 = [2 \ \ 1 \ \ 5 \ \ 1] \quad \text{and} \quad v_5 + v_6 = \begin{bmatrix} 4 \\ 4 \\ 3 \end{bmatrix}. \quad \blacksquare$$

Elements in $\mathbb{R}^n$ are also often called **vectors**. We add them just as if they were row vectors:

$$(x_1, x_2, \ldots, x_n) + (y_1, y_2, \ldots, y_n) = (x_1 + y_1, x_2 + y_2, \ldots, x_n + y_n).$$

Before listing properties of addition we give a little more notation. Let **0** represent the $m \times n$ matrix all entries of which are 0. [Context will always make plain what size this matrix is.] For **A** in $\mathfrak{M}_{m,n}$ the matrix $-\mathbf{A}$, called the **negative of A**, is obtained by negating each entry in **A**. Thus if $\mathbf{A} = [a_{ij}]$, then $-\mathbf{A} = [-a_{ij}]$; equivalently, $(-\mathbf{A})[i,j] = -\mathbf{A}[i,j]$.

Theorem	For all **A**, **B** and **C** in $\mathfrak{M}_{m,n}$

  (a) $\mathbf{A} + (\mathbf{B} + \mathbf{C}) = (\mathbf{A} + \mathbf{B}) + \mathbf{C}$ [associative law]
  (b) $\mathbf{A} + \mathbf{B} = \mathbf{B} + \mathbf{A}$ [commutative law]
  (c) $\mathbf{A} + \mathbf{0} = \mathbf{0} + \mathbf{A} = \mathbf{A}$ [additive identity]
  (d) $\mathbf{A} + (-\mathbf{A}) = (-\mathbf{A}) + \mathbf{A} = \mathbf{0}$ [additive inverses]

Proof. These properties of matrix addition are reflections of corresponding properties of addition of real numbers and are easy to check. We check (a) and leave the rest to Exercise 12.

Say $\mathbf{A} = [a_{ij}]$, $\mathbf{B} = [b_{ij}]$ and $\mathbf{C} = [c_{ij}]$. The (i, j) entry of $\mathbf{B} + \mathbf{C}$ is $b_{ij} + c_{ij}$, so the (i, j) entry of $\mathbf{A} + (\mathbf{B} + \mathbf{C})$ is $a_{ij} + (b_{ij} + c_{ij})$. Similarly, the (i, j) entry of $(\mathbf{A} + \mathbf{B}) + \mathbf{C}$ is $(a_{ij} + b_{ij}) + c_{ij}$. Since addition of real numbers is associative, corresponding entries of $\mathbf{A} + (\mathbf{B} + \mathbf{C})$ and $(\mathbf{A} + \mathbf{B}) + \mathbf{C}$ are equal, so the matrices are equal. ∎

Since addition of matrices is associative, we can write $\mathbf{A} + \mathbf{B} + \mathbf{C}$ without causing ambiguity.

Matrices can be multiplied by real numbers, which in this context are often called **scalars**. Given **A** in $\mathfrak{M}_{m,n}$ and c in $\mathbb{R}$, $c\mathbf{A}$ is the $m \times n$ matrix whose (i, j) entry is ca_{ij}; thus $(c\mathbf{A})[i,j] = c\mathbf{A}[i,j]$. This multiplication is called **scalar multiplication** and $c\mathbf{A}$ is called the **scalar product**.

EXAMPLE 3   (a) If

$$\mathbf{A} = \begin{bmatrix} 2 & 1 & -3 \\ -1 & 0 & 4 \end{bmatrix},$$

then

$$2\mathbf{A} = \begin{bmatrix} 4 & 2 & -6 \\ -2 & 0 & 8 \end{bmatrix} \quad \text{and} \quad -7\mathbf{A} = \begin{bmatrix} -14 & -7 & 21 \\ 7 & 0 & -28 \end{bmatrix}.$$

(b) In general, the scalar product $(-1)\mathbf{A}$ is the negative $-\mathbf{A}$ of $\mathbf{A}$. ∎

We end this section by describing a matrix that is useful for studying finite relations, digraphs and graphs. We first consider a finite digraph G with vertex set $V(G)$. Let $v_1, v_2, \ldots, v_n$ be a list of the vertices in $V(G)$. The **adjacency matrix** is the $n \times n$ matrix $\mathbf{M}$ such that each entry $\mathbf{M}[i,j]$ is the number of edges from v_i to v_j. Thus $\mathbf{M}[i,j] = 0$ if there is no edge from v_i to v_j and $\mathbf{M}[i,j]$ is a positive integer otherwise.

EXAMPLE 4   (a) The digraph in Figure 1 has adjacency matrix

$$\mathbf{M} = \begin{bmatrix} 0 & 1 & 0 & 0 \\ 1 & 0 & 0 & 0 \\ 1 & 2 & 0 & 0 \\ 3 & 0 & 0 & 1 \end{bmatrix}.$$

Note that the matrix $\mathbf{M}$ contains all the information about the digraph. It tells us that there are four vertices, and it tells us how many edges connect each pair of vertices.

(b) Don't look for it now, but here is the adjacency matrix for a digraph in §3.2:

$$\mathbf{M} = \begin{bmatrix} \mathbf{0} & 1 & 0 & 1 \\ 0 & \mathbf{0} & 0 & 1 \\ 1 & 1 & \mathbf{0} & 0 \\ 0 & 1 & 1 & \mathbf{1} \end{bmatrix}.$$

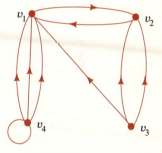

FIGURE 1

Let's see how much information we can glean from the matrix alone. There are four vertices because the matrix is 4×4. There are eight edges because the sum of the entries is 8. Since all the entries are 0's and 1's, there are no multiple edges. There is one loop since there is one 1 on the **main diagonal** [in boldface]. Now look at Figure 2 in §3.2 for the digraph, where we have mentally relabeled the vertices w, x, y, z as v_1, v_2, v_3, v_4. For example, the loop is at $z = v_4$. ∎

Recall that every relation R on a finite set S corresponds to a finite digraph G with no multiple edges. Hence it also corresponds to a matrix $\mathbf{M}_R$ of 0's and 1's. Since the vertex set of G is the set S, if $|S| = n$, then the matrix is $n \times n$.

EXAMPLE 5 We return to the relations of Example 6 of §3.1.

(a) The relation R_1 on $\{0, 1, 2, 3\}$ is defined by $m \leq n$. The matrix for R_1 is

$$\mathbf{M}_{R_1} = \begin{bmatrix} 1 & 1 & 1 & 1 \\ 0 & 1 & 1 & 1 \\ 0 & 0 & 1 & 1 \\ 0 & 0 & 0 & 1 \end{bmatrix}.$$

We have again mentally relabeled the set of vertices; this time 0, 1, 2, 3 correspond to v_1, v_2, v_3, v_4.

(b) The relation R_2 on $S = \{1, 2, 3, 4, 5\}$ is defined by requiring $m - n$ to be even. If we keep 1, 2, 3, 4, 5 in their usual order, the matrix for R_2 is

$$\mathbf{M}_{R_2} = \begin{bmatrix} 1 & 0 & 1 & 0 & 1 \\ 0 & 1 & 0 & 1 & 0 \\ 1 & 0 & 1 & 0 & 1 \\ 0 & 1 & 0 & 1 & 0 \\ 1 & 0 & 1 & 0 & 1 \end{bmatrix}.$$

If we reorder S as 1, 3, 5, 2, 4, we obtain

$$\begin{bmatrix} 1 & 1 & 1 & 0 & 0 \\ 1 & 1 & 1 & 0 & 0 \\ 1 & 1 & 1 & 0 & 0 \\ 0 & 0 & 0 & 1 & 1 \\ 0 & 0 & 0 & 1 & 1 \end{bmatrix}.$$

From this matrix it is clear that the first three elements of S [namely, 1, 3 and 5] are all related to each other and the last two are related to each other. ∎

The matrix for the converse $R^{\leftarrow}$ of a relation R is the transpose of the matrix for R. In symbols, $\mathbf{M}_{R^{\leftarrow}} = \mathbf{M}_R^T$.

EXAMPLE 6 (a) The matrix for $R_1^{\leftarrow}$, where R_1 is in Example 5, is

$$\begin{bmatrix} 1 & 0 & 0 & 0 \\ 1 & 1 & 0 & 0 \\ 1 & 1 & 1 & 0 \\ 1 & 1 & 1 & 1 \end{bmatrix}.$$

In other words, this is the matrix for the relation defined by $m \geq n$.

(b) Consider the relation R_2 in Example 5. Once the order of the set S is fixed, the matrix for $R_2^{\leftarrow}$ is the same as for R_2 because $R_2 = R_2^{\leftarrow}$ [since R_2 is symmetric]. ∎

In general, a relation R is symmetric if and only if its matrix $\mathbf{M}_R$ equals its transpose. In fact, matrices $\mathbf{M}$ such that $\mathbf{M}^T = \mathbf{M}$ are called **symmetric**, so we have: A relation is symmetric if and only if its matrix is. A symmetric matrix whose entries are nonnegative integers can represent a graph as well as a digraph.

EXAMPLE 7   Figure 2 is a picture of the graph obtained from Figure 1 by ignoring the arrows. Its matrix is

$$\mathbf{M} = \begin{bmatrix} \mathbf{0} & 2 & 1 & 3 \\ 2 & \mathbf{0} & 2 & 0 \\ 1 & 2 & \mathbf{0} & 0 \\ 3 & 0 & 0 & \mathbf{1} \end{bmatrix}.$$

Note that $\mathbf{M}^T = \mathbf{M}$ as expected. As before, the graph is completely determined by the matrix, and a lot of information about the graph can be read off from the matrix. For example, we can see the one loop from the boldface diagonal. Counting edges is a bit trickier now, however, because each nonloop appears twice in the matrix. The number of edges is now the sum of the entries on and below the diagonal. ∎

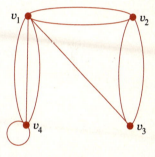

FIGURE 2

EXERCISES 3.3

1. Consider the matrix

$$\mathbf{A} = \begin{bmatrix} 1 & -2 & 5 \\ 3 & -2 & 3 \\ 2 & 0 & 1 \end{bmatrix}.$$

Evaluate

(a) a_{11} (b) a_{13} (c) a_{31} (d) $\sum_{i=1}^{3} a_{ii}$

2. Consider the matrix

$$\mathbf{B} = \begin{bmatrix} 1 & 2 & -2 & 1 \\ 3 & 0 & 1 & 2 \\ 2 & -1 & 4 & 1 \\ 0 & -3 & 1 & 3 \end{bmatrix}.$$

Evaluate

(a) b_{12}
(b) b_{21}
(c) b_{23}
(d) $\sum_{i=1}^{4} b_{ii}$

3. Consider the matrices

$$\mathbf{A} = \begin{bmatrix} -1 & 0 & 2 \\ 1 & 3 & -2 \\ 4 & 2 & 3 \end{bmatrix}, \quad \mathbf{B} = \begin{bmatrix} 6 & 8 & 5 \\ 4 & -2 & 7 \\ 3 & 1 & 2 \end{bmatrix}, \quad \mathbf{C} = \begin{bmatrix} 1 & 3 \\ 2 & -4 \\ 5 & -2 \end{bmatrix}.$$

Calculate the following when they exist.

(a) $\mathbf{A}^T$
(b) $\mathbf{C}^T$
(c) $\mathbf{A} + \mathbf{B}$
(d) $\mathbf{A} + \mathbf{C}$
(e) $(\mathbf{A} + \mathbf{B})^T$
(f) $\mathbf{A}^T + \mathbf{B}^T$
(g) $\mathbf{B} + \mathbf{B}^T$
(h) $\mathbf{C} + \mathbf{C}^T$
(i) $(\mathbf{A} + \mathbf{A}) + \mathbf{B}$

4. Consider the following elements in $\mathbb{R}^3$:

$$\mathbf{v}_1 = (1, 0, 0), \quad \mathbf{v}_2 = (0, -1, 1), \quad \mathbf{v}_3 = (1, 0, -1).$$

Find

(a) $\mathbf{v}_1 + \mathbf{v}_2$
(b) $\mathbf{v}_1 + \mathbf{v}_3$
(c) $\mathbf{v}_3 + \mathbf{v}_2$
(d) $(\mathbf{v}_1 + \mathbf{v}_2) + \mathbf{v}_1$

5. Let $\mathbf{A} = [a_{ij}]$ and $\mathbf{B} = [b_{ij}]$ be matrices in $\mathfrak{M}_{4,3}$ defined by $a_{ij} = (-1)^{i+j}$ and $b_{ij} = i + j$. Find the following matrices when they exist.

(a) $\mathbf{A}^T$
(b) $\mathbf{A} + \mathbf{B}$
(c) $\mathbf{A}^T + \mathbf{B}$
(d) $\mathbf{A}^T + \mathbf{B}^T$
(e) $(\mathbf{A} + \mathbf{B})^T$
(f) $\mathbf{A} + \mathbf{A}$

6. Let $\mathbf{A}$ and $\mathbf{B}$ be matrices in $\mathfrak{M}_{3,3}$ defined by $A[i, j] = ij$ and $B[i, j] = i + j^2$.

(a) Find $\mathbf{A} + \mathbf{B}$.

(b) Calculate $\sum_{i=1}^{3} A[i, i]$.

(c) Calculate $\sum_{i=1}^{3} \left(\sum_{j=1}^{3} B[i, j] \right)$ and $\sum_{j=1}^{3} \left(\sum_{i=1}^{3} B[i, j] \right)$.

(d) Calculate $\sum_{i=1}^{2} \left(\sum_{j=2}^{3} B[i, j] \right)$ and $\sum_{j=1}^{2} \left(\sum_{i=2}^{3} B[i, j] \right)$.

(e) Calculate $\prod_{i=1}^{2} \left(\sum_{j=2}^{3} A[i, j] \right)$.

(f) Does $\mathbf{A}$ equal its transpose $\mathbf{A}^T$?

(g) Does $\mathbf{B}$ equal its transpose $\mathbf{B}^T$?

7. (a) List the 3×3 matrices whose rows are the row vectors

$$[1 \ \ 0 \ \ 0], \quad [0 \ \ 1 \ \ 0] \quad \text{and} \quad [0 \ \ 0 \ \ 1].$$

(b) Which matrices in part (a) are equal to their transposes?

8. In this exercise, **A** and **B** represent matrices. True or false?

(a) $(\mathbf{A}^T)^T = \mathbf{A}$ for all **A**.

(b) If $\mathbf{A}^T = \mathbf{B}^T$, then $\mathbf{A} = \mathbf{B}$.

(c) If $\mathbf{A} = \mathbf{A}^T$, then **A** is a square matrix.

(d) If **A** and **B** are the same size, then $(\mathbf{A} + \mathbf{B})^T = \mathbf{A}^T + \mathbf{B}^T$.

9. For each $n \in \mathbb{N}$, let

$$\mathbf{A}_n = \begin{bmatrix} 1 & n \\ 0 & 1 \end{bmatrix} \quad \text{and} \quad \mathbf{B}_n = \begin{bmatrix} 1 & (-1)^n \\ -1 & 1 \end{bmatrix}.$$

(a) Give $\mathbf{A}_n^T$ for all $n \in \mathbb{N}$.   (b) Find $\{n \in \mathbb{N} : \mathbf{A}_n^T = \mathbf{A}_n\}$.

(c) Find $\{n \in \mathbb{N} : \mathbf{B}_n^T = \mathbf{B}_n\}$.   (d) Find $\{n \in \mathbb{N} : \mathbf{B}_n = \mathbf{B}_0\}$.

10. For **A** and **B** in $\mathfrak{M}_{m,n}$ let $\mathbf{A} - \mathbf{B} = \mathbf{A} + (-\mathbf{B})$. Show that

(a) $(\mathbf{A} - \mathbf{B}) + \mathbf{B} = \mathbf{A}$

(b) $-(\mathbf{A} - \mathbf{B}) = \mathbf{B} - \mathbf{A}$

(c) $(\mathbf{A} - \mathbf{B}) - \mathbf{C} \neq \mathbf{A} - (\mathbf{B} - \mathbf{C})$ in general

11. Consider **A**, **B** in $\mathfrak{M}_{m,n}$ and a, b, c in $\mathbb{R}$. Show that

(a) $c(a\mathbf{A} + b\mathbf{B}) = (ca)\mathbf{A} + (cb)\mathbf{B}$

(b) $-a\mathbf{A} = (-a)\mathbf{A} = a(-\mathbf{A})$

(c) $(a\mathbf{A})^T = a\mathbf{A}^T$

12. Prove (b), (c) and (d) of the theorem.

13. Give the matrices for the digraphs in Figure 3.

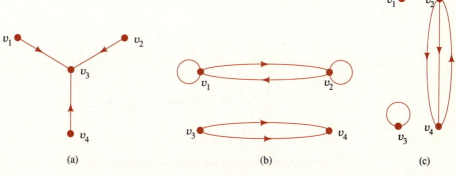

FIGURE 3

14. Write matrices for the graphs in Figure 4.

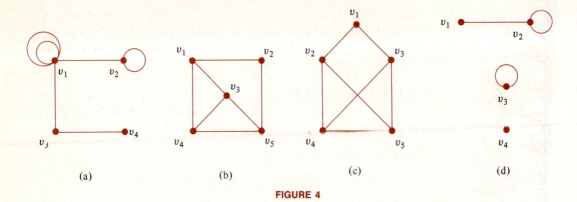

FIGURE 4

15. For each matrix in Figure 5, draw a digraph having the matrix.

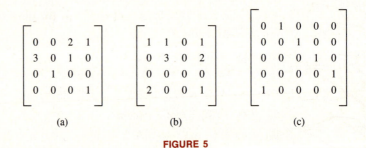

FIGURE 5

16. For each matrix in Figure 6, draw a graph having the matrix.

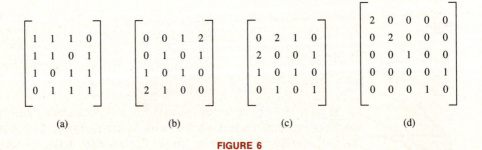

FIGURE 6

17. Give a matrix for each of the relations in Exercise 1 of § 3.1.

18. Draw a digraph having the matrix in Figure 6(b).

19. Give a matrix for each of the relations in Exercise 2 of § 3.1.

§ 3.4 Multiplication of Matrices

As we saw in the preceding section, addition and scalar multiplication of matrices are straightforward. However, the definition that we are about to make of the product of two matrices may appear peculiar and un-natural. The standard linear algebra explanation for choosing this definition shows that matrices correspond to certain functions, called linear transformations; then multiplication of matrices corresponds to composition of the linear transformations. A motivation can also be given in terms of systems of linear equations. A treatment along either of these lines would take us too far into linear algebra. We will draw our motivation instead from graphs and digraphs.

EXAMPLE 1 Consider the digraph in Figure 1(a). Its adjacency matrix is

$$\mathbf{M} = \begin{bmatrix} 1 & 2 & 1 & 1 \\ 0 & 2 & 0 & 0 \\ 1 & 0 & 0 & 1 \\ 0 & 1 & 0 & 0 \end{bmatrix}.$$

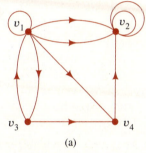
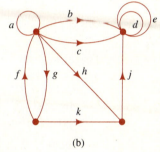

(a)   (b)

FIGURE 1

Observe that the (i, j)-entry of $\mathbf{M}$ gives the number of paths of length 1 from v_i to v_j. Let's count the number of paths of length 2 in the digraph. The edges are labeled in Figure 1(b). By inspection we see that the paths of length 2 from v_1 to v_2 are ab, ac, bd, be, cd, ce and hj. So there are seven such paths. A similar inspection could be applied to find the number of paths of length 2 from any v_i to any v_j.

There must be a better method for counting paths than just by inspection, especially when we deal with larger digraphs. Let's count again the paths of length 2 from v_1 to v_2. Such a path passes through v_1, v_2, v_3 or v_4 on the way, so we can count the number of paths from v_1 to v_2 through v_1, the number through v_2, through v_3 and through v_4 and add these numbers to get the total. Now to get the number of paths with vertex sequence $v_1 v_1 v_2$, for example, we count the edges from v_1 to v_1 [i.e., loops] and the edges from v_1 to v_2 and multiply these numbers to get $1 \cdot 2 = 2$. The corresponding paths are ab and ac. The numbers we multiply, $M[1, 1]$ and $M[1, 2]$, are given in the matrix M. As one more illustration, to get the number of paths with vertex sequence $v_1 v_2 v_2$ we count the edges from v_1 to v_2 and from v_2 to v_2 and multiply to get $M[1, 2] \cdot M[2, 2] = 2 \cdot 2 = 4$. The corresponding paths are bd, be, cd and ce. The general situation for paths from v_1 to v_2 is illustrated in Table 1. The total number of paths of length 2 from v_1 to v_2 is thus

$$M[1, 1] \cdot M[1, 2] + M[1, 2] \cdot M[2, 2] + M[1, 3] \cdot M[3, 2]$$

$$+ M[1, 4] \cdot M[4, 2] = 2 + 4 + 0 + 1 = 7.$$

TABLE 1

Vertex v_i	Number of edges from v_1 to v_i	Number of edges from v_i to v_2	Number of paths with vertex sequence $v_1 v_i v_2$
v_1	$M[1, 1] = 1$	$M[1, 2] = 2$	$M[1, 1] \cdot M[1, 2] = 1 \cdot 2 = 2$
v_2	$M[1, 2] = 2$	$M[2, 2] = 2$	$M[1, 2] \cdot M[2, 2] = 2 \cdot 2 = 4$
v_3	$M[1, 3] = 1$	$M[3, 2] = 0$	$M[1, 3] \cdot M[3, 2] = 1 \cdot 0 = 0$
v_4	$M[1, 4] = 1$	$M[4, 2] = 1$	$M[1, 4] \cdot M[4, 2] = 1 \cdot 1 = 1$

This turns out to be the entry in the first row and second column of a matrix called the "product matrix" $MM = M^2$. It is the sum of products of entries from the first row of M and the second column of M.

In general, the (i, j)-entry of the matrix M^2 is the number of paths of length 2 from v_i to v_j. It turns out for our example that

$$M^2 = \begin{bmatrix} 2 & 7 & 1 & 2 \\ 0 & 4 & 0 & 0 \\ 1 & 3 & 1 & 1 \\ 0 & 2 & 0 & 0 \end{bmatrix}.$$

The entries in the product $M^2M = M^3$ tell us, similarly, the number of paths of length 3 connecting vertices, etc. These matrices are easy to compute [see Exercises 8 and 11] using methods that we are about to describe. ■

Two matrices A and B can be multiplied to get a product matrix provided that the number of columns of A equals the number of rows of B. Consider an $m \times n$ matrix A and an $n \times p$ matrix B. The **product** AB is the $m \times p$ matrix defined by

$$c_{ik} = \sum_{j=1}^{n} a_{ij} b_{jk} \qquad \text{for } 1 \leq i \leq m \quad \text{and} \quad 1 \leq k \leq p.$$

In subscript-free notation,

$$(\mathbf{AB})[i, k] = \sum_{j=1}^{n} \mathbf{A}[i, j]\mathbf{B}[j, k].$$

Schematically, the (i, k)-entry of $\mathbf{AB}$ is obtained by multiplying terms of the ith row of $\mathbf{A}$ by corresponding terms of the kth column of $\mathbf{B}$ and summing. See Figure 2. One can calculate c_{ik} by mentally lifting the ith row of $\mathbf{A}$, rotating it clockwise by $90°$, placing it on top of the kth column of $\mathbf{B}$, and then summing the products of the corresponding terms:

$$c_{ik} = a_{i1}b_{1k} + a_{i2}b_{2k} + \cdots + a_{in}b_{nk}.$$

$$
\begin{bmatrix}
a_{11} & a_{12} & \dots & a_{1n} \\
 & & & \\
a_{i1} & a_{i2} & \cdots & a_{in} \\
 & & & \\
a_{m1} & a_{m2} & & a_{mn}
\end{bmatrix}
\begin{bmatrix}
b_{11} & b_{12} & \dots & b_{1k} & \dots & b_{1p} \\
b_{21} & b_{22} & \dots & b_{2k} & \dots & b_{2p} \\
 & & & & & \\
b_{n1} & b_{n2} & \dots & b_{nk} & \dots & b_{np}
\end{bmatrix}
=
\begin{bmatrix}
c_{11} & c_{12} & \dots & c_{1p} \\
c_{21} & c_{22} & \dots & c_{2p} \\
 & & c_{ik} & \\
c_{m1} & c_{m2} & \dots & c_{mp}
\end{bmatrix}
$$

$$\mathbf{A} \qquad\qquad \mathbf{B} \qquad\qquad \mathbf{AB} = \mathbf{C}$$

FIGURE 2

For this calculation to make sense, the rows of $\mathbf{A}$ and the columns of $\mathbf{B}$ must have the same number of entries. If $\mathbf{A}$ is $m \times n$ and $\mathbf{B}$ is $r \times p$, the matrix product $\mathbf{AB}$ is only defined if $n = r$, in which case $\mathbf{AB}$ is an $m \times p$ matrix.

EXAMPLE 2 Consider matrices and vectors

$$\mathbf{A} = \begin{bmatrix} 3 & -1 \\ -2 & 4 \end{bmatrix}, \qquad \mathbf{B} = \begin{bmatrix} -1 & 0 & 3 \\ 2 & 1 & -5 \end{bmatrix},$$

$$\mathbf{v}_1 = \begin{bmatrix} 2 & -3 & 4 \end{bmatrix} \quad \text{and} \quad \mathbf{v}_2 = \begin{bmatrix} 1 \\ -3 \end{bmatrix}.$$

(a) To calculate $\mathbf{AB}$, we begin by mentally rotating the first row of $\mathbf{A}$ and then placing it over the first, second, and third columns of $\mathbf{B}$ in turn. These three computations give the first row of $\mathbf{AB}$:

$$\mathbf{AB} = \begin{bmatrix} -3 - 2 & 0 - 1 & 9 + 5 \end{bmatrix} = \begin{bmatrix} -5 & -1 & 14 \end{bmatrix}.$$

Using the second row of $\mathbf{A}$ in the same way, we obtain the second row of $\mathbf{AB}$:

$$\mathbf{AB} = \begin{bmatrix} -5 & -1 & 14 \\ 10 & 4 & -26 \end{bmatrix}.$$

(b) The product **BA** is not defined, since **B** is a 2×3 matrix, **A** is a 2×2 matrix and $3 \neq 2$. Furthermore, our schematic procedure breaks down, since the rows of **B** have three terms and the columns of **A** have two terms; so it is not clear how we would mentally place the rows of **B** on top of the columns of **A**.

(c) We have

$$\mathbf{A}^2 = \mathbf{AA} = \begin{bmatrix} 3 & -1 \\ -2 & 4 \end{bmatrix}\begin{bmatrix} 3 & -1 \\ -2 & 4 \end{bmatrix} = \begin{bmatrix} 11 & -7 \\ -14 & 18 \end{bmatrix}.$$

(d) We have

$$\mathbf{A}\mathbf{v}_2 = \begin{bmatrix} 3 & -1 \\ -2 & 4 \end{bmatrix}\begin{bmatrix} 1 \\ -3 \end{bmatrix} = \begin{bmatrix} 6 \\ -14 \end{bmatrix}.$$

(e) Neither $\mathbf{B}\mathbf{v}_1$ nor $\mathbf{v}_1\mathbf{B}$ is defined. But $\mathbf{v}_1\mathbf{B}^T$ and $\mathbf{B}\mathbf{v}_1^T$ are:

$$\mathbf{v}_1\mathbf{B}^T = \begin{bmatrix} 2 & -3 & 4 \end{bmatrix}\begin{bmatrix} -1 & 2 \\ 0 & 1 \\ 3 & -5 \end{bmatrix} = \begin{bmatrix} 10 & -19 \end{bmatrix}$$

and

$$\mathbf{B}\mathbf{v}_1^T = \begin{bmatrix} -1 & 0 & 3 \\ 2 & 1 & -5 \end{bmatrix}\begin{bmatrix} 2 \\ -3 \\ 4 \end{bmatrix} = \begin{bmatrix} 10 \\ -19 \end{bmatrix}.$$

The similarity of these two products is not an accident, as noted in Exercise 19. ■

Just as in Example 1, powers of adjacency matrices can be used to count paths in [undirected] graphs.

EXAMPLE 3 Consider the graph with adjacency matrix

$$\mathbf{M} = \begin{bmatrix} 1 & 2 & 2 & 1 \\ 2 & 2 & 0 & 1 \\ 2 & 0 & 0 & 1 \\ 1 & 1 & 1 & 0 \end{bmatrix}.$$

The graph can be obtained from Figure 1 by removing the arrows. We obtain

$$\mathbf{M}^2 = \begin{bmatrix} 10 & 7 & 3 & 5 \\ 7 & 9 & 5 & 4 \\ 3 & 5 & 5 & 2 \\ 5 & 4 & 2 & 3 \end{bmatrix}.$$

Thus we see that there are 10 paths of length 2 from v_1 to itself, 7 paths of length 2 from v_1 to v_2, etc. To get the number of paths of length 3, we would use

$$\mathbf{M}^3 = \begin{bmatrix} 35 & 39 & 25 & 20 \\ 39 & 36 & 18 & 21 \\ 25 & 18 & 8 & 13 \\ 20 & 21 & 13 & 11 \end{bmatrix}.$$

It is clear that counting paths of length 3 in this graph by inspection would be tedious and that errors would be hard to avoid. ■

EXAMPLE 4   Consider

$$\mathbf{A} = \begin{bmatrix} 3 & -1 \\ -2 & 4 \end{bmatrix}, \qquad \mathbf{B} = \begin{bmatrix} 1 & 2 \\ -3 & 1 \end{bmatrix}, \qquad \mathbf{C} = \begin{bmatrix} 1 & 0 \\ 2 & 3 \end{bmatrix}.$$

(a) We have

$$\mathbf{AB} = \begin{bmatrix} 3 & -1 \\ -2 & 4 \end{bmatrix} \begin{bmatrix} 1 & 2 \\ -3 & 1 \end{bmatrix} = \begin{bmatrix} 6 & 5 \\ -14 & 0 \end{bmatrix}$$

and

$$\mathbf{BA} = \begin{bmatrix} 1 & 2 \\ -3 & 1 \end{bmatrix} \begin{bmatrix} 3 & -1 \\ -2 & 4 \end{bmatrix} = \begin{bmatrix} -1 & 7 \\ -11 & 7 \end{bmatrix}.$$

This example shows that multiplication of matrices is not commutative! Even when both **AB** and **BA** are defined, they may or may not be equal.

(b) Since

$$\mathbf{AB} = \begin{bmatrix} 6 & 5 \\ -14 & 0 \end{bmatrix},$$

we have

$$(\mathbf{AB})\mathbf{C} = \begin{bmatrix} 6 & 5 \\ -14 & 0 \end{bmatrix} \begin{bmatrix} 1 & 0 \\ 2 & 3 \end{bmatrix} = \begin{bmatrix} 16 & 15 \\ -14 & 0 \end{bmatrix}.$$

On the other hand,

$$\mathbf{BC} = \begin{bmatrix} 1 & 2 \\ -3 & 1 \end{bmatrix} \begin{bmatrix} 1 & 0 \\ 2 & 3 \end{bmatrix} = \begin{bmatrix} 5 & 6 \\ -1 & 3 \end{bmatrix}$$

and so

$$\mathbf{A}(\mathbf{BC}) = \begin{bmatrix} 3 & -1 \\ -2 & 4 \end{bmatrix} \begin{bmatrix} 5 & 6 \\ -1 & 3 \end{bmatrix} = \begin{bmatrix} 16 & 15 \\ -14 & 0 \end{bmatrix}.$$

These rather different computations show that $(\mathbf{AB})\mathbf{C} = \mathbf{A}(\mathbf{BC})$ in this case. Equality here is *not* an accident, as we will see later in this section.

■

EXAMPLE 5   Consider the $n \times n$ matrix

$$\mathbf{I} = \begin{bmatrix} 1 & 0 & 0 & \cdots & 0 \\ 0 & 1 & 0 & \cdots & 0 \\ 0 & 0 & 1 & \cdots & 0 \\ \vdots & \vdots & \vdots & & \vdots \\ 0 & 0 & 0 & \cdots & 1 \end{bmatrix};$$

thus $\mathbf{I}[i, i] = 1$ for $i = 1, 2, \ldots, n$ and $\mathbf{I}[i, j] = 0$ for $i \neq j$. This special matrix is called the $n \times n$ **identity matrix**. Whenever we wish to specify its size explicitly we will denote it by $\mathbf{I}_n$. For example,

$$\mathbf{I}_2 = \begin{bmatrix} 1 & 0 \\ 0 & 1 \end{bmatrix} \quad \text{and} \quad \mathbf{I}_4 = \begin{bmatrix} 1 & 0 & 0 & 0 \\ 0 & 1 & 0 & 0 \\ 0 & 0 & 1 & 0 \\ 0 & 0 & 0 & 1 \end{bmatrix}.$$

Now consider any $m \times n$ matrix $\mathbf{A}$. Then the product $\mathbf{AI}_n$ is defined and

$$(\mathbf{AI}_n)[i, k] = \sum_{j=1}^{n} \mathbf{A}[i, j]\mathbf{I}_n[j, k]$$

for $1 \leq i \leq m$ and $1 \leq k \leq n$. If $j \neq k$, we have $\mathbf{I}_n[j, k] = 0$, so this sum collapses to one term, $\mathbf{A}[i, k]\mathbf{I}_n[k, k] = \mathbf{A}[i, k]$. This collapsing occurs for all i and k, and hence

(1) $\mathbf{AI}_n = \mathbf{A}$ for all $\mathbf{A} \in \mathfrak{M}_{m,n}$.

Next consider an $n \times p$ matrix $\mathbf{B}$. Then $\mathbf{I}_n\mathbf{B}$ is defined and

$$(\mathbf{I}_n\mathbf{B})[i, k] = \sum_{j=1}^{n} \mathbf{I}_n[i, j]\mathbf{B}[j, k] = \mathbf{B}[i, k].$$

Thus

(2) $\mathbf{I}_n\mathbf{B} = \mathbf{B}$ for all $\mathbf{B} \in \mathfrak{M}_{n,p}$.

Both assertions (1) and (2) apply to $n \times n$ matrices, so

(3) $\mathbf{AI}_n = \mathbf{I}_n\mathbf{A} = \mathbf{A}$ for all $\mathbf{A} \in \mathfrak{M}_{n,n}$. ∎

Consider a fixed n in $\mathbb{P}$. We claim that the set $\mathbb{R}^n$, the set $\mathfrak{M}_{n,1}$ of column vectors and the set $\mathfrak{M}_{1,n}$ of row vectors are in one-to-one correspondence with each other. In fact,

$$f(x_1, x_2, \ldots, x_n) = [x_1 \quad x_2 \quad \cdots \quad x_n]$$

defines a one-to-one correspondence f between $\mathbb{R}^n$ and $\mathfrak{M}_{1,n}$, and TRANS$(\mathbf{A}) = \mathbf{A}^T$ provides a one-to-one correspondence TRANS between $\mathfrak{M}_{1,n}$ and $\mathfrak{M}_{n,1}$. Composition of these correspondences provides a one-to-one correspondence between $\mathbb{R}^n$ and $\mathfrak{M}_{n,1}$.

Multiplication of matrices is associative [that is, $(\mathbf{AB})\mathbf{C} = \mathbf{A}(\mathbf{BC})$ whenever either side makes sense], as we illustrated in Example 4. At the computational level, this fact is rather mysterious; see Exercise 22. The mystery vanishes when we relate matrices to linear transformations, since the composition of functions is associative. However, we have not explained the connection between matrices and linear transformations, so we simply state the general associative law here without proof.

<table>
<tr><td>*Associative Law*
for Matrices</td><td>If $\mathbf{A}$ is an $m \times n$ matrix, $\mathbf{B}$ is an $n \times p$ matrix, and $\mathbf{C}$ is a $p \times q$ matrix, then $(\mathbf{AB})\mathbf{C} = \mathbf{A}(\mathbf{BC})$.</td></tr>
</table>

Since multiplication of matrices is associative, we can write $\mathbf{ABC}$ without ambiguity. Also, powers such as $\mathbf{A}^3 = \mathbf{AAA}$ are unambiguous. We stress again that although *multiplication of matrices is associative* it is *not commutative*. $\mathbf{AB}$ need not equal $\mathbf{BA}$ even if both products are defined, as we saw in Example 4(a). Examples of other laws of arithmetic not satisfied by matrices appear in Exercises 16 and 20(b).

Two $n \times n$ matrices $\mathbf{A}$ and $\mathbf{B}$ are said to be inverses to each other provided that $\mathbf{AB} = \mathbf{BA} = \mathbf{I}_n$. A matrix $\mathbf{A}$ can have only one inverse. To see this, suppose that $\mathbf{B}$ and $\mathbf{C}$ are both inverses to $\mathbf{A}$. Then $\mathbf{BA} = \mathbf{I}_n$ and $\mathbf{AC} = \mathbf{I}_n$, so $\mathbf{B} = \mathbf{BI}_n = \mathbf{B}(\mathbf{AC}) = (\mathbf{BA})\mathbf{C} = \mathbf{I}_n\mathbf{C} = \mathbf{C}$. A matrix $\mathbf{A}$ that has an inverse is called **invertible** and its unique **inverse** is written $\mathbf{A}^{-1}$. There are several techniques for finding inverses of matrices, and many calculators can find them, too. We will not pursue this topic, although Exercise 14 gives the formula for 2×2 matrices.

EXERCISES 3.4

1. Let

$$\mathbf{A} = \begin{bmatrix} 1 & 2 & 4 \\ 3 & 0 & 2 \end{bmatrix} \quad \text{and} \quad \mathbf{B} = \begin{bmatrix} 2 & 1 \\ -1 & 0 \\ -2 & 3 \end{bmatrix}.$$

Find the following when they exist.

(a) $\mathbf{AB}$ (b) $\mathbf{BA}$ (c) $\mathbf{ABA}$

(d) $\mathbf{A} + \mathbf{B}^T$ (e) $3\mathbf{A}^T - 2\mathbf{B}$ (f) $(\mathbf{AB})^2$

2. Let

$$\mathbf{C} = \begin{bmatrix} 1 \\ 0 \\ 1 \end{bmatrix}$$

and let $\mathbf{A}$ and $\mathbf{B}$ be as in Exercise 1. Find the following when they exist.

(a) **AC**

(b) **BC**

(c) **C²**

(d) **C^T C**

(e) **CC^T**

(f) **73C**

3. Let

$$\mathbf{A} = \begin{bmatrix} 3 & -4 & 3 & 1 \\ 2 & 0 & 1 & -2 \\ -1 & 1 & 2 & 0 \end{bmatrix} \quad \text{and} \quad \mathbf{B} = \begin{bmatrix} -1 & 1 & 0 \\ 1 & 2 & 1 \\ 0 & 1 & -1 \end{bmatrix}.$$

Find the following when they exist.

(a) **A²**

(b) **B²**

(c) **AB**

(d) **BA**

(e) **BA^T**

(f) **A^T B**

4. Let **A** and **B** be as in Exercise 3, and let $\mathbf{v} = \begin{bmatrix} -2 & 1 & -1 \end{bmatrix}$. Find the following when they exist.

(a) **vA**

(b) **vB**

(c) **Bv^T**

(d) **(vB)^T**

(e) **5(vB)^T − 3Bv^T**

5. (a) Calculate both **(AB)C** and **A(BC)** for

$$\mathbf{A} = \begin{bmatrix} -1 & 4 \\ 2 & 5 \end{bmatrix}, \quad \mathbf{B} = \begin{bmatrix} 1 & 1 \\ 0 & 1 \end{bmatrix} \quad \text{and} \quad \mathbf{C} = \begin{bmatrix} 2 & -1 \\ 1 & 3 \end{bmatrix}.$$

(b) Calculate both **B(AC)** and **(BA)C**.

6. Let **A**, **B** and **C** be as in Exercise 5. Calculate

(a) both **AB** and **BA**

(b) both **AC** and **CA**

(c) **A²**

7. Let

$$\mathbf{A} = \begin{bmatrix} 3 & -1 \\ 2 & 1 \\ -2 & 4 \end{bmatrix}, \quad \mathbf{B} = \begin{bmatrix} 1 & 2 \\ 0 & 1 \end{bmatrix} \quad \text{and} \quad \mathbf{C} = \begin{bmatrix} -1 & 3 \\ 2 & 1 \end{bmatrix}.$$

(a) Calculate **A(BC)** and **(AB)C**.

(b) Calculate **A(B²)** and **(AB)B**.

8. Show that $\mathbf{M}^2 = \begin{bmatrix} 2 & 7 & 1 & 2 \\ 0 & 4 & 0 & 0 \\ 1 & 3 & 1 & 1 \\ 0 & 2 & 0 & 0 \end{bmatrix}$, where **M** is the adjacency matrix in Example 1.

9. Use **M²** in Exercise 8 to find the number of paths of length 2

(a) from v_1 to itself

(b) from v_1 to v_3

(c) from v_1 to v_4

(d) from v_2 to v_1

10. By inspection, list the paths of length 2 described in Exercise 9. Use the labeling of Figure 1(b).

11. (a) Calculate $\mathbf{M}^3$ for the adjacency matrix in Example 1.

(b) Find the number of paths of length 3 from v_3 to v_2.

(c) List the paths of length 3 from v_3 to v_2 using the labeling of Figure 1(b).

12. This exercise refers to the graph in Example 3.

(a) Draw the graph; just remove the arrows from Figure 1(a). Label the edges as in Figure 1(b).

(b) How many paths of length 2 are there from v_3 to itself?

(c) List the paths from v_3 to itself of length 2.

(d) How many paths of length 3 are there from v_3 to itself?

(e) List the paths from v_3 to itself of length 3.

13. Repeat parts (a) to (d) of Exercise 12 for the vertex v_2.

14. Show that a 2×2 matrix $\mathbf{A} = \begin{bmatrix} a & b \\ c & d \end{bmatrix}$ has an inverse if and only if $ad - bc \neq 0$, in which case the inverse is

$$\mathbf{A}^{-1} = \frac{1}{ad - bc} \begin{bmatrix} d & -b \\ -c & a \end{bmatrix}.$$

Hint: Try to solve $\begin{bmatrix} a & b \\ c & d \end{bmatrix} \begin{bmatrix} x & y \\ z & w \end{bmatrix} = \begin{bmatrix} 1 & 0 \\ 0 & 1 \end{bmatrix}$ for x, y, z, w.

15. Use Exercise 14 to determine which of the following matrices have inverses. Find the inverses when they exist and check your answers.

(a) $\mathbf{I} = \begin{bmatrix} 1 & 0 \\ 0 & 1 \end{bmatrix}$ (b) $\mathbf{A} = \begin{bmatrix} 1 & 1 \\ 0 & 1 \end{bmatrix}$ (c) $\mathbf{B} = \begin{bmatrix} 1 & 1 \\ 1 & 1 \end{bmatrix}$

(d) $\mathbf{C} = \begin{bmatrix} 2 & -3 \\ 5 & 8 \end{bmatrix}$ (e) $\mathbf{D} = \begin{bmatrix} 0 & 1 \\ 1 & 0 \end{bmatrix}$

16. Find 2×2 matrices that show that $(\mathbf{A} + \mathbf{B})(\mathbf{A} - \mathbf{B}) = \mathbf{A}^2 - \mathbf{B}^2$ does not generally hold.

17. Let $\mathbf{A} = \begin{bmatrix} 1 & 0 \\ 1 & 1 \end{bmatrix}$.

(a) Calculate $\mathbf{A}^n$ for $n = 1, 2, 3, 4$.

(b) Guess a general formula for $\mathbf{A}^n$ and check that your guess satisfies $\mathbf{A}^n \cdot \mathbf{A} = \mathbf{A}^{n+1}$ in general.

18. Consider $\mathbf{A}$, $\mathbf{B}$ in $\mathfrak{M}_{m,n}$ and a in $\mathbb{R}$. Show that $(a\mathbf{A})\mathbf{B} = a(\mathbf{A}\mathbf{B}) = \mathbf{A}(a\mathbf{B})$.

19. Show that if $\mathbf{A}$ is an $m \times n$ matrix and $\mathbf{B}$ is an $n \times p$ matrix, then $(\mathbf{A}\mathbf{B})^T = \mathbf{B}^T\mathbf{A}^T$. Note that both sides of the equality represent $p \times m$ matrices.

20. (a) Prove the cancellation law for $\mathfrak{M}_{m,n}$ under addition; i.e., prove that if $\mathbf{A}$, $\mathbf{B}$, $\mathbf{C}$ are in $\mathfrak{M}_{m,n}$ and $\mathbf{A} + \mathbf{C} = \mathbf{B} + \mathbf{C}$, then $\mathbf{A} = \mathbf{B}$.

(b) Show that the cancellation law for $\mathfrak{M}_{n,n}$ under multiplication fails; i.e., show that $\mathbf{A}\mathbf{C} = \mathbf{B}\mathbf{C}$ need not imply $\mathbf{A} = \mathbf{B}$ even when $\mathbf{C} \neq \mathbf{0}$.

21. (a) Let

$$\mathbf{A} = \begin{bmatrix} a & 0 \\ 0 & a \end{bmatrix}$$

for some fixed a in $\mathbb{R}$. Show that $\mathbf{AB} = \mathbf{BA}$ for all $\mathbf{B}$ in $\mathfrak{M}_{2,2}$.

(b) Consider a fixed matrix $\mathbf{A}$ in $\mathfrak{M}_{2,2}$ that satisfies $\mathbf{AB} = \mathbf{BA}$ for all $\mathbf{B} \in \mathfrak{M}_{2,2}$. Show that

$$\mathbf{A} = \begin{bmatrix} a & 0 \\ 0 & a \end{bmatrix} \quad \text{for some} \quad a \in \mathbb{R}.$$

Hint: Write $\mathbf{A} = \begin{bmatrix} a & b \\ c & d \end{bmatrix}$ and try $\mathbf{B} = \begin{bmatrix} 1 & 0 \\ 0 & 0 \end{bmatrix}$ and $\begin{bmatrix} 0 & 1 \\ 0 & 0 \end{bmatrix}$.

22. (a) Show directly that $\mathbf{A(BC)} = \mathbf{(AB)C}$ for matrices $\mathbf{A}$, $\mathbf{B}$ and $\mathbf{C}$ in $\mathfrak{M}_{2,2}$.

(b) Did you enjoy part (a)? If yes, give a direct proof of the general associative law for matrices.

23. (a) Let $\mathbf{A}$ and $\mathbf{B}$ be $m \times n$ matrices and let $\mathbf{C}$ be an $n \times p$ matrix. Show that the distributive law holds: $\mathbf{(A + B)C} = \mathbf{AC} + \mathbf{BC}$.

(b) Verify the distributive law $\mathbf{A(B + C)} = \mathbf{AB} + \mathbf{AC}$. First specify the sizes of the matrices for which this makes sense.

§ 3.5   Equivalence Relations and Partitions

In this section we study equivalence relations, which are relations that group together elements that have similar characteristics or share some property. Equivalence relations occur throughout mathematics and other fields, even though they are not always formally identified as such.

EXAMPLE 1   (a) Let S be a set of marbles. We might regard marbles s and t as equivalent if they have the same color, in which case we might write $s \sim t$. Note that the relation $\sim$ satisfies three properties:

(R) $s \sim s$ for all marbles s.
(S)   If $s \sim t$, then $t \sim s$.
(T)   If $s \sim t$ and $t \sim u$, then $s \sim u$.

These are nearly obvious; for example, (T) asserts that if marbles s and t have the same color and t and u have the same color, then s and u have the same color. Note also that we can break S up into disjoint subsets so that elements belong to the same subset if and only if they are equivalent, i.e., if and only if they have the same color.

(b) For the same set S of marbles, we might regard marbles s and t as equivalent if they are of the same size, and write $s \approx t$ in this case. All the comments in part (a) apply to $\approx$, with obvious changes.   ■

Let S be any set and suppose that we have a relation $\sim$ on S; i.e., for each pair (x, y) in $S \times S$ we know whether $x \sim y$ or not. The relation $\sim$ is called an **equivalence relation** provided that it satisfies the reflexive, symmetric and transitive laws:

(R) $s \sim s$ for every $s \in S$.
(S)   If $s \sim t$, then $t \sim s$.
(T)   If $s \sim t$ and $t \sim u$, then $s \sim u$.

If $s \sim t$, we say that s and t are **equivalent**; depending on the circumstances, we might also say that s and t are **similar** or **congruent** or **isomorphic**. Other notations sometimes used for equivalence relations are $s \approx t$, $s \cong t$, $s \equiv t$ and $s \leftrightarrow t$. All these notations are intended to convey the idea that s and t have equal [or equivalent] status, a reasonable view because of the symmetry law.

EXAMPLE 2   Triangles T_1 and T_2 in the plane are said to be **similar**, and we write $T_1 \approx T_2$, if their angles can be put in one-to-one correspondence so that corresponding angles are equal. If the corresponding sides are also equal, we say that the triangles are **congruent**, and we write $T_1 \cong T_2$. In Figure 1, $T_1 \cong T_2$, $T_1 \approx T_3$ and $T_2 \approx T_3$, but T_3 is not congruent to T_1 or to T_2. Both $\approx$ and $\cong$ are equivalence relations on the set of all triangles in the plane. All the laws (R), (S) and (T) are evident for these relations. ■

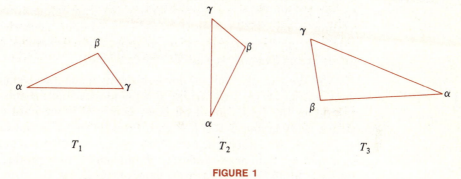

FIGURE 1

We normally use a notation like $\sim$ or $\equiv$ only when we have an equivalence relation. Of course, the use of such notation does not automatically guarantee that we have an equivalence relation.

EXAMPLE 3   (a) For $m, n \in \mathbb{Z}$, define $m \sim n$ in case $m - n$ is odd. The relation $\sim$ is symmetric but is highly nonreflexive and nontransitive. In fact,

$$m \nsim m \quad \text{for all} \quad m \in \mathbb{Z}$$

and

$$m \sim n \quad \text{and} \quad n \sim p \quad \text{always imply} \quad m \nsim p.$$

(b) Consider the set S of all functions mapping $[0, 1]$ into $\mathbb{R}$. Define $f \sim g$ provided $|f(x) - g(x)| \leq 1$ for all $x \in [0, 1]$. One could say that $f \sim g$ in case f and g are "close enough," or "approximately equal." The relation $\sim$ is reflexive and symmetric on S, but it fails to be transitive. For example, if $f(x) = 0$, $g(x) = x$ and $h(x) = 2x$ for $x \in [0, 1]$, then $f \sim g$ and $g \sim h$ but $f \nsim h$. ∎

EXAMPLE 4   (a) Let G be a graph. Define $\sim$ on $V(G)$ by $v \sim w$ if v and w are connected by an edge. Thus $\sim$ is the adjacency relation. It is reflexive only if there is a loop at each vertex. We can build a new relation $\simeq$ on $V(G)$ that *is* reflexive by defining $v \simeq w$ if and only if either $v = w$ or $v \sim w$. This relation $\simeq$ is reflexive and symmetric, but it still need not be transitive. Think of an example of vertices u, v, w with $u \simeq v$ and $v \simeq w$, but with $u \neq w$ and with no edge from u to w.

(b) §3.2 we defined the reachable relation R on $V(G)$ by $(v, w) \in R$ if there is a path of length at least 1 from v to w. This R is what we get if we try to make a transitive relation out of the adjacency relation $\sim$. The new relation R is transitive and symmetric, but if G has isolated vertices without loops, then it is not reflexive. We can make an equivalence relation $\cong$ from R with the trick from part (a). Define $v \cong w$ in case either $v = w$ or $(v, w) \in R$. ∎

EXAMPLE 5   (a) Consider a machine that accepts input strings in Σ^* for some alphabet Σ and generates output strings. We can define an equivalence relation $\sim$ on Σ^* by letting $w_1 \sim w_2$ if the machine generates the same output string for either w_1 or w_2 as input. To see whether or not two words are equivalent, we can ask the machine.

(b) We can also talk about equivalent machines. Define the relations $\approx_1, \approx_2, \approx_3, \ldots$ by writing $B \approx_k C$ for machines B and C if B and C produce the same output for every choice of input word of length k. Define $\approx$ by letting $B \approx C$ if $B \approx_k C$ for all $k \in \mathbb{P}$. Then all the relations $\approx_k$ and $\approx$ are equivalence relations on the set of machines, and two machines are equivalent under $\approx$ if and only if they produce the same response to all input words with letters from Σ. ∎

EXAMPLE 6   Let $\mathscr{S}$ be a family of sets, and for $S, T \in \mathscr{S}$ define $S \sim T$ if there is a one-to-one function mapping S onto T. The relation $\sim$ is an equivalence relation on $\mathscr{S}$. Indeed:

(R) $S \sim S$ because the identity function $1_S \colon S \to S$ is a one-to-one mapping of S onto S.

(S)   If $S \sim T$, then there is a one-to-one mapping f of S onto T. Its inverse f^{-1} is a one-to-one mapping of T onto S, so $T \sim S$.

(T)   If $S \sim T$ and $T \sim U$, then there are one-to-one correspondences $f \colon S \to T$ and $g \colon T \to U$. It is easy to check that $g \circ f$ is a one-to-one correspondence of S onto U. This also follows from Exercise 9 of §1.4 in conjunction with the theorem of that section. In any event, we conclude that $S \sim U$.

Observe that if S is finite, then $S \sim T$ if and only if S and T are the same size. If S is infinite, then $S \sim T$ for some of the infinite sets T in $\mathscr{S}$ but probably not all of them, because not all infinite sets are equivalent to each other. This last assertion is not obvious. A brief glimpse at this fascinating story is available in the final section of the book, §13.3. ∎

In Example 1 we explicitly observed that the set S of marbles can be viewed as a disjoint union of subsets, where two marbles belong to the same subset if and only if they are equivalent. The original collection is cut up into disjoint subsets, consisting of objects that are equivalent to each other. In fact, a similar phenomenon occurs in each of the examples of this section, though it may not be so obvious in some cases. There is a technical term for such a family of sets. A **partition** of a nonempty set S is a collection of nonempty subsets that are disjoint and whose union is S.

EXAMPLE 7   (a) Let f be a function from a set S onto a set T. Then the set $\{f^{\leftarrow}(y):y \in T\}$ of all inverse images $f^{\leftarrow}(y)$ partitions S. In the first place, each $f^{\leftarrow}(y)$ is nonempty, since f maps *onto* T. Every x in S is in exactly one subset of the form $f^{\leftarrow}(y)$, namely the set $f^{\leftarrow}(f(x))$, which consists of all s in S with $f(s) = f(x)$. If $y \neq z$, then we have $f^{\leftarrow}(y) \cap f^{\leftarrow}(z) = \varnothing$. Also, the union $\bigcup_{y \in T} f^{\leftarrow}(y)$ of all the sets $f^{\leftarrow}(y)$ is S, so $\{f^{\leftarrow}(y):y \in T\}$ partitions S. We saw this sort of partition in Example 7 of §1.4.

(b) Let's return to the set S of marbles in Example 1(a). If we let C be the set of colors of the marbles and define the function $f: S \to C$ by $f(s) =$ "the color of s" for each s in S, then the partition $\{f^{\leftarrow}(c):c \in C\}$ is exactly the partition of S mentioned in Example 1. That is, two marbles have the same image under f and hence belong to the same subset if and only if they have the same color. The function $g: S \to \mathbb{R}$ given by $g(s) =$ "the diameter of s" puts two marbles in the same set $g^{\leftarrow}(d)$ if and only if they have the same size. The connection between equivalence relations and partitions given by inverse images is a general phenomenon, as we will see shortly in Theorem 2. ∎

Consider again an equivalence relation $\sim$ on a set S. For each $s \in S$ we define

$$[s] = \{t \in S : s \sim t\}.$$

The set $[s]$ is called the **equivalence class** containing s. For us, "class" and "set" are synonymous, so that $[s]$ could have been called an "equivalence set," but it never is. The set of all equivalence classes of S is denoted by

[S], i.e., [S] = {[s] : s ∈ S}. We will sometimes attach subscripts to [s] or [S] to clarify exactly which of several possible equivalence relations is being used.

EXAMPLE 8   (a) In the marble setting of Example 1(a) the equivalence class [s] of a given marble s is the set of all marbles that are the same color as s; this includes s itself. The equivalence classes are {blue marbles}, {red marbles}, {green marbles}, etc.

(b) Consider the equivalence relation $\cong$ on the set V of vertices of a graph, built from the reachable relation in Example 4(b). Two vertices are equivalent precisely if they belong to the same connected part of the graph. [In fact, this is how we will define "connected" in §6.2.] For example, the equivalence classes for the graph in Figure 2 are $\{v_1, v_6, v_8\}$, $\{v_2, v_4, v_{10}\}$ and $\{v_3, v_5, v_7, v_9, v_{11}, v_{12}\}$. If a graph is connected, then the only equivalence class is the set V itself. ■

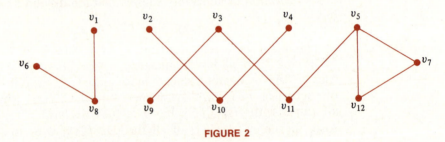

FIGURE 2

The equivalence classes in Example 8 also form a partition of the underlying set. Before we give Theorem 1, which states that something like this always happens, we prove a lemma that gives the key facts.

Lemma   Let $\sim$ be an equivalence relation on a set S. For s and t in S the following assertions are logically equivalent:

(i) $s \sim t$;

(ii) $[s] = [t]$;

(iii) $[s] \cap [t] \neq \emptyset$.

Proof. "Logically equivalent" means that all three assertions are true or else all are false; if any one is true, they all are. We prove (i) ⇒ (ii), (ii) ⇒ (iii) and (iii) ⇒ (i).

$(s \sim t) \Rightarrow ([s] = [t])$. Suppose that $s \sim t$, and consider $s' \in [s]$. Then $s \sim s'$. By symmetry $t \sim s$. Since $t \sim s$ and $s \sim s'$, transitivity shows that

$t \sim s'$. Thus $s' \in [t]$. We've shown that every s' in $[s]$ belongs to $[t]$, and hence $[s] \subseteq [t]$. Similarly, $[t] \subseteq [s]$.

$([s] = [t]) \Rightarrow ([s] \cap [t] \ne \varnothing)$ is obvious, since each set $[s]$ is non-empty [why?].

$([s] \cap [t] \ne \varnothing) \Rightarrow (s \sim t)$. Select u in $[s] \cap [t]$. Then $s \sim u$ and $t \sim u$. By symmetry $u \sim t$. Since $s \sim u$ and $u \sim t$, we have $s \sim t$ by transitivity.

∎

Theorem 1

> (a) If $\sim$ is an equivalence relation on a nonempty set S, then $[S]$ is a partition of S.
>
> (b) Conversely, if $\{A_i : i \in I\}$ is a partition of S, then the sets A_i are the equivalence classes corresponding to some equivalence relation on S.

Proof. (a) To show that $[S]$ partitions S, we need to show

$$(1) \qquad \bigcup_{s \in S} [s] = S$$

and

$$(2) \qquad \text{for} \quad s, t \in S \quad \text{either} \quad [s] = [t] \quad \text{or} \quad [s] \cap [t] = \varnothing.$$

Clearly, $[s] \subseteq S$ for each s in S, so that $\bigcup_{s \in S} [s] \subseteq S$. Given any s in S, we have $s \sim s$, so $s \in [s]$; hence $S \subseteq \bigcup_{s \in S} [s]$. Therefore, (1) holds.

Assertion (2) is logically equivalent to

$$(3) \qquad \text{if} \quad [s] \cap [t] \ne \varnothing \quad \text{then} \quad [s] = [t],$$

which follows from the lemma.

(b) Given a partition $\{A_i : i \in I\}$ of S, define the relation $\sim$ on S by $s \sim t$ if s and t belong to the same set A_i. Properties (R), (S) and (T) are obvious, so $\sim$ is an equivalence relation on S. Given a nonempty set A_i we have $A_i = [s]$ for all $s \in A_i$, so the partition consists precisely of all equivalence classes $[s]$. ∎

Sometimes an equivalence relation is defined in terms of a function on the underlying space. In a sense, which we will make precise in Theorem 2, every equivalence relation arises in this way.

EXAMPLE 9   Define $\sim$ on $\mathbb{N} \times \mathbb{N}$ by $(m, n) \sim (j, k)$ provided that $m^2 + n^2 = j^2 + k^2$. It is easy to show directly that $\sim$ is an equivalence relation. We take a slightly different approach. Define $f : \mathbb{N} \times \mathbb{N} \to \mathbb{N}$ by the rule

$$f(m, n) = m^2 + n^2.$$

Then ordered pairs are equivalent exactly when they have equal images under f. The equivalence classes are simply the nonempty sets $f^{\leftarrow}(r)$, where $r \in \mathbb{N}$. Some of the sets $f^{\leftarrow}(r)$, like $f^{\leftarrow}(3)$, are empty, but this does no harm. ∎

Theorem 2

(a) Let S be a nonempty set. Let f be a function with domain S, and define $s_1 \sim s_2$ if $f(s_1) = f(s_2)$. Then $\sim$ is an equivalence relation on S, and the equivalence classes are the nonempty sets $f^{\leftarrow}(t)$, where t is in the codomain T of f.

(b) Every equivalence relation $\sim$ on a set S is determined by a suitable function with domain S, as in part (a).

Proof. We check that $\sim$ is an equivalence relation.

(R) $f(s) = f(s)$, so $s \sim s$ for all $s \in S$.

(S)   If $f(s_1) = f(s_2)$, then $f(s_2) = f(s_1)$, so $s_1 \sim s_2$ implies $s_2 \sim s_1$.

(T)   If $f(s_1) = f(s_2)$ and $f(s_2) = f(s_3)$, then $f(s_1) = f(s_3)$, so $\sim$ is transitive.

The statement about equivalent classes is just the definition of $f^{\leftarrow}(t)$.

To prove (b), we define the function $v: S \to [S]$, called the **natural mapping** of S onto $[S]$, by

$$v(s) = [s] \qquad \text{for} \quad s \in S.$$

[That's not a v; it's a lowercase Greek nu, for natural.] By the lemma before Theorem 1, $s \sim t$ if and only if $[s] = [t]$, so $s \sim t$ if and only if $v(s) = v(t)$. That is, $\sim$ is the equivalence relation determined by v. Note that v maps S onto $[S]$ and $v^{\leftarrow}([s]) = [s]$ for every s in S. ∎

EXAMPLE 10

(a) If S is our familiar set of marbles and $f(s)$ is the color of s, then $v(s)$ is the class $[s]$ of all marbles the same color as s. We could think of $v(s)$ as a bag of marbles, with v the function that puts each marble in its proper bag. With this fanciful interpretation, $[S]$ is a collection of bags, each with at least one marble in it. The number of bags is the number of colors used.

Switching to the function g for which $g(s)$ is the diameter of s could give a new v and a partition $[S]$ consisting of new bags, one for each possible size of marble.

(b) Define the function $f: \mathbb{R} \times \mathbb{R} \to \mathbb{R}$ by $f(x, y) = x^2 + y^2$. Then f gives an equivalence relation $\sim$ defined by $(x, y) \sim (z, w)$ in case $x^2 + y^2 = z^2 + w^2$. The equivalence classes are the circles in the plane $\mathbb{R} \times \mathbb{R}$ centered at $(0, 0)$, because $x^2 + y^2 = z^2 + w^2$ means $\sqrt{x^2 + y^2} = \sqrt{z^2 + w^2}$, i.e.,

(x, y) and (z, w) are the same distance from $(0, 0)$. Thus $[\mathbb{R} \times \mathbb{R}]$ consists of these circles, including the set $\{(0, 0)\}$ [the circle of radius 0]. The function v maps each point (x, y) to the circle it lives on. There is a one-to-one correspondence between the set of circles and the set of values of f. ■

EXAMPLE 11   The set $\mathbb{Q}$ of rational numbers consists of numbers of the form $\dfrac{m}{n}$ with $m, n \in \mathbb{Z}$ and $n \neq 0$. Each rational number can be written in lots of ways; for instance,

$$\frac{2}{3} = \frac{4}{6} = \frac{8}{12}, \quad -5 = \frac{-5}{1} = \frac{-10}{2} \quad \text{and} \quad 0 = \frac{0}{1} = \frac{0}{73}.$$

We can view the members of $\mathbb{Q}$ as equivalence classes of pairs of integers, so that $\dfrac{m}{n}$ corresponds to the class of (m, n). Here's how.

We will want $(m, n) \sim (p, q)$ in case $\dfrac{m}{n} = \dfrac{p}{q}$, so for n and q not 0 we set

$$(m, n) \sim (p, q) \quad \text{in case } m \cdot q = n \cdot p,$$

a definition that just involves multiplying integers. No fractions here. It is easy to check [Exercise 13] that $\sim$ is an equivalence relation, and we see that (m, n) and (p, q) are equivalent in case the numerical ratio of m to n is the ratio of p to q [hence the term "rational"]. The class $[(2, 3)] = [(4, 6)] = [(8, 12)]$ corresponds to $\frac{2}{3}$, and we could even think of $\frac{2}{3}$ as another name for this class. ■

Theorem 2(b) says that every equivalence relation is determined by a function, and Theorem 2(a) says that the equivalence classes match up with the possible values of the function. The mapping θ defined by $\theta([s]) = f(s)$ is a one-to-one correspondence between the set $[S]$ of equivalence classes and the set $f(S)$ of values of f.

Defining functions, such as θ just now, on sets of equivalence classes can be tricky. One always has to check that the function definition does not really depend on which representative of the class is used. In the case of θ there is no problem, because if $[s] = [t]$, then $f(s) = f(t)$, so we get the same value of $\theta([s])$ whether we think of $[s]$ as $[s]$ or as $[t]$. The next example shows how one can go wrong, though. To ask whether a function f is **well-defined** is to ask whether each value $f(x)$ depends only on what x *is* and not just on what name we've given it. Does the rule defining f give the same value to $f(x)$ and to $f(y)$ if $x = y$?

EXAMPLE 12

(a) We can define the function $f: \mathbb{Q} \to \mathbb{Q}$ by $f\left(\dfrac{m}{n}\right) = \dfrac{m^2}{n^2}$, because

if $\dfrac{m}{n} = \dfrac{p}{q}$, then $\dfrac{m^2}{n^2} = \dfrac{p^2}{q^2}$. In the notation of Example 11, $\dfrac{m}{n}$ corresponds to the equivalence class $[(m, n)]$, and if $(m, n) \sim (p, q)$, then $m \cdot q = n \cdot p$, $m^2 \cdot q^2 = n^2 \cdot p^2$, and thus $(m^2, n^2) \sim (p^2, q^2)$.

(b) We can't define the function $g: \mathbb{Q} \to \mathbb{Q}$ by $g\left(\dfrac{m}{n}\right) = m + n$. If we could, we would want $g\left(\dfrac{1}{2}\right) = 1 + 2 = 3$ but also $g\left(\dfrac{2}{4}\right) = 2 + 4 = 6$. The trouble comes because we have two different names, $\frac{1}{2}$ and $\frac{2}{4}$, for the same object, and our rule for g is based on the name, not on the object itself. When we look at the problem in terms of equivalence classes, we see that $[(1, 2)] = [(2, 4)]$. We would have no trouble defining $g(m, n) = m + n$, but this g does not respect the equivalence relation, since $(m, n) \sim (p, q)$ does not imply $m + n = p + q$. Our original g was not well-defined. ■

EXERCISES 3.5

1. Which of the following describe equivalence relations? For those that are not equivalence relations, specify which of (R), (S) and (T) fail, and illustrate the failures with examples.

   (a) $L_1 \| L_2$ for straight lines in the plane if L_1 and L_2 are the same or are parallel.

   (b) $L_1 \perp L_2$ for straight lines in the plane if L_1 and L_2 are perpendicular.

   (c) $p_1 \sim p_2$ for Americans if p_1 and p_2 live in the same state.

   (d) $p_1 \approx p_2$ for Americans if p_1 and p_2 live in the same state or in neighboring states.

   (e) $p_1 \approx p_2$ for people if p_1 and p_2 have a parent in common.

   (f) $p_1 \cong p_2$ for people if p_1 and p_2 have the same mother.

2. For each example of an equivalence relation in Exercise 1, describe the members of some equivalence class.

3. Let S be a set. Is equality, i.e., "=," an equivalence relation?

4. (a) Define the relation $\equiv$ on $\mathbb{Z}$ by $m \equiv n$ in case $m - n$ is even. Is $\equiv$ an equivalence relation? Explain.

   (b) Repeat part (a) for the relation $\approx$ defined by $m \approx n$ in case $|m - n| \le 1$.

5. If G and H are both graphs with vertex set $\{1, 2, \ldots, n\}$ we say that G is isomorphic to H, and write $G \simeq H$, in case there is a way to label the vertices of G so that it becomes H. For example, for $n = 3$,

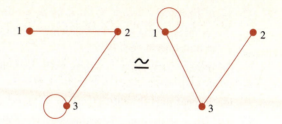

by relabeling $f(1) = 2$, $f(2) = 3$ and $f(3) = 1$.

(a) Give a picture of another graph isomorphic to these two.

(b) Show that $\simeq$ is an equivalence relation on the set of all graphs with vertex set $\{1, 2, \ldots, n\}$.

6. Can you think of situations in the real world where you'd use the term "equivalent" and where a natural equivalence relation is involved?

7. Define the relation $\approx$ on $\mathbb{Z}$ by $m \approx n$ in case $m^2 = n^2$.

(a) Show that $\approx$ is an equivalence relation on $\mathbb{Z}$.

(b) Describe the equivalence classes for $\approx$. How many are there?

8. Verify the claims in Example 3(a).

9. Consider the functions g and h mapping $\mathbb{Z}$ into $\mathbb{N}$ defined by $g(n) = |n|$ and $h(n) = 1 + (-1)^n$.

(a) Describe the sets in the partition $\{g^\leftarrow(k) : k \in \mathbb{N}\}$ of $\mathbb{Z}$. How many sets are there?

(b) Describe the sets in the partition $\{h^\leftarrow(k) : k \in \mathbb{N}\}$ of $\mathbb{Z}$. How many sets are there?

10. On the set $\mathbb{N} \times \mathbb{N}$ define $(m, n) \sim (k, l)$ if $m + l = n + k$.

(a) Show that $\sim$ is an equivalence relation on $\mathbb{N} \times \mathbb{N}$.

(b) Draw a sketch of $\mathbb{N} \times \mathbb{N}$ that shows the equivalence classes.

11. Let Σ be an alphabet, and for w_1 and w_2 in Σ^* define $w_1 \sim w_2$ if length$(w_1) =$ length(w_2). Explain why $\sim$ is an equivalence relation, and describe the equivalence classes.

12. Let P be a set of computer programs and regard programs p_1 and p_2 as equivalent if they always produce the same outputs for given inputs. Is this an equivalence relation? Explain.

13. Consider $\mathbb{Z} \times \mathbb{P}$ and define $(m, n) \sim (p, q)$ if $mq = np$.

(a) Show that $\sim$ is an equivalence relation on $\mathbb{Z} \times \mathbb{P}$.

(b) Show that $\sim$ is the equivalence relation corresponding to the function $\mathbb{Z} \times \mathbb{P} \to \mathbb{Q}$ given by $f(m, n) = m/n$; see Theorem 2(a).

14. In the proof of Theorem 2(b), we obtained the equality $v^\leftarrow([s]) = [s]$. Does this mean that the function v has an inverse and that the inverse of v is the identity function on $[S]$? Discuss.

15. As in Exercise 7 define $\approx$ on $\mathbb{Z}$ by $m \approx n$ in case $m^2 = n^2$.

(a) What is wrong with the following "definition" of $\leq$ on $[\mathbb{Z}]$? Let $[m] \leq [n]$ if and only if $m \leq n$.

(b) What, if anything, is wrong with the following "definition" of a function $f: [\mathbb{Z}] \rightarrow \mathbb{Z}$? Let $f([m]) = m^2 + m + 1$.

(c) Repeat part (b) with $g([m]) = m^4 + m^2 + 1$.

(d) What, if anything, is wrong with the following "definition" of the operation $\oplus$ on $[\mathbb{Z}]$? Let $[m] \oplus [n] = [m + n]$.

16. Which of the following expressions are well-defined definitions of functions on $\mathbb{Q}^+ = \left\{ \dfrac{m}{n} : m, n \in \mathbb{P} \right\}$?

(a) $f\left(\dfrac{m}{n} \right) = \dfrac{n}{m}$

(b) $g\left(\dfrac{m}{n} \right) = m^2 + n^2$

(c) $h\left(\dfrac{m}{n} \right) = \dfrac{m^2 + n^2}{mn}$

17. Let $\sim$ be the relation in Example 3(b), given by $f \sim g$ in case $|f(x) - g(x)| \leq 1$ for all $x \in [0, 1]$. Define $\approx$ by $f \approx g$ in case there is a chain $f = f_1, f_2, \ldots,$ $f_n = g$ with $f_1 \sim f_2, f_2 \sim f_3, \ldots, f_{n-1} \sim f_n$.

(a) Show that $\approx$ is an equivalence relation on the set S.

(b) Describe the equivalence class of the identically zero function $z(x) = 0$.

18. Let S be the set of all sequences (s_n) of real numbers and define $(s_n) \approx (t_n)$ if $\{n \in \mathbb{N} : s_n \neq t_n\}$ is finite. Show that $\approx$ is an equivalence relation on S.

§ 3.6   The Division Algorithm and $\mathbb{Z}(p)$

This section is devoted to those equivalence relations on $\mathbb{Z}$ that are tied to the algebraic operations $+$ and $\cdot$. The key to defining the relations is integer division, so we start by examining what division in $\mathbb{Z}$ actually means.

When we divide 6 by 3 to get 2, there's no mystery; $6 \div 3 = 2$ is just another way of saying $6 = 2 \cdot 3$. When we divide 7 by 3, though, it doesn't "come out even." There is no integer q with $7 = q \cdot 3$. The best we can do is to get two 3's out of 7, with a remainder of 1; $7 = 2 \cdot 3 + 1$. In general, when we try to divide an integer n by a nonzero integer p this sort of outcome is all we can expect. The following theorem, which is surely familiar, says that we can always get a quotient and remainder, and there's only one possible answer.

The Division Algorithm

Let $p \in \mathbb{P}$. For each integer n there are unique integers q and r satisfying

$$n = p \cdot q + r \quad \text{and} \quad 0 \leq r < p.$$

The numbers q and r are called the **quotient** and **remainder**, respectively, **when** n **is divided by** p. For example, if $p = 7$ and $n = 31$, then $31 = 7 \cdot 4 + 3$, so $q = 4$ and $r = 3$.

It may seem odd to call this theorem an algorithm, since the statement doesn't explain a procedure for finding either q or r. The name for the theorem is traditional, however, and in most applications the actual method of computation is unimportant.

We will not stop to prove the Division Algorithm now. The uniqueness of q and r is not too hard to show [Exercise 19], and their existence can be proved fairly quickly using a nonconstructive argument. In §4.1 we will develop an algorithm that produces q and r given n and p, which of course yields a constructive proof of the Division Algorithm. In any case, we all believe it.

So how do we compute q and r? Given a calculator, we can add or multiply integers to get integers, but when we divide, we typically get some decimal expression, rather than q and r. No problem. We can easily get q and r from decimal output, as follows.

Rewrite the conditions $n = p \cdot q + r$ and $0 \le r < p$ as

$$\frac{n}{p} = q + \frac{r}{p} \quad \text{with} \quad 0 \le \frac{r}{p} < 1.$$

The new version says that q is the **integer part** of n/p, denoted by $\lfloor n/p \rfloor$ using the **floor function** notation in which $\lfloor x \rfloor$ stands for the greatest integer less than or equal to x. The quantity r/p is the **fractional part** of n/p, i.e., the number to the right of the decimal point in n/p. So to get q we calculate $\lfloor n/p \rfloor$, and then we compute $r = (n/p - \lfloor n/p \rfloor) \cdot p$.

EXAMPLE 1   (a) Take $n = 31$ and $p = 7$. A pocket calculator gives $31 \div 7 \approx 4.429$. Thus $q = \lfloor 4.429 \rfloor = 4$ and $r \approx (4.429 - 4) \cdot 7 = .429 \cdot 7 = 3.003$, so $r = 3$. We check that $31 = 4 \cdot 7 + 3$.

(b) Now consider $p = 7$ and $n = -31$. Then $-31/7 \approx -4.429$. Does this mean that $q = -4$ and $r = -3$? No, because r must be nonnegative. Recall that q is always the largest integer less than or equal to n/m. In our present case, $q = -5$ since $-5 < -4.429 < -4$ and so $r = -31 - (-5) \cdot 7 = 4$. ∎

Some calculators and most computer languages will do these manipulations for us, so we can ask for q and r directly, using the two built-in integer functions DIV and MOD. The definitions are

$$n \,\text{DIV}\, p = \left\lfloor \frac{n}{p} \right\rfloor \quad \text{and} \quad n \,\text{MOD}\, p = \left(\frac{n}{p} - n \,\text{DIV}\, p \right) \cdot p,$$

so that

$$n = (n \,\text{DIV}\, p) \cdot p + n \,\text{MOD}\, p \quad \text{and} \quad 0 \le n \,\text{MOD}\, p < p.$$

We will suppose that DIV and MOD satisfy these conditions even if n is negative, but one should double-check that assumption in practice, as definitions for the negative case vary. Exercise 18 shows how to use functions DIV and MOD that work for $n \in \mathbb{N}$ to handle $n \le 0$ as well.

The integers n DIV p and n MOD p are the unique q and r guaranteed by the Division Algorithm. Given a positive integer p, n MOD p is the remainder obtained when n is divided by p and is called the **remainder MOD p**. MOD p is a function of n even though we write n MOD p instead of (MOD p)(n); the variable n has appeared in unusual places before, as with $|n|$ and $n!$. The values of MOD p are in the set $\mathbb{Z}(p) = \{0, 1, \dots, p-1\}$, and so MOD $p: \mathbb{Z} \to \mathbb{Z}(p)$. In fact, MOD p maps $\mathbb{Z}$ *onto* $\mathbb{Z}(p)$ because n MOD $p = n$ for $n \in \mathbb{Z}(p)$.

The sets $\mathbb{Z}(p)$ and the mappings MOD p for p in $\mathbb{P}$ will be especially important to us in Chapter 12. They play an important role, as well, in a host of applications ranging from signal transmission to hashing to random number generators to fast computer graphics.

EXAMPLE 2   (a) We have 31 MOD 7 = 3, and 31 DIV 7 = 4. Also, (-31) MOD 7 = 4, and (-31) DIV 7 = -5. Note that (-31) MOD 7 $\ne -(31$ MOD 7$)$ and (-31) DIV 7 $\ne -(31$ DIV 7$)$, so we must be careful about writing $-n$ MOD p or $-n$ DIV p.

(b) n MOD 2 is 0 if n is even and 1 if n is odd.

(c) n MOD 10 is the last decimal digit of n. ■

We are now ready to look at the equivalence relations on $\mathbb{Z}$ that are associated with arithmetic. Given p in $\mathbb{P}$, we regard integers m and n as equivalent mod p if they have the same remainders when divided by p. Thus for $m, n \in \mathbb{Z}$ we define

$$m \equiv n \ (\text{mod } p) \quad \text{in case} \quad m \text{ MOD } p = n \text{ MOD } p.$$

The relation "$\equiv$ (mod p)" is called **congruence mod p**. The expression $m \equiv n$ (mod p) is read as "m [is] congruent to n mod p [or modulo p]." To avoid confusing "mod p" with "MOD p," remember that MOD p is the name for a *function* [like FACT] whereas mod p is *part* of the notation for a *relation*.

Congruence mod p is the equivalence relation on $\mathbb{Z}$ determined by the function MOD $p: \mathbb{Z} \to \mathbb{Z}(p)$ in the manner described by Theorem 2 of § 3.5. Thus

(R) $m \equiv m$ (mod p) for all $m \in \mathbb{Z}$.

(S)   If $m \equiv n$ (mod p), then $n \equiv m$ (mod p).

(T)   If $m \equiv n$ (mod p) and $n \equiv r$ (mod p), then $m \equiv r$ (mod p).

The next theorem shows that the definition above is consistent with the one given in § 3.1.

Theorem 1	Let p be in $\mathbb{P}$. For $m, n \in \mathbb{Z}$ we have $$m \equiv n \,(\text{mod } p) \quad \text{if and only if} \quad m - n \text{ is a multiple of } p.$$

Proof. We use

$$m = (m \,\text{DIV}\, p) \cdot p + m \,\text{MOD}\, p \quad \text{and} \quad n = (n \,\text{DIV}\, p) \cdot p + n \,\text{MOD}\, p.$$

If $m \equiv n \,(\text{mod } p)$, then we have $m \,\text{MOD}\, p = n \,\text{MOD}\, p$, and therefore $m - n = [m \,\text{DIV}\, p - n \,\text{DIV}\, p] \cdot p$ is a multiple of p. Conversely, suppose that $m - n$ is a multiple of p. Then

$$m \,\text{MOD}\, p - n \,\text{MOD}\, p = (m - n) + [n \,\text{DIV}\, p - m \,\text{DIV}\, p] \cdot p$$

is also a multiple of p. Since $m \,\text{MOD}\, p$ and $n \,\text{MOD}\, p$ are both in $\{0, 1, \ldots, p - 1\}$, their difference is at most $p - 1$. Thus they must be equal and $m \equiv n \,(\text{mod } p)$. ∎

Since $m - (m \,\text{MOD}\, p) = (m \,\text{DIV}\, p) \cdot p$, we have $m \,\text{MOD}\, p \equiv m \,(\text{mod } p)$. This is where the MOD notation came from historically. Since distinct elements of $\mathbb{Z}(p)$ cannot be congruent mod p, we see that $r = m \,\text{MOD}\, p$ is the unique number in $\mathbb{Z}(p)$ for which $r \equiv m \,(\text{mod } p)$.

The equivalence class of n with respect to the equivalence relation $\equiv \,(\text{mod } p)$ is called its **congruence class mod p** and denoted by $[n]_p$, or sometimes just by $[n]$ if p is understood. Thus

$$[n]_p = \{m \in \mathbb{Z} : m \equiv n \,(\text{mod } p)\}.$$

The case $p = 1$ is quite special and somewhat boring [Exercise 13], so unless we explicitly say otherwise we will be thinking of $p \geq 2$ in what follows. Many of the arguments will still work for $p = 1$, though.

EXAMPLE 3   (a) Two integers are congruent mod 2 if they are both even or if they are both odd, i.e., if both have the same remainder when divided by 2. The congruence classes are $[0]_2 = [2]_2 = \cdots = \{n \in \mathbb{Z} : n \text{ is even}\}$ and $[1]_2 = [-3]_2 = [73]_2 = \{n \in \mathbb{Z} : n \text{ is odd}\}$.

(b) The integers that are multiples of 5, namely

$$\ldots, -25, -20, -15, -10, -5, 0, 5, 10, 15, 20, 25, \ldots,$$

are all congruent to each other mod 5 since the difference between any two numbers on this list is a multiple of 5. These numbers all have remainders 0 when divided by 5.

If we add 1 to each member of this list, we get a new list:

$$\ldots, -24, -19, -14, -9, -4, 1, 6, 11, 16, 21, 26, \ldots.$$

The *differences* between numbers haven't changed, so the differences are all still multiples of 5. For instance,

$$21 - (-14) = (20 + 1) - (-15 + 1) = 20 + 1 - (-15) - 1$$
$$= 20 - (-15) = 35.$$

Thus the numbers on the new list are also congruent to each other mod 5. They all have remainder 1 when divided by 5.

The integers

$$\ldots, -23, -18, -13, -8, -3, 2, 7, 12, 17, 22, 27, \ldots$$

form another congruence class. So do the integers

$$\ldots, -22, -17, -12, -7, -2, 3, 8, 13, 18, 23, 28, \ldots$$

and the integers

$$\ldots, -21, -16, -11, -6, -1, 4, 9, 14, 19, 24, 29, \ldots .$$

Each integer belongs to exactly one of these five classes, i.e., the classes form a partition of $\mathbb{Z}$, and each of the classes contains exactly one of the numbers 0, 1, 2, 3, 4. We could list the classes as $[0]_5$, $[1]_5$, $[2]_5$, $[3]_5$ and $[4]_5$. ■

Our next theorem shows the link between congruence mod p and the arithmetic in $\mathbb{Z}$.

Theorem 2

Let $m, m', n, n' \in \mathbb{Z}$ and let $p \in \mathbb{P}$. If $m' \equiv m \,(\mathrm{mod}\ p)$ and $n' \equiv n \,(\mathrm{mod}\ p)$, then

$$m' + n' \equiv m + n \,(\mathrm{mod}\ p) \quad \text{and} \quad m' \cdot n' \equiv m \cdot n \,(\mathrm{mod}\ p).$$

Proof. By hypothesis, $m' = m + k \cdot p$ and $n' = n + l \cdot p$ for some k, $l \in \mathbb{Z}$. Thus

$$m' + n' = m + n + (k + l) \cdot p \equiv m + n \,(\mathrm{mod}\ p) \quad \text{and}$$
$$m' \cdot n' = m \cdot n + (k \cdot n + m \cdot l + k \cdot p \cdot l) \cdot p \equiv m \cdot n \,(\mathrm{mod}\ p). \quad ■$$

Taking $m' = m \,\mathrm{MOD}\, p$ and $n' = n \,\mathrm{MOD}\, p$ in Theorem 2 gives the following useful consequence.

Corollary

Let $m, n \in \mathbb{Z}$ and $p \in \mathbb{P}$. Then

(a) $m \,\mathrm{MOD}\, p + n \,\mathrm{MOD}\, p \equiv m + n \,(\mathrm{mod}\ p)$.

(b) $(m \,\mathrm{MOD}\, p) \cdot (n \,\mathrm{MOD}\, p) \equiv m \cdot n \,(\mathrm{mod}\ p)$.

We can use this corollary to help us transfer the arithmetic of $\mathbb{Z}$ over to $\mathbb{Z}(p)$. First, we define two operations, $+_p$ and $*_p$, on $\mathbb{Z}(p)$ by

$$a +_p b = (a + b)\,\text{MOD}\,p \quad \text{and} \quad a *_p b = (a \cdot b)\,\text{MOD}\,p$$

for $a, b \in \mathbb{Z}(p)$. Since $m\,\text{MOD}\,p \in \mathbb{Z}(p)$ for all m, $a +_p b$ and $a *_p b$ are in $\mathbb{Z}(p)$.

EXAMPLE 4   (a) A very simple but important case is $\mathbb{Z}(2)$. The addition and multiplication tables for $\mathbb{Z}(2)$ are given in Figure 1.

$+_2$	0	1		$*_2$	0	1
0	0	1		0	0	0
1	1	0		1	0	1

$\mathbb{Z}(2)$

FIGURE 1

(b) For $p = 6$ and $\mathbb{Z}(6) = \{0, 1, 2, 3, 4, 5\}$ we have $4 +_6 5 = 3$, since $(4 + 5)\,\text{MOD}\,6 = 9\,\text{MOD}\,6 = 3$. Similarly, $4 *_6 4 = 4$, since $4 \cdot 4 \equiv 4 \pmod 6$. The complete addition and multiplication tables for $\mathbb{Z}(6)$ are given in Figure 2. Notice that the product of nonzero elements under $*_6$ can be 0.

$+_6$	0	1	2	3	4	5		$*_6$	0	1	2	3	4	5
0	0	1	2	3	4	5		0	0	0	0	0	0	0
1	1	2	3	4	5	0		1	0	1	2	3	4	5
2	2	3	4	5	0	1		2	0	2	4	0	2	4
3	3	4	5	0	1	2		3	0	3	0	3	0	3
4	4	5	0	1	2	3		4	0	4	2	0	4	2
5	5	0	1	2	3	4		5	0	5	4	3	2	1

$\mathbb{Z}(6)$

FIGURE 2

(c) Figure 3 gives the tables for $\mathbb{Z}(5)$.

$+_5$	0	1	2	3	4		$*_5$	0	1	2	3	4
0	0	1	2	3	4		0	0	0	0	0	0
1	1	2	3	4	0		1	0	1	2	3	4
2	2	3	4	0	1		2	0	2	4	1	3
3	3	4	0	1	2		3	0	3	1	4	2
4	4	0	1	2	3		4	0	4	3	2	1

$\mathbb{Z}(5)$

FIGURE 3   ■

The new operations $+_p$ and $*_p$ are consistent with the old operations $+$ and $\cdot$ on $\mathbb{Z}$.

Theorem 3

> Let $m, n \in \mathbb{Z}$ and $p \in \mathbb{P}$. Then
>
> (a) $(m + n) \,\text{MOD}\, p = (m \,\text{MOD}\, p) +_p (n \,\text{MOD}\, p)$.
> (b) $(m \cdot n) \,\text{MOD}\, p = (m \,\text{MOD}\, p) *_p (n \,\text{MOD}\, p)$.

Proof. (a) According to Theorem 2 or its corollary, $m + n \equiv m \,\text{MOD}\, p + n \,\text{MOD}\, p \pmod{p}$. This means that

$$(m + n) \,\text{MOD}\, p = (m \,\text{MOD}\, p + n \,\text{MOD}\, p) \,\text{MOD}\, p,$$

which is $(m \,\text{MOD}\, p) +_p (n \,\text{MOD}\, p)$ by definition.

The proof of (b) is similar. ∎

Thus the function $\text{MOD}\, p$ carries sums in $\mathbb{Z}$ to sums [under $+_p$] in $\mathbb{Z}(p)$ and products in $\mathbb{Z}$ to products [under $*_p$] in $\mathbb{Z}(p)$.

EXAMPLE 5

(a) $(6 + 3) \,\text{MOD}\, 2 = 9 \,\text{MOD}\, 2 = 1$, but also $6 \,\text{MOD}\, 2 +_2 3 \,\text{MOD}\, 2 = 0 +_2 1 = 1$. In fact, $(\text{even} + \text{odd}) \,\text{MOD}\, 2 = \text{odd} \,\text{MOD}\, 2 = 1 = 0 +_2 1 = \text{even} \,\text{MOD}\, 2 +_2 \text{odd} \,\text{MOD}\, 2$ in general.

(b) $(8 \cdot 3) \,\text{MOD}\, 6 = 24 \,\text{MOD}\, 6 = 0$, and $8 \,\text{MOD}\, 6 *_6 3 \,\text{MOD}\, 6 = 2 *_6 3 = 0$, too. ∎

Theorem 3 also lets us show that $+_p$ and $*_p$ satisfy some familiar algebraic laws.

Theorem 4

> Let $p \in \mathbb{P}$ and let $m, n, r \in \mathbb{Z}(p)$. Then
>
> (a) $m +_p n = n +_p m$ and $m *_p n = n *_p m$.
> (b) $(m +_p n) +_p r = m +_p (n +_p r)$ and
> $(m *_p n) *_p r = m *_p (n *_p r)$.
> (c) $(m +_p n) *_p r = (m *_p r) +_p (n *_p r)$.

Proof. We show the distributive law (c). The other proofs are similar. [See Exercise 21.]

Since $(m + n) \cdot r = m \cdot r + n \cdot r$ in $\mathbb{Z}$, we have $(m + n) \cdot r \,\text{MOD}\, p = (m \cdot r + n \cdot r) \,\text{MOD}\, p$. By Theorem 3

$$((m + n) \cdot r) \,\text{MOD}\, p = (m + n) \,\text{MOD}\, p *_p (r \,\text{MOD}\, p)$$

$$= (m \,\text{MOD}\, p +_p n \,\text{MOD}\, p) *_p (r \,\text{MOD}\, p).$$

Since m, n and r are already in $\mathbb{Z}(p)$, $m \, \text{MOD} \, p = m$, $n \, \text{MOD} \, p = n$ and $r \, \text{MOD} \, p = r$, so this equation just says that

$$((m + n) \cdot r) \, \text{MOD} \, p = (m +_p n) *_p r.$$

Similarly,

$$(m \cdot r + n \cdot r) \, \text{MOD} \, p = (m *_p r) +_p (n *_p r). \quad \blacksquare$$

The set $\mathbb{Z}(p)$ with its operations $+_p$ and $*_p$ acts, in a way, like a finite model of $\mathbb{Z}$. We need to be a little careful, however, because although Theorem 4 shows that many of the laws of arithmetic hold in $\mathbb{Z}(p)$, cancellation may not work as expected. We saw that $3 *_6 5 = 3 *_6 3 = 3 *_6 1 = 3$, but $5 \neq 3 \neq 1$ in $\mathbb{Z}(6)$. Moreover, $3 *_6 2 = 0$ with $3 \neq 0$ and $2 \neq 0$. Chapter 12 contains a thorough account of $\mathbb{Z}(p)$ as an algebraic structure, and explains why $\mathbb{Z}(5)$ and $\mathbb{Z}(7)$ are nicer than $\mathbb{Z}(6)$.

EXAMPLE 6   (a) We could try to define operations $+$ and $\cdot$ on the collection $[\mathbb{Z}]_p$ of congruence classes $[m]_p$. The natural candidates are

$$[m]_p + [n]_p = [m + n]_p \quad \text{and} \quad [m]_p \cdot [n]_p = [m \cdot n]_p.$$

We know from §3.5 that we need to be careful about definitions on sets of equivalence classes. To be sure that $+$ and $\cdot$ are well-defined on $[\mathbb{Z}]_p$ we need to check that if $[m]_p = [m']_p$ and $[n]_p = [n']_p$ then

(1) $[m + n]_p = [m' + n']_p$ and
(2) $[m \cdot n]_p = [m' \cdot n']_p$.

Now $[m]_p = [m']_p$ if and only if $m' \equiv m \pmod{p}$, and similarly, $[n]_p = [n']_p$ means $n' \equiv n \pmod{p}$. Condition (1) translates to $(m' + n') \equiv (m + n) \pmod{p}$, which follows from Theorem 2. The proof of (2) is similar. Thus our new operations $+$ and $\cdot$ are well-defined on $[\mathbb{Z}]_p$.

(b) For instance, $[3]_6 + [5]_6 = [8]_6 = [2]_6$ and $[3]_5 \cdot [2]_5 = [6]_5 = [1]_5$. Notice for comparison that $3 +_6 5 = 2$ and $3 *_5 2 = 1$. Our operations on $[\mathbb{Z}]_6$ and $[\mathbb{Z}]_5$ look a lot like our operations on $\mathbb{Z}(6)$ and $\mathbb{Z}(5)$. [See Exercise 20 for the full story.]

If we write EVEN for $[0]_2$ and ODD for $[1]_2$ in $[\mathbb{Z}]_2$, then EVEN + ODD = ODD, ODD $\cdot$ ODD = ODD, etc.

(c) Let's try to define $f : [\mathbb{Z}]_6 \to \mathbb{Z}$ by the rule $f([m]_6) = m^2$. For example, $f([2]_6) = 2^2 = 4$, $f([3]_6) = 3^2 = 9$, $f([8]_6) = 64 = $ OOPS! The trouble is that $[8]_6 = [2]_6$, but $4 \neq 64$. Too bad. Our f is not well-defined. $\blacksquare$

EXERCISES 3.6

1. Use any method to find q and r as in the Division Algorithm for the following values of n and m.

   (a) $n = 20, m = 3$ (b) $n = 20, m = 4$

   (c) $n = -20, m = 3$ (d) $n = -20, m = 4$

   (e) $n = 371{,}246, m = 65$ (f) $n = -371{,}246, m = 65$

2. Find n DIV m and n MOD m for the following values of n and m.

   (a) $n = 20, m = 3$ (b) $n = 20, m = 4$

   (c) $n = -20, m = 3$ (d) $n = -20, m = 4$

   (e) $n = 371{,}246, m = 65$ (f) $n = -371{,}246, m = 65$

3. List three integers that are congruent mod 4 to each of the following.

   (a) 0   (b) 1   (c) 2   (d) 3   (e) 4

4. (a) List all equivalence classes of $\mathbb{Z}$ for the equivalence relation congruence mod 4.

   (b) How many different equivalence classes of $\mathbb{Z}$ are there with respect to congruence mod 73?

5. For each of the following integers m find the unique integer r in $\{0, 1, 2, 3\}$ such that $m \equiv r \pmod 4$.

   (a) 17   (b) 7   (c) -7 (d) 2   (e) -88

6. Calculate

   (a) $4 +_7 4$ (b) $5 +_7 6$

   (c) $4 *_7 4$ (d) $0 +_7 k$ for any $k \in \mathbb{Z}(7)$

   (e) $1 *_7 k$ for any $k \in \mathbb{Z}(7)$

7. (a) Calculate $6 +_{10} 7$ and $6 *_{10} 7$.

   (b) Describe in words $m +_{10} k$ for any $m, k \in \mathbb{Z}(10)$.

   (c) Do the same for $m *_{10} k$.

8. (a) List the elements in the sets A_0, A_1 and A_2 defined by

   $$A_k = \{m \in \mathbb{Z} : -10 \le m \le 10 \quad \text{and} \quad m \equiv k \pmod 3\}.$$

   (b) What is A_3? A_4? A_{73}?

9. Give the complete addition and multiplication tables for $\mathbb{Z}(4)$.

10. Use Figure 2 to solve the following equations for x in $\mathbb{Z}(6)$.

    (a) $1 +_6 x = 0$ (b) $2 +_6 x = 0$

    (c) $3 +_6 x = 0$ (d) $4 +_6 x = 0$

    (e) $5 +_6 x = 0$

11. Use Figure 3 to solve the following equations for x in $\mathbb{Z}(5)$.

    (a) $1 *_5 x = 1$ (b) $2 *_5 x = 1$

    (c) $3 *_5 x = 1$ (d) $4 *_5 x = 1$

12. For m, n in $\mathbb{N}$ define $m \sim n$ if $m^2 - n^2$ is a multiple of 3.

    (a) Show that $\sim$ is an equivalence relation on $\mathbb{N}$.

    (b) List four elements in the equivalence class $[0]$.

    (c) List four elements in the equivalence class $[1]$.

    (d) Do you think there are any more equivalence classes?

13. The definition of $m \equiv n \pmod{p}$ makes sense even if $p = 1$.

    (a) Describe this equivalence relation for $p = 1$ and the corresponding equivalence classes in $\mathbb{Z}$.

    (b) What meaning can you attach to m DIV 1 and m MOD 1?

    (c) What does Theorem 3 say if $p = 1$?

14. (a) Prove that if $m, n \in \mathbb{Z}$ and $m \equiv n \pmod{p}$, then $m^2 \equiv n^2 \pmod{p}$.

    (b) Is the function $f: [\mathbb{Z}]_p \to [\mathbb{Z}]_p$ given by $f([n]_p) = [n^2]_p$ well-defined? Explain.

    (c) Repeat part (b) for the function $g: [\mathbb{Z}]_6 \to [\mathbb{Z}]_{12}$ given by $g([n]_6) = [n^2]_{12}$.

    (d) Repeat part (b) for the function $h: [\mathbb{Z}]_6 \to [\mathbb{Z}]_{12}$ given by $h([n]_6) = [n^3]_{12}$.

15. (a) Show that the four-digit number $n = abcd$ is divisible by 9 if and only if the sum of the digits $a + b + c + d$ is divisible by 9.

    (b) Is the statement in part (a) valid for every n in $\mathbb{P}$ regardless of the number of digits? Explain.

16. (a) Show that four-digit number $n = abcd$ is divisible by 2 if and only if the last digit d is.

    (b) Show that $n = abcd$ is divisible by 5 if and only if d is.

17. Show that the four-digit number $n = abcd$ is divisible by 11 if and only if $a - b + c - d$ is.

18. (a) Show that if n MOD $p = 0$, then $(-n)$ MOD $p = 0$ and $(-n)$ DIV $p = -(n$ DIV $p)$.

    (b) Show that if $0 < n$ MOD $p < p$, then $(-n)$ MOD $p = p - n$ MOD p and $(-n)$ DIV $p = -(n$ DIV $p) - 1$.

    This exercise shows that one can easily compute n DIV p and n MOD p for negative n with DIV p and MOD p functions that are only defined for $n \in \mathbb{N}$.

19. Show that q and r are unique in the Division Algorithm. That is, show that if $p, q, r, q', r' \in \mathbb{Z}$ with $0 \neq p$, and if

    $$q \cdot p + r = q' \cdot p + r', \quad 0 \leq r < p \quad \text{and} \quad 0 \leq r' < p,$$

    then $q = q'$ and $r = r'$.

20. Let $p \geq 2$. Define the mapping $\theta: \mathbb{Z}(p) \to [\mathbb{Z}]_p$ by $\theta(m) = [m]_p$.

    (a) Show that θ is a one-to-one correspondence of $\mathbb{Z}(p)$ onto $[\mathbb{Z}]_p$.

    (b) Show that $\theta(m +_p n) = \theta(m) + \theta(n)$ and $\theta(m *_p n) = \theta(m) \cdot \theta(n)$, where the operations $+$ and $\cdot$ on $[\mathbb{Z}]_p$ are as defined in Example 6(a).

21. (a) Verify the commutative law for $+_p$ in Theorem 4(a).

   (b) Verify the associative law for $+_p$ in Theorem 4(b).

   (c) Observe that the proofs of the commutative and associative laws for $*_p$ are almost identical to those for $+_p$.

CHAPTER HIGHLIGHTS

For the items listed below:

   (a) Satisfy yourself that you can define and use each concept and notation.

   (b) Give at least one reason why the item was included in this chapter.

   (c) Think of at least one example of each concept and at least one situation in which each fact would be useful.

The purpose of this review is to tie each of the ideas to as many other ideas and to as many concrete examples as possible. That way, when an example or concept is brought to mind, everything tied to it will also be called up for possible use.

CONCEPTS AND NOTATION

binary relation on S or from S to T

   reflexive, antireflexive, symmetric, antisymmetric, transitive [for relations on S]

   converse relation

   function as relation

equivalence relation

   equivalence class $[s]$

   partition $[S]$

   natural mapping v of s to $[s]$

   congruence mod p, $\equiv (\text{mod } p)$, $[n]_p$

quotient, DIV p, remainder, MOD p

$\mathbb{Z}(p)$, $+_p$, $*_p$

digraph or graph [undirected]

   vertex, edge, loop, parallel edges

   initial, terminal vertex, end point of edge

   path

      length

      closed path, cycle

acyclic path, acyclic digraph
adjacency relation
reachable relation
picture
of a digraph or graph
of a relation
matrix
transpose, sum, product, scalar multiple, negative, inverse
square, symmetric
special matrices **0**, **I**
adjacency matrix of a graph or digraph
matrix of a relation

FACTS

Equivalence relations and partitions are two views of the same concept.
Every equivalence relation is of form $a \sim b$ if and only if $f(a) = f(b)$ for some function f.
Matrix addition is associative and commutative.
Matrix multiplication is associative but not commutative.
Powers of the adjacency matrix count paths in digraphs and graphs.
Definitions of functions on equivalence classes must be independent of choice of representatives.
Division Algorithm $[n = p \cdot q + r, 0 \leq r < p]$.
The unique x in $\mathbb{Z}(p)$ with $x \equiv m \pmod{p}$ is m MOD p.
If $m \equiv m' \pmod{p}$ and $n \equiv n' \pmod{p}$, then $m + n \equiv m' + n' \pmod{p}$ and $m \cdot n \equiv m' \cdot n' \pmod{p}$.
Operations $+_p$ and $*_p$ on $\mathbb{Z}(p)$ mimic $+$ and $\cdot$ on $\mathbb{Z}$ except for cancellation with respect to $*_p$.

INDUCTION AND RECURSION

4

The theme of this chapter can be described by the statement, "If we start out right and if nothing can go wrong, then we will always be right." The context to which we apply this observation will be a list of propositions that we are proving, a succession of steps that we are executing in an algorithm, or a sequence of values that we are computing. In each case, we wish to be sure that the results we get are always correct.

In §4.1 we look at algorithm segments, called "while loops," that repeat a sequence of steps as long as a specified condition is met. The next section introduces mathematical induction, a fundamental tool for proving sequences of propositions. Section 4.3 introduces recursive definition of sequences and the calculation of terms from previous terms. Then §4.4 gives explicit methods that apply to sequences given by some common forms of recurrence equations. The principles of induction introduced in §4.2 apply only to a special kind of recursive formulation, so in §4.5 we extend those methods to more general recursive settings. The chapter concludes with a discussion of greatest common divisors and the recursive Euclidean algorithm for computing them. Section 4.6 includes applications to solving congruences.

§ 4.1   Loop Invariants

In the last section of Chapter 3 we claimed that for every pair of integers m and n with $m \geq 0$ and $n > 0$ there are integers q and r, the quotient and remainder, such that $m = q \cdot n + r$ and $0 \leq r < n$. We will prove that claim in this section by first giving an algorithm that allegedly constructs q and r, and then by proving that the algorithm really does what it should. The algorithm we present has a special feature, called a "while loop," that makes the algorithm easy to understand and helps us to give a proof of its correctness. The main point of this section is to understand while loops and some associated notation. In the next section we will see how the logic associated with these loops can be used to develop mathematical induction, one of the most powerful proof techniques in mathematics.

Here is the basic outline of our algorithm for division.

Guess values of q and r so that $m = q \cdot n + r$.

Keep improving the guesses, without losing the property that $m = q \cdot n + r$, until eventually we stop, with $0 \leq r < n$.

Actually, our initial guess will not be very imaginative: $q = 0$ and $r = m$ certainly work. This q is probably too small, and the r probably too large. Once we are started, $m = q \cdot n + r$ will remain true if we increase q by 1 and decrease r by n, because $q \cdot n + r = (q + 1) \cdot n + (r - n)$. If we make these changes enough times, we hope to get $r < n$, as desired.

We can refine this rough algorithm into one that will actually compute q and r with the required properties. Here is the final version, written out in the style of a computer program. Don't worry; it is not necessary to know programming to follow the ideas.

Division Algorithm

```
{Input: integers m ≥ 0 and n > 0.}
{Output: integers q and r with q·n + r = m and 0 ≤ r < n.}
begin
{Initialize.}
q := 0
r := m
{Do the real work.}
while r ≥ n do
    q := q + 1
    r := r - n
end ■
```

Here and from now on we will use the notational convention that statements in braces $\{\cdots\}$ are comments, not part of the algorithm itself. The sequence of lines in the algorithm is a recipe. It says:

1. Set the value of q to be 0. [The notation $a := b$ means "set, or define, the value of a to equal the value of b."]
2. Set the value of r to be m.
3. Check to see if $r \geq n$.
    If $r \geq n$, then:
        Increase the value of q by 1,
        Decrease the value of r by n and
        Go back to 3.
    Otherwise, i.e., if $r < n$,
        Go on to 4.
4. Stop.

The initial guesses for q and r are likely to be wrong, of course, but repetition of step 3 is meant to improve them until they are right.

The while loop here is step 3. In general, a **while loop** is a sequence of steps in an algorithm or program that has the form

> while g do
> > S

and is interpreted to mean

> (∗) Check if g is true.
> > If g is true, then do whatever S says to do, and after that go back to (∗).
> > Otherwise, skip over S and go on to whatever follows the loop in the main program.

Here g is some proposition, for instance "$r \geq n$," called the **guard** of the loop, and S is a sequence of steps, called the **body** of the loop. We could let S contain instructions to jump out of the loop to some other place in the program, but there is no real loss of generality if we just think of S as some program segment that we execute, after which we go back to test g again. Of course, S may change the truth value of g; that's the whole idea. We call an execution of S in the loop a **pass** through the loop, or an **iteration**, and we say that the loop **terminates** or **exits** if at some stage g is false and the program continues past the loop.

EXAMPLE 1   What a loop actually does may depend on the status of various quantities at the time the program or algorithm arrives at the loop. Figure 1 shows some simple loops.

while $n < 5$ do	while $n \neq 8$ do	while $A \neq \emptyset$ do
$n := n + 1$	print n^2	choose x in A
print n^2	$n := n + 2$	remove x from A
(a)	(b)	(c)

FIGURE 1

If $n = 0$ when the program enters the loop in Figure 1(a), the loop replaces n by 1, prints 1, replaces n by 2, prints 4, ..., replaces n by 5, prints 25, observes that now the guard $n < 5$ is false, and goes on to the next part of the program, whatever that is. If $n = 4$ at the start of the loop, we just print 25 before exiting. If $n = 6$ initially, we print nothing at all, since the guard $6 < 5$ is false immediately.

If $n = 0$ initially, the loop of Figure 1(b) prints 0, 4, 16, 36 and then exits. Wiith an input value of $n = 1$, however, this loop prints 1, 9, 25,

49, 81, 121, etc. It *never* exits, because although the values of n keep changing, the guard $n \neq 8$ is always true. As this example shows, loops are not required to terminate. Figure 1(c) gives a nonnumeric example. If A is some set, for instance the set of edges of a graph, the loop just keeps throwing elements out of A until there are none left. If A is finite, the loop terminates; otherwise, it doesn't. ∎

In the case of the Division Algorithm we set the initial values of q and r before we enter the loop. Then S instructs us to increase q by 1 and decrease r by n. Figure 2 shows the successive values of q, r and the proposition $r \geq n$ during the execution of this algorithm with inputs $m = 17$ and $n = 7$.

$m = 17, \ n = 7$	q	r	$r \geq n$
Initially	0	17	True
After the first pass	1	10	True
After the second pass	2	3	False

FIGURE 2

The algorithm does not go through the while loop a third time, since the guard $r \geq n$ is false after the second pass through. The final values are $q = 2, r = 3$. Since $17 = 2 \cdot 7 + 3$, the algorithm produces the correct result for the given input data.

Our main concerns for the Division Algorithm are that:

1. It stops after a while, and
2. When it stops, it gives the right answers.

Let us first see why the algorithm stops. The action is all in the while loop, since the initialization presents no problems. Each pass through the loop, i.e., each execution of the sequence $q := q + 1$ and $r := r - n$, takes a fixed amount of time, so we just need to show that the algorithm only makes a finite number of passes through the loop. The key is what happens to r. At first, $r = m \geq 0$. Each execution of the body of the loop reduces the value of r by n, so the successive values of r are m, $m - n$, $m - 2n$, etc. These integers keep decreasing, since $n > 0$. Sooner or later we reach a k with $r = m - k \cdot n < n$. Then the guard $r \geq n$ is false, we exit from the loop, and we quit.

How do we know we'll reach such a k? The set $\mathbb{N}$ of natural numbers has the following important property:

Well-Ordering Principle

Every nonempty subset of $\mathbb{N}$ has a smallest element.

This principle implies, in particular, that decreasing sequences in $\mathbb{N}$ do not go on forever; every decreasing sequence $a > b > c > \cdots$ in $\mathbb{N}$ is finite in length, because no member of the sequence can come after the smallest member. Hence every infinite decreasing sequence in $\mathbb{Z}$ must eventually have negative terms.

In our case, the sequence $m, m - n, m - 2n, \ldots$ must stop at some value $m - k \cdot n \geq 0$ such that $m - (k + 1) \cdot n < 0$, i.e., such that $m - k \cdot n < n$.

We now know that the Division Algorithm stops, but why does it always give the right answers? To study this question, we introduce a new idea. We say that a proposition p is an **invariant** of the loop

> while g do
> S

in case it satisfies the following condition.

If p and g are true before we do S, then p is true after we finish S.

The following theorem gives one reason why loop invariants are useful.

Loop Invariant Theorem

Suppose that p is an invariant of the loop "while g do S," and that p is true on entry into the loop. Then p is true after each iteration of the loop. If the loop terminates, it does so with p true and g false.

Cause P has nothing to do with the loop?

Proof. The theorem is really fairly obvious. By hypothesis, p is true at the beginning, i.e., after 0 iterations of the loop. Suppose, if possible, that p is false after some iteration. Then there is a first time that p is false, say after the nth pass. Now $n \geq 1$, and p was true at the end of the $(n - 1)$th pass, and hence at the start of the nth pass. So was g, or we wouldn't even have made the nth iteration. Because p is a loop invariant, it must also be true after the nth pass, contrary to the choice of n. This contradiction shows that it can never happen that p is false at the end of a pass through the loop, so it must always be true.

If the loop terminates [because g is false] it does so at the end of a pass, so p must be true at that time. ∎

The idea that we used in this proof could be called the Principle of the Smallest Criminal—if there's a bad guy, there's a smallest bad guy. It is a variant of "There's a first time for everything," which is really just another way of stating the Well-Ordering Principle. In our proof we looked at $\{n \in \mathbb{N} : p$ is false after the nth pass$\}$. Supposing this set to be nonempty made it have a smallest member and led eventually to a contradiction.

Division Algorithm

{Input: integers $m \geq 0$ and $n > 0$.}
{Output: integers q and r with $q \cdot n + r = m$ and $0 \leq r < n$.}
begin
  $q := 0$
  $r := m$
  {$q \cdot n + r = m$ and $r \geq 0$.}
  while $r \geq n$ do
    $q := q + 1$
    $r := r - n$
end

while "r" changes here,
$r \geq n$ is True, az is $q \cdot n + r = m$.
Since $n > 0$, then $r > 0$
$+ r$

FIGURE 3

How does this theorem help us validate the Division Algorithm? Figure 3 shows the algorithm again, for reference, with a new comment added just before the while loop. We next show that the statement "$q \cdot n + r = m$ and $r \geq 0$" is a loop invariant. It is surely true before we enter the while loop, because we have set up q and r that way. Moreover, if $q \cdot n + r = m$ and $r \geq n$, and if we execute the body of the loop to get new values r' and q', then *(cause r & q change, & loop*

$$q' \cdot n + r' = (q + 1) \cdot n + (r - n) = q \cdot n + r = m$$

and

$$r' = r - n \geq 0,$$

so the statement remains true after the pass through the loop. We already know that the loop terminates. The Loop Invariant Theorem tells us that when it does stop, $m = q \cdot n + r$ and $r \geq 0$ are true, and $r \geq n$ is false, i.e., $r < n$. These are exactly the conditions we want the output to satisfy. We have shown the following.

Theorem | Given integers m and n with $m \geq 0$ and $n > 0$, the Division Algorithm constructs integers q and r with $m = q \cdot n + r$ and $0 \leq r < n$.

Why did we choose the loop invariant we did? We wanted to have $m = q \cdot n + r$ and $0 \leq r < n$. Of course "$m = q \cdot n + r$" and "$0 \leq r$" are each invariants of our loop, but neither by itself is good enough. We want both, so we list both. We can't use "$r < n$" as an invariant for the loop we set up, but it's a natural negation for a guard "$r \geq n$," so it will hold when the guard fails, on exit from the loop.

Loop invariants can be used to guide the development of algorithms, as well as to help verify their correctness. We illustrate the method with the Division Algorithm.

At the beginning of this section we first saw a rough outline of the algorithm. A slightly refined version might have looked like this.

begin
initialize q and r so that $m = q \cdot n + r$
while $r \geq n$ do
    make progress toward stopping, and
    keep $m = q \cdot n + r$
end.

The next version could be somewhat more explicit.

begin
let $q := 0$
choose r so that $m = q \cdot n + r$
while $r \geq n$ do
    increase q
    keep $m = q \cdot n + r$ {so r will decrease}
end

At this stage it is clear that r has to be m initially, and that if q is increased by 1 in the body of the loop then r has to decrease by n. Finally, $r \geq 0$ comes from insisting that $m \geq 0$ in the input. Developing the algorithm this way makes it easy to show, using the invariant, that the algorithm gives the correct result.

This may be the place to point out that the Division Algorithm we have given is much slower than the pencil and paper division method we learned as children. One reason the childhood method and its computerized relatives are a lot faster is that they take advantage of the representation of the input data in decimal or binary form; some organizational work has already been done. The algorithm we have given could also be greatly speeded up by an improved strategy for guessing new values for q and r. Since we just wanted to know that the right q and r *could* be computed, though, we deliberately kept the account uncomplicated.

Of course just listing a proposition p in braces doesn't make it an invariant, any more than writing down a program makes it compute what it's supposed to. In order to use the Loop Invariant Theorem, one needs to check that p does satisfy the definition of an invariant.

EXAMPLE 2   (a) We defined $n! = 1 \cdot 2 \cdots n$ in § 1.5. If $n > 1$, then $n! = (n - 1)! \cdot n$, a fact that lets us compute the value of $n!$ with a simple algorithm. Figure 4 shows one such algorithm.

The alleged loop invariant is "FACT $= m!$". This equation is certainly true initially, and if FACT $= m!$ at the start of the while loop, then

(new FACT) = (old FACT) $\cdot$ (new m) = $m! \cdot (m + 1) = (m + 1)! =$ (new m)!,

```
{Input:  integer n > 0}
{Output:  integer FACT with FACT = n!}
begin
m := 1
FACT := 1
{FACT = m!}
while m < n do
    m := m + 1
    FACT := FACT · m
end
```

FIGURE 4

as required at the end of the loop. The loop exits with $m = n$, and hence with FACT $= n!$. Once we had the idea of using FACT $= m!$ as a loop invariant, this algorithm practically wrote itself.

With the convention that $0! = 1$, we could initialize the algorithm with $m := 0$ instead of $m := 1$ and still produce the correct output.

Notice one apparently unavoidable feature of this algorithm; even if we just want 150!, we must compute $1!, 2!, \ldots, 149!$ first.

(b) Figure 5 shows another algorithm for computing $n!$, this time building in the factors in reverse order, from the top down.

```
{Input:  integer n ≥ 0}
{Output:  integer REV with REV = n!}
begin
REV := 1
m := n
{REV · m! = n!}
while m > 0 do
    REV := REV · m
    m := m − 1
end
```

FIGURE 5

Again, the loop invariant is easy to check. Initially, it is obviously true, and if REV $\cdot m! = n!$ at the start of a pass through the loop, then

$$(\text{new REV}) \cdot (\text{new } m)! = (\text{REV} \cdot m) \cdot (m - 1)! = \text{REV} \cdot m! = n!$$

at the end of the pass. The guard makes the exit at the right time, with $m = 1$, so the algorithm produces the correct output, even if $n = 0$.

This algorithm does not compute $1!, 2!, \ldots$ along the way to $n!$, but what it does is just as bad, or worse. We still must compute a lot of partial products, in this instance pretty big ones from the very beginning, before we are done. ■

The examples we have just seen begin to show the value of loop invariants. In the next section we will use the Loop Invariant Theorem to develop the method of proof by mathematical induction, one of the most important techniques in this book. Later chapters will show us how the use of loop invariants can help us to understand the workings of some fairly complicated algorithms, as well as to verify that they produce the correct results.

Both the Division Algorithm and Example 2(a), the first of our algorithms to compute $n!$, had variables that increased by 1 on each pass through the loop. In the case of the Division Algorithm, we did not know in advance how many passes we would make through the loop, but in Example 2(a) we knew that m would take the values $1, 2, 3, \ldots, n$ in that order, and that the algorithm would then stop. There is a convenient notation to describe this kind of predictable incrementation.

The algorithm instruction "**for** $k = m$ **to** n **do** S" tells us to substitute $m, m + 1, \ldots, n$, for k, in that order, and to do S each time. Thus the segment

FACT $:= 1$
for $k = 1$ to n do
　　FACT $:=$ FACT $\cdot k$

produces FACT $= n!$ at the end.

Using a while loop, we could rewrite a segment "for $k = m$ to n do S" as

$k := m$
while $k \leq n$ do
　　S
　　$k := k + 1,$

with one major and obvious warning: the segment S cannot change the value of k. If S changes k, we should replace k in the algorithm by some other letter, not m or n, that S does not change. See Exercise 13 for what can go wrong otherwise.

Some programming languages contain a construction "**for** $k = n$ **downto** m **do** S," meaning substitute $n, n - 1, \ldots, m$ in decreasing order for k, and do S each time. Our second factorial algorithm could have been written using this device.

EXAMPLE 3　　(a) Here is an algorithm to compute the sum

$$\sum_{k=1}^{n} k^2 = 1 + 4 + 9 + \cdots + n^2,$$

where n is a given positive integer.

$$s := 0$$
for $k = 1$ to n do
$$\quad s := s + k^2$$

(b) More generally, the algorithm

$$s := 0$$
for $k = m$ to n do
$$\quad s := s + a_k$$

computes $\displaystyle\sum_{k=m}^{n} a_k = a_m + \cdots + a_n.$ ■

EXERCISES 4.1

1. List the first five values of x for each of the following algorithm segments.

(a) $x := 0$
    while $0 \leq x$ do
      $x := 2x + 3$

(b) $x := 1$
    while $0 \leq x$ do
      $x := 2x + 3$

(c) $x := 1$
    while $0 \leq x$ do
      $x := 2x - 1$

2. Find an integer b so that the last number printed is 6.

(a) $n := 0$
    while $n < b$ do
      print n
      $n := n + 1$

(b) $n := 0$
    while $n < b$ do
      $n := n + 1$
      print n

3. (a) List the values of m and n during the execution of the following algorithm, using the format of Figure 2.

```
begin
m := 0
n := 0
while n ≠ 4 do
    m := m + 2·n + 1
    n := n + 1
end
```

(b) Modify the algorithm of part (a) so that $m = 17^2$ when it stops.

4. Interchange the lines "$n := n + 1$" and "print n^2" in Figure 1(a).

(a) List the values printed if $n = 0$ at the start of the loop.

(b) Do the same for $n = 4$.

5. Interchange the lines "print n^2" and "$n := n + 2$" in Figure 1(b).

(a) List the values printed if $n = 0$ at the start of the loop.

(b) Do the same for $n = 1$.

6. (a) Write the segment

   for $i = 1$ to 17 do
   $\quad k := k + 2i$

   using a while loop.

   (b) Repeat part (a) with

   for $k = 8$ downto 1 do
   $\quad i := i + 2k$

7. Show that the following are loop invariants for the loop

   while $1 \leq m$ do
   $\quad m := m + 1$
   $\quad n := n + 1$

   (a) $m + n$ is even.   (b) $m + n$ is odd.

8. Show that the following are loop invariants for the loop

   while $1 \leq m$ do
   $\quad m := 2m$
   $\quad n := 3n$

   (a) $n^2 \geq m^3$.   (b) $2m^6 < n^4$.

9. Consider the loop

   while $j \geq n$ do
   $\quad i := i + 2$
   $\quad j := j + 1$

   where i and j are integers.

   (a) Is $i < j^2$ a loop invariant if $n = 1$? Explain.

   (b) Is $i < j^2$ a loop invariant if $n = 0$? Explain.

   (c) Is $i \leq j^2$ a loop invariant if $n = 0$? Explain.

   (d) Is $i \geq j^2$ a loop invariant if $n = 0$? Explain.

10. Consider the loop

   while $k \geq 1$ do
   $\quad k := 2k$

   (a) Is $k^2 \equiv 1 \pmod{3}$ a loop invariant? Explain.

   (b) Is $k^2 \equiv 1 \pmod{4}$ a loop invariant? Explain.

11. For which values of the integer b does each of the following algorithms terminate?

(a) begin	(b) begin	(c) begin
$k := b$	$k := b$	$k := b$
while $k < 5$ do	while $k \neq 5$ do	while $k < 5$ do
$k := 2k - 1$	$k := 2k - 1$	$k := 2k + 1$
end	end	end

12. Suppose that the loop body in Example 2(a) is changed to

$$\text{FACT} := \text{FACT} \cdot m$$
$$m := m + 1.$$

How should the initialization and guard be changed in order to get the same output, FACT $= n!$? The loop invariant may change.

13. Do the two algorithms shown produce the same output? Explain.

Algorithm A

```
begin
k := 1
while k ≤ 4 do
  k := k²
  print k
  k := k + 1
end
```

Algorithm B

```
begin
for k = 1 to 4 do
  k := k²
  print k
end
```

14. Do the two algorithms shown produce the same output? Explain.

Algorithm C

```
begin
m := 1
n := 1
while 1 ≤ m ≤ 3 do
  while 1 ≤ n ≤ 3 do
    m := 2m
    n := n + 1
    print m
end
```

Algorithm D

```
begin
m := 1
n := 1
while 1 ≤ m ≤ 3 and 1 ≤ n ≤ 3 do
  m := 2m
  n := n + 1
  print m
end
```

15. Here's a primitive pseudo-random number generator that generates numbers in $\{0, 1, 2, \ldots, 72\}$:

```
begin
r := c
while r > 0 do
  r := 31 · r MOD 73
end
```

For example, if $c = 3$, we get $3, 20, 36, 21, 67, 33, \ldots$. Which of the following are loop invariants?

(a) $r < 73$.   (b) $r \equiv 0 \pmod 5$.   (c) $r = 0$.

16. (a) Show that $S = I^2$ is an invariant of the loop

>   while $1 \leq I$ do
> $S := S + 2I + 1$
> $I := I + 1$.

(b) Show that $S = I^2 + 1$ is also an invariant of the loop in part (a).

(c) What is the 73rd number printed by the following? Explain.

>   begin
> $S := 1$
> $I := 1$
>   while $1 \leq I$ do
>   print S
> $S := S + 2I + 1$
> $I := I + 1$
>   end

(d) The same question as in part (c), but with $S := 2$ initially.

17. Which of these sets of integers does the Well-Ordering Principle say have smallest elements? Explain.

(a) $\mathbb{P}$

(b) $\mathbb{Z}$

(c) $\{n \in \mathbb{P} : n^2 > 17\}$

(d) $\{n \in \mathbb{P} : n^2 < 0\}$

(e) $\{n \in \mathbb{Z} : n^2 > 17\}$

(f) $\{n^2 : n \in \mathbb{P} \text{ and } n! > 80^n\}$

18. (a) What would we get if we tried to apply the algorithm in Figure 4 with $n = -3$ or with $n = -73$? Are these sensible outputs?

(b) Answer the same question for the algorithm in Figure 5.

19. Consider the following loop, with $a, b \in \mathbb{P}$. [Recall that $m = (m \text{ DIV } n) \cdot n + (m \text{ MOD } n)$ with $0 \leq (m \text{ MOD } n) < n$.]

>   while $r > 0$ do
> $a := b$
> $b := r$
> $r := a \text{ MOD } b$

Which of the following are invariants of the loop? Explain.

(a) a, b and r are multiples of 5.

(b) a is a multiple of 5.

(c) $r < b$.

(d) $r \leq 0$.

20. (a) Is $5^k < k!$ an invariant of the following loop?

>   while $4 \leq k$ do
> $k := k + 1$

(b) Can you conclude that $5^k < k!$ for all $k \geq 4$?

21. Here is an algorithm that factors an integer as a product of an odd integer and a power of 2.

> {input: positive integer n}
> {output: nonnegative integers k and m with m odd and $m \cdot 2^k = n$}
> begin
> $m := n$
> $k := 0$
> while m is even do
> $$m := \frac{m}{2} \text{ and } k := k + 1$$
> end

(a) Show that $m \cdot 2^k = n$ is a loop invariant.

(b) Explain why the algorithm terminates, and show that m is odd on exit from the loop.

22. Suppose that p and q are both invariants of the loop

> while g do S.

(a) Is $p \land q$ an invariant of the loop? Explain.

(b) Is $p \lor q$ an invariant of the loop? Explain.

23. The following algorithm gives a fast method for raising a number to a power.

> {input: number a and $n \in \mathbb{N}$}
> {output: $p = a^n$}
> begin
> $p := 1$
> $q := a$
> $i := n$
> while $i > 0$ do
>   if i is odd, then $p := p \cdot q$
>   {Do the next two steps whether or not i is odd.}
> $q := q \cdot q$
> $i := i$ DIV 2
> end

(a) Give a table similar to Figure 2 showing successive values of p, q and i with input $a = 2$ and $n = 11$.

(b) Verify that $q^i \cdot p = a^n$ is a loop invariant and that $p = a^n$ on exit from the loop.

24. (a) Consider the following algorithm with inputs $n \in \mathbb{P}$, $x \in \mathbb{R}$.

> begin
> $m := n$
> $y := x$
> while $m \neq 0$ do
>   body
> end

Suppose that "body" is chosen so that "$x^n = x^m \cdot y$ and $m \geq 0$" is an invariant of the while loop. What is the value of y at the time the algorithm terminates, if it ever does?

(b) Is it possible to choose z so that

while $m \neq 0$ do
$\quad m := m - 1$
$\quad y := z$

has "$x^n - x^m \cdot y$ and $m \geq 0$" as an invariant? Explain.

§ 4.2   Mathematical Induction

This section develops a framework for proving lots of propositions all at once. We introduce the main idea, which follows from the Well-Ordering Principle for $\mathbb{N}$, as a natural consequence of the Loop Invariant Theorem of §4.1.

EXAMPLE 1   We claim that $37^{500} - 37^{100}$ is a multiple of 10. One way to check this would be to compute $37^{500} - 37^{100}$ and, if the last digit is 0, proclaim victory. The trouble with this plan is that 37^{500} is pretty big—it has over 780 decimal digits—so we might be a while computing it. Since we aren't claiming to know the exact value of $37^{500} - 37^{100}$, but just claiming that it's a multiple of 10, maybe there's a simpler way.

Since 37 is odd, so are 37^{500} and 37^{100}, so at least we know that $37^{500} - 37^{100}$ is a multiple of 2. To establish our claim, we just need to show that it's also a multiple of 5. We notice that $37^{500} = (37^{100})^5$, so we're interested in $(37^{100})^5 - 37^{100}$. Maybe $n^5 - n$ is *always* a multiple of 5 for $n \in \mathbb{P}$. If it is, then we can get our claim by letting $n = 37^{100}$.

Experiments with small numbers look promising. For instance, we have $1^5 - 1 = 0, 2^5 - 2 = 30, 3^5 - 3 = 240$ and even $17^5 - 17 = 1{,}419{,}840$. So there is hope.

Now here comes the trick. We construct a simple loop, in effect an elementary machine whose job it is to verify that $n^5 - n$ is a multiple of 5 for every n in $\mathbb{P}$, starting with $n = 1$ and running at least until $n = 37^{100}$. We will also give the machine some help. Figure 1 shows the loop. Its output is irrelevant; what will matter is its invariant.

begin
$\quad n := 1$
while $n < 37^{100}$ do
$\quad\quad$ if $n^5 - n$ is a multiple of 5 then
$\quad\quad\quad n := n + 1$
end

FIGURE 1

The while loop simply checks that $n^5 - n$ is a multiple of 5 and, if it is true, moves on to the next value of n. We have already checked that $1^5 - 1$ is a multiple of 5, so the alleged loop invariant is true when we arrive at the loop. We claim that the algorithm terminates [with $n = 37^{100}$], and that "$n^5 - n$ is a multiple of 5" is a loop invariant. If these claims are true, then of course $37^{500} - 37^{100}$ is a multiple of 5 by the Loop Invariant Theorem.

Consider a pass through the loop, say with $n = k < 37^{100}$. If $k^5 - k$ is *not* a multiple of 5, then the body of the loop does nothing, and the algorithm just repeats the loop forever with n stuck at k. On the other hand, if $k^5 - k$ *is* a multiple of 5, then the "if" condition is satisfied, the loop body increases the value of n to $k + 1$ and the algorithm goes back to check if the guard $n < 37^{100}$ is still true. To make sure that the algorithm terminates, we want to be sure that at each iteration the "if" condition is true.

Here is where we give the algorithm some help. If it is making the $(k + 1)$th pass through the loop, then $k^5 - k$ must have been a multiple of 5. We give a little proof now to show that this fact forces $(k + 1)^5 - (k + 1)$ also to be a multiple of 5. The reason [using some algebra] is that

$$(k + 1)^5 - (k + 1) = k^5 + 5k^4 + 10k^3 + 10k^2 + 5k + 1 - k - 1$$

$$= (k^5 - k) + 5(k^4 + 2k^3 + 2k^2 + k).$$

If $k^5 - k$ is a multiple of 5, then since the second term is obviously a multiple of 5, $(k + 1)^5 - (k + 1)$ must also be a multiple of 5. What this all means is that we can tell the algorithm not to bother checking the "if" condition each time; $n^5 - n$ will always be a multiple of 5, because it was the time before.

This argument is enough to show that the condition forms a loop invariant. Moreover, during each pass through the loop the condition checked for "if" holds, so n increases by 1 each time. Eventually, $n = 37^{100}$ and the loop terminates, as we claimed.

This loop we have devised may look like a stupid algorithm. To check that $37^{500} - 37^{100}$ is a multiple of 5 we seem to go to all the work of looking at $1^5 - 1, 2^5 - 2, 3^5 - 3, \ldots, (37^{100} - 1)^5 - (37^{100} - 1)$ first. That really *would* be stupid. But notice that as a matter of fact the only one we checked outright was $1^5 - 1$. For the rest, we just gave a short algebraic argument that *if* $k^5 - k$ is a multiple of 5, then so is $(k + 1)^5 - (k + 1)$. The algorithm was never intended to be run, but was just meant to give us a proof that $37^{500} - 37^{100}$ is a multiple of 5, without computing either of the huge powers. In fact, a tiny change in the algorithm shows, with the same proof, that 5 divides $73^{5000} - 73^{1000}$. Indeed, we can show that $n^5 - n$ is a multiple of 5 for *every* $n \in \mathbb{P}$ by applying exactly the same arguments to the loop in Figure 2, which does not terminate. ■

$$n := 1$$
$$\text{while } 1 \le n \text{ do}$$
$$\quad \text{if } n^5 - n \text{ is a multiple of 5 then}$$
$$\quad\quad n: = n + 1$$

FIGURE 2

Suppose, more generally, that we have a finite list of propositions, say $p(m)$, $p(m + 1)$, ..., $p(n)$, where m, $m + 1$, ..., n are successive integers. In Example 1, for instance, we had $p(k) = $ "$k^5 - k$ is a multiple of 5" for $k = 1, 2, ..., 37^{100}$. Suppose that we also know, as we did in Example 1, that:

(B) $p(m)$ is true and
(I) $p(k + 1)$ is true whenever $p(k)$ is true and $m \le k < n$.

We claim that all of the propositions $p(m)$, $p(m + 1)$, ..., $p(n)$ must then be true. The argument is just like the one we went through in Example 1. We construct the loop in Figure 3.

$$k := m$$
$$\{p(k) \text{ is true}\}$$
$$\text{while } m \le k < n \text{ do}$$
$$\quad \text{if } p(k) \text{ is true then}$$
$$\quad\quad k := k + 1$$

FIGURE 3

By (I) the statement "$p(k)$ is true" is a loop invariant, and the loop terminates with $k = n$. Of course the condition that $p(k) \Rightarrow p(k + 1)$ is the reason that $p(k)$ is true at the end of each pass. In terms of the Well-Ordering Principle, we are just observing that if $p(k)$ ever failed there would be a first time when it did, and the conditions (B) and (I) prevent every time from being that first bad time.

We have proved the following important fact.

Principle of Finite Mathematical Induction

> Let $p(m)$, $p(m + 1)$, ..., $p(n)$ be a finite sequence of propositions. If
>
> (B) $p(m)$ is true and
> (I) $p(k + 1)$ is true whenever $p(k)$ is true and $m \le k < n$,
>
> then all the propositions are true.

An infinite version is available, too. Just replace the condition "while $m \le k < n$" by "while $m \le k$" to get the following.

Principle of Mathematical Induction

> Let $p(m)$, $p(m + 1)$, ... be a sequence of propositions. If
>
> > (B) $p(m)$ is true and
> > (I) $p(k + 1)$ is true whenever $p(k)$ is true and $m \le k$,
>
> then all the propositions are true.

Condition (B) in each of these principles of induction is called the **basis**, and (I) is the **inductive step**. Given a list of propositions, these principles help us to organize a proof that all of the propositions are true. The basis is usually easy to check; the inductive step is sometimes quite a bit more complicated to verify.

The principles tell us that *if* we can show (B) and (I), then we are done, but they do not help us show either condition. Of course, if the $p(k)$'s are not all true, the principles cannot show that they are. Either (B) or (I) must fail in such a case.

EXAMPLE 2 (a) For each $n \in \mathbb{P}$ let $p(n)$ be the proposition "$n! > 2^n$," which we proved for all $n \ge 4$ in § 1.6 just by staring hard at it. To give a proof by induction we verify $p(n)$ for $n = 4$, i.e., check that $4! > 2^4$, and then show

(I) If $4 \le k$ and if $k! > 2^k$, then $(k + 1)! > 2^{k+1}$.

The proof of (I) is straightforward:

$$
\begin{aligned}
(k + 1)! &= k! \cdot (k + 1) \\
&> 2^k \cdot (k + 1) && \text{[by the inductive assumption } k! > 2^k\text{]} \\
&\ge 2^k \cdot 2 && \text{[since } k + 1 \ge 5 > 2\text{]} \\
&= 2^{k+1}.
\end{aligned}
$$

Since we have checked the basis and the inductive step, $p(n)$ is true for every integer $n \ge 4$ by induction.

(b) The fact that

$$1 + 2 + \cdots + n = \frac{n(n + 1)}{2} \qquad \text{for all} \quad n \in \mathbb{P}$$

is useful to know. It can be proved by an averaging argument, but also by induction.

Let $p(n)$ be "$\sum_{i=1}^{n} i = n(n + 1)/2$." Then $p(1)$ is "$\sum_{i=1}^{1} i = 1(1 + 1)/2$," which is true.

Assume inductively that $p(k)$ is true for some $k \in \mathbb{P}$, i.e., that $\sum_{i=1}^{k} i = k(k+1)/2$. We want to show that this implies that $p(k+1)$ is true. Now

$$\sum_{i=1}^{k+1} i = \left(\sum_{i=1}^{k} i \right) + (k+1) \qquad [\text{definition of } \Sigma \text{ notation}]$$

$$= \frac{k(k+1)}{2} + (k+1) \qquad [\text{assumption that } p(k) \text{ is true}]$$

$$= \left[\frac{k}{2} + 1 \right](k+1) \qquad [\text{factor out } k+1]$$

$$= \left[\frac{k+2}{2} \right](k+1)$$

$$= \frac{((k+1)+1)(k+1)}{2},$$

so $p(k+1)$ holds. By induction, $p(n)$ is true for every n in $\mathbb{P}$.

(c) We can use induction to establish the formula for the sum of the terms in a geometric series:

$$\sum_{i=0}^{n} r^i = \frac{r^{n+1} - 1}{r - 1} \qquad \text{if } r \neq 0, r \neq 1 \text{ and } n \in \mathbb{N}.$$

Let $p(n)$ be "$\sum_{i=0}^{n} r^i = \frac{r^{n+1} - 1}{r - 1}$." Then $p(0)$ is

$$r^0 = \frac{r^1 - 1}{r - 1},$$

which is true because $r^0 = 1$. Thus the basis for induction is true.

We prove the inductive step $p(k) \Rightarrow p(k+1)$ as follows:

$$\sum_{i=0}^{k+1} r^i = \left(\sum_{i=0}^{k} r^i \right) + r^{k+1}$$

$$= \frac{r^{k+1} - 1}{r - 1} + r^{k+1} \qquad [\text{by the inductive assumption } p(k)]$$

$$= \frac{r^{k+1} - 1 + r^{k+2} - r^{k+1}}{r - 1} \qquad [\text{algebra}]$$

$$= \frac{r^{k+2} - 1}{r - 1} \qquad [\text{more algebra}]$$

$$= \frac{r^{(k+1)+1} - 1}{r - 1},$$

i.e., $p(k + 1)$ holds. By induction, $p(n)$ is true for every $n \in \mathbb{N}$.

(d) We prove that all numbers of the form $8^n - 2^n$ are divisible by 6. More precisely, we show that $8^n - 2^n$ is divisible by 6 for each $n \in \mathbb{P}$. Our nth proposition is

$$p(n) = \text{"}8^n - 2^n \text{ is divisible by 6."}$$

The basis for the induction, $p(1)$, is clearly true, since $8^1 - 2^1 = 6$. For the inductive step, assume that $p(k)$ is true. Our task is to use this assumption somehow to establish $p(k + 1)$:

$$8^{k+1} - 2^{k+1} \text{ is divisible by 6.}$$

Thus we would like to write $8^{k+1} - 2^{k+1}$ somehow in terms of $8^k - 2^k$, in such a way that any remaining terms are easily seen to be divisible by 6. A little trick is to write $8^{k+1} - 2^{k+1}$ as $8(8^k - 2^k)$ plus appropriate correction terms:

$$8^{k+1} - 2^{k+1} = 8(8^k - 2^k) + 8 \cdot 2^k - 2^{k+1}$$
$$= 8(8^k - 2^k) + 8 \cdot 2^k - 2 \cdot 2^k = 8(8^k - 2^k) + 6 \cdot 2^k.$$

Now $8^k - 2^k$ is divisible by 6 by assumption $p(k)$, and $6 \cdot 2^k$ is obviously a multiple of 6, so the same is true of $8^{k+1} - 2^{k+1}$. We have shown that the inductive step is valid, so our proof is complete by the Principle of Mathematical Induction. ■

It is worth emphasizing that prior to the last sentence in each of these proofs, we did *not* prove "$p(k + 1)$ is true." We merely proved an implication: "if $p(k)$ is true, then $p(k + 1)$ is true." In a sense we proved an infinite number of assertions, namely: $p(1)$; if $p(1)$ is true, then $p(2)$ is true; if $p(2)$ is true, then $p(3)$ is true; if $p(3)$ is true, then $p(4)$ is true; etc. Then we applied mathematical induction to conclude $p(1)$ is true; $p(2)$ is true; $p(3)$ is true; $p(4)$ is true; etc.

Note also that when we use the induction principles we don't need to write any loops. We used loops to justify the principles, but when we apply the principles we only need to verify conditions (B) and (I).

EXAMPLE 3 Here are some poor examples of the use of induction.

(a) For $n \in \mathbb{P}$ let $p(n)$ be "$n^2 \leq 100$." Then we can check directly that $p(1), p(2), \ldots, p(10)$ are true. If we try to prove by induction that $p(n)$ is true for all $n \in \mathbb{P}$, or even for $1 \leq n \leq 20$, we will fail. We can show the basis (B), but the inductive step (I) must not be provable, because we know, for example, that $p(11)$ is false.

(b) For $n \in \mathbb{N}$ let $r(n)$ be "$n = n + 5$." Then $r(0)$ is "$0 = 5$," so (B) is false here, but in fact $r(k) \Rightarrow r(k + 1)$, because *if* $k = k + 5$, then $k + 1 = (k + 1) + 5$. It is not enough just to verify that (I) holds; we also need (B).

(c) For $m \in \mathbb{P}$ let $s(m)$ be "$m \cdot (m + 1)$ is even." We could certainly prove $s(m)$ true for all m in $\mathbb{P}$ using induction [think of how such a proof would go], but we don't need such an elaborate proof. Either m or $m + 1$ is even, so their product is surely even in any case. Sometimes it is as easy to prove directly that all propositions in a sequence are true as it is to invoke induction. ■

EXAMPLE 4   (a) An application of induction often starts with a guessing game. Suppose, for instance, that the sequence $(s_0, s_1, s_2, \ldots)$ satisfies the conditions $s_0 = a$ and $s_n = 2s_{n-1} + b$ for some constants a and b and all $n \in \mathbb{P}$. Can we find a formula to describe s_n? We compute the first few terms of the sequence to see if there is a pattern. Figure 4 shows the results.

n	s_n
0	a
1	$2a + b$
2	$2(2a + b) + b = 4a + 3b$
3	$2(4a + 3b) + b = 8a + 7b$
4	$2(8a + 7b) + b = 16a + 15b$

FIGURE 4

It looks as if maybe $s_n = 2^n a + (2^n - 1)b$ in general. [This kind of guessing a general fact from a few observations is what is called "inductive reasoning" in the sciences. The use of the word "inductive" in that context is quite different from the mathematical usage.] Now we have a list of propositions; for $n \in \mathbb{N}$, $p(n)$ is "$s_n = 2^n a + (2^n - 1)b$." It turns out [Exercise 5] that we have made a good guess, and one can prove by *mathematical* induction that $p(n)$ is true for each n in $\mathbb{N}$.

(b) As we are looking for a pattern, it sometimes helps to examine closely how a particular case follows from the one before. Then we may be able to see how to construct an argument for the inductive step.

Let $\mathscr{P}(S)$ be the power set of some finite set S. If S has n elements, then $\mathscr{P}(S)$ has 2^n members. This proposition was shown to be plausible in Example 2 of §1.1. We prove it now, by induction.

We verified this assertion earlier for $n = 0, 1, 2$ and 3. In particular, the case $n = 0$ establishes the basis for induction. Before proving the inductive step, let's experiment a little and compare $\mathscr{P}(S)$ for $S = \{a, b\}$ and for $S = \{a, b, c\}$. Note that

$$\mathcal{P}(\{a, b, c\}) = \{\emptyset, \{a\}, \{b\}, \{a, b\}, \{c\}, \{a, c\}, \{b, c\}, \{a, b, c\}\}.$$

The first four sets make up $\mathcal{P}(\{a, b\})$; each of the remaining sets consists of a set in $\mathcal{P}(\{a, b\})$ with c added to it. This is why $\mathcal{P}(\{a, b, c\})$ has twice as many sets as $\mathcal{P}(\{a, b\})$. This argument looks as if it generalizes: every time an element is added to S, the size of $\mathcal{P}(S)$ doubles.

To prove the inductive step, we assume the proposition is valid for n. We consider a set S with $n + 1$ elements; for convenience we use $S = \{1, 2, 3, \ldots, n, n + 1\}$. Let $T = \{1, 2, 3, \ldots, n\}$. The sets in $\mathcal{P}(T)$ are simply the subsets of S that do not contain $n + 1$. By the assumption for n, $\mathcal{P}(T)$ contains exactly 2^n sets. Each remaining subset of S contains the number $n + 1$, so it is the union of a set in $\mathcal{P}(T)$ with the one-element set $\{n + 1\}$. That is, $\mathcal{P}(S)$ has another 2^n sets that are not subsets of T. It follows that $\mathcal{P}(S)$ has $2^n + 2^n = 2^{n+1}$ members. This completes the inductive step, and hence the proposition is true for all n, by mathematical induction. ■

The method of mathematical induction applies to situations, like the one in this example, in which:

1. We know the answer in the beginning.
2. We know how to determine the answer at one stage from the answer at the previous stage.
3. We have a guess at the general answer.

Of course, if our guess is wrong, that's too bad; we won't be able to prove it is right with this method, or with any other. But if our guess is correct, then mathematical induction often gives us a framework for confirming the guess with a proof.

Sometimes it only makes sense to talk about a finite sequence of propositions $p(n)$. For instance, we might have an iterative algorithm that we know terminates after a while, and we might want to know that some condition holds while it is running. Verifying invariants for while loops is an example of this sort of problem. If we can write our condition as $p(k)$, in terms of some variable k that increases steadily during the execution, say $k = m, m + 1, \ldots, N$, then we may be able to use the Principle of Finite Mathematical Induction to prove that $p(k)$ is true for each allowed value of k. Verifying an invariant of a segment "for $i = m$ to N do S" amounts to checking the inductive step for this principle.

This section has covered the basic ideas of mathematical induction and given us the tools to deal with a great number of common situations. As we will see in §4.5, there are other forms of induction that we can use to handle many problems in which the principles from this section do not naturally apply.

EXERCISES 4.2

1. Prove

$$\sum_{i=1}^{n} i^2 = 1 + 4 + 9 + \cdots + n^2 = \frac{n(n+1)(2n+1)}{6} \qquad \text{for} \quad n \in \mathbb{P}.$$

2. Prove

$$4 + 10 + 16 + \cdots + (6n - 2) = n(3n + 1) \qquad \text{for all} \quad n \in \mathbb{P}.$$

3. Show each of the following.
   (a) $37^{100} - 37^{20}$ is a multiple of 10.
   (b) $37^{20} - 37^4$ is a multiple of 10.
   (c) $37^{500} - 37^4$ is a multiple of 10.
   (d) $37^4 - 1$ is a multiple of 10.
   (e) $37^{500} - 1$ is a multiple of 10.

4. Prove

$$\frac{1}{1 \cdot 5} + \frac{1}{5 \cdot 9} + \frac{1}{9 \cdot 13} + \cdots + \frac{1}{(4n-3)(4n+1)} = \frac{n}{4n+1} \qquad \text{for} \quad n \in \mathbb{P}.$$

5. Show by induction that if $s_0 = a$ and $s_n = 2s_{n-1} + b$ for $n \in \mathbb{P}$, then $s_n = 2^n a + (2^n - 1)b$ for every $n \in \mathbb{N}$.

6. Consider the following procedure.

   ```
   begin
   S := 1
   while 1 ≤ S do
       print S
       S := S + 2√S + 1
   end
   ```

   (a) List the first four printed values of S.
   (b) Use mathematical induction to show that the value of S is always an integer. [It is easier to prove the stronger statement that the value of S is always the square of an integer; in fact, $S = n^2$ at the start of the nth pass through the loop.]

7. Prove that $11^n - 4^n$ is divisible by 7 for all $n \in \mathbb{P}$.

8. (a) Choose m and $p(k)$ in the segment

   ```
   k := m
   while m ≤ k do
       if p(k) is true then
           k := k + 1
   ```

   so that proving $p(k)$ an invariant of the loop would show that $2^n < n!$ for all integers $n \geq 4$.

(b) Verify that your $p(k)$ in part (a) is an invariant of the loop.

(c) The proposition $p(k) = $ "$8^k < k!$" is an invariant of this loop. Does it follow that $8^n < n!$ for all $n \geq 4$? Explain.

9. (a) Show that $\sum_{i=0}^{k} 2^i = 2^{k+1} - 1$ is an invariant of the loop in the algorithm

    begin
    $k := 0$
    while $0 \leq k$ do
    $k := k + 1$
    end.

    (b) Repeat part (a) for the invariant $\sum_{i=0}^{k} 2^i = 2^{k+1}$.

    (c) Can you use part (a) to prove that $\sum_{i=0}^{k} 2^i = 2^{k+1} - 1$ for every $k \in \mathbb{N}$? Explain.

    (d) Can you use part (b) to prove that $\sum_{i=0}^{k} 2^i = 2^{k+1}$ for every $k \in \mathbb{N}$? Explain.

10. Prove that $n^2 > n + 1$ for $n \geq 2$.

11. (a) Calculate $1 + 3 + \cdots + (2n - 1)$ for a few values of n, and then guess a general formula for this sum.

    (b) Prove the formula obtained in part (a) by induction.

12. For which $n \in \mathbb{P}$ does the inequality $4n \leq n^2 - 7$ hold? Explain.

13. Consider the proposition $p(n) = $ "$n^2 + 5n + 1$ is even."

    (a) Prove that the truth of $p(k)$ implies the truth of $p(k + 1)$, for each $k \in \mathbb{P}$.

    (b) For which values of n is $p(n)$ actually true? What is the moral of this exercise?

14. Prove $(2n + 1) + (2n + 3) + (2n + 5) + \cdots + (4n - 1) = 3n^2$ for $n \in \mathbb{P}$. The sum can also be written $\sum_{i=n}^{2n-1} (2i + 1)$.

15. Prove that $5^n - 4n - 1$ is divisible by 16 for $n \in \mathbb{P}$.

16. Prove $1^3 + 2^3 + \cdots + n^3 = (1 + 2 + \cdots + n)^2$, i.e., $\sum_{i=1}^{n} i^3 = \left(\sum_{i=1}^{n} i \right)^2$ for $n \in \mathbb{P}$. *Hint:* Use the identity in Example 2(b).

17. Prove that

$$\frac{1}{n+1} + \frac{1}{n+2} + \cdots + \frac{1}{2n} = 1 - \frac{1}{2} + \frac{1}{3} - \frac{1}{4} + \cdots + \frac{1}{2n-1} - \frac{1}{2n}$$

for $n \in \mathbb{P}$. For $n = 1$ this says that $\frac{1}{2} = 1 - \frac{1}{2}$ and for $n = 2$ this says that $\frac{1}{3} + \frac{1}{4} = 1 - \frac{1}{2} + \frac{1}{3} - \frac{1}{4}$.

18. For $n \in \mathbb{P}$, prove

(a) $\displaystyle\sum_{i=1}^{n} \frac{1}{\sqrt{i}} \geq \sqrt{n}$ (b) $\displaystyle\sum_{i=1}^{n} \frac{1}{\sqrt{i}} \leq 2\sqrt{n} - 1$

19. Prove that $5^{n+1} + 2 \cdot 3^n + 1$ is divisible by 8 for $n \in \mathbb{N}$.

20. Prove that $8^{n+2} + 9^{2n+1}$ is divisible by 73 for $n \in \mathbb{N}$.

21. This exercise requires a little knowledge of trigonometric identities. Prove that $|\sin nx| \leq n|\sin x|$ for all $x \in \mathbb{R}$ and all $n \in \mathbb{P}$.

§ 4.3   Recursive Definitions

The values of the terms in a sequence may be given explicitly by formulas such as $s_n = n^3 - 73n$ or by descriptions like "let t_n be the weight of the nth edge on the path." Terms may also be defined sometimes by descriptions that involve other terms which come before them in the sequence.

We say that a sequence is defined **recursively** provided that:

(B)   Some finite set of values, usually the first one or first few, are specified.

(R)   The remaining values of the sequence are defined in terms of previous values of the sequence. [A formula that gives such a definition is called a **recurrence formula** or **relation**.]

The requirement (B) gives the **basis** or starting point for the definition. The remainder of the sequence is determined by using the relation (R) repeatedly [i.e., recurrently].

EXAMPLE 1   (a) We can define the familiar sequence FACT recursively by

(B)   FACT$(0) = 1$,
(R)   FACT$(n + 1) = (n + 1) \cdot$ FACT(n) for $n \in \mathbb{N}$.

Condition (R) lets us calculate FACT(1), then FACT(2), then FACT(3), etc., and we quickly see that FACT$(n) = n!$ for the first several values of n. This equation can be proved for all n by mathematical induction. By (B), FACT$(0) = 1 = 0!$. Assuming inductively that FACT$(m) = m!$ for some $m \in \mathbb{N}$ and using (R), we get FACT$(m + 1) = (m + 1) \cdot$ FACT$(m) = (m + 1) \cdot m! = (m + 1)!$ for the inductive step. Since we understand the sequence $n!$ already, the recursive definition above may seem silly, but we will try to convince you in part (b) that recursive definitions of even simple sequences are useful.

(b) Consider the sequence SUM$(n) = \displaystyle\sum_{i=0}^{n} \frac{1}{i!}$. To write a computer program that calculates the values of SUM for large values of *n*, one could use the following recursive definition:

(B) $\text{SUM}(0) = 1$,

(R) $\text{SUM}(n + 1) = \text{SUM}(n) + \dfrac{1}{(n + 1)!}$.

The added term in (R) is the reciprocal of $(n + 1)!$, so FACT$(n + 1)$ will be needed as the program progresses. At each *n*, one could instruct the program to calculate FACT$(n + 1)$ from scratch or one could store a large number of these values. Clearly, it would be more efficient to alternately calculate FACT$(n + 1)$ and SUM$(n + 1)$ using the recursive definition in part (a) for FACT and the recursive definition above for SUM.

(c) Define the sequence SEQ as follows:

(B) $\text{SEQ}(0) = 1$,
(R) $\text{SEQ}(n + 1) = (n + 1)/\text{SEQ}(n)$ for $n \in \mathbb{N}$.

With $n = 0$, (R) gives $\text{SEQ}(1) = 1/1 = 1$. Then with $n = 1$, we find $\text{SEQ}(2) = 2/1 = 2$. Continuing in this way, we see that the first few terms are 1, 1, 2, 3/2, 8/3, 15/8, 16/5, 35/16. It is by no means apparent what a general formula for SEQ(n) might be. It is evident that SEQ(73) exists, but it would take considerable calculation to find it.   ∎

In Example 1, how did we *know* that SEQ(73) exists? Our certainty is based on the belief that recursive definitions do indeed define sequences on all of $\mathbb{N}$, unless some step leads to an illegal computation such as division by 0. We could prove by induction that the recursive definition in Example 1 defines a sequence, but we will use the Well-Ordering Principle instead. Let

$$S = \{n \in \mathbb{N} : \text{SEQ}(n) = 0 \text{ or } \text{SEQ}(n) \text{ is not defined}\}.$$

We want to show that S is empty. If not, it has a smallest member, say *m*. By (B), $m \neq 0$, so $m - 1 \in \mathbb{N}$. Since *m* is the smallest bad guy, $\text{SEQ}(m - 1) \neq 0$ and $\text{SEQ}(m - 1)$ is defined. But then (R) defines $\text{SEQ}(m) = m/\text{SEQ}(m - 1) \neq 0$, contrary to $m \in S$. Thus S must be empty.

This proof that SEQ(n) is defined for each *n* uses the fact that SEQ$(n + 1)$ only depends on SEQ(n). Recursive definitions allow a term to depend on other terms besides the one just before it. In such cases either the Well-Ordering Principle or an enhanced version of the Principle of Mathematical Induction that we take up in §4.5 can be used to prove that the sequences are well defined.

The values of the terms in a recursively defined sequence can be calculated in more than one way. An **iterative calculation** finds s_n by computing all of the values $s_1, s_2, \ldots, s_{n-1}$ first, so they are available to use in computing s_n. We had in mind an iterative calculation of the sequences in Example 1. To calculate FACT(73), for instance, we would first calculate FACT(k) for $k = 1, 2, \ldots, 72$, even though we might have no interest in these preliminary values themselves.

In the case of FACT there really seems to be no better alternative. Sometimes, though, there is a more clever way to calculate a given value of s_n. A **recursive calculation** finds the value of s_n by looking to see which terms s_n depends on, then which terms those terms depend on, and so on. It may turn out that the value of s_n only depends on the values of a relatively small set of its predecessors, in which case the other previous terms can be ignored.

EXAMPLE 2   (a) The integer part of a real number a, denoted by $\lfloor a \rfloor$ as in §3.6, is the largest integer m such that $m \leq a$. Define the sequence T by

(B) $T(1) = 1$,
(R) $T(n) = 2 \cdot T(\lfloor n/2 \rfloor)$ for $n \geq 2$.

Then

$$T(73) = 2 \cdot T(\lfloor 73/2 \rfloor) = 2 \cdot T(36) = 2 \cdot 2 \cdot T(18) = 2 \cdot 2 \cdot 2 \cdot T(9)$$
$$= 2 \cdot 2 \cdot 2 \cdot 2 \cdot T(4) = 2 \cdot 2 \cdot 2 \cdot 2 \cdot 2 \cdot T(2) = 2 \cdot 2 \cdot 2 \cdot 2 \cdot 2 \cdot 2 \cdot T(1) = 2^6.$$

For this calculation we only need the values of $T(36)$, $T(18)$, $T(9)$, $T(4)$, $T(2)$, and $T(1)$, and we have no need to compute the other 66 values of $T(n)$ that precede $T(73)$.

This sequence can be described in another way as follows [Exercise 19 of §4.5]:

$$T(n) \text{ is the largest integer } 2^k \text{ with } 2^k \leq n.$$

Using this description we could compute $T(73)$ by looking at the list of powers of 2 less than 73 and taking the largest one.

(b) A slight change in (R) from part (a) gives the sequence Q with

(B) $Q(1) = 1$,
(R) $Q(n) = 2 \cdot Q(\lfloor n/2 \rfloor) + n$ for $n \geq 2$.

Now the general term is not so clear, but we can still find $Q(73)$ recursively from $Q(36)$, $Q(18)$, $\ldots$, $Q(2)$, $Q(1)$. ■

EXAMPLE 3   (a) The **Fibonacci sequence** is defined as follows:

(B)   FIB$(0) = $ FIB$(1) = 1$,
(R)   FIB$(n) = $ FIB$(n - 1) + $ FIB$(n - 2)$ for $n \geq 2$.

Note that the recurrence formula makes no sense for $n = 1$, so FIB(1) had to be defined separately in the basis. The first few terms of this sequence are

$$1, 1, 2, 3, 5, 8, 13, 21, 34, 55, 89.$$

(b) Here is an easy way to define the sequence

$$0, 0, 1, 1, 2, 2, 3, 3, \ldots.$$

(B) $\text{SEQ}(0) = \text{SEQ}(1) = 0,$

(R) $\text{SEQ}(n) = 1 + \text{SEQ}(n - 2)$ for $n \geq 2$.

Compare Exercise 11 of §1.5. ∎

EXAMPLE 4   Let $\Sigma = \{a, b\}$.

(a) We are interested in the number s_n of words of length n that do not have consecutive a's, i.e., that do not contain the string aa. Let's write A_n for the set of words in Σ^n having no consecutive a's. Then $A_0 = \{\lambda\}$, $A_1 = \Sigma$ and $A_2 = \Sigma^2 \backslash \{aa\}$, so $s_0 = 1$, $s_1 = 2$ and $s_2 = 4 - 1 = 3$. To get a recurrence formula for s_n, we consider $n \geq 2$ and count the number of words in A_n in terms of shorter words. If a word in A_n ends in b, it can be preceded by any word in A_{n-1}. So s_{n-1} words in A_n end in b. If a word in A_n ends in a, the last two letters must be ba and this string can be preceded by any word in A_{n-2}. So s_{n-2} words in A_n end in a. Thus $s_n = s_{n-1} + s_{n-2}$ for $n \geq 2$. This is the recurrence relation for the Fibonacci sequence, but note that the basis is different: $s_1 = 2$, while $\text{FIB}(1) = 1$. In fact, $s_n = \text{FIB}(n + 1)$ for $n \in \mathbb{N}$ [Exercise 13].

(b) Since Σ^n has 2^n words in it, there are $2^n - s_n = 2^n - \text{FIB}(n + 1)$ words of length n that contain consecutive a's. ∎

EXAMPLE 5   Let $\Sigma = \{a, b, c\}$, let B_n be the set of words in Σ^n with an even number of a's, and let t_n denote the number of words in B_n. Then $B_0 = \{\lambda\}$, $B_1 = \{b, c\}$ and $B_2 = \{aa, bb, bc, cb, cc\}$, so $t_0 = 1$, $t_1 = 2$ and $t_2 = 5$. We count the number of words in B_n by looking at the last letter. If a word in B_n ends in b, it can be preceded by any word in B_{n-1}. So t_{n-1} words in B_n end in b. Similarly, t_{n-1} words in B_n end in c. If a word in B_n ends in a it must be preceded by a word in Σ^{n-1} with an *odd* number of a's. Since Σ^{n-1} has 3^{n-1} words, $3^{n-1} - t_{n-1}$ of them must have an odd number of a's. Hence $3^{n-1} - t_{n-1}$ words in B_n end in a. Thus

$$t_n = t_{n-1} + t_{n-1} + (3^{n-1} - t_{n-1}) = 3^{n-1} + t_{n-1}$$

for $n \geq 1$. Hence $t_3 = 3^2 + t_2 = 9 + 5 = 14$, $t_4 = 3^3 + t_3 = 27 + 14 = 41$, etc.

In this case, it's relatively easy to find an explicit formula for t_n. First note that

$$t_n = 3^{n-1} + t_{n-1} = 3^{n-1} + 3^{n-2} + t_{n-2} = \cdots$$

$$= 3^{n-1} + 3^{n-2} + \cdots + 3^0 + t_0 = 1 + \sum_{k=0}^{n-1} 3^k.$$

If those three dots make you nervous, you can supply a proof by induction. Now we apply Example 2(c) of §4.2 to obtain

$$t_n = 1 + \frac{3^n - 1}{3 - 1} = 1 + \frac{3^n - 1}{2} = \frac{3^n + 1}{2}.$$

Our mental process works for $n \geq 1$, but the formula works for $n = 0$, too. The formula for t_n looks right: about half of the words in Σ^n use an even number of a's.

Notice the technique that we used to get the formula for t_n. We wrote t_n in terms of t_{n-1}, then t_{n-2}, then $t_{n-3}, \ldots$ and collected the left-over terms to see if they suggested a formula. They looked like $3^{n-1} + 3^{n-2} + 3^{n-3} + \cdots + 3^{n-k} + t_{n-k}$, so we guessed that carrying the process backward far enough would give $t_n = 3^{n-1} + \cdots + 3^0 + t_0$.

We illustrated one of the tricks of the trade, but it really doesn't matter how we arrive at the formula as long as we can show that it's correct. We could have asked mother for the answer or simply made an inspired guess.

To prove that our guess is correct, it is enough to prove that it satisfies the recurrence conditions that define t_n, namely, $t_0 = 1$ and $t_n = 3^{n-1} + t_{n-1}$. We simply check:

$$\frac{3^0 + 1}{2} = 1 \quad \text{and} \quad \frac{3^n + 1}{2} = 3^{n-1} + \frac{3^{n-1} + 1}{2} \qquad \text{for} \quad n \in \mathbb{P}.$$

This method of proof is legitimate because the recurrence conditions define the sequence t_n uniquely. ■

EXAMPLE 6   (a) Define the sequence S by

(B) $S(0) = 0$, $S(1) = 1$,
(R) $S(n) = S(\lfloor n/2 \rfloor) + S(\lfloor n/5 \rfloor)$ for $n \geq 2$.

It makes sense to calculate the values of S recursively, rather than iteratively. For instance,

$$S(73) = S(36) + S(14) = [S(18) + S(7)] + [S(7) + S(2)]$$
$$= S(18) + 2S(7) + S(2)$$
$$= S(9) + S(3) + 2[S(3) + S(1)] + S(1) + S(0)$$
$$= \cdots = 8S(1) + 6S(0) = 8.$$

The calculation of $S(73)$ involves the values of $S(36)$, $S(18)$, $S(14)$, $S(9)$, $S(7)$, $S(4)$, $S(3)$, $S(2)$, $S(1)$ and $S(0)$, but that's still better than finding all values of $S(k)$ for $k = 1, \ldots, 72$. One can show more generally that the value of $S(n)$ in this example only depends on the values of the terms $S(m)$ for m of the form $\lfloor n/2^a 5^b \rfloor$.

(b) Recursive calculation requires storage space for the intermediate values that have been called for but not yet computed. It may be possible, though, to keep the number of storage slots fairly small. For example, the recursive calculation of FACT(6) goes like this:

$$\text{FACT}(6) = 6 \cdot \text{FACT}(5) = 30 \cdot \text{FACT}(4) = 120 \cdot \text{FACT}(3) = \cdots$$

Only one address is needed for the intermediate [unknown] value FACT(k) for $k < 6$. Similarly,

$$\text{FIB}(6) = \text{FIB}(5) + \text{FIB}(4) = (\text{FIB}(4) + \text{FIB}(3)) + \text{FIB}(4)$$
$$= 2 \cdot \text{FIB}(4) + \text{FIB}(3)$$
$$= 3 \cdot \text{FIB}(3) + 2 \cdot \text{FIB}(2)$$
$$= 5 \cdot \text{FIB}(2) + 3 \cdot \text{FIB}(1)$$
$$= 8 \cdot \text{FIB}(1) + 5 \cdot \text{FIB}(0)$$

only requires two intermediate addresses. ■

EXAMPLE 7   Figure 1 shows a graph and its adjacency matrix **M**. As we observed in §3.3, the nth power $\mathbf{M}^n$ of **M** gives the exact number of paths of length n connecting each pair of vertices.

$$\mathbf{M} = \begin{bmatrix} 1 & 1 \\ 1 & 0 \end{bmatrix}$$

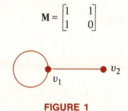

FIGURE 1

We can compute the first few powers directly.

$$\mathbf{M}^1 = \begin{bmatrix} 1 & 1 \\ 1 & 0 \end{bmatrix}, \quad \mathbf{M}^2 = \begin{bmatrix} 2 & 1 \\ 1 & 1 \end{bmatrix}, \quad \mathbf{M}^3 = \begin{bmatrix} 3 & 2 \\ 2 & 1 \end{bmatrix},$$
$$\mathbf{M}^4 = \begin{bmatrix} 5 & 3 \\ 3 & 2 \end{bmatrix}, \quad \mathbf{M}^5 = \begin{bmatrix} 8 & 5 \\ 5 & 3 \end{bmatrix}.$$

The numbers look familiar. In fact [Exercise 12],

$$\mathbf{M}^n = \begin{bmatrix} \text{FIB}(n) & \text{FIB}(n-1) \\ \text{FIB}(n-1) & \text{FIB}(n-2) \end{bmatrix} \quad \text{for} \quad n \geq 2.$$

Thus this simple graph leads to matrices whose entries are defined by the recursively defined Fibonacci sequence in Example 3. ■

Of course, we can give recursive definitions even if the sequence is not real-valued.

EXAMPLE 8   Let S be a set and let f be a function from S into S. We define

(B) $f^{(0)} = 1_S$ [the identity function on S],
(R) $f^{(n+1)} = f^{(n)} \circ f$.

Thus

(1) $\qquad\qquad f^{(1)} = f, \quad f^{(2)} = f \circ f, \quad f^{(3)} = f \circ f \circ f, \quad$ etc.

In other words,

(2) $\qquad\qquad\qquad f^{(n)} = f \circ f \circ \cdots \circ f \qquad$ [n times].

$f^{(n)}$ is simply the composite of the function f, n times. The recursive definition is more precise than the "etc." in (1) or the three dots in (2).   ∎

As in Example 8, we will often use recursive definitions to give concise definitions of objects that we already understand quite well.

EXAMPLE 9   Let a be a nonzero real number. Define the powers of a by

(B) $a^0 = 1$,
(R) $a^{n+1} = a^n \cdot a$ for $n \in \mathbb{N}$.

Equivalently,

(B) $\text{POW}(0) = 1$,
(R) $\text{POW}(n+1) = \text{POW}(n) \cdot a$ for $n \in \mathbb{N}$.   ∎

EXAMPLE 10   Let $(a_j)_{j \in \mathbb{P}}$ be a sequence of real numbers. We can define the product notation by

(B) $\displaystyle\prod_{j=1}^{1} a_j = a_1$,

(R) $\displaystyle\prod_{j=1}^{n+1} a_j = a_{n+1} \cdot \prod_{j=1}^{n} a_j$ for $n \geq 1$.

Equivalently,

(B) $\text{PROD}(1) = a_1$,
(R) $\text{PROD}(n+1) = a_{n+1} \cdot \text{PROD}(n)$ for $n \geq 1$.

These recursive definitions start at $n = 1$. An alternative is to define the "empty product" to be 1, i.e.,

(B) $\displaystyle\prod_{j=1}^{0} a_j = 1$ [which looks peculiar], or

   (B) PROD$(0) = 1$.

Then the same recursive relation (R) as before serves to define the remaining terms of the sequence. ∎

EXERCISES 4.3

1. We recursively define $s_0 = 1$ and $s_{n+1} = 2/s_n$ for $n \in \mathbb{N}$.

  (a) List the first few terms of the sequence.

  (b) What is the set of values of s?

2. We recursively define SEQ$(0) = 0$ and SEQ$(n + 1) = 1/[1 + $ SEQ$(n)]$ for $n \in \mathbb{N}$. Calculate SEQ(n) for $n = 1, 2, 3, 4$ and 6.

3. Consider the sequence $(1, 3, 9, 27, 81, \ldots)$.

  (a) Give a formula for the nth term SEQ(n) where SEQ$(0) = 1$.

  (b) Give a recursive definition for the sequence SEQ.

4. (a) Give a recursive definition for the sequence $(2, 2^2, (2^2)^2, ((2^2)^2)^2, \ldots)$, i.e., $(2, 4, 16, 256, \ldots)$.

  (b) Give a recursive definition for the sequence $(2, 2^2, 2^{(2^2)}, 2^{(2^{(2^2)})}, \ldots)$, i.e., $(2, 4, 16, 65536, \ldots)$.

5. Is the following a recursive definition for a sequence SEQ? Explain.

    (B) SEQ$(0) = 1$,
    (R) SEQ$(n + 1) = $ SEQ$(n)/[100 - n]$.

6. (a) Calculate SEQ(9) where SEQ is as in Example 1(c).

  (b) Calculate FIB(11) where FIB is as in Example 3(a).

7. Let $\Sigma = \{a, b, c\}$ and let s_n denote the number of words of length n that do not have consecutive a's.

  (a) Calculate s_0, s_1 and s_2.

  (b) Find a recurrence formula for s_n.

  (c) Calculate s_3 and s_4.

8. Let $\Sigma = \{a, b\}$ and let s_n denote the number of words of length n that do not contain the string ab.

  (a) Calculate s_0, s_1, s_2 and s_3.

  (b) Find a formula for s_n and prove that it is correct.

9. Let $\Sigma = \{a, b\}$ and let t_n denote the number of words of length n with an even number of a's.

  (a) Calculate t_0, t_1, t_2 and t_3.

(b) Find a formula for t_n and prove that it is correct.

(c) Does your formula for t_n work for $n = 0$?

10. Consider the sequence defined by

    (B)   SEQ(0) = 1, SEQ(1) = 0,

    (R)   SEQ(n) = SEQ($n - 2$)   for $n \geq 2$.

(a) List the first few terms of this sequence.

(b) What is the set of values of this sequence?

11. We recursively define $a_0 = a_1 = 1$ and $a_n = a_{n-1} + 2a_{n-2}$ for $n \geq 2$.

(a) Calculate a_6 recursively.

(b) Prove that all the terms a_n are odd.

12. Prove that the sequence $\mathbf{M}_1, \mathbf{M}_2, \ldots$ of matrices defined by

$$\mathbf{M}_1 = \begin{bmatrix} 1 & 1 \\ 1 & 0 \end{bmatrix}, \quad \mathbf{M}_2 = \begin{bmatrix} 2 & 1 \\ 1 & 1 \end{bmatrix} \quad \text{and} \quad \mathbf{M}_n = \begin{bmatrix} \text{FIB}(n) & \text{FIB}(n-1) \\ \text{FIB}(n-1) & \text{FIB}(n-2) \end{bmatrix}$$

for $n \geq 2$ satisfies $\mathbf{M}_{n+1} = \mathbf{M}_1 \cdot \mathbf{M}_n$ for $n \in \mathbb{P}$.

13. Show that if the sequence s_n satisfies $s_0 = 1$, $s_1 = 2$ and $s_n = s_{n-1} + s_{n-2}$ for $n \geq 2$, then $s_n = \text{FIB}(n + 1)$ for $n \in \mathbb{N}$. *Hint:* Consider the smallest member of $S = \{n \in \mathbb{N} : s_n \neq \text{FIB}(n + 1)\}$.

14. Recursively define $b_0 = b_1 = 1$ and $b_n = 2b_{n-1} + b_{n-2}$ for $n \geq 2$.

(a) Calculate b_5 iteratively.

(b) Explain why all the terms b_n are odd. *Hint:* Consider the first one that is even.

15. Let SEQ(0) = 1 and SEQ(n) = $\sum_{i=0}^{n-1} \text{SEQ}(i)$ for $n \geq 1$. This is actually a simple, familiar sequence. What is it?

16. We recursively define $a_0 = 0$, $a_1 = 1$, $a_2 = 2$ and $a_n = a_{n-1} - a_{n-2} + a_{n-3}$ for $n \geq 3$.

(a) List the first few terms of the sequence until the pattern is clear.

(b) What is the set of values of the sequence?

17. The process of assigning n children to n classroom seats can be broken down into (1) choosing a child for the first seat and (2) assigning the other $n - 1$ children to the remaining seats. Let $A(n)$ be the number of different assignments of n children to n seats.

(a) Write a recursive definition of the sequence A.

(b) Calculate $A(6)$ recursively.

(c) Is the sequence A familiar?

18. Consider the process of assigning $2n$ children to n cars in a playground train so that two children go in each car. First choose two children for the front car [there are $2n(2n - 1)/2$ ways to do this, as we will see in Chapter 5].

Then distribute the rest of the children to the remaining $n - 1$ cars. Let $B(n)$ be the number of ways to assign $2n$ children to n cars.

(a) Write a recursive definition of the sequence B.

(b) Calculate $B(3)$ recursively.

(c) Calculate $B(5)$ iteratively.

(d) Give an explicit formula for $B(n)$.

19. Consider the sequence FOO defined by

$$(B) \quad \text{FOO}(0) = 1, \text{FOO}(1) = 1$$

$$(R) \quad \text{FOO}(n) = \frac{10 \cdot \text{FOO}(n - 1) + 100}{\text{FOO}(n - 2)} \quad \text{for } n \geq 2.$$

(a) What is the set of values of FOO?

(b) Repeat part (a) for the sequence GOO defined by

$$(B) \quad \text{GOO}(0) = 1, \text{GOO}(1) = 2$$

$$(R) \quad \text{GOO}(n) = \frac{10 \cdot \text{GOO}(n - 1) + 100}{\text{GOO}(n - 2)} \quad \text{for } n \geq 2.$$

We learned about these entertaining sequences from our colleague Ivan Niven.

20. Let $(a_1, a_2, \ldots)$ be a sequence of real numbers.

(a) Give a recursive definition for $\text{SUM}(n) = \sum_{j=1}^{n} a_j$ for $n \geq 1$.

(b) Revise your recursive definition for $\text{SUM}(n)$ by starting with $n = 0$. What is the "empty sum"?

21. Let $(A_1, A_2, \ldots)$ be a sequence of subsets of some set S.

(a) Give a recursive definition for $\bigcup_{j=1}^{n} A_j$.

(b) How would you define the "empty union"?

(c) Give a recursive definition for $\bigcap_{j=1}^{n} A_j$.

(d) How would you define the "empty intersection"?

22. Let $(A_1, A_2, \ldots)$ be a sequence of subsets of some set S. Define

$$(B) \quad \text{SYM}(1) = A_1,$$
$$(R) \quad \text{SYM}(n + 1) = A_{n+1} \oplus \text{SYM}(n) \quad \text{for } n \geq 1.$$

Recall that $\oplus$ denotes symmetric difference. It turns out that an element x in S belongs to $\text{SYM}(n)$ if and only if the set $\{k : x \in A_k \text{ and } k \leq n\}$ has an odd number of elements. Prove this fact by mathematical induction.

§ 4.4   Recurrence Relations

Sequences that appear in mathematics and science are frequently defined by recurrences, rather than by formulas. A variety of techniques have been developed to obtain explicit formulas for the values in such cases. In this section we give the complete answer for sequences defined by recurrences of the form

$$s_n = as_{n-1} + bs_{n-2},$$

where a and b are constants, and we get substantial information about sequences that satisfy recurrences of the form

$$s_{2n} = 2 \cdot s_n + f(n)$$

for known functions f.

We begin by considering

$$s_n = as_{n-1} + bs_{n-2},$$

assuming that the two initial values s_0 and s_1 are known. The cases where $a = 0$ or $b = 0$ are especially easy to deal with.

If $b = 0$, so that $s_n = as_{n-1}$ for $n \geq 1$, then $s_1 = as_0$, $s_2 = as_1 = a^2 s_0$, etc. A simple induction argument shows that $s_n = a^n s_0$ for all $n \in \mathbb{N}$.

Now suppose that $a = 0$. Then $s_2 = bs_0$, $s_4 = bs_2 = b^2 s_0$, etc., so that $s_{2n} = b^n s_0$ for all $n \in \mathbb{N}$. Similarly, $s_3 = bs_1$, $s_5 = b^2 s_1$, etc., so that $s_{2n+1} = b^n s_1$ for all $n \in \mathbb{N}$.

EXAMPLE 1   (a) Consider the recurrence relation $s_n = 3s_{n-1}$ with $s_0 = 5$. Here $a = 3$, so $s_n = 5 \cdot 3^n$ for $n \in \mathbb{N}$.

(b) Consider the recurrence relation $s_n = 3s_{n-2}$ with $s_0 = 5$ and $s_1 = 2$. Here $b = 3$, so $s_{2n} = 5 \cdot 3^n$ and $s_{2n+1} = 2 \cdot 3^n$ for $n \in \mathbb{N}$. ■

From now on we will assume that $a \neq 0$ and $b \neq 0$. It will be convenient to ignore the specified values of s_0 and s_1 until later. From the special cases we have examined it seems reasonable to hope that some solutions have the form $s_n = cr^n$ for a nonzero constant c. This hope, if true, would force

$$r^n = ar^{n-1} + br^{n-2}.$$

Dividing by r^{n-2} would then give $r^2 = ar + b$, or $r^2 - ar - b = 0$. In other words, if $s_n = cr^n$ for all n, then r must be a solution of the quadratic equation $x^2 - ax - b = 0$, which is called the **characteristic equation** of the recurrence relation. The characteristic equation has either one or two solutions, which we will assume from now on are real numbers.

Theorem 1 | Consider a recurrence relation of the form

$$s_n = as_{n-1} + bs_{n-2}$$

with characteristic equation

$$x^2 - ax - b = 0,$$

where a and b are nonzero constants.

(a) If the characteristic equation has two different solutions r_1 and r_2, then

$(*)$ $\qquad\qquad s_n = c_1 r_1^n + c_2 r_2^n$

for certain constants c_1 and c_2. If s_0 and s_1 are specfied, the constants can be determined by setting $n = 0$ and $n = 1$ in $(*)$ and solving the two equations for c_1 and c_2.

(b) If the characteristic equation has only one solution r, then

$(**)$ $\qquad\qquad s_n = c_1 r^n + c_2 \cdot n \cdot r^n$

for certain constants c_1 and c_2. As in (a), c_1 and c_2 can be determined if s_0 and s_1 are specified.

Warning. This fine theorem only applies to recurrence relations of the form $s_n = as_{n-1} + bs_{n-2}$.

EXAMPLE 2 Consider the recurrence relation $s_n = s_{n-1} + 2s_{n-2}$ where $s_0 = s_1 = 3$. Here $a = 1$ and $b = 2$. The characteristic equation $x^2 - x - 2 = 0$ has solutions $r_1 = 2$ and $r_2 = -1$ since $x^2 - x - 2 = (x-2)(x+1)$, so part (a) of the theorem applies. By Theorem 1,

$$s_n = c_1 \cdot 2^n + c_2 \cdot (-1)^n$$

for constants c_1 and c_2. By setting $n = 0$ and $n = 1$ we find that

$$s_0 = c_1 \cdot 2^0 + c_2 \cdot (-1)^0 \quad \text{and} \quad s_1 = c_1 \cdot 2^1 + c_2 \cdot (-1)^1,$$

i.e., that

$$3 = c_1 + c_2 \quad \text{and} \quad 3 = 2c_1 - c_2.$$

Solving this system of two equations gives $c_1 = 2$ and $c_2 = 1$. We conclude that

$$s_n = 2 \cdot 2^n + 1 \cdot (-1)^n = 2^{n+1} + (-1)^n \quad \text{for} \quad n \in \mathbb{N}. \quad \blacksquare$$

EXAMPLE 3   Consider again the Fibonacci sequence in §4.3. Writing s_n for FIB(n), we have $s_0 = s_1 = 1$ and $s_n = s_{n-1} + s_{n-2}$ for $n \geq 2$. Here $a = b = 1$, so we solve $x^2 - x - 1 = 0$. By the quadratic formula, the equation has two solutions:

$$r_1 = \frac{1 + \sqrt{5}}{2} \quad \text{and} \quad r_2 = \frac{1 - \sqrt{5}}{2}.$$

Thus part (a) of the theorem applies, so

$$s_n = c_1 \left(\frac{1 + \sqrt{5}}{2} \right)^n + c_2 \left(\frac{1 - \sqrt{5}}{2} \right)^n \quad \text{for} \quad n \in \mathbb{N}.$$

While solving for c_1 and c_2, it's convenient to retain the notation r_1 and r_2. Setting $n = 0$ and $n = 1$ gives

$$1 = c_1 + c_2 \quad \text{and} \quad 1 = c_1 r_1 + c_2 r_2.$$

If we replace c_2 by $1 - c_1$ in the second equation we get $1 = c_1 r_1 + (1 - c_1) r_2$, so $1 - r_2 = c_1(r_1 - r_2)$ and

$$c_1 = \frac{1 - r_2}{r_1 - r_2}.$$

Since $r_1 + r_2 = 1$ and $r_1 - r_2 = \sqrt{5}$, we conclude that $c_1 = r_1/\sqrt{5}$. Now $c_2 = 1 - c_1 = (\sqrt{5} - r_1)/\sqrt{5} = -r_2/\sqrt{5}$. Finally,

$$s_n = c_1 r_1^n + c_2 r_2^n = \frac{r_1}{\sqrt{5}} r_1^n - \frac{r_2}{\sqrt{5}} r_2^n = \frac{1}{\sqrt{5}} (r_1^{n+1} - r_2^{n+1})$$

and so

$$\text{FIB}(n) = s_n = \frac{1}{\sqrt{5}} \left[\left(\frac{1 + \sqrt{5}}{2} \right)^{n+1} - \left(\frac{1 - \sqrt{5}}{2} \right)^{n+1} \right].$$

This description for FIB(n) is hardly what one could have expected when we started. FIB(n) is an integer, of course, whereas the expression with the radicals certainly doesn't look much like an integer. The formula is correct, though. Try it with a calculator for a couple of values of n to be convinced. By using the formula, we can compute any value of FIB(n) without having to find the previous values, and we can also estimate the rate at which FIB(n) grows as n increases. Since

$$\frac{\sqrt{5} - 1}{2} < 0.62,$$

the absolute value of

$$\frac{1}{\sqrt{5}} \cdot \left(\frac{1 - \sqrt{5}}{2} \right)^{n+1}$$

is less than 0.5 for every positive n, and thus FIB(n) is the closest integer to

$$\frac{1}{\sqrt{5}} \cdot \left(\frac{1 + \sqrt{5}}{2}\right)^{n+1}. \quad \blacksquare$$

EXAMPLE 4   Consider (s_n) defined by $s_0 = 1$, $s_1 = -3$ and $s_n = 6s_{n-1} - 9s_{n-2}$ for $n \geq 2$. Here the characteristic equation is $x^2 - 6x + 9 = 0$, which has exactly one solution, namely $r = 3$. By part (b) of the theorem,

$$s_n = c_1 \cdot 3^n + c_2 \cdot n \cdot 3^n \qquad \text{for} \quad n \in \mathbb{N}.$$

Setting $n = 0$ and $n = 1$ we get

$$s_0 = c_1 \cdot 3^0 + 0 \quad \text{and} \quad s_1 = c_1 \cdot 3^1 + c_2 \cdot 3^1$$

or

$$1 = c_1 \quad \text{and} \quad -3 = 3c_1 + 3c_2.$$

So $c_1 = 1$ and $c_2 = -2$. Therefore,

$$s_n = 3^n - 2 \cdot n \cdot 3^n \qquad \text{for} \quad n \in \mathbb{N}. \quad \blacksquare$$

Proof of Theorem 1.   (a) No matter what s_0 and s_1 are, the equations

$$s_0 = c_1 + c_2 \quad \text{and} \quad s_1 = c_1 r_1 + c_2 r_2$$

can be solved for c_1 and c_2, since $r_1 \neq r_2$. The original sequence (s_n) is determined by the values s_0 and s_1 and the recurrence condition $s_n = as_{n-1} + bs_{n-2}$, so it suffices to show that the sequence defined by (∗) also satisfies this recurrence condition. Since $x = r_1$ satisfies $x^2 = ax + b$, we have $r_1^n = ar_1^{n-1} + br_1^{n-2}$, so the sequence (r_1^n) satisfies the condition $s_n = as_{n-1} + bs_{n-2}$. So does (r_2^n). It is now easy to check that the sequence defined by (∗) also satisfies the recurrence condition:

$$as_{n-1} + bs_{n-2} = a(c_1 r_1^{n-1} + c_2 r_2^{n-1}) + b(c_1 r_1^{n-2} + c_2 r_2^{n-2})$$

$$= c_1[ar_1^{n-1} + br_1^{n-2}] + c_2[ar_2^{n-1} + br_2^{n-2}]$$

$$= c_1 r_1^n + c_2 r_2^n = s_n.$$

(b) If r is the only solution of the characteristic equation, then the characteristic equation has the form $(x - r)^2 = 0$. Thus $x^2 - 2rx + r^2 = x^2 - ax - b$, so $a = 2r$ and $b = -r^2$. The recurrence relation can now be written

$$s_n = 2rs_{n-1} - r^2 s_{n-2}.$$

Putting $n = 0$ and $n = 1$ in (∗∗) gives the equations

$$s_0 = c_1 \quad \text{and} \quad s_1 = c_1 r + c_2 r.$$

Since $r \neq 0$, these equations have the solutions $c_1 = s_0$ and $c_2 = -s_0 + s_1/r$. As in our proof of part (a), it suffices to show that any sequence defined by (**) satisfies $s_n = 2rs_{n-1} - r^2 s_{n-2}$. But

$$2rs_{n-1} - r^2 s_{n-2}$$
$$= 2r[c_1 r^{n-1} + c_2(n-1)r^{n-1}] - r^2[c_1 r^{n-2} + c_2(n-2)r^{n-2}]$$
$$= 2c_1 r^n + 2c_2(n-1)r^n - c_1 r^n - c_2(n-2)r^n$$
$$= c_1 r^n + c_2 \cdot n \cdot r^n = s_n. \quad \blacksquare$$

The proof of the theorem is still valid if the roots of the characteristic equation are not real, so the theorem is true in that case as well. Finding the values of the terms with formula (*) will then involve complex arithmetic, but the answers will of course all be real if a, b, s_0 and s_1 are real. This situation is analogous to the calculation of the Fibonacci numbers, which are integers, using $\sqrt{5}$, which is not.

Theorem 1 applies only to recurrences of the form $s_n = as_{n-1} + bs_{n-2}$, but since recurrences like this come up fairly frequently it is a useful tool. Another important kind of recurrence arises from estimating running times of what are called **divide-and-conquer** algorithms. The general algorithm of this type splits its input problem into two or more pieces, solves the pieces separately and then puts the results together for the final answer.

A simple example of a divide-and-conquer algorithm is the process of finding the largest number in a set of numbers by breaking the set into two batches of approximately equal size, finding the largest number in each batch and then comparing those two to get the largest number overall. Merge sorting of a set of numbers by sorting the two batches separately and then fitting the sorted lists together is another example.

If $T(n)$ is the time such an algorithm takes to process an input of size n, then the structure of the algorithm leads to a recurrence of the form

$$T(n) = T\left(\frac{n}{2}\right) + T\left(\frac{n}{2}\right) + F(n),$$

where the two $T(n/2)$'s give the time it takes to process the two batches separately, and $F(n)$ is the time to fit the two results together. Of course, this equation only makes sense if n is even.

Our next theorem gives information about recurrences such as this, of the form

$$s_{2n} = 2 \cdot s_n + f(n).$$

We have stated the theorem in a fairly general form, with one very important special case highlighted. Exercise 17 gives two variations on the theme.

Theorem 2

Let (s_n) be a sequence that satisfies a recurrence relation of the form

$$s_{2n} = 2 \cdot s_n + f(n) \qquad \text{for} \quad n \in \mathbb{P}.$$

Then

$$s_{2^m} = 2^m \cdot \left[s_1 + \frac{1}{2} \sum_{i=0}^{m-1} \frac{f(2^i)}{2^i} \right] \qquad \text{for} \quad m \in \mathbb{N}.$$

In particular, if

$$s_{2n} = 2 \cdot s_n + A + B \cdot n$$

for constants A and B, then

$$s_{2^m} = 2^m \cdot s_1 + (2^m - 1) \cdot A + \frac{B}{2} \cdot 2^m \cdot m.$$

Thus if $n = 2^m$ in this case we have

$$s_n = n s_1 + (n - 1)A + \frac{B}{2} \cdot n \cdot \log_2 n.$$

Before we discuss the proof, let's see how we can use the theorem.

EXAMPLE 5

(a) Finding the largest member of a set by dividing and conquering leads to the recurrence

$$T(2n) = 2T(n) + A,$$

where the constant A is the time it takes to compare the winners from the two halves of the set. According to Theorem 2 with $B = 0$,

$$T(2^m) = 2^m \cdot T(1) + (2^m - 1) \cdot A.$$

Here $T(1)$ is the time it takes to find the biggest element in a set with one element. It seems reasonable to regard $T(1)$ as an overhead or handling charge for examining each element, whereas A is the cost of an individual comparison. If $n = 2^m$ we get

$$T(n) = n \cdot T(1) + (n - 1) \cdot A.$$

We examine n elements and make $n - 1$ comparisons to find the largest one. In this case, dividing and conquering gives no essential improvement over simply running through the elements one at a time, at each step keeping the largest number seen so far.

(b) Sorting a set by dividing and conquering gives a recurrence

$$T(2n) = 2 \cdot T(n) + B \cdot n,$$

since the time to fit together the two ordered halves into one set is proportional to the sizes of the two sets being merged. Theorem 2 with $A = 0$ says that

$$T(2^m) = 2^m \cdot T(1) + \frac{B}{2} \cdot 2^m \cdot m.$$

If $n = 2^m$, then

$$T(n) = n \cdot T(1) + \frac{B}{2} \cdot n \cdot \log_2 n = O(n \cdot \log_2 n).$$

For large n the dominant cost comes from the merging, not from examining individual elements. This algorithm is actually a reasonably efficient one for many applications.   ■

Proof of Theorem 2. We verify that if

(∗) $$S_{2^m} = 2^m \cdot \left[s_1 + \frac{1}{2} \sum_{i=0}^{m-1} \frac{f(2^i)}{2^i} \right]$$

for some $m \in \mathbb{N}$, then

$$S_{2^{m+1}} = 2^{m+1} \cdot \left[s_1 + \frac{1}{2} \sum_{i=0}^{m} \frac{f(2^i)}{2^i} \right].$$

Since $s_{2^0} = 2^0 \cdot s_1$, because there are no terms in the summation if $m = 0$, it will follow by induction on m that (∗) holds for every $m \in \mathbb{N}$. [Though it's not necessary to check the case $m = 1$ separately, the reader may find it comforting.] Here are the details:

$$S_{2^{m+1}} = S_{2 \cdot 2^m} = 2 \cdot s_{2^m} + f(2^m) \qquad \text{[recurrence]}$$

$$= 2^{m+1} \cdot \left[s_1 + \frac{1}{2} \sum_{i=0}^{m-1} \frac{f(2^i)}{2^i} \right] + f(2^m) \qquad \text{[by (∗)]}$$

$$= 2^{m+1} \cdot \left[s_1 + \frac{1}{2} \sum_{i=0}^{m-1} \frac{f(2^i)}{2^i} + \frac{1}{2 \cdot 2^m} f(2^m) \right]$$

$$= 2^{m+1} \cdot \left[s_1 + \frac{1}{2} \sum_{i=0}^{m} \frac{f(2^i)}{2^i} \right].$$

The special case $f(n) = A + Bn$ gives a summation that we can compute fairly easily, but it is just as simple to verify directly [Exercise 16] that

$$S_{2^m} = 2^m \cdot s_1 + (2^m - 1) \cdot A + \frac{B}{2} \cdot 2^m \cdot m$$

if (s_n) satisfies the recurrence $s_{2n} = 2 \cdot s_n + A + Bn$.   ■

This proof is technically correct, but unilluminating. How did we get the formula for s_{2^m} in the first place? Easy. We just used the recurrence to calculate the first few terms s_2, s_4, s_8, s_{16} and then guessed the pattern.

The theorem only tells us the values of s_n in cases in which n is a power of 2. That leaves some enormous gaps in our knowledge. It is often the case, however, that the sequence s is **monotone**, that is, $s_k \le s_n$ whenever $k \le n$. Timing estimates for algorithms generally have this property, for example. If s is monotone and if we know a function g such that $s_n \le g(n)$ whenever n is a power of 2, then for a general integer k we have $2^{m-1} < k \le 2^m$ for some m, so $s_{2^{m-1}} \le s_k \le s_{2^m} \le g(2^m)$.

We often just want to bound the values of s_n, typically by concluding that $s_n = O(h(n))$ for some function h. Replacing "$=$" by "$\le$" in the proof of Theorem 2 shows that if

$$s_{2n} \le 2 \cdot s_n + f(n) \qquad \text{for} \quad n \in \mathbb{P}$$

then

$$s_{2^m} \le 2^m \cdot \left[s_1 + \frac{1}{2} \sum_{i=0}^{m-1} \frac{f(2^i)}{2^i} \right] \qquad \text{for} \quad m \in \mathbb{N}.$$

In particular, if $s_{2n} \le 2 \cdot s_n + A$, then

$$s_{2^m} \le 2^m \cdot s_1 + (2^m - 1) \cdot A - O(2^m).$$

If in addition s is monotone and $2^{m-1} < k \le 2^m$, then $s_k \le s_{2^m} \le C \cdot 2^m$ for some C [for instance, $C = s_1 + A$], and so $s_k \le C \cdot 2 \cdot 2^{m-1} \le C \cdot 2 \cdot k$. That is, $s_k = O(k)$.

The inequality $s_{2n} \le 2 \cdot s_n + B \cdot n$ leads similarly to $s_{2^m} = O(2^m \cdot m)$ and to $s_k = O(k \cdot \log_2 k)$. We were actually a little imprecise when we estimated the time for the merge sort in Example 5(b). If one of the two lists is exhausted early, we can just toss the remainder of the other list in without further action. What we *really* have in that example is the inequality $T(2n) \le 2T(n) + Bn$, which of course still gives $T(n) = O(n \cdot \log_2 n)$.

EXERCISES 4.4

1. Give an explicit formula for s_n where $s_0 = 3$ and $s_n = -2s_{n-1}$ for $n \ge 1$.

2. (a) Give an explicit formula for $s_n = 4s_{n-2}$ where $s_0 = s_1 = 1$.

   (b) Repeat part (a) for $s_0 = 1$ and $s_1 = 2$.

3. Prove that if $s_n = as_{n-1}$ for $n \ge 1$ and $a \ne 0$, then $s_n = a^n \cdot s_0$ for $n \in \mathbb{N}$.

4. Verify that the sequence given by $s_n = 2^{n+1} + (-1)^n$ in Example 2 satisfies the conditions $s_0 = s_1 = 3$ and $s_n = s_{n-1} + 2s_{n-2}$ for $n \ge 2$.

5. Verify that the sequence $s_n = 3^n - 2 \cdot n \cdot 3^n$ in Example 4 satisfies the conditions $s_0 = 1$, $s_1 = -3$ and $s_n = 6s_{n-1} - 9s_{n-2}$ for $n \ge 2$.

6. Use the formula for FIB(n) in Example 3 and a hand calculator to verify that FIB(5) = 8.

7. Give an explicit formula for s_n where $s_0 = 3$, $s_1 = 6$ and $s_n = s_{n-1} + 2s_{n-2}$ for $n \geq 2$. *Hint:* Imitate Example 2, but note that now $s_1 = 6$.

8. Repeat Exercise 7 with $s_0 = 3$ and $s_1 = -3$.

9. Give an explicit formula for the sequence in Example 4 of §4.3: $s_0 = 1$, $s_1 = 2$ and $s_n = s_{n-1} + s_{n-2}$ for $n \geq 2$. *Hint:* Use Example 3.

10. Consider the sequence s_n where $s_0 = 2$, $s_1 = 1$ and $s_n = s_{n-1} + s_{n-2}$ for $n \geq 2$.

    (a) Calculate s_n for $n = 2, 3, 4, 5$ and 6.

    (b) Give an explicit formula for s_n.

11. In each of the following cases give an explicit formula for s_n.

    (a) $s_0 = 2$, $s_1 = -1$ and $s_n = -s_{n-1} + 6s_{n-2}$ for $n \geq 2$.

    (b) $s_0 = 2$ and $s_n = 5 \cdot s_{n-1}$ for $n \geq 1$.

    (c) $s_0 = 1$, $s_1 = 8$ and $s_n = 4s_{n-1} - 4s_{n-2}$ for $n \geq 2$.

    (d) $s_0 = c$, $s_1 = d$ and $s_n = 5s_{n-1} - 6s_{n-2}$ for $n \geq 2$. Here c and d are unspecified constants.

    (e) $s_0 = 1$, $s_1 = 4$ and $s_n = s_{n-2}$ for $n \geq 2$.

    (f) $s_0 = 1$, $s_1 = 2$ and $s_n = 3 \cdot s_{n-2}$ for $n \geq 2$.

    (g) $s_0 = 1$, $s_1 = -3$ and $s_n = -2s_{n-1} + 3s_{n-2}$ for $n \geq 2$.

    (h) $s_0 = 1$, $s_1 = 2$ and $s_n = -2s_{n-1} + 3s_{n-2}$ for $n \geq 2$.

12. Recall that if $s_n = bs_{n-2}$ for $n \geq 2$, then $s_{2n} = b^n s_0$ and $s_{2n+1} = b^n s_1$ for $n \in \mathbb{N}$. Show that Theorem 1 holds for $a = 0$ and $b > 0$, and reconcile this assertion with the preceding sentence. That is, specify r_1, r_2, c_1 and c_2 in terms of b, s_0 and s_1.

13. In each of the following cases give an explicit formula for s_{2m}.

    (a) $s_{2n} = 2s_n + 3$, $s_1 = 1$ (b) $s_{2n} = 2s_n$, $s_1 = 3$

    (c) $s_{2n} = 2s_n + 5n$, $s_1 = 0$ (d) $s_{2n} = 2s_n + 3 + 5n$, $s_1 = 2$

    (e) $s_{2n} = 2s_n - 7$, $s_1 = 1$ (f) $s_{2n} = 2s_n - 7$, $s_1 = 5$

    (g) $s_{2n} = 2s_n - n$, $s_1 = 3$ (h) $s_{2n} = 2s_n + 5 - 7n$, $s_1 = 0$.

14. Suppose that the sequence (s_n) satisfies the given inequality and that $s_1 = 7$. Give your best estimate for how large s_{2m} can be.

    (a) $s_{2n} \leq 2s_n + 1$ (b) $s_{2n} \leq 2(s_n + n)$

15. Suppose that the sequence (s_n) satisfies the recurrence

$$s_{2n} = 2s_n + n^2.$$

Give a formula for s_{2m} and an argument that your formula is correct. *Suggestion:* Apply Theorem 2 or guess a formula and check that it is correct.

16. Verify the formula in Theorem 2 for s_{2m} in the case $f(n) = A + B \cdot n$.

17. Theorem 2 does not specifically apply to the following two recurrences, but the ideas carry over.

   (a) Given that $t_{2n} = b \cdot t_n + f(n)$ for some constant b and function f, find a formula for t_{2m} in terms of b, t_1 and the values of f.

   (b) Given that $t_{3n} = 3t_n + f(n)$, find a formula for t_{3m}.

§ 4.5 More Induction

The principle of mathematical induction studied in §4.2 is often called the First Principle of Mathematical Induction. We restate it here in a slightly different form.

First Principle of Mathematical Induction

> Let m be an integer and let $p(n)$ be a sequence of propositions defined on the set $\{n \in \mathbb{Z} : n \geq m\}$. If
>
>   (B) $p(m)$ is true and
>   (I)   for $k > m$, $p(k)$ is true whenever $p(k - 1)$ is true,
>
> then $p(n)$ is true for every $n \geq m$.

In the inductive step (I), each proposition is true provided that the proposition immediately preceding it is true. To use this principle as a framework for constructing a proof, we need to check that $p(m)$ is true and that each proposition is true *assuming that the proposition just before it is true*. It is this right to assume the immediately previous case that makes the method of proof by induction so powerful. It turns out that in fact we are permitted to assume *all* previous cases. This apparently stronger assertion is a consequence of the following principle, whose proof we discuss at the end of this section.

Second Principle of Mathematical Induction

> Let m be an integer and let $p(n)$ be a sequence of propositions defined on the set $\{n \in \mathbb{Z} : n \geq m\}$. If
>
>   (B) $p(m)$ is true and
>   (I)   for $k > m$, $p(k)$ is true whenever $p(m), \ldots, p(k - 1)$ are all true,
>
> then $p(n)$ is true for every $n \geq m$.

To verify (I) for $k = m + 1$, one shows that $p(m)$ implies $p(m + 1)$. To verify (I) for $k = m + 2$, one shows that $p(m)$ and $p(m + 1)$ together imply $p(m + 2)$. And so on. To verify (I) in general one considers a $k > m$, assumes that the propositions $p(n)$ are true for $m \leq n < k$, and shows that $p(k)$ is true. The Second Principle of Mathematical Induction is the appropriate version to use when the truths of the propositions follow from predecessors other than the immediate predecessors.

EXAMPLE 1   We show that every integer $n \geq 2$ can be written as a product of primes. Note that if n is prime, the "product of primes" is simply the number n by itself.

For $n \geq 2$ let $p(n)$ be the proposition "n can be written as a product of primes." Observe that the First Principle of Mathematical Induction is really unsuitable here. The lone fact that the integer 1,311,819, say, happens to be a product of primes is of no help in showing that 1,311,820 is also a product of primes. We apply the Second Principle. Clearly, $p(2)$ is true, since 2 is a prime.

Consider $k > 2$ and assume that $p(n)$ is true for all n satisfying $2 \leq n < k$. We need to show that then $p(k)$ is true. If k is prime, then $p(k)$ is clearly true. Otherwise, k can be written as a product $i \cdot j$ where i and j are integers greater than 1. Thus $2 \leq i < k$ and $2 \leq j < k$. Since both $p(i)$ and $p(j)$ are assumed to be true, we can write i and j as products of primes. Then $k = i \cdot j$ is also a product of primes. We have checked the basis and induction step for the Second Principle of Mathematical Induction, so we infer that all the propositions $p(n)$ are true. ■

Often the general proof of the inductive step (I) does not work for the first few values of k. In such a case, these first few values of k need to be checked separately, so they may serve as part of the basis. We restate the Second Principle of Mathematical Induction in a more general version that applies in such situations.

Second Principle of Mathematical Induction

Let m be an integer, let $p(n)$ be a sequence of propositions defined on the set $\{n \in \mathbb{Z} : n \geq m\}$, and let l be a nonnegative integer. If

(B) $p(m), \ldots, p(m + l)$ are all true and
(I)   for $k > m + l$, $p(k)$ is true whenever $p(m), \ldots, p(k - 1)$ are true,

then $p(n)$ is true for all $n \geq m$.

If $l = 0$, this is our original version of the Second Principle.

In §4.3 we saw that many sequences are defined recursively using earlier terms other than the immediate predecessors. The Second Principle is the natural form of induction for proving results about such sequences.

EXAMPLE 2   (a) In Exercise 14 of §4.3 we recursively defined $b_0 = b_1 = 1$ and $b_n = 2b_{n-1} + b_{n-2}$ for $n \geq 2$. In part (b), we asked for an explanation of why all b_n's are odd integers. A proof at that point would require use of the Well-Ordering Principle. It is more straightforward to apply the Second Principle of Mathematical Induction, as follows.

The nth proposition $p(n)$ is "b_n is odd." In the inductive step we will use the relation $b_k = 2b_{k-1} + b_{k-2}$, so we'll need $k \geq 2$. Hence we'll check the cases $n = 0$ and 1 separately. Thus we will use the Second Principle with $m = 0$ and $l = 1$.

(B)   The propositions $p(0)$ and $p(1)$ are obviously true, since $b_0 = b_1 = 1$.

(I)   Consider $k \geq 2$ and assume that b_n is odd for all n satisfying $0 \leq n < k$. In particular, b_{k-2} is odd. Clearly, $2b_{k-1}$ is even, and so $b_k = 2b_{k-1} + b_{k-2}$ is the sum of an even and an odd integer. Thus b_k is odd. It follows from the Second Principle of Mathematical Induction that all b_n's are odd.

Note that in this proof the oddness of b_k followed from the oddness of b_{k-2}.

(b) For the sequence above, we prove that $b_n < 6b_{n-2}$ for $n \geq 4$. Direct computation shows that $b_2 = 3$, $b_3 = 7$, $b_4 = 17$ and $b_5 = 41$. For $n = 4$ the inequality says that $b_4 < 6b_2$ or $17 < 6 \cdot 3$, and for $n = 5$ it says that $b_5 < 6b_3$ or $41 < 6 \cdot 7$; these are true. We now consider $k \geq 6$ and assume that

$$b_n < 6b_{n-2} \qquad \text{for} \quad 4 \leq n < k.$$

Since both $k - 1$ and $k - 2$ are at least 4, we have $b_{k-1} < 6b_{k-3}$ and $b_{k-2} < 6b_{k-4}$ by assumption. Hence

$$
\begin{aligned}
b_k &= 2b_{k-1} + b_{k-2} && [\text{definition of } b_k] \\
&< 2(6b_{k-3}) + 6b_{k-4} && [\text{inductive assumption}] \\
&= 6[2b_{k-3} + b_{k-4}] && [\text{algebra}] \\
&= 6b_{k-2} && [\text{definition of } b_{k-2}].
\end{aligned}
$$

Hence by the Second Principle of Mathematical Induction the inequality holds for all $n \geq 4$.

Note that we checked the assertion for $n = 4$ and $n = 5$ before going on to the inductive step. Thus we applied the Second Principle with $m = 4$ and $l = 1$. Why did we check the inequality for $n = 5$, as well as for $n = 4$, before proceeding with the inductive step? Before we wrote up the proof, we had observed that in the inductive step we were going to need to use $b_n < 6b_{n-2}$ for $n = k - 2$, so we would need $k - 2 \geq 4$ or $k \geq 6$. In other words, the inductive step wouldn't work for $n = 5$: $b_5 = 2b_4 + b_3$, but b_3 isn't less than $6b_1$. ■

EXAMPLE 3   We recursively define $a_0 = a_1 = a_2 = 1$ and $a_n = a_{n-2} + a_{n-3}$ for $n \geq 3$. The first few terms of the sequence are 1, 1, 1, 2, 2, 3, 4, 5, 7, 9, 12, 16, 21, 28, 37, 49. We prove that $a_n \leq (\frac{4}{3})^n$ for all $n \in \mathbb{N}$. This inequality is clear for $n = 0$, 1 and 2. So we consider $k \geq 3$ and assume that $a_n < (\frac{4}{3})^n$ for $0 \leq n < k$. In particular, $a_{k-2} \leq (\frac{4}{3})^{k-2}$ and $a_{k-3} \leq (\frac{4}{3})^{k-3}$. Thus we have

$$a_k = a_{k-2} + a_{k-3} \leq \left(\frac{4}{3}\right)^{k-2} + \left(\frac{4}{3}\right)^{k-3} = \left(\frac{4}{3}\right)^k \cdot \left[\left(\frac{3}{4}\right)^2 + \left(\frac{3}{4}\right)^3\right].$$

Since $(\frac{3}{4})^2 + (\frac{3}{4})^3 = \frac{63}{64} < 1$, we conclude that $a_k \leq (\frac{4}{3})^k$. This establishes the inductive step. Hence we infer from the Second Principle of Mathematical Induction [with $m = 0$ and $l = 2$] that $a_n \leq (\frac{4}{3})^n$ for all $n \in \mathbb{N}$.

In this proof we were lucky that $(\frac{3}{4})^2 + (\frac{3}{4})^3 < 1$. If this inequality hadn't held, we would have had to find another proof, prove something else, or abandon the problem. Induction gives us a framework for proofs, but it doesn't provide the details, which are determined by the particular problem at hand. ■

We have already stated the First Principle of Mathematical Induction for finite sequences in § 4.2. Both versions of the Second Principle can also be stated for finite sequences. The changes are simple. Suppose that the propositions $p(n)$ are defined for $m \leq n \leq m^*$. Then the first version of the Second Principle can be stated as follows. If

(B) $p(m)$ is true and
(I)   for $m < k \leq m^*$, $p(k)$ is true whenever $p(m), \ldots, p(k-1)$ are all true,

then $p(n)$ is true for all n satisfying $m \leq n \leq m^*$.

We return to the infinite principles of induction and end this section by discussing the logical relationship between the two principles and explaining why we regard both as valid for constructing proofs.

It turns out that each of the two principles implies the other, in the sense that if we accept either as valid, then the other is also valid. It is clear that the Second Principle implies the First Principle since, if we are allowed to assume all previous cases, then we are surely allowed to assume the immediately preceding case. A rigorous proof can be given by showing that (B) and (I) of the Second Principle are consequences of (B) and (I) of the First Principle.

It is perhaps more surprising that the First Principle implies the Second. A proof can be given using the propositions

$$q(n) = p(m) \wedge \cdots \wedge p(n) \qquad \text{for} \quad n \geq m$$

and showing that if the sequence $p(n)$ satisfies (B) and (I) of the Second Principle, then the sequence $q(n)$ satisfies (B) and (I) of the First Principle. Therefore, every $q(n)$ will be true by the First Principle, so that every $p(n)$ will also be true.

The equivalence of the two principles is of less concern to us than an assurance that they are valid rules. For this we rely on the Well-Ordering Principle stated in §4.1

Proof of the Second Principle.   Assume

(B) $p(m), \ldots, p(m + l)$ are all true and

(I)   for $k > m + l$, $p(k)$ is true whenever $p(m), \ldots, p(k - 1)$ are true, but that $p(n)$ is false for some $n \geq m$. Then the set

$$S = \{n \in \mathbb{Z} : n \geq m \text{ and } p(n) \text{ is false}\}$$

is nonempty. By the Well-Ordering Principle, S has a smallest element n_0. In view of (B) we must have $n_0 > m + l$. Since $p(n)$ is true for $m \leq n < n_0$, $p(n_0)$ is also true by (I). This contradicts the fact that n_0 belongs to S. It follows that if (B) and (I) hold, then every $p(n)$ is true.   ∎

A similar proof can be given for the First Principle but, since the principles are equivalent, it is not needed.

EXERCISES 4.5

Some of the exercises for this section require only the First Principle of Mathematical Induction and are included to provide extra practice.

1. Prove $3 + 11 + \cdots + (8n - 5) = 4n^2 - n$ for $n \in \mathbb{P}$.

2. For $n \in \mathbb{P}$, prove

   (a) $1 \cdot 2 + 2 \cdot 3 + \cdots + n(n + 1) = \frac{1}{3}n(n + 1)(n + 2)$

   (b) $\dfrac{1}{1 \cdot 2} + \dfrac{1}{2 \cdot 3} + \cdots + \dfrac{1}{n(n + 1)} = \dfrac{n}{n + 1}$

3. Prove that $n^5 - n$ is divisible by 10 for all $n \in \mathbb{P}$.

4. (a) Calculate b_6 for the sequence (b_n) in Example 2.

   (b) Use the recursive definition of (a_n) in Example 3 to calculate a_9.

5. Is the First Principle of Mathematical Induction adequate to prove the fact in Exercise 11(b) of §4.3? Explain.

6. Recursively define $a_0 = 1$, $a_1 = 2$ and $a_n = \dfrac{a_{n-1}^2}{a_{n-2}}$ for $n \geq 2$.

   (a) Calculate the first few terms of the sequence.

   (b) Using part (a), guess the general formula for a_n.

   (c) Prove the guess in part (b).

7. Recursively define $a_0 = a_1 = 1$ and $a_n = \dfrac{a_{n-1}^2 + a_{n-2}}{a_{n-1} + a_{n-2}}$ for $n \geq 2$.

   Repeat Exercise 6 for this sequence.

8. Recursively define $a_0 = 1$, $a_1 = 2$ and $a_n = \dfrac{a_{n-1}^2 - 1}{a_{n-2}}$ for $n \geq 2$.

   Repeat Exercise 6 for this sequence.

9. Recursively define $a_0 = 0$, $a_1 = 1$ and $a_n = \frac{1}{4}(a_{n-1} - a_{n-2} + 3)^2$ for $n \geq 2$.
   Repeat Exercise 6 for this sequence.

10. Recursively define $a_0 = 1$, $a_1 = 2$, $a_2 = 3$ and $a_n = a_{n-2} + 2a_{n-3}$ for $n \geq 3$.

    (a) Calculate a_n for $n = 3, 4, 5, 6, 7$.

    (b) Prove that $a_n > (\frac{3}{2})^n$ for all $n \geq 1$.

11. Recursively define $a_0 = a_1 = a_2 = 1$ and $a_n = a_{n-1} + a_{n-2} + a_{n-3}$ for $n \geq 3$.

    (a) Calculate the first few terms of the sequence.

    (b) Prove that all the a_n's are odd.

    (c) Prove that $a_n \leq 2^{n-1}$ for all $n \geq 1$.

12. Recursively define $a_0 = 1$, $a_1 = 3$, $a_2 = 5$ and $a_n = 3a_{n-2} + 2a_{n-3}$ for $n \geq 3$.

    (a) Calculate a_n for $n = 3, 4, 5, 6, 7$.

    (b) Prove that $a_n > 2^n$ for $n \geq 1$.

    (c) Prove that $a_n < 2^{n+1}$ for $n \geq 1$.

    (d) Prove that $a_n = 2a_{n-1} + (-1)^{n-1}$ for $n \geq 1$.

13. Recursively define $b_0 = b_1 = b_2 = 1$ and $b_n = b_{n-1} + b_{n-3}$ for $n \geq 3$.

    (a) Calculate b_n for $n = 3, 4, 5, 6$.

    (b) Show that $b_n \geq 2b_{n-2}$ for $n \geq 3$.

    (c) Prove the inequality $b_n \geq (\sqrt{2})^{n-2}$ for $n \geq 2$.

14. For the sequence in Exercise 13, show that $b_n \leq (\frac{3}{2})^{n-1}$ for $n \geq 1$.

15. Recursively define $\text{SEQ}(0) = 0$, $\text{SEQ}(1) = 1$ and

$$\text{SEQ}(n) = \frac{1}{n} * \text{SEQ}(n-1) + \frac{n-1}{n} * \text{SEQ}(n-2)$$

   for $n \geq 2$. Prove that $0 \leq \text{SEQ}(n) \leq 1$ for all $n \in \mathbb{N}$.

16. As in Exercise 15 of §4.3, let $\text{SEQ}(0) = 1$ and $\text{SEQ}(n) = \sum_{i=0}^{n-1} \text{SEQ}(i)$ for $n \geq 1$.
    Prove that $\text{SEQ}(n) = 2^{n-1}$ for $n \geq 1$.

17. Recall the Fibonacci sequence in Example 3 of § 4.3:

   (B)   FIB(0) = FIB(1) = 1,
   (R)   FIB(n) = FIB(n − 1) + FIB(n − 2)   for n ≥ 2.

   Prove that

   $$\text{FIB}(n) = 1 + \sum_{k=0}^{n-2} \text{FIB}(k) \qquad \text{for } n \geq 2.$$

18. The **Lucas sequence** is defined as follows:

   (B)   LUC(1) = 1 and LUC(2) = 3,
   (R)   LUC(n) = LUC(n − 1) + LUC(n − 2)   for n ≥ 3.

   (a) List the first eight terms of the Lucas sequence.

   (b) Prove that LUC(n) = FIB(n) + FIB(n − 2) for n ≥ 2, where FIB is the Fibonacci sequence defined in Exercise 17.

19. Let the sequence T be defined as in Example 2(a) of § 4.3 by

   (B)   T(1) = 1,
   (R)   T(n) = 2 · T(⌊n/2⌋)   for n ≥ 2.

   Show that T(n) is the largest integer of form 2^k with $2^k \leq n$. [That is, $T(n) = 2^{\lfloor \log n \rfloor}$ where the logarithm is to the base 2.]

20. (a) Show that if T is defined as in Exercise 19, then T(n) is O(n).

   (b) Show that if the sequence Q is defined as in Example 2(b) of § 4.3 by

   (B)   Q(1) = 1,
   (R)   Q(n) = 2 · Q(⌊n/2⌋) + n   for n ≥ 2,

   then Q(n) is $O(n^2)$.

   (c) Show that, in fact, Q(n) is $O(n \log_2 n)$ for Q as in part (b).

21. Show that if S is defined as in Example 6 of § 4.3 by

   (B)   S(0) = 0, S(1) = 1,
   (R)   S(n) = S(⌊n/2⌋) + S(⌊n/5⌋)   for n ≥ 2,

   then S(n) is O(n).

§ 4.6   The Euclidean Algorithm

In this section we develop an algorithm that computes greatest common divisors, and we show how to modify the algorithm to solve $x \cdot m \equiv a \pmod{n}$ for the unknown x.

Recall that an integer d is a **divisor** of the integer m in case m is a multiple of d, i.e., in case $m = d \cdot k$ for some $k \in \mathbb{Z}$. In this case we also say that d **divides** m. Because $0 = d \cdot 0$, every number is a divisor of 0, but 0 is only a divisor of itself. The divisors of $-m$ and m are always the same [for example, the divisors of -6 and 6 are $-6, -3, -2, -1, 1, 2, 3$ and 6]. Moreover, d is a divisor of m if and only if $-d$ is. So we will normally restrict our attention to nonnegative divisors d of nonnegative integers m. If $m > 0$ and $m = d \cdot k$ with $d, k \in \mathbb{Z}$, then $d = m/k$, so $|d| = m/|k| \le m$. Thus all the divisors of m are between $-m$ and m.

A **common divisor** of m and n is an integer that is a divisor of both m and n. Observe that 1 and   1 are always common divisors of m and n. If m and n are not both 0, they can only have a finite number of common divisors. In this case, we define the largest of them to be the **greatest common divisor** or **gcd** of m and n. We denote it by $\mathbf{gcd(m, n)}$.

EXAMPLE 1   (a) The divisors of 12 are $\pm 1, \pm 2, \pm 3, \pm 4, \pm 6$ and ± 12, which we find by factoring 12 into its prime factors $2 \cdot 2 \cdot 3$ and then packaging the factors in all possible combinations. The divisors of $45 = 3 \cdot 3 \cdot 5$ are $\pm 1, \pm 3, \pm 5, \pm 9, \pm 15$ and ± 45. The common divisors of 12 and 45 are thus $-3, -1, 1$ and 3, of which 3 is the largest. Thus $\gcd(12, 45) = 3$.

(b) The common divisors of $12 = 2 \cdot 2 \cdot 3$ and $30 = 2 \cdot 3 \cdot 5$ are $-6, -3, -2, -1, 1, 2, 3$ and 6, so $\gcd(12, 30) = 6$.

(c) The common divisors of $20 = 2 \cdot 2 \cdot 5$ and $150 = 2 \cdot 3 \cdot 5 \cdot 5$ are $-10, -5, -2, -1, 1, 2, 5$ and 10. Hence $\gcd(20, 150) = 10$.

(d) Since the common divisors of $20 = 2 \cdot 2 \cdot 5$ and $63 = 3 \cdot 3 \cdot 7$ are -1 and 1, we have $\gcd(20, 63) = 1$.

(e) Since the common divisors of a and 0 are simply the divisors of a, if $a > 0$ then $\gcd(a, 0) = a = \gcd(-a, 0)$. ∎

These examples illustrate a general phenomenon: the common divisors of m and n all divide their greatest common divisor. We could give a somewhat complicated proof of this fact now using the Division Algorithm of §3.6; instead we will get it as a corollary to Theorem 3.

Example 1 suggests one way to find the gcd of m and n in general: factor each integer into prime factors, find the largest power of each prime that divides both m and n, and multiply all these prime powers together to get $\gcd(m, n)$. Although this approach is sound in theory, it has two weaknesses in practice. First, if m and n are large integers—for practical applications today, m and n may have more than 100 decimal digits—then factoring them is enormously time consuming, requiring unrealistic computing facilities. Second, most of the work is wasted finding factors that don't get used. For instance, we can see with no calculation at all that m and $m + 1$ have no common divisors other than 1 and -1 [any common divisor must divide their difference, 1] so $\gcd(m, m + 1) = 1$, and the factors of m and $m + 1$ are irrelevant. The case $\gcd(m, n) = 1$ is an important case for the applications, and it is fairly likely to occur if m and n are chosen more or less at random.

Fortunately, there is a way to compute gcd's reasonably rapidly without factoring, using just addition, subtraction, and the functions DIV and MOD from §3.6. Our next goal is to develop an algorithm that does so. The key to the algorithm is the following fact.

Proposition

If m and n are integers with $n \neq 0$, then the common divisors of m and n are the same as the common divisors of n and m MOD n. Hence

$$\gcd(m, n) = \gcd(n, m \text{ MOD } n).$$

Proof. We have $m = n \cdot (m \text{ DIV } n) + m \text{ MOD } n$. This equation holds even if $n = 1$, since $m \text{ DIV } 1 = m$ and $m \text{ MOD } 1 = 0$. If n and $m \text{ MOD } n$ are multiples of d, then so is m, since both $n \cdot (m \text{ DIV } n)$ and $m \text{ MOD } n$ are multiples of d. In the other direction, if m and n are multiples of e, then since $m \text{ MOD } n = m - n \cdot (m \text{ DIV } n)$, $m \text{ MOD } n$ is also a multiple of e.

Since the pairs (m, n) and $(n, m \text{ MOD } n)$ have the same sets of common divisors, the two gcd's must be the same. ∎

EXAMPLE 2   (a) We have $\gcd(45, 12) = \gcd(12, 45 \text{ MOD } 12) = \gcd(12, 9) = \gcd(9, 12 \text{ MOD } 9) = \gcd(9, 3) = \gcd(3, 9 \text{ MOD } 3) = \gcd(3, 0) = 3$.

(b) Also, $\gcd(20, 63) = \gcd(63, 20 \text{ MOD } 63) = \gcd(63, 20) = \gcd(20, 63 \text{ MOD } 20) = \gcd(20, 3) = \gcd(3, 20 \text{ MOD } 3) = \gcd(3, 2) = \gcd(2, 3 \text{ MOD } 2) = \gcd(2, 1) = \gcd(1, 2 \text{ MOD } 1) = \gcd(1, 0) = 1$.

(c) As a third example, $\gcd(12, 6) = \gcd(6, 12 \text{ MOD } 6) = \gcd(6, 0) = 6$. ∎

These examples suggest a general strategy for finding $\gcd(m, n)$: Replace m and n by n and $m \text{ MOD } n$ and try again. We hope to reduce in this way to a case where we know the answer. Here is the resulting algorithm, which was already known to Euclid over 2000 years ago and may be older than that.

Algorithm GCD

{input: $m, n \in \mathbb{N}$, not both 0}
{output: $d = \gcd(m, n)$}
{auxiliary variables: integers a and b}
$a := m; b := n$
{The pairs (a, b) and (m, n) have the same gcd.}
while $b \neq 0$ do
$\quad (a, b) := (b, a \text{ MOD } b)$
$d := a$ ∎

EXAMPLE 3 Figure 1 lists the successive values of (a, b) when GCD is applied to the numbers in Example 2. The output gcd's are shown in red. [At the end of the middle column, recall that $2 \text{ MOD } 1 = 0$.]

(a, b)	(a, b)	(a, b)
(45, 12)	(20, 63)	(12, 6)
(12, 9)	(63, 20)	(6, 0)
(9, 3)	(20, 3)	
(3, 0)	(3, 2)	
	(2, 1)	
	(1, 0)	
$m = 45, \quad n = 12$	$m = 20, \quad n = 63$	$m = 12, \quad n = 6$

FIGURE 1   ■

Theorem 1 | Algorithm GCD computes the gcd of the input integers m and n.

> **Proof.** We need to check that the algorithm terminates, and that when it does d is the gcd of m and n.
>
> If $n = 0$, the algorithm sets $a := m$ and $b := 0$, does not execute the while loop and terminates with $d = a = m$. Since $\gcd(m, 0) = m$ for $m \in \mathbb{P}$, the algorithm terminates correctly in this case. Thus suppose that $n > 0$.
>
> As long as b is positive, the while loop replaces b by $a \text{ MOD } b$. Since $0 \le a \text{ MOD } b < b$, the new value of b is smaller than the old value. A decreasing sequence of positive integers must terminate, so b is eventually 0 and the algorithm terminates.
>
> The Proposition shows that the statement "$\gcd(a, b) = \gcd(m, n)$" is an invariant of the while loop. When the loop terminates, $a > b = 0$, so $(a, b) = (d, 0)$. Since $d = \gcd(d, 0)$, we have $d = \gcd(m, n)$, as required.   ■

The argument that the algorithm terminates shows that it makes at most n passes through the while loop. In fact, the algorithm is much faster than that.

Theorem 2 | For input integers $m > n \ge 0$ algorithm GCD makes at most $2 \log_2 (m + n)$ passes through the loop.

> **Proof.** We will show in general that if $a \ge b$, then

$$(*) \qquad\qquad b + a \text{ MOD } b < \frac{2}{3} \cdot (a + b).$$

This inequality means that the value of $a + b$ decreases by a factor of at least 2/3 with each pass through the loop

> while $b \neq 0$ do
> $\quad (a, b) := (b, a \bmod b).$

At first, $a + b = m + n$, and after k passes through the loop we have $a + b \leq (\frac{2}{3})^k \cdot (m + n)$. Now $a + b \geq 1 + 0$ always, so $1 \leq (\frac{2}{3})^l \cdot (m + n)$ after the last, lth, pass through the loop. Hence $m + n \geq (\frac{3}{2})^l$, so $\log_2 (m + n) \geq l \cdot \log_2 (\frac{3}{2}) > \frac{1}{2} l$ and thus $l < 2 \cdot \log_2 (m + n)$, as desired.

It only remains to prove $(*)$, which we rewrite as

$$3b + 3 \cdot (a \bmod b) < 2a + 2b$$

or $b + 3 \cdot (a \bmod b) < 2a$. Since $a = b \cdot (a \operatorname{DIV} b) + a \bmod b$, this is equivalent to

$$b + 3 \cdot (a \bmod b) < 2b \cdot (a \operatorname{DIV} b) + 2 \cdot (a \bmod b)$$

or

$(**)$ $\qquad\qquad\qquad b + a \bmod b < 2b \cdot (a \operatorname{DIV} b).$

Since $a \geq b$, we have $a \operatorname{DIV} b \geq 1$. Moreover, $a \bmod b < b$, and thus

$$b + a \bmod b < 2b \leq 2b \cdot (a \operatorname{DIV} b),$$

proving $(**)$. ■

In fact, GCD typically terminates after far fewer than $2 \log_2 (m + n)$ passes through the loop, though there are inputs m and n that do require at least $\log_2 (m + n)$ passes [see Exercise 17]. Moreover, the numbers a and b get smaller with each pass, so that even starting with very large inputs one quickly gets manageable numbers. Note also that the hypothesis that $m > n$ is not a serious constraint; if $n > m$, then GCD replaces (m, n) by $(n, m \bmod n) = (n, m)$ on the first pass.

Why would anyone be interested in finding gcd's of large numbers anyway? Aside from purely mathematical problems, there are everyday applications to fast implementation of "infinite-precision" computer arithmetic and to public-key cryptosystems for secure transmission of data. In § 12.8 we will discuss the use of gcd's in solving Chinese remainder problems, which play a role in the design of fast arithmetic algorithms.

The solution of these problems depends on solving congruences of the form

$$m \cdot x \equiv a \pmod{n}$$

for unknown x, given a, m and n. If we could simply divide a by m, then $x = a/m$ would solve the congruence. Unfortunately, with typical integers m and a it's not likely that a will be a multiple of m, so this method usually won't work. Fortunately, we don't need to have $m \cdot x = a$ anyway, but just want $m \cdot x \equiv a \pmod{n}$, and there is another approach that does work. It is based on the properties of arithmetic mod n that we developed in Theorems 2 and 3 of § 3.6.

Dividing by m is like multiplying by $1/m$. If we can somehow find an integer s with $m \cdot s \equiv 1 \pmod{n}$, then s will act like $1/m \pmod{n}$, and we will have $m \cdot s \cdot a \equiv 1 \cdot a \equiv a \pmod{n}$, so that $x = s \cdot a$ [which looks like $\frac{1}{m} \cdot a \pmod{n}$] will solve the congruence, no matter what a is. Finding s is the key, and it turns out that a modification of GCD will give us s if it exists.

It can happen that there is no solution to $m \cdot x \equiv a \pmod{n}$. Even if a solution exists for a particular lucky value of a, there may be no s with $m \cdot s \equiv 1 \pmod{n}$. For example, $4 \cdot x \equiv 2 \pmod{6}$ has the solution $x = 2$, but $4 \cdot x \equiv 1 \pmod{6}$ has no solution at all, because $4 \cdot x \equiv 1 \pmod{6}$ is another way of saying $4x = 1 + 6k$ for some integer k, but $4x - 6k$ is even, so it can't equal 1.

This last argument shows more generally that if $\gcd(m, n) = d > 1$ then $m \cdot x = a \pmod{n}$ has no solution unless a is a multiple of d. [Exercise 15(c) shows that this necessary condition is also sufficient.] If we expect to be able to solve $m \cdot x \equiv a \pmod{n}$ for arbitrary values of a, then we must have $\gcd(m, n) = 1$. Integers m and n with $\gcd(m, n) = 1$ are said to be **relatively prime**; they have no common prime factors.

We will modify the algorithm GCD so that it produces not only the integer $d = \gcd(m, n)$ but also integers s and t with $d = s \cdot m + t \cdot n$. Then $d \equiv s \cdot m \pmod{n}$. If $d = 1$, then s is what we want, and if $d \neq 1$, then we will have learned that m and n are not relatively prime and that we cannot expect to solve $m \cdot x \equiv a \pmod{n}$ in general. To make the new algorithm clearer, we first slightly rewrite GCD, using DIV instead of MOD. Here is the result.

Algorithm **GCD$^+$**

{input: $m, n \in \mathbb{N}$, not both 0}
{output: $d = \gcd(m, n)$}
{auxiliary variables: integers a, b and q}
$a := m;\ b := n$
while $b \neq 0$ do
  $q := a$ DIV b
  $(a, b) := (b, a - q \cdot b)$
$d := a$ ■

EXAMPLE 4   Figure 2 gives a table of successive values of a, b, q and $-q \cdot b$ for $m = 135$ and $n = 40$. The output d is in red. We have labeled b as a_{next} to remind ourselves that each b becomes a in the next pass. In fact, since the b's just become a's, we could delete the b column from the table without losing information.

a	$b = a_{next}$	q	$-q \cdot b$
135	40	3	-120
40	15	2	-30
15	10	1	-10
10	5	2	-10
5	0		

FIGURE 2   ■

In the general case, just as in Example 4, if we set $a_0 = m$ and $b_0 = a_1 = n$, then GCD$^+$ builds sequences $a_0, a_1, \ldots, a_l, a_{l+1} = 0, b_0, b_1, \ldots, b_l$ and $q_1, q_2, \ldots, q_l$. If a_{i-1} and b_{i-1} are the values of a and b at the start of the ith pass through the loop and a_i and b_i the values at the end, then $a_i = b_{i-1}, q_i = a_{i-1}$ DIV a_i, and $a_{i+1} = b_i = a_{i-1} - q_i \cdot b_{i-1} = a_{i-1} - q_i \cdot a_i$. That is, the finite sequences (a_i) and (q_i) satisfy

$$q_i = a_{i-1} \text{ DIV } a_i \quad \text{and} \quad a_{i+1} = a_{i-1} - q_i \cdot a_i$$

for $i = 1, \ldots, l$. Moreover, $a_l = \gcd(m, n)$.

Now we are ready to compute s and t with $\gcd(m, n) = s \cdot m + t \cdot n$. We will construct sequences $s_0, s_1, \ldots, s_l$ and $t_0, t_1, \ldots, t_l$ such that

$$a_i = s_i \cdot m + t_i \cdot n \quad \text{for} \quad i = 0, 1, \ldots, l.$$

Taking $i = l$ will give $\gcd(m, n) = a_l = s_l \cdot m + t_l \cdot n$, which is what we want.

To start, we want $m = a_0 = s_0 \cdot m + t_0 \cdot n$; set $s_0 = 1$ and $t_0 = 0$. Next we want $n = a_1 = s_1 \cdot m + t_1 \cdot n$; set $s_1 = 0$ and $t_1 = 1$. From now on, we exploit the recurrence $a_{i+1} = a_{i-1} - q_i \cdot a_i$. If we already have

$$a_{i-1} = s_{i-1} \cdot m + t_{i-1} \cdot n \quad \text{and} \quad a_i = s_i \cdot m + t_i \cdot n,$$

then

$$a_{i+1} = a_{i-1} - q_i \cdot a_i = [s_{i-1} - q_i \cdot s_i] \cdot m + [t_{i-1} - q_i \cdot t_i] \cdot n.$$

Thus if we set

$$s_{i+1} = s_{i-1} - q_i \cdot s_i \quad \text{and} \quad t_{i+1} = t_{i-1} - q_i \cdot t_i$$

we get $a_{i+1} = s_{i+1} \cdot m + t_{i+1} \cdot n$, as desired.

EXAMPLE 5   Figure 3 shows the successive values of a_i, q_i, s_i and t_i for the case $m = 135$, $n = 40$ of Example 4. Sure enough, $5 = 3 \cdot 135 + (-10) \cdot 40$, as claimed.

i	a_i	q_i	s_i	t_i
0	135		1	0
1	40	3	0	1
2	15	2	1	−3
3	10	1	−2	7
4	5	2	3	−10
5	0			

FIGURE 3   ■

To recast the recursive definition of the sequences (s_i) and (t_i) in forms suitable for a while loop, we introduce new variables s' and t', corresponding to s_{i+1} and t_{i+1}. While we are at it, we let $a' = b$ as well. Here is the algorithm that results.

The Euclidean Algorithm

{input: $m, n \in \mathbb{N}$, not both 0}
{output: $d = \gcd(m, n)$, integers s and t with $d = s \cdot m + t \cdot n$}
{auxiliary variables: integers q, a, a', s, s', t, t'}
$a := m;\ a' := n;\ s := 1;\ s' := 0;\ t := 0;\ t' := 1$
{$a = s \cdot m + t \cdot n$ and $a' = s' \cdot m + t' \cdot n$}
while $a' \neq 0$ do
   $q := a\, \text{DIV}\, a'$
   $(a, a') := (a', a - q \cdot a')$
   $(s, s') := (s', s - q \cdot s')$
   $(t, t') := (t', t - q \cdot t')$
$d := a$ ■

We have simply added two more lines to the loop, without affecting the value of a, so this algorithm makes exactly as many passes through the loop as GCD did. The argument above shows that the equations $a = s \cdot m + t \cdot n$ and $a' = s' \cdot m + t' \cdot n$ are loop invariants, so $d = a = s \cdot m + t \cdot n$ when the algorithm terminates, which it does after at most $2 \log_2 (m + n)$ iterations. We have shown the following.

Theorem 3

> For input integers $m > n \geq 0$, the Euclidean algorithm produces $d = \gcd(m, n)$ and integers s and t with $d = s \cdot m + t \cdot n$, using $O(\log_2 m)$ arithmetic operations of the form $-, \cdot$ and DIV

Corollary

> The greatest common divisor $\gcd(m, n)$ of m and n is a multiple of all the common divisors of m and n.

Proof. If c is a common divisor of m and n, then $m = k \cdot c$ and $n = l \cdot c$ for integers k and l. Thus $\gcd(m, n) = s \cdot m + t \cdot n = [s \cdot k + t \cdot l] \cdot c$, so $\gcd(m, n)$ is a multiple of c. ∎

The actual running time for the Euclidean algorithm will depend on how fast the arithmetic operations are. For large m and n the operation DIV is the bottleneck. In our application to solving the congruence $m \cdot s \equiv 1 \pmod{n}$ we only need s and not t, so we can omit the line of the algorithm that calculates t, with corresponding savings in time. Moreover, since we only need s MOD n in that case, we can replace $s' - q \cdot s$ by $(s' - q \cdot s)$ MOD n in the computation and keep all the numbers s and s' smaller than n.

We made the comment in §3.6 that, although $+_p$ and $*_p$ behave on $\mathbb{Z}(p) = \{0, 1, \ldots, p - 1\}$ like $+$ and $\cdot$ on $\mathbb{Z}$, one must be careful about cancellation in $\mathbb{Z}(p)$. We have seen in this section that if m and p are relatively prime, then we can find an s in $\mathbb{Z}$ so that $m \cdot s \equiv 1 \pmod{p}$. If $m \in \mathbb{Z}(p)$ then $m *_p (s \text{ MOD } p) = 1$ in $\mathbb{Z}(p)$, so that s MOD p acts like $1/m$ in $\mathbb{Z}(p)$ and $a *_p (s \text{ MOD } p)$ is a/m in $\mathbb{Z}(p)$. Thus division by m is possible in $\mathbb{Z}(p)$ if $\gcd(m, p) = 1$. In case p is prime, $\gcd(m, p) = 1$ for every nonzero m in $\mathbb{Z}(p)$, so division by all nonzero members of $\mathbb{Z}(p)$ is legal.

Cancellation of a factor on both sides of an equation is the same as division of both sides by the canceled factor. Thus if $\gcd(m, p) = 1$ and $a *_p m = b *_p m$ in $\mathbb{Z}(p)$, then $a = b$, because $a = a *_p m *_p (s \text{ MOD } p) = b *_p m *_p (s \text{ MOD } p) = b$, where s MOD p acts like $1/m$ in $\mathbb{Z}(p)$. In terms of congruences, if m and p are relatively prime and if $a \cdot m \equiv b \cdot m \pmod{p}$, then $a \equiv b \pmod{p}$.

EXAMPLE 6   (a) Since 5 is prime, each of the congruences $m \cdot x \equiv 1 \pmod{5}$, with $m = 1, 2, 3, 4$, has a solution. These congruences are equivalent to the equations $m *_5 x = 1$ in $\mathbb{Z}(5)$, which were solved in Exercise 11 of §3.6 using a table. With such small numbers, the Euclidean algorithm was not needed. The solutions are $1 \cdot 1 \equiv 1 \pmod{5}$, $2 \cdot 3 \equiv 1 \pmod{5}$, $3 \cdot 2 \equiv 1 \pmod{5}$ and $4 \cdot 4 \equiv 1 \pmod{5}$.

(b) Let's solve $10x \equiv 1 \pmod{37}$. Since 37 is prime, we know there is a solution, but creating a table would be inconvenient. We apply the Euclidean algorithm with $m = 10$ and $n = 37$ to obtain $s = -11$ and $t = 3$, so that $(-11) \cdot 10 + 3 \cdot 37 = 1$. Then $10 \cdot (-11) \equiv 1 \pmod{37}$, so $x = -11$ is a solution of $10x \equiv 1 \pmod{37}$. Any integer y with $y \equiv -11 \pmod{37}$ is also a solution, so -11 MOD $37 = -11 + 37 = 26$ is a solution in $\mathbb{Z}(37) = \{0, 1, 2, \ldots, 36\}$. Notice that the value of t is completely irrelevant here.

(c) Consider the equation $m \cdot x \equiv 1 \pmod{15}$. We regard m as fixed and seek a solution x in $\mathbb{Z}(15)$. If m and 15 are not relatively prime, there is no solution. In particular, there is no solution if m is a multiple of 3 or 5.

Let's solve $13x \equiv 1 \pmod{15}$, which has a solution since $\gcd(13, 15) = 1$. We apply the Euclidean algorithm with $m = 13$ and $n = 15$ to obtain $s = 7$ and $t = -6$, so that $7 \cdot 13 + (-6) \cdot 15 = 1$. [Again, we don't care about t.] Thus $13 \cdot 7 \equiv 1 \pmod{15}$, i.e., $x = 7$ is a solution of $13x \equiv 1 \pmod{15}$.

(d) The congruence $8x \equiv 6 \pmod{15}$ is equivalent to the congruence $4x \equiv 3 \pmod{15}$; we can cancel a factor of 2 on both sides because 2 is relatively prime to 15. Doubters can use the Euclidean algorithm [or inspection] to observe that $2 \cdot 8 \equiv 1 \pmod{15}$, so if $8x \equiv 6 \pmod{15}$, then $4x \equiv (8 \cdot 2) \cdot 4x \equiv 8 \cdot 8x \equiv 8 \cdot 6 \equiv (8 \cdot 2) \cdot 3 \equiv 3 \pmod{15}$. Multiplication by 8 is division by 2 (mod 15).

To solve $4x \equiv 3 \pmod{15}$, we first solve $4y \equiv 1 \pmod{15}$ by the Euclidean algorithm or by inspection. The solutions are the integers y with $y \equiv 4 \pmod{15}$, so $x \equiv 3y \equiv 12 \pmod{15}$ gives the solutions of $4x \equiv 3 \pmod{15}$ and hence of $8x \equiv 6 \pmod{15}$. ∎

Now that we know how to solve $m \cdot x \equiv a \pmod{n}$ we can solve the simultaneous congruences

$$\begin{cases} x \equiv a_1 \pmod{n_1} \\ x \equiv a_2 \pmod{n_2}. \end{cases}$$

The general solution to $x \equiv a_1 \pmod{n_1}$ is $x = a_1 + n_1 \cdot y$ for some integer y, so we want to solve

$$a_1 + n_1 \cdot y \equiv a_2 \pmod{n_2}$$

or

$$n_1 \cdot y \equiv a_2 - a_1 \pmod{n_2}.$$

If $\gcd(n_1, n_2) = 1$, we can use the Euclidean algorithm to get s with $n_1 \cdot s \equiv 1 \pmod{n_2}$. Then $y = s \cdot (a_2 - a_1)$ satisfies $n_1 \cdot y \equiv a_2 - a_1 \pmod{n_2}$, so $x = a_1 + n_1 \cdot s \cdot (a_2 - a_1)$ solves both original congruences. This approach is the key to solving the Chinese remainder problems in §12.8.

EXAMPLE 7   We illustrate this method by solving the simultaneous congruences

$$\begin{cases} x \equiv 1 \pmod{13} \\ x \equiv 4 \pmod{15}. \end{cases}$$

For any integer y, $1 + 13y$ is a solution of $x \equiv 1 \pmod{13}$, so we solve $1 + 13y \equiv 4 \pmod{15}$ or $13y \equiv 3 \pmod{15}$. First we need to solve $13s \equiv 1 \pmod{15}$, but we did this in Example 6(c) and found that $13 \cdot 7 \equiv 1 \pmod{15}$. It follows that $13 \cdot 7 \cdot 3 \equiv 3 \pmod{15}$. Thus $y = 21$ is a solution of $13y \equiv 3 \pmod{15}$. Finally, $x = 1 + 13y = 1 + 13 \cdot 21 = 274$ satisfies both $x \equiv 1 \pmod{13}$ and $x \equiv 4 \pmod{15}$.

Any number congruent to 274 modulo $13 \cdot 15 = 195$ will also solve these two congruences, so we can find a solution x satisfying $0 \le x < 194$. It is $274 - 195 = 79$. ∎

EXERCISES 4.6

Notice that many of the answers are easy to check, once found.

1. Use any method to find gcd(m, n) for the following pairs.

   (a) $m = 20, n = 20$ 　　　　　　　　(b) $m = 20, n = 10$

   (c) $m = 20, n = 1$ 　　　　　　　　　(d) $m = 20, n = 0$

   (e) $m = 20, n = 72$ 　　　　　　　　(f) $m = 20, n = -20$

   (g) $m = 120, n = 162$ 　　　　　　　(h) $m = 20, n = 27$

2. Repeat Exercise 1 for the following pairs.

   (a) $m = 17, n = 34$ 　　　　　　　　(b) $m = 17, n = 72$

   (c) $m = 17, n = 850$ 　　　　　　　(d) $m = 170, n = 850$

   (e) $m = 289, n = 850$ 　　　　　　　(f) $m = 2890, n = 850$

3. List the pairs (a, b) that arise when algorithm GCD is applied to the numbers m and n, and find gcd(m, n).

   (a) $m = 20, n = 14$ 　　　　　　　　(b) $m = 20, n = 7$

   (c) $m = 20, n = 30$ 　　　　　　　　(d) $m = 2000, n = 987$

4. Repeat Exercise 3 for the following.

   (a) $m = 30, n = 30$ 　　　　　　　　(b) $m = 30, n = 10$

   (c) $m = 30, n = 60$ 　　　　　　　　(d) $m = 3000, n = 999$

5. Use the Euclidean algorithm to find gcd(m, n) and integers s and t with gcd(m, n) $= s \cdot m + t \cdot n$ for the following. *Suggestion:* Make tables like the one in Figure 3.

   (a) $m = 20, n = 14$ 　　　　　　　　(b) $m = 72, n = 17$

   (c) $m = 20, n = 30$ 　　　　　　　　(d) $m = 320, n = 30$

6. Repeat Exercise 5 for the following.

   (a) $m = 14{,}259, n = 3521$ 　　　　(b) $m = 8359, n = 9373$

7. For each value of m solve $m \cdot x \equiv 1 \pmod{26}$ with $0 \le x < 26$, or explain why no solution exists.

   (a) $m = 5$ 　　　　　　　　　　　　(b) $m = 11$

   (c) $m = 4$ 　　　　　　　　　　　　(d) $m = 9$

   (e) $m = 17$ 　　　　　　　　　　　(f) $m = 13$

8. Repeat Exercise 7 for the congruence $m \cdot x \equiv 1 \pmod{24}$ with $0 \le x < 24$.

9. Solve for x.

   (a) $8x \equiv 1 \pmod{13}$ 　　　　　　(b) $8x \equiv 4 \pmod{13}$

   (c) $99x \equiv 1 \pmod{13}$ 　　　　　(d) $99x \equiv 5 \pmod{13}$

10. Solve for x.

   (a) $2000x \equiv 1 \pmod{643}$ 　　　(b) $643x \equiv 1 \pmod{2000}$

   (c) $1647x \equiv 1 \pmod{788}$ 　　　(d) $788x \equiv 24 \pmod{1647}$

11. Use any method to solve for x with $0 \leq x < 13 \cdot 99$. *Hint:* Cancellation can save some work on (a) and (b).

    (a) $\begin{cases} x \equiv 8 \ (\text{mod } 13) \\ x \equiv 0 \ (\text{mod } 99) \end{cases}$
    (b) $\begin{cases} x \equiv 0 \ (\text{mod } 13) \\ x \equiv 65 \ (\text{mod } 99) \end{cases}$

    (c) $\begin{cases} x \equiv 8 \ (\text{mod } 13) \\ x \equiv 65 \ (\text{mod } 99) \end{cases}$

12. Use any method to solve for x with $0 \leq x < 1300$.

    (a) $\begin{cases} x \equiv 1 \ (\text{mod } 13) \\ x \equiv 1 \ (\text{mod } 99) \end{cases}$
    (b) $\begin{cases} x \equiv -1 \ (\text{mod } 13) \\ x \equiv -1 \ (\text{mod } 99) \end{cases}$

    (c) $\begin{cases} x \equiv -2 \ (\text{mod } 13) \\ x \equiv -2 \ (\text{mod } 99) \end{cases}$
    (d) $\begin{cases} x \equiv 10 \ (\text{mod } 13) \\ x \equiv 96 \ (\text{mod } 99) \end{cases}$

13. Show that the equations $a = s \cdot m + t \cdot n$ and $a' = s' \cdot m + t' \cdot n$ are invariants for the loop in the Euclidean algorithm no matter how q is defined in the loop. [Thus making a mistake in calculating or guessing q does no permanent harm.]

14. Consider integers m and n, not both 0. Show that $\gcd(m, n)$ is the *smallest* positive integer that can be written as $a \cdot m + b \cdot n$ for integers a and b.

15. Suppose that $d = \gcd(m, n) = s \cdot m + t \cdot n$ for some integers s and t.

    (a) Show that m/d and n/d are relatively prime.

    (b) Show that if $d = s' \cdot m + t' \cdot n$ for $s', t' \in \mathbb{Z}$, then $s' = s + k \cdot n/d$ for some $k \in \mathbb{Z}$.

    (c) Show that if a is a multiple of d, then $mx \equiv a \ (\text{mod } n)$ has a solution. *Hint:* $ms \equiv d \ (\text{mod } n)$.

16. Suppose that $d = \gcd(m, n) = s \cdot m + t \cdot n$ for some $s, t \in \mathbb{Z}$ where $n \geq 1$.

    (a) Show that if $s' = s + k \cdot n/d$ and $t' = t - k \cdot m/d$, then $d = s' \cdot m + t' \cdot n$.

    (b) Show that there are $s', t' \in \mathbb{Z}$ such that $d = s' \cdot m + t' \cdot n$ and $0 \leq s' < n/d$.

17. (a) Show that with $m = \text{FIB}(l + 1)$ and $n = \text{FIB}(l)$ as inputs and $l \geq 1$, algorithm GCD makes exactly l passes through the loop. The Fibonacci sequence is defined in §4.3, Example 3. *Hint:* Use induction on l.

    (b) Show that $k \geq \log_2 \text{FIB}(k + 2)$ for $k \geq 3$.

    (c) Show that if $l \geq 3$, $m = \text{FIB}(l + 1)$ and $n = \text{FIB}(l)$, then GCD makes at least $\log_2 (m + n)$ passes through the loop.

CHAPTER HIGHLIGHTS

As usual:

   (a) Satisfy yourself that you can define each concept and can describe each method.

(b) Give at least one reason why the item was included in this chapter.

(c) Think of at least one example of each concept and at least one situation in which each fact or method would be useful.

CONCEPTS AND NOTATION

while loop

   guard, body, pass = iteration, terminate = exit

   invariant

for loop, for ... downto loop

mathematical induction

   basis, inductive step

recursive definition of sequence

   basis, recurrence formula

   iterative, recursive calculation

   Fibonacci sequence

characteristic equation

divide-and-conquer recurrence

divisor, common divisor, gcd

relatively prime integers

FACTS AND PRINCIPLES

Loop Invariant Theorem.

Well-Ordering Principle for $\mathbb{N}$.

First and Second Principles of [Finite] Mathematical Induction.
   [Principles are logically equivalent.]

Theorem 2 of §4.4 on recurrences of form $s_{2n} = 2s_n + f(n)$.

If p is prime, division by nonzero elements is legal in $\mathbb{Z}(p)$.

METHODS

Use of loop invariants to develop algorithms and verify correctness.

Solution of $s_n = as_{n-1} + bs_{n-2}$ with characteristic equation.

Euclidean algorithm to compute $\gcd(m, n)$ and integers s, t with
   $\gcd(m, n) = s \cdot m + t \cdot n$.

Application of the Euclidean algorithm to congruences

$$m \cdot x \equiv a \ (\text{mod } n)$$

   and systems

$$\begin{cases} x \equiv a_1 \ (\text{mod } n_1) \\ x \equiv a_2 \ (\text{mod } n_2) \end{cases}.$$

COUNTING

5

One major goal of this chapter is to develop methods for counting large finite sets without actually listing their elements. The theory that results has applications to computing probabilities in finite sample spaces, which we introduce in § 5.2. The chapter concludes with a discussion of the Pigeon-Hole Principle, a proof technique that applies to finite sets.

§ 5.1   Basic Counting Techniques

We begin with some counting rules with which you are probably more or less familiar. As in § 1.2, we write $|S|$ for the number of elements in the finite set S.

Union Rules

> Let S and T be finite sets.
>
> (a) If S and T are disjoint, i.e., if $S \cap T = \emptyset$, then $|S \cup T| = |S| + |T|$.
>
> (b) In general, $|S \cup T| = |S| + |T| - |S \cap T|$.

The intuitive reason (b) holds is that in counting $S \cup T$ by calculating $|S| + |T|$, elements in $S \cap T$ are counted twice, so $|S \cap T|$ needs to be subtracted from the sum $|S| + |T|$ to obtain $|S \cup T|$. Assertion (b) is a consequence of (a), as follows. Applying (a) twice gives

$$|S \cup T| = |S| + |T \backslash S| \quad \text{and} \quad |T| = |T \backslash S| + |S \cap T|.$$

Thus

$$|S \cup T| + |S \cap T| = |S| + |T \backslash S| + |S \cap T| = |S| + |T|$$

which implies (b).

EXAMPLE 1 How many integers in $S = \{1, 2, 3, \ldots, 1000\}$ are divisible by 3 or 5 or both? We let

$$D_3 = \{n \in S : n \text{ is divisible by } 3\}$$

and

$$D_5 = \{n \in S : n \text{ is divisible by } 5\}.$$

We seek the number of elements in $D_3 \cup D_5$, which is not obvious. But $|D_3|$ is easily seen to be $\lfloor 1000/3 \rfloor = 1000 \text{ DIV } 3 = 333$. Doubters should note that

$$D_3 = \{3m : 1 \le m \le 333\}.$$

Similarly, $|D_5| = 200$. Since $D_3 \cap D_5 = \{n \in S : n \text{ is divisible by } 15\}$ and $1000/15$ is $66\frac{2}{3}$, $|D_3 \cap D_5|$ equals 66. By the Union Rule (b),

$$|D_3 \cup D_5| = |D_3| + |D_5| - |D_3 \cap D_5| = 333 + 200 - 66 = 467. \quad \blacksquare$$

For finite sets S and T we have $|S \times T| = |S| \cdot |T|$, since

$$S \times T = \{(s, t) : s \in S \text{ and } t \in T\}$$

and for each of the $|S|$ choices of s in S there are $|T|$ choices of t in T to make up the ordered pair (s, t). The identity $|S \times T| = |S| \cdot |T|$ is illustrated in Figures 7 and 8 of §1.2. A similar equality holds for the product of more than two sets.

Product Rules

(a) For finite sets $S_1, S_2, \ldots, S_k$ we have

$$|S_1 \times S_2 \times \cdots \times S_k| = \prod_{j=1}^{k} |S_j|.$$

(b) More generally, suppose that a set of ordered k-tuples $(s_1, s_2, \ldots, s_k)$ has the following structure. There are n_1 possible choices of s_1. Given s_1 there are n_2 possible choices of s_2, given s_1 and s_2 there are n_3 possible choices of s_3, and in general, given $s_1, s_2, \ldots, s_{j-1}$ there are n_j choices of s_j. Then the set has $n_1 n_2 \cdots n_k$ elements.

In practice we will often use Product Rule (b), but almost never with the forbidding formalism suggested in its statement.

EXAMPLE 2 (a) We calculate the number of ways of selecting five cards with replacement from a deck of 52 cards. Thus we are counting ordered 5-tuples consisting of cards from the deck. **With replacement** means that each card is returned to the deck before the next card is drawn. The set of ways of selecting five cards with replacement is in one-to-one correspondence with $D \times D \times D \times D \times D = D^5$, where D is the 52-element set of cards. Thus by Product Rule (a), the set has 52^5 elements.

This problem can also be solved by Product Rule (b). There are 52 ways of selecting the first card. After selecting a few cards that have been returned to the deck, there are still 52 ways of selecting the next card. So there are $52 \cdot 52 \cdot 52 \cdot 52 \cdot 52$ ways of selecting five cards with replacement.

(b) Now we calculate the number of ways of selecting five cards without replacement from a deck of 52 cards. **Without replacement** means that once a card is drawn, it is not returned to the deck. This time Product Rule (a) does *not* apply, since not all ordered 5-tuples in D^5 are allowed. Specifically, ordered 5-tuples with cards repeated are forbidden. But Product Rule (b) does apply. The first card can be selected in 52 ways. Once it is selected the second card can be selected in 51 ways. The third card can be selected in 50 ways, the fourth in 49 ways and the fifth in 48 ways. So five cards can be selected without replacement in $52 \cdot 51 \cdot 50 \cdot 49 \cdot 48$ ways.

So far we have only counted ordered 5-tuples of cards, not 5-card subsets. We will return to the subset question in Example 10. ■

EXAMPLE 3   (a) Let $\Sigma = \{a, b, c, d, e, f, g\}$. The number of words in Σ^* having length 5 is $7^5 = 16{,}807$, i.e., $|\Sigma^5| = 16{,}807$. This is seen by applying either of the Product Rules, just as in Example 2(a). The number of words in Σ^5 that have no letters repeated is $7 \cdot 6 \cdot 5 \cdot 4 \cdot 3 = 2520$. This is because the first letter can be selected in 7 ways, then the second letter can be selected in 6 ways, etc.

(b) Let $\Sigma = \{a, b, c, d\}$. The number of words in Σ^2 without repetitions of letters is $4 \cdot 3 = 12$ by Product Rule (b). We can illustrate this by a picture called a tree; see Figure 1. Each path from the start corresponds to a word in Σ^2 without repetitions. For example, the path ending at $*$ corresponds to the word bc. One can imagine a similar but very large tree for the computation in part (a). In fact, one can imagine such a tree for any situation to which Product Rule (b) applies. ■

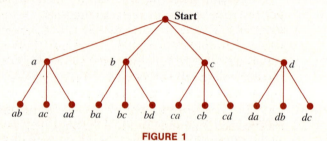

FIGURE 1

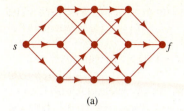
(a)

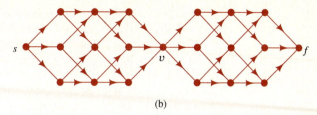
(b)

FIGURE 2

EXAMPLE 4   The number of paths from s to f in Figure 2(a) is $3 \cdot 2 \cdot 2 \cdot 1 = 12$ since there are 3 choices for the first edge and then 2 for the second and third edges. After selecting the first three edges, the fourth edge is forced.

A similar computation can be used to count the number of paths from s to f in Figure 2(b). Alternatively, notice that from above there are 12 paths from s to v and 12 paths from v to f, so there are $12 \cdot 12 = 144$ paths from s to f. ■

EXAMPLE 5   (a) Let S and T be finite sets. We will count the number of functions $f : S \to T$. Here it is convenient to write

$$S = \{s_1, s_2, \ldots, s_m\} \quad \text{and} \quad T = \{t_1, t_2, \ldots, t_n\}$$

so that $|S| = m$ and $|T| = n$. A function $f : S \to T$ can be obtained by specifying $f(s_1)$ to be one of the n elements in T, then specifying $f(s_2)$ to be one of the n elements in T, etc. This process leads to $n^m = n \cdot n \cdots n$ [m times] different results, each of which specifies a different function. We conclude that there are n^m functions mapping S into T.

(b) In part (a) we determined $|\text{FUN}(S, T)|$ for the set $\text{FUN}(S, T)$ of all functions from S into T. Some people write T^S in place of $\text{FUN}(S, T)$. This is strange-looking notation, but it does allow one to write

$$|T^S| = |T|^{|S|},$$

giving a "power rule" analogous to our union rules and product rules. ■

Consider a nonempty finite set S with n elements, and consider a positive integer $r \leq n$. An r-**permutation** of S is a sequence of r distinct elements of S. That is, an r-**permutation** is a one-to-one mapping σ [lowercase Greek sigma] of the set $\{1, 2, \ldots, r\}$ into S. An r-permutation σ is completely described by the ordered r-tuple $(\sigma(1), \sigma(2), \ldots, \sigma(r))$, and we will sometimes use the r-tuple as a notation for the r-permutation itself. An r-permutation can be obtained by assigning 1 to any of the n elements in S. Then 2 can be assigned to any of the $n - 1$ remaining elements, etc. Hence by Product Rule (b), the set S has $n(n - 1)(n - 2) \cdots$ r-permutations, where the product consists of exactly r factors. The last factor turns out to be $n - r + 1$. We sometimes abbreviate the product as $P(n, r)$. Thus S has exactly

$$P(n, r) = n(n - 1)(n - 2) \cdots (n - r + 1) = \prod_{j=0}^{r-1} (n - j)$$

r-permutations. We call the n-permutations simply **permutations**. The set S has exactly $P(n, n) = n!$ permutations. Note that $P(n, r) \cdot (n - r)! = n!$ and so

$$P(n, r) = \frac{n!}{(n - r)!} \qquad \text{for} \quad 1 \le r \le n.$$

It is also convenient to decree that $P(n, 0) = 1$, the unique 0-permutation being the "empty permutation."

EXAMPLE 6   (a) The five cards drawn without replacement in Example 2(b) correspond to 5-permutations of the 52-element set of cards. So the number of such drawings is $P(52, 5) = 52 \cdot 51 \cdot 50 \cdot 49 \cdot 48$.

(b) Let Σ be the seven-letter alphabet in Example 3(a). The words in Σ^5 that have no letters repeated are 5-permutations of Σ. There are $P(7, 5) = 7 \cdot 6 \cdot 5 \cdot 4 \cdot 3$ such 5-permutations. Note that the empty word λ is the empty permutation of Σ.

(c) The two-letter words without repetitions in the tree of Example 3(b) are 2-permutations of the 4-element set Σ. There are $P(4, 2) = 4 \cdot 3 = 12$ of them. ■

EXAMPLE 7   The number of different orderings of a deck of 52 cards is $52! \approx 8.07 \cdot 10^{67}$, which is enormous. Some very interesting theorems about card shuffling were established as recently as the 1980's. They deal with questions like: How many shuffles are needed to get a reasonably well mixed deck of cards? The methods are mostly theoretical: Not even the largest computers can store 52! items and mindlessly verify results case by case. ■

In counting problems where order matters, r-permutations are clearly relevant. Often order is irrelevant, in which case the ability to count sets becomes important. We already know that a set S with n elements has 2^n subsets altogether. For $0 \le r \le n$ let $\binom{n}{r}$ be the number of r-element subsets of S. The number $\binom{n}{r}$, called a **binomial coefficient**, is read "n choose r" and is sometimes called the number of **combinations** of n things taken r at a time. Binomial coefficients get their name from the Binomial Theorem, which will be discussed in § 5.3.

Theorem | For $0 \leq r \leq n$ we have

$$\binom{n}{r} = \frac{n!}{(n-r)! \, r!}.$$

Proof. Let S be a set with n elements. Consider the process of choosing an r-permutation in two steps: first choose an r-element subset of S $\left[\text{in one of the } \binom{n}{r} \text{ ways} \right]$, and then arrange it in order [in one of $r!$ ways]. Altogether, there are $P(n, r)$ possible outcomes, so Product Rule (b) applies to give

$$P(n, r) = \binom{n}{r} \cdot r!.$$

Hence

$$\binom{n}{r} = \frac{P(n, r)}{r!} = \frac{n!}{(n-r)! \cdot r!}. \quad \blacksquare$$

The trick we used here, applying the Product Rule when we know the product and want one of the factors, is worth remembering.

EXAMPLE 8 We count the number of strings of 0's and 1's having length n that contain exactly r 1's. This is equivalent to counting the number of functions from $\{1, 2, \ldots, n\}$ to $\{0, 1\}$ that take the value 1 exactly r times. In other words, we need to count the number of characteristic functions χ_A where $|A| = r$. This is exactly the number of r-element subsets of $\{1, 2, \ldots, n\}$, i.e., $\binom{n}{r}$. $\blacksquare$

EXAMPLE 9 Consider a graph with no loops that is **complete** in the sense that each pair of distinct vertices has exactly one edge connecting them. If the graph has n vertices, how many edges does it have? Let's assume that $n \geq 2$. Each edge determines a 2-element subset of the set V of vertices and, conversely, each 2-element subset of V determines an edge. In other words, the set of edges is in one-to-one correspondence with the set of 2-element subsets of V. Hence there are

$$\binom{n}{2} = \frac{n!}{(n-2)! \, 2!} = \frac{n(n-1)}{2}$$

edges of the graph. $\blacksquare$

An excellent way to illustrate the techniques of this section is to calculate the numbers of various kinds of poker hands. A deck of cards consists of four suits called clubs, diamonds, hearts and spades. Each suit consists of thirteen cards with values A, 2, 3, 4, 5, 6, 7, 8, 9, 10, J, Q, K. Here A stands for ace, J for jack, Q for queen and K for king. There are four cards of each value, one from each suit. A **poker hand** is a set of 5 cards from a 52-card deck of cards. The order in which the cards are chosen is irrelevant. A **straight** consists of five cards whose values form a consecutive sequence such as 8, 9, 10, J, Q. The ace A can be at the bottom of a sequence A, 2, 3, 4, 5 or at the top of a sequence 10, J, Q, K, A. Poker hands are classified into disjoint sets as follows, listed in reverse order of their likelihood.

Royal flush 10, J, Q, K, A all in the same suit.

Straight flush A straight all in the same suit that is not a royal flush.

Four of a kind Four cards in the hand have the same value: for example, four 3's and a 9.

Full house Three cards of one value and two cards of another value: for example, three jacks and two 8's.

Flush Five cards all in the same suit, but not a royal or straight flush.

Straight A straight that is not a royal or straight flush.

Three of a kind Three cards of one value, a fourth card of a second value and a fifth card of a third value.

Two pairs Two cards of one value, two more cards of a second value and the remaining card a third value: for example, two queens, two 4's and a 7.

One pair Two cards of one value, but not classified above: for example, two kings, a jack, a 9 and a 6.

Nothing None of the above.

EXAMPLE 10 (a) There are $\binom{52}{5}$ poker hands. Note that

$$\binom{52}{5} = \frac{52 \cdot 51 \cdot 50 \cdot 49 \cdot 48}{5 \cdot 4 \cdot 3 \cdot 2 \cdot 1} = 52 \cdot 17 \cdot 10 \cdot 49 \cdot 6 = 2,598,960.$$

(b) How many poker hands are full houses? Let's call a hand consisting of three jacks and two 8's a full house of type (J, 8), with similar notation for other types of full houses. Order matters, since hands of type (8, J) have three 8's and two jacks. Also, types like (J, J) and (8, 8) are impossible. So types of full houses correspond to 2-permutations of the set of possible values of cards; hence there are $13 \cdot 12$ different types of full houses.

Now we count the number of full houses of each type, say type (J, 8). There are $\binom{4}{3} = 4$ ways to choose three jacks from four jacks, and there are then $\binom{4}{2} = 6$ ways to select two 8's from four 8's. Thus there are $4 \cdot 6 = 24$ hands of type (J, 8). This argument works for all $13 \cdot 12$ types of hands and so there are $13 \cdot 12 \cdot 24 = 3744$ full houses.

(c) How many poker hands are two pairs? Let's say that a hand with two pairs is of type $\{Q, 4\}$ if it has two queens and two 4's. This time we have used set notation because order does not matter: hands of type $\{4, Q\}$ are hands of type $\{Q, 4\}$ and we don't want to count them twice. There are $\binom{13}{2}$ types of hands. For each type, say $\{Q, 4\}$, there are $\binom{4}{2}$ ways of choosing two queens, $\binom{4}{2}$ ways of choosing two 4's and $52 - 8 = 44$ ways of choosing the fifth card. Hence there are

$$\binom{13}{2} \cdot \binom{4}{2} \cdot \binom{4}{2} \cdot 44 = 123{,}552$$

poker hands consisting of two pairs.

(d) How many poker hands are straights? First we count all possible straights even if they are royal or straight flushes. Let's call a straight consisting of the values 8, 9, 10, J, Q a straight of type Q. In general, the type of a straight is the highest value in the straight. Since any of the values 5, 6, 7, 8, 9, 10, J, Q, K, A can be the highest value in a straight, there are 10 types of straights. Given a type of straight, there are 4 choices for each of the 5 values. So there are 4^5 straights of each type and $10 \cdot 4^5 = 10{,}240$ straights altogether. There are 4 royal flushes and 36 straight flushes and so there are 10,200 straights that are not of these exotic varieties.

(e) You are asked to count the remaining kinds of poker hands in Exercise 11, for which all answers are given. ■

EXERCISES 5.1

1. Calculate

(a) $\binom{8}{3}$

(b) $\binom{8}{0}$

(c) $\binom{8}{5}$

(d) $\binom{52}{50}$

(e) $\binom{52}{52}$

(f) $\binom{52}{1}$

2. Let $A = \{1, 2, 3, 4, 5, 6, 7, 8, 9, 10\}$ and $B = \{2, 3, 5, 7, 11, 13, 17, 19\}$.

(a) Determine the sizes of the sets $A \cup B$, $A \cap B$ and $A \oplus B$.

(b) How many subsets of A are there?

(c) How many 4-element subsets of A are there?

(d) How many 4-element subsets of A consist of 3 even and 1 odd number?

3. Among 150 people, 45 swim, 40 bike and 50 jog. Also, 32 people jog but don't bike, 27 people jog and swim, and 10 people do all three.

   (a) How many people jog but don't swim and don't bike?

   (b) If 21 people bike and swim, how many do none of the three activities?

4. A certain class consists of 12 men and 16 women. How many committees can be chosen from this class consisting of

   (a) seven people?

   (b) three men and four women?

   (c) seven women or seven men?

5. (a) How many committees consisting of 4 people can be chosen from 9 people?

   (b) Redo part (a) if there are two people, Ann and Bob, who will not serve on the same committee.

6. How many committees consisting of 4 men and 4 women can be chosen from a group of 8 men and 6 women?

7. Let $S = \{a, b, c, d\}$ and $T = \{1, 2, 3, 4, 5, 6, 7\}$.

   (a) How many one-to-one functions are there from T into S?

   (b) How many one-to-one functions are there from S into T?

   (c) How many functions are there from S into T?

8. Let $P = \{1, 2, 3, 4, 5, 6, 7, 8, 9\}$ and $Q = \{A, B, C, D, E\}$.

   (a) How many 4-element subsets of P are there?

   (b) How many permutations, i.e., 5-permutations, of Q are there?

   (c) How many license plates are there consisting of three letters from Q followed by two numbers from P? Repetition is allowed; for example, DAD 88 is allowed.

9. Cards are drawn from a deck of 52 cards with replacement.

   (a) In how many ways can ten cards be drawn so that the tenth card is not a repetition?

   (b) In how many ways can ten cards be drawn so that the tenth card is a repetition?

10. Let Σ be the alphabet $\{a, b, c, d, e\}$ and let $\Sigma^k = \{w \in \Sigma^* : \text{length}(w) = k\}$. How many elements are there in each of the following sets?

    (a) Σ^k, for each $k \in \mathbb{N}$

    (b) $\{w \in \Sigma^3 : \text{no letter in } w \text{ is used more than once}\}$

    (c) $\{w \in \Sigma^4 : \text{the letter } c \text{ occurs in } w \text{ exactly once}\}$

    (d) $\{w \in \Sigma^4 : \text{the letter } c \text{ occurs in } w \text{ at least once}\}$

11. Count the number of poker hands of the following kinds:

    (a) four of a kind

    (b) flush [but not straight or royal flush]

(c) three of a kind

(d) one pair

12. (a) In how many ways can the letters a, b, c, d, e, f be arranged so that the letters a and b are adjacent?

(b) In how many ways can the letters a, b, c, d, e, f be arranged so that the letters a and b are not adjacent?

(c) In how many ways can the letters a, b, c, d, e, f be arranged so that the letters a and b are adjacent but a and c are not.

13. (a) Give the matrix for a complete graph with n vertices; see Example 9.

(b) Use the matrix in part (a) to count the number of edges of the graph. *Hint:* How many entries in the matrix are equal to 1?

14. (a) Consider a complete graph G with n vertices, $n \geq 3$. Find the number of cycles in G of length n. Cycles are defined in § 6.1.

(b) How many cycles are there in a complete graph with 5 vertices?

15. Consider a complete graph with n vertices, $n \geq 4$.

(a) Find the number of paths of length 3.

(b) Find the number of paths of length 3 whose vertex sequences consist of distinct vertices.

(c) Find the number of simple paths of length 3 consisting of distinct edges.

§ 5.2 Elementary Probability

In the preceding section we calculated the number of poker hands and the numbers of various kinds of poker hands. These numbers may not be all that fascinating, but poker players *are* interested in the fraction of poker hands that are flushes, full houses, etc. Why? Because if all poker hands are equally likely, then these fractions represent the likelihood or probability of getting one of these good hands. We return to poker hands after establishing some notation and terminology.

The underlying structure in probability is a set, called a **sample space**, consisting of **outcomes** that might result from an experiment, a game of chance, a survey, etc. It is traditional to denote a generic sample space by big omega Ω [a sort of Greek O, for **O**utcome]. Generic possible outcomes in Ω are denoted by little omegas ω or other Greek letters. Subsets of Ω are called **events**. A **probability on** Ω is a function P that assigns to each event $E \subseteq \Omega$ a number $P(E)$ in $[0, 1]$ and that satisfies

(P$_1$) $P(\Omega) = 1$,

(P$_2$) $P(E \cup F) = P(E) + P(F)$ for disjoint events E and F.

For now we think of Ω as a finite set.

The idea is that $P(E)$, the **probability of** E, should be a measure of the likelihood or the chance that an outcome in E occurs. We want $P(E) = 0$ to mean that there is really no chance the outcome of our experiment will be in E and $P(E) > 0$ to mean that there is some chance that an outcome in E occurs. We then surely want $P(\Omega) > 0$, and it is the standard convention to take $P(\Omega) = 1$, as we have done in (P$_1$). Condition (P$_2$) reflects the way we think chances work in real life; the chance of pulling a red marble or a green marble from a sack is the chance of pulling a red one *plus* the chance of pulling a green one. After we look at some examples, we will show that conditions (P$_1$) and (P$_2$) by themselves are enough to give us a workable mathematical model of real-world probability.

Often, as with poker hands, it is reasonable to assume that all the outcomes are **equally likely**. From (P$_2$) and (P$_1$) we have

$$\sum_{\omega \in \Omega} P(\{\omega\}) = P(\Omega) = 1,$$

so in this case $P(\{\omega\}) = \dfrac{1}{|\Omega|}$ for $\omega \in \Omega$. Using (P$_2$) again, we see that

$$P(E) = \frac{|E|}{|\Omega|} \qquad \text{for } E \subseteq \Omega \quad \text{when outcomes are equally likely.}$$

Our first example uses calculations from Example 10 in §5.1.

EXAMPLE 1   (a) In poker the set Ω of all possible outcomes is the set of all poker hands. Thus $|\Omega| = 2{,}598{,}960$. A typical event is

$$H = \{\omega \in \Omega : \omega \text{ is a full house}\}.$$

In everyday language the event "She was dealt a full house" corresponds exactly to the statement: the poker hand ω belongs to H. The probability of being dealt a full house is

$$P(H) = \frac{|H|}{|\Omega|} = \frac{3744}{2{,}598{,}960} \approx .00144$$

or about 1 in 700. Real poker players see full houses much more often than this because real poker games allow players to selectively exchange cards after the original deal. The value $P(H)$ above is the probability of being dealt a full house without taking advantage of the fancy rules.

(b) Another important event in poker is

$$F = \{\omega \in \Omega : \omega \text{ is a flush}\}.$$

We might abbreviate this by writing $F = $ "ω is a flush" or simply $F = $ "flush." From Exercise 11 of §5.1, we have $|F| = 5108$, so

$$P(F) = \frac{|F|}{|\Omega|} = \frac{5108}{2,598,960} \approx .00197$$

or about 1 in 500.

(c) The probability of obtaining a full house or flush is $P(H \cup F)$. No poker hand is both a full house and a flush, so the events H and F are disjoint. Hence

$$P(H \cup F) = P(H) + P(F) \approx .00341.$$

This illustrates the fundamental axiom (P_2) for a probability: the probability of the union of disjoint events is the sum of their probabilities. All axiom (P_1) tells us is that if a poker hand is dealt, then the probability that a poker hand is dealt is 1. Yawn! ■

Consider again any probability P on a sample space Ω. Since Ω consists of *all* possible outcomes, the probability that nothing happens surely should be zero. That is, $P(\varnothing) = 0$. This also follows easily from axiom (P_2):

$$P(\varnothing) = P(\varnothing \cup \varnothing) = P(\varnothing) + P(\varnothing) \quad \text{implies} \quad P(\varnothing) = 0.$$

The probability that an event fails should be 1 minus the probability of the event. That is, $P(E^c) = 1 - P(E)$ for $E \subseteq \Omega$. Again this is a consequence of the axioms, since

$$P(E) + P(E^c) = P(E \cup E^c) = P(\Omega) = 1.$$

We have already established parts (a) and (b) of the next theorem.

Theorem | Let P be a probability on a sample space Ω.

(a) $P(\varnothing) = 0$.

(b) $P(E^c) = 1 - P(E)$ for events E.

(c) $P(E \cup F) = P(E) + P(F) - P(E \cap F)$ for events E and F whether they are disjoint or not.

(d) $P(E_1 \cup E_2 \cup \cdots \cup E_m) = \sum_{k=1}^{m} P(E_k) = P(E_1) + P(E_2) + \cdots + P(E_m)$ for pairwise disjoint events $E_1, E_2, \ldots, E_m$.

By **pairwise disjoint** we mean that $E_j \cap E_k = \varnothing$ for $j \neq k$.

Proof. (c) The argument is essentially the same as for Union Rule (b) in §5.1. We have $E \cup F = E \cup (F \backslash E)$ and $F = (F \backslash E) \cup (E \cap F)$ with both unions disjoint. Hence $P(E \cup F) = P(E) + P(F \backslash E)$ and $P(F) = P(F \backslash E) + P(E \cap F)$, so $P(E \cup F) = P(E) + P(F) - P(E \cap F)$.

(d) This is an easy induction argument. The result is true for $m = 2$ by axiom (P$_2$). Assume that the identity holds for m sets and that E_1, $E_2, \ldots, E_{m+1}$ are pairwise disjoint. Then $E_1 \cup E_2 \cup \cdots \cup E_m$ is disjoint from E_{m+1}, so

$$P(E_1 \cup E_2 \cup \cdots \cup E_{m+1}) = P(E_1 \cup E_2 \cup \cdots \cup E_m) + P(E_{m+1})$$

$$= P(E_1) + P(E_2) + \cdots + P(E_m) + P(E_{m+1}). \quad \blacksquare$$

EXAMPLE 2   In Example 4 of §5.1 we counted 12 paths from s to f in Figure 2(a). Note that each such path has length 4. Suppose that a set of four edges is selected at random. What is the probability that the four edges are the edges of a path from s to f?

Here Ω consists of all the 4-element subsets of the set of all edges. There are 18 edges in all, so $|\Omega| = \binom{18}{4} = 3060$. Only 12 of these sets give paths from s to f, so the answer to the question is $12/3060 \approx .000392$. $\blacksquare$

Time for an easy example.

EXAMPLE 3   The usual die [plural is dice] in Figure 1(a) has six sides with 1, 2, 3, 4, 5 and 6 dots on them, respectively. When the die is tossed, one of the six numbers appears at the top. If all six are equally likely, we say that the die is **fair**. We will assume that our dice are fair. Let $\Omega = \{1, 2, 3, 4, 5, 6\}$ be the set of possible outcomes. Since the die is fair, $P(k) = \frac{1}{6}$ for each $k \in \Omega$; here we are writing $P(k)$ in place of $P(\{k\})$. If E is the event "k is even," i.e., if $E = \{2, 4, 6\}$, then $P(E) = \frac{1}{6} + \frac{1}{6} + \frac{1}{6} = \frac{1}{2}$. If F is the

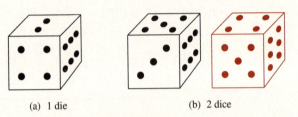

(a) 1 die   (b) 2 dice

FIGURE 1

event "four or five," i.e., if $F = \{4, 5\}$, then $P(F) = \frac{1}{3}$. Also,

$$P(E \cup F) = P(\{2, 4, 5, 6\}) = \frac{2}{3} \quad \text{and} \quad P(E \cap F) = P(\{4\}) = \frac{1}{6}.$$

Of course, $P(E \cup F) = P(E) + P(F) - P(E \cap F)$. ∎

The last example was too easy. Here is one that is just right.

EXAMPLE 4 We now consider tossing two fair dice, one black and one red. Usually the outcome of interest is the sum of the two values appearing at the tops of the dice; for example, the sum is 9 for the dice in Figure 1(b). For the first time it is not absolutely clear what the sample space Ω should be. In fact, the choice is up to us, but some choices may be easier to work with than others. Since the outcomes of interest are the sums, it is tempting to set Ω equal to the 11-element set consisting of 2, 3, 4, 5, 6, 7, 8, 9, 10, 11 and 12. The trouble is that these outcomes are not equally likely. While we could use their probabilities [determined below], it is easier to let Ω consist of all ordered pairs of values on the two dice:

$$\Omega = \{(k, l) : 1 \le k \le 6, \ 1 \le l \le 6\}.$$

Here the first entry, k, is the value on the black die and the second entry, l, is the value on the red die. The outcome shown in Figure 1(b) corresponds to the pair (5, 4). It seems reasonable, and is justified in §9.1, that these 36 outcomes are equally likely. They are listed in Figure 2(a). The events of special interest are "sum is k," for $k = 2, 3, 4, \ldots, 12$. For example,

$$P(\text{sum is 9}) = P(\{(3, 6), (4, 5), (5, 4), (6, 3)\}) = \frac{4}{36} = \frac{1}{9}.$$

This and the other values of $P(\text{sum is } k)$ are given in Figure 2(b). Note that the outcomes (4, 5) and (5, 4) are really different. On the other hand,

(1, 1)	(1, 2)	(1, 3)	(1, 4)	(1, 5)	(1, 6)
(2, 1)	(2, 2)	(2, 3)	(2, 4)	(2, 5)	(2, 6)
(3, 1)	(3, 2)	(3, 3)	(3, 4)	(3, 5)	(3, 6)
(4, 1)	(4, 2)	(4, 3)	(4, 4)	(4, 5)	(4, 6)
(5, 1)	(5, 2)	(5, 3)	(5, 4)	(5, 5)	(5, 6)
(6, 1)	(6, 2)	(6, 3)	(6, 4)	(6, 5)	(6, 6)

(a)

Sum	P (Sum)	Sum
2	1/36	12
3	2/36 = 1/18	11
4	3/36 = 1/12	10
5	4/36 = 1/9	9
6	5/36	8
7	6/36 = 1/6	

(b)

FIGURE 2

as we can see from Figure 2(a), there is only one way to get two fives, namely (5, 5). Thus

$$P(\text{sum is } 10) = P(\{(4, 6), (5, 5), (6, 4)\}) = \frac{3}{36} = \frac{1}{12}.$$

Figure 2(a) can be used to solve any probabilistic question about our two dice. Either Figure 2(a) or 2(b) can be used to handle questions only involving sums.

(a) What is the probability that the sum of the values on the dice is greater than 7? This event consists of the ordered pairs below the dashed line in Figure 2(a). There are 15 such pairs, so the answer is $\frac{15}{36} = \frac{5}{12}$. We can also use the table in Figure 2(b) to get

$$P(\text{sum} > 7) = \sum_{k=8}^{12} P(\text{sum is } k) = \frac{5}{36} + \frac{4}{36} + \frac{3}{36} + \frac{2}{36} + \frac{1}{36} = \frac{15}{36}.$$

(b) What is the probability that the number on the black die divides the number on the red die? That is, what is $P(E)$ where $E = \{(k, l) : k | l\}$? This time Figure 2(b) is no help, but we can list the elements of E with or without using Figure 2(a):

$$E = \{(1, 1), (1, 2), (1, 3), (1, 4), (1, 5), (1, 6), (2, 2), (2, 4), (2, 6), (3, 3),$$

$$(3, 6), (4, 4), (5, 5), (6, 6)\}.$$

Hence $P(E) = \frac{14}{36}$. ∎

EXAMPLE 5   (a) A tossed coin is said to be **fair** if the probability of heads is $\frac{1}{2}$. If H signifies heads and T signifies tails, then $P(\text{H}) = P(\text{T}) = \frac{1}{2}$. We want to toss the coin several times, say n times. The set of possible outcomes corresponds to the set Ω of all n-tuples of H's and T's. For example, (T, H, H, T, H) corresponds to $n = 5$ tosses where the first and fourth tosses are tails and the other three are heads. As in Example 4, it's reasonable to assume that all 2^n n-tuples in Ω are equally likely.

For $r = 0, 1, \ldots, n$, we calculate $P(\text{exactly } r \text{ of the tosses are heads})$. The number of n-tuples with r heads is exactly the number of strings of 0's and 1's having length n and r 1's. This is $\binom{n}{r}$, as shown in Example 8 of § 5.1. Hence

$$P(\text{exactly } r \text{ of the tosses are heads}) = \frac{\binom{n}{r}}{2^n}.$$

(b) If we tossed a fair coin ten times we would be surprised if we got 8 or more heads or if we got 8 or more tails. Should we be? We find

$$P(\text{number of heads is} \geq 3 \text{ and} \leq 7) = \sum_{r=3}^{7} P(\text{number of heads is } r)$$

$$= \frac{1}{2^{10}}\left[\binom{10}{3} + \binom{10}{4} + \binom{10}{5} + \binom{10}{6} + \binom{10}{7}\right]$$

$$= \frac{1}{1024}[120 + 210 + 252 + 210 + 120] = \frac{912}{1024} \approx .891.$$

Therefore, there is more than a 10 percent chance of getting 8 or more heads or 8 or more tails; we conclude that it is only mildly surprising when this happens.

(c) The computation in part (b) would get out of hand if we tossed the coin 100 times and wanted to calculate

$$P(\text{number of heads is} \geq 36 \text{ and} \leq 64),$$

say. In §9.4 we will indicate how such numbers can be estimated. We will also see that the significance of these estimates goes far beyond the coin-tossing problem. Incidentally,

$$P(\text{number of heads is} \geq 36 \text{ and} \leq 64) \approx .996,$$

so it really would be surprising if we obtained more than 64 or fewer than 36 heads. ∎

Our discussion of "surprising" events in the preceding example was clearly subjective. However, in many statistical analyses events with probabilities less than .05 are regarded as surprising or unexpected while events with probabilities greater than .05 are not.

Sample spaces are often infinite. In this case the axioms need to be modified. First we give some examples.

EXAMPLE 6 (a) Consider the experiment of tossing a fair coin until a head is obtained. Our sample space will be $\Omega = \mathbb{P} = \{1, 2, 3, \dots\}$ where outcome k corresponds to k tosses. From Example 5 we see that

$$P(1) = \text{probability of head on first toss} = \frac{1}{2},$$

$$P(2) = \text{probability of a tail, then a head} = \frac{1}{2^2},$$

$$P(3) = \text{probability of two tails, then a head} = \frac{1}{2^3},$$

etc. In general, $P(k) = 1/2^k$. For finite $E \subseteq \Omega$ we define $P(E) = \sum_{k \in E} \frac{1}{2^k}$.

For example, the probability of getting a head by tossing the coin fewer than six times is

$$P(\{1, 2, 3, 4, 5\}) = \frac{1}{2} + \frac{1}{4} + \frac{1}{8} + \frac{1}{16} + \frac{1}{32} = \frac{31}{32} \approx .969.$$

We haven't verified that P satisfies the axioms for a probability, but it does. If you've ever seen an infinite series you know that

$$P(\Omega) = \sum_{k \in \mathbb{P}} P(k) = \sum_{k=1}^{\infty} \frac{1}{2^k} = 1.$$

If you haven't, this is just shorthand for the statement

$$\sum_{k=1}^{n} \frac{1}{2^k} \text{ is very close to 1 for very large } n.$$

More generally, the "sum" $\sum_{k \in E} \frac{1}{2^k}$ makes sense for every $E \subseteq \Omega$ and we define $P(E)$ to be this sum. For example, the probability that an odd number of tosses are needed is

$$P(\{1, 3, 5, 7, 9, \ldots\}) = \sum_{k \text{ odd}} \frac{1}{2^k},$$

which turns out to be $\frac{2}{3}$.

(b) People talk about random numbers in $\Omega = [0, 1)$. Computers claim that they can produce random numbers in $[0, 1)$. What do they mean? If we ignore the fact that people and computers *really* only work with finite sets, we have to admit that Ω has an infinite number of outcomes that we want to be equally likely. Then $P(\omega)$ must be 0 for all ω in $[0, 1)$, and the definition $P(E) = \sum_{\omega \in E} P(\omega)$ for $E \subseteq \Omega$ leads to nonsense.

All is not lost, but we cannot base our probability P on outcomes ω alone. We'd like to have $P([0, \frac{1}{2})) = \frac{1}{2}$, $P([\frac{3}{4}, 1)) = \frac{1}{4}$, etc. It turns out that there is a probability P defined on *some* subsets of $[0, 1)$, called events, such that $P([a, b)) = b - a$ whenever $[a, b) \subseteq [0, 1)$. This P satisfies the axioms for a probability and more:

$$P\left(\bigcup_{k=1}^{\infty} E_k\right) = \sum_{k=1}^{\infty} P(E_k)$$

for pairwise disjoint sequences of events $E_1, E_2, \ldots$ in $[0, 1)$.

(c) There are many useful probabilities P on $\Omega = [0, 1)$. The probability in part (b) is the one where $P([a, b)) = b - a$ for $[a, b) \subseteq [0, 1)$. There are also many probabilities on $\mathbb{P}$ in addition to the one described in part (a).   ■

Here are the modifications for a probability P on an infinite sample space Ω. As hinted in Example 6(b), in general only certain subsets of Ω are regarded as events. Also, (P$_2$) is strengthened to

$$(\mathrm{P}'_2) \qquad P\left(\bigcup_{k=1}^{\infty} E_k\right) = \sum_{k=1}^{\infty} P(E_k) \qquad \begin{array}{l}\text{for pairwise disjoint sequences}\\\text{of events in } \Omega.\end{array}$$

EXERCISES 5.2

Whenever choices or selections are "at random" the possible outcomes are assumed to be equally likely.

1. An integer in $\{1, 2, 3, \ldots, 25\}$ is selected at random. Find the probability that the number is

   (a) divisible by 3   (b) divisible by 5

   (c) a prime

2. A letter of the alphabet is selected at random. What is the probability that it is a vowel $[a, e, i, o$ or $u]$?

3. A four-letter word is selected at random from Σ^4 where $\Sigma = \{a, b, c, d, e\}$.

   (a) What is the probability that the letters in the word are distinct?

   (b) What is the probability that there are no vowels in the word?

   (c) What is the probability that the word begins with a vowel?

4. A five-letter word is selected at random from Σ^5 where $\Sigma = \{a, b, c\}$. Repeat Exercise 3.

5. An urn contains three red and four black balls. A set of three balls is removed at random from the urn [without replacement]. Give the probabilities that the three balls are

   (a) all red   (b) all black

   (c) one red and two black   (d) two red and one black

   (e) Sum the answers to parts (a) to (d).

6. An urn has three red and two black balls. Two balls are removed at random [without replacement]. What is the probability that the two balls are

   (a) both red?   (b) both black?   (c) different colors?

7. Suppose that an experiment leads to events A, B and C with the following probabilities: $P(A) = .5$, $P(B) = .8$, $P(A \cap B) = .4$. Find

   (a) $P(B^c)$ (b) $P(A \cup B)$ (c) $P(A^c \cup B^c)$

8. Suppose that an experiment leads to events A, B and C with the following probabilities: $P(A) = .6$ and $P(B) = .7$. Show that $P(A \cap B) \geq .3$.

9. A five card poker hand is dealt. Find the probability of getting

   (a) four of a kind   (b) three of a kind

(c) a [nonexotic] straight   (d) two pairs

(e) one pair

Hint: Use Example 10 and Exercise 11 of §5.1.

10. A poker hand is dealt.

    (a) what's the probability of getting a hand better than one pair? "Better" here means any other special hand listed in §5.1.

    (b) What's the probability of getting a pair of jacks or better? The only pairs that are "better" than a pair of jacks are a pair of queens, a pair of kings, and a pair of aces.

11. A black die and a red die are tossed as in Example 4. What's the probability that

    (a) the sum of the values is even?

    (b) the number on the red die is bigger than the number on the black die?

    (c) the number on the red die is twice the number on the black die?

12. Two dice are tossed as in Exercise 11. What's the probability that

    (a) the maximum of the numbers on the dice is 4?

    (b) the minimum of the numbers on the dice is 4?

    (c) the product of the numbers on the dice is 4?

13. Let P be a probability on a sample space Ω. For events E_1, E_2 and E_3, show that

$$P(E_1 \cup E_2 \cup E_3) = P(E_1) + P(E_2) + P(E_3) - P(E_1 \cap E_2) - P(E_1 \cap E_3)$$
$$- P(E_2 \cap E_3) + P(E_1 \cap E_2 \cap E_3).$$

14. Let P be a probability on a sample space Ω. Show that if E and F are events and $E \subseteq F$, then $P(E) \le P(F)$.

15. A fair coin is tossed six times. Find the probabilities of getting

    (a) no heads   (b) one head   (c) two heads

    (d) three heads   (e) more than three heads

16. A fair coin is tossed until a head is obtained. What is the probability that the coin was tossed at least four times?

17. A fair coin is tossed n times. Show that the probability of getting an even number of heads is $\frac{1}{2}$.

18. The probability of my winning the first game of backgammon is .5, the second game .4, and both games .3. What is the probability that I will lose both games?

19. A set of four numbers is selected at random from $S = \{1, 2, 3, 4, 5, 6, 7, 8\}$ [no replacement]. What is the probability that

    (a) exactly two of them are even?

    (b) none of them is even?

(c) exactly one of them is even?

(d) exactly three of them are even?

(e) all four of them are even?

20. (a) A student answers a three-question true-false test at random. What is the probability that she will get at least two-thirds of the questions correct?

(b) Repeat part (a) for a six-question test.

(c) Repeat part (a) for a nine-question test.

21. A computer program selects an integer in $\{k : 1 \leq k \leq 1{,}000{,}000\}$ at random and prints the result. This is repeated one million times. What's the probability that the value $k = 1$ appears in the printout at least once? *Hints:*

(a) A number is selected at random from $\{1, 2, 3\}$ three times. What's the probability that 1 was selected at least once? [First find the probability that 1 is not selected.]

(b) A number is selected at random from $\{1, 2, 3, 4\}$ four times. What's the probability that 1 was selected at least once?

(c) A number is selected at random from $\{1, 2, \ldots, n\}$ n times. What's the probability that 1 was selected at least once?

(d) Set $n = 1{,}000{,}000$.

§ 5.3   Inclusion–Exclusion Principle and Binomial Methods

This section contains extensions of the counting methods we introduced in §5.1. The Inclusion–Exclusion Principle generalizes the Union Rule, to count unions of more than two sets. The Binomial Theorem, one of the basic facts of algebra, is closely related to counting. Our third topic, putting objects in boxes, shows yet another application for binomial coefficients.

It is often easy to count elements in an intersection of sets, where the key connective is "and." On the other hand, a direct count of the elements in a union of sets is often difficult. The Inclusion–Exclusion Principle will tell us the size of a union in terms of the sizes of various intersections.

Let $A_1, A_2, \ldots, A_n$ be finite sets. For $n = 2$, Union Rule (b) of §5.1 states that

$$|A_1 \cup A_2| = |A_1| + |A_2| - |A_1 \cap A_2|.$$

For $n = 3$, the Inclusion–Exclusion Principle below will assert that

$$|A_1 \cup A_2 \cup A_3| = |A_1| + |A_2| + |A_3|$$
$$- \{|A_1 \cap A_2| + |A_1 \cap A_3| + |A_2 \cap A_3|\} + |A_1 \cap A_2 \cap A_3|$$

and for $n = 4$ it will say that

$$|A_1 \cup A_2 \cup A_3 \cup A_4| = |A_1| + |A_2| + |A_3| + |A_4|$$
$$- \{|A_1 \cap A_2| + |A_1 \cap A_3| + |A_1 \cap A_4| + |A_2 \cap A_3| + |A_2 \cap A_4| + |A_3 \cap A_4|\}$$
$$+ \{|A_1 \cap A_2 \cap A_3| + |A_1 \cap A_2 \cap A_4| + |A_1 \cap A_3 \cap A_4| + |A_2 \cap A_3 \cap A_4|\}$$
$$- |A_1 \cap A_2 \cap A_3 \cap A_4|.$$

Here is a statement of the general principle in words. A version with symbols is offered in Exercise 6.

Inclusion–Exclusion Principle

> To calculate the size of $A_1 \cup A_2 \cup \cdots \cup A_n$, calculate the sizes of all possible intersections of sets from $\{A_1, A_2, \ldots, A_n\}$, add the results obtained by intersecting an odd number of the sets, and then subtract the results obtained by intersecting an even number of the sets.

In terms of the phrase "inclusion–exclusion," include or add the sizes of the sets, then exclude or subtract the sizes of all intersections of two sets, then include or add the sizes of all intersections of three sets, etc.

EXAMPLE 1 We count the number of integers in $S = \{1, 2, 3, \ldots, 2000\}$ that are divisible by 9, 11, 13 or 15. For each $k \in \mathbb{P}$, we let $D_k = \{n \in S : n$ is divisible by $k\}$ and we seek $|D_9 \cup D_{11} \cup D_{13} \cup D_{15}|$. Note that $|D_k|$ is the largest integer $\leq 2000/k$, i.e., $|D_k| = \lfloor 2000/k \rfloor$, so we can compute

$$|D_9| = 222, \qquad |D_{11}| = 181, \qquad |D_{13}| = 153, \qquad |D_{15}| = 133,$$

$$|D_9 \cap D_{11}| = |D_{99}| = 20, \qquad\qquad |D_9 \cap D_{13}| = |D_{117}| = 17,$$

$$|D_9 \cap D_{15}| = |D_{45}| = 44, \qquad\qquad |D_{11} \cap D_{13}| = |D_{143}| = 13,$$

$$|D_{11} \cap D_{15}| = |D_{165}| = 12, \qquad\qquad |D_{13} \cap D_{15}| = |D_{195}| = 10,$$

$$|D_9 \cap D_{11} \cap D_{13}| = |D_{1287}| = 1, \qquad |D_9 \cap D_{11} \cap D_{15}| = |D_{495}| = 4,$$

$$|D_9 \cap D_{13} \cap D_{15}| = |D_{585}| = 3, \qquad |D_{11} \cap D_{13} \cap D_{15}| = |D_{2145}| = 0,$$

$$|D_9 \cap D_{11} \cap D_{13} \cap D_{15}| = |D_{6435}| = 0.$$

Note that $D_9 \cap D_{15} = D_{45}$ [not D_{135}] since $\mathrm{lcm}(9, 15) = 45$; similar care is needed in dealing with $D_9 \cap D_{11} \cap D_{15}$, $D_9 \cap D_{13} \cap D_{15}$, etc. Now by the Inclusion–Exclusion Principle, we have

$$|D_9 \cup D_{11} \cup D_{13} \cup D_{15}| = 222 + 181 + 153 + 133$$

$$- (20 + 17 + 44 + 13 + 12 + 10) + (1 + 4 + 3 + 0) - 0 = 581. \blacksquare$$

The Inclusion–Exclusion Principle is ideally suited to situations in which:

(a) We just want the size of $A_1 \cup \cdots \cup A_n$, not a listing of its elements, and

(b) Multiple intersections are fairly easy to count.

Example 1 showed just such a problem.

If we want an actual list of the members of $A_1 \cup \cdots \cup A_n$, then we can use an iterative algorithm to list A_1 first, then $A_2 \backslash A_1$, then $A_3 \backslash (A_1 \cup A_2)$, and so on, or we can list A_1 and then recursively list $(A_2 \backslash A_1) \cup \cdots \cup (A_n \backslash A_1)$. Look again at Example 1 to see the sort of calculations that either of these methods would require to produce a list of the elements of $D_9 \cup D_{11} \cup D_{13} \cup D_{15}$. One would certainly want machine help and a friendly data structure. If we just want a *count*, though, and not a list, the Inclusion–Exclusion Principle makes the job fairly painless.

EXAMPLE 2   A number is selected at random from $T = \{1000, 1001, \ldots, 9999\}$. We find the probability that the number has at least one digit that is 0, at least one that is 1 and at least one that is 2. For example, 1072 and 2101 are two such numbers. It is easier to count numbers that exclude certain digits, so we deal with complements. That is, for $k = 0$, 1 and 2 we let

$$A_k = \{n \in T : n \text{ has no digit equal to } k\}.$$

Then each A_k^c consists of those n in T that have at least one digit equal to k, so $A_0^c \cap A_1^c \cap A_2^c$ consists of those n in T that have at least one 0, one 1 and one 2 among their digits. This is exactly the set whose size we are after.

Since $A_0^c \cap A_1^c \cap A_2^c = (A_0 \cup A_1 \cup A_2)^c$ by DeMorgan's law, we will first calculate $|A_0 \cup A_1 \cup A_2|$ using the Inclusion–Exclusion Principle. By Product Rule (a) in §5.1 we have $|A_1| = 8 \cdot 9 \cdot 9 \cdot 9$, since there are 8 choices for the first digit, which cannot be 0 or 1, and 9 choices for the other digits. Similar computations yield

$$|A_0| = 9 \cdot 9 \cdot 9 \cdot 9 = 6561, \qquad |A_1| = |A_2| = 8 \cdot 9 \cdot 9 \cdot 9 = 5832,$$

$$|A_0 \cap A_1| = |A_0 \cap A_2| = 8 \cdot 8 \cdot 8 \cdot 8 = 4096,$$

$$|A_1 \cap A_2| = 7 \cdot 8 \cdot 8 \cdot 8 = 3584,$$

$$|A_0 \cap A_1 \cap A_2| = 7 \cdot 7 \cdot 7 \cdot 7 = 2401.$$

By the Inclusion–Exclusion Principle,

$$|A_0 \cup A_1 \cup A_2| = 6561 + 5832 + 5832$$

$$- (4096 + 4096 + 3584) + 2401 = 8850,$$

so

$$\left|(A_0 \cup A_1 \cup A_2)^c\right| = |T| - |A_0 \cup A_1 \cup A_2| = 9000 - 8850 = 150.$$

There are 150 integers in T whose digits include at least one 0, 1 and 2.
Hence the probability of this event is $\dfrac{150}{|T|} = \dfrac{150}{9000} = \dfrac{1}{60}$. ■

An Explanation of the Inclusion–Exclusion Principle. The main barrier to proving the general principle is the notation [cf. Exercise 6]. The principle can be proved by induction on n. We show how the result for $n = 2$ leads to the result for $n = 3$. Using the $n = 2$ case we have

(1) $$|A \cup B \cup C| = |A \cup B| + |C| - |(A \cup B) \cap C|$$

and

(2) $$|A \cup B| = |A| + |B| - |A \cap B|.$$

Applying the distributive law for unions and intersections [rule 3b in Table 1 of §1.2], we also obtain

(3) $$\begin{aligned}|(A \cup B) \cap C| &= |(A \cap C) \cup (B \cap C)| \\ &= |A \cap C| + |B \cap C| - |A \cap B \cap C|.\end{aligned}$$

Substitution of (2) and (3) into (1) yields

(4) $$\begin{aligned}|A \cup B \cup C| = {} &|A| + |B| - |A \cap B| + |C| - |A \cap C| \\ &- |B \cap C| + |A \cap B \cap C|,\end{aligned}$$

which is the principle for $n = 3$. ■

The next theorem is probably familiar from algebra. It has many applications, and because $a + b$ is a binomial it explains why we called $\dbinom{n}{r}$ a "binomial coefficient."

Binomial Theorem

For real numbers a and b and for $n \in \mathbb{N}$, we have

$$(a + b)^n = \sum_{r=0}^{n} \binom{n}{r} a^r b^{n-r}.$$

Proof. The theorem can be proved by induction, using the recurrence relation

$$\binom{n+1}{r} = \binom{n}{r-1} + \binom{n}{r} \qquad \text{for} \quad 1 \le r \le n;$$

see Exercise 12. This relation can in turn be proved by algebraic manipulation, but let us give a set-theoretic explanation instead, in the spirit of counting.

There are $\binom{n+1}{r}$ r-element subsets of $\{1, 2, \ldots, n, n+1\}$. We separate them into two classes. There are $\binom{n}{r}$ subsets that contain only members of $\{1, 2, \ldots, n\}$. The remaining subsets each consist of the number $n + 1$ and some $r - 1$ members of $\{1, 2, \ldots, n\}$. Since there are $\binom{n}{r-1}$ ways to choose the elements that aren't $n + 1$, there are $\binom{n}{r-1}$ subsets of this type. Hence there are exactly $\binom{n}{r} + \binom{n}{r-1}$ r-element subsets of $\{1, 2, \ldots, n, n+1\}$, so that

$$\binom{n}{r} + \binom{n}{r-1} = \binom{n+1}{r},$$

as claimed.

The *real* reason, though, that $\binom{n}{r}$ is the coefficient of $a^r b^{n-r}$ in $(a + b)^n$ is that it counts the terms with r a's in them that arise when we multiply out $(a + b)^n$. Look at

$$(a + b)^2 = (a + b)(a + b) = aa + (ab + ba) + bb \quad \text{and}$$

$$(a + b)^3 = (a + b)(a + b)(a + b)$$

$$= aaa + (aab + aba + baa) + (abb + bab + bba) + bbb.$$

There are $3 = \binom{3}{2}$ terms with two a's in $(a + b)^3$, corresponding to the choices

$$(\boldsymbol{a} + b)(\boldsymbol{a} + b)(a + \boldsymbol{b}), \quad (\boldsymbol{a} + b)(a + \boldsymbol{b})(\boldsymbol{a} + b) \quad \text{and} \quad (a + \boldsymbol{b})(\boldsymbol{a} + b)(\boldsymbol{a} + b).$$

In the general case of $(a + b)^n$, each choice of r factors $(a + b)$ to take the a's from produces a term with r a's [and so with $n - r$ b's]. There are $\binom{n}{r}$ possible choices, so there are $\binom{n}{r}$ such terms. ∎

Sometimes the Binomial Theorem is useful for computing the value of a sum, such as $\sum_{r=0}^{n} (-1)^r \binom{n}{r}$ [Exercise 10], $\sum_{r=0}^{n} \binom{n}{r}$ [Exercise 13] or $\sum_{r=0}^{n} \binom{n}{r} 2^r$ [Exercise 14]. At other times we can use it to compute specific coefficients, often the first few, in powers that we don't want to write out in detail.

EXAMPLE 3   The first few terms of $(2 + x)^{100}$ are

$$2^{100} + \binom{100}{1}2^{99}x + \binom{100}{2}2^{98}x^2 + \cdots$$

$$= 2^{100} + 100 \cdot 2^{99}x + 4950 \cdot 2^{98}x^2 + \cdots.$$

If x is very near 0, it is tempting to say that since $x^3, x^4, \ldots$ are tiny we can pretty much ignore the terms after x^2. Beware! The coefficients in the middle, such as $\binom{100}{50}$, can get pretty big, and collectively the discarded terms can add up to a lot. In our case, if $x = 0.01$, we make about a 1.4 percent error in $(2.01)^{100}$ by dropping the higher powers, but with $x = 0.1$ our "approximate" answer would be less than 5 percent of the correct value. ■

Our next counting principle can be applied in a variety of settings. We offer it in a form that is easy to remember.

Placing Objects in Boxes

> There are $\binom{n + k - 1}{k - 1}$ ways to place n identical objects into k distinguishable boxes.

Proof. The proof is both elegant and illuminating; we illustrate it for the case $n = 5$ and $k = 4$. We let five 0's represent the objects and then we add three 1's to serve as dividers among the four boxes. We claim that there is a one-to-one correspondence between the strings consisting of five 0's and three 1's and the ways to place the five 0's into four boxes. Specifically, a given string corresponds to the placement of the 0's before the first 1 into the first box, the 0's between the first and second 1 into the second box, the 0's between the second and third 1 into the third box, and the 0's after the third 1 into the fourth box. For example,

$$0\ 0\ 1\ 1\ 0\ 0\ 0\ 1 \ \longrightarrow \ 0\ 0\ \big\|\ 0\ 0\ 0\ \big|\ \longrightarrow$$

$$\text{box}\quad\text{box}\quad\text{box}\quad\text{box}$$
$$1\qquad 2\qquad 3\qquad 4$$

In this instance, boxes 2 and 4 are empty because there are no 0's between the first and second dividers and there are no 0's after the last divider.

More examples:

$$1\ 0\ 0\ 1\ 0\ 0\ 1\ 0 \;\longrightarrow\; \boxed{\ \ }\ \boxed{0\ 0}\ \boxed{0\ 0}\ \boxed{0}\ ;$$

$$0\ 0\ 0\ 1\ 1\ 1\ 0\ 0 \;\longrightarrow\; \boxed{\begin{array}{c}0\ 0\\ 0\end{array}}\ \boxed{\ \ }\ \boxed{\ \ }\ \boxed{0\ 0}\ .$$

There are $\binom{8}{3}$ strings having five 0's and three 1's. Since $\binom{8}{3} = \binom{5+4-1}{4-1}$, this establishes the result for $n = 5$ and $k = 4$.

In the general case, we consider strings of n 0's and $k - 1$ 1's. The 0's correspond to objects and the 1's to dividers. There are $\binom{n+k-1}{k-1}$ such strings and, as above, there is a one-to-one correspondence between these strings and the placing of n 0's into k boxes. ∎

EXAMPLE 4   (a) In how many ways can ten red marbles be placed into five distinguishable bags? Here $n = 10$, $k = 5$ and the answer is

$$\binom{10+5-1}{5-1} = \binom{14}{4} = 1001.$$

(b) In how many ways can ten red marbles be placed into five indistinguishable bags? This is much harder. You should be aware that counting problems can get difficult quickly. Here one would like to apply part (a) somehow. Even with the methods of the next section, however, there is no natural way to do so. We abandon this problem. Any solution we are aware of involves the consideration of several cases. ∎

Sometimes problems need to be manipulated before it is clear how our principles apply.

EXAMPLE 5   How many numbers in $\{1, 2, 3, \ldots, 100000\}$ have the property that the sum of their digits is 7? We can ignore the very last number, 100000, and we can assume that all the numbers have five digits, by placing zeros in front if necessary. So, for example, we replace 1 by 00001 and 73 by 00073. Our question is now: How many strings of five digits have the property that the sum of their digits is 7? We can associate each such string with the placement of seven balls in five boxes; for example,

$$0\ 0\ 1\ 4\ 2 \;\longrightarrow\; \boxed{\ \ }\ \boxed{\ \ }\ \boxed{0}\ \boxed{\begin{array}{c}0\ 0\\ 0\ 0\end{array}}\ \boxed{0\ 0}\ ;$$

$$3\ 0\ 1\ 2\ 1 \;\longrightarrow\; \boxed{\begin{array}{c}0\ 0\\ 0\end{array}}\ \boxed{\ \ }\ \boxed{0}\ \boxed{0\ 0}\ \boxed{0}\ .$$

There are $\binom{11}{4} = 330$ such placements, so there are 330 numbers with the desired property.   ■

We now give a little different interpretation of the "Placing Objects in Boxes" principle. Consider the k boxes first and assume that each box contains an unlimited supply of objects labeled according to which box they are in. Applying the principle in reverse, we see that there are $\binom{n + k - 1}{k - 1}$ ways to remove n objects from the k boxes. In other words,

> The number of ways to select a set of n objects of k distinguishable types, allowing repetitions, is
> $$\binom{n + k - 1}{k - 1}.$$

EXAMPLE 6   In how many ways can ten coins be selected from an unlimited supply of pennies, nickels, dimes and quarters? This is tailor-made for the principle just stated. Let $n = 10$ [for the ten coins] and $k = 4$ [for the four types of coins]. Then the answer is

$$\binom{10 + 4 - 1}{4 - 1} = \binom{13}{3} = 286.$$

The new interpretation can be avoided as follows. The problem is equivalent to counting ordered 4-tuples of nonnegative integers whose sum is 10. For example, $(5, 3, 0, 2)$ corresponds to the selection of 5 pennies, 3 nickels and 2 quarters. Counting these ordered 4-tuples is equivalent to counting the ways of placing ten indistinguishable objects into 4 boxes and this can be done in $\binom{13}{3}$ ways.   ■

EXERCISES 5.3

1. Among 200 people, 150 either swim or jog or both. If 85 swim and 60 swim and jog, how many jog?

2. Let $S = \{100, 101, 102, \ldots, 999\}$, so that $|S| = 900$.

   (a) How many numbers in S have at least one digit that is a 3 or a 7? Examples: 300, 707, 736, 103, 997.

(b) How many numbers in S have at least one digit that is a 3 *and* at least one digit that is a 7? Examples: 736 and 377 but not 300, 707, 103, 997.

3. An integer is selected at random from $\{1, 2, 3, \ldots, 1000\}$. What is the probability that it is divisible by 4, 5 or 6?

4. An investor has 7 $1000 bills to distribute by mail among 3 mutual funds.

   (a) In how many ways can she invest her money?

   (b) In how many ways can she invest her money if each fund must get at least $1000?

5. An integer is selected at random from $\{1, 2, 3, \ldots, 1000\}$. What is the probability that the integer is

   (a) divisible by 7?   (b) divisible by 11?

   (c) not divisible by 7 or 11?

   (d) divisible by 7 or 11 but *not* both?

6. Consider finite sets $\{A_1, A_2, \ldots, A_n\}$. Let $\mathscr{P}_+(n)$ be the set of nonempty subsets I of $\{1, 2, \ldots, n\}$. Show that the Inclusion–Exclusion Principle says that

$$\left| \bigcup_{i=1}^{n} A_i \right| = \sum_{I \in \mathscr{P}_+(n)} (-1)^{|I|+1} \cdot \left| \bigcap_{i \in I} A_i \right|.$$

7. Twelve identical letters are to be placed into four mailboxes.

   (a) In how many ways can this be done?

   (b) How many ways are possible if each mailbox must receive at least two letters?

8. How many different mixes of candy are possible if a mix consists of 10 pieces of candy and if there are 4 different kinds of candy available in unlimited quantities?

9. Use the binomial theorem to expand the following:

   (a) $(x + 2y)^4$ (b) $(x - y)^6$ (c) $(3x + 1)^4$ (d) $(x + 2)^5$

10. Prove that $\displaystyle\sum_{r=0}^{n} (-1)^r \binom{n}{r} = 0$ for $n \in \mathbb{P}$. *Hint:* Use the binomial theorem.

11. (a) Show that $\dbinom{n}{r} = \dbinom{n}{n-r}$ for $0 \le r \le n$.

    (b) Give a set-theoretic interpretation of the identities in part (a).

12. (a) Prove the binomial theorem.

    (b) Prove $\dbinom{n+1}{r} = \dbinom{n}{r-1} + \dbinom{n}{r}$ for $1 \le r \le n$ algebraically.

13. Prove that $2^n = \displaystyle\sum_{r=0}^{n} \binom{n}{r}$

    (a) by setting $a = b = 1$ in the binomial theorem.

    (b) by counting subsets of an n-element set.

    (c) by induction using the recurrence relation in Exercise 12(b).

14. Prove that $\displaystyle\sum_{r=0}^{n} \binom{n}{r} 2^r = 3^n$ for $n \in \mathbb{P}$.

15. (a) Verify that $\displaystyle\sum_{k=m}^{n} \binom{k}{m} = \binom{n+1}{m+1}$ for some small values of m and n, such as $m = 3$ and $n = 5$.

(b) Prove the identity by induction on n for $n \geq m$.

(c) Prove the identity by counting the $(m+1)$-element subsets of $\{1, 2, \ldots, n+1\}$. *Hint:* How many of these sets are there whose largest element is $k + 1$? What can k be?

16. In the proof of the "Placing Objects in Boxes" principle, we set up a one-to-one correspondence between strings of five 0's and three 1's and placements of five 0's in four boxes.

(a) Give the placements that correspond to the following strings:

$$1\ 0\ 1\ 0\ 1\ 0\ 0\ 0, \quad 0\ 1\ 0\ 0\ 1\ 0\ 0\ 1,$$

$$1\ 0\ 0\ 0\ 0\ 0\ 1\ 1, \quad 1\ 1\ 1\ 0\ 0\ 0\ 0\ 0.$$

(b) Give the strings that correspond to the following placements:

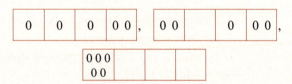

17. (a) How many ways are there to put 14 objects in 3 boxes with at least 8 objects in one box?

(b) How many ways are there to put 14 objects in 3 boxes with no more than 7 objects in any box?

(c) How many numbers between 0 and 999 have the sum of their digits equal to 20? *Hint:* Each digit must be at least 2; part (b) applies.

§ 5.4   Counting and Partitions

Recall that a partition of a set S is a collection of disjoint subsets whose union is the set S itself. This section will focus on problems that are associated with partitions.

Our first elementary observation concerns special partitions in which all the subsets are the same size. It is a consequence of Product Rule (b) or the Union Rule, and it is really just a formalization of common sense. To count the number of sheep in a flock, simply count the legs and divide by 4. As another example, if a box contains 30 marbles of different colors and if there are 6 marbles of each color, then there must be 5 colors of marbles. This observation is the special case of the next lemma in which

A is the set of marbles, B is the set of colors, ψ is the function that maps each marble to its color, and $r = 6$.

Counting
Lemma

If $\psi: A \rightarrow B$ maps the finite set A onto B and if all the sets
$$\psi^{\leftarrow}(b) = \{a \in A : \psi(a) = b\}$$
for $b \in B$ have the same number of elements, say r, then
$$|B| = \frac{|A|}{r}.$$

Proof. The set A is the union of the disjoint sets $\psi^{\leftarrow}(b)$, so the Union Rule gives
$$|A| = \sum_{b \in B} |\psi^{\leftarrow}(b)| = \sum_{b \in B} r = r \cdot |B|.$$

Therefore, $|B| = |A|/r$.

Alternatively, think of choosing one of the $|A|$ members of A in two stages, by first choosing $b \in B$ in one of $|B|$ ways and then choosing one of the r a's for which $\psi(a) = b$. Again, $|A| = |B| \cdot r$. ■

EXAMPLE 1   This problem begins with a hat containing slips of paper on which letters are written. The slips are drawn from the hat one at a time and placed in a row as they are drawn, and the question is: How many different words can arise in this way? Here we regard two words as different if there is at least one position in which they have different letters.

If all the letters in the hat are different, the problem is easy. For example, if the hat contains the ten letters I, M, P, O, R, T, A, N, C, E, then there are 10! different words that can be drawn, most of which are gibberish and only one of which is IMPORTANCE.

If the hat contains the eight letters E, E, N, N, N, O, S, S, however, then there are 8! ways to pull them from the hat, but more than one way can lead to the same word. To analyze this, let's put subscripts on the duplicate letters, so that the hat contains $E_1, E_2, N_1, N_2, N_3, O, S_1$ and S_2. There are 8! permutations of these subscripted letters. If ψ is the function that maps each such permutation to itself but with the subscripts erased, then the image of ψ is the set of distinguishable words using E, E, N, N, N, O, S, S. For example,
$$\psi(N_3 \, O \, N_1 \, S_2 \, E_1 \, N_2 \, S_1 \, E_2) = \text{NONSENSE}.$$

How many of the permutations map onto NONSENSE? There are 3! different orders in which N_1, N_2, N_3 can appear, 2! different orders for S_1 and S_2, and 2! different orders for E_1 and E_2. So there are $3! \cdot 2! \cdot 2! = 24$ different permutations that give NONSENSE. That is, the inverse image

$\psi^{\leftarrow}$(NONSENSE) has 24 elements. Similarly, there are 24 permutations that give SNEENONS, or any other word in these letters. The Counting Lemma says that the total number of distinguishable words is

$$\frac{8!}{3! \cdot 2! \cdot 2!} = 1680.$$

Thus if we draw slips from the hat at random the probability of getting NONSENSE is 1/1680 [and hence the probability of getting gibberish is 1679/1680]. ∎

The argument in this example can be used to prove the following general fact.

A Counting
Principle

Suppose that a set of n objects is partitioned into subsets of k different types, with $n_1, n_2, \ldots, n_k$ members, so that $n = n_1 + n_2 + \cdots + n_k$. Regard two permutations of the set as distinguishable in case their entries in at least one position are of different types. Then there are

$$\frac{n!}{n_1! \, n_2! \cdots n_k!}$$

distinguishable permutations of the set.

EXAMPLE 2   Let $\Sigma = \{a, b, c\}$. The number of words in Σ^* having length 10 using 4 a's, 3 b's and 3 c's is

$$\frac{10!}{4! \, 3! \, 3!} = 4200.$$

The number of words using 5 a's, 3 b's and 2 c's is

$$\frac{10!}{5! \, 3! \, 2!} = 2520$$

and the number using 5 a's and 5 b's is

$$\frac{10!}{5! \, 5!} = 252.$$

For comparison, note that Σ^{10} has $3^{10} = 59{,}049$ words. ∎

We have just been counting permutations distinguished by a partition, but now we will start counting the partitions themselves. An **ordered partition** of a set S is a sequence $(A_1, A_2, \ldots, A_k)$ whose members A_1,

$A_2, \ldots, A_k$ form a partition of S. The A_i's themselves are not assumed to be ordered internally, but the order in which they appear in the list does matter.

EXAMPLE 3 Let $S = \{1, 2, 3, 4, 5, 6, 7, 8\}$.

(a) Here are some ordered partitions of S:

$(\{1, 3, 5\}, \{2, 4, 6, 7, 8\})$, $(\{2, 4, 6, 7, 8\}, \{1, 3, 5\})$,

$(\{3, 6\}, \{2, 5, 8\}, \{1, 4, 7\})$, $(\{1\}, \{2, 4, 6, 8\}, \{3, 5, 7\})$,

$(\{1, 6\}, \{2, 5, 8\}, \{3, 4\}, \{7\})$ and $(\{6, 1\}, \{2, 5, 8\}, \{4, 3\}, \{7\})$.

The last two are the same, since $\{1, 6\} = \{6, 1\}$ and $\{3, 4\} = \{4, 3\}$, but all of the others are distinct.

(b) We count the number of ordered partitions of S of the form (A, B, C, D) where $|A| = 2$, $|B| = 3$, $|C| = 1$ and $|D| = 2$. This problem can be viewed as a question of counting distinguishable permutations of 8 objects of 4 types, that is, as a letters-in-a-hat problem. Here's how.

Consider an ordered partition (A, B, C, D) of $\{1, 2, \ldots, 8\}$ into sets with 2, 3, 1 and 2 members, respectively. For instance, take $(\{5, 8\}, \{1, 3, 6\}, \{2\}, \{4, 7\})$. Then 5 and 8 are of type A, while 1, 3 and 6 are of type B, etc. If we list the types of $1, 2, 3, \ldots, 8$ in order, we get the sequence B, C, B, D, A, B, D, A which completely describes the partition as long as we know how to decode it. If instead of calling the blocks A, B, C, D we label them E, N, O and S, then $(\{5, 8\}, \{1, 3, 6\}, \{2\}, \{4, 7\})$ corresponds to our old friend N, O, N, S, E, N, S, E.

This argument shows that the ordered partitions of $\{1, 2, \ldots, 8\}$ into blocks of sizes 2, 3, 1 and 2 correspond one-to-one with the distinguishable permutations of A, A, B, B, B, C, D, D, so the number of ordered partitions is $\dfrac{8!}{2! \cdot 3! \cdot 1! \cdot 2!} = 1680$. ∎

The argument in Example 3(b) easily generalizes to establish the following.

Counting
Ordered
Partitions

If a set has n elements and if $n_1 + n_2 + \cdots + n_k = n$, then there are

$$\frac{n!}{n_1! \, n_2! \cdots n_k!}$$

ordered partitions $(A_1, A_2, \ldots, A_k)$ of the set with $|A_i| = n_i$ for $i = 1, 2, \ldots, k$.

The number given by the formula can also be written as

$$\binom{n}{n_1} \cdot \binom{n-n_1}{n_2} \cdots \binom{n-n_1-\cdots-n_{k-1}}{n_k},$$

a form that is natural if we think of choosing the n_1 members to go into A_1 from the full set of n elements, then the n_2 members for A_2 from the remaining $n - n_1$, and so on, choosing the members for each A_i from $(A_1 \cup \cdots \cup A_{i-1})^c$. The last factor is $\binom{n_k}{n_k}$, so it can be omitted. One possible advantage of the factored form, when n is very large and $n!$ is enormous, is that the binomial coefficient factors are a lot smaller and can perhaps be calculated readily.

EXAMPLE 4   In how many ways can three disjoint committees be formed from twenty people if they must have 3, 5 and 7 people, respectively? This problem is equivalent to counting ordered partitions (A, B, C, D) of the set of twenty people where $|A| = 3$, $|B| = 5$, $|C| = 7$ and $|D| = 5$. The set D corresponds to the people with no committee assignment. There are

$$\frac{20!}{3!\,5!\,7!\,5!} = \binom{20}{3} \cdot \binom{17}{5} \cdot \binom{12}{7} \cdot \binom{5}{5} = 1140 \cdot 6188 \cdot 792 \cdot 1 \approx 5.587 \cdot 10^9$$

possible ways to form such committees. Note that although $|D| = |B|$, the committee B and the set D play different roles; the ordering of the partition is significant and we are not just interested in ways of breaking the big set up into a 3-element subset, a couple of 5-element subsets and a 7-element subset. ■

EXAMPLE 5   (a) A **bridge deal** is an ordered partition of 52 cards involving four sets with 13 cards each. Thus there are

$$\frac{52!}{13!\,13!\,13!\,13!} = \frac{52!}{(13!)^4} \approx 5.3645 \cdot 10^{28}$$

bridge deals.

(b) Given a bridge deal, we find the probability that each hand of 13 cards contains one ace. First we deal out the aces; this can be done in $4! = 24$ ways. Just as in part (a), the remaining cards can be partitioned in $48!/(12!)^4$ ways. So $24 \cdot 48!/(12!)^4$ of the bridge deals yield one ace in each hand. The probability of such a deal is

$$24 \frac{48!}{(12!)^4} \cdot \frac{(13!)^4}{52!} = \frac{24 \cdot 13^4}{49 \cdot 50 \cdot 51 \cdot 52} \approx .1055.$$

(c) A single **bridge hand** consists of 13 cards drawn from a 52-card deck. There are $\binom{52}{13} \approx 6.394 \cdot 10^{11}$ bridge hands. We say that a bridge hand has distribution $n_1 - n_2 - n_3 - n_4$ where $n_1 \geq n_2 \geq n_3 \geq n_4$ and $n_1 + n_2 + n_3 + n_4 = 13$ if there are n_1 cards of some suit, n_2 cards of a second suit, n_3 cards of a third suit and n_4 cards of the remaining suit. To illustrate, we count the number of bridge hands having 4–3–3–3 distribution. There are

$$\binom{13}{4}\binom{13}{3}\binom{13}{3}\binom{13}{3}$$

ways to choose 4 clubs and three of each of the other suits. We get the same result if we replace clubs by one of the other suits. So we conclude that there are

$$4\binom{13}{4}\binom{13}{3}^3 \approx 6.6906 \cdot 10^{10}$$

bridge hands having 4–3–3–3 distribution. The probability that a bridge hand will have 4–3–3–3 distribution is

$$\frac{6.6906}{63.94} \approx .1046. \quad \blacksquare$$

Sometimes problems reduce to counting unordered partitions. In such cases, count ordered partitions first and then divide by suitable numbers to take into account the lack of order.

EXAMPLE 6   (a) In how many ways can twelve students be divided into three groups, with four students in each group, so that one group studies topic T_1, one studies topic T_2 and one studies topic T_3? Here order matters: if we permuted groups of students, the students would be studying different topics. So we count ordered partitions, of which there are

$$\frac{12!}{4!\,4!\,4!} = \binom{12}{4}\cdot\binom{8}{4} = 495 \cdot 70 = 34{,}650.$$

(b) In how many ways can twelve students be divided into three study groups, with four students in each group, but with each group studying the same topic? Now we wish to count unordered partitions, since we regard partitions like (A, B, C) and (B, A, C) as equivalent. They correspond to the same partition of the twelve students into three equal groups. From part (a), there are 34,650 ordered partitions. If we map each ordered partition (A, B, C) to the unordered partition $\psi((A, B, C)) = \{A, B, C\}$, we find that $\psi^{\leftarrow}(\{A, B, C\})$ has $3! = 6$ elements, namely $(A, B, C), (A, C, B), (B, A, C), (B, C, A), (C, A, B)$ and (C, B, A). So by the

Counting Lemma there are $34,650/6 = 5775$ unordered partitions of the desired type. Hence the answer to our question is 5775. ∎

EXAMPLE 7   (a) In how many ways can nineteen students be divided into five groups, two groups of five and three groups of three, so that each group studies a different topic? As in Example 6(a), we count ordered partitions, of which there are

$$\frac{19!}{5!\,5!\,3!\,3!\,3!} \approx 3.911 \cdot 10^{10}.$$

(b) In how many ways can the students in part (a) be divided if all five groups are to study the same topic? In part (a) we counted all ordered partitions (A, B, C, D, E) where $|A| = |B| = 5$ and $|C| = |D| = |E| = 3$. If A and B are permuted and C, D, E are permuted, we will get the same study groups, but we cannot permute groups of different sizes like A and D. To count unordered partitions we let $\psi((A, B, C, D, E)) = (\{A, B\}, \{C, D, E\})$. Each inverse image $\psi^{\leftarrow}(\{A, B\}, \{C, D, E\})$ has $2! \cdot 3!$ elements [such as (B, A, C, E, D)] and so, by the Counting Lemma, there are

$$\frac{19!}{5!\,5!\,3!\,3!\,3!} \cdot \frac{1}{2!\,3!} \approx 3.26 \cdot 10^9$$

unordered partitions of students into study groups. ∎

EXAMPLE 8   In how many ways can we split a group of 12 contestants into 4 teams of 3 contestants each? More generally, what if we want to divide $3n$ contestants among n teams of 3? We are asking for unordered partitions. There are

$$\frac{(3n)!}{(3!) \cdots (3!)} = \frac{(3n)!}{6^n}$$

ordered partitions $(A_1, A_2, \ldots, A_n)$ for which each A_i has 3 members. Any permutation of the n sets gives the same unordered partition, so there are $\dfrac{(3n)!}{6^n \cdot n!}$ unordered partitions of $3n$ elements into n sets with 3 elements each. Note that this number is

$$\frac{3n(3n-1)(3n-2)}{6n} \cdot \frac{(3n-3)(3n-4)(3n-5)}{6(n-1)} \cdots \frac{6 \cdot 5 \cdot 4}{6 \cdot 2} \cdot \frac{3 \cdot 2 \cdot 1}{6 \cdot 1}$$

$$= \frac{(3n-1)(3n-2)}{2} \cdot \frac{(3n-4)(3n-5)}{2} \cdots \frac{5 \cdot 4}{2} \cdot \frac{2 \cdot 1}{2}$$

$$= \binom{3n-1}{2} \cdot \binom{3n-4}{2} \cdots \binom{5}{2} \cdot \binom{2}{2}.$$

This factorization suggests another way to solve the problem. List the contestants in some order. Pick the first contestant and choose 2 more for her team, in one of $\binom{3n-1}{2}$ ways. Now pick the next unchosen contestant and choose 2 more to go with him, in one of $\binom{3n-4}{2}$ ways. And so on. [Or use induction.]

In case there are just 12 contestants, there are

$$\binom{11}{2}\cdot\binom{8}{2}\cdot\binom{5}{2}\cdot\binom{2}{2} = 15,400$$

ways to choose the teams. ■

The principles in this section and in §5.3 may appear to be a bag of tricks needed to work the problems we set for ourselves. To some extent the subject of counting is like that, although we prefer to think of our techniques as tools rather than tricks. They *are* applicable in a variety of common situations, but won't handle every problem that comes up. The thought processes that we demonstrated in the proofs of the principles are as valuable as the principles themselves, since the same sort of analysis can often be used on problems where the ready-made tools don't apply.

EXERCISES 5.4

1. From a total of 15 people, 3 committees consisting of 3, 4 and 5 people, respectively, are to be chosen.

   (a) How many such sets of committees are possible if no person may serve on more than one committee?

   (b) How many such sets of committees are possible if there is no restriction on the number of committees on which a person may serve?

2. Compare.

   (a) $\binom{7}{2}\cdot\binom{5}{2}$ and $\dfrac{7!}{2!\cdot 2!\cdot 3!}$.

   (b) $\binom{12}{3}\cdot\binom{9}{4}$ and $\dfrac{12!}{3!\cdot 4!\cdot 5!}$.

   (c) $\binom{n}{k}\cdot\binom{n-k}{r}$ and $\dfrac{n!}{k!\cdot r!\cdot(n-k-r)!}$.

3. Three disjoint study groups are to be selected from a class of thirteen students, with each group studying the same topic. In how many ways can this be done if the groups are to have

   (a) 5, 3 and 2 students?   (b) 4, 3 and 3 students?   (c) 3 students each?

4. How many different signals can be created by lining up nine flags in a vertical column if 3 of them are white, 2 are red and 4 are blue?

5. Let S be the set of all sequences of 0's, 1's and 2's of length ten. For example, S contains 0 2 1 1 0 1 2 2 0 1.

   (a) How many elements are in S?

   (b) How many sequences in S have exactly five 0's and five 1's?

   (c) How many sequences in S have exactly three 0's and seven 1's?

   (d) How many sequences in S have exactly three 0's?

   (e) How many sequences in S have exactly three 0's, four 1's and three 2's?

   (f) How many sequences in S have at least one 0, at least one 1 and at least one 2?

6. Find the number of permutations that can be formed from all the letters of the following words.

   (a) FLORIDA   (b) CALIFORNIA

   (c) MISSISSIPPI   (d) OHIO

7. (a) How many 4-digit numbers can be formed using only the digits 3, 4, 5, 6 and 7?

   (b) How many of the numbers in part (a) have some digit repeated?

   (c) How many of the numbers in part (a) are even?

   (d) How many of the numbers in part (b) are bigger that 5000?

8. Let $\Sigma = \{a, b, c\}$. If $n_1 \geq n_2 \geq n_3$ and $n_1 + n_2 + n_3 = 6$, we call a word in Σ^6 of type $n_1-n_2-n_3$ if one of the letters appears in the word n_1 times, another letter appears n_2 times and the other letter appears n_3 times. For example, $accabc$ is of type 3–2–1 and $caccca$ is of type 4–2–0. The number of words in Σ^6 of each type is

Type	6–0–0	5–1–0	4–2–0	4–1–1	3–3–0	3–2–1	2–2–2
Number	3	36	90	90	60	360	90

   Verify this assertion for three of the types.

9. In how many ways can $2n$ elements be partitioned into two sets with n elements each?

10. The English alphabet consists of 21 consonants and 5 vowels. The vowels are a, e, i, o, u.

   (a) Prove that no matter how the letters of the English alphabet are listed in order [e.g., z u v a r q l g h $\cdots$] there must be 4 consecutive consonants.

   (b) Give a list to show that there need not be 5 consecutive consonants.

(c) Suppose now that the letters of the English alphabet are put in a circular array; for example,

Prove that there must be 5 consecutive consonants in such an array.

11. Two of the 12 contestants, Ann and Bob, want to be on the same team in Example 8.

   (a) How many ways are there to choose the teams so that Ann and Bob are together?

   (b) If all ways of choosing teams are equally likely, what is the probability that Ann and Bob are chosen for the same team?

12. A basketball tournament has 16 teams. How many ways are there to match up the teams in 8 pairs?

13. (a) For how many integers between 1000 and 9999 is the sum of the digits exactly 9? Examples: 1431, 5121, 9000, 4320.

   (b) How many of the integers counted in part (a) have all nonzero digits?

14. Consider finite sets satisfying $\chi_A + \chi_B = \chi_C + \chi_D$. Show that

$$|A| + |B| = |C| + |D|.$$

15. How many equivalence relations are there on $\{0, 1, 2, 3\}$? *Hint:* Count unordered partitions. Why does this solve the problem?

16. A hat contains the letters B, B, B, E, O, O, P, P. What is the probability that when they are drawn out of the hat in sequence the letters will spell B E B O P B O P?

§ 5.5   **Pigeon-Hole Principle**

The usual Pigeon-Hole Principle asserts that if m objects are placed in n boxes or pigeon-holes and if $m > n$, then some box will receive more than one object. Here is a slight generalization of this fact.

<table>
<tr><td>*Pigeon-Hole*
Principle</td><td>If a finite set S is partitioned into k sets, then at least one of the sets has $|S|/k$ or more elements.</td></tr>
</table>

Proof. Say the sets in the partition are $A_1, \ldots, A_k$. Then the average value of $|A_i|$ is $\frac{1}{k} \cdot (|A_1| + \cdots + |A_k|) = \frac{1}{k} \cdot |S|$, so the largest A_i has at least this many elements. ∎

We will often apply this principle when the partition is given by a function. The principle can then be stated as follows.

<table>
<tr><td>*Pigeon-Hole*
Principle</td><td>Consider a function $f: S \to T$ where S and T are finite sets satisfying $|S| > r \cdot |T|$. Then at least one of the sets $f^{\leftarrow}(t)$ has more than r elements.</td></tr>
</table>

Proof. The family $\{ f^{\leftarrow}(t) : t \in T \}$ partitions S into k sets with $k \leq |T|$. By the principle just proved, some set $f^{\leftarrow}(t)$ has at least $|S|/k$ members. Since $|S|/k \geq |S|/|T| > r$ by hypothesis, such a set $f^{\leftarrow}(t)$ has more than r elements. ∎

When $r = 1$, this principle tells us that if $f: S \to T$ and $|S| > |T|$, then at least one of the sets $f^{\leftarrow}(t)$ has more than one element. It is remarkable how often this simple observation is helpful in problem solving.

EXAMPLE 1 Given three integers, there must be two of them whose sum is even, because either two of the integers are even or else two of them are odd, and in either case their sum must be even. Here is a tighter argument. Let S be the set of three integers. Then MOD 2: $S \to \{0, 1\}$ and by the Pigeon-Hole Principle one of the sets $(\text{MOD } 2)^{\leftarrow}(0)$ or $(\text{MOD } 2)^{\leftarrow}(1)$ has more than one element. That is, either S contains two [or more] even integers or else S contains two [or more] odd integers. ∎

EXAMPLE 2 We show that if $a_1, a_2, \ldots, a_p$ are integers, not necessarily distinct, then some of them add up to a number that is a multiple of p. Consider the function MOD $p: \mathbb{Z} \to \mathbb{Z}(p)$ applied to the set

$$S = \{0, a_1, a_1 + a_2, a_1 + a_2 + a_3, \ldots, a_1 + a_2 + a_3 + \cdots + a_p\}.$$

Since $|S| = p + 1 > p = |\mathbb{Z}(p)|$, the Pigeon-Hole Principle shows that two distinct numbers m and n in S have the same image in $\mathbb{Z}(p)$. Since m MOD $p = n$ MOD p, we have $m \equiv n \pmod{p}$ and so $n - m$ and $m - n$ are multiples of p. One of these differences has the form $a_k + a_{k+1} + \cdots + a_l$, i.e., it's a sum of integers from our list that is a multiple of p.

Note that we have proved more than we claimed; some *consecutive* batch of a_i's adds up to a multiple of p. The result just proved is sharp, in the sense that we can give integers $a_1, a_2, \ldots, a_{p-1}$ for which no nonempty subset has sum that is a multiple of p. Simply let $a_j = 1$ for $j = 1, 2, \ldots, p-1$. ∎

EXAMPLE 3   Let A be some fixed 10-element subset of $\{1, 2, 3, \ldots, 50\}$. We show that A possesses two different 5-element subsets, the sums of whose elements are equal. Let $\mathscr{S}$ be the family of 5-element subsets B of A. For each B in $\mathscr{S}$, let $f(B)$ be the sum of the numbers in B. Note that we must have $f(B) \geq 1 + 2 + 3 + 4 + 5 = 15$ and $f(B) \leq 50 + 49 + 48 + 47 + 46 = 240$, so that $f: \mathscr{S} \to T$ where $T = \{15, 16, 17, \ldots, 240\}$. Since $|T| = 226$ and $|\mathscr{S}| = \binom{10}{5} = 252$, the Pigeon-Hole Principle shows that $\mathscr{S}$ contains different sets with the same image under f, i.e., different sets the sums of whose elements are equal. ∎

Some applications of the Pigeon-Hole Principle require considerable ingenuity.

EXAMPLE 4   Here we show that if $a_1, a_2, \ldots, a_{n^2+1}$ is a sequence of $n^2 + 1$ distinct real numbers, then there is a subsequence with $n + 1$ terms that is either increasing or decreasing. This means that there exist subscripts $s(1) < s(2) < \cdots < s(n+1)$, so that either

$$a_{s(1)} < a_{s(2)} < \cdots < a_{s(n+1)}$$

or

$$a_{s(1)} > a_{s(2)} > \cdots > a_{s(n+1)}.$$

For each j in $\{1, 2, \ldots, n^2 + 1\}$, let $\text{INC}(j)$ be the length of the longest increasing subsequence stopping at a_j and $\text{DEC}(j)$ be the length of the longest decreasing subsequence stopping at a_j. Then define $f(j) = (\text{INC}(j), \text{DEC}(j))$. For example, suppose that $n = 3$ and the original sequence is given by

a_1	a_2	a_3	a_4	a_5	a_6	a_7	a_8	a_9	a_{10}
11	3	15	8	6	12	17	2	7	1.

Here $a_5 = 6$, $\text{INC}(5) = 2$ since a_2, a_5 is the longest increasing subsequence stopping at a_5, and $\text{DEC}(5) = 3$ since a_1, a_4, a_5 and a_3, a_4, a_5 [i.e., 11, 8, 6 and 15, 8, 6] are longest decreasing subsequences stopping at a_5. Similarly, $\text{INC}(6) = 3$ and $\text{DEC}(6) = 2$, so $f(5) = (2, 3)$ and $f(6) = (3, 2)$. Indeed, in this example

$$f(1) = (1, 1) \qquad f(2) = (1, 2) \qquad f(3) = (2, 1)$$
$$f(4) = (2, 2) \qquad f(5) = (2, 3) \qquad f(6) = (3, 2)$$
$$f(7) = (4, 1) \qquad f(8) = (1, 4) \qquad f(9) = (3, 3)$$
$$f(10) = (1, 5).$$

This particular example has increasing subsequences of length 4 such as a_2, a_5, a_6, a_7 [note INC(7) = 4], and also decreasing subsequences of length 4 such as a_1, a_4, a_5, a_8 [note DEC(8) = 4]. Since DEC(10) = 5, it even has a decreasing subsequence of length 5. Note that f is one-to-one in this example, so that f cannot map the 10-element set $\{1, 2, 3, \ldots, 10\}$ into the 9-element set $\{1, 2, 3\} \times \{1, 2, 3\}$. In other words, the one-to-oneness of f alone forces at least one INC(j) or DEC(j) to exceed 3, and this in turn forces our sequence to have an increasing or decreasing subsequence of length 4.

To prove the general result we first prove directly that f must *always* be one-to-one. Consider j, k in $\{1, 2, 3, \ldots, n^2 + 1\}$ with $j < k$. If $a_j < a_k$, then INC(j) < INC(k) since a_k could be attached to the longest increasing sequence ending at a_j to get a longer increasing sequence ending at a_k. Similarly, if $a_j > a_k$, then DEC(j) < DEC(k). In either case the *ordered pairs* $f(j)$ and $f(k)$ cannot be equal, i.e., $f(j) \neq f(k)$. Since f is one-to-one, the Pigeon-Hole Principle says that f cannot map

$$\{1, 2, 3, \ldots, n^2 + 1\} \quad \text{into} \quad \{1, 2, \ldots, n\} \times \{1, 2, \ldots, n\},$$

so there is either a j such that INC(j) $\geq n + 1$ or one such that DEC(j) $\geq n + 1$. Hence the original sequence has an increasing or decreasing subsequence with $n + 1$ terms. ∎

EXAMPLE 5   You probably know that the decimal expansions of rational numbers repeat themselves, but you may never have seen a proof. For example,

$$\frac{29}{54} = .5370370370370370370370370\cdots;$$

check this *by long division* before proceeding further! The general fact can be seen as a consequence of the Pigeon-Hole Principle, as follows.

We may assume that the given rational number has the form m/n where $0 < m < n$. Let us analyze the steps in the algorithm for long division. When we divide m by n we obtain $.d_1 d_2 d_3 \cdots$ where

$$10 \cdot m = n \cdot d_1 + r_1 \qquad 0 \leq r_1 < n$$
$$10 \cdot r_1 = n \cdot d_2 + r_2 \qquad 0 \leq r_2 < n$$
$$10 \cdot r_2 = n \cdot d_3 + r_3 \qquad 0 \leq r_3 < n$$

etc. so that $10 \cdot r_j = n \cdot d_{j+1} + r_{j+1}$ where $0 \le r_{j+1} < n$. That is,

$$r_{j+1} = (10 \cdot r_j) \,\mathrm{MOD}\, n \quad \text{and} \quad d_{j+1} = (10 \cdot r_j) \,\mathrm{DIV}\, n.$$

Figure 1 shows the details for our sample division. The remainders r_j all take their values in $\{0, 1, 2, \ldots, n-1\} = \mathbb{Z}(n)$. By the Pigeon-Hole Principle, after a while the values must repeat. In fact, two of the numbers $r_1, r_2, \ldots, r_{n+1}$ must be equal. Hence there are k and l in $\{1, 2, \ldots, n+1\}$ with $k < l$ and $r_k = r_l$. Let $p = l - k$, so that $r_k = r_{k+p}$. [In our carefully selected example, k can be 1 and p can be 3.] We will show that the sequences of r_i's and d_i's repeat every p terms beginning with $i = k + 1$.

$. d_1 d_2 d_3 d_4 d_5 \,\cdots$	$= .53703 \,\cdots$	
$54 \overline{\smash{)}290}$	$10 \cdot 29 = 10m$	
270	$54 \cdot 5 = nd_1$	
200	$10 \cdot 20 = 10r_1$	$r_1 = 20$
162	$54 \cdot 3 = nd_2$	
380	$10 \cdot 38 = 10r_2$	$r_2 = 38$
378	$54 \cdot 7 = nd_3$	
20	$10 \cdot 2 = 10r_3$	$r_3 = 2$
0	$54 \cdot 0 = nd_4$	
200	$10 \cdot 20 = 10r_4$	$r_4 = 20$
162	$54 \cdot 3 = nd_5$	
380	$10 \cdot 38 = 10r_5$	$r_5 = 38$
378	etc.	

FIGURE 1

First, we show by induction that the remainders repeat:

$(*)$ $\qquad\qquad\qquad r_j = r_{j+p} \qquad$ for $\quad j \ge k.$

We have $r_k = r_{k+p}$ by the choice of k and p. Assume inductively that $r_j = r_{j+p}$ for some j. Then $r_{j+1} = (10 \cdot r_j) \,\mathrm{MOD}\, n = (10 \cdot r_{j+p}) \,\mathrm{MOD}\, n = r_{j+p+1}$. Thus $(*)$ holds for $j \ge k$ by induction on j.

Now for $j > k$ $(*)$ implies that

$$d_j = (10 \cdot r_{j-1}) \,\mathrm{DIV}\, n = (10 \cdot r_{j+p-1}) \,\mathrm{DIV}\, n = d_{j+p}.$$

Thus the d_j's also repeat in cycles of length p and we have

$$d_{k+1} = d_{k+p+1}, \quad d_{k+2} = d_{k+p+2}, \quad \ldots, \quad d_{k+p} = d_{k+2p}$$

so that

$$d_{k+1} d_{k+2} \cdots d_{k+p} = d_{k+p+1} d_{k+p+2} \cdots d_{k+2p}.$$

In fact, this whole block repeats indefinitely. In other words, the decimal expansion of m/n is a repeating expansion.   ■

The next example doesn't exactly apply the Pigeon-Hole Principle but it is a pigeon-hole problem in spirit.

EXAMPLE 6   Consider nine nonnegative real numbers $a_1, a_2, a_3, \ldots, a_9$ with sum 90.

(a) We show that there must be three of the numbers having sum at least 30. This is easy because

$$90 = (a_1 + a_2 + a_3) + (a_4 + a_5 + a_6) + (a_7 + a_8 + a_9),$$

so at least one of the sums in parentheses must be at least 30.

(b) We show that there must be four of the numbers having sum at least 40. There are several ways to do this, but none of them is quite as simple as the method in part (a). Our first approach is to note that the sum of all the numbers in Figure 2 is 360 since each row sums to 90. Hence one of the nine columns must have sum at least $360/9 = 40$.

a_1	a_2	a_3	a_4	a_5	a_6	a_7	a_8	a_9
a_2	a_3	a_4	a_5	a_6	a_7	a_8	a_9	a_1
a_3	a_4	a_5	a_6	a_7	a_8	a_9	a_1	a_2
a_4	a_5	a_6	a_7	a_8	a_9	a_1	a_2	a_3

FIGURE 2

Our second approach is to use part (a) to select three of the numbers having sum $s \geq 30$. One of the remaining six numbers must have value at least $\frac{1}{6}$ of their sum $90 - s$. Adding this to the selected three gives four numbers with sum at least

$$s + \frac{1}{6}(90 - s) = 15 + \frac{5}{6}s \geq 15 + \frac{5}{6} \cdot 30 = 40.$$

Our third approach is to note that we may as well assume that $a_1 \geq a_2 \geq \cdots \geq a_9$. Then it is clear that $a_1 + a_2 + a_3 + a_4$ is the largest sum using four of the numbers and our task is relatively concrete: to show that $a_1 + a_2 + a_3 + a_4 \geq 40$. Moreover, this suggests showing

(∗) $$a_1 + a_2 + \cdots + a_n \geq 10n$$

for $1 \leq n \leq 9$. We can do this by a finite induction, i.e., by noting that (∗) holds for $n = 1$ and showing that if (∗) holds for n with $1 \leq n < 9$, then (∗) holds for $n + 1$. We will adapt the method of our second approach. Assume that (∗) holds for n, and let $s = a_1 + a_2 + \cdots + a_n$. Since a_{n+1} is the largest of the remaining $9 - n$ numbers, we have

$a_{n+1} \geq (90 - s)/(9 - n)$. Hence

$$a_1 + a_2 + \cdots + a_n + a_{n+1} = s + a_{n+1} \geq s + \frac{90 - s}{9 - n}$$

$$= s + \frac{90}{9 - n} - \frac{s}{9 - n} = s\left(1 - \frac{1}{9 - n}\right) + \frac{90}{9 - n}$$

$$\geq 10n\left(1 - \frac{1}{9 - n}\right) + \frac{90}{9 - n} \qquad \text{[inductive assumption]}$$

$$= 10n + \frac{90 - 10n}{9 - n} = 10n + 10 = 10(n + 1).$$

This finite induction argument shows that (∗) holds for $1 \leq n \leq 9$.

For this particular problem, the first and last approaches are far superior because they generalize in an obvious way without further tricks. ∎

The Pigeon-Hole Principle can be generalized to allow the sets A_i to overlap.

Generalized
Pigeon-Hole
Principle

> Let $A_1, \ldots, A_k$ be subsets of the finite set S such that each element of S is in at least t of the sets A_i. Then the average number of elements in the A_i's is at least $t \cdot |S|/k$.

Proof. We use the interesting and powerful technique of counting a set of pairs in two ways. Let P be the set of all pairs (s, A_i) with $s \in A_i$. We can count P by counting the pairs for each s in S and then adding up the numbers. We get

$$|P| = \sum_{s \in S} [\text{number of pairs } (s, A_i) \text{ with } s \in A_i]$$

$$= \sum_{s \in S} [\text{number of } A_i\text{'s with } s \in A_i]$$

$$\geq \sum_{s \in S} t \qquad \text{[by assumption]}$$

$$= t \cdot |S|.$$

We can also count P by counting the pairs for each A_i and adding the numbers. Thus

$$|P| = \sum_{i=1}^{k} [\text{number of pairs } (s, A_i) \text{ with } s \in A_i]$$

$$= \sum_{i=1}^{k} [\text{number of } s\text{'s with } s \in A_i] = \sum_{i=1}^{k} |A_i|.$$

Putting these two results together, we get

$$\sum_{i=1}^{k} |A_i| \geq t \cdot |S|,$$

so the average of the $|A_i|$'s, namely

$$\frac{1}{k} \cdot \sum_{i=1}^{k} |A_i|,$$

is at least $t \cdot |S|/k$.

Our proof also shows that if each s is in exactly t of the sets A_i, then the average of the $|A_i|$'s is exactly $t \cdot |S|/k$. The ordinary Pigeon-Hole Principle is the special case $t = 1$. ■

EXAMPLE 7   A roulette wheel is divided into 36 sectors with numbers $1, 2, 3, \ldots, 36$ in some unspecified order. [We are omitting sections with 0 and 00 that are included in Las Vegas and give the house the edge in gambling.] The average value of a sector is 18.5. If this isn't obvious, observe that

$$1 + 2 + \cdots + 36 = 666$$

by Example 2(b) of §4.2 and that $666/36 = 18.5$.

(a) There are 36 pairs of consecutive sectors. We show that the average of the sums of the numbers on these pairs of consecutive sectors is 37. Imagine that each sector is replaced by a bag containing as many marbles as the number on the sector. For each pair of consecutive sectors, consider the set of marbles from the two bags. Each marble belongs to exactly two of these sets and there are 36 sets in all. Since the total number of marbles is 666, the remark at the end of the proof of the Generalized Pigeon-Hole Principle shows that the average number of marbles in these sets is $2 \cdot 666/36 = 37$. The number of marbles in each set corresponds to the sum of the numbers on the two consecutive sectors.

The fact that each marble belongs to exactly two of the sets is essential to the argument above. Suppose, for example, that we omit one pair of consecutive sectors from consideration. If the numbers on these two sectors are, say, 5 and 17, then the average of the sums of the numbers on the remaining 35 pairs of sectors would be

$$\frac{2 \cdot 666 - 22}{35} \approx 37.43.$$

(b) Some consecutive pair of sectors must have sum at least 38. To see this, group the 36 sectors into 18 disjoint consecutive pairs. The average sum of the 18 pairs is 37. Some sum exceeds 37 or else each pair has sum exactly 37. In the latter case, shift clockwise one sector, which changes each sum. Some new sum will have to be greater than 37.

(c) Part (a) easily generalizes. Let t be an integer with $2 \le t \le 36$. There are 36 blocks of t consecutive sectors and the average of the sums of the numbers on these blocks is $18.5 \cdot t$. Imagine the marbles in part (a) again. Each block corresponds to the set of marbles associated to the t sectors of the block. Since each marble belongs to exactly t of these sets, the average number of marbles in these sets is

$$\frac{t \cdot 666}{36} = 18.5 \cdot t.$$

As before, the number of marbles in each set is the sum of the numbers on the sectors in the corresponding block. ∎

EXERCISES 5.5

Most, but not all, of the following exercises involve the Pigeon-Hole Principle. They also provide more practice using the techniques from §§ 5.1–5.4. The exercises are not all equally difficult, and some may require extra ingenuity.

1. (a) Given any four integers, explain why two of them must be congruent mod 3.

   (b) Prove that if $a_1, a_2, \ldots, a_{p+1}$ are integers, then two of them must be congruent mod p.

2. (a) A sack contains 50 marbles of four different colors. Explain why there are at least 13 marbles of the same color.

   (b) If exactly 8 of the marbles are red, explain why there are at least 14 of the same color.

3. Suppose that 73 marbles are placed in eight boxes.

   (a) Show that some box contains at least 10 marbles.

   (b) Show that if two of the boxes are empty, then some box contains at least 13 marbles.

4. (a) Let B be a 12-element subset of $\{1, 2, 3, 4, 5, 6\} \times \{1, 2, 3, 4, 5, 6\}$. Show that B contains two different ordered pairs, the sums of whose entries are equal.

   (b) How many times can a pair of dice be tossed without obtaining the same sum twice?

5. Let A be a 10-element subset of $\{1, 2, 3, \ldots, 50\}$. Show that A possesses two different 4-element subsets, the sums of whose elements are equal.

6. Let S be a 3-element set of integers. Show that S has two different nonempty subsets such that the sums of the numbers in each of the subsets are congruent mod 6.

7. Let A be a subset of $\{1, 2, 3, \ldots, 149, 150\}$ consisting of 25 numbers. Show that there are two disjoint pairs of numbers from A having the same sum [for example, $\{3, 89\}$ and $\{41, 51\}$ have the same sum, namely 92].

8. For the following sequences, find an increasing or decreasing subsequence of length 5 if you can.

   (a) 4, 3, 2, 1, 8, 7, 6, 5, 12, 11, 10, 9, 16, 15, 14, 13

   (b) 17, 13, 14, 15, 16, 9, 10, 11, 12, 5, 6, 7, 8, 1, 2, 3, 4

   (c) 10, 6, 2, 14, 3, 17, 12, 8, 7, 16, 13, 11, 9, 15, 4, 1, 5

9. Find the decimal expansions for 1/7, 2/7, 3/7, 4/7, 5/7 and 6/7 and compare them.

10. (a) Show that if ten nonnegative integers have sum 101, there must be three with sum at least 31.

    (b) Prove a generalization of part (a): If $1 \leq k \leq n$ and if n nonnegative integers have sum m, there must be k with sum at least _____.

11. In this problem the twenty-four numbers 1, 2, 3, 4, $\ldots$, 24 are permuted in some way, say $(n_1, n_2, n_3, n_4, \ldots, n_{24})$.

    (a) Show that there must be four consecutive numbers in the permutation that are less than 20, i.e., at most 19.

    (b) Show that $n_1 + n_2 + n_3 + \cdots + n_{24} = 300$.

    (c) Show that there must be three consecutive numbers in the permutation with sum at least 38.

    (d) Show that there must be five consecutive numbers in the permutation with sum at least 61.

12. Consider the roulette wheel in Example 7.

    (a) Use Example 7(c) to show that there are four consecutive sectors with sum at least 74.

    (b) Show that, in fact, 74 in part (a) can be improved to 75.

    (c) Use Example 7(c) to show that there are five consecutive sectors with sum at least 93.

    (d) Show that, in fact, 93 in part (c) can be improved to 95.

13. Let n_1, n_2 and n_3 be distinct positive integers. Show that at least one of n_1, n_2, n_3, $n_1 + n_2$, $n_2 + n_3$ or $n_1 + n_2 + n_3$ is divisible by 3. *Hint:* Map $\{n_1, n_1 + n_2, n_1 + n_2 + n_3\}$ to $\mathbb{Z}(3)$ using $f = \text{MOD } 3$.

14. A club has six men and nine women members. A committee of five is selected at random. Find the probability that

    (a) there are two men and three women on the committee.

    (b) there are at least one man and at least one woman on the committee.

    (c) the committee consists of only men or else consists of only women.

15. Six-digit numbers are to be formed using the integers in the set

$$A = \{1, 2, 3, 4, 5, 6, 7, 8\}.$$

(a) How many such numbers can be formed if repetitions are allowed?

(b) In part (a), how many of the numbers contain at least one 3 and at least one 5?

(c) How many six-digit numbers can be formed if each digit in A can be used at most once?

(d) How many six-digit numbers can be formed that consist of one 2, two 4's and three 5's?

16. How many divisors are there of 6000? *Hint:* $6000 = 2^4 \cdot 3 \cdot 5^3$ and every divisor has the form $2^m 3^n 5^r$ where $m \le 4$, $n \le 1$ and $r \le 3$.

17. Consider n in $\mathbb{P}$ and let S be a subset of $\{1, 2, \ldots, 2n\}$ consisting of $n + 1$ numbers.

(a) Show that S contains two numbers that are relatively prime.

(b) Show that S contains two numbers such that one of them divides the other.

(c) Show that part (a) can fail if S has only n elements.

(d) Show that part (b) can fail if S has only n elements.

18. (a) Consider a subset A of $\{0, 1, 2, \ldots, p\}$ such that $|A| > \frac{1}{2}p + 1$. Show that A contains two different numbers whose sum is p.

(b) For $p = 6$, find A with $|A| = \frac{1}{2}p + 1$ not satisfying the conclusion in part (a).

(c) For $p = 7$, find A with $|A| = \frac{1}{2}(p - 1) + 1$ not satisfying the conclusion in part (a).

19. A class of 21 students wants to form 7 study groups so that each student belongs to exactly 2 study groups.

(a) Show that the average size of the study groups would have to be 6.

(b) Indicate how the students might be assigned to the study groups so that each group has exactly 6 students.

CHAPTER HIGHLIGHTS

To check your understanding of the material in this chapter, follow our usual suggestions for review. Think always of examples.

CONCEPTS

selection with/without replacement

r-permutation, permutation, combination

sample space, event, outcome, probability [on Ω]

$|S \cup T| = |S| + |T| - |S \cap T|$ for finite sets.

$|S_1 \times S_2 \times \cdots \times S_n| = \prod_{k=1}^{n} |S_k|.$

$P(E^c) = 1 - P(E).$

$P(E \cup F) = P(E) + P(F) - P(E \cap F).$

$\dbinom{n}{r} = \dfrac{n!}{(n-r)!\,r!}.$

Binomial Theorem: $(a + b)^n = \sum_{r=0}^{n} \dbinom{n}{r} a^r b^{n-r}.$

Counting Lemma.

Formula $\dfrac{n!}{n_1!\,n_2! \cdots n_k!}$ for counting indistinguishable permutations or ordered partitions.

Formula $\dbinom{n+k-1}{k-1}$ for ways to place n objects in k boxes.

Inclusion–Exclusion Principle.

Pigeon-Hole Principle, Generalized Pigeon-Hole Principle.

METHODS

Counting a set of pairs in two different ways.

Numerous clever ideas illustrated in the examples.

INTRODUCTION TO GRAPHS AND TREES

6

We have already seen graphs and digraphs in Chapter 3, where they gave us ways to picture relations. In this chapter we develop the fundamental theory of the graphs and digraphs themselves. Section 6.1 introduces the main ideas and terminology, and §§ 6.2 and 6.5 discuss paths with special properties. The remainder of the chapter is devoted to trees, which we will study further in Chapter 7. Trees are graphs that can also be thought of as digraphs in a natural way. Section 6.4 is devoted to rooted trees, which arise commonly as data structures. The final section of the chapter treats graphs whose edges have weights, and contains two algorithms for constructing spanning trees of smallest weight.

§ 6.1   Graphs

In § 3.2 we introduced the idea of a digraph and of a graph. We now pursue a more detailed study of graphs. Recall from § 3.2 that a **path** of **length** n is a sequence $e_1 e_2 \cdots e_n$ of edges together with a sequence $v_1 v_2 \cdots v_{n+1}$ of vertices, where $\gamma(e_i) = \{v_i, v_{i+1}\}$ for $i = 1, 2, \ldots, n$. Here γ is the function that specifies the vertices of each edge. The vertices v_i and v_{i+1} can be equal, in which case e_i is a **loop**. If $v_{n+1} = v_1$, the path is said to be **closed**.

A closed path can consist just of going out and back along the same sequence of edges, for example of $efggfe$. The most important closed paths in what follows, though, will be the ones in which no edge is repeated. A path is called **simple** if all of its edges are different. Thus a simple path cannot use any edge twice, although it may go through the same vertex more than once. A closed simple path with vertex sequence $x_1 \cdots x_n x_1$, $n \geq 1$, is called a **cycle** if the vertices $x_1, \ldots, x_n$ are distinct. A graph that contains no cycles is called **acyclic**. We will see soon that a graph is acyclic if and only if it contains no simple closed paths.

A path is **acyclic** if the "subgraph" consisting of the vertices and edges of the path is acyclic. In general, a graph H is a **subgraph** of a graph G if $V(H) \subseteq V(G)$, $E(H) \subseteq E(G)$, and the function γ for G defined on $E(G)$ agrees with the γ for H on $E(H)$. If G has no parallel edges and if we think of $E(G)$ as a set of one- or two-element subsets of $V(G)$, the condition on γ follows from $E(H) \subseteq E(G)$. It follows from the definition that if H is a subgraph of G and if G is acyclic, then so is H.

EXAMPLE 1   (a) Consider the graph pictured in Figure 1(a). The path ee with vertex sequence $x_1 x_2 x_1$ is a closed path, but it is not a cycle because it is not simple. Neither is the path ee with vertex sequence $x_2 x_1 x_2$. This graph is acyclic.

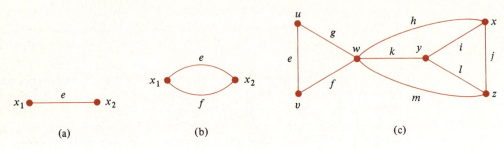

(a)   (b)   (c)

FIGURE 1

(b) The path ef with vertex sequence $x_1 x_2 x_1$ in the graph of Figure 1(b) is a cycle. So is the path ef with vertex sequence $x_2 x_1 x_2$.

(c) In the graph of Figure 1(c) the path $efhikg$ of length 6 with vertex sequence $uvwxywu$ is closed and simple but is not a cycle because the first six vertices u, v, w, x, y and w are not all different. The path with vertex sequence $uwvwuvu$ also fails to be a cycle. The graph as a whole is not acyclic, and neither are these two paths, since $uvwu$ is a cycle in both of their subgraphs. ∎

A path $e_1 \cdots e_n$ in G with vertex sequence $x_1 \cdots x_{n+1}$ consisting of distinct vertices must surely be simple since no two edges in it can have the same set of endpoints. Example 1(a) shows, however, that a closed path with $x_1, \ldots, x_n$ distinct need not be simple. That bad example, which is a closed path of length 2, is essentially the only one there is, as the following shows.

Proposition 1   Every closed path $e_1 \cdots e_n$ of length at least 3 with $x_1, \ldots, x_n$ distinct is a cycle.

Proof. We only need to show that $e_1, \ldots, e_n$ are different. Since $x_1, \ldots, x_n$ are distinct, the path $e_1 \cdots e_{n-1}$ is simple. That is, $e_1, \ldots, e_{n-1}$

are all different. But $\gamma(e_n) = \{x_n, x_1\}$ and $\gamma(e_i) = \{x_i, x_{i+1}\}$ for $i < n$. Since $n \geq 3$, $e_n \neq e_i$ for $i < n$, so the path is simple. ■

For paths that are not closed, distinctness of vertices can be characterized another way.

Proposition 2 | A path has all vertices distinct if and only if it is simple and acyclic.

Proof. Consider first a path with distinct vertices. It must be simple, as we observed earlier. The subgraph consisting of the vertices and edges of the path could be redrawn as a straight line with vertices on it, so it's clearly acyclic.

For the converse, suppose that a simple path has vertex sequence $x_1 \cdots x_{n+1}$ with some repeated vertices. Consider two such vertices, say x_i and x_j with $i < j$ and with the difference $j - i$ as small as possible. Then $x_i, x_{i+1}, \ldots, x_{j-1}$ are all distinct, so the simple path $x_i x_{i+1} \cdots x_{j-1} x_j$ is a cycle [even if $j = i + 1$] and the original path contains a cycle. ■

It follows from Proposition 2 that every simple closed path contains a cycle, and in fact the proof shows that such a cycle can be built out of successive edges in the path. As another consequence of Proposition 2, if a graph is acyclic, then it contains no cycles and hence contains no simple closed paths. We will look more closely at such graphs in the later sections on trees.

In the language of §3.2, the next theorem says that cycles have no bearing on the reachable relation.

Theorem 1 | If u and v are distinct vertices of a graph G and if there is a path in G from u to v, then there is a simple acyclic path from u to v.

Proof. Among all paths from u to v in G choose one of smallest length, say with vertex sequence $x_1 \cdots x_{n+1}$, $x_1 = u$ and $x_{n+1} = v$. As noted in Proposition 2, this path is simple and acyclic provided that the vertices $x_1, \ldots, x_{n+1}$ are distinct. But otherwise we would have $x_i = x_j$ for some i and j with $1 \leq i < j \leq n + 1$. Then the path $x_i x_{i+1} \cdots x_j$ from x_i to x_j would be closed [see Figure 2 for an illustration] and the path $x_1 \cdots x_i x_{j+1} \cdots x_{n+1}$ obtained by omitting this part would still go from u to v. Since $x_1 \cdots x_n x_{n+1}$ has smallest length, this shorter path is impossible. So the vertices are distinct and the shortest path is simple and acyclic. ■

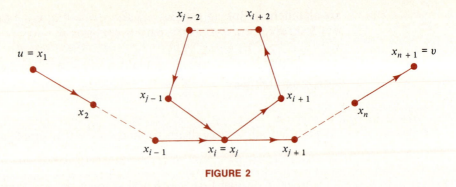

FIGURE 2

Corollary | If e is an edge in a simple closed path in G, then e belongs to some cycle.

Proof. If e is a loop we are done, so suppose that e joins distinct vertices u and v. Remove e from G. Since the given path through u and v is closed, even with e gone there is a path from u to v; just go the long way around. According to Theorem 1, there is a simple acyclic path from u to v not including e. Joining the ends of such a path with e forms a cycle through u and v. ■

We will have other occasions, especially in § 6.3, to remove an edge e from a graph G to get a new graph $G\backslash\{e\}$. The graph $G\backslash\{e\}$ is the subgraph of G with $V(G\backslash\{e\}) = V(G)$ and $E(G\backslash\{e\}) = E(G)\backslash\{e\}$.

We will need the following fact when we study trees in § 6.3.

Theorem 2 | If u and v are distinct vertices of the acyclic graph G, then there is at most one simple path in G from u to v.

Proof. We suppose the theorem is false, and among pairs of vertices joined by two different simple paths choose a pair (u, v) with the shortest possible path length from u to v.

Consider two simple paths from u to v, one of which is as short as possible. If the two paths had no vertices but u and v in common, then going from u to v along one path and returning from v to u along the other would yield a cycle, contrary to the fact that G is acyclic. Thus the paths must both go through at least one other vertex, say w. Then the paths from u to v form two different simple paths from u to w or else two different paths from w to v. But w is closer to u and v than they are to each other, contradicting the choice of the pair (u, v). ■

EXAMPLE 2 Consider the graph in Figure 1(c). The path $x\,w\,y\,z\,x$ formed by going out from x to w on one path and back to x on the other is a cycle. So is $w\,u\,v\,w$, formed by going on from w to u and then back to w another way.

The simple paths $u\,w\,y\,x\,w\,v$ and $u\,w\,y\,z\,w\,v$ go from u to v. The cycle $y\,x\,w\,z\,y$ is made out of parts of these paths, one part from y to w and the other back from w to y. ■

It frequently happens that two graphs are "essentially the same," even though they differ in the names of their edges and vertices. General graph-theoretic statements that can be made about one of the graphs are then equally true of the other. To make these ideas mathematically precise, we introduce the concept of isomorphism. Generally speaking, two sets with some mathematical structure are said to be **isomorphic** [pronounced *eye-so-MOR-fik*] if there exists a one-to-one correspondence between them that preserves [i.e., is compatible with] the structure. Recall that if G and H are graphs without parallel edges, then we consider edges to be one- or two-element sets of vertices. An **isomorphism** of G onto H is a one-to-one correspondence $\alpha: V(G) \to V(H)$ such that $\{u, v\}$ is an edge of G if and only if $\{\alpha(u), \alpha(v)\}$ is an edge of H. Two graphs G and H are **isomorphic**, written $G \simeq H$, if there is an isomorphism α of one onto the other, in which case the inverse correspondence α^{-1} is also an isomorphism.

For graphs with parallel edges the situation is slightly more complicated: we require two one-to-one correspondences $\alpha: V(G) \to V(H)$ and $\beta: E(G) \to E(H)$ such that an edge e of $E(G)$ joins vertices u and v in $V(G)$ if and only if the corresponding edge $\beta(e)$ joins $\alpha(u)$ and $\alpha(v)$. Thus two graphs are isomorphic if and only if they have the same picture except for the labels on edges and vertices. This observation is mostly useful as a way of verifying an alleged isomorphism by actually drawing matching pictures of the two graphs.

EXAMPLE 3 The correspondence α with $\alpha(t) = t'$, $\alpha(u) = u'$, $\ldots$, $\alpha(z) = z'$ is an isomorphism between the graphs pictured in Figures 3(a) and 3(b). The graphs shown in Figures 3(c) and 3(d) are also isomorphic to each other, but not to the graphs in parts (a) and (b). ■

To tell the graphs of Figures 3(a) and 3(c) apart we can simply count vertices. Isomorphic graphs have the same number of vertices and the same number of edges. These two numbers are examples of **isomorphism invariants** for graphs. Other examples include the number of loops and number of simple paths of a given length.

It is often useful to count the number of edges attached to a particular vertex. To get the right count we need to treat loops differently from edges with two distinct vertices. We define deg(v), the **degree** of the vertex v, to be the number of 2-vertex edges with v as a vertex plus twice

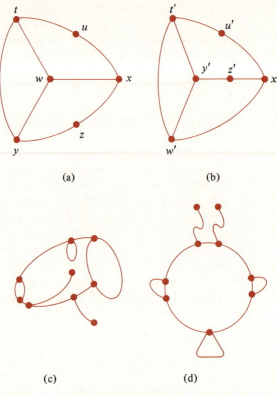

(a) (b)

(c) (d)

FIGURE 3

the number of loops with v as vertex. The number $D_k(G)$ of vertices of degree k in G is an isomorphism invariant, as is the **degree sequence** $(D_0(G), D_1(G), D_2(G), \ldots)$.

EXAMPLE 4 (a) The graphs shown in Figures 3(a) and 3(b) each have degree sequence $(0, 0, 2, 4, 0, 0, \ldots)$. Those in Figures 3(c) and 3(d) have degree sequence $(0, 2, 0, 6, 1, 0, 0, \ldots)$.

(b) All four of the graphs in Figure 4 have eight vertices of degree 3 and no others. It turns out that $H_1 \simeq H_2 \simeq H_3$, but that none of these three is isomorphic to H_4. Having the same degree sequence does not guarantee isomorphism. ■

Graphs in which all vertices have the same degree, such as those in Figure 4, are called **regular** graphs. As the example shows, regular graphs with the same number of vertices need not be isomorphic. Graphs without

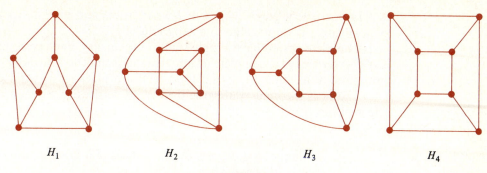

H_1 H_2 H_3 H_4

FIGURE 4

loops or multiple edges and in which every vertex is joined to every other by an edge are called **complete** graphs. A complete graph with n vertices has all vertices of degree $n - 1$, so such a graph is regular. All complete graphs with n vertices are isomorphic to each other, so we use the symbol K_n for any of them.

EXAMPLE 5 Figure 5(a) shows the first five complete graphs. The graph in Figure 5(b) has four vertices, each of degree 3, but is not complete. ■

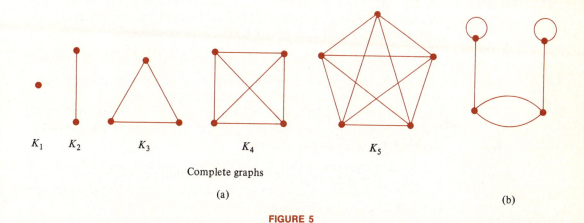

K_1 K_2 K_3 K_4 K_5

Complete graphs

(a)

(b)

FIGURE 5

A complete graph K_n contains subgraphs isomorphic to the graphs K_m for $m = 1, 2, \ldots, n$. Such a subgraph can be obtained by selecting any m of the n vertices and using all the edges in K_n joining them. Thus K_5 contains $\binom{5}{2} = 10$ subgraphs isomorphic to K_2, $\binom{5}{3} = 10$ subgraphs isomorphic to K_3 [i.e., triangles], and $\binom{5}{4} = 5$ subgraphs isomorphic to

K_4. In fact, every graph with n or fewer vertices and with no loops or parallel edges is isomorphic to a subgraph of K_n.

Complete graphs have a high degree of symmetry. Each permutation α of the vertices of a complete graph gives an isomorphism of the graph onto itself, since both $\{u, v\}$ and $\{\alpha(u), \alpha(v)\}$ are edges whenever $u \neq v$.

The next theorem relates the degrees of vertices to the number of edges of the graph.

Theorem 3

> (a) The sum of the degrees of the vertices of a graph is twice the number of edges. That is,
>
> $$\sum_{v \in V(G)} \deg(v) = 2 \cdot |E(G)|.$$
>
> (b) $D_1(G) + 2D_2(G) + 3D_3(G) + 4D_4(G) + \cdots = 2 \cdot |E(G)|.$

Proof. (a) Each edge, whether a loop or not, contributes 2 to the degree sum.

(b) The total degree sum contribution from the $D_k(G)$ vertices of degree k is $k \cdot D_k(G)$. ■

EXAMPLE 6

(a) The graph of Figure 3(c) [and its isomorph in Figure 3(d)] has vertices of degrees 1, 1, 3, 3, 3, 3, 3, 3 and 4, and has twelve edges. The degree sequence is $(0, 2, 0, 6, 1, 0, 0, 0, \ldots)$. Sure enough,

$$1 + 1 + 3 + 3 + 3 + 3 + 3 + 3 + 4 = 2 \cdot 12 = 2 + 2 \cdot 0 + 3 \cdot 6 + 4 \cdot 1.$$

(b) The graphs in Figure 4 have eight vertices of degree 3 and twelve edges. Their degree sequence is $(0, 0, 0, 8, 0, 0, 0, \ldots)$, and

$$2 \cdot 12 = 0 + 2 \cdot 0 + 3 \cdot 8$$

confirms Theorem 3(b) in this case.

(c) The complete graph K_n has n vertices, each of degree $n - 1$, and has $n(n - 1)/2$ edges. ■

EXERCISES 6.1

1. For the graph in Figure 6(a), give the vertex sequence of a shortest path connecting the following pairs of vertices and give its length.

(a) s and v

(b) s and z

(c) u and y

(d) v and w

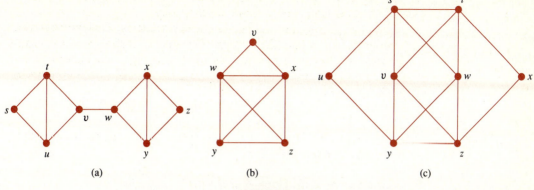

FIGURE 6

2. For each pair of vertices in Exercise 1, give the vertex sequence of a longest path connecting them that repeats no edges. Is there a longest path connecting them if edges can be repeated?

3. True or False. "True" means "true in all circumstances under consideration." Consider a graph.

   (a) If there is an edge from a vertex u to a vertex v, then there is an edge from v to u.

   (b) If there is an edge from a vertex u to a vertex v and an edge from v to a vertex w, then there is an edge from u to w.

4. Repeat Exercise 3 with the word "edge" replaced by "path" everywhere.

5. Repeat Exercise 3 with the word "edge" replaced by "path of even length" everywhere.

6. (a) Confirm Theorem 3(a) for each graph in Figure 6 by calculating

   (i) the sum of the degrees of all of the vertices;

   (ii) the number of edges.

   (b) Can a graph have an odd number of vertices of odd degree?

7. Give an example of a graph with vertices x, y and z with all three of the following properties:

   (i) there is a cycle using vertices x and y;

   (ii) there is a cycle using vertices y and z;

   (iii) no cycle uses vertices x and z.

8. Suppose that a cycle contains a loop. What is its length? Can a cycle contain two loops?

9. (a) Give a table of the function γ for the graph G pictured in Figure 7.

   (b) List the edges of this graph, considered as subsets of $V(G)$. For example, $a = \{w, x\}$.

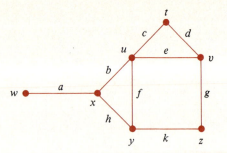

FIGURE 7

10. Draw a picture of the graph G with $V(G) = \{x, y, z, w\}$, $E(G) = \{a, b, c, d, f, g, h\}$ and γ given by the table:

e	a	b	c	d	f	g	h
$\gamma(e)$	$\{x, y\}$	$\{x, y\}$	$\{w, x\}$	$\{w, y\}$	$\{y, z\}$	$\{y, z\}$	$\{w, z\}$

11. In each part of this exercise two paths are given that join a pair of points in the graph of Figure 7. Use the idea of the proof of Theorem 2 to construct cycles out of sequences of edges in the two paths.

   (a) $e\,b\,a$ and $g\,k\,h\,a$ (b) $a\,b\,c\,d\,g\,k$ and $a\,h$

   (c) $e\,b\,a$ and $d\,c\,f\,h\,a$

12. (a) Draw pictures of all fourteen graphs with three vertices and three edges. "All" here means that every such graph is isomorphic to one of the fourteen, and no two of the fourteen are isomorphic.

   (b) Draw pictures of all graphs with four vertices and four edges that have no loops or parallel edges.

   (c) List the four graphs in parts (a) and (b) that are regular.

13. (a) Draw pictures of all five of the regular graphs with four vertices, each vertex of degree 2.

   (b) Draw pictures of all of the regular graphs with four vertices, each of degree 3, and with no loops or parallel edges.

   (c) Draw pictures of all of the regular graphs with five vertices, each of degree 3.

14. Suppose that a graph H is isomorphic to the graph G of Figure 7.

   (a) How many vertices of degree 1 does H have?

   (b) Give the degree sequence of H.

   (c) How many different isomorphisms are there of G onto G? Explain.

   (d) How many isomorphisms are there of G onto H?

15. Which, if any, of the pairs of graphs shown in Figure 8 are isomorphic? Justify your answer by describing an isomorphism or explaining why one does not exist.

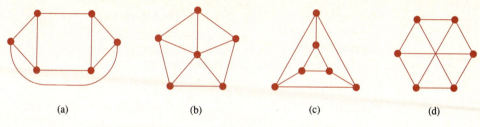

(a)　　　　　　　(b)　　　　　　　(c)　　　　　　　(d)

FIGURE 8

16. Describe an isomorphism between the graphs shown in Figure 9.

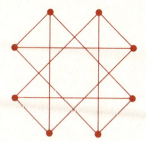 

FIGURE 9

17. Consider the complete graph K_8 with vertices $v_1, v_2, \ldots, v_8$.

    (a) How many subgraphs of K_8 are isomorphic to K_5?

    (b) How many simple paths with 3 or fewer edges are there from v_1 to v_2?

    (c) How many simple paths with 3 or fewer edges are there altogether in K_8?

18. (a) A graph with 21 edges has 7 vertices of degree 1, 3 of degree 2, 7 of degree 3 and the rest of degree 4. How many vertices does it have?

    (b) How would your answer to part (a) change if the graph also had 6 vertices of degree 0?

19. Which of the following are degree sequences of graphs? In each case either draw a graph with the given degree sequence or explain why no such graph exists.

    (a) $(1, 1, 0, 3, 1, 0, 0, \ldots)$　　　　(b) $(4, 1, 0, 3, 1, 0, 0, \ldots)$

    (c) $(0, 1, 0, 2, 1, 0, 0, \ldots)$　　　　(d) $(0, 0, 2, 2, 1, 0, 0, \ldots)$

    (e) $(0, 0, 1, 2, 1, 0, 0, \ldots)$　　　　(f) $(0, 1, 0, 1, 1, 1, 0, \ldots)$

    (g) $(0, 0, 0, 4, 0, 0, 0, \ldots)$　　　　(h) $(0, 0, 0, 0, 5, 0, 0, \ldots)$

20. Show that a path in a graph G is a cycle if and only if it is possible to assign directions to the edges of G so that the path is a [directed] cycle in the resulting digraph.

21. Show that every finite graph in which each vertex has degree at least 2 contains a cycle.

22. Show that every graph with n vertices and at least n edges contains a cycle. *Hint:* Use induction on n and Exercise 21.

23. Show that

$$2|E(G)| - |V(G)| = -D_0(G) + D_2(G) + 2D_3(G) + \cdots + (k-1)D_k(G) + \cdots$$

24. (a) Let $\mathscr{S}$ be a set of graphs. Show that isomorphism $\sim$ is an equivalence relation on $\mathscr{S}$.

    (b) How many equivalence classes are there if $\mathscr{S}$ consists of the four graphs in Figure 4?

§ 6.2 Edge Traversal Problems

One of the oldest problems involving graphs is the Königsberg bridge problem: Is it possible to take a walk in the town shown in Figure 1(a) crossing each bridge exactly once and returning home? The Swiss mathematician Leonhard Euler [pronounced *OIL-er*] solved this problem in 1736. He constructed the graph shown in Figure 1(b), replacing the land areas by vertices and the bridges joining them by edges. The question then became: Is there a closed path in this graph that uses each edge exactly once? We call such a path an **Euler circuit** of the graph. More generally, a simple path that contains all edges of a graph G is called an **Euler path** of G. Euler showed that there is no Euler circuit for the Königsberg bridge graph, in which all vertices have odd degree, by establishing the following elementary fact.

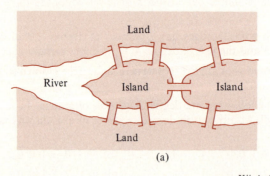

(a)   (b)

Königsberg graph

FIGURE 1

Theorem 1 | A graph that has an Euler circuit must have all vertices of even degree.

Proof. Start at some vertex on the circuit and follow the circuit from vertex to vertex, erasing each edge as you go along it. When you go through a vertex you erase one edge going in and one going out, or else you erase a loop. Either way, the erasure reduces the degree of the vertex by 2. Eventually, every edge gets erased and all vertices have degree 0. So all vertices must have had even degree to begin with. ■

Corollary

> A graph G that has an Euler path has either two vertices of odd degree or no vertices of odd degree.

Proof. Suppose that G has an Euler path starting at u and ending at v. If $u = v$, the path is closed and Theorem 1 says all vertices have even degree. If $u \neq v$, create a new edge e joining u and v. The new graph $G \cup \{e\}$ has an Euler circuit consisting of the Euler path for G followed by e, so all vertices of $G \cup \{e\}$ have even degree. Remove e. Then u and v are the only vertices of $G = (G \cup \{e\}) \backslash \{e\}$ of odd degree. ■

EXAMPLE 1 The graph shown in Figure 2(a) has no Euler circuit, since u and v have odd degree, but the path $b\,a\,c\,d\,g\,f\,e$ is an Euler path. The graph in Figure 2(b) has all vertices of even degree and in fact has an Euler circuit. The graph in Figure 2(c) has all vertices of even degree but has no Euler circuit, for the obvious reason that the graph is disconnected into two subgraphs that are not connected to each other. Each of the subgraphs, however, has its own Euler circuit. ■

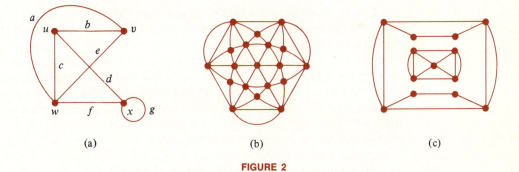

(a)   (b)   (c)

FIGURE 2

Theorem 1 shows that the even-degree condition is necessary for the existence of an Euler circuit. Euler's major contribution to the problem was his proof that, except for the sort of obvious trouble we ran into in Figure 2(c), the condition is also sufficient to guarantee an Euler circuit.

We need some terminology to describe the exceptional cases. A graph is **connected** if each pair of distinct vertices is joined by a path in the graph. The graphs in Figures 2(a) and 2(b) are connected, but the one in Figure 2(c) is not. A connected subgraph of a graph G that is not contained in a larger connected subgraph of G is called a **component** of G. The component containing a given vertex v consists of v together with all vertices and edges on paths starting at v.

EXAMPLE 2   (a) The graphs of Figures 2(a) and 2(b) are connected. In these cases the graph has just one component, namely the graph itself.

(b) The graph of Figure 2(c) has two components, the one drawn on the outside and the one on the inside. Another picture of this graph is shown in Figure 3(a). In this picture there is no "inside" component, but of course there are still two components.

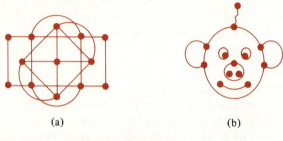

(a)   (b)

FIGURE 3

(c) The graph of Figure 3(b) has seven components, two of which are isolated vertices. ■

Euler's theorem states that a finite connected graph in which every vertex has even degree has an Euler circuit. To really understand this theorem we should be able to find a proof, or develop an algorithm or procedure that would always produce an Euler circuit. Indeed, these two approaches are intimately connected. A full understanding of a proof often leads to an algorithm, and behind every algorithm there's a proof. Here's a simple explanation of Euler's theorem, which we illustrate using Figure 4. Start with any vertex, say w, and any edge connected to it, say a. The other vertex, x in this case, has even degree and has been used an odd number of times [once] and so there is an unused edge leaving x. Pick one, say b. Continue in this way. The process won't stop until the starting vertex w is reached, since whenever any other vertex is reached, only an odd number of its edges have been used. In

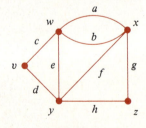

FIGURE 4

our example, this algorithm might start out with edges $a\,b\,e$ and vertices $w\,x\,w\,y$. At y we can choose any of three edges: d, f or h. If we select f, the rest of the process is determined. We end up with the Euler circuit $a\,b\,e\,f\,g\,h\,d\,c$ with vertex sequence $w\,x\,w\,y\,x\,z\,y\,v\,w$.

Simple, wasn't it? Well, it is too simple. What would have happened if, when we first reached vertex y, we had chosen edge d? After choosing edge c, we'd have been trapped at vertex w and our path $a\,b\,e\,d\,c$ would have missed edges f, g and h. Our explanation and our algorithm must be too simple. In our example it is clear that edge d should have been avoided when we first reached vertex y, but why? What general principle should have warned us to avoid this choice? Think about it. We will continue this discussion after we give a nonconstructive proof of Euler's theorem.

Theorem 2 (Euler's theorem)

A finite connected graph in which every vertex has even degree has an Euler circuit.

Proof. Suppose that G is a finite connected graph with every vertex of even degree. If G has just one vertex the theorem is trivial, so assume $|V(G)| \geq 2$. Since G is connected, each vertex has degree at least 1, and hence at least 2 since all degrees are even.

Let $v_1\,v_2 \cdots v_n$ be the vertex sequence of a path of longest possible length with distinct edges. Since $E(G)$ is finite, such a path must exist. We claim that $v_n = v_1$. To show this, we consider the graph G' obtained by removing the edges of this path, and see how the degrees of the vertices are affected. The vertex v_n may or may not appear in the list $v_2, \ldots, v_{n-1}$. If it does, for each such appearance the degree of v_n in G' is reduced by 2, by 1 for the edge going in and by 1 for the edge going out. Since the removal of the last edge on the path reduces the degree of v_n by 1, if v_n were not v_1 the degree of v_n in the graph G' would be odd. In that case, we could add an unused edge with vertex v_n to the given path and obtain a longer path of distinct edges, contrary to our choice of longest path. Hence $v_n = v_1$.

We next show that our longest path visits every vertex of G. If not, then since G is connected there is a path from some nonvisited vertex to a vertex in $\{v_1, \ldots, v_n\}$. The first edge of this path is in G'. The last edge e of this path that is in G' must connect to some vertex in $\{v_1, \ldots, v_n\}$, say v_i. If we now go around the longest path from v_i to v_i and then add the edge e, we obtain a longer path with distinct edges, again contrary to our choice of longest path.

To complete the proof, we show that our longest path uses every edge of G and hence is an Euler circuit. In fact, if e is an edge that's missed, it must have some v_i as a vertex [two v_i's unless e is a loop]. As in the preceding paragraph, the edge e can be attached to the longest path to get a longer path and a contradiction.   ■

Corollary | A finite connected graph that has exactly two vertices of odd degree has an Euler path.

Proof. Say u and v have odd degree. Create a new edge e joining them. Then $G \cup \{e\}$ has all vertices of even degree and so has an Euler circuit by Theorem 2. Remove e again. What remains of the circuit is an Euler path for G.   ■

We will show in §8.1, using essentially the same proof we have given here, that the digraph analog of Theorem 2 is also true. This theorem tells us when a graph has an Euler circuit, but it does not tell us how to find one. We would like an algorithm that finds an Euler circuit or path, one edge at a time, rather like the vague procedure we applied to Figure 4. Let's take another look at the Euler circuit we found for Figure 4, which we reproduce in Figure 5(a). As we select edges let's remove them from the graph and consider the subgraphs so obtained. Our path started out with edges $a\,b\,e$; Figure 5(b) shows the graph with these edges removed. In our successful search for an Euler circuit we next selected f, and we noted that if we had selected d we were doomed. Figure 5(c)

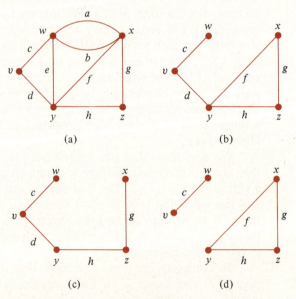

FIGURE 5

shows the graph if f is also removed, while Figure 5(d) shows the graph if instead d is removed. There is a difference: removal of d disconnected the graph while removal of f did not. This is the clue for an algorithm that works. At each vertex, FLEURY'S algorithm instructs us to select, if possible, an edge whose removal will not disconnect the graph. If this is not possible, there is exactly one edge available. [This is a nontrivial fact that requires proof. This is where mathematics comes in.] We select it; then remove it and the vertex from the graph.

We next formalize the algorithm. When the algorithm terminates, the sequence ES is the edge sequence of an Euler path or circuit and VS is its vertex sequence.

FLEURY'S *Algorithm*

Step 1. Start at any vertex v of odd degree if there is one. Otherwise, start at any vertex v. Let $VS = v$ and let $ES = \lambda$ [the empty sequence].

Step 2. If there is no edge remaining at v, stop.

Step 3. If there is exactly one edge remaining at v, say e from v to w, then remove e from $E(G)$ and v from $V(G)$ and go to Step 5.

Step 4. If there is more than one edge remaining at v, choose such an edge, say e from v to w, whose removal will not disconnect the graph; then remove e from $E(G)$.

Step 5. Add w to the end of VS, add e to the end of ES, replace v by w and go to Step 2. ■

Before discussing why this algorithm works, we give an example to illustrate its use.

EXAMPLE 3 Consider the graph of Figure 6(a). This graph does not have an Euler circuit, but it does have an Euler path joining the vertices z and y of odd degree. We start from z; i.e., let $v = z$ in Step 1. Thus $VS = z$ and $ES = \lambda$. The only edge at z is i. So we go to Step 3, choose $e = i$ and $w = y$, remove i from $E(G)$, remove z from $V(G)$ and go to Step 5. Then $VS = zy$ and $ES = i$. The new G with both i and z deleted is shown in Figure 6(b). Let $v = y$ and return to Step 2. There are three edges at v, so Step 4 applies.

Now e can be f, g or h. Let's take $e = f$ in Step 4. Step 5 gives $VS = zyx$, $ES = if$, $v = x$, and the new G shown in Figure 6(c). Return to Step 2. Again there are three edges at v and Step 4 applies.

Now e can be either a or d, but not h since removing h would disconnect the graph. Let's choose $e = a$; Step 5 gives $VS = zyxr$, $ES = ifa$ and $v = r$.

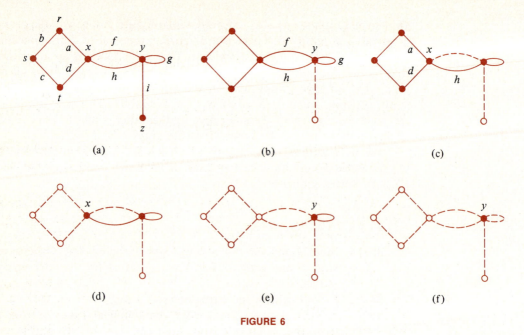

(a) (b) (c)

(d) (e) (f)

FIGURE 6

The next three moves are forced. Each leads to Step 3 and the removal of a vertex as well as an edge. We arrive at $VS = z\,y\,x\,r\,s\,t\,x$, $ES = i\,f\,a\,b\,c\,d$, $v = x$, and the graph shown in Figure 6(d).

The remaining two moves are also forced and lead to the graphs in Figures 6(e) and 6(f). The final vertex and edge sequences are $VS = z\,y\,x\,r\,s\,t\,x\,y\,y$ and $ES = i\,f\,a\,b\,c\,d\,h\,g$. ∎

To prove that FLEURY'S algorithm works, we need the next theorem. It will also be useful in our treatment of trees in §6.3.

Theorem 3   Let e be an edge of a connected graph G. The following are equivalent:

    (a) $G \backslash \{e\}$ is connected.
    (b) e is an edge of some cycle in G.
    (c) e is an edge of some simple closed path in G.

Proof. First note that if e is a loop, then $G \backslash \{e\}$ is connected, while e is a cycle all by itself. Since cycles are simple closed paths, the theorem holds in this case, and we may assume that e is not a loop. Thus e connects distinct vertices u and v. If f is another edge connecting u and v, then clearly $G \backslash \{e\}$ is connected and ef is a cycle containing e. So the theorem also holds in this case. Hence we may assume that e is the unique edge connecting u and v.

(a) $\Rightarrow$ (b). Suppose that $G\backslash\{e\}$ is connected. By Theorem 1 of §6.1 there is a simple acyclic path $x_1 x_2 \cdots x_m$ with $u = x_1$ and $x_m = v$. Since there is no edge from u to v in $G\backslash\{e\}$, we have $x_2 \neq v$, and so $m \geq 3$. As noted in Proposition 2 of §6.1, the vertices $x_1, x_2, \ldots, x_m$ are distinct, so $x_1 x_2 \cdots x_m u$ is a cycle in G containing the edge e.

(b) $\Leftrightarrow$ (c). Obviously, (b) $\Rightarrow$ (c) since cycles are simple closed paths, and (c) $\Rightarrow$ (b) even if G is not connected, by the Corollary to Theorem 1 of §6.1.

(b) $\Rightarrow$ (a). The edge e is one path between u and v, while the rest of the cycle containing e gives an alternate route. It is still possible to get from any vertex of G to any other, even if the bridge is out at e, just by replacing e by the alternative if necessary. ■

Proof That **FLEURY'S** *Algorithm Works.* We consider a finite connected graph all of whose vertices have even degree, and we show that FLEURY'S algorithm produces an Euler circuit. Modifications showing that we get an Euler path in the case of two vertices of odd degree are straightforward.

Each pass through the loop from Step 2 through Step 5 removes an edge from G and adds it to ES in such a way that the edges in ES form a path. Since G has only a finite number of edges to begin with, the algorithm must stop—or break down—sooner or later, and no edge can appear more than once in the path determined by ES. We need to show that the algorithm does not break down, and that when it stops ES contains every edge of G.

The only possible breakdown might occur at Step 4. How do we know that there is an edge we can remove without disconnecting the graph? Say G' is the current value of G and v' the current value of v, and suppose that there is more than one edge at v' in G'. We claim that in fact we can choose *any* of the edges at v', with at most one exception. Since the edges in ES form a path, there are two cases to consider: either G' has two vertices of odd degree, one of which is v' and the other the vertex v_0 chosen in Step 1, or else $v' = v_0$ and G' has no vertices of odd degree.

In the first case, the corollary to Theorem 2 says that G' has an Euler path from v' to v_0. Every time this path returns to v' it completes a simple closed path that includes two of the edges at v', so except perhaps for the edge that the path uses the last time it leaves v', all edges at v' belong to simple closed paths in G'. In the second case, Theorem 2 says that G' has an Euler circuit, so a similar argument shows that every edge at v' belongs to a simple closed path of G'. By Theorem 3, removing an edge from a simple closed path cannot disconnect G'. Because there is at most one bad edge at v', if the first one we try doesn't belong to a simple closed path, then any other choice of an edge at v' will surely work.

Why are there no edges left at the end? At the start the graph is connected, when we execute Step 4 the graph remains connected, and when we execute Step 3 the graph also remains connected, because after we remove the edge e we also remove the isolated vertex v that is left. So at all times the graph is connected. When we're forced to stop because there is no edge left at v' there must not be any other vertices left in G' either. Hence G' has no edges whatever and ES contains every edge of G. ■

Most of the operations in FLEURY'S algorithm, such as adding or removing edges, take a fixed amount of time independent of how many vertices G has. The operation that takes longer is the test for connectedness of $G\backslash\{e\}$. We will show in §8.4 that there is a connectedness test based on DIJKSTRA'S algorithm that runs in time $O(|V(G)|^2)$. The proof that FLEURY'S algorithm works shows that we only need to test one edge at Step 4. Since the algorithm makes one pass through the Step $2, \ldots,$ Step 5 loop for each edge in G and then stops, the total time required by FLEURY'S algorithm is $O(|V(G)|^2 \cdot |E(G)|)$.

EXERCISES 6.2

1. Which of the graphs in Figure 7 have Euler circuits? Give the vertex sequence of an Euler circuit in each case in which one exists.

2. Use FLEURY'S algorithm to find an Euler circuit in the graph of Figure 7(b).

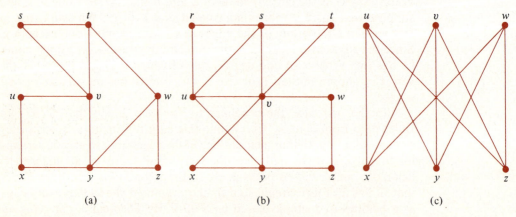

(a)   (b)   (c)

FIGURE 7

3. Apply FLEURY'S algorithm to the graph of Figure 7(a) until it breaks down. Start at vertex w.

4. Repeat Exercise 3 for the graph in Figure 7(c).

5. Consider the graph shown in Figure 8(a).

   (a) Describe an Euler path for this graph or explain why there isn't one.

   (b) Describe an Euler circuit for this graph or explain why there isn't one.

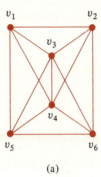

(a)

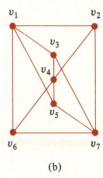

(b)

FIGURE 8

6. Repeat Exercise 5 for the graph of Figure 8(b).

7. Is it possible for an insect to crawl along the edges of a cube so as to travel along each edge exactly once? Explain.

8. Apply FLEURY'S algorithm as in Example 3 to get an Euler path for the graph of Figure 2(a). Sketch the intermediate graphs obtained in the application of the algorithm, as was done in Figure 6.

9. Construct a graph with vertex set $\{0, 1\}^3$ and with an edge between vertices v and w if v and w differ in exactly two coordinates.

   (a) How many components does the graph have?

   (b) How many vertices does the graph have of each degree?

   (c) Does the graph have an Euler circuit?

10. Answer the same questions as in Exercise 9 for the graph with vertex set $\{0, 1\}^3$ and with an edge between v and w if v and w differ in two or three coordinates.

11. (a) Show that if a connected graph G has exactly $2k$ vertices of odd degree and $k > 0$, then $E(G)$ is the disjoint union of the edge sets of k simple paths. *Hint:* Add more edges, as in the proof of the corollary to Theorem 2.

    (b) Find two disjoint simple paths whose edge set union is $E(G)$ for the Königsberg graph in Figure 1.

    (c) Do the same for the graph in Figure 8(b) of Exercise 6.

12. Which complete graphs K_n have Euler circuits?

13. An old puzzle presents a house with five rooms and sixteen doors, as shown in Figure 9. The problem is to figure out how to walk around and through the house so as to go through each door exactly once.

(a) Is such a walk possible? Explain.

(b) How does your answer change if the door joining the two large rooms is sealed shut?

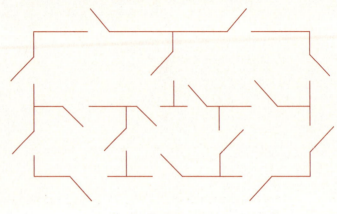

FIGURE 9

§ 6.3 Trees

In this section we study the graphs that are acyclic and connected; they are called **trees**. Since they are acyclic, trees have no parallel edges or loops. We already saw one tree in Figure 1 of § 5.1. Here are some more examples.

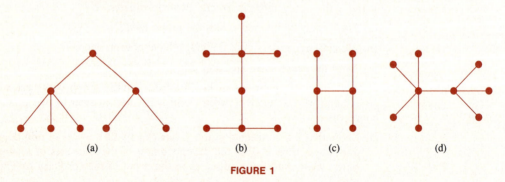

(a)   (b)   (c)   (d)

FIGURE 1

EXAMPLE 1 Figure 1 contains pictures of some trees. The ones in Figures 1(a) and 1(b) are isomorphic. Their pictures are different but the essential structures [vertices and edges] are the same. They share all graph-theoretic properties such as the numbers of vertices and edges, the number of

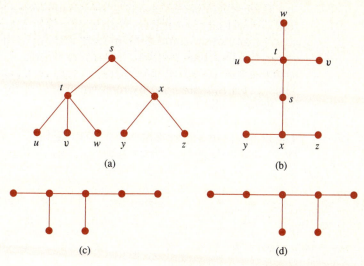

FIGURE 2

vertices of each degree, etc. To make this clear, we have redrawn them in Figures 2(a) and 2(b) and labeled corresponding vertices. The trees in Figures 2(c) and 2(d) are also isomorphic. ∎

EXAMPLE 2   The eleven trees with seven vertices are pictured in Figure 3. In other words, every tree with seven vertices is isomorphic to one drawn in Figure 3 and no two of the trees in Figure 3 are isomorphic to each other. ∎

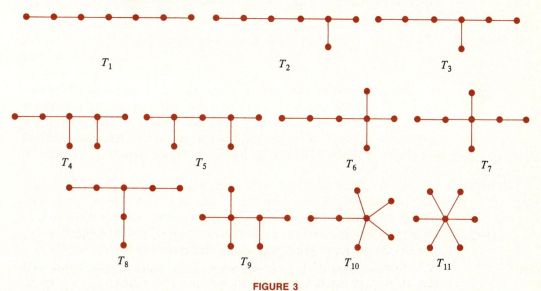

FIGURE 3

Given a connected graph G, we are interested in minimal subgraphs that connect all the vertices. Such a subgraph must be acyclic since one edge of any cycle could be removed without losing the connectedness property [Theorem 3 of §6.2]. In other words, such a subgraph T is a **spanning tree**: it's a tree that includes every vertex of G, i.e., $V(T) = V(G)$. Thus T is a tree obtained by removing some of the edges of G, perhaps, but keeping all of the vertices.

EXAMPLE 3   The graph H in Figure 4 has over 300 spanning trees, of which four have been sketched. They all have 6 edges. ∎

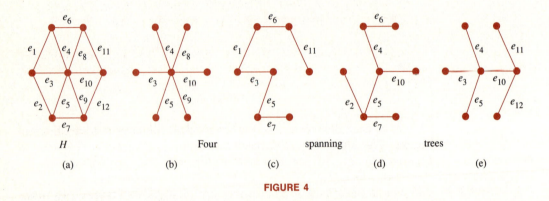

FIGURE 4

The next theorem guarantees that spanning trees exist for all finite connected graphs. In §6.6 we will see how to construct them.

Theorem 1 | Every finite connected graph G has a spanning tree.

Proof. Consider a connected subgraph G' of G that uses all the vertices of G and has as few edges as possible. Suppose that G' contains a cycle, say one involving the edge e. By Theorem 3 of §6.2, $G'\backslash\{e\}$ is a connected subgraph of G that has fewer edges than G' has, contradicting the choice of G'. So G' has no cycles. Since it's connected, G' is a tree. ∎

EXAMPLE 4   (a) We illustrate Theorem 1 using the connected graph in Figure 5(a). Note that e_1 does not belong to a cycle and that $G\backslash\{e_1\}$ is disconnected: no path in $G\backslash\{e_1\}$ connects v to the other vertices. Similarly, e_5 belongs to no cycle and $G\backslash\{e_5\}$ is disconnected. The remaining edges belong to cycles. Removal of any *one* of them will not disconnect G.

(b) Note that $G\backslash\{e_{10}\}$ is still connected but has cycles. If we also remove e_8, the resulting graph still has a cycle, namely $e_2 e_3 e_4$. But if

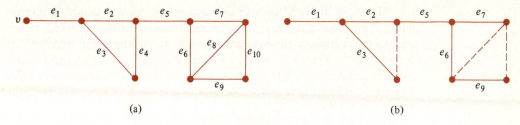

FIGURE 5

we then remove one of the edges in the cycle, say e_4, we obtain an acyclic connected subgraph, i.e., a spanning tree. See Figure 5(b). Clearly, several different spanning trees can be obtained in this way. ∎

In characterizing trees we lose nothing by restricting our attention to graphs with no loops or parallel edges. Our first characterizations hold even if the graph is infinite.

Theorem 2

Let G be a graph with more than one vertex, no loops and no parallel edges. The following are equivalent:

(a) G is a tree.
(b) Each pair of distinct vertices is connected by exactly one simple path.
(c) G is connected, but will not be if any edge is removed.
(d) G is acyclic, but will not be if any edge is added.

Proof. This proof consists of four short proofs.

(a) ⇒ (b). Suppose that G is a tree, so that G is connected and acyclic. By Theorem 1 of §6.1, each pair of vertices is connected by at least one simple path, and by Theorem 2 of §6.1, there is just one simple path.

(b) ⇒ (c). If (b) holds, G is clearly connected. Let $e = \{u, v\}$ be an edge of G and assume that $G\backslash\{e\}$ is still connected. Note that $u \neq v$ since G has no loops. By Theorem 1 of §6.1 there is a simple path in $G\backslash\{e\}$ from u to v. Since this path and the one-edge path e are different simple paths in G from u to v, we contradict (b).

(c) ⇒ (d). Suppose that (c) holds. If G had a cycle, we could remove an edge from G and retain connectedness by Theorem 3 of §6.2. So G is acyclic. Now consider an edge e not in the graph G and let G' denote the graph G with this new edge adjoined. Since $G'\backslash\{e\} = G$ is connected, we apply Theorem 3 of §6.2 to G' to conclude that e belongs to some cycle of G'. In other words, adding e to G destroys acyclicity.

(d) $\Rightarrow$ (a). If (d) holds and G is not a tree, then G is not connected. Then there exist distinct vertices u and v that are not connected by any paths in G. Consider the new edge $e = \{u, v\}$. According to our assumption (d), $G \cup \{e\}$ has a cycle and e must be part of it. The rest of the cycle is a path in G that connects u and v. This contradicts our choice of u and v. Hence G is connected and G is a tree. ∎

In order to appreciate Theorem 2, draw a tree or look at a tree in one of the Figures 1 through 4 and observe that it possesses all the properties (a) to (d) in Theorem 2. Then draw or look at a nontree and observe that it possesses none of the properties (a) to (d).

We need two lemmas for our characterization of finite trees. For a tree, vertices of degree one are called **leaves** [the singular is **leaf**].

EXAMPLE 5 Of the trees in Figure 3: T_1 has two leaves; T_2, T_3 and T_8 have three leaves; T_4, T_5, T_6 and T_7 have four leaves; T_9 and T_{10} have five leaves; and T_{11} has six leaves. ∎

Lemma 1 | A finite tree with at least one edge has at least two leaves.

Proof. Consider a longest simple acyclic path, say $v_1 v_2 \cdots v_n$. Then $v_1 \neq v_n$ and both v_1 and v_n are leaves. ∎

Lemma 2 | A tree with n vertices has exactly $n - 1$ edges.

Proof. We apply induction. For $n = 2$ the lemma is clear. Assume that the result is true for some n, and consider a tree T with $n + 1$ vertices. By Lemma 1, T has a leaf v_0. Let T_0 be the graph obtained by removing v_0 and the edge attached to v_0. Then T_0 is a tree, as is easily checked, and has n vertices. By the inductive assumption T_0 has $n - 1$ edges, so T has n edges. ∎

Theorem 3 | Let G be a finite graph with n vertices, no loops and no parallel edges. The following are equivalent:

(a) G is a tree.
(b) G is acyclic and has $n - 1$ edges.
(c) G is connected and has $n - 1$ edges.

In other words, any two of the properties "connectedness," "acyclicity" and "having $n - 1$ edges" imply the third one.

Proof. The theorem is obvious for $n = 1$, so we assume that $n \geq 2$. Both (a) ⇒ (b) and (a) ⇒ (c) follow from Lemma 2.

(b) ⇒ (a). Assume that (b) holds but that G is not a tree. Then (d) of Theorem 2 cannot hold. Since G is acyclic we can evidently add some edge and retain acyclicity. Now add as many edges as possible and still retain acyclicity. The graph G' so obtained will satisfy Theorem 2(d), so G' will be a tree. Since G' has n vertices and at least n edges, this contradicts Lemma 2. Thus G is a tree.

(c) ⇒ (a). Assume that (c) holds but that G is not a tree. By Theorem 1, G has a spanning tree T, which must have fewer than $n - 1$ edges. This contradicts Lemma 2, so G is a tree. ■

An acyclic graph, whether or not it is connected, is sometimes called a **forest**. Clearly, the connected components of a forest are trees. It is possible to generalize Theorem 3 to characterize forests as well as trees; see Exercise 9.

EXERCISES 6.3

1. Find all trees with six vertices.

2. The trees in Figure 6 have seven vertices. Specify which tree in Figure 3 each is isomorphic to.

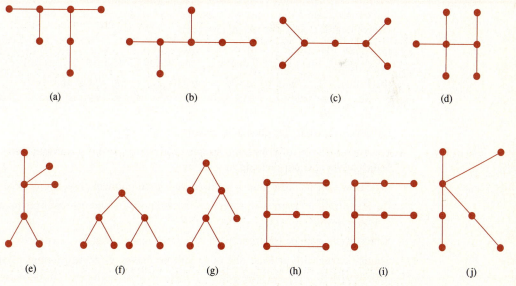

(a)   (b)   (c)   (d)

(e)   (f)   (g)   (h)   (i)   (j)

FIGURE 6

3. Count the number of spanning trees in the graphs of Figure 7.

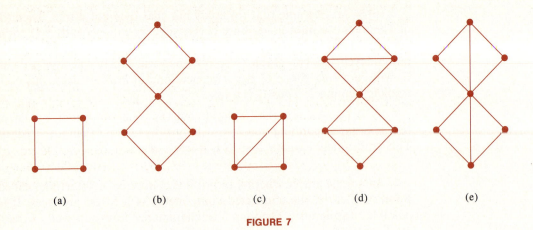

(a)   (b)   (c)   (d)   (e)

FIGURE 7

4. Find two nonisomorphic spanning trees of $K_{3,3}$ drawn in Figure 7 of §6.5.

5. Consider a tree with n vertices. It has exactly $n - 1$ edges [Lemma 2], so the sum of the degrees of its vertices is $2n - 2$ [Theorem 3 of §6.1].

(a) A certain tree has two vertices of degree 4, one vertex of degree 3 and one vertex of degree 2. If the other vertices have degree 1, how many vertices are there in the graph? *Hint:* If the tree has n vertices, $n - 4$ of them will have to have degree 1.

(b) Draw a tree as described in part (a).

6. Repeat Exercise 5 for a tree with two vertices of degree 5, three of degree 3, two of degree 2 and the rest of degree 1.

7. (a) Show that there is a tree with six vertices of degree 1, one vertex of degree 2, one vertex of degree 3, one vertex of degree 5 and no others.

(b) For $n \geq 2$, consider n positive integers $d_1, \ldots, d_n$ whose sum is $2n - 2$. Show that there is a tree with n vertices whose vertices have degrees $d_1, \ldots, d_n$.

(c) Show that part (a) illustrates part (b), where $n = 9$.

8. Draw pictures of all connected graphs with 4 edges and 4 vertices. Don't forget loops and parallel edges.

9. (a) Show that a forest with n vertices and m components has $n - m$ edges.

(b) Show that a graph with n vertices, m components and $n - m$ edges must be a forest.

10. Show that a connected graph with n vertices has at least $n - 1$ edges.

11. Sketch a tree with at least one edge and no leaves. *Hint:* See Lemma 1 to Theorem 3.

§ 6.4   Rooted Trees

A **rooted tree** is a tree with one vertex, called its **root**, singled out. This concept is simple, but amazingly useful. Besides having important applications to data structures, rooted trees also help us to organize and to visualize relationships in a wide variety of settings. In this section we look at several examples of rooted trees and at some of their properties.

Rooted trees are commonly drawn with their roots at the top, just upside-down from the trees in the woods. Our first examples include some typical applications, with accompanying diagrams.

EXAMPLE 1   (a) A set such as a list of numbers or an alphabetized file can be conveniently organized by a special type of rooted tree, called a **binary search tree**. Figure 1 shows an example of such a tree that holds a set of client records, organized in alphabetical order.

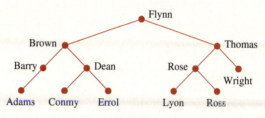

FIGURE 1

To locate the records for Conmy, say, we compare Conmy alphabetically with the root, Flynn. Conmy comes first, so we take the left branch and compare Conmy with Brown. Conmy comes second, so we take the right branch. Then go left at Dean to find Conmy. If we had wanted the records for Dean, we would have stopped when we got to Dean's vertex. What makes the search procedure work and gives the tree its "binary" name is the fact that at each vertex [or **node**, as they are frequently called in this setting] there are at most two edges downward, at most one to the left and at most one to the right. It is easy to see where the records for a client named Romanelli should go: to the right of Flynn, left of Thomas, left of Rose and right of Lyon, and hence at a new vertex below and to the right of Lyon.

The chief advantage to organizing data with a binary search tree is that one only needs to make a few comparisons in order to locate the correct address, even when the total number of records is large. The idea as we have described it here is quite primitive. The method is so important that a number of schemes have been devised for creating and updating search trees to keep the average length of a search path relatively small, on the order of $\log_2 n$ where n is the number of vertices in the tree.

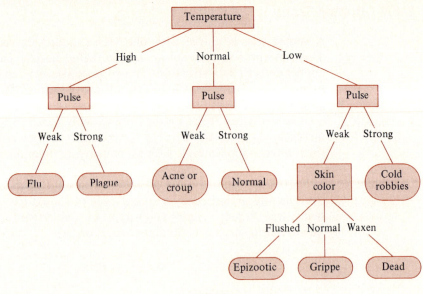

FIGURE 2

(b) Data structures for diagnosis or identification can frequently be viewed as rooted trees. Figure 2 shows the idea. To use such a data structure we start at the top and proceed from vertex to vertex, taking an appropriate branch in each case to match the symptoms of the patient. The final leaf on the path gives the name of the most likely condition or conditions for the given symptoms. The same sort of rooted tree structure is the basis for the key system used in field guides for identifying mushrooms, birds, wildflowers and the like. In the case of the client records in (a) there was a natural left-to-right way to arrange the labels on the nodes. In this example and the next, the order in which the vertices are listed has no special significance.

(c) The chains of command of an organization can often be represented by a rooted tree. Part of the hierarchy of a university is indicated in Figure 3. ■

In these examples we have given the vertices of the trees fancy names or labels, such as "Ross" or "epizootic," rather than u, v, w, etc., to convey additional information. In fact, the *name* of a vertex and its *label* do not have to be the same, and we could quite reasonably have several vertices with different names but the same label. For instance, the label of a vertex might be a dollar figure associated with it. When we speak of a **labeled tree** we will mean a tree that has some additional information attached to its vertices. In practical applications we can think of the label of a vertex as being the information that is stored at the vertex. We will see

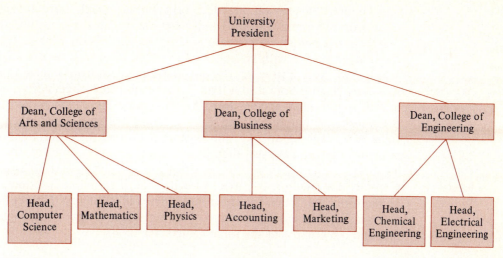

FIGURE 3

shortly that it may also be possible to choose the vertex names themselves to convey information about the locations of the vertices in the graph.

There is a natural way to view a rooted tree as a digraph. Simply pick up the tree by its root and let gravity direct the edges downward. Figure 4(a) shows a tree with root r directed in this way. It is common to leave the arrows off the edges, with the agreement that they all point downward. We can think of the rooted tree in Figure 4(b) as either an undirected tree with distinguished vertex r or as the digraph in Figure 4(a); there is no essential difference between the two points of view.

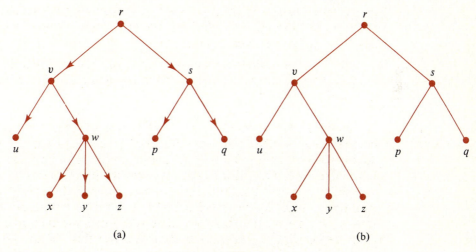

FIGURE 4

To make the gravitational definition mathematically respectable, we recall from Theorem 2 of §6.3 that whenever r and v are vertices of a tree there is a unique simple path joining r to v. If r is the root of T and if e is an edge of T with vertices u and w, then either u is on the unique simple path from r to w or w is on the simple path from r to u. In the first case, direct e from u to w, and in the second case direct e from w to u. This is just what gravity does. We will denote by T_r the rooted tree made from the undirected tree T by choosing the root r, and when we think of T_r as a digraph we will consider it to have the natural digraph structure that we have just described.

EXAMPLE 2   Consider the [undirected] tree in Figure 5(a). If we select v, x and z to be the roots, we obtain the three rooted trees illustrated in Figures 5(b), 5(c) and 5(d). The exact placement of the vertices is unimportant; Figures 5(b) and 5(b') represent the same rooted tree.

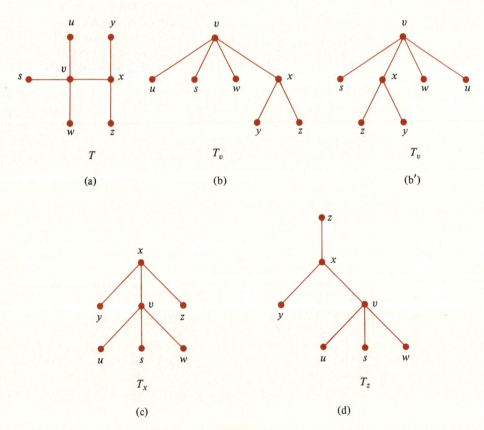

FIGURE 5

Note that (v, w) is an edge in Figure 5(d) because $\{v, w\}$ is an edge of the unique simple path in Figure 5(a) from z to w. On the other hand, (w, v) is not an edge of the rooted tree T_z; even though $\{w, v\}$ is an edge of the original tree, it is not an edge of the unique simple path from z to v. Similar remarks apply to all of the other edges. ■

The terms used to describe various parts of a tree are a curious mixture derived both from the trees in the woods and from family trees. As before, the vertices of degree 1 are called **leaves**; there is one exception: occasionally [as in Figure 5(d)], the root will have degree 1 but we will not call it a leaf. Viewing a tree as a digraph, we see that its root is the only vertex that is not a terminal vertex of an edge, while the leaves are the ones that are not initial vertices. The remaining vertices are sometimes called "branch nodes" or "interior nodes" and the leaves are sometimes called "terminal nodes." We adopt the convention that if (v, w) is an edge of a rooted tree, then v is the **parent** of w and w is a **child** of v. Every vertex except the root has exactly one parent. A parent may have several **children**. More generally, w is a **descendant** of v provided that $w \neq v$ and v is a vertex of the unique simple path from r to w. Finally, for any vertex v the **subtree with root** v is precisely the tree T_v consisting of v, all its descendants and all the directed edges connecting them. Whenever v is a leaf, the subtree with root v is a trivial one-vertex tree.

EXAMPLE 3   Consider the rooted tree in Figure 4 redrawn in Figure 6. There are six leaves. The parent v has two children, u and w, and five descendants: u, w, x, y and z. All the vertices except r itself are descendants of r. The whole tree itself is clearly a subtree rooted at r, and there are six trivial subtrees consisting of leaves. The interesting subtrees are given in Figure 6. ■

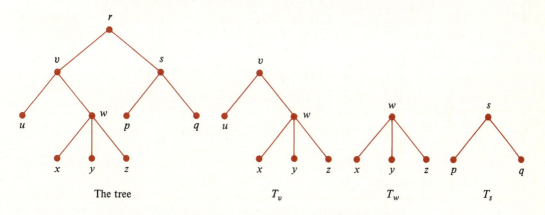

The tree T_v T_w T_s

Subtrees of the tree

FIGURE 6

A rooted tree is a **binary tree** in case each node has at most two children: a left child, a right child, both a left child and a right child, or no children at all. This sort of right–left distinction was important in the binary search tree of Example 1(a). Similarly, for $m > 2$ an **m–ary tree** is one in which the children of each parent are labeled with distinct members of $\{1, 2, \ldots, m\}$. A parent is not required to have a complete set of m children; we say that the ith child is **absent** in case no child is labeled with i. In digraph terminology an m–ary tree has outdeg$(v) \leq m$ for all vertices v. An m–ary tree [or binary tree] is a **regular m–ary tree** if outdeg$(v) = m$ or 0 for all vertices v.

The **level number** of a vertex v is the length of the unique simple path from the root to v. In particular, the root itself has level number 0. The **height** of a rooted tree is the largest level number of a vertex. Only leaves can have their level numbers equal to the height of the tree. A regular m–ary tree is said to be a **full m–ary tree** if all the leaves have the same level number, namely the height of the tree.

EXAMPLE 4   (a) The rooted tree in Figures 4 and 6 is a 3–ary tree and, in fact, is an m–ary tree for $m \geq 3$. It is not a regular 3–ary tree since vertices r, v and s have outdegree 2. Vertices v and s have level number 1, vertices u, w, p and q have level number 2, and the leaves x, y and z have level number 3. The height of the tree is 3.

(b) The labeled tree in Figure 8(a) below is a full regular binary tree of height 3. The labeled tree in Figure 8(b) is a regular 3–ary tree of height 3. It is not a full 3–ary tree since one leaf has level number 1 and five leaves have level number 2. ■

EXAMPLE 5   Consider a full m–ary tree of height h. There are m vertices at level 1. Each parent at level 1 has m children, so there are m^2 vertices at level 2. A simple induction shows that because the tree is full it has m^l vertices at level l for each $l \leq h$. Thus it has $1 + m + m^2 + \cdots + m^h$ vertices in all. Since

$$(m - 1)(1 + m + m^2 + \cdots + m^h) = m^{h+1} - 1,$$

as one can check by multiplying and canceling, we have

$$1 + m + m^2 + \cdots + m^h = \frac{m^{h+1} - 1}{m - 1}.$$

Note that the same tree has $p = (m^h - 1)/(m - 1)$ parents and $t = m^h$ leaves. ■

Example 1(a), the binary search tree, depended on having alphabetical ordering to compare client names. The same idea can be applied more generally; all we need is some ordering or listing to tell us which of two given objects comes first. [The technical term for this kind of listing, which we will discuss further in § 11.2, is **linear order**.] Whenever we want to, we can decide on an order in which to list the children of a given parent in a rooted tree. If we order the children of *every* parent in the tree, we obtain what we call an **ordered rooted tree**. When we draw such a tree, we draw the children in order from left to right. It is convenient to use the notation $v < w$ to mean that v comes before w in the order, even in cases in which v and w are not numbers.

EXAMPLE 6   (a) If we view Figure 5(b) as an ordered rooted tree, then the children of v are ordered: $u < s < w < x$. The children of x are ordered: $y < z$. Figure 5(b') is the picture of a different ordered rooted tree, since $s < x < w < u$ and $z < y$.

(b) As soon as we draw a rooted tree it looks like an ordered rooted tree, even if we do not care about the order structure. For example, the important structure in Figure 3 is the rooted tree structure. The order of the "children" is not important. The head of the computer science department precedes the head of the mathematics department simply because we chose to list the departments in alphabetical order.

(c) A binary tree or, more generally, an m–ary tree is in a natural way an ordered tree, but there is a difference between the two ideas. In a binary tree the right child will be the first child if there is no left child. In the 3–ary tree of Figure 6 the children w, s and q may have the label 3, even though their parents only have two children. ■

EXAMPLE 7   (a) Consider an alphabet Σ, ordered in some way. We make Σ^* into a rooted tree as follows. The empty word λ will serve as the root. For any word w in Σ^*, its set of children is

$$\{wx : x \in \Sigma\}.$$

Since Σ is ordered, we can order each set of children to obtain an ordered rooted tree Σ_λ^* by letting $wx < wy$ in case $x < y$ in Σ.

(b) Let $\Sigma = \{a, b\}$ where $a < b$. Each vertex has two children. For instance, the children of $a\,b\,b\,a$ are $a\,b\,b\,a\,a$ and $a\,b\,b\,a\,b$. Part of the infinite ordered rooted tree Σ_λ^* is drawn in Figure 7 on the next page.

(c) Figure 1 of § 5.1 showed part of the tree $\{a, b, c, d\}_\lambda^*$ with the natural ordering determined by $a < b < c < d$. ■

There are a variety of ways to name [or label] the vertices of an ordered rooted tree so that the names describe their locations. One such scheme resembles Example 7: vertices of an m–ary tree can be named using words from Σ^* where $\Sigma = \mathbb{Z}(m) = \{0, 1, \ldots, m - 1\}$. The ordered children of the root have names from $\{0, 1, 2, \ldots, m - 1\}$. If a vertex is

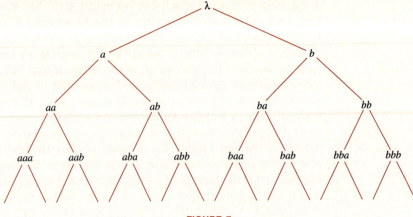

FIGURE 7

named by the word *w*, then its ordered children are named using *w*0, *w*1, *w*2, etc. The name of a vertex tells us the exact location of the vertex in the tree. For example, a vertex named 1021 would be the number two child of the vertex named 102, which, in turn, would be the number three child of the vertex named 10, etc. The level of the vertex is the length of its name; a vertex named 1021 is at level 4.

EXAMPLE 8   All of the vertices in Figure 8(a) except the root are named in this way. In Figure 8(b) we have only named the leaves. The names of the other vertices should be clear. ■

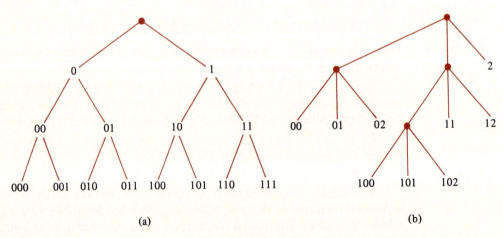

(a)   (b)

FIGURE 8

Trees have been used for a long time to organize or store information. In old mathematics textbooks the individual paragraphs were frequently labeled by using numbers, i.e., words in Σ^* where $\Sigma = \{1, 2, 3, \ldots\}$. Decimal points were used to set off the letters of the words in Σ^*. Thus 3.4.1.2 would refer to the second paragraph of the first subsection of the fourth section of Chapter 3, while 3.4.12 would refer to the twelfth subsection of the fourth section of Chapter 3. This scheme is not very pretty, so modern authors usually avoid it, but it has some real advantages that carry over to present-day uses of trees. One can always insert new paragraphs or sections without disrupting the numbering system. With a little care, paragraphs and sections can be deleted without causing trouble, especially if one doesn't mind gaps. Also a label, such as 3.4.12, specifies exactly where the subsection or paragraph fits into the book. Although the book itself is printed linearly as one long string of characters, the numbering scheme lets us visualize a tree structure associated with it. In contrast, one famous mathematics book has theorems numbered from 1 to 460. All the label "Theorem 303" tells us is that this theorem probably appears about two-thirds of the way through the book.

Of course, being able to locate a node in a tree quickly is most useful if we know which node we want to reach. If we are storing patient records with a tree indexed by last names, and if we need to look at every node, for example to determine which patients are taking a particular drug, the tree structure is not much help. In §7.3 we will investigate search schemes for systematically going through all the nodes in a rooted tree.

EXERCISES 6.4

1. (a) For the tree in Figure 4, draw a rooted tree with new root v.

   (b) What is the level number of the vertex r?

   (c) What is the height of the tree?

2. Create a binary search tree of height 4 for the usual English alphabet $\{a, b, \ldots, z\}$ with its usual order.

3. (a) For each rooted tree in Figure 5, give the level numbers of the vertices and the height of the tree.

   (b) Which of the trees in Figure 5 are regular m–ary for some m?

4. Discuss why ordinary family trees are not rooted trees.

5. (a) There are seven different types of rooted trees of height 2 in which each node has at most two children. Draw one tree of each type.

   (b) To which of the types in part (a) do the regular binary trees of height 2 belong?

(c) Which of the trees in part (a) are full binary trees?

(d) How many different types of binary trees are there of height 2?

6. (a) Repeat Exercise 5(a) for the seven types of rooted trees of height 3 in which each node that is not a leaf has two children.

(b) How many different types of regular binary trees are there of height 3?

7. For each n draw a binary search tree whose nodes are $1, 2, 3, \ldots, n$ and whose height is as small as possible.

(a) $n = 7$ (b) $n = 15$ (c) $n = 4$ (d) $n = 6$

8. A **2–3 tree** is a rooted tree such that each interior node has either two or three children and all paths from the root to the leaves have the same length. There are seven different types of 2–3 trees of height 2. Draw one tree of each type. [2–3 trees yield data structures that are comparatively easy to update.]

9. (a) Draw full m–ary trees of height h for $m = 2$, $h = 2$; $m = 2$, $h = 3$; and $m = 3$, $h = 2$.

(b) Which trees in part (a) have m^h leaves?

10. Consider a full binary tree T of height h.

(a) How many leaves does T have?

(b) How many vertices does T have?

11. Consider a full m–ary tree with p parents and t leaves. Show that $t = (m - 1)p + 1$ no matter what the height is.

12. Give some real-life examples of information storage that can be viewed as labeled trees.

13. Let $\Sigma = \{a, b\}$ and consider the rooted tree Σ_λ^*; see Example 7. Describe the set of vertices at level k. How big is this set?

14. Draw part of the rooted tree Σ_λ^* where $\Sigma = \{a, b, c\}$ and $a < b < c$ as usual.

The next two exercises illustrate some of the problems associated with updating binary search trees.

15. (a) Suppose that in Example 1(a) client Rose moves away. How might we naturally rearrange the tree to delete the records of Rose without disturbing the rest of the tree too much?

(b) Repeat part (a) with Brown moving instead of Rose.

16. (a) Suppose that in Example 1(a) a new client, Smith, must be added to the binary search tree. Show how to do so without increasing the height of the tree. Try not to disturb the rest of the tree more than necessary.

(b) Suppose that three new clients, Smith1, Smith2 and Smith3, must be added to the tree in Example 1(a). Show how to do so without increasing the height of the binary search tree.

(c) What happens in part (b) if there are four new clients?

§ 6.5 Vertex Traversal Problems

Euler's theorem in §6.2 tells us which graphs have closed paths using each edge exactly once, and FLEURY'S algorithm gives a way to construct the paths when they exist. In contrast, much less is known about graphs with paths that use each vertex exactly once. The Irish mathematician Sir William Hamilton was one of the first to study such graphs, and at one time even marketed a puzzle based on the problem.

A path is called a **Hamilton path** if it visits every vertex of the graph exactly once. A closed path that visits every vertex of the graph exactly once, except for the last vertex which duplicates the first one, is called a **Hamilton circuit**. A graph with a Hamilton circuit is called a **Hamiltonian graph**. A Hamilton path must be simple, and by Proposition 1 of §6.1, if G has at least 3 vertices, then a Hamilton circuit of G must be a cycle.

EXAMPLE 1   (a) The graph shown in Figure 1(a) has Hamilton circuit $v\,w\,x\,y\,z\,v$.

(b) Adding more edges can't hurt, so the graph K_5 of Figure 1(b) is also Hamiltonian. In fact, every complete graph K_n for $n \geq 3$ is Hamiltonian; we can go from vertex to vertex in any order we please.

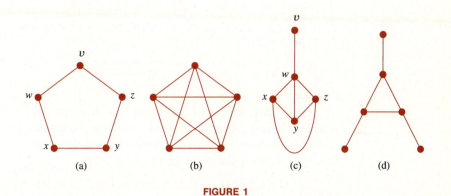

FIGURE 1

(c) The graph of Figure 1(c) has the Hamilton path $v\,w\,x\,y\,z$ but has no Hamilton circuit since no cycle goes through v.

(d) The graph of Figure 1(d) has no Hamilton path. ∎

With Euler's theorem, the theory of Euler circuits is very nice and complete. What can be proved about Hamilton circuits? Under certain conditions, graphs will have so many edges compared to the number of vertices that they must have Hamilton circuits. But the graph in Figure 1(a) has very few edges, yet it has a Hamilton circuit. And the graph in Figure 1(c) has lots of edges but no Hamilton circuit. It turns out that there is no known simple characterization of those connected graphs possessing Hamilton circuits. The concept of Hamilton circuit seems very close to that of Euler circuit, yet the theory of Hamilton circuits is vastly more complicated. In particular, no efficient algorithm is known for finding Hamilton circuits. The problem is a special case of the Traveling Salesperson Problem. Here one begins with a graph whose edges are assigned **weights** that may represent mileage, cost, computer time or some other quantity that we wish to minimize. In Figure 2 the weights might represent mileage between cities on a traveling salesperson's route. The goal is to find the shortest round trip that visits each city exactly once. That is, the goal is to find a Hamilton circuit minimizing the sum of the weights of the edges. A nice algorithm solving this problem would also be able to find Hamilton circuits in an unweighted graph, since we could always assign weight 1 to each edge.

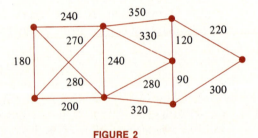

FIGURE 2

Obviously, a Hamiltonian graph with n vertices must have at least n edges. This necessary condition may not be sufficient, as Figures 1(c) and 1(d) illustrate. Of course, loops and parallel edges are of no use. The following gives a simple sufficient condition.

Theorem 1 | If the graph G has no loops or parallel edges, if $|V(G)| = n \geq 3$ and if $\deg(v) \geq n/2$ for each vertex v of G, then G is Hamiltonian.

EXAMPLE 2   (a) The graph K_5 in Figure 1(b) has $\deg(v) = 4$ for each v and has $|V(G)| = 5$, so it satisfies the condition of Theorem 1.

(b) Each of the graphs in Figure 3 has $|V(G)|/2 = 5/2$ and has a vertex of degree 2. They do not satisfy the hypotheses of Theorem 1, but are nevertheless Hamiltonian. ■

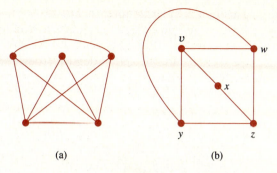

(a)   (b)

FIGURE 3

Theorem 1 imposes a uniform condition on all the vertices. Our next theorem requires only that there be enough edges somewhere in the graph. We will establish both of these sufficient conditions as consequences of Theorem 3, which gives a criterion in terms of degrees of pairs of vertices.

Theorem 2   A graph with n vertices and with no loops or parallel edges that has at least $\frac{1}{2}(n-1)(n-2)+2$ edges is Hamiltonian.

EXAMPLE 3   (a) The Hamiltonian graph of Figure 3(a) has $n = 5$, which gives $\frac{1}{2}(n-1)(n-2)+2 = 8$. It has 8 edges, so it satisfies the hypotheses and the conclusion of Theorem 2.

(b) The Hamiltonian graph of Figure 3(b) also has $n = 5$, so $\frac{1}{2}(n-1)(n-2)+2 = 8$, but it has only 7 edges. It fails to satisfy the hypotheses of Theorem 2 as well as Theorem 1. If there were no vertex in the middle, we would have K_4 with $n = 4$, so $\frac{1}{2}(n-1)(n-2)+2 = 5$, and the 6 edges would be more than enough. As it stands, the graph satisfies the hypotheses of the next theorem.   ■

Theorem 3   Suppose that the graph G has no loops or parallel edges and that $|V(G)| = n \geq 3$. If

$$\deg(v) + \deg(w) \geq n$$

for each pair of vertices v and w that are not connected by an edge, then G is Hamiltonian.

EXAMPLE 4   For the graph in Figure 3(b), $n = 5$. There are three pairs of distinct vertices that are not connected by an edge. We verify the hypotheses of

Theorem 3 by examining them:

$$\text{for } (v, z), \qquad \deg(v) + \deg(z) = 3 + 3 = 6 \geq 5;$$

$$\text{for } (w, x), \qquad \deg(w) + \deg(x) = 3 + 2 = 5 \geq 5;$$

$$\text{for } (x, y), \qquad \deg(x) + \deg(y) = 2 + 3 = 5 \geq 5. \quad \blacksquare$$

Proof of Theorem 3. Suppose that the theorem is false for some n, and let G be a counterexample with $|E(G)|$ as large as possible. Now G is a subgraph of the Hamiltonian graph K_n. Adjoining to G an edge from K_n would give a graph that still satisfies the degree condition but has more than $|E(G)|$ edges. By the choice of G, any such graph would have a Hamilton circuit. This means that G must already have a Hamilton *path*, say with vertex sequence $v_1 v_2 \cdots v_n$. Since G has no Hamilton circuit, v_1 and v_n are not connected by an edge in G, and so we have $\deg(v_1) + \deg(v_n) \geq n$.

Define subsets S_1 and S_n of $\{2, \ldots, n\}$ by

$$S_1 = \{i : \{v_1, v_i\} \in E(G)\} \quad \text{and} \quad S_n = \{i : \{v_{i-1}, v_n\} \in E(G)\}.$$

Then $|S_1| = \deg(v_1)$ and $|S_n| = \deg(v_n)$. Since $|S_1| + |S_n| \geq n$ and $S_1 \cup S_n$ has at most $n - 1$ elements, $S_1 \cap S_n$ must be nonempty. Thus there is an i for which both $\{v_1, v_i\}$ and $\{v_{i-1}, v_n\}$ are edges of G. Then [see Figure 4] *the path $v_1 \cdots v_{i-1} v_n \cdots v_i v_1$ is a Hamilton circuit in G*, contradicting the choice of G as a counterexample. $\blacksquare$

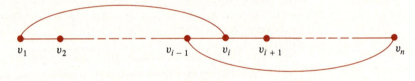

FIGURE 4

Our first two sufficient conditions follow easily from Theorem 3.

Proofs of Theorems 1 and 2. Suppose that G has no loops or parallel edges and $|V(G)| = n \geq 3$.

If $\deg(v) \geq n/2$ for each v, then $\deg(v) + \deg(w) \geq n$ for any v and w whether joined by an edge or not, so the hypothesis of Theorem 3 is satisfied and G is Hamiltonian.

Suppose that

$$|E(G)| \geq \frac{1}{2}(n-1)(n-2) + 2 = \binom{n-1}{2} + 2,$$

and consider vertices u and v with $\{u, v\} \notin E(G)$. Remove from G the vertices u and v and all edges with u or v as a vertex. Since $\{u, v\} \notin E(G)$ we have removed $\deg(u) + \deg(v)$ edges and 2 vertices. The graph G' which is left is a subgraph of K_{n-2}, so

$$\binom{n-2}{2} = |E(K_{n-2})| \geq |E(G')| \geq \binom{n-1}{2} + 2 - \deg(u) - \deg(v).$$

Hence

$$\deg(u) + \deg(v) \geq \binom{n-1}{2} - \binom{n-2}{2} + 2$$

$$= \tfrac{1}{2}(n-1)(n-2) - \tfrac{1}{2}(n-2)(n-3) + 2$$

$$= \tfrac{1}{2}(n-2)[(n-1) - (n-3)] + 2$$

$$= \tfrac{1}{2}(n-2)[2] + 2 = n.$$

Again, G satisfies the hypothesis of Theorem 3. ∎

Theorems 1, 2 and 3 are somewhat unsatisfactory in two ways. Not only are their sufficient conditions not necessary, the theorems give no guidance for finding a Hamilton circuit when one is guaranteed to exist. As we mentioned earlier, as of this writing no efficient algorithm is known for finding Hamilton paths or circuits. On the positive side, a Hamiltonian graph must certainly be connected, so all three theorems give sufficient conditions for a graph to be connected.

EXAMPLE 5   A **Gray code** of length n is a list of all 2^n distinct strings of n binary digits in such a way that adjacent strings differ in exactly one digit and so that the last string differs from the first string in exactly one digit. For example, $0\,0$, $0\,1$, $1\,1$, $1\,0$ is a Gray code of length 2.

We can view the construction of a Gray code as a graph-theoretic problem. Let $V(G)$ be the set $\{0, 1\}^n$ of binary n-tuples, and join u and v by an edge if u and v differ in exactly one digit. A Gray code of length n is, in effect, a Hamilton circuit of the graph G. Figure 5(a) shows the

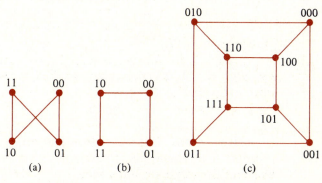

FIGURE 5

graph G for $n = 2$. Figure 5(b) shows the same graph redrawn. This graph has two Hamilton circuits, one in each direction, which shows that there are two [essentially equivalent] Gray codes of length 2. Figure 5(c) shows the graph for $n = 3$. There are 12 Gray codes of length 3. Figure 6 indicates the Hamilton circuit corresponding to one such code. Thus

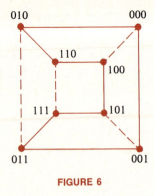

$$000, \quad 001, \quad 011, \quad 111,$$
$$101, \quad 100, \quad 110, \quad 010$$

FIGURE 6

is a Gray code of length 3.

Gray codes can be used to label the individual processors in a hypercube array. [The square and cube in Figure 5 are hypercubes of dimension 2 and 3.] Using such a labeling scheme, two processors are connected if and only if their labels differ in just one bit. ∎

The vertices in the graphs we constructed in Example 5 can be partitioned into two sets, those with an even number of 1's and those with an odd number, so that each edge joins a member of one set to a member of the other. We conclude this section with some observations about Hamilton circuits in graphs with this sort of partition.

A graph G is called **bipartite** if $V(G)$ is the union of two disjoint nonempty subsets V_1 and V_2 such that every edge of G joins a vertex of V_1 to a vertex of V_2. A graph is called a **complete bipartite** graph if, in addition, every vertex of V_1 is joined to every vertex of V_2 by a unique edge.

EXAMPLE 6 The graphs shown in Figure 7 are all bipartite. All but the one in Figure 7(b) are complete bipartite graphs. ∎

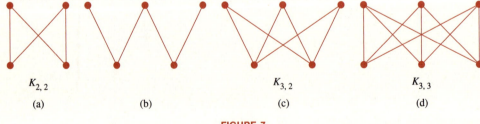

$K_{2,2}$ $K_{3,2}$ $K_{3,3}$

(a)   (b)   (c)   (d)

FIGURE 7

Given m and n, the complete bipartite graphs with $|V_1| = m$ and $|V_2| = n$ are all isomorphic to each other; we denote them by $K_{m,n}$. Note that $K_{m,n}$ and $K_{n,m}$ are isomorphic.

Theorem 4 Let G be a bipartite graph with partition $V(G) = V_1 \cup V_2$. If G has a Hamilton circuit, then $|V_1| = |V_2|$. If G has a Hamilton path, then the numbers $|V_1|$ and $|V_2|$ differ by at most 1. For complete bipartite graphs with at least 3 vertices the converse statements are also true.

Proof. The vertices on a path in G alternately belong to V_1 and V_2. If $x_1 x_2 \cdots x_n x_1$ is a closed path that goes through each vertex once, then $x_1, x_3, x_5, \ldots$ must belong to one of the sets, say V_1. Since $\{x_n, x_1\}$ is an edge, n must be even and $x_2, x_4, \ldots, x_n$ all belong to V_2. So $|V_1| = |V_2|$. Similar remarks apply to a nonclosed path $x_1 x_2 \cdots x_n$, except that n might be odd, in which case one of V_1 and V_2 will have an extra vertex.

Now suppose that $G = K_{m,n}$. If $m = n$, we can simply go back and forth from V_1 to V_2, since edges exist to take us wherever we want. If $m = n + 1$, we should start in V_1 to get a Hamilton path. ∎

A computer scientist we know tells the story of how he once spent over two weeks on a computer searching for a Hamilton path in a bipartite graph with 42 vertices before he realized that the graph violated the condition of Theorem 4. The story has two messages: (1) people do have practical applications for bipartite graphs and Hamilton paths, and (2) *thought should precede computation.*

EXERCISES 6.5

1. (a) Explain why the graph in Figure 1(c) has no Hamilton circuit.

   (b) Explain why the graph in Figure 1(d) has no Hamilton path.

2. For each graph in Figure 8 give a Hamilton circuit or explain why none exists.

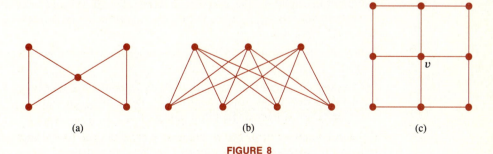

(a)   (b)   (c)

FIGURE 8

3. Consider the graph shown in Figure 9(a).

(a) Is this a Hamiltonian graph?

(b) Is this a complete graph?

(c) Is this a bipartite graph?

(d) Is this a complete bipartite graph?

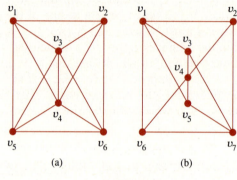

(a)   (b)

FIGURE 9

4. Answer the same questions as in Exercise 3 for the graph in Figure 9(b).

5. (a) How many Hamilton circuits does the graph $K_{n,n}$ have for $n \geq 2$? [Count circuits as different if they have different starting points or vertex sequences.]

(b) How many Hamilton paths does $K_{n,n-1}$ have for $n \geq 2$?

(c) Which complete bipartite graphs $K_{m,n}$ have Euler paths?

6. Redraw the graphs in Figure 5 and mark each of the subsets V_1 and V_2 of the bipartite partition of $V(G)$.

7. Arrange eight 0's and 1's in a circle so that each 3-digit binary number occurs as a string of 3 consecutive symbols somewhere in the circle. *Hint:* Find a Hamilton circuit in the graph with vertex set $\{0, 1\}^3$ and with an edge between vertices (v_1, v_2, v_3) and (w_1, w_2, w_3) whenever $(v_1, v_2) = (w_2, w_3)$ or $(v_2, v_3) = (w_1, w_2)$.

8. Give two other examples of Gray codes of length 3 besides the one in Example 5.

9. Consider the graph that has vertex set $\{0, 1\}^3$ and an edge between vertices whenever they differ in two coordinates. Does the graph have a Hamilton circuit? Does it have a Hamilton path?

10. Repeat Exercise 9 for the graph that has vertex set $\{0, 1\}^3$ and an edge between vertices if they differ in two or three coordinates.

11. For $n \geq 4$ build the graph K_n^+ from the complete graph K_{n-1} by adding one more vertex in the middle of an edge of K_{n-1}. [Figure 3(b) shows K_5^+.]

    (a) Show that K_n^+ does not satisfy the condition of Theorem 2.

    (b) Use Theorem 3 to show that K_n^+ is nevertheless Hamiltonian.

12. For $n \geq 4$ build a graph K_n^{++} from the complete graph K_{n-1} by adding one more vertex and an edge from the new vertex to a vertex of K_{n-1}. [Figure 1(c) shows K_5^{++}.] Show that K_n^{++} is not Hamiltonian. Observe that K_n^{++} has n vertices and $\frac{1}{2}(n-1)(n-2) + 1$ edges. This example shows that the number of edges required in Theorem 2 cannot be decreased.

13. The **complement** of a graph G is the graph with vertex set $V(G)$ and with an edge between distinct vertices v and w if G does *not* have an edge joining v and w.

    (a) Draw the complement of the graph of Figure 3(b).

    (b) How many components does the complement in part (a) have?

    (c) Show that if G is not connected, then its complement is connected.

    (d) Give an example of a graph that is isomorphic to its complement.

    (e) Is the converse to the statement in part (c) true?

14. Suppose that the graph G is regular of degree $k \geq 1$ [i.e., each vertex has degree k] and has at least $2k + 2$ vertices. Show that the complement of G is Hamiltonian. *Hint:* Use Theorem 1.

15. Show that Gray codes of length n always exist. *Hint:* Use induction on n and consider the graph G_n in which a Hamilton circuit corresponds to a Gray code of length n, as described in Example 5.

16. Explain why none of the theorems in this section can be used to solve Exercise 15.

§ 6.6   Minimum Spanning Trees

The theorems that characterize trees suggest two methods for finding a spanning tree of a finite connected graph. Using the idea in the proof of Theorem 1 of §6.3, we could just remove edges one after another without destroying connectedness, i.e., remove edges that belong to cycles, until we are forced to stop. If G has n vertices and more than $2n$ edges, this procedure will examine and throw out more than half of the edges. It might be faster, if we could do it, to build up a spanning tree by choosing its $n - 1$ edges one at a time so that at each stage the subgraph of chosen edges is acyclic. The algorithms in this section all build trees in this second way.

Our first algorithm starts from an initially chosen vertex v. If the given graph is connected, then the algorithm produces a spanning tree for it. Otherwise, the algorithm gives a spanning tree for the connected component of the graph that contains v.

Algorithm **TREE(v)**

{input: vertex v of the finite graph G}
{output: a set E of edges of a spanning tree for the component of G
    that contains v}
{auxiliary variable: a list V of visited vertices; ultimately $V = V(G)$}
Let $V := \{v\}$ and $E := \varnothing$.
While there are edges of G joining vertices in V to vertices that are
    not in V,
    choose such an edge $\{u, w\}$ joining u in V to w not in V,
    put w in V and put $\{u, w\}$ in E. ■

To get a spanning forest for G, we just keep growing trees.

Algorithm **FOREST**

{input: a finite graph G}
{output: a set EE of edges of a spanning forest for G}
Set $VV := \varnothing$ and $EE := \varnothing$.
While $VV \neq V(G)$
    choose $v \in V(G) \backslash VV$
    do TREE(v) {to get a vertex set V and edge set E for a tree spanning
        the component containing v}
    put v in VV and put the members of E in EE. ■

EXAMPLE 1 We can illustrate the operation of TREE and FOREST on the graph
shown in Figure 1(a). Figure 2 shows the steps in TREE(1) and in TREE(2);
then FOREST puts them together. Where choices are available we have
chosen in increasing numerical order and exhausted all edges from a
given vertex before going on to the next vertex. Other choice schemes
are possible, of course. Figure 1(b) shows the two trees that are grown
in the spanning forest, with vertices labeled according to whether they
are in the component of 1 or the component of 2. ■

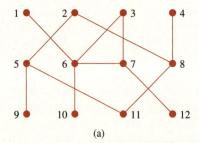

(a)

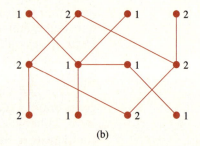

(b)

FIGURE 1

TREE(1)
$V := \{1\}; \quad E := \emptyset.$
Choose edge $\{1, 6\}$.
  $V := \{1, 6\}; \quad E := \{\{1, 6\}\}.$
Choose edges $\{6, 3\}, \{6, 7\}, \{6, 10\}$ [order doesn't matter here].
  $V := \{1, 6, 3, 7, 10\}; \quad E := \{\{1, 6\}, \{6, 3\}, \{6, 7\}, \{6, 10\}\}.$
Choose edge $\{7, 12\}$ [not $\{3, 7\}$ since 3 and 7 are both in V].
  $V := \{1, 6, 3, 7, 10, 12\}; \quad E := \{\{1, 6\}, \{6, 3\}, \{6, 7\}, \{6, 10\}, \{7, 12\}\}.$

TREE(2)
$V := \{2\}; \quad E := \emptyset.$
Choose edges $\{2, 5\}, \{2, 8\}$.
  $V := \{2, 5, 8\}; \quad E := \{\{2, 5\}, \{2, 8\}\}.$
Choose edges $\{5, 9\}, \{5, 11\}$.
  $V := \{2, 5, 8, 9, 11\}; \quad E := \{\{2, 5\}, \{2, 8\}, \{5, 9\}, \{5, 11\}\}.$
Choose edge $\{8, 4\}$.
  $V := \{2, 5, 8, 9, 11, 4\}; \quad E := \{\{2, 5\}, \{2, 8\}, \{5, 9\}, \{5, 11\}, \{8, 4\}\}.$

FOREST
Put together the two lists E from TREE(1) and TREE(2) to form EE.

FIGURE 2

Theorem 1

> TREE(v) produces a spanning tree for the component of G containing v. Hence FOREST produces a spanning forest for G.

Proof. Here is TREE(v) again for reference.

TREE(v)
Let $V := \{v\}$ and $E := \emptyset.$
While there are edges of G joining vertices in V to vertices that are not in V, choose such an edge $\{u, w\}$ joining u in V to w not in V, put w in V and put $\{u, w\}$ in E.

Each pass through the while loop increases the size of V, so the algorithm must stop eventually. The statement "V and E are the vertices and edges of a tree with v as vertex" is clearly true when we first enter the loop. We show that it is an invariant of the loop, so that when the algorithm stops it stops with a tree containing v. Theorem 3 of §6.3 implies that attaching a new vertex to a tree with a new edge yields a tree. Since this is precisely what the while loop does, the statement is a loop invariant. When the algorithm stops there can't still be vertices in the component of v that are not in V, for if there were the guard on the loop would still be true. Thus TREE(v) performs as advertised. ∎

The time that TREE(v) takes depends on the scheme for making choices and on how the list of available edges is maintained. If we mark each vertex when we choose it and at the same time tell its neighbors that it's marked, then each vertex and each edge is handled only once. If G is connected, then TREE(v) builds a spanning tree in time $O(|V(G)| + |E(G)|)$. The same argument shows that in the general case FOREST also runs in time $O(|V(G)| + |E(G)|)$.

The number of components of G is the number of trees in a spanning forest. It is easy to keep track of the components as we build trees with FOREST, by using a function C that assigns the value u to each vertex w in the list V produced by TREE(u). Then each pass through the while loop in FOREST produces a different value of C, which is shared by all the vertices in the component for that pass. The labels 1 and 2 in Figure 1(b) were assigned using such a function C. To modify the algorithm, simply set $C(v) := v$ at the start of TREE(v) and add the line "Set $C(w) := C(u)$" after putting the edge $\{u, w\}$ in E.

FOREST can be used to test connectivity of a graph; just check whether the algorithm produces more than one tree. FOREST can also be used to give a relatively fast test for the presence of cycles in a graph. If G is acyclic, then the spanning forest produced is just G itself; otherwise, $|EE| < |E(G)|$ at the conclusion of the algorithm.

The question of finding spanning trees is especially interesting if the edges are **weighted**, i.e., if each edge e of G is assigned a nonnegative number $W(e)$. The **weight** $W(H)$ of a subgraph H of G is simply the sum of the weights of the edges of H. The problem is to find a **minimum spanning tree**, i.e., a spanning tree whose weight is less than or equal to that of any other. If a graph G is not weighted and if we assign each edge the weight 1, then all spanning trees are minimum, since they all have weight $|V(G)| - 1$.

Our first algorithm builds a minimum spanning tree for a weighted graph G whose edges $e_1, \ldots, e_m$ have been initially sorted so that

$$W(e_1) \le W(e_2) \le \cdots \le W(e_m).$$

The algorithm proceeds one by one through the list of edges of G, beginning with the smallest weights, choosing edges that do not introduce cycles. When the algorithm stops, the set E is supposed to be the set of edges in a minimum spanning tree for G. The notation $E \cup \{e_j\}$ in the statement of the algorithm stands for the subgraph whose edge set is $E \cup \{e_j\}$ and whose vertex set is $V(G)$.

KRUSKAL'S *Algorithm*

{input: a finite weighted connected graph G with edges listed in order
　　of increasing weight}
{output: a set E of edges of a minimum spanning tree for G}
Set $E := \varnothing$.
For $j = 1$ to $|E(G)|$
　if $E \cup \{e_j\}$ is acyclic put e_j in E.   ∎

EXAMPLE 2   Figure 3(a) shows a weighted graph with the weights indicated next to the edges. Figure 3(b) shows one possible way to number the edges of the graph so that the weights form a nondecreasing sequence, i.e., with $W(e_i) \leq W(e_j)$ whenever $i < j$.

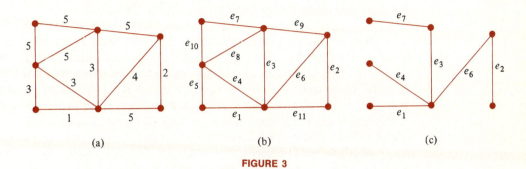

(a)   (b)   (c)

FIGURE 3

Applying KRUSKAL'S algorithm to this weighted graph gives the spanning tree T with edges $e_1, e_2, e_3, e_4, e_6, e_7$ sketched in Figure 3(c). Edge e_5 was rejected because $e_1 e_4 e_5$ would form a cycle. Edges e_8 through e_{11} were rejected for similar reasons. The spanning tree T has weight 18. ■

Theorem 2	KRUSKAL'S algorithm produces a minimum spanning tree.

Proof. We show first that the statement "E is contained in a minimum spanning tree of G" is a loop invariant. This statement is clearly true initially when E is empty. Suppose that it is true at the start of the jth pass through the for loop, so that E is contained in some minimum spanning tree T. If $E \cup \{e_j\}$ is not acyclic, then E doesn't change, so we may suppose that $E \cup \{e_j\}$ is acyclic. We want to find a minimum spanning tree T^* that contains $E \cup \{e_j\}$. If e_j is in T, we can take $T^* = T$. Thus suppose that e_j is not in T.

Then e_j belongs to some cycle C in $T \cup \{e_j\}$ by Theorem 3 of §6.3. Since $E \cup \{e_j\}$ is acyclic, C must contain some edge f in $T \setminus (E \cup \{e_j\})$. Form $T^* := (T \cup C) \setminus \{f\} = (T \cup \{e_j\}) \setminus \{f\}$. Then T^* is connected, spans G and has $|V(G)| - 1$ edges. Hence, by Theorem 3 of §6.3, T^* is a spanning tree for G containing $E \cup \{e_j\}$. Because $E \cup \{f\} \subseteq T$, $E \cup \{f\}$ is acyclic. Since f has not yet been picked to be adjoined to E, it must be that e_j has first chance, i.e., that $W(e_j) \leq W(f)$. Since $W(T^*) = W(T) + W(e_j) - W(f) \leq W(T)$, T^* is a minimum spanning tree, as desired.

Since E is always contained in a minimum spanning tree, it only remains to show that the graph with edge set E and vertex set $V(G)$ is connected when the algorithm stops. Let v and w be two vertices of G. Since G is connected, there is a path from u to v in G. If some edge f on that path is not in E, then $E \cup \{f\}$ contains a cycle [else f would have been chosen in its turn], so f can be replaced in the path by the part of the cycle that's in E. Making necessary replacements in this way, we obtain a path from u to v lying entirely in E. ∎

Note, by the way, that Kruskal's algorithm works even if G has loops or parallel edges. It never chooses loops, and it will select the first edge listed in a collection of parallel edges. It is not even necessary for G to be connected in order to apply KRUSKAL'S algorithm. In the general case the algorithm produces a **minimum spanning forest** made up of minimum spanning trees for the various components of G.

If G has n vertices, KRUSKAL'S algorithm can't produce more than $n - 1$ edges in E. The algorithm could be programmed to stop when $|E| = n - 1$, but it might still need to examine every edge in G before it stopped.

Each edge examined requires a test to see if e_j belongs to a cycle. The algorithm FOREST can be applied to the graph $E \cup \{e_j\}$ to test whether it contains a cycle. The idea in FOREST can be applied in another way to give an acyclicity check, using the observations in Theorem 2 of §6.3. Suppose that G' is the graph with $V(G') = V(G)$ and $E(G') = E$ when the algorithm is examining e_j. If we know which components of G' the endpoints of e_j lie in, then we can add e_j to E if they lie in different components and reject e_j otherwise.

This test is quick, provided that we keep track of the components. At the start, each component consists of a single vertex, and it's easy to update the component list after accepting e_j; the components of the endpoints of e_j just merge into a single component. The resulting version of KRUSKAL'S algorithm runs in time $O(|E(G)| \cdot \log_2 |E(G)|)$, including the time that it takes to sort $E(G)$ initially. For complete details see, for example, the accounts of KRUSKAL'S algorithm in *Data Structures and Algorithms* by Aho, Hopcroft and Ullmann, *Introduction to Algorithms* by Cormen, Leiserson and Rivest or *Data Structures and Network Algorithms* by Tarjan.

In the case of the graph in Example 2 it would have been quicker to delete a few bad edges from G than it was to build T up one edge at a time. There is a general algorithm that works by deleting edges: given a connected graph with the edges listed in increasing order of weight, go through the list starting at the big end, throwing out an edge if and only if it belongs to a cycle in the current subgraph of G. The subgraphs that arise during the operation of this algorithm are all connected, and the algorithm only stops when it reaches an acyclic graph, so the final result is a spanning tree for G. It is, in fact, a minimum spanning tree [Exercise 14]. Indeed, it's the same tree that KRUSKAL'S algorithm produces. If

$|E(G)| < 2|V(G)| - 1$ this procedure may take less time than KRUSKAL'S algorithm, but of course if G has so few edges then both algorithms work quite quickly.

KRUSKAL'S algorithm makes sure that the subgraph being built is always acyclic, while the deletion procedure we have just described keeps all subgraphs connected. Both algorithms are greedy, in the sense that they always choose the smallest edge to add or the largest to delete. Fortunately, greed pays off this time.

The algorithm we next describe is doubly greedy; it makes minimum choices while simultaneously keeping the subgraph both acyclic and connected. Moreover, it does not require the edges of G to be sorted initially. The procedure works just like TREE(v), but takes weights into account. It grows a tree T inside G, with $V(T) = V$ and $E(T) = E$. At each stage the algorithm looks for an edge of smallest weight that joins a vertex in T to some new vertex outside T. Then it adds such an edge and vertex to T and repeats the process.

PRIM'S *Algorithm*

{input: a finite weighted connected graph G [with edges listed in any order]}
{output: a set E of edges of a minimum spanning tree for G}
Set $E := \varnothing$.
Choose w in $V(G)$ and set $V := \{w\}$.
While $V \neq V(G)$
   choose an edge $\{u, v\}$ in $E(G)$ of smallest possible weight with $u \in V$ and $v \in V(G) \backslash V$
   put $\{u, v\}$ in E and put v in V. ■

EXAMPLE 3   We apply PRIM'S algorithm to the weighted graph shown in Figure 4(a). Since there are choices possible at several stages of the execution, the resulting tree is not uniquely determined. The solid edges in Figure 4(b) show one possible outcome, with the edges labeled a, b, c, d, e in the order in which they are chosen. The dashed edges b' and d' are alternate

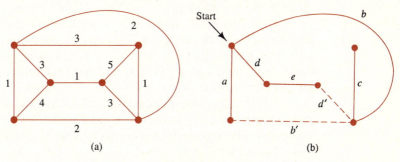

(a)   (b)

FIGURE 4

choices. Note that KRUSKAL'S algorithm would have chosen edges c and e before b. ■

Theorem 3

PRIM'S algorithm produces a minimum spanning tree for a connected weighted graph.

Proof. The proof of Theorem 1, that TREE(v) works, shows that PRIM'S algorithm terminates and yields a spanning tree for G; all that is new here is greed. The statement "T is contained in a minimum spanning tree of G" is surely true at the start, when T is just a single vertex. We claim that this statement is an invariant of the while loop.

Suppose that at the beginning of some pass through the while loop T is contained in the minimum spanning tree T^* of G. Suppose that the algorithm now chooses the edge $\{u, v\}$. If $\{u, v\} \in E(T^*)$, then the new T is still contained in T^*, which is wonderful. Suppose not. Because T^* is a spanning tree, there is a path in T^* from u to v. Now $u \in V$ and $v \notin V$, so there must be some edge in the path that joins a vertex z in V to a vertex w in $V(G)\backslash V$. Since PRIM'S algorithm chose $\{u, v\}$ instead of $\{z, w\}$, we have $W(u, v) \le W(z, w)$. Take $\{z, w\}$ out of $E(T^*)$ and replace it with $\{u, v\}$. The new graph T^{**} is still connected, so it's a tree by Theorem 3 of §6.3. Since $W(T^{**}) \le W(T^*)$, T^{**} is also a minimum spanning tree, and T^{**} contains the new T. At the end of the loop, T is still contained in some minimum spanning tree, as we wanted to show.
■

PRIM'S algorithm makes $n - 1$ passes through the while loop for a graph G with n vertices. Each pass involves choosing a smallest edge subject to a specified condition. A stupid implementation could require looking through all the edges of $E(G)$ to find the right edge. A more clever implementation would keep a record for each vertex x in $V(G)\backslash V$ of the vertex u in V with smallest value $W(u, x)$, and would also store the corresponding value of $W(u, x)$. The algorithm could then simply run through the list of vertices x in $V(G)\backslash V$, find the smallest $W(u, x)$, add $\{u, x\}$ to E and add x to V. Then it could check for each y in $V(G)\backslash V$ whether x is now the vertex in V closest to y, and if so, update the record for y. The time to find the closest x to V and then update records is just $O(n)$, so PRIM'S algorithm with the implementation we have just described runs in time $O(n^2)$.

PRIM'S algorithm can easily be modified [Exercise 13] to produce a minimum spanning forest for a graph, whether or not the graph is connected. The algorithm as it stands will break down if the given graph is not connected.

As a final note, we observe that the weight of a minimum spanning tree helps provide a lower bound for the Traveling Salesperson Problem we mentioned in §6.5. Suppose that the path $C = e_1 e_2 \cdots e_n$ is a solution for the Traveling Salesperson Problem on G, i.e., a Hamilton circuit of smallest possible weight. Then $e_2 \cdots e_n$ visits each vertex just once, so it's a spanning tree for G. If M is the weight of a minimum spanning tree for G [a number that we can compute using KRUSKAL'S or PRIM'S algorithm], then

$$M \leq W(e_2) + \cdots + W(e_n) = W(C) - W(e_1).$$

Hence $W(C) \geq M +$ (smallest edge weight in G). If we can find, by any method, some Hamilton circuit C of G with weight close to $M +$ (smallest edge weight), we should probably take it and not spend time trying to do better.

EXERCISES 6.6

1. (a) Apply TREE(1) to the graph in Figure 5(a). Draw a picture of the tree that results, and label the edges with a, b, c, d, e, f in the order in which they are chosen. Use the choice scheme of Example 1.

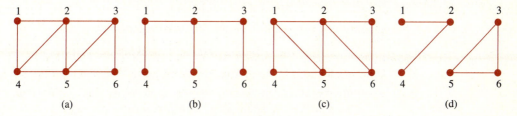

FIGURE 5

(b) Repeat part (a) for the graph in Figure 5(b).

(c) Repeat part (a) for the graph in Figure 5(c).

(d) Repeat part (a) for the graph in Figure 5(d).

2. Apply FOREST to the graph in Figure 6.

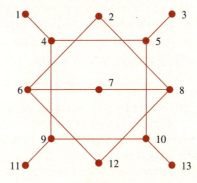

FIGURE 6

3. Figure 7(a) shows a weighted graph and Figures 7(b) and 7(c) show two different ways to label its edges with $a, b, \ldots, n$ in order of increasing weight.

   (a) Apply KRUSKAL'S algorithm to the graph with the edges ordered as in Figure 7(b). Draw the resulting minimum spanning tree and give its weight.

   (b) Repeat part (a) with the ordering in Figure 7(c).

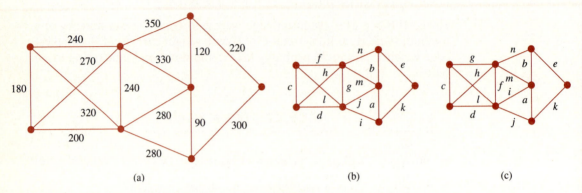

(a)   (b)   (c)

FIGURE 7

4. Suppose that the graph in Figure 7(b) is weighted so that

   $$W(a) \geq W(b) \geq \cdots \geq W(n).$$

   Draw a minimum spanning tree for the graph.

5. (a) Apply PRIM'S algorithm to the graph in Figure 7(a), starting at the vertex of degree 4. Draw the resulting minimum spanning tree and label the edges alphabetically in the order chosen.

   (b) What is the weight of a minimum spanning tree for this graph?

   (c) How many different answers are there to part (a)?

6. Suppose that the graph in Figure 5(a) of §6.3 is weighted so that

   $$W(e_1) > W(e_2) > \cdots > W(e_{10}).$$

   (a) List the edges in a minimum spanning tree for this graph in the order in which KRUSKAL'S algorithm would choose them.

   (b) Repeat part (a) for PRIM'S algorithm, starting at the upper right vertex.

7. Repeat Exercise 6 with weights

   $$W(e_1) < W(e_2) < \cdots < W(e_{10}).$$

8. (a) Use KRUSKAL'S algorithm to find a minimum spanning tree of the graph in Figure 8(a). Label the edges in alphabetical order as you choose them. Give the weight of the minimum spanning tree.

   (b) Repeat part (a) with PRIM'S algorithm, starting at the lower middle vertex.

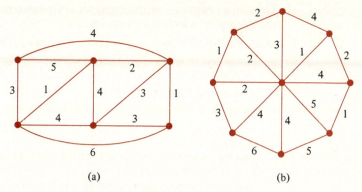

(a)   (b)

FIGURE 8

9. (a) Repeat Exercise 8(a) for the graph in Figure 8(b).

(b) Repeat Exercise 8(b) for the graph in Figure 8(b), starting at the top vertex.

10. (a) Find all spanning trees of the graph in Figure 9.

(b) Which edges belong to every spanning tree?

(c) For a general finite connected graph, characterize the edges that belong to every spanning tree. Prove your assertion.

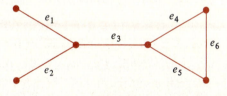

FIGURE 9

11. An oil company wants to connect the cities in the mileage chart below by pipelines going directly between cities. What is the minimum number of miles of pipeline needed?

	Des Moines	Milwaukee	Minneapolis	Omaha	Pierre	Winnipeg
Bismarck	670	758	427	581	211	369
Des Moines		361	252	132	492	680
Milwaukee			332	493	690	759
Minneapolis				357	394	431
Omaha					391	650
Pierre						521

12. Does every edge of a finite connected graph with no loops belong to some spanning tree? Justify your answer.

13. Modify PRIM'S algorithm to produce a minimum spanning forest.

14. (a) Show that if H is a subgraph of the weighted graph G that contains a minimum spanning tree of G, then every minimum spanning tree of H is a minimum spanning tree of G.

    (b) Show that the edge-deletion algorithm described after KRUSKAL'S algorithm produces a minimum spanning tree. *Hint:* Show that the set of edges remaining after each deletion contains a minimum spanning tree for G. *Outline:* Suppose that the set E of edges remaining when e is about to be deleted contains a minimum spanning tree for G. Apply part (a) to the minimum spanning tree K *produced for* E by KRUSKAL'S algorithm. Let C be a cycle in E containing e. Argue that every edge f in $C \backslash (K \cup \{e\})$ precedes e on the edge list and forms a cycle with some edges in K that also precede e. Get a path in $K \backslash \{e\}$ that joins the ends of e and conclude that $e \notin K$.

15. Let G be a finite weighted connected graph in which different edges have different weights. Show that G has exactly one minimum spanning tree. *Hint:* Assume that G has more than one minimum spanning tree. Consider the edge of smallest weight that belongs to some but not all minimum spanning trees.

CHAPTER HIGHLIGHTS

As always, one of the best ways to use this material for review is to follow the suggestions at the end of Chapter 1. Ask yourself: What does it mean? Why is it here? How can I use it? Keep thinking of examples. Though there are lots of items on the lists below, there are not really as many new ideas to master as the lists would suggest. Have courage.

CONCEPTS

path
    closed, simple, cycle, acyclic
    Euler path, circuit
    Hamilton path, circuit
isomorphism, invariant
degree, degree sequence
graph
    regular, complete, bipartite, complete bipartite
connected, component
tree, leaf, forest
spanning tree, spanning forest

rooted tree
  root, parent, child, descendant, subtree with root v
  binary search tree, labeled tree
  binary [m–ary] rooted tree
    regular, full
  level number, height
  ordered rooted tree
weight of edge, path, subgraph
minimum spanning tree, forest
Gray code

FACTS

A path has all vertices distinct if and only if it is simple and acyclic.

If there is a path between distinct vertices then there is a simple acyclic path between them.

There is at most one simple path between two vertices in an acyclic graph or digraph. There is exactly one simple path between two vertices in a tree.

If e is an edge of a connected graph G, then e belongs to some cycle if and only if $G\backslash\{e\}$ is connected. Thus an algorithm that checks connectedness can test for cycles.

The following statements are equivalent for a graph G with $n \geq 1$ vertices and no loops:
  (a) G is a tree.
  (b) G is connected, but won't be if an edge is removed.
  (c) G is acyclic, but won't be if an edge is added.
  (d) G is acyclic and has $n - 1$ edges [as many as possible].
  (e) G is connected and has $n - 1$ edges [as few as possible].

Choosing a root gives a tree a natural directed structure.

$$\sum_{v \in V(G)} \deg(v) = 2 \cdot |E(G)|.$$

A graph has an Euler circuit if and only if it is connected and all vertices have even degree. Euler paths exist if at most two vertices have odd degree.

If a graph has no loops or parallel edges, and if $|V(G)| = n \geq 3$, then G is Hamiltonian if any of the following is true:
  (a) $\deg(v) \geq n/2$ for each vertex v [high degrees].
  (b) $|E(G)| \geq \frac{1}{2}(n - 1)(n - 2) + 2$ [lots of edges].
  (c) $\deg(v) + \deg(w) \geq n$ whenever v and w are not connected by an edge.

Theorem 4 of §6.5 gives information on Hamilton paths in bipartite graphs.

ALGORITHMS

FLEURY'S algorithm to construct an Euler circuit of a graph.

TREE(v) and FOREST to build a spanning forest or find components of a graph in time $O(|V(G)| + |E(G)|)$.

KRUSKAL'S and PRIM'S algorithms to construct minimum spanning trees [or forests] for weighted graphs.

7

RECURSION, TREES AND ALGORITHMS

One of the main goals of this chapter is an understanding of how recursive algorithms work and how they can be verified. Before we discuss algorithms, we look at the general topic of recursive definition, which we touched on briefly for sequences in Chapter 4, and we consider a recursive generalization of mathematical induction. Recursive algorithms can often be thought of as working their way downward through some sort of tree. To give emphasis to that view we have chosen examples of algorithms that explicitly examine trees or that naturally give rise to trees. The idea of a sorted labeling of a digraph, which our algorithms produce in § 7.3, will play a role later, in Chapter 8. The last two sections of this chapter present applications of recursive methods to algebraic notation and to prefix codes, which are used for file compression as well as for designing data structures.

§ 7.1 General Recursion

The recursive definitions in § 4.3 allowed us to define sequences by giving their first few terms, together with a recipe for getting later terms from earlier ones. In our treatment of tree algorithms it will be convenient to have a more general version of recursion, which we consider now. There are three main themes to the section: recursively defined sets, a generalization of induction, and recursively defined functions. These topics are closely linked, as we will see.

Roughly speaking, a set of objects is defined recursively in case it is built up by a process in which some elements are put in at the beginning and others are added later because of elements that are already members of the club. Of course, such a description is too fuzzy to be useful. We will be more precise in a moment, but first here are some examples of recursively defined sets.

EXAMPLE 1   (a) We build $\mathbb{N}$ recursively by

(B) $0 \in \mathbb{N}$;
(R)   If $n \in \mathbb{N}$, then $n + 1 \in \mathbb{N}$.

Condition (B) puts 0 in $\mathbb{N}$ and (R) gives a recipe for generating new members of $\mathbb{N}$ from old ones. By invoking (B) and then repeatedly using (R) we get $0, 1, 2, 3, \ldots$ in $\mathbb{N}$.

(b) The conditions

(B) $1 \in S$;
(R)   If $n \in S$, then $2n \in S$

give members of a subset S of $\mathbb{P}$ that contains 1 [by (B)], 2 [by (R), since $1 \in S$], 4 [since $2 \in S$], 8, 16, and so on. It is not hard to see, by induction, that S contains $\{2^m : m \in \mathbb{N}\}$. Moreover, the numbers of the form 2^m are the only ones that are forced to belong to S by (B) and (R) [see Exercise 1]. It seems reasonable to say that (B) and (R) define the set $\{2^m : m \in \mathbb{N}\}$. ■

EXAMPLE 2   Consider an alphabet Σ. The rules

(B) $\lambda \in \Sigma^*$;
(R)   If $w \in \Sigma^*$ and $x \in \Sigma$, then $wx \in \Sigma^*$.

give a recursive description of Σ^*. The empty word λ is in Σ^* by decree, and repeated applications of (R) let us build longer words. For instance, if Σ is the English alphabet the words $\lambda, b, bi, big, \ldots, bigwor$, $bigword$ are in the set described by (B) and (R). ■

A **recursive definition of a set** S consists of two parts: a **basis** of the form

(B) $X \subseteq S$,

where X is some specified set, and a **recursive clause**

(R)   If s is determined from members of S by following certain rules, then $s \in S$.

The particular rules in (R) will be specified, of course, as they were in Examples 1 and 2, and they can be anything that makes sense. As in Example 2, they can refer to objects that are not in S itself.

Our implicit understanding is always that an element beongs to S only if it is required to by (B) and (R). Thus in Example 1(b) only the powers of 2 are *required* to be in S, so S consists only of powers of 2.

The conditions (B) and (R) allow us to build S up by spreading out from X in layers. Define the chain of sets $S_0 \subseteq S_1 \subseteq S_2 \subseteq \cdots$ by

$$S_0 = X, \quad \text{and}$$

$$S_{n+1} = \text{everything in } S_n \text{ or constructed from members of } S_n$$
$$\text{by using the rules in (R)}.$$

Then S_n consists of the members of S that can be built from X by n or fewer applications of rules in (R), and $S = \displaystyle\bigcup_{n=0}^{\infty} S_n$.

EXAMPLE 3   (a) For $S = \mathbb{N}$ as in Example 1(a), $S_0 = \{0\}$, $S_1 = \{0, 1\}, \ldots, S_n = \{0, 1, \ldots, n\}, \ldots$.

(b) For $S = \{2^m : m \in \mathbb{N}\}$ as defined in Example 1(b), $S_0 = \{1\}$, $S_1 = \{1, 2\}, \ldots, S_n = \{1, 2, 4, \ldots, 2^n\}$.

(c) For $\Sigma = \{a, b\}$ the definition of Σ^* in Example 2 leads to $\Sigma_0^* = \{\lambda\}$, $\Sigma_1^* = \{\lambda, a, b\}$, $\Sigma_2^* = \{\lambda, a, b, aa, ab, ba, bb\}$, etc., so $\Sigma_n^* = \{\lambda\} \cup \Sigma \cup \Sigma^2 \cup \cdots \cup \Sigma^n$ in our old notation.

(d) Let $\Sigma = \{a, b\}$ and define S by

(B) $\Sigma \subseteq S$;
(R)   If $w \in S$, then $awb \in S$.

Then $S_0 = \Sigma = \{a, b\}$, $S_1 = \{a, b, aab, abb\}$, $S_2 = S_1 \cup \{aaabb, aabbb\}$, etc., and S itself consists of the words in Σ^* of the form $a \cdots ab \cdots b$ with one more or one fewer a than the number of b's.   ∎

EXAMPLE 4   We can define the class of finite trees recursively. For convenience, let us say that the graph G' is **obtained from** the graph G **by attaching** v **as a leaf** in case

(a) $V(G') = V(G) \cup \{v\} \neq V(G)$, and

(b) $E(G') = E(G) \cup \{e\}$, where the edge e joins v to a vertex of G.

Then the class of trees is defined by:

(B)   Every graph with one vertex and no edges is a [trivial] tree;
(R)   If T is a tree and if T' is obtained from T by attaching a leaf, then T' is a tree.

We can think of this recursive definition as building up trees by adding leaves. Figure 1 shows a typical construction sequence.

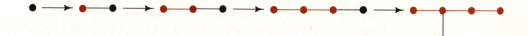

FIGURE 1

We previously defined trees to be connected acylic graphs. To show that the new recursive definition coincides with the old definition, we would need to check that

(1) (B) and (R) only produce connected acyclic graphs, and
(2) Every finite connected acyclic graph can be constructed using only (B) and (R).

It is fairly easy to see that (1) is true, because (B) gives trees and if T is connected and acyclic, so is any T' constructed from T by (R). In effect, we are arguing by induction that if S_n consists of trees, so does S_{n+1}.

To show (2) is a little harder. Imagine that there are [finite] trees that we can't construct with (B) and (R), and suppose that T is such a tree with as few vertices as possible. By (B), T has more than one vertex. Then T has at least two leaves by Lemma 1 in §6.3. Prune a leaf of T to get a new tree T''. Then T'' is constructible from (B) and (R), by the minimal choice of T. But T is obtainable from T'' by (R); just attach the leaf again. Hence T itself is constructible, which is a contradiction to the way it was chosen. ∎

EXAMPLE 5

(a) We can mimic the recursive definition in Example 4 to obtain the class of [finite] rooted trees.

(B)   A graph with one vertex v and no edges is a [trivial] rooted tree with root v;
(R)   If T is a rooted tree with root r and T' is obtained by attaching a leaf to T, then T' is a rooted tree with root r.

As in Example 4, we see that this definition gives nothing but rooted trees. The second argument in Example 4 shows that every rooted tree is constructible using (B) and (R); since every nontrivial tree has at least two leaves, we can prune a leaf that's not the root.

(b) Here is another way to describe the class of rooted trees recursively. We will define a class $\mathscr{R}$ of ordered pairs (T, r) in which T is a tree and r is a vertex of T; r is called the **root** of the tree T. For convenience, say that (T_1, r_1) and (T_2, r_2) are **disjoint** in case T_1 and T_2 have no vertices in common. If the pairs $(T_1, r_1), \ldots, (T_k, r_k)$ are disjoint, we will say that T is **obtained by hanging** $(T_1, r_1), \ldots, (T_k, r_k)$ **from** r in case:

(1) r is not a vertex of any T_i,
(2) $V(T) = V(T_1) \cup \cdots \cup V(T_k) \cup \{r\}$, and
(3) $E(T) = E(T_1) \cup \cdots \cup E(T_k) \cup \{e_1, \ldots, e_k\}$, where the edge e_i joins r to r_i.

Figure 2 shows the tree obtained by hanging (T_1, r_1), (T_2, r_2) and (T_3, r_3) from r.

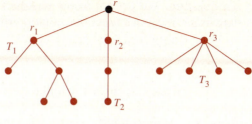

FIGURE 2

Here is the definition of $\mathcal{R}$:

(B)   If T is a graph with one vertex v and no edges, then $(T, v) \in \mathcal{R}$;

(R)   If $(T_1, r_1), \ldots, (T_k, r_k)$ are disjoint members of $\mathcal{R}$ and if (T, r) is obtained by hanging $(T_1, r_1), \ldots, (T_k, r_k)$ from r, then $(T, r) \in \mathcal{R}$.

As in (a), we have no difficulty in showing that $\mathcal{R}$ consists of pairs (T, r) in which T is a tree with vertex r, that is, a tree rooted at r. To see that every rooted tree arises from (B) and (R), imagine a rooted tree T that doesn't, with $|V(T)|$ as small as possible. Because T doesn't come from (B), it has more than one vertex. Consider the children of the root r of T. The subtrees that have these children as their roots are smaller than T, so they're obtainable by (B) and (R). Since T is the result of hanging these subtrees from r, (T, r) is in $\mathcal{R}$ after all, contrary to assumption.   ■

In our examples so far we know which sets we are trying to define, and the recursive definitions are aimed directly at them. One can also give perfectly acceptable recursive definitions without knowing for sure what they define.

EXAMPLE 6   (a) We define a set A of integers by

(B) $1 \in A$;

(R)   If $n \in A$, then $3n \in A$, and if $2n + 1 \in A$, then $n \in A$.

These conditions describe A unambiguously, but it is not completely clear which integers are in A. A little experimentation suggests that A contains no n with $n \equiv 2 \pmod 3$, and in fact [Exercise 9] that is the case. More experimentation suggests that perhaps $A = \{n \in \mathbb{P} : n \not\equiv 2 \pmod 3\}$, i.e., $A = \{1, 3, 4, 6, 7, 9, 10, 12, 13, \ldots\}$. It is not clear how we could verify such a guess. The chain 1, 3, 9, 27, 13, 39, 19, 57, 171, 85, 255, 127, 63, 31, 15, 7, which is the shortest route to showing that 7 is in A, gives no hint at a general argument. Several hours of study have not completely settled the question. We do not know whether our guess is correct, although experts on recursive functions could perhaps tell us the answer.

(b) Consider the set S defined by

(B) $1 \in S$;

(R)   If $n \in S$, then $2n \in S$, and if $3n + 1 \in S$ with n odd, then $n \in S$.

As of this writing, *nobody* knows whether or not $S = \mathbb{P}$, although many people have looked at the question.   ■

Recursively defined classes support a kind of generalized induction. Suppose that the class S is defined recursively by a set X, given in (B), and a set of rules for producing new members, given in (R). Suppose also that each s in S has an associated proposition $p(s)$.

Generalized Principle of Induction

In this setting, if

(b) $p(s)$ is true for every $s \in X$, and

(r) $p(s)$ is true whenever s is produced from members t of S for which $p(t)$ is true,

then $p(s)$ is true for every $s \in S$.

Proof. Let $T = \{s \in S : p(s) \text{ is true}\}$, and recall the sequence $X = S_0 \subseteq S_1 \subseteq S_2 \subseteq \cdots$ defined earlier. By (b), $X = S_0 \subseteq T$. If $S \not\subseteq T$, then there is a smallest $n \in \mathbb{P}$ for which $S_n \not\subseteq T$, and there is some $s \in S_n \backslash T$. Because $S_{n-1} \subseteq T$, $s \notin S_{n-1}$, so s must be constructed from members of S_{n-1} using rules in (R). But then (r) says that $p(s)$ is true, contrary to the choice of s. Thus $S \subseteq T$, as claimed.   ■

EXAMPLE 7   (a) If S is $\mathbb{N}$, given by

(B) $0 \in S$;

(R) $n \in S$ implies $n + 1 \in S$,

then (b) and (r) become

(b) $p(0)$ is true;

(r) $p(n + 1)$ is true whenever $p(n)$ is true,

so the generalized principle in this case is just the ordinary principle of mathematical induction.

(b) For the class $\mathscr{R}$ of finite rooted trees, defined as in Example 5(b), we get the following principle:

Suppose that

(b) $p(T)$ is true for every 1-vertex rooted tree T, and
(r) $p(T)$ is true whenever p is true for every subtree hanging from the root of T.

Then $p(T)$ is true for every rooted tree T.

(c) The recursive definition of the class of trees in Example 4 yields the following.

Suppose that $p(T)$ is true for every 1-vertex graph T with no edges, and suppose that $p(T')$ is true whenever $p(T)$ is true and T' is obtained by attaching a leaf to T. Then $p(T)$ is true for every tree T.

(d) We apply the principle in (c) to give another proof that a tree with n vertices has $n - 1$ edges; compare Lemma 2 in §6.3. Let $p(T)$ be "$|E(T)| = |V(T)| - 1$." If T has one vertex and no edges, then $p(T)$ is true. If $p(T)$ is true for some tree T and we attach a leaf to T, then both sides of the equation increase by 1, so $p(T')$ is still true. The generalized principle implies that $p(T)$ is true for every tree T. ∎

In general, if the class S is recursively defined, it can happen that a member of S_n can be constructed from members of S_{n-1} in more than one way. There are often advantages to working with recursive definitions in which new members of S are produced from other members in only one way. We will call such recursive definitions **uniquely determined**.

EXAMPLE 8   Most of the recursive definitions given so far have been uniquely determined. Here we discuss the exceptions.

(a) The recursive definition of the class of finite trees in Example 4 is not uniquely determined. In fact, the tree constructed in Figure 1 can also be constructed as indicated in Figure 3.

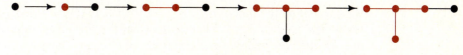

FIGURE 3

(b) The first recursive definition of the class of finite rooted trees in Example 5 is also not uniquely determined. See Exercise 7. On the other hand, the "hanging from the root" definition in Example 5(b) is uniquely determined. A rooted tree is uniquely determined by the subtrees that hang from its root, each of the subtrees is uniquely determined in the same way, and so on.

(c) The definitions in Example 6 are not uniquely determined. For example, 1 is in A by (B), and also by using (R) twice to get 3 and then 1 in A. Similarly, 1 reappears in S of Example 6(b) via the sequence 1, 2, 4, 1. ∎

If the set S is recursively defined by (B) and (R), we may be able to define functions f on S recursively, using the following procedure:

    (1) Define $f(s)$ explicitly for s in the set X described in (B);

    (2) Give a recipe for defining $f(s)$ in terms of values of $f(t)$ for t's in S that come before s.

To say what we mean by "come before" we need to recall our chain $X = S_0 \subseteq S_1 \subseteq S_2 \subseteq \cdots$ for building up S. For each s in S there is a first value of n for which $s \in S_n$. We view all the members of S_{n-1} as coming before such an s, although, of course, the rules in (R) may produce s from just a few of them. If the recursive definition of S is uniquely determined, the recipe in step (2) will be unambiguous. Otherwise, one must be sure that the definition is well-defined. That is, one must check that the value of $f(s)$ does not depend on how s is constructed from t's that come before s.

EXAMPLE 9

    (a) Let $\mathbb{N}$ be given by

    (B) $0 \in \mathbb{N}$;

    (R) $n \in \mathbb{N}$ implies $n + 1 \in \mathbb{N}$.

As we saw earlier, $S_m = \{0, \ldots, m\}$, so n appears for the first time in S_n and the numbers that come before n are $0, \ldots, n - 1$. To define a function f on $\mathbb{N}$ we can define $f(0)$ and then explain how to get $f(n)$ from $f(0), \ldots, f(n - 1)$.

    (b) Define $\mathbb{N}$ recursively by

    (B) $\{0, 1\} \subseteq \mathbb{N}$;

    (R) $n \in \mathbb{N}$ implies $n + 1 \in \mathbb{N}$.

We can define the function f on $\mathbb{N}$ by $f(0) = 1$, $f(1) = 1$ and $f(n + 1) = f(n) + f(n - 1)$ for $n \geq 1$. Then f describes the Fibonacci sequence. With this particular recursive definition of $\mathbb{N}$ we have $\{0, 1\} = S_0 \subseteq \{0, 1, 2\} = S_1 \subseteq \{0, 1, 2, 3\} = S_2 \subseteq \cdots$, so whenever $n + 1 \in S_m$ we have n and $n - 1$ in S_{m-1}. ∎

EXAMPLE 10

    (a) The set Σ^* was defined recursively in Example 2. Define the **length** function l on Σ^* by:

    (1) $l(\lambda) = 0$;

    (2) If $w \in \Sigma^*$ and $x \in \Sigma$, then $l(wx) = l(w) + 1$.

Then $l(w)$ is the number of letters in w.

(b) We could also have defined Σ^* as follows:

(B) $\{\lambda\} \cup \Sigma \subseteq \Sigma^*$;
(R)   If $w, u \in \Sigma^*$, then $wu \in \Sigma^*$.

Then length could have been defined by:

(1) $l'(\lambda) = 0$, $l'(x) = 1$ for $x \in \Sigma$;
(2) If $w, u \in \Sigma^*$, then $l'(wu) = l'(w) + l'(u)$.

Or could it? The potential difficulty comes because this new recursive definition of Σ^* is not uniquely determined. It may be possible to construct a given word as wu in more than one way. For instance, if $\Sigma = \{a, b\}$ then $a, ab, (ab)a, ((ab)a)b$ is one path to $abab$, but $a, ab, (ab)(ab)$ is another, and $a, b, ba, (ba)b, a((ba)b)$ is still another. How do we know that our "definition" of l' gives the same value of $l'(abab)$ for all of these paths?

It turns out [Exercise 13] that l' is well-defined, and in fact it is the function l that we saw in (a) above.

(c) Let's try to define a **depth** function d on Σ^* by:

(1) $d(\lambda) = 0$, $d(x) = 1$ for $x \in \Sigma$;
(2) If $w, u \in \Sigma^*$, then $d(wu) = \max\{d(w), d(u)\} + 1$.

This time we fail. If d were well-defined, we would have $d(a) = 1$, $d(ab) = 1 + 1 = 2$, $d((ab)a) = 2 + 1 = 3$ and $d(((ab)a)b) = 3 + 1 = 4$, but also $d((ab)(ab)) = 2 + 1 = 3$. Thus $d(abab)$ would not be given unambiguously. ■

Examples 10(b) and 10(c) illustrate the potential difficulty in defining functions recursively when the recursive definition is not uniquely determined. A uniquely determined recursive definition is preferable for defining functions recursively. Thus the definition of Σ^* that we used in Example 10(a) is preferable to the one in Example 10(b). Similarly, the definition of rooted trees in Example 5(b) has advantages over that in Example 5(a). Sometimes we don't have a choice: We know of no uniquely determined recursive definition for finite trees, but the definition in Example 4 is still useful.

EXAMPLE 11 We wish to recursively define the height of finite rooted trees so that the height is the length of a longest simple path from the root to a leaf. For this purpose, our recursive definition of rooted trees in Example 5(a) is of little use, since the addition of a leaf need not increase the height of the tree. However, the recursive definition in Example 5(b) is ideal for the task.

(a) Recall that the class $\mathscr{R}$ of finite rooted trees is defined by:

(B)   If T is a graph with one vertex v and no edges, then $(T, v) \in \mathscr{R}$;

(R)   If $(T_1, r_1), \ldots, (T_k, r_k)$ are disjoint members of $\mathscr{R}$ and if (T, r) is obtained by hanging $(T_1, r_1), \ldots, (T_k, r_k)$ from r, then $(T, r) \in \mathscr{R}$.

The **height** of a member of $\mathscr{R}$ is defined by:

(1) Trivial one-vertex rooted trees have height 0;

(2) If (T, r) is defined as in (R), and if the trees $(T_1, r_1), \ldots, (T_k, r_k)$ have heights $h_1, \ldots, h_k$, then the height of (T, r) is $1 + \max\{h_1, \ldots, h_k\}$.

(b) Other concepts for finite rooted trees can be defined recursively. For example, for an integer m greater than 1, the class of **m−ary rooted trees** is defined by:

(B)   A trivial one-vertex rooted tree is an m−ary rooted tree;

(R)   A rooted tree obtained by hanging at most m m−ary rooted trees from the root is an m−ary rooted tree.

(c) We use these recursive definitions to prove that an m−ary tree of height h has at most m^h leaves. This statement is clear for a trivial tree since $m^0 = 1$. Consider an m−ary tree (T, r), defined as in part (b), with the property that each subtree (T_i, r_i) has height h_i. Since (T, r) is an m−ary tree, each (T_i, r_i) is an m−ary tree, so it has at most m^{h_i} leaves. Let $h^* = \max\{h_1, \ldots, h_k\}$. Then the number of leaves of (T, r) is bounded by

$$m^{h_1} + \cdots + m^{h_k} \le m^{h^*} + \cdots + m^{h^*} = k \cdot m^{h^*}.$$

Since the root has at most m subtrees, $k \le m$, and so (T, r) has at most $m \cdot m^{h^*} = m^{h^*+1}$ leaves. Since $h^* + 1$ equals the height h of (T, r), we are done. ■

EXERCISES 7.1

1. Let S be the recursively defined set in Example 1(b).

   (a) Show that $2^m \in S$ for all $m \in \mathbb{N}$.

   (b) Show that if n is in S, then n has the form 2^m for some $m \in \mathbb{N}$.

   (c) Conclude that $S = \{2^m : m \in \mathbb{N}\}$.

2. Use the definition in Example 2 to show that the following objects are in Σ^*, where Σ is the usual English alphabet.

   (a) *cat*   (b) *math*   (c) *zzpq*   (d) *aint*

3. (a) Describe the subset S of $\mathbb{N} \times \mathbb{N}$ recursively defined by

   (B) $(0, 0) \in S$;

   (R) If $(m, n) \in S$ with $m < n$, then $(m + 1, n) \in S$, and if $(m, m) \in S$ then $(0, m + 1) \in S$.

   (b) Use the recursive definition to show that $(1, 2) \in S$.

   (c) Is the recursive definition uniquely determined?

4. (a) Describe the subset T of $\mathbb{N} \times \mathbb{N}$ recursively defined by

   (B) $(0, 0) \in T$;

   (R) If $(m, n) \in T$, then $(m, n + 1) \in T$ and $(m + 1, n + 1) \in T$.

   (b) Use the the recursive definition to show that $(3, 5) \in T$.

   (c) Is the recursive definition uniquely determined?

5. Let $\Sigma = \{a, b\}$ and let S be the set of words in Σ^* in which all the a's precede all the b's. For example, aab, $abbb$, a, b, and even λ belong to S, but bab and ba do not.

   (a) Give a recursive definition for the set S.

   (b) Use your recursive definition to show that $abbb \in S$.

   (c) Use your recursive definition to show that $aab \in S$.

   (d) Is your recursive definition uniquely determined?

6. Let $\Sigma = \{a, b\}$ and let T be the set of words in Σ^* that have exactly one a.

   (a) Give a recursive definition for the set T.

   (b) Use your recursive definition to show that $bbab \in T$.

   (c) Is your recursive definition uniquely determined?

7. (a) Describe two distinct constructions of the rooted tree in Figure 4, using the recursive definition in Example 5(a).

   (b) Describe the construction of the rooted tree in Figure 4, using the definition in Example 5(b).

8. Verify (2) of Example 4 by showing that S_n contains all trees with at most $n + 1$ vertices.

FIGURE 4

9. Let A be the recursively defined set in Example 6(a). Show that if n is in A, then $n \equiv 0 \pmod 3$ or $n \equiv 1 \pmod 3$.

10. Show that 4, 6, 10 and 12 are in the set A defined in Example 6(a).

11. (a) Show that the set S defined in Example 6(b) includes 1, 2, 3, 4, 5 and 6.

    (b) Show that 7 is in S.

12. (a) Give a recursive definition for the class of regular m–ary trees.

    (b) Do the same for full m–ary trees.

13. Define Σ^* as in Example 10(b) and let l' be as in that example. Show that $l'(w)$ is the number of letters in w for every $w \in \Sigma^*$. *Hint:* Use generalized induction with $X = \{\lambda\} \cup \Sigma$.

14. Let Σ^* be as defined in Example 2. We define the **reversal** $\overleftarrow{w}$ of a word w in Σ^* recursively as follows:

   (B) $\overleftarrow{\lambda} = \lambda$;
   (R)   If $\overleftarrow{w}$ has been defined and $x \in \Sigma$, then $\overleftarrow{wx} = x\overleftarrow{w}$.

   This is another well-defined definition.

   (a) Prove that $\overleftarrow{x} = x$ for all $x \in \Sigma$.

   (b) Use this definition to find the reversal of *cab*.

   (c) Use this definition to find the reversal of *abbaa*.

   (d) If w_1 and w_2 are in Σ^*, what is $\overleftarrow{w_1 w_2}$ in terms of $\overleftarrow{w_1}$ and $\overleftarrow{w_2}$? What is $\overleftarrow{\overleftarrow{w_1}}$?

15. Here is a recursive definition for a subset S of $\mathbb{N} \times \mathbb{N}$:

   (B) $(0, 0) \in S$;
   (R)   If $(m, n) \in S$, then $(m + 2, n + 3) \in S$.

   (a) List four members of S.

   (b) Prove that if $(m, n) \in S$, then 5 divides $m + n$.

   (c) Is the converse to the assertion in part (b) true?

16. Here is a recursive definition for another subset T of $\mathbb{N} \times \mathbb{N}$:

   (B) $(0, 0) \in T$;
   (R)   If $(m, n) \in T$, then each of $(m + 1, n)$, $(m + 1, n + 1)$ and $(m + 1, n + 2)$ is in T.

   (a) List six members of T.

   (b) Prove that $2m \geq n$ for all $(m, n) \in T$.

   (c) Is this recursive definition uniquely determined?

17. Consider the following recursive definition for a subset A of $\mathbb{N} \times \mathbb{N}$:

   (B) $(0, 0) \in A$;
   (R)   If $(m, n) \in A$, then $(m + 1, n)$ and $(m, n + 1)$ are in A.

   (a) Show that $A = \mathbb{N} \times \mathbb{N}$.

   (b) Let $p(m, n)$ be a proposition-valued function on $\mathbb{N} \times \mathbb{N}$. Use part (a) to devise a general recursive procedure for proving $p(m, n)$ true for all m and n.

18. Let $\Sigma = \{a, b\}$ and let B be the subset of Σ^* defined recursively as follows:

   (B) a and b are B;
   (R)   If $w \in B$, then abw and baw are in B.

   (a) List six members of B.

   (b) Prove that if $w \in B$, then length(w) is odd.

   (c) Is the converse to the assertion in part (b) true?

   (d) Is this recursive definition uniquely determined?

19. Let Σ be a finite alphabet. For words w in Σ^*, the reversal $\overleftarrow{w}$ is defined in Exercise 14.

(a) Prove length(w) = length($\overleftarrow{w}$) for all $w \in \Sigma^*$.

(b) Prove $\overleftarrow{w_1 w_2} = \overleftarrow{w_2}\,\overleftarrow{w_1}$ for all $w_1, w_2 \in \Sigma^*$.

§ 7.2   Recursive Algorithms

This section describes the structure of a general recursive algorithm. It also discusses how to prove that a recursive algorithm does what it is claimed to do. The presentation is a continuation of the account of recursion begun in the last section.

For both generalized induction and the recursive definition of functions we start out with a recursively defined set S. Our goal is to show that the subset of S for which $p(s)$ is true or $f(s)$ is well-defined is the whole set S. Another important question we can ask about a recursively defined set S is, "How can we tell whether or not something is a member of S?"

EXAMPLE 1   As in Example 1(b) of §7.1, define $S = \{2^m : m \in \mathbb{N}\}$ recursively by

(B) $1 \in S$;
(R) $2k \in S$ whenever $k \in S$.

Then $S \subseteq \mathbb{P}$. Given $n \in \mathbb{P}$, we wish to test whether or not $n \in S$, and if it turns out that $n \in S$ we would like to find m with $n = 2^m$.

Suppose that n is odd. If $n = 1$, then $n \in S$ by (B), and $n = 2^0$. If $n > 1$, then (R) can't force n to be in S, so $n \notin S$. We know the answer quickly in the odd case.

If n is even, say $n = 2k$, then $n \in S$ if $k \in S$, by (R), but otherwise (R) won't put n in S. So we need to test $k = n/2$ to see if $k \in S$. If $k \in S$ with $k = 2^a$, then n will be in S with $n = 2^{a+1}$, and if $k \notin S$ then $n \notin S$.

This analysis leads us to design the following algorithm.

TEST(n; b, m)
{input: $n \in \mathbb{P}$}
{output: b = true if $n \in S$, false if $n \notin S$
  $m = \log_2 n$ if $n \in S$, $-\infty$ if $n \notin S$}
If n is odd then
  if $n = 1$ then
    $b := $ true; $m := 0$
  else
    $b := $ false; $m := -\infty$
else {so n is even}
  TEST($n/2$; b', m')
    {b' is true if and only if $n/2 \in S$, in which case $n/2 = 2^{m'}$}
    $b := b'$; $m := m' + 1$.

As usual, the statements in braces $\{\cdot\cdot\cdot\}$ are comments, and not part of the algorithm. We will say more about such comments shortly. Our way of listing the input and output of TEST in its name, separated by a semicolon, is a useful notation. It helps us follow algorithms, such as this one, that call themselves as subroutines.

The way that TEST works is that if $n = 1$ it reports success, if n is odd and not 1 it reports failure, and if n is even it checks out $n/2$. For example, with input $n = 4$, TEST(4;) invokes TEST(2;), which invokes TEST(1;), which succeeds. We get TEST(1; true, 0), TEST(2; true, 1), and hence TEST(4; true, 2). These outputs reflect the facts that $1 = 2^0 \in S$, $2 = 2^1 \in S$ and $4 = 2^2 \in S$. With input $n = 6$, TEST(6;) invokes TEST(3;), which reports failure. In our notation, TEST(3; false, $-\infty$) leads to TEST(6; false, $-\infty$). ■

The membership-testing algorithm TEST in Example 1 has the interesting feature that if the original input n is even, then TEST asks itself about $n/2$. Algorithms that call themselves are termed **recursive algorithms**.

TEST doesn't stop calling itself until the recursive calls reach an odd number, so even though 4 is in S, TEST(4;) doesn't know at first that b = true; it calls TEST (2;) which calls TEST(1;), which finally reports values for b and m that are then passed back through the calling chain. Here are the executions of TEST for inputs 4 and 6.

TEST (4;)	TEST(6;)
TEST(2;)	TEST(3;)
TEST(1;)	$b := $ false; $m := -\infty$
$b := $ true; $m := 0$	$b := $ false; $m := -\infty + 1 = -\infty$
$b := $ true; $m := 0 + 1 = 1$	
$b := $ true; $m := 1 + 1 = 2$	

EXAMPLE 2   The class of trees was defined in Example 4 of §7.1 by

(B)   Trivial [1-vertex] graphs are trees;
(R)   If T is obtained by attaching a leaf to a tree, then T is a tree.

We can devise a recursive algorithm TESTTREE for testing graphs to see if they are trees.

TESTTREE(G; b)
{input: finite graph G}
{output: b = true if G is a tree, false if not}
$b := $ false
If G has one vertex and no edges then
    $b := $ true

> else
>   if G has a leaf then
>   prune a leaf and its associated edge, to get G'
>   TESTTREE(G'; b')
> $\{b' = \text{true if } G' \text{ is a tree, false if it's not}\}$
> $b := b'.$

If G really *is* a tree, this algorithm will prune edges and vertices off it until all that's left is a single vertex, and then it will report success, i.e., $b = \text{true}$, back up the recursive ladder. Figure 1 shows a chain G, G', G'', ... of inputs considered in one such recursion.

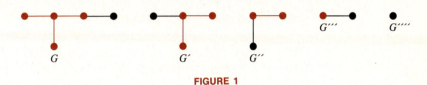

FIGURE 1

Figure 2 shows a chain for an input H that is not a tree. In this case TESTTREE(H''';) sets $b := \text{false}$ and then cannot either apply the case for a trivial tree or recur to a smaller graph. It simply reports $b = \text{false}$ and then TESTTREE(H'';), TESTTREE(H';) and TESTTREE(H;) accept that verdict. ■

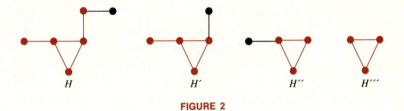

FIGURE 2

To help motivate the next example, consider the following questions that probably did not arise in algebra: What is an acceptable formula? What is it that makes $(x + y)(x - y)$ look good and makes $(x + -(^4/y$ look worthless? Why can a computer make sense of the first expression but not the second? An answer leads to the notion of a **well-formed formula** or **wff**.

EXAMPLE 3   (a) Here is a definition of wff's for algebra.

      (B) Numerical constants and variables, such as x, y, z, are wff's;

      (R) If f and g are wff's, so are $(f + g)$, $(f - g)$, (fg), (f/g) and (f^g).

Being variables, x and y are wff's. Therefore, both $(x + y)$ and $(x - y)$ are wff's. We can hence conclude that $((x + y)(x - y))$ is a wff. The definition isn't entirely satisfactory, since the outside parentheses here seem extraneous. However, without them the square $(((x + y)(x - y))^2)$ would look like $((x + y)(x - y)^2)$, and these two expressions have different meanings. The problem is that in algebra we traditionally allow the omission of parentheses in some circumstances. Taking all the exceptional cases into account, we would be led to a complicated definition. Note also that our definition does not exclude division by 0. Thus $(0/0)$ is a wff even though we would not assign a numerical value to this expression.

(b) In computer science, the symbol $*$ is often used for multiplication and $^\wedge$ is used for exponentiation [$a^\wedge b$ means a^b]. With this notation, the definition of wff's can be rewritten as:

(B) Numerical constants and variables are wff's;
(R) If f and g are wff's, so are $(f + g)$, $(f - g)$, $(f * g)$, (f/g) and $(f^\wedge g)$.

For example,

$$(((((X + Y)^\wedge 2) - (2 * (X * Y))) - (X^\wedge 2)) - (Y^\wedge 2))$$

is a wff.

(c) In §7.4 we will discuss Polish notation, which is a parenthesis-free notation. The preceding examples and related exercises may help you appreciate its value.

(d) How can we test a string of symbols to see if it's a wff? The recursive definition gives us an idea for an algorithm. Suppose that we have some way to recognize numerical constants and variables, and that our input string is made out of allowable symbols. The question is whether or not the symbols are put together in a meaningful way. Here's an outline of a test algorithm.

TESTWFF(w; b)
{input: string w of allowable symbols}
{output: b = true if w is a wff, false otherwise}
$b := $ false
if w is a numerical constant or a variable then
    $b := $ true
else
    look at strings f and g for which w is of the form $(f + g)$, $(f - g)$,
        $(f * g)$, (f/g) or $(f^\wedge g)$, and for each such f and g
    TESTWFF(f; b')
    TESTWFF(g; b'')
    $b := b' \wedge b''$ {so b = true if and only if b' and b'' are true}

If the algorithm finds an f and g that *are* wff's, it can skip testing other possible f's and g's. The reason is that condition (R) only allows new wff's to be constructed in a single way. A formula $(f * g)$ cannot be also of the form (h/k) for wff's h and k or of the form $(f' * g')$ for wff's f' and g' with $f' \neq f$ or $g' \neq g$. Writing out the details for the else branch would be tedious but would lead to no surprises. It is easy to believe that one could write a computer program to recognize wff's. ∎

A computer needs to be able to read strings of input to see if they contain instructions or other meaningful expressions. Designing a compiler amounts to designing an instruction language and a recognition algorithm so that the computer can "parse," i.e., make sense out of, input strings. The recent development of the theory of formal languages has been greatly motivated by parsing questions.

Recursive algorithms are useful for a lot more than just testing membership in sets. We will look at some other illustrations in this section, and we will have important examples in this and later chapters as well. The basic feature of a recursive algorithm is that it calls itself, but of course that may not be much help unless the recursive calls are somehow closer to termination than the original invocation. To make sure that an algorithm terminates we must know that chains of recursive calls cannot go on forever. We need some **base cases** for which the result is produced without further recursive calls, and we need to guarantee that the inputs for the recursive calls always get closer to the base cases somehow. A typical recursive algorithm with a chance of terminating has the following rough form.

RECUR(I; O)
{input: I; output: O}
If I is a base case then
   $O :=$ whatever it should be
else
     find some I_k's closer to base cases than I is
     RECUR(I_k; O_k) for each k
     use the O_k's somehow to determine O.

EXAMPLE 4 Figure 3 shows six very simple algorithms. All of them look alike for the first few lines. In each instance the input n is assumed to be a positive integer, and the base case is $n = 1$. Do these algorithms terminate, and if they do, how is the output r related to the input n?

We can experiment a little. With $n = 1$, FOO gives $r := 1$. With $n = 2$ it calls FOO(1;) to get $s = 1$, so $r := 1 + 1 = 2$. With $n = 3$ it calls FOO(2;) to get $s = 2$, as we just saw, so $r := 2 + 1 = 3$. In fact, it looks as if $r = n$ every time. In any case, since the chain n, $n - 1$, $n - 2, \ldots$ can't go on forever without reaching the base case, FOO must terminate. We will come back in a moment to the question of how one might prove that $r = n$ for every n.

| FOO(*n*; *r*)
If *n* = 1 then
  r := 1
else
  FOO(*n* − 1; *s*)
  r := *s* + 1. | GOO(*n*; *r*)
If *n* = 1 then
  r := 1
else
  GOO(*n* − 1; *s*)
  r := *n* ∗ *s*. | BOO(*n*; *r*)
If *n* = 1 then
  r := 1
else
  BOO(*n* + 1; *s*)
  r := *s* + 1. |
| MOO(*n*; *r*)
If *n* = 1 then
  r := 1
else
  MOO(*n*/2; *s*)
  r := *s* + 1. | TOO(*n*; *r*)
If *n* = 1 then
  r := 1
else
  TOO(*n* DIV 2; *s*)
  r := *s* + 1. | ZOO(*n*; *r*)
If *n* = 1 then
  r := 1
else
  ZOO(*n* DIV 2; *s*)
  r := 2 ∗ *s*. |

FIGURE 3

Algorithm GOO also terminates, for the same reason as FOO does. Here are the first few values of r: for $n = 1$, $r = 1$; for $n = 2$, $n − 1 = 1$, so $s = 1$ and $r = 2 ∗ 1 = 2$; for $n = 3$, $n − 1 = 2$, so $s = 2$ and $r = 3 ∗ 2 = 6$. In fact, it looks as if we'll always get $r = n!$, so GOO gives us a recursive way to compute factorials.

Algorithm BOO never terminates. For instance, BOO(2;) calls BOO(3;), which calls BOO(4;), etc. The recursive calls don't get closer to the base case $n = 1$.

Algorithm MOO fails for a different reason. Even though $n/2$ is closer to 1 than n is, $n/2$ may not be an integer, so MOO($n/2$;) may not make sense. If we decide to allow noninteger inputs, then MOO need not terminate; for instance, the chain 3, 3/2, 3/4, 3/8, ... goes on forever.

The MOO problem is fixed in TOO, since n DIV 2 is surely an integer, and if $n > 1$, then $n > n$ DIV $2 \geq 1$. Table 1 shows the first few values of r for TOO, as well as for the other algorithms in Figure 3.

TABLE 1

n	FOO	GOO	BOO	MOO	TOO	ZOO
			Value of *r*			
1	1	1	1	1	1	1
2	2	2		2	2	2
3	3	6			2	2
4	4	24		3	3	4
5	5	120			3	4
6	6	720			3	4
7	7	5040			3	4

Algorithm ZOO terminates, just as TOO does, but produces a different output. ∎

Once we know that a recursive algorithm terminates, we still need to know what its output values are. If the algorithm is intended to produce a particular result, we must verify that it does what it is supposed to do. In the notation of our generic algorithm RECUR(I; O), we need to check that:

(b) If I is a base case, then O has the correct value;
(r) If each O_k has the correct value in the recursive calls RECUR(I_k; O_k), then O has the correct value for I.

These conditions look like the ones we must check in a proof by generalized induction, and of course there is a connection. Recursively define the set S of allowable inputs for RECUR as follows. For (B) let X be the set of base case inputs. Agree that I is **obtained from** $\{I_1, \ldots, I_m\}$ in case the else branch of RECUR(I; O) can compute O from the outputs $O_1, \ldots, O_m$ of RECUR(I_1; O_1), $\ldots$, RECUR(I_m; O_m). Let the recursive clause be

(R) If I is obtained from $I_1, \ldots, I_m$ in S, then $I \in S$.

Then S is precisely the set of inputs for which RECUR can compute outputs. Setting $p(I) = $ "RECUR gives the correct value for input I" defines a set of propositions on S, and (b) and (r) become:

(b') $p(I)$ is true for all $I \in X$;
(r') $p(I)$ is true whenever I is obtained from members I_k of S for which $p(I_k)$ is true.

The Generalized Principle of Induction says that these conditions guarantee that $p(I)$ is true for every I in S.

EXAMPLE 5

(a) To show that GOO produces $n!$ we simply need to verify that:

(b) $1! = 1$, and
(r) $n! = n * (n-1)!$.

These equations are true by definition of $n!$.

(b) We claim that the algorithm TEST of Example 1 produces $b = $ true and $m = \log_2 n$ when the input n is a power of 2 and gives $b = $ false, $m = -\infty$ otherwise. Here is the algorithm again, for reference.

$\text{TEST}(n; b, m)$
If n is odd then
    if $n = 1$ then
        $b := \text{true}; m := 0$
    else
        $b := \text{false}; m := -\infty$
else
    $\text{TEST}(n/2; b', m')$
    $b := b'; m := m' + 1.$

What are the base cases and the allowable inputs? We certainly want the set S of allowable inputs to be $\mathbb{P}$. If we just permitted $n = 1$ as a base case, then since (R) only lets $2s$ be obtained from s, S would just consist of the powers of 2. Instead, the algorithm treats all odd members of $\mathbb{P}$ as base cases.

To check that the algorithm is correct, i.e., that it gives the answer we claim, we just need to verify:

(b) If $n = 1$ then $n = 2^0$, and if n is odd and greater than 1, then n is not a power of 2;

(r) If $n/2 = 2^{m'}$, then $n = 2^{m'+1}$.

Both (b) and (r) are clearly true.

While we are at it, we can estimate the time, $T(n)$, that TEST takes for an input n. The base case verification and disposition in the odd case take some constant time, say C. If n is even, the else branch takes $T(n/2)$ for the call to $\text{TEST}(n/2; \quad)$ plus some time D to compute b and m from b' and m'. Thus

$$(*) \qquad\qquad T(n) \le C + T(n/2) + D.$$

Now $n = 2^m \cdot k$ with k odd and $m > 0$. Hence

$$T(n) = T(2^m \cdot k) \le T(2^{m-1} \cdot k) + C + D \le T(2^{m-2} \cdot k) + C + D + C + D$$

$$\le \cdots$$

$$\le T(k) + m \cdot (C + D) \le C + m \cdot (C + D)$$

$$\le 2m \cdot (C + D).$$

Since $\dfrac{n}{k} = 2^m$ we have

$$m = \log_2\left(\frac{n}{k}\right) \le \log_2 n, \quad \text{and so} \quad T(n) \le 2(C + D) \cdot \log_2 n.$$

Thus $T(n) = O(\log_2 n)$. The three dots in the argument are harmless; once we guess the inequality $T(n) \le 2(C + D) \cdot \log_2 n$, we can prove it inductively, by using $(*)$.

This algorithm is more efficient for large n than any of the divide-and-conquer algorithms covered by Theorem 2 of §4.4. Note that (∗) gives

$$T(2n) \leq T(n) + (C + D) \qquad \text{for all } n.$$

This is a sharper estimate than any inequality of the form

$$s_{2n} \leq 2s_n + f(n)$$

because of the coefficient 2. However, an inequality version of Exercise 17 of §4.4, with $b = 1$ and $f(n) = C + D$, could be applied to obtain

$$T(2^m) \leq C + m \cdot (C + D) \qquad \text{for } m > 0.$$

(c) The base cases for TESTTREE in Example 2 include not only the graphs with one vertex but also all finite graphs without leaves. For any such graph the algorithm computes b without recurring further.

(d) Similarly, the base cases for TESTWFF in Example 3 include all strings that are *not* of form $(f + g)$, $(f - g)$, $(f * g)$, (f/g) or $(f^\wedge g)$. ∎

At the expense of added storage for intermediate results we can convert while loops to recursive procedures. For instance, the recursive segment

```
RECUR(  ;  )
if γ if false then
    do nothing
else
    S
    RECUR(  ;  )
```

produces the same effect as the loop

```
while γ do
    S.
```

EXAMPLE 6   We presented the Euclidean Algorithm of §4.6 using a while loop, but recursion seems more natural. The key observation that $\gcd(m, n) = \gcd(n, m \text{ MOD } n)$ if $n \neq 0$ gives the following recursive algorithm.

Algorithm **EUCLID**$(m, n; d)$

{input: $m, n \in \mathbb{N}$, not both 0}
{output: $d = \gcd(m, n)$}
If $n = 0$ then
    $d := m$
else
    EUCLID$(n, m \text{ MOD } n; d')$
    $d := d'$. ∎

A minor addition gives an algorithm that computes integers s and t with $d = sm + tn$.

Algorithm **EUCLID$^+$(*m*, *n*; *d*, *s*, *t*)**

If $n = 0$ then
    $d := m; s := 1; t := 0$
else
    EUCLID$^+$(n, m MOD n; d', s', t')
    $d := d'; s := t'; t := s' - t' \cdot (m \text{ DIV } n)$. ■

The running times for these algorithms are essentially the same as for the iterative versions in §4.6.

For reference we repeat the conditions a recursive algorithm must satisfy.

VERIFICATION CONDITIONS FOR RECURSIVE ALGORITHMS

   (a) The algorithm must terminate. In particular, recursive calls must make progress toward base cases.
   (b) The algorithm must give the correct results in the base cases.
   (c) The algorithm must give the correct result if all recursive calls produce correct results.

EXERCISES 7.2

1. (a) Illustrate the execution of the algorithm TEST for input 20.

   (b) Do the same for input 8.

2. Illustrate the algorithm TESTTREE for the graphs in Figure 4. You may draw pictures like those in Figures 1 and 2.

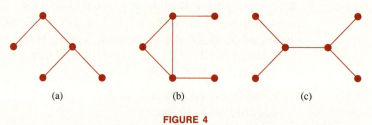

        (a)   (b)   (c)

FIGURE 4

3. Add enough parentheses to the following algebraic expressions so that they are wff's as defined in Example 3.

(a) $x + y + z$

(b) $x + y/z$

(c) xyz

(d) $(x + y)^{x+y}$

4. Add enough parentheses to the following algebraic expressions so that they are wff's as defined in Example 3.

(a) $X + Y + Z$

(b) $X * (Y + Z)$

(c) $X^2 + 2 * X + 1$

(d) $X + Y/Z - Z * X$

5. Use the recursive definition of wff in Example 3 to show that the following are wff's.

(a) $((x^2) + (y^2))$

(b) $(((X^2) + (Y^2))^2)$

(c) $((X + Y) * (X - Y))$

6. (a) Let GOOF be the algorithm GOO with the assignment $r := 1$ changed to $r := 0$ [and the recursive call to GOOF$(n - 1; s)$]. What is the output r if the input is a positive integer?

(b) What happens if the assignment $r := 1$ is unchanged but the assignment $r := n * s$ is changed to $r := (n - 1) * s$?

7. Supply the lines for $n = 8$ and $n = 9$ in Table 1.

8. Verify that the algorithm TOO produces $k + 1$ whenever $2^k \le n < 2^{k+1}$. *Hint:* Use induction on k.

9. Verify that the algorithm ZOO produces 2^k whenever $2^k \le n < 2^{k+1}$.

10. Show how the algorithm EUCLID$^+$ computes $d = \gcd(80, 35)$ and finds s and t so that $d = 80s + 35t$.

11. Repeat Exercise 10 for $d = \gcd(108, 30)$.

12. Repeat Exercise 10 for $d = \gcd(56, 21)$.

13. (a) Verify the algorithm EUCLID in Example 6.

(b) Verify the algorithm EUCLID$^+$ in Example 6.

14. We recursively define the **depth** of a wff in algebra as follows; see Example 3.

(B) Numerical constants and variables have depth 0;

(R) If depth(f) and depth(g) have been defined, then each of $(f + g)$, $(f - g)$, $(f * g)$, (f/g) and (f^g) has depth equal to $1 + \max\{\text{depth}(f), \text{depth}(g)\}$.

This turns out to be a well-defined definition; compare Example 10 of §7.1. Calculate the depth of the following algebraic expressions:

(a) $((x^2) + (y^2))$

(b) $(((X^2) + (Y^2))^2)$

(c) $((X + Y) * (X - Y))$

(d) $(((((X + Y)^2) - (2 * (X * Y))) - (X^2)) - (Y^2))$

(e) $(((x + (x + y)) + z) - y)$

(f) $(((X * Y)/X) - (Y^4))$

15. The following is a uniquely determined recursive definition of wff's for the propositional calculus.

    (B)   Variables, such as p, q, r, are wff's;

    (R)   If P and Q are wff's, so are $(P \vee Q)$, $(P \wedge Q)$, $(P \rightarrow Q)$, $(P \leftrightarrow Q)$ and $\neg P$.

Note that we do not require parentheses when we negate a proposition. Consequently, the negation symbol $\neg$ always negates the shortest subexpression following it that is a wff. In practice, we tend to omit the outside parentheses, and for the sake of readability, brackets [ ] and braces { } may be used for parentheses. Show that the following are wff's in the propositional calculus.

(a) $\neg(p \vee q)$ (b) $(\neg p \wedge \neg q)$

(c) $((p \leftrightarrow q) \rightarrow ((r \rightarrow p) \vee q))$

16. Modify the definition in Exercise 15 so that the "exclusive or" connective $\oplus$ is allowable.

17. Throughout this exercise, let p and q be *fixed* propositions. We recursively define the family $\mathscr{F}$ of compound propositions using only p, q, $\wedge$ and $\vee$ as follows:

    (B) $p, q \in \mathscr{F}$;

    (R)   If $P, Q \in \mathscr{F}$, then $(P \wedge Q)$ and $(P \vee Q)$ are in $\mathscr{F}$.

(a) Use this definition to verify that $(p \wedge (p \vee q))$ is in $\mathscr{F}$.

(b) Prove that if p and q are false, then all the propositions in $\mathscr{F}$ are false.

(c) Show that $p \rightarrow q$ is not logically equivalent to any proposition in $\mathscr{F}$. This verifies the unproved claim in the answer to Exercise 17(c) of §2.4.

§ 7.3   Depth-First Search Algorithms

A tree traversal algorithm is an algorithm for listing [or visiting or searching] all the vertices of a finite ordered rooted tree. The three most common such algorithms provide preorder listing, inorder listing [for binary trees *only*] and postorder listing. All three are recursive algorithms.

    In the **preorder listing**, the root is listed first and the subtrees are listed in order of their roots. Because of the way we draw ordered rooted trees, we will refer to the order as left to right. In this algorithm the root gets listed first, while the listing of the descendants is farmed out to the children. A base case occurs when the input vertex has no children, i.e., when it is a leaf. In this case, the first step adds the leaf to the list and the for loop is vacuously completed, so the list has only one entry, namely v.

Algorithm PREORDER(*v*)

{input: a finite ordered rooted tree with root *v*}
{output: a list $L(v)$ of all the vertices of the tree, in which parents
    always come before their children}
Put *v* on the list $L(v)$.
For each child *w* of *v* (taken from left to right)
    PREORDER(*w*) {to get a list $L(w)$ of *w* and its descendants}
    attach the resulting list $L(w)$ to the end of $L(v)$. ∎

In the **postorder listing**, the subtrees are listed in order first, and then
the root is listed at the end. Again a base case occurs when the input is
a leaf. In this case, the first step is vacuously completed, and again the
list has only one entry, *v*.

Algorithm POSTORDER(*v*)

{input: a finite ordered rooted tree with root *v*}
{output: a list $L(v)$ of all the vertices of the tree, in which parents
    always appear after their children}
Begin with an empty list; $L(v) = \lambda$.
For each child *w* of *v* (taken from left to right)
    POSTORDER(*w*) {to get a list $L(w)$ of *w* and its descendants}
    attach the resulting list $L(w)$ to the end of the list $L(v)$ obtained
      so far.
Put *v* on the end of $L(v)$. ∎

EXAMPLE 1 We apply the algorithms PREORDER and POSTORDER to the tree *T*
in Figure 1. Since an understanding of all the algorithms in this section
depends on a solid understanding of these simple ones, we will explain
every step in this example.

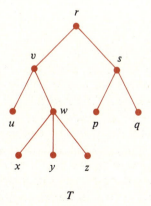

T

FIGURE 1

(a) The PREORDER algorithm proceeds as follows:

List r. $L(r) = r$.
 Go to r's first child v. List v. $L(v) = v$.
  Go to v's first child u. List u. **$L(u) = u$.**
  $\{u$ has no children.$\}$ $L(v) = v\,u$.
  Go to v's next child w. List w. $L(w) = w$.
   Go to w's first child x. List x. **$L(x) = x$.**
   $\{x$ has no children.$\}$ $L(w) = w\,x$.
   Go to w's next child y. List y. **$L(y) = y$.**
   $\{y$ has no children.$\}$ $L(w) = w\,x\,y$.
   Go to w's next child z. List z. **$L(z) = z$.**
   $\{z$ has no children.$\}$ **$L(w) = w\,x\,y\,z$.**
  $\{$All of w's children have now been dealt with.$\}$
  $L(v) = v\,u\,w\,x\,y\,z$.
 $\{$All of v's children have now been dealt with.$\}$
 $L(r) = r\,v\,u\,w\,x\,y\,z$.
 Go to r's next child s. List s. $L(s) = s$.
  Go to s's first child p. List p. **$L(p) = p$.**
  $\{p$ has no children.$\}$ $L(s) = s\,p$.
  Go to s's next child q. List q. **$L(q) = q$.**
  $\{q$ has no children.$\}$ **$L(s) = s\,p\,q$.**
 $\{$All of s's children have now been dealt with.$\}$
 $L(r) = r\,v\,u\,w\,x\,y\,z\,s\,p\,q$.
$\{$All of r's children have now been dealt with.$\}$
Stop.

The list obtained is $r\,v\,u\,w\,x\,y\,z\,s\,p\,q$. In the PREORDER algorithm, each vertex is listed when it is first visited, so this listing can be obtained from the picture of T as illustrated in Figure 2. Just follow the dashed line, listing each vertex the first time it is reached.

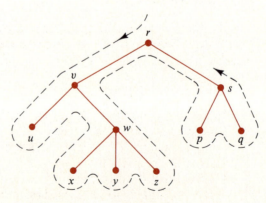

FIGURE 2

(b) The POSTORDER algorithm does the listing of the parents after their children are listed:

Visit r. $L(r) = \lambda$.
  Visit r's first child v. $L(v) = \lambda$.
    Visit v's first child u. $L(u) = \lambda$.
      {u has no children.} List u. $\boldsymbol{L(u) = u}$; $L(v) = u$.
    Visit v's second child w. $L(w) = \lambda$.
      Visit w's first child x. $L(x) = \lambda$.
        {x has no children.} List x. $\boldsymbol{L(x) = x}$; $L(w) = x$.
      Visit w's second child y. $L(y) = \lambda$.
        {y has no children.} List y. $\boldsymbol{L(y) = y}$; $L(w) = x\,y$.
      Visit w's third child z. $L(z) = \lambda$.
        {z has no children.} List z. $\boldsymbol{L(z) = z}$; $L(w) = x\,y\,z$.
      {All of w's children have been visited.}
      List w. $\boldsymbol{L(w) = x\,y\,z\,w}$; $L(v) = u\,x\,y\,z\,w$.
    {All of v's children have been visited.}
    List v. $\boldsymbol{L(v) = u\,x\,y\,z\,w\,v}$; $L(r) = u\,x\,y\,z\,w\,v$.
  Visit r's second child s. $L(s) = \lambda$.
    Visit s's first child p. $L(p) = \lambda$.
      {p has no children.} List p. $\boldsymbol{L(p) = p}$; $L(s) = p$.
    Visit s's second child q. $L(q) = \lambda$.
      {q has no children.} List q. $\boldsymbol{L(q) = q}$; $L(s) = p\,q$.
    {All of s's children have been visited.}
    List s. $\boldsymbol{L(s) = p\,q\,s}$; $L(r) = u\,x\,y\,z\,w\,v\,p\,q\,s$.
  {All of r's children have been visited.}
  List r. $\boldsymbol{L(r) = u\,x\,y\,z\,w\,v\,p\,q\,s\,r}$.
  Stop.

The list obtained is $u\,x\,y\,z\,w\,v\,p\,q\,s\,r$. In the POSTORDER algorithm, each vertex is listed when it is last visited, so this listing can be obtained from the picture of T in Figure 2, provided that each vertex is listed the *last* time it is visited. ∎

For *binary* ordered rooted trees a third kind of listing is available. The **inorder listing** puts the root v in between the lists for the subtrees rooted at its left and right children.

Algorithm INORDER(v)

{input: a finite ordered binary tree with root v}
{output: a list $L(v)$ of the vertices of the tree, in which left children appear before their parents, and right children appear after them}
Start with an empty list; $L(v) = \lambda$.
If v has a left child w then
  INORDER(w) {to get a list $L(w)$ of w and its descendants}
  attach $L(w)$ to $L(v)$.

Add v at the end of $L(v)$.
If v has a right child u then
    INORDER(u) {to get $L(u)$}
    attach $L(u)$ to the end of $L(v)$. ■

If the input tree is not regular, then some vertices have just one child. Since each such child must be designated as either a left or a right child, the algorithm still works. As with PREORDER and POSTORDER, the list for a childless vertex consists of just the vertex itself.

EXAMPLE 2   (a) Consider the binary tree in Figure 3. Algorithm INORDER proceeds as follows:

Visit r, w, and v, then list v. $L(w) = \boldsymbol{L(v) = v}$.
    Return to w, list w. $L(w) = vw$.
    Then visit x and y; list y. $L(x) = \boldsymbol{L(y) = y}$.
      Return to x, list x. $L(x) = yx$.
      Then visit z and list z. $\boldsymbol{L(z) = z}$; $\boldsymbol{L(x) = yxz}$; $\boldsymbol{L(w) = vwyxz}$.
Return to r, list r. $L(r) = vwyxzr$.
Then visit u and t; list t. $L(u) = \boldsymbol{L(t) = t}$.
    Return to u, list u. $L(u) = tu$.
    Then visit s and p; list p. $L(s) = \boldsymbol{L(p) = p}$.
      Return to s, list s. $L(s) = ps$.
      Then visit q and list q. $\boldsymbol{L(q) = q}$; $\boldsymbol{L(s) = psq}$; $\boldsymbol{L(u) = tupsq}$;
      $\boldsymbol{L(r) = vwyxzrtupsq}$.
Return to s, u and r, and stop.

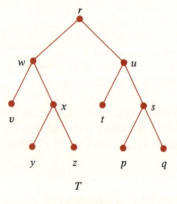

T

FIGURE 3

The final list is $vwyxzrtupsq$. In this example each parent has both left and right children. If some of the children had been missing the algorithm would still have performed successfully. For example, if y had been missing, the list $L(x)$ would have been just $x\,z$, and if the whole x, y, z branch had been missing, $L(w)$ would have been just $v\,w$.

(b) Consider the labeled tree in Figure 8(a) of §6.4. The inorder listing of the subtree with root 00 is 000, 00, 001; the other subtrees are also easy to list. The inorder listing of the entire tree is

$$000, 00, 001, 0, 010, 01, 011, \text{root}, 100, 10, 101, 1, 110, 11, 111. \quad \blacksquare$$

If we are given a listing of a tree, can we reconstruct the tree from the listing? In general, the answer is no. Different trees can have the same listing.

EXAMPLE 3   Consider again the tree T in Figure 3. Figure 4 gives two more binary trees having the same inorder listing. The binary tree in Figure 5(a) has the same preorder listing as T, and the tree in Figure 5(b) has the same postorder listing as T. Exercise 8 asks you to verify these assertions. $\blacksquare$

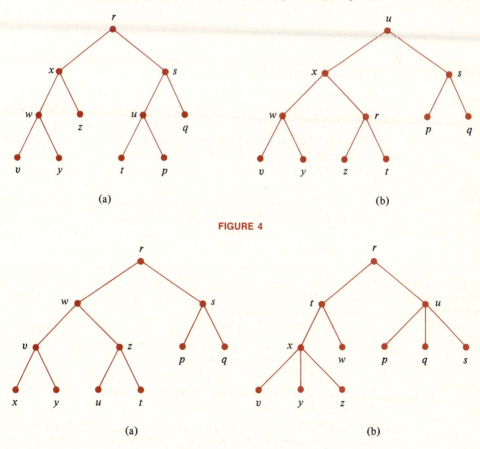

(a)   (b)

FIGURE 4

(a)   (b)

FIGURE 5

In spite of these examples, there are important situations under which trees can be recovered from their listings. Since our applications will be to Polish notation, and since the ideas are easier to grasp in that setting, we will return to this matter at the end of the next section.

We can analyze how long the three listing algorithms take by using a method of "charges." Let t_1 be the time that it takes to label a vertex, and let t_2 be the time that it takes to attach a list of vertices to another list. [We can assume that the attachment time is constant, by using a linked-list representation for our lists of vertices.] The idea now is to think of the total time for the various operations in the execution of the algorithm as being charged to the individual vertices and edges of the tree. We charge a vertex v with the time t_1 that it takes to label it, and we charge an edge (v, w) with the time t_2 that it takes to attach the list $L(w)$ obtained by the call to w to make it part of the list of $L(v)$. Then the time for every operation gets charged somewhere, every vertex and every edge gets charged, and nobody gets charged twice. By adding up the charges we find that the total time is $t_1|V(T)| + t_2|E(T)|$, which is $O(\max\{|V(T)|, |E(T)|\})$. Since $|E(T)| = |V(T)| - 1$ for the tree T, the time is $O(|V(T)|)$.

The three listing algorithms above are all examples of what are called **depth-first search** or **backtrack** algorithms. Each of them goes as far as it can by following the edges of the tree away from the root, then it backs up a bit and again goes as far as it can, and so on. The same idea can be useful in settings in which we have no obvious tree to start with. We will illustrate with an algorithm that labels the vertices of a digraph G with integers in a special way. Before we get into the details of the general case, we look at how such an algorithm might work on an ordered rooted tree.

The list of vertices given by algorithm POSTORDER provides a natural way to label the vertices of the tree; simply give the kth vertex on the list the label k. This labeling has the property that parents have larger labels than their children, and hence have larger labels than any of their descendants. For such a labeling, called a **sorted labeling**, the leaves tend to have small labels and the root has the largest label. Algorithm POSTORDER can be converted into an algorithm LABELTREE to produce a sorted labeling.

Before we write out the new algorithm, let's think again about how POSTORDER(r) arranges vertices on the output list. The vertices that precede a given vertex w on the list include not only w's own descendants but also all the vertices on branches of the tree that have been labeled by the time w's turn comes. If we are to label w and its descendants by their positions in the list, then we need to know which labels have been used up when we get to w. If $1, 2, \ldots, k$ have been used, then the labels for the subtree rooted at w should start with $k + 1$. To keep track of available labels, our algorithm has an extra input variable that says where the labels start, and its output gives the last label used so far. [Later, in algorithm TREESORT, we will see a way to avoid carrying these variables along so explicitly.] Remember, LABELTREE is really just POSTORDER, except that instead of listing each vertex in turn, the algorithm gives each a label.

Algorithm **LABELTREE(v, k; n)**

{input: finite rooted tree with root v, integer $k \geq 0$}
{output: integer n and a sorted labeling of the tree with the labels
   $k + 1, \ldots, n$}
$n := k$.
For w a child of v (children taken in some definite order)
   LABELTREE(w, n; m) {labels descendants of w with $n + 1, \ldots$,
      $m - 1$, labels w itself with m}
   $n := m$ {sets the starting point for the next child's labels}.
   Increase n by 1.
   Label v with n. ■

To label the whole tree rooted at r, starting with the label 1, we call LABELTREE(r, 0;   ).

Observe that the ordering of the children is irrelevant if all that is wanted is a sorted labeling of the tree. That is, it doesn't matter in which order the children are assigned their labeling tasks, as long as two children don't try to use the same labels.

EXAMPLE 4   Here is how **LABELTREE** would deal with our tree of Example 1.

LABELTREE(r, 0;   )
  LABELTREE(v, 0;   )
    LABELTREE(u, 0;   ); label **u** with **1**
    LABELTREE(w, 1;   )
      LABELTREE(x, 1;   ); label **x** with **2**
      LABELTREE(y, 2;   ); label **y** with **3**
      LABELTREE(z, 3;   ); label **z** with **4**
    Label **w** with **5**
  Label **v** with **6**
  LABELTREE(s, 6;   )
    LABELTREE(p, 6;   ); label **p** with **7**
    LABELTREE(q, 7;   ); label **q** with **8**
  Label **s** with **9**
Label **r** with **10**
Stop ■

Sorted labelings can be defined and studied in any digraph. Consider a finite digraph G and a subset L of $V(G)$. A **sorted labeling** of L is a numbering of its vertices with $1, 2, \ldots, |L|$ such that whenever there is a path in G from vertex i to vertex j, then $i > j$.

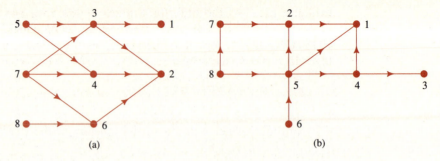

FIGURE 6

EXAMPLE 5   Figure 6 shows two digraphs whose vertex sets have sorted labelings. Notice that if $i > j$, there does not have to be a path from i to j; we simply require that $i > j$ *if* there is such a path. ■

A digraph whose vertex set has a sorted labeling cannot possibly contain a cycle: if the vertices of a cycle were labeled $ij \cdots ki$ we would have $i > j > \cdots > k > i$. If we restrict ourselves to acyclic digraphs, i.e., to ones that contain no cycles, then a sorted labeling of $V(G)$ is always possible. We prove this fact by exhibiting an algorithm, LABEL, that constructs one. The "child" and "descendant" terminology for the tree case gets changed into the language of "successors" and "accessible" for a general digraph, but the idea of the algorithm is the same. We use the notation SUCC(v) for the set of immediate successors of v, those vertices w for which there is an edge from v to w, and let ACC(v) be the set of vertices w for which there is a path from v to w, together with v itself. More generally, if $L \subseteq V(G)$, we let ACC(L) be the union of the sets ACC(v) for v in L.

EXAMPLE 6   The successor and accessible sets for the graph in Figure 6(a) are given in Figure 7. ■

Vertex	1	2	3	4	5	6	7	8
SUCC( )	∅	∅	{1}	{2}	{3,4}	{2}	{3,4,6}	{6}
ACC( )	{1}	{2}	{1,2,3}	{2,4}	{1,2,3,4,5}	{2,6}	{1,2,3,4,6,7}	{2,6,8}

FIGURE 7

There are two reasons why algorithm LABEL will be more complicated than LABELTREE. In a rooted tree, all vertices are accessible from the root, so the algorithm starts at the root and works down. However, there may not be a unique vertex in an acyclic digraph from which all other vertices are accessible. For example, see either of the digraphs in Figure 6. Therefore, LABEL will need subroutines that label sets like ACC(v). The other difference is that the children of a parent in a rooted tree have disjoint sets of descendants, so none of the children ever try to label some other child's descendant. On the other hand, a vertex in an acyclic digraph can be accessible from different successors of the same vertex. For example, vertex 1 in Figure 6(b) is accessible from both successors of 8. Thus the algorithm will need to keep track of vertices already labeled and avoid trying to label them again.

Algorithm LABEL first chooses a vertex v, labels all the vertices accessible from v [including v] and then calls itself recursively to handle the vertices that don't have labels yet. For clarity, we give the procedure for labeling the vertices accessible from v as a separate algorithm, which we call TREESORT for a reason that will become clear.

TREESORT is also recursive. When we apply TREESORT to a rooted tree, starting at the root r, it asks a child of r to label its descendants and then it recursively labels the remaining descendants of r and r itself. So in this case TREESORT works rather like LABELTREE(r). By writing the recursion in a slightly different form we have avoided some of the clumsiness of the extra variables k and n in LABELTREE, but the order of traversal for the two algorithms is exactly the same.

When either LABEL or TREESORT is called recursively, some of the vertices may already have been assigned labels. The algorithms will have to be passed that information, so that they won't try to label those vertices again. For convenience in verifying the algorithms' correctness, we make one more definition. Say that a subset L of $V(G)$ is **labeled** in case it has a sorted labeling and $L = $ ACC(L). Here is the first algorithm.

Algorithm **TREESORT(v, L)**

{input: finite acyclic digraph G, labeled subset L of $V(G)$, vertex
$v \in V(G) \backslash L$}
{output: $L \cup$ ACC(v) as a labeled set, with a sorted labeling that agrees
with the given labeling on L}
If SUCC(v) $\subseteq L$ {so $L \cup$ ACC(v) $= L \cup \{v\}$} then
label L as before and label v with $|L| + 1$
else
choose $w \in$ SUCC(v)$\backslash L$
TREESORT(w, L)
{get $L \cup$ ACC(w) labeled to agree with L labels}
TREESORT(v, $L \cup$ ACC(w))
{get $L \cup$ ACC(w) $\cup$ ACC(v) $= L \cup$ ACC(v) labeled to agree with
$L \cup$ ACC(w) labels}. ■

The first instruction takes care of the base case when all the vertices in ACC(v) except v are already labeled. Note that this base case holds vacuously if v has no successors at all. As with POSTORDER [where successors are called children], the else instruction refers the labeling task to the successors, but with an extra complication because already labeled vertices need to be dealt with. After the first chosen successor w_1 and the vertices accessible from w_1 are all labeled, TREESORT($v, L \cup$ ACC(w_1)) leads to one of two possibilities. If all of v's successors are in $L \cup$ ACC(w_1), then v is the only unlabeled vertex in $L \cup$ ACC(v), so it gets labeled. Otherwise, another successor w_2 of v is chosen; w_2 and the vertices accessible from w_2 *that were not previously labeled* are given labels. And so on.

EXAMPLE 7 We illustrate TREESORT on the digraph G of Figure 6(a), which is redrawn in Figure 8(a) with letter names for the vertices. As usual, arbitrary choices are made in alphabetic preference.

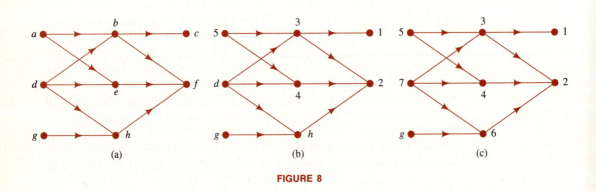

FIGURE 8

(a) First we use TREESORT(v, L) with $v = a$ and $L = \varnothing$ to label ACC(a).

> Start with a, choose its successor b, choose its successor c, label **c** with **1**.
> Return to b, choose its successor f, and label f with **2**.
> Return to b and label **b** with **3**.
> Return to a, choose its successor e which has no unlabeled successors, label e with **4**.
> Return to a and label **a** with **5**.

Figure 8(b) shows the labeling of ACC(a). If the execution of the algorithm just described is clear, skip to part (b). If not, here's the detailed explanation specifying the subroutines.

TREESORT(a, $\varnothing$)
Choose successor b of a {since a has unlabeled successors}
 TREESORT(b, $\varnothing$)
  Choose successor c of b {since b has unlabeled successors}
   TREESORT(c, $\varnothing$)
   Label c with **1** {since c has no successors at all}
  TREESORT(b, $\{c\}$)
  Choose successor f of b {f is still unlabeled}
   Label f with **2** {since f has no successors}
  Label b with **3** {since b now has no unlabeled successors}
   {ACC(b) = $\{c, f, b\}$ is now labeled}
 TREESORT(a, $\{c, f, b\}$)
 Choose successor e of a {a still has an unlabeled successor}
  TREESORT(e, $\{c, f, b\}$)
  Label e with **4** {the successor f is labeled already}
 TREESORT(a, $\{c, f, b, e\}$)
 Label a with **5** {a now has no unlabeled successors}

(b) We next use **TREESORT**(v, L) with $v = d$ and $L = \{a, b, c, e, f\}$ to label $L \cup$ ACC(d). See Figure 8(b).

> Start with d, choose its unlabeled successor h, then label h with **6** since h has no unlabeled successors.
> Return to d and label d with **7** since d now has no unlabeled successors.

Figure 8(c) shows the labeling of $\{a, b, c, e, f\} \cup$ ACC(d).

(c) Finally, we use **TREESORT**(v, L) with $v = g$ and $L = \{a, b, c, d, e, f, h\}$ to label $V(G) = L \cup$ ACC(g). See Figure 8(c). This will be easy.

> Start with g and label g with **8** since g has no unlabeled successors.

Figure 6(a) shows the labeling of G. We would, of course, obtain different labelings if we used different alphabetic preferences. See Exercises 13. ∎

In Example 7 the algorithm performed depth-first searches following the paths indicated by the dashed curves in Figure 9. Each vertex gets labeled the last time it is visited. In each search, the set of vertices and edges visited forms a tree if the directions on the edges are ignored. Thus the algorithm sorts within a tree that it finds inside G; this is why we called it TREESORT. In §6.6 we saw the algorithm TREE that worked in a similar way to build spanning trees inside connected graphs.

To verify that TREESORT(v, L) labels $L \cup$ ACC(v) correctly, we need to check that it gives the right result in the base case SUCC(v) $\subseteq L$

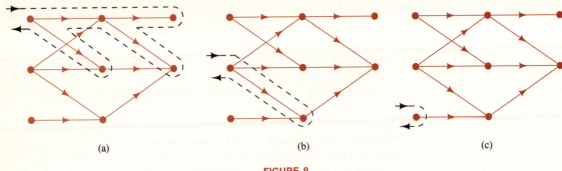

(a)   (b)   (c)

FIGURE 9

and that if its recursive calls TREESORT(w, L) and TREESORT($v, L \cup$ ACC(w)) give the right results, so does TREESORT(v, L). We should also check that the recursive calls are valid and are to cases closer to the base case than (v, L) is.

Suppose that SUCC(v) $\subseteq L$. Except for v itself, every vertex accessible from v is in ACC(w) for some successor w of v. Since all the w's are in L and $L =$ ACC(L), all the vertices accessible from the w's are also in L. Hence, in this case, ACC($L \cup \{v\}$) = ACC(L) $\cup$ ACC(v) = $L \cup \{v\}$. Since TREESORT labels L as before with $1, \ldots, |L|$ and labels v with $|L| + 1$, it makes $L \cup \{v\}$ a labeled set as claimed.

If the recursive calls give the right results, i.e., if the bracketed comments after them are correct, then $L \cup$ ACC(v) = $L \cup$ ACC(w) $\cup$ ACC(v) is a labeled set whose labeling agrees with the one given on L, so the output for TREESORT(v, L) is correct.

We can use $|$ACC(v)$\backslash L|$, the number of unlabeled vertices in ACC(v), as a measure of distance from a base case. If $w \in$ SUCC(v)$\backslash L$, then ACC(w) $\subset$ ACC(v) and $|$ACC(v)$\backslash L| > |$ACC(w)$\backslash L|$. Moreover, since $w \in$ ACC(v)$\backslash L$ we have $|$ACC(v)$\backslash L| > |$ACC(v)$\backslash (L \cup$ ACC(w))$|$. Thus the recursive calls are closer to base cases than (v, L) is. [This is the point at which we need the graph to be acyclic.] Thus TREESORT does perform as claimed.

By the time we finished Example 7 we had labeled the entire acyclic digraph G. We repeatedly applied TREESORT to unlabeled vertices until all the vertices were labeled, i.e., we executed algorithm LABEL(L) below with $L = \varnothing$.

Algorithm **LABEL(G, L)**

{input: finite acyclic digraph G, labeled subset L of $V(G)$}
{output: $V(G)$ with a sorted labeling that agrees with that on L}
If $L = V(G)$ then
    stop

else
    choose $v \in V(G) \backslash L$
    TREESORT(v, L)
      {get $L \cup$ ACC(v) labeled to agree with L labels}
    LABEL($G, L \cup$ ACC(v))
      {label what's left of $V(G)$ and retain the labels on $L \cup$ ACC(v)}. ∎

The verification that LABEL performs correctly is similar to the argument for TREESORT. The appropriate measure of distance from the base case $L = V(G)$ is $|V(G) \backslash L|$, the number of vertices not yet labeled.

EXAMPLE 8   As already noted, if we apply LABEL with $L = \varnothing$ to the digraph in Figure 8(a) we obtain the sorted labeling in Figure 6(a). We summarize.

Choose a
Label ACC(a) = $\{c, b, f, e, a\}$ with TREESORT($a, \varnothing$) {Example 7(a)}
Recur to $L = \{c, b, f, e, a\}$
  Choose d
  Label $L \cup$ ACC(d) = $L \cup \{h, d\}$ with TREESORT(d, L) {Example 7(b)}
  Recur to $L = \{c, b, f, e, a, h, d\}$
    Choose g
    Label $L \cup$ ACC(g) = $L \cup \{g\} = V(G)$ {Example 7(c)}
    Recur to $L = V(G)$
    Stop ∎

Although our examples have all been connected, algorithms TREESORT and LABEL would have worked as well for digraphs with more than one component.

The method of charges gives an estimate for how long TREESORT takes. Charge each vertex with the time it takes to label it. Assign charges to the edges in two different ways: If the edge (u, w) is not chosen, because w is already in L when the choice could be made, charge (u, w) with the time it takes to remove w from the list of available choices in SUCC(u); if w *is* chosen in SUCC(u), charge (u, w) with the time it takes to recur to w and later get back to u. For example, in the graph of Figure 8(a) the edge (e, f) is not chosen; i.e., TREESORT does not recur from e to f, because f is already labeled when the algorithm gets to e. On the other hand, we see that the dashed curve goes down and up the edge (a, e), down for the call to TREESORT at e and back up to report the result to TREESORT at a.

Since LABEL simply keeps applying TREESORT and updating L, these same charges apply to LABEL as well, and show that the time to label $V(G)$ starting with $L = \varnothing$ is $O(|V(G)| + |E(G)|)$. One might object, and say that since only the edges (u, w) for $w \in$ SUCC(V)$\backslash L$ get used, they are the only ones that take time. In fact, though, each edge in G has to be considered, if only briefly, to determine whether it is or is not available for use.

EXERCISES 7.3

1. Use the "dashed-line" idea of Figure 2 to give the preorder and postorder listings of the vertices of the tree in Figure 4(a).

2. Repeat Exercise 1 for Figure 5(a).

3. Repeat Exercise 1 for Figure 5(b).

4. Give the inorder listing of the vertices of the labeled tree in Figure 10.

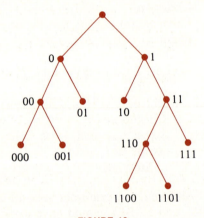

FIGURE 10

5. (a) Use the **PREORDER** algorithm to list the vertices of the tree T in Figure 3. Give the values of $L(w)$ and $L(u)$ that the algorithm produces.

   (b) Repeat part (a) using the **POSTORDER** algorithm.

6. List the successors and accessible sets for the graph in Figure 6(b); use the format of Figure 7.

7. (a) Use the **POSTORDER** algorithm to list the vertices in Figure 4(b).

   (b) Repeat part (a) using the **PREORDER** algorithm.

   (c) Repeat part (a) using the **INORDER** algorithm.

8. Verify the statements in Example 3.

9. (a) Apply algorithm TREESORT$(a, \emptyset)$ to the digraph of Figure 11(a). At each choice point in the algorithm use alphabetic preference to choose successors of the vertex. Draw a dashed curve to show the search pattern and show the labels on the vertices, as in Figure 8(c).

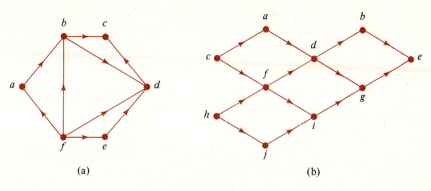

(a)   (b)

FIGURE 11

(b) List the steps for part (a) in the less detailed manner of Example 7(a).

(c) Apply algorithm LABEL to the digraph of Figure 11(a) and draw a picture that shows the resulting labels on the vertices.

10. (a) Repeat Exercise 9(a) for the digraph of Figure 11(b), using the algorithm TREESORT($a, \varnothing$).

    (b) Repeat Exercise 9(c) for this digraph.

11. Apply TREESORT(v, L) to the digraph in Figure 12 with

    (a) $v = a, L = \varnothing$.   (b) $v = d, L = \{a, b, c\}$.

    (c) $v = h, L = \{a, b, c, d, e, f, g\}$.

12. Use Exercise 11 and LABEL to give a sorted labeling of the digraph in Figure 12.

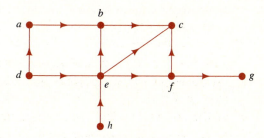

FIGURE 12

13. (a) The digraph in Figure 8(a) is redrawn in Figure 13(a) with different letter names on the vertices. Use algorithm LABEL and the usual alphabetic preference to give a sorted labeling of the digraph.

    (b) Repeat part (a) using the digraph of Figure 13(b).

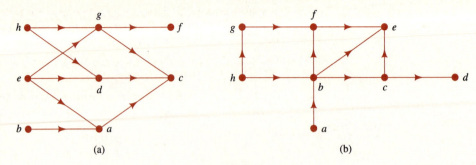

FIGURE 13

14. Algorithm TREESORT resembles POSTORDER in that it labels each vertex the last time it visits it. Here is an algorithm BUSHSORT that tries to imitate PREORDER.

> *Algorithm* **BUSHSORT(v, L)**
>
> {input: finite acyclic digraph G, subset L of $V(G)$ labeled somehow, vertex $v \in V(G)\backslash L$}
> {output: $L \cup$ ACC(v) labeled somehow to agree with the labeling on L}
> Label v with $|L| + 1$ {when we first come to v}.
> Replace L by $L \cup \{v\}$.
> For $w \in$ SUCC(v)$\backslash L$ (successors taken in some order)
>   BUSHSORT(w, L)
>   replace L by $L \cup$ ACC(w).

(a) Apply BUSHSORT(a, $\varnothing$) to label the graph in Figure 14.

(b) Is your labeling in part (a) a sorted labeling for the digraph? Is it the reverse of a sorted labeling, i.e., do vertices have smaller numbers than their successors? Explain.

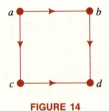

FIGURE 14

15. (a) What are the base cases for algorithm LABELTREE?

(b) What is the output n when LABELTREE(v, k;   ) is called at a base case?

(c) Give a measure of the distance from a base case for algorithm LABELTREE and show that the recursive calls at (w, n) in the algorithm are closer to base cases than the case (v, k) that calls them.

16. (a) Give an example of an acyclic digraph without parallel edges that has 4 vertices and $4(4 - 1)/2 = 6$ edges. How many sorted labelings does your example have?

(b) Show that every acyclic digraph without parallel edges that has n vertices and $n(n - 1)/2$ edges has exactly one sorted labeling.

17. Show that an acyclic digraph without parallel edges that has n vertices cannot have more than $n(n - 1)/2$ edges. [Hence LABEL takes time no more than $O(n^2)$.]

§ 7.4   Polish Notation

Preorder, postorder and inorder listing give ways to list the vertices of an ordered rooted tree. If the vertices have labels such as numbers, addition signs, multiplication signs and the like, the list itself may have a meaningful interpretation. For instance, using ordinary algebraic notation, the list $4 * 3 \div 2$ determines the number 6. The list $4 + 3 * 2$ seems ambiguous; is it the number 14 or 10? Polish notation, which we describe below, is a method for defining algebraic expressions without parentheses, using lists obtained from trees. It is important that the lists completely determine the corresponding labeled trees and their associated expressions. After we discuss Polish notation we prove that under even quite general conditions the lists do determine the trees uniquely.

Polish notation can be used to write expressions that involve objects from some system [of numbers, or matrices, or propositions in the propositional calculus, etc.] and certain operations on the objects. The operations are usually, but not always, **binary**, i.e., ones that combine two objects, or **unary**, ones that act on only one object. Examples of binary operations are $+, *, \wedge, \rightarrow$; the operation $\neg$ is unary. The corresponding ordered rooted trees have leaves labeled with objects from the system [such as numbers] or by variables representing objects from the system [such as x]. The other vertices are labeled by the operations.

EXAMPLE 1   The algebraic expression

$$((x - 4)\wedge 2) * ((y + 2)/3)$$

is represented by the tree in Figure 1(a). This expression uses several familiar binary operations on $\mathbb{R}$: $+, -, *, /, \wedge$. Recall that $*$ represents multiplication and that $\wedge$ represents exponentiation: $a \wedge b$ means a^b. Thus our expression is equivalent to

$$(x - 4)^2 \left(\frac{y + 2}{3} \right).$$

Note that the tree is an *ordered* tree; if x and 4 were interchanged, for example, the tree would represent a different algebraic expression.

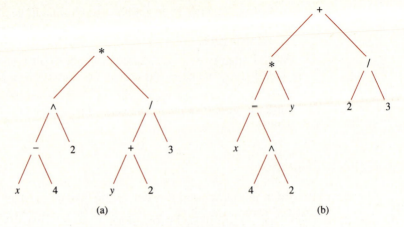

FIGURE 1

It is clear that the ordered rooted tree determines the algebraic expression. Note that the inorder listing of the vertices yields $x - 4 \wedge 2 * y + 2/3$, and this is exactly the original expression *except for the parentheses.* Moreover, the parentheses are crucial since this expression determines neither the tree nor the original algebraic expression. This listing could just as well come from the algebraic expression

$$((x - (4 \wedge 2)) * y) + (2/3)$$

whose tree is drawn in Figure 1(b). This algebraic expression is equivalent to $(x - 16)y + \frac{2}{3}$, a far cry from

$$(x - 4)^2 \left(\frac{y + 2}{3} \right).$$

Let us return to our original algebraic expression given by Figure 1(a). Preorder listing yields

$$* \wedge - x\,4\,2\,/ + y\,2\,3$$

and postorder listing yields

$$x\,4 - 2 \wedge y\,2 + 3\,/\,*.$$

It turns out that each of these listings uniquely determines the tree, and hence the original algebraic expression. Thus these expressions are unambiguous *without parentheses.* This extremely useful observation was made by the Polish logician Łukasiewicz. The preorder listing is known now as **Polish notation** or **prefix notation**. The postorder listing is known as **reverse Polish notation** or **postfix notation**. Our usual algebraic notation, with the necessary parentheses, is known as **infix** notation. ■

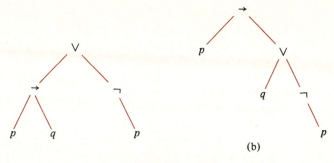

(b)

FIGURE 2

EXAMPLE 2   Consider the compound proposition $(p \to q) \vee (\neg p)$. We treat the binary operations $\to$ and $\vee$ as before. However, $\neg$ is a 1-ary or unary operation. We decree that its child is a right child since the operation precedes the proposition that it operates on. The corresponding binary tree in Figure 2(a) can be traversed in all three ways. The preorder is $\vee \to p\,q\,\neg\,p$ and the postorder is $p\,q \to p\,\neg\,\vee$. The inorder is the original expression *without* parentheses. Another tree with the same inorder expression is drawn in Figure 2(b). As in Example 1, the preorder and postorder listings determine the tree and the original compound proposition. ■

As we illustrated in Example 3 of §7.3, in general the preorder or postorder list of vertices will not determine a tree, even a regular binary tree. More information is needed. It turns out that if we are provided with the level of each vertex, the preorder or postorder list determines the tree. We will not pursue this. It also turns out that if we know how many children each vertex has, the tree is determined by its preorder or postorder list.

EXAMPLE 3   We illustrate the last sentence by beginning with the expression

$$x\,4 - 2\,{}^{\wedge}\,y\,2 + 3\,/\,*$$

in postfix notation. Each vertex $*$, $\wedge$, $-$, $/$ and $+$ represents a binary operation and so has two children. The other vertices have no children, so we know exactly how many children each vertex has.

Let us recover the tree, but instead of drawing subtrees we'll determine their corresponding subexpressions. To reconstruct the tree we recall that in the postorder listing each vertex is immediately preceded by lists of the subtrees rooted at its children. Our task is to recognize those subtree lists in the given expression.

Starting from the left, the first binary operation, $-$, must have two children, namely the leaves x and 4. The subtree rooted at $-$ has post-order listing $x\,4\,-$, corresponding to the ordinary [infix] algebraic expression $x - 4$. We replace $x\,4\,-$ by $(x - 4)$ to get the modified sequence $(x - 4)\,2\,\hat{}\,y\,2 + 3\,/\,*$. The next binary operation, $\hat{}$, has two subtrees hanging from it. One is the subtree $(x - 4)$ that we just found, and the other is the leaf 2. The corresponding [infix] expression is $((x - 4)\hat{}\,2)$, leading to the new sequence $((x - 4)\hat{}\,2)y\,2 + 3\,/\,*$. Now the operation $+$ acts on its own immediately preceding subtrees, namely the leaves y and 2 to give $(y + 2)$ and we obtain $((x - 4)\hat{}\,2)(y + 2)\,3\,/\,*$. Then $/$ acts on its two immediate predecessors $(y + 2)$ and 3 to give $((y + 2)/3)$ and the sequence $((x - 4)\hat{}\,2)((y + 2)/3)\,*$. Finally, we have one operation, $*$, preceded by two expressions representing subtrees, so we obtain

$$((x - 4)\hat{}\,2) * ((y + 2)/3).$$

We have recovered the expression for the tree in Figure 1(a).

We briefly summarize the procedure of the last paragraph:

$$\boldsymbol{x\,4 - 2\,\hat{}\,y\,2 + 3\,/\,*}$$

$$\boldsymbol{(x - 4)\,2\,\hat{}\,y\,2 + 3\,/\,*}$$

$$\boldsymbol{((x - 4)\hat{}\,2)\,y\,2 + 3\,/\,*}$$

$$\boldsymbol{((x - 4)\hat{}\,2)(y + 2)\,3\,/\,*}$$

$$\boldsymbol{((x - 4)\hat{}\,2)((y + 2)/3)\,*}$$

$$((x - 4)\hat{}\,2) * ((y + 2)/3).$$

There is another way to view this procedure, which will serve as a model for the proof of the theorem after Example 6. Working left to right, we can pick off the children of each vertex as follows: each binary operation gets the two nearest available children in the stack of children to its left. A child is "available" if it has not already been assigned to a parent. The other vertices get no children.

Vertex	x	4	$-$	2	$\hat{}$	y	2	$+$	3	$/$	$*$
Available children		x	$x, 4$	$-$	$-, 2$	$\hat{}$	$\hat{}, y$	$\hat{}, y, 2$	$\hat{}, +$	$\hat{}, +, 3$	$\hat{}, /$
Assigned children			$x, 4$		$-, 2$			$y, 2$		$+, 3$	$\hat{}, /$

■

EXAMPLE 4　The same method works for a compound proposition in reverse Polish notation, but when we meet the unary operation $\neg$, it acts on just the immediately preceding subexpression. For example,

$$\boldsymbol{p\,q} \to p\,q \wedge \neg \vee$$

$$(p \to q)\,\boldsymbol{p\,q} \wedge \neg \vee$$

$$(p \to q)\,(\boldsymbol{p \wedge q})\,\neg \vee \qquad [\neg \text{ just acts on } (p \wedge q)]$$

$$(p \to q)\,(\boldsymbol{\neg(p \wedge q)}) \vee$$

$$(p \to q) \vee (\neg(p \wedge q)).$$

The reader can draw a tree representing this compound proposition. Alternatively, we can pick off the children working from left to right as in Example 3:

Vertex	p	q	$\to$	p	q	$\wedge$	$\neg$	$\vee$
Available children		p	p, q	$\to$	$\to, p$	$\to, p, q$	$\to, \wedge$	$\to, \neg$
Assigned children			p, q			p, q	$\wedge$	$\to, \neg$

Not all strings of operations and symbols lead to meaningful expressions.

EXAMPLE 5   Suppose it is alleged that

$$y + 2x * {}^{\wedge}4 \quad \text{and} \quad q \neg p\,q \vee \wedge \to$$

are in reverse Polish notation. The first one is hopeless right away, since $+$ is not preceded by two expressions. The second one breaks down when we attempt to decode [i.e., parse] it as in Example 4:

$$\boldsymbol{q\,\neg}\,p\,q \vee \wedge \to$$

$$(\boldsymbol{\neg q})\,\boldsymbol{p\,q} \vee \wedge \to$$

$$(\neg q)\,(\boldsymbol{p \vee q}) \wedge \to$$

$$((\neg q) \wedge (p \vee q)) \to.$$

Unfortunately, the operation $\to$ has only one subexpression preceding it. We conclude that neither of the strings of symbols given above represents a meaningful expression.   ■

Just as we did with ordinary algebraic expressions in §7.2, we can recursively define what we mean by well-formed formulas [wff's] for Polish and reverse Polish notation. We give one example by defining **wff's for reverse Polish notation** of algebraic expressions:

(B)   Numerical constants and variables are wff's;

(R)   If f and g are wff's, so are $f\,g\,+,\ f\,g\,-,\ f\,g\,*,\ f\,g\,/$ and $f\,g\,{}^{\wedge}$.

EXAMPLE 6   We show that $x\, 2 \wedge y - x\, y * / $ is a wff. All the variables and constants are wff's by (B). Then $x\, 2 \wedge$ is a wff by (R). Hence $x\, 2 \wedge y -$ is a wff, where we use (R) with $f = x\, 2 \wedge$ and $g = y$. Similarly, $x\, y *$ is a wff by (R). Finally, the entire expression is a wff by (R), since it has the form $f\, g /$ where $f = x\, 2 \wedge y -$ and $g = x\, y *$. ∎

We end the section by proving the theorem that shows that expressions in Polish and reverse Polish notation uniquely determine the original expression. The proof is based on the second algorithm illustrated in Examples 3 and 4.

Theorem

> Let T be a finite ordered rooted tree whose vertices have been listed by a preorder listing or a postorder listing. Suppose that the number of children of each vertex is known. Then the tree is determined; that is, the tree can be recovered from the listing.

Proof. We consider only the case of postorder listing. We are given the postordered list $v_1 v_2 \cdots v_n$ of T and the numbers $c_1, \ldots, c_n$ of children of $v_1, \ldots, v_n$. We'll show that for each vertex v_m of T the set $S(v_m)$ of children of v_m and its order are uniquely determined.

Consider some vertex v_m. Then v_m is the root of the subtree T_m consisting of v_m and its descendants. When Algorithm POSTORDER lists v_m, the subtrees of T_m have already been listed and their lists immediately precede v_m, in the order determined by the order of $S(v_m)$. Moreover, the algorithm does not insert later entries into the list of T_m, so the list of T_m appears in the complete list $v_1 v_2 \cdots v_n$ of T as an unbroken string with v_m at the right end.

Since v_1 has no predecessors, v_1 is a leaf. Thus $S(v_1) = \varnothing$, and its order is vacuously determined. Assume inductively that for each k with $k < m$ the set $S(v_k)$ and its order are determined, and consider v_m. [We are using the second principle of induction here on the finite set $\{1, 2, \ldots, n\}$.] Then the set $U(m) = \{v_k : k < m\} \backslash \bigcup_{k < m} S(v_k)$ is determined. It consists of the vertices v_k to the left of v_m whose parents are not to the left of v_m. The children of v_m are the members of $U(m)$ that are in the tree T_m. Since the T_m list is immediately to the left of v_m, the children of v_m are the c_m members of $U(m)$ farthest to the right. Since $U(m)$ is determined and c_m is given, the set $S(v_m)$ of children of v_m is determined. Moreover, its order is the order of appearance in the list $v_1 v_2 \cdots v_n$.

By induction, each ordered set $S(v_m)$ of children is determined by the postordered list and the sequence $c_1, c_2, \ldots, c_n$. The root of the tree is, of course, the last vertex v_n. Thus the complete structure of the ordered rooted tree is determined. ∎

EXAMPLE 7 The list $u\,x\,y\,z\,w\,v\,p\,q\,s\,r$ and sequence $0, 0, 0, 0, 3, 2, 0, 0, 2, 2$ give the following sets.

v_k	u	x	y	z	w	v	p	q	s	r
$U(k)$	$\varnothing$	$\{u\}$	$\{u, x\}$	$\{u, x, y\}$	$\{u, x, y, z\}$	$\{u, w\}$	$\{v\}$	$\{v, p\}$	$\{v, p, q\}$	$\{v, s\}$
$S(v_k)$	$\varnothing$	$\varnothing$	$\varnothing$	$\varnothing$	$\{x, y, z\}$	$\{u, w\}$	$\varnothing$	$\varnothing$	$\{p, q\}$	$\{v, s\}$

The ordered sets $S(v_k)$ can be assembled recursively into the tree T of Figure 1 of §7.3. ■

EXERCISES 7.4

1. Write the algebraic expression given by Figure 1(b) in reverse Polish and in Polish notation.

2. For the ordered rooted tree in Figure 3(a), write the corresponding algebraic expression in reverse Polish notation and also in the usual infix algebraic notation.

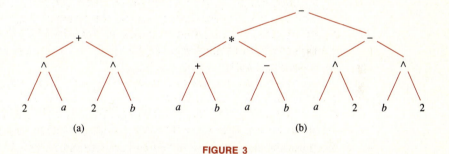

(a)   (b)

FIGURE 3

3. (a) For the ordered rooted tree in Figure 3(b), write the corresponding algebraic expression in Polish notation and also in the usual infix algebraic notation.

   (b) Simplify the algebraic expression obtained in part (a) and then draw the corresponding tree.

4. Calculate the following expressions given in reverse Polish notation.

   (a) $3\,3\,4\,5\,1 - * + +$ (b) $3\,3 + 4 + 5 * 1 -$

   (c) $3\,3\,4 + 5 * 1 - +$

5. Calculate the following expressions given in reverse Polish notation.

(a) $6\,3\,/\,3 + 7\,3 - *$

(b) $3\,2\,\char`\^\,4\,2\,\char`\^\, + 5\,/\,2\,*$

6. Calculate the following expressions in Polish notation.

(a) $- *3\,\char`\^5\,2\,2$

(b) $\char`\^ *3\,5 - 2\,2$

(c) $- \char`\^ *3\,5\,2\,2$

(d) $/\,*2 + 2\,5\,\char`\^\, + 3\,4\,2$

(e) $* + /\,6\,3\,3 - 7\,3$

7. Write the following algebraic expressions in reverse Polish notation.

(a) $(3x - 4)^2$

(b) $(a + 2b)/(a - 2b)$

(c) $x - x^2 + x^3 - x^4$

8. Write the algebraic expressions in Exercise 7 in Polish notation.

9. (a) Write the algebraic expressions $a(bc)$ and $(ab)c$ in reverse Polish notation.

(b) Do the same for $a(b + c)$ and $ab + ac$.

(c) What do the associative and distributive laws look like in reverse Polish notation?

10. Write the expression $x\,y + 2\,\char`\^\,x\,y - 2\,\char`\^\, - x\,y\,*\,/$ in the usual infix algebraic notation and simplify.

11. Consider the compound proposition represented by Figure 2(b).

(a) Write the proposition in the usual infix notation [with parentheses].

(b) Write the proposition in reverse Polish and in Polish notation.

12. The following compound propositions are given in Polish notation. Draw the corresponding rooted trees, and rewrite the expressions in the usual infix notation.

(a) $\leftrightarrow \neg \wedge \neg p \neg q \vee p q$

(b) $\leftrightarrow \wedge p q \neg \rightarrow p \neg q$

[These are laws from Table 1 of §2.2.]

13. Repeat Exercise 12 for the following.

(a) $\rightarrow \wedge p \rightarrow p q q$

(b) $\rightarrow \wedge \wedge \rightarrow p q \rightarrow r s \vee p r \vee q s$

14. Write the following compound propositions in reverse Polish notation.

(a) $[(p \rightarrow q) \wedge (q \rightarrow r)] \rightarrow (p \rightarrow r)$

(b) $[(p \vee q) \wedge \neg p] \rightarrow q$

15. Illustrate the ambiguity of "parenthesis-free infix notation" by writing the following pairs of expressions in infix notation without parentheses.

(a) $(a/b) + c$ and $a/(b + c)$

(b) $a + (b^3 + c)$ and $(a + b)^3 + c$

16. Use the recursive definition for wff's for reverse Polish notation to show that the following are wff's.

(a) $3\,x\,2\,\char`\^\,*$

(b) $x\,y + 1\,x\,/\,1\,y\,/\, + *$

(c) $4\,x\,2\,\char`\^\,y\,z + 2\,\char`\^\,/\,-$

17. (a) Define wff's for Polish notation for algebraic expressions.

(b) Use the definition in part (a) to show that $\char`\^ + x\,/\,4\,x\,2$ is a wff.

18. Let $S_1 = x_1\,2\,\char`\^$ and $S_{n+1} = S_n\,x_{n+1}\,2\,\char`\^\, +$ for $n \geq 1$. Here $x_1, x_2, \ldots$ represent variables.

(a) Show that each S_n is a wff for reverse Polish notation. *Hint:* Use induction.

(b) What does S_n look like in the usual infix notation?

19. (a) Define wff's for reverse Polish notation for the propositional calculus; see the definition just before Example 6.

(b) Use the definition in part (a) to show that $p\,q \neg \wedge \neg p\,q \neg \rightarrow \vee$ is a wff.

(c) Define wff's for Polish notation for the propositional calculus.

(d) Use the definition in part (c) to show that $\vee \neg \wedge p \neg q \rightarrow p \neg q$ is a wff.

20. (a) Draw the tree with postorder vertex sequence $s\,t\,v\,y\,r\,z\,w\,u\,x\,q$ and number of children sequence 0, 0, 0, 2, 2, 0, 0, 0, 2, 3.

(b) Is there a tree with $s\,t\,v\,y\,r\,z\,w\,u\,x\,q$ as preorder vertex sequence and number of children sequence 0, 0, 0, 2, 2, 0, 0, 0, 2, 3? Explain.

§ 7.5   Weighted Trees

A **weighted tree** is a finite rooted tree in which each leaf is assigned a nonnegative real number, called the **weight** of the leaf. In this section we discuss general weighted trees and we give applications to prefix codes and sorted lists.

To establish some notation, we assume that our weighted tree T has t leaves whose weights are $w_1, w_2, \ldots, w_t$. We lose no generality if we also assume that $w_1 \le w_2 \le \cdots \le w_t$. It will be convenient to label the leaves by their weights, so we will often refer to the leaf by referring to its weight. Let $l_1, l_2, \ldots, l_t$ denote the corresponding level numbers of the leaves, so that l_i is the length of the path from the root to the leaf w_i. The **weight** of the tree T is the number

$$W(T) = \sum_{i=1}^{t} w_i l_i,$$

in which the weights of the leaves are multiplied by their level numbers.

EXAMPLE 1   (a) The six leaves of the weighted tree in Figure 1(a) have weights 2, 4, 6, 7, 7 and 9. Thus $w_1 = 2$, $w_2 = 4$, $w_3 = 6$, $w_4 = 7$, $w_5 = 7$ and $w_6 = 9$. There are two leaves labeled 7, and it does not matter which we regard as w_4 and which we regard as w_5. For definiteness, we let w_4 represent the leaf labeled 7 at level 2. Then the level numbers are $l_1 = 3$, $l_2 = 1$, $l_3 = 3$, $l_4 = 2$, $l_5 = 1$ and $l_6 = 2$. Hence

$$W(T) = \sum_{i=1}^{6} w_i l_i = 2 \cdot 3 + 4 \cdot 1 + 6 \cdot 3 + 7 \cdot 2 + 7 \cdot 1 + 9 \cdot 2 = 67.$$

(b) The same six weights can be placed on a binary tree as in Figure 1(b), for instance. Now the level numbers are $l_1 = 3$, $l_2 = 3$, $l_3 = 2$, $l_4 = 3$, $l_5 = 3$ and $l_6 = 2$, so

$$W(T) = 2 \cdot 3 + 4 \cdot 3 + 6 \cdot 2 + 7 \cdot 3 + 7 \cdot 3 + 9 \cdot 2 = 90.$$

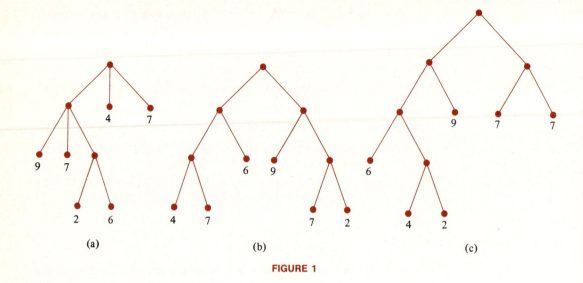

FIGURE 1

(c) Figure 1(c) shows another binary tree with these weights. Its weight is

$$W(T) = 2 \cdot 4 + 4 \cdot 4 + 6 \cdot 3 + 7 \cdot 2 + 7 \cdot 2 + 9 \cdot 2 = 88.$$

The total weight is less than in part (b), because the heavier leaves are near the root and the lighter ones are farther away. Later in this section we will discuss an algorithm for obtaining a binary tree with minimum weight for any specified sequence of weights $w_1, w_2, \ldots, w_t$. ■

EXAMPLE 2   (a) As we will explain later in this section, certain sets of binary numbers can serve as codes. An example of such a set is {00, 01, 100, 1010, 1011, 11}. These numbers are the labels of the leaves in the binary tree of Figure 2(a). This set could serve as a code for the letters in an alphabet Σ that has six letters. Suppose that we know how frequently each letter in Σ is used in sending messages. In Figure 2(b) we have placed a weight at each leaf that signifies the percentage of code symbols using that leaf. For example, the letter coded 00 appears 25 percent of the time, the letter coded 1010 appears 20 percent of the time, etc. Since the length of each code symbol as a word in 0's and 1's is exactly its level in the binary tree, the average length of a code message using 100 letters from Σ will just be the weight of the weighted tree, in this case

$$25 \cdot 2 + 10 \cdot 2 + 10 \cdot 3 + 20 \cdot 4 + 15 \cdot 4 + 20 \cdot 2 = 280.$$

This weight measures the efficiency of the code. As we will see in Example 7, there are more efficient codes for this example, i.e., for the set of frequencies 10, 10, 15, 20, 20, 25.

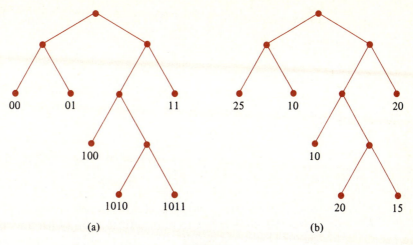

FIGURE 2

(b) Instead of standing for letters in the alphabet Σ, the code words in (a) can be thought of as addresses for the leaves in the binary search tree in Figure 2(a). Suppose that the time it takes to reach a leaf and look up the record stored there is proportional to the number of decision [branch] nodes on the path to the leaf. Then, to minimize the average lookup time, the most frequently looked up records should be placed at leaves near the root, with the less popular ones farther away. If the weights in Figure 2(b) show the percentages of lookups of records in the corresponding locations, then the average lookup time is proportional to the weight of the weighted tree. ■

EXAMPLE 3 Consider a collection of sorted lists, say $L_1, L_2, \ldots, L_n$. For example, each L_i could be an alphabetically sorted mailing list of clients or a pile of exam papers arranged in increasing order of grades. To illustrate the ideas involved, let's suppose that each list is a set of real numbers arranged by the usual order $\leq$. Suppose that we can merge lists two at a time to produce new lists. Our problem is to determine how to merge the n lists most efficiently to produce a single sorted list.

Two lists are merged by comparing the first numbers of both lists and selecting the smaller of the two [either one if they are equal]. The selected number is removed and becomes the first member of the merged list, and the process is repeated for the two lists that remain. The next number selected is placed second on the merged list, and so on. The process ends when one of the lists is empty.

For instance, to merge 4, 8, 9 and 3, 6, 10: compare 3 and 4, choose **3**, and reduce to lists 4, 8, 9 and 6, 10; compare 4 and 6, choose **4**, and reduce to 8, 9 and 6, 10; compare 8 and 6, choose **6**, and reduce to 8, 9

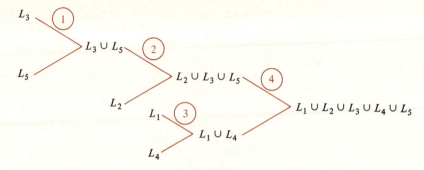

FIGURE 3

and 10; compare 8 and 10, choose **8**, and reduce to 9 and 10; compare 9 and 10, choose **9**, and reduce to the empty list and 10; choose **10**. The merged list is 3, 4, 6, 8, 9, 10. In this example 5 comparisons were required. If the lists contain j and k elements, respectively, then in general the process must end after $j + k - 1$ or fewer comparisons. The goal is to merge $L_1, L_2, \ldots, L_n$ in pairs while minimizing the worst case number of comparisons involved.

Suppose, for example, that we have five lists L_1, L_2, L_3, L_4, L_5 with 15, 22, 31, 34, 42 items and suppose they are merged as indicated in Figure 3. There are four merges indicated by the circled numbers. The first merge involves at most $|L_3| + |L_5| - 1 = 72$ comparisons. The second merge involves at most $|L_2| + |L_3| + |L_5| - 1 = 94$ comparisons. The third and fourth merges involve at most $|L_1| + |L_4| - 1 = 48$ and $|L_1| + |L_2| + |L_3| + |L_4| + |L_5| - 1 = 143$ comparisons. The entire process involves at most 357 comparisons. This number isn't very illuminating by itself, but note that

$$357 = 2 \cdot |L_1| + 2 \cdot |L_2| + 3 \cdot |L_3| + 2 \cdot |L_4| + 3 \cdot |L_5| - 4.$$

This is just 4 less than the weight of the tree in Figure 4. Note the intimate connection between Figures 3 and 4. No matter how we merge the

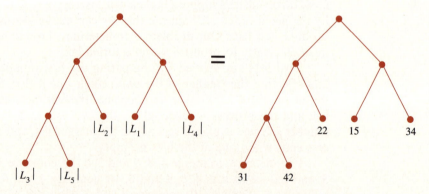

FIGURE 4

five lists in pairs, there will be 4 merges. A computation like the one above shows that the merge will involve at most $W(T) - 4$ comparisons, where T is the tree corresponding to the merge. So finding a merge that minimizes the worst case number of comparisons is equivalent to finding a binary tree with weights 15, 22, 31, 34, 42 having minimal weight. We return to this problem in Example 8.

The merging of n lists in pairs involves $n - 1$ merges. In general, a merge of n lists will involve at most $W(T) - (n - 1)$ comparisons, where T is the weighted tree corresponding to the merge. ■

Examples 2 and 3 suggest the following general problem. We are given a list $L = (w_1, \ldots, w_t)$ of at least two nonnegative numbers, and we want to construct a binary weighted tree T with the members of L as weights so that $W(T)$ is as small as possible. We call such a tree T an **optimal binary tree** for the weights $w_1, \ldots, w_t$. The following recursive algorithm solves the problem by producing an optimal binary tree $T(L)$.

Algorithm HUFFMAN(*L*)

{input: list $L = (w_1, \ldots, w_t)$ of nonnegative numbers, $t \geq 2$}
{output: optimal binary tree $T(L)$ for L}
If $t = 2$ let $T(L)$ be a tree with 2 leaves, of weights w_1 and w_2.
Otherwise, find the two smallest members of L, say u and v,
    let L' be the list obtained from L by removing u and v and inserting
        $u + v$,
    do HUFFMAN(L') to get $T(L')$, and
    form $T(L)$ from $T(L')$ by replacing a leaf of weight $u + v$ in $T(L')$
        by a subtree with two leaves, of weights u and v. ■

This algorithm ultimately reduces the problem to the base case of finding optimal binary trees with two leaves, which is trivial to solve. We will show shortly that HUFFMAN(L) always produces an optimal binary tree. First we look at some examples of how it works and we apply the algorithm to the problems that originally motivated our looking at optimal trees.

EXAMPLE 4 Consider weights 2, 4, 6, 7, 7, 9. First the algorithm repeatedly combines the smallest two weights to obtain shorter and shorter weight sequences. Here is the recursive chain:

HUFFMAN(**2, 4**, 6, 7, 7, 9) replaces 2 and 4 by $2 + 4$ and calls
   HUFFMAN(**6, 6**, 7, 7, 9), which replaces 6 and 6 by 12 and calls
      HUFFMAN(**7, 7**, 9, 12), which calls
         HUFFMAN(**9, 12**, 14), which calls
           HUFFMAN(14, 21), which builds the first tree
              $T(14, 21)$ in Figure 5.

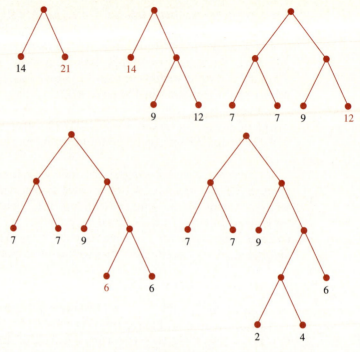

FIGURE 5

Now each of the previous recursive calls constructs its tree and passes it back up the calling chain. Figure 5 shows the full sequence of trees $T(14, 21)$, $T(9, 12, 14)$, ..., $T(2, 4, 6, 7, 7, 9)$. Note, for example, that the third tree $T(7, 7, 9, 12)$ is obtained from the second tree $T(9, 12, 14)$ by replacing the leaf of weight $\mathbf{14} = 7 + 7$ by a subtree with two leaves of weight 7 each. The final weighted tree is essentially the tree drawn in Figure 1(c), so that tree is an optimal binary tree. As noted in Example 1(c), it has weight 88. ■

EXAMPLE 5 Let us find an optimal binary tree with weights 2, 3, 5, 7, 10, 13, 19. We repeatedly combine the smallest two weights to obtain the weight sequences

$$\mathbf{2, 3}, 5, 7, 10, 13, 19 \to \mathbf{5, 5}, 7, 10, 13, 19$$

$$\to \mathbf{7, 10}, 10, 13, 19 \to \mathbf{10, 13}, 17, 19 \to \mathbf{17, 19}, 23 \to 23, 36.$$

Then we use HUFFMAN'S algorithm to build the optimal binary trees in Figure 6. After the fourth tree is obtained, either leaf of weight 10 could have been replaced by the subtree with weights 5 and 5. Thus the last two trees could have been as drawn in Figure 7. Either way, the final tree

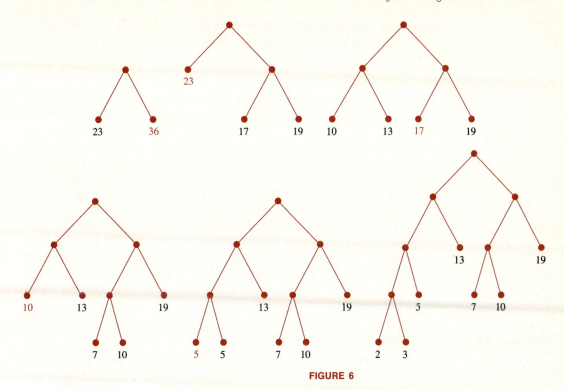

FIGURE 6

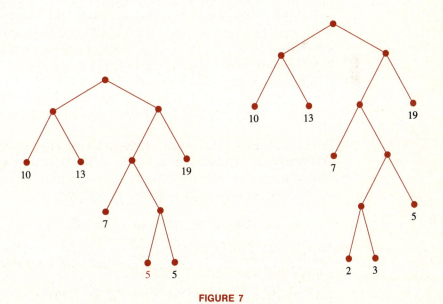

FIGURE 7

has weight 150 [Exercise 2]. Note that the optimal tree is by no means unique; the one in Figure 6 has height 4, while the one in Figure 7 has height 5. ■

We now return to codes using binary numbers. To avoid ambiguity, we must impose some restrictions; for example, a code should not use all three of 10, 01 and 0110, since a string such as 011010 would be ambiguous. If we think of strings of 0's and 1's as labels for vertices in a binary tree, each label gives instructions for finding its vertex from the root: go left on 0, right on 1. Motivated in part by the search tree in Example 2(b), we impose the condition that no code symbol can be the initial string of another code symbol. Thus, for instance, 01 and 0110 cannot both be code symbols. In terms of the binary tree, this condition means that no vertex labeled with a symbol can lie below another such vertex in the tree. We also impose one more condition on the code: The binary tree whose leaves are labeled by the code words must be regular, so that every nonleaf has two children. A code meeting these two conditions is called a **prefix code**.

EXAMPLE 6   The set {00, 01, 100, 1010, 1011, 11} is a prefix code. It is the set of leaves for the labeled binary tree in Figure 2(a), which we have redrawn in Figure 8(a) with all vertices but the root labeled. Every string of 0's and 1's of length 4 begins with one of these code symbols, since every path of length 4 from the root in the full binary tree in Figure 8(b) runs into one of the code vertices. This means that we can attempt to decode any string of 0's and 1's by proceeding from left to right in the string, finding the first substring that is a code symbol, then the next substring after that, and so on. This procedure either uses up the whole string or it leave at most three 0's and 1's undecoded at the end.

For example, consider the string

$$1\,1\,1\,0\,1\,0\,1\,1\,0\,1\,1\,0\,0\,0\,1\,0\,0\,1\,1\,1\,1\,1\,0\,0\,1\,0.$$

We visit vertex 1, then vertex 11. Since vertex 11 is a leaf, we record 11 and return to the root. We next visit vertices 1, 10, 101 and 1010. Since 1010 is a leaf, we record 1010 and return again to the root. Proceeding in this way, we obtain the sequence of code symbols

$$11, \quad 1010, \quad 11, \quad 01, \quad 100, \quad 01, \quad 00, \quad 11, \quad 11, \quad 100$$

and have 10 left over. This scheme for decoding arbitrary string of 0's and 1's will work for any code with the property that every path from the root in a full binary tree runs into a unique code vertex. Prefix codes have this property by their definition. ■

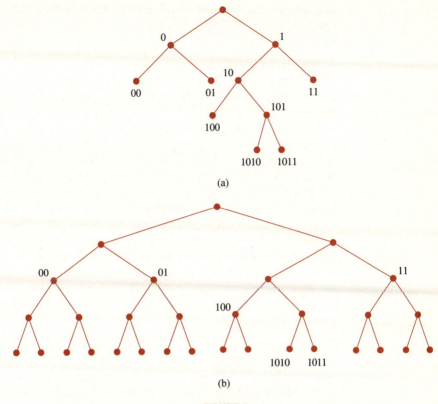

(a)

(b)

FIGURE 8

EXAMPLE 7   (a) We now solve the problem suggested by Example 2(a); that is, we find a prefix code for the set of frequencies 10, 10, 15, 20, 20, 25 that is as efficient as possible. We want to minimize the average length of a code message using 100 letters from Σ. Thus all we need is an optimal binary tree for these weights. Using the procedure illustrated in Examples 4 and 5, we obtain the weighted tree in Figure 9(a). We label this tree with binary digits in Figure 9(b). Then {00, 01, 10, 110, 1110, 1111} will be a most efficient code for Σ provided that we match the letters of Σ to code symbols so that the frequencies of the letters are given by Figure 9(a). With this code, the average length of a code message using 100 letters from Σ is

$$20\cdot 2 + 20\cdot 2 + 25\cdot 2 + 15\cdot 3 + 10\cdot 4 + 10\cdot 4 = 255,$$

an improvement over the average length 280 obtained in Example 2.

(b) The solution we have just obtained also gives a most efficient binary search tree for looking up records whose lookup frequencies are 10, 10, 15, 20, 20, 25, as in Example 2(b). ∎

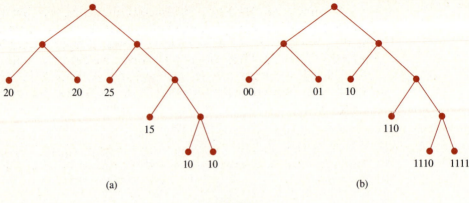

(a)   (b)

FIGURE 9

EXAMPLE 8 We complete the discussion on sorted lists begun in Example 3. There we saw that we needed an optimal binary tree with weights 15, 22, 31, 34, 42. Using the procedure in Examples 4 and 5, we obtain the tree in Figure 10. This tree has weight 325. The corresponding merge in pairs given in Figure 11 will require at most $325 - 4 = 321$ comparisons. ∎

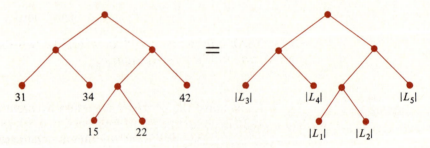

FIGURE 10

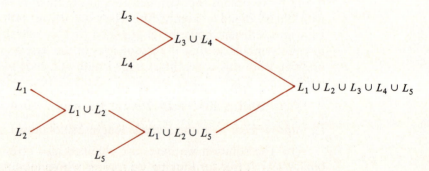

FIGURE 11

To show that HUFFMAN's algorithm works, we first prove a lemma that tells us that in optimal binary trees the heavy leaves are near the root. The lemma and its corollary are quite straightforward if all the weights are distinct [Exercise 14]. However, we need the more general case. Even if we begin with distinct weights, HUFFMAN'S algorithm may lead to the case where the weights are not all distinct, as occurred in Example 5.

Lemma

> Let T be an optimal binary tree for weights $w_1, w_2, \ldots, w_t$. For $i = 1, 2, \ldots, t$, let l_i denote the level of w_i. If $w_j < w_k$, then $l_j \geq l_k$.

Proof. Assume that $w_j < w_k$ and $l_j < l_k$ for some j and k. Let T' be the tree obtained by interchanging the weights w_j and w_k. In calculating $W(T)$ the leaves w_j and w_k contribute $w_j l_j + w_k l_k$, while in calculating $W(T')$ they contribute $w_j l_k + w_k l_j$. Since the other leaves contribute the same to both $W(T)$ and $W(T')$, we have

$$W(T) - W(T') = w_j l_j + w_k l_k - w_j l_k - w_k l_j = (w_k - w_j)(l_k - l_j) > 0.$$

Hence $W(T') < W(T)$ and T is not an optimal binary tree, contrary to our hypothesis. ∎

Corollary

> There is an optimal binary tree T for which the two smallest weights w_1 and w_2 are both at the lowest level l.

Proof. There are at least two leaves at the lowest level, say w_j and w_k. If $w_1 < w_j$, then $l_1 \geq l_j = l$ by the lemma, so $l_1 = l$ and w_1 is at level l. If $w_1 = w_j$, then conceivably $l_1 < l_j$, but we can interchange w_1 and w_j without changing the total weight of T. Similarly, by interchanging w_2 and w_k if necessary, we may suppose that w_2 is at level l. ∎

The following result shows that HUFFMAN(L) works.

Theorem

> Suppose that $0 \leq w_1 \leq w_2 \leq \cdots \leq w_t$. Let T' be an optimal binary tree with weights $w_1 + w_2, w_3, \ldots, w_t$ and let T be the weighted binary tree obtained from T' by replacing a leaf of weight $w_1 + w_2$ by a subtree with two leaves having weights w_1 and w_2. Then T is an optimal binary tree with weights $w_1, w_2, \ldots, w_t$.

Proof. Since there are only finitely many binary trees with t leaves, there must be an optimal binary tree T_0 with weights $w_1, w_2, \ldots, w_t$.

Our task is to show that $W(T) = W(T_0)$. By the corollary of the lemma, we may suppose that the weights w_1 and w_2 for T_0 are at the same level. The total weight of T_0 won't change if weights at the same level are interchanged. Thus we may assume that w_1 and w_2 are children of the same parent p. These three vertices form a little subtree T_p with p as root.

Now let T'_0 be the tree with weights $w_1 + w_2, w_3, \ldots, w_t$ obtained from T_0 by replacing the subtree T_p by a leaf $\bar{p}$ of weight $w_1 + w_2$. Let l be the level of the vertex p and observe that in calculating $W(T_0)$ the subtree T_p contributes $w_1(l + 1) + w_2(l + 1)$, while in calculating $W(T'_0)$, the vertex $\bar{p}$ with weight $w_1 + w_2$ contributes $(w_1 + w_2)l$. Thus

$$W(T_0) = W(T'_0) + w_1 + w_2.$$

The same argument shows that

$$W(T) = W(T') + w_1 + w_2.$$

Since T' is optimal for the weights $w_1 + w_2, w_3, \ldots, w_t$, we have $W(T') \le W(T'_0)$, so

$$W(T) = W(T') + w_1 + w_2 \le W(T'_0) + w_1 + w_2 = W(T_0).$$

Of course, $W(T_0) \le W(T)$, since T_0 is optimal for the weights $w_1, w_2, \ldots, w_t$, so $W(T) = W(T_0)$, as desired. That is, T is an optimal binary tree with weights $w_1, w_2, \ldots, w_t$. ∎

HUFFMAN(L) leads recursively to $t - 1$ choices of parents in $T(L)$. Each choice requires a search for the two smallest members of the current list, which can be done in time $O(t)$ by just running through the list. So the total operation of the algorithm takes time $O(t^2)$.

There are at least two ways to speed up the algorithm. It is possible to find the two smallest members in time $O(\log_2 t)$, using a binary tree as a data structure for the list L. Alternatively, there are algorithms that can initially sort L into nondecreasing order in time $O(t \log_2 t)$. Then the smallest elements are simply the first two on the list, and after we remove them we can maintain the nondecreasing order on the new list by inserting their sum at the appropriate place, just as we did in Examples 4 and 5. The correct insertion point can be found in time $O(\log_2 t)$, so this scheme, too, works in time $O(t \log_2 t)$.

Minimum weight prefix codes, called **Huffman codes**, are of considerable practical interest because of their applications to efficient message transmission and to the design of search tree data structures, as described in Examples 2(a) and 2(b). In both of these settings the actual frequencies of message symbols or record lookups are determined by experience, and may change over time. In the search tree setting, the number of code symbols may even change as records are added or deleted. The problem of dynamically modifying a Huffman code to reflect changing circumstances is challenging and interesting, but unfortunately beyond the scope of this book.

EXERCISES 7.5

1. (a) Calculate the weights of all the trees in Figure 5.

   (b) Calculate the weights of all the trees in Figure 6.

2. Calculate the weights of the two trees in Figures 6 and 7 with weights 2, 3, 5, 7, 10, 13, 19.

3. Construct an optimal binary tree for the following sets of weights and compute the weight of the optimal tree.

   (a) $\{1, 3, 4, 6, 9, 13\}$ (b) $\{1, 3, 5, 6, 10, 13, 16\}$

   (c) $\{2, 4, 5, 8, 13, 15, 18, 25\}$ (d) $\{1, 1, 2, 3, 5, 8, 13, 21, 34\}$

4. Find an optimal binary tree for the weights 10, 10, 15, 20, 20, 25 and compare your answer with Figure 9(a).

5. Which of the following sets of sequence are prefix codes? If the set is a prefix code, construct a binary tree whose leaves represent this binary code. Otherwise, explain why the set is not a prefix code.

   (a) $\{000, 001, 01, 10, 11\}$ (b) $\{00, 01, 110, 101, 0111\}$

   (c) $\{00, 0100, 0101, 011, 100, 101, 11\}$

6. Here is a prefix code: $\{00, 010, 0110, 0111, 10, 11\}$.

   (a) Construct a binary tree whose leaves represent this binary code.

   (b) Decode the string

   $$0\,0\,1\,0\,0\,0\,0\,1\,1\,0\,0\,1\,0\,0\,0\,1\,0\,0\,1\,1\,1\,1\,1\,0\,1\,1\,0$$

   if $00 = A$, $10 = D$, $11 = E$, $010 = H$, $0110 = M$ and 0111 represents the apostrophe '. You will obtain the very short poem titled "Fleas."

   (c) Decode $0\,1\,0\,1\,1\,0\,1\,1\,0\,0\,0\,1\,0\,1\,1\,0\,1\,1\,0\,1\,1\,0\,1\,1\,0\,0\,0\,1\,0$.

   (d) Decode the following soap opera. $1\,0\,0\,0\,1\,0\,0\,1\,0\,0\,0\,1\,0\,0\,1\,1\,0\,0\,0\,1\,0\,0\,0\,0$ $1\,1\,0$. $0\,1\,0\,1\,1\,0\,1\,1\,0\,0\,0\,1\,0\,1\,1\,0\,1\,1\,0\,0\,0\,0\,1\,1\,0\,0\,0\,1\,0$. $0\,1\,1\,0\,0\,0\,0\,1\,1\,0\,0\,0\,1$ $0\,1\,1\,1\,0\,0\,0\,1\,0\,1\,0\,1\,1\,0\,0\,1\,0$.

7. Suppose that we are given a fictitious alphabet Σ of seven letters a, b, c, d, e, f and g with the following frequencies per 100 letters: $a-11$, $b-20$, $c-4$, $d-22$, $e-14$, $f-8$, $g-21$.

   (a) Design an optimal binary prefix code for this alphabet.

   (b) What is the average length of a code message using 100 letters from Σ?

8. Repeat Exercise 7 for the frequencies: $a-25$, $b-2$, $c-15$, $d-10$, $e-38$, $f-4$, $g-6$.

9. (a) Show that the code $\{000, 001, 10, 110, 111\}$ satisfies all the requirements of a prefix code, except that the corresponding binary tree is not regular.

(b) Show that some strings of binary digits are meaningless for this code.

(c) Show that $\{00, 01, 10, 110, 111\}$ is a prefix code, and compare its binary tree with that of part (a).

10. Repeat Exercise 7 for the frequencies: $a-31$, $d-31$, $e-12$, $h-6$, $m-20$.

11. Let L_1, L_2, L_3, L_4 be sorted lists having 23, 31, 61 and 73 elements, respectively. How many comparisons at most are needed if the lists are merged as indicated?

(a)

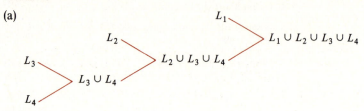

(b)

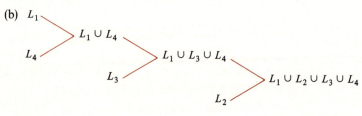

(c)

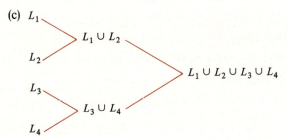

(d)

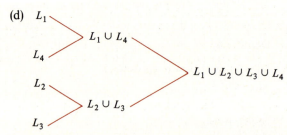

(e) How should the four lists be merged so that the total number of comparisons is a minimum? It is not sufficient simply to examine parts (a) to (d) since there are other ways to merge the lists.

12. Let $L_1, L_2, L_3, L_4, L_5, L_6$ be sorted lists having $5, 6, 9, 22, 29, 34$ elements, respectively.

(a) Show how the six lists should be merged so that the total number of comparisons can be kept to a minimum.

(b) How many comparisons might be needed in your procedure?

13. Repeat Exercise 12 for seven lists having $2, 5, 8, 12, 16, 22, 24$ elements, respectively.

14. Let T be an optimal binary tree whose weights satisfy $w_1 < w_2 < \cdots < w_t$. Show that the corresponding level numbers satisfy

$$l_1 \geq l_2 \geq l_3 \geq \cdots \geq l_t.$$

15. Look at Exercise 1 again, and note that whenever a vertex of weight $w_1 + w_2$ in a tree T' is replaced by a subtree with weights w_1 and w_2, then the weight increases by $w_1 + w_2$. That is, the new tree T has weight $W(T') + w_1 + w_2$, just as in the proof of the theorem.

CHAPTER HIGHLIGHTS

For suggestions on how to use this material, see the Highlights at the end of Chapter 1.

CONCEPTS AND NOTATION

recursive definition of a set, of a function
　　basis, recursive clause
　　uniquely determined
wff for algebra, for reverse Polish notation
recursive algorithm
　　base case
　　input obtained from other inputs
preorder, postorder, inorder listing
sorted labeling of a rooted tree, of an acyclic digraph
SUCC(V), ACC(V)
Polish, reverse Polish, infix notation
binary, unary operation
weighted tree [weights at leaves], weight of a tree
optimal binary tree
　　merge of lists
　　prefix code
Huffman code = minimum weight prefix code

The classes of finite trees and finite rooted trees can be defined recursively.

Recursive algorithms can test membership in sets defined recursively and compute values of functions defined recursively.

Generalized Principle of Induction.

Verification Conditions for Recursive Algorithms.

An ordered rooted tree cannot always be recovered from its preorder, inorder or postorder listing.

It *can* be recovered from its preorder or postorder listing given knowledge of how many children each vertex has.

METHODS AND ALGORITHMS

Depth-first search to traverse a rooted tree.

PREORDER, POSTORDER and INORDER to list vertices of an ordered rooted tree.

Method of charges for estimating running time of an algorithm.

LABELTREE to give a sorted labeling of a rooted tree.

TREESORT and LABEL to give a sorted labeling of an acyclic digraph in time $O(|V(G)| + |E(G)|)$.

Use of binary weighted trees to determine efficient merging patterns and efficient prefix codes.

HUFFMAN'S algorithm to find an optimal binary tree with given weights.

DIGRAPHS

8

This chapter is devoted to directed graphs, which were introduced in Chapter 3. The first section is primarily concerned with acyclic digraphs. It contains an account of sorted labelings that is independent of the treatment in Chapter 7, and it includes a digraph version of Euler's theorem. The remainder of the chapter deals with weighted digraphs, the primary focus being on paths of smallest weight between pairs of vertices. Section 8.2 introduces weighted digraphs and also contains a discussion of scheduling networks, in which the important paths are the ones of largest weight. Although this chapter contains a few references to sorted labelings and to § 7.3, it can be read independently of Chapter 7.

§ 8.1 Digraphs

Closed paths and cycles in graphs and digraphs were introduced in § 3.2. In this section we study acyclicity in digraphs and we give a version of Euler's theorem for digraphs.

We first observe that in some sense cycles in paths are redundant.

Theorem 1

> If u and v are different vertices of a digraph G, and if there is a path in G from u to v, then there is an acyclic path from u to v.

The proof is exactly the same as that of Theorem 1 in § 6.1. The only difference now is that the edges of the paths are directed.

Corollary 1

> If there is a closed path from v to v, then there is a cycle from v to v.

Proof. If there is an edge e of the graph from v to itself, then the one-element sequence e is a cycle from v to v. Otherwise, there is a closed path from v to v having the form $v x_2 \cdots x_n v$ where $x_n \neq v$. Then by Theorem 1 there is an acyclic path from v to x_n. Tacking on the last edge from x_n to v gives the desired cycle. ■

Corollary 2 | A path is acyclic if and only if all its vertices are distinct.

Proof. If a path has no repeated vertex, then it is surely acyclic. If a path has a repeated vertex, then it contains a closed path, so by Corollary 1 it contains a cycle. ■

We next introduce special kinds of vertices that do not arise in [undirected] graphs. Call a vertex of a digraph a **sink** if it is not an initial vertex of any edge. Sinks correspond to points with no arrows leading away from them. Points with no arrows leading into them are also special. We call a vertex of a digraph a **source** if it is not a terminal vertex of any edge.

EXAMPLE 1 Consider the digraph drawn in Figure 1. Vertices v and y are sinks, while t and z are sources. This digraph is acyclic. The next theorem shows that it is not an accident that there is at least one sink and at least one source. ■

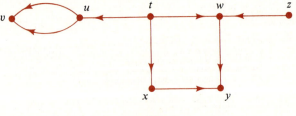

FIGURE 1

Theorem 2 | Every finite acyclic digraph has at least one sink and at least one source.

We give three proofs for sinks, since analogous facts about sources can be verified by reversing all the arrows. See Exercise 13.

First Proof. Since the digraph is acyclic, every path in it is acyclic. Since the digraph is finite, the path lengths are bounded and there must be a path of largest length, say $v_1 v_2 \cdots v_n$. Then v_n must be a sink. [Of course, if the digraph has no edges at all, every vertex is a sink.] ■

This proof is short and elegant, but it doesn't tell us how to find v_n or any other sink. Our next argument is constructive.

Second Proof. Choose any vertex v_1. If v_1 is a sink, we are done. If not, there is an edge from v_1 to some v_2. If v_2 is a sink, we are done. If not, etc. We obtain in this way a sequence $v_1, v_2, v_3, \ldots$ such that $v_1 v_2 \cdots v_k$ is a path for each k. As in the first proof, such paths cannot be arbitrarily long, so at some stage we reach a sink. ■

Third Proof. Omit this proof unless you have studied §7.3. Algorithm LABEL provides a sorted labeling for a finite acyclic digraph. This means that if there is a path from a vertex labeled i to one labeled j, then $i > j$. In particular, the vertex with the smallest label cannot be the initial vertex of any edge, so it must be a sink. ■

We will give an algorithm, based on the construction in the second proof, that returns a sink when it is applied to a finite acyclic digraph G. The algorithm uses **immediate successor** sets SUCC(v), defined by **SUCC(v)** $= \{u \in V(G): \text{there is an edge from } v \text{ to } u\}$. These sets also appeared in algorithm TREESORT, which is a subroutine for algorithm LABEL mentioned in the third proof above. Note that a vertex v is a sink if and only if SUCC(v) is the empty set. These data sets SUCC(v) would be supplied in the description of the digraph G when the algorithm is carried out.

***Algorithm* SINK**

{input: a finite acyclic digraph G}
{output: SINK(G), a sink of G}
Choose a vertex v in $V(G)$.
Let SINK$(G) := v$.
While SUCC$(v) \neq \varnothing$
   choose u in SUCC(v)
   let SINK$(G) := u$
   let $v := u$. ■

EXAMPLE 2 Consider the acyclic digraph G shown in Figure 1. The immediate successor sets are SUCC$(t) = \{u, w, x\}$, SUCC$(u) = \{v\}$, SUCC$(v) = \varnothing$, SUCC$(w) = \{y\}$, SUCC$(x) = \{y\}$, SUCC$(y) = \varnothing$ and SUCC$(z) = \{w\}$. One possible sequence of choices using algorithm SINK on G is t, w, y. Others starting with t are t, x, y and t, u, v. A different first choice could lead to z, w, y. We could even get lucky and choose a sink first time. In any case the value SINK returns is either v or y. ■

The time algorithm SINK takes is proportional to the number of vertices it chooses before it gets to a sink, so for a digraph with n vertices the algorithm runs in time $O(n)$.

If a digraph with n vertices can have its vertices numbered from 1 to n in such a way that $i > j$ whenever there is a path from vertex i to vertex j, then such a labeling of its vertices is called a **sorted labeling**. Algorithm LABEL in §7.3 shows that an acyclic digraph always has a sorted labeling. We now give an independent proof using sinks. This proof will lead to another algorithm, which will not be as efficient as LABEL.

Theorem 3	A finite acyclic digraph has a sorted labeling.

Proof. We use induction on the number n of vertices, and note that the assertion is obvious for $n = 1$. Assume inductively that acyclic digraphs with fewer than n vertices can be labeled as described, and consider an acyclic digraph G with n vertices. By Theorem 2, G has a sink, say s. Give s the number 1. Form a new graph H with $V(H) = V(G)\backslash\{s\}$ and with edges all those edges of G that do not have s as a vertex. Since G has no cycles, H has no cycles. Since H has only $n-1$ vertices, it has a sorted labeling by the inductive assumption. Let's increase the value of each label by 1, so that the vertices of H are numbered $2, 3, \ldots, n$. The vertex s is numbered 1, so every vertex in $V(G)$ has a number between 1 and n.

Now suppose there is a path in G from vertex i to vertex j. If the path lies entirely in H, then $i > j$ since H is properly numbered. Otherwise, some vertex along the path is s, and since s is a sink it must be the last vertex, vertex j. But then $j = 1$, so $i > j$ in this case too. Hence G, with n vertices, has a sorted labeling. The Principle of Mathematical Induction now shows that the theorem holds for all n. ∎

The idea in the proof of Theorem 3 can be developed into a procedure for constructing sorted labelings for acyclic digraphs.

Algorithm NUMBERING VERTICES

{input: a finite acyclic digraph G with n vertices}
{output: a sorted labeling of the vertices of G}
Let $V := V(G)$ and $E := E(G)$.
While $V \neq \varnothing$
   let H be the digraph with vertex set V and edge set E
   apply SINK to H {to get a sink of H}
   label SINK(H) with $n - |V| + 1$
   remove SINK(H) from V and all edges attached to it from E. ∎

Each pass through the while loop removes one vertex from the original set of vertices $V(G)$, so the algorithm must stop, and when it stops
each vertex is labeled. The labels used are $1, 2, \ldots, n$. As written, the
algorithm calls SINK as a subroutine. You may find it instructive to
apply this algorithm to number the graph of Figure 1. Also see Exercise
14 for a procedure that begins numbering with n.

If we estimate the running time of NUMBERING VERTICES, assuming that it calls SINK to find sinks as needed, the best we can do is
$O(n^2)$ where n is the number of vertices of G. This algorithm is inefficient
because the SINK subroutine may examine the same vertex over and
over. Algorithm LABEL in §7.3 is more efficient, although harder to
execute by hand.

EXAMPLE 3   Consider the digraph in Figure 2. The algorithm NUMBERING VER-
TICES might begin with vertex v_1 every time it calls SINK. It would
find the sink v_n the first time, label and remove it. The next time it would
find v_{n-1}, label it and remove it. And so on. The number of vertices that
it would examine is

$$n + (n - 1) + \cdots + 2 + 1 = \tfrac{1}{2}n(n + 1) > \tfrac{1}{2}n^2.$$

In contrast, suppose that algorithm LABEL in §7.3 began with vertex v_1.
Instead of starting over each time, once it reached v_n it would backtrack
to v_{n-1}, label it, backtrack to v_{n-2}, label it, etc. for a total of $2n$ vertex
examinations. ■

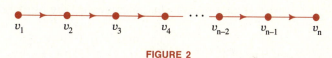

FIGURE 2

The second proof of Theorem 2 consisted of constructing a path
from a given vertex v to a sink. We call a vertex u **reachable from** v in
G if there is a path of length at least 1 in G from v to u, and we define

$$\boldsymbol{R(v)} = \{u \in V(G) : u \text{ is reachable from } v\}.$$

Then $R(v) = \varnothing$ if and only if v is a sink, and the theorem's second proof
showed in effect that in an acyclic digraph each nonempty set $R(v)$ contains at least one sink. The notion of reachability has already appeared
in a slightly different context. A pair (v, w) of vertices belongs to the reachable relation, as defined in §3.2, if and only if the vertex w is reachable
from v. Also, in §7.3 we used the set $\text{ACC}(v)$ of vertices accessible from
v, i.e., the set $R(v)$ together with the vertex v itself.

EXAMPLE 4   (a) For the digraph in Figure 2 we have

$$R(v_k) = \{v_j : k < j \le n\} \qquad \text{for} \quad k = 1, 2, \ldots, n.$$

This is even correct for $k = n$, since both sets are empty in this case.

(b) For the digraph in Figure 1, we have $R(t) = \{u, v, w, x, y\}$, $R(u) = \{v\}$, $R(v) = \varnothing$, $R(w) = \{y\}$, $R(x) = \{y\}$, $R(y) = \varnothing$ and $R(z) = \{w, y\}$. ∎

Even if G is not acyclic, the sets $R(v)$ may be important. As we shall see in §11.4, determining all sets $R(v)$ amounts to finding the transitive closure of a certain relation. In §8.4 we will study algorithms for finding the $R(v)$'s, as well as for answering other graph-theoretic questions.

We next give a digraph version of Euler's theorem of §6.2. Since a digraph can be viewed as a graph, the notion of **degree** of a vertex still makes sense. To take into account which way the edges are directed we refine the idea. For a vertex v of a digraph G, the **indegree** of v is the number of edges of G with v as terminal vertex, and the **outdegree** of v is the number with v as initial vertex. With obvious notation, we have

$$\text{indeg}(v) + \text{outdeg}(v) = \text{deg}(v)$$

for all vertices v. A loop gets counted twice here, once going out and once coming in.

EXAMPLE 5   (a) Consider again the acyclic digraph in Figure 1. The sinks v and y have outdegree 0, while the sources t and z have indegree 0. Other in- and outdegrees are easy to read off. For example, $\text{indeg}(u) = 1$ and $\text{outdeg}(u) = 2$. The sum of all indegrees is just the number of edges, namely 8, which is also the sum of the outdegrees.

(b) The digraphs in Figure 3 have the following rather special property: at each vertex the indegree and the outdegree are equal. ∎

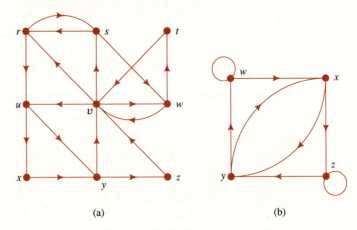

(a)   (b)

FIGURE 3

The proof of the next theorem is very like that for the undirected case in Theorems 1 and 2 of §6.2. Nevertheless, we have written out the proofs for the reader's convenience and also because we will apply this theorem in §12.1.

*Theorem 4
(Euler's
Theorem
for Digraphs)*

> Suppose that a finite digraph G is connected when viewed as a graph. There is a closed [directed] path in G using all the edges of G if and only if $\text{indeg}(v) = \text{outdeg}(v)$ for every vertex v.

Proof. Suppose that such a closed path exists. We start at some vertex and follow the path, erasing each edge in order. As we go through a vertex we erase one edge going in and one going out, or else we erase a loop. Either way, the erasure reduces both indegree and the outdegree by 1. Eventually all edges are erased, so all indegrees and outdegrees are equal to 0. Hence, at each vertex, the indegree and outdegree must have been equal at the start.

Now suppose that the in- and outdegree are equal at each vertex. We may assume that there is more than one vertex, since otherwise the theorem is trivial. Then every vertex has positive in- and outdegree. Let $v_1 v_2 \cdots v_n$ be the vertex sequence of a directed path of longest possible length with distinct edges. Let's see how the in- and outdegrees of v_n are affected if the edges of this path are removed from the digraph. Each time v_n is encountered in the list $v_2, \ldots, v_{n-1}$ the in- and outdegree are both reduced by 1, while the last encounter with v_n reduces the indegree by 1 but not the outdegree. If $v_n \neq v_1$, there must be an unused edge with initial vertex v_n that could be added to our longest path to get an even longer path with distinct edges. This contradiction shows that $v_n = v_1$.

We next explain why our longest path visits every vertex of G. If not, then since G is connected as an undirected graph there is an *undirected* path from some unvisited vertex to a vertex in $\{v_1, \ldots, v_n\}$. The first edge in this path is not part of our longest path. The last edge e of this path that is not part of our longest path must connect to some vertex in $\{v_1, \ldots, v_n\}$, say v_i. Of course, this edge is actually directed. If v_i is an initial vertex of e, we go around the longest path from v_i to v_i and then add the edge e to obtain a longer path with distinct edges, a contradiction. If v_i is a terminal vertex of e, we can start with e and then go around the longest path from v_i to v_i to again obtain a longer path with distinct edges and reach a contradiction.

Finally, we argue that the longest path uses all the edges of G. If some edge is missed, consider one of its vertices, v_i. Just as in the preceding paragraph, the edge can be attached to the beginning or end of the longest path from v_i to v_i to get a longer path, which is again a contradiction. ∎

EXAMPLE 6   (a) The digraphs in Figure 3 each have a closed path using all of their edges [Exercise 9].

(b) A **de Bruijn sequence of order** n is a circular arrangement of 2^n 0's and 1's so that each string of length n appears exactly once as n consecutive digits. Figure 4(a) is such an arrangement for $n = 4$.

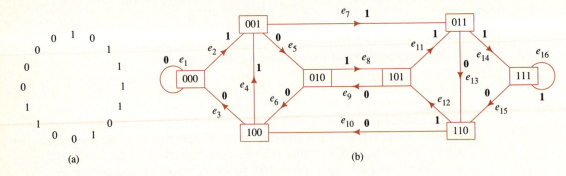

FIGURE 4

We obtained the arrangement in Figure 4(a) using the digraph in Figure 4(b), which needs some explanation. The set of $2^{n-1} = 8$ vertices of the digraph consists of all the strings of 0's and 1's of length $n - 1 = 3$. A directed edge connects two such strings provided the last two digits of the initial vertex agree with the first two digits of the terminal vertex. We label each edge with the last digit of the terminal vertex. Thus edges e_1, e_3, e_5, e_6, etc. are labeled 0 and e_2, e_4, etc. are labeled 1. Put differently, the label of the edge of a path of length one gives the last digit of the terminal vertex of the path. It follows that the labels of the two edges of a path of length two give the last two digits of the terminal vertex of the path in the same order. Similarly, the labels of the edges of a path of length three give all the digits of the terminal vertex in the same order.

Now observe that each vertex of the digraph has indegree 2 and outdegree 2. Since the digraph is connected, Euler's theorem guarantees that there is a closed path that uses each edge exactly once. We claim that the labels of these $2^n = 16$ edges provide a de Bruijn sequence of order $n = 4$. For example, the labels of the edges of the closed path

$$e_2, e_5, e_8, e_{11}, e_{14}, e_{16}, e_{15}, e_{12}, e_9, e_6, e_4, e_7, e_{13}, e_{10}, e_3, e_1$$

give the circular array in Figure 4(a) starting at the top and proceeding clockwise. Since there are only 16 different consecutive sequences of digits in the circular array, it suffices to show that any sequence $d_1 d_2 d_3 d_4$ of 0's and 1's does appear in the circular arrangement. Since $d_1 d_2 d_3$ is the initial vertex for an edge labeled 0 and also for an edge labeled 1, some edge in the path has initial vertex $d_1 d_2 d_3$ and is labeled d_4. As noted in the preceding paragraph, the three edges preceding it are labeled d_1, d_2 and d_3, in that order. So the labels of the four consecutive edges are d_1, d_2, d_3 and d_4, as claimed. ∎

Note that if a finite digraph is not connected but satisfies the degree conditions of Euler's theorem, the theorem still applies to its connected components.

EXERCISES 8.1

1. Find the sinks and sources for the digraph in Figure 5.

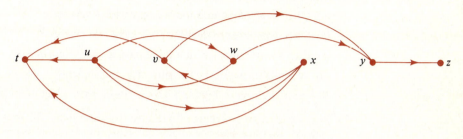

FIGURE 5

2. (a) Give the immediate successor sets succ(v) for all vertices in the digraph shown in Figure 5.

   (b) What value for SINK(G) does algorithm SINK give with an initial choice of vertex w?

   (c) What sinks of G are in R(x)?

3. Consider the digraph G pictured in Figure 6.

   (a) Find R(v) for each vertex v in V(G).

   (b) Find all sinks of G.

   (c) Is G acyclic?

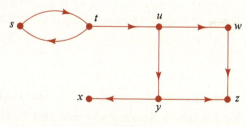

FIGURE 6

4. Does algorithm SINK work on digraphs that have cycles? Explain.

5. Consider a digraph G with the following immediate successor sets:
   succ(r) = {s, u}, succ(s) = ∅, succ(t) = {r, w}, succ(u) = ∅,
   succ(w) = {r, t, x, y}, succ(x) = ∅, succ(y) = {w, z} and succ(z) = ∅.

   (a) Draw a picture of such a digraph.

   (b) Do these sets succ(v) determine E(G) uniquely? Explain.

   (c) Find all sinks in G.

   (d) Find paths from the vertex w to three different sinks in the digraph.

6. Give two sorted labelings for the acyclic digraph in Figure 1.

7. Give two sorted labelings for the acyclic digraph in Figure 5.

8. A **tournament** is a digraph in which every two vertices have exactly one edge between them. [Think of (x, y) as an edge provided that x defeats y.]

   (a) Give an example of a tournament with 4 vertices.

   (b) Show that a tournament cannot have two sinks.

   (c) Can a tournament with a cycle have a sink? Explain.

   (d) Would you like to be the sink of a tournament?

9. For each digraph in Figure 3, give a closed path using all of its edges.

10. Use the digraph in Figure 4(b) to give two essentially different de Bruijn sequences that are also essentially different from the one in Figure 4(a). [We regard two circular sequences as essentially the same if one can be obtained by rotating or reversing the other.]

11. (a) Create a digraph, similar to the one in Figure 4(b), that can be used to find de Bruijn sequences of order 3.

   (b) Use your digraph to draw two de Bruijn sequences of order 3.

12. (a) Explain how the discussion in Example 6(b) needs to be modified to show that de Bruijn sequences exist of all orders $n \geq 3$.

   (b) It is easy to draw a de Bruijn sequence of order 2. Do so.

13. The **reverse** of a digraph G is the digraph $\hat{G}$ obtained by reversing all the arrows of G. That is, $V(\hat{G}) = V(G)$, $E(\hat{G}) = E(G)$ and if $\gamma(e) = (x, y)$, then $\hat{\gamma}(e) = (y, x)$. Use $\hat{G}$ and Theorem 2 to show that if G is acyclic [and finite], then G has a source.

14. (a) Modify algorithm NUMBERING VERTICES by using sources instead of sinks to produce an algorithm that numbers $V(G)$ in decreasing order.

   (b) Use your algorithm to number the digraph of Figure 1.

15. (a) Suppose that a finite acyclic digraph has just one sink. Show that there is a path to the sink from each vertex.

   (b) What is the corresponding statement for sources?

16. Let G be a digraph and define the relation $\sim$ on $V(G)$ by $x \sim y$ if $x = y$ or if x is reachable from y and y is reachable from x.

   (a) Show that $\sim$ is an equivalence relation.

   (b) Find the equivalence classes for the digraph pictured in Figure 6.

   (c) Describe the relation $\sim$ in the case that G is acyclic.

17. (a) Show that in Theorem 1 and Corollary 1 the path without repeated vertices can be constructed from edges of the given path. Thus every closed path contains at least one cycle.

   (b) Show that if u and v are vertices of a digraph and if there is a path from u to v, then there is a path from u to v in which no edge is repeated. [Consider the case $u = v$, as well as $u \neq v$.]

18. Let G be a digraph.

(a) Show that if u is reachable from v, then $R(u) \subseteq R(v)$.

(b) Give an alternative proof of Theorem 2 by choosing v in $V(G)$ with $|R(v)|$ as small as possible.

(c) Does your proof in part (b) lead to a useful constructive procedure? Explain.

§ 8.2   Weighted Digraphs

In many applications of digraphs one wants to know if a given vertex v is reachable from another vertex u, that is, if it is possible to get to v from u by following arrows. For instance, suppose that each vertex represents a state a machine can be in such as FETCH, DEFER or EXECUTE, and there is an edge from s to t whenever the machine can change from state s to state t in response to some input. If the machine is in state u, can it later be in state v? The answer is "yes" if and only if the digraph contains a path from u to v.

Now suppose that there is a cost associated with each transition from one state to another, i.e., with each edge in the digraph. Such a cost might be monetary, might be a measure of the time involved to carry out the change, or might have some other meaning. We could now ask for a path from u to v with the smallest total associated cost, obtained by adding all the costs for the edges in the path.

If all edges cost the same amount, the cheapest path is simply the shortest. In general, however, edge costs might differ. A digraph with no parallel edges is called **weighted** if each edge has an associated number, called its **weight**. In a given application it might better be called "cost" or "length" or "capacity" or have some other interpretation. Weights are normally assumed to be nonnegative, but many of the results about weighted digraphs are true without such a limitation. We can describe the weighting of a digraph G with a function W from $E(G)$ to $\mathbb{R}$ where $W(e)$ is the weight of the edge e. The **weight** of a path $e_1 e_2 \cdots e_m$ in G is then the sum $\sum_{i=1}^{m} W(e_i)$. Since a weighted digraph has no parallel edges, we may suppose that $E(G) \subseteq V(G) \times V(G)$, and write $W(u, v)$ for the weight of an edge (u, v) from u to v.

EXAMPLE 1   (a) The digraph shown in Figure 1 is stolen from Figure 1 of §3.2, where it described a rat and some cages. It could just as well describe a machine with states A, B, C and D, and the number next to an arrow [the weight of the edge] could be the number of microseconds necessary to get from its initial state to its terminal state. With that interpretation it takes .3 microsecond to go from state C to state B, .9 microsecond to go from D to A by way of C, .8 microsecond to go directly from D to A

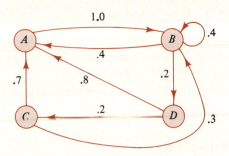

FIGURE 1

and .4 microsecond to stay in state B in response to an input. What is the shortest time to get from D to B?

(b) Figure 2 shows a more complicated example. In this case the shortest paths $svxf$ and $swxf$ from s to f have weights $6 + 7 + 4 = 17$ and $3 + 7 + 4 = 14$, respectively, but the longer path $swvyxzf$ has weight $3 + 2 + 1 + 3 + 1 + 3 = 13$, which is less than either of these. Thus length is not directly related to weight. This example also shows a path swv from s to v that has smaller weight than the edge from s to v.

This digraph has a cycle $wvyw$. Clearly, the whole cycle cannot be part of a path of minimum weight, but pieces of it can be. For instance, wvy is the path of smallest weight from w to y and the edge yw is the path of smallest weight from y to w. ■

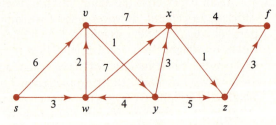

FIGURE 2

If we wish, we can display the weight function W in a tabular form by labeling rows and columns of an array with the members of $V(G)$ and entering the value of $W(u, v)$ at the intersection of row u and column v.

EXAMPLE 2   The array for the digraph in Figure 2 is given in Figure 3(a). The numbers appear in locations corresponding to edges of the digraph. The table in Figure 3(a) contains enough information to let us reconstruct the weighted digraph, since from the table we know just where the edges go and what their weights are. Figure 3(b) is a tabulation of the weight function W^*, where $W^*(u, v)$ is the smallest weight of a path from u to v if such a path exists. ■

W	s	v	w	x	y	z	f
s		6	3				
v				7	1		
w	2			7			
x					1	4	
y		4	3			5	
z							3
f							

(a)

W*	s	v	w	x	y	z	f
s		5	3	9	6	10	13
v	7		5	4	1	5	8
w	2	7		6	3	7	10
x						1	4
y	6	4	3	7		4	7
z							3
f							

(b)

FIGURE 3

We will call the smallest weight of a path [of length at least 1] from u to v the **min-weight** from u to v, and we will generally denote it by $W^*(u, v)$, as in Example 2. We also call a path from u to v that has this weight a **min-path**.

It is no real restriction to suppose that our weighted digraphs have no loops [why?], so we could just decide not to worry about $W(u, u)$, or we might define $W(u, u) := 0$ for all vertices u. It turns out, though, that we will learn more useful information by another choice, based on the following idea.

Consider vertices u and v with no edge from u to v in G. We can create a fictitious new edge from u to v of enormous weight, so big that the edge would never get chosen in a min-path if a real path from u to v were available. Suppose that we create such fictitious edges wherever edges are missing in G. If we ever find that the weight of a path in the enlarged graph is enormous, we will know that the path includes at least one fictitious edge, so it can't be a path in the original digraph G. A convenient notation is to write $W(u, v) = \infty$ if there is no edge from u to v in G; $W^*(u, v) = \infty$ means that there is no path from u to v. The operating rules for the symbol ∞ are: $\infty + x = x + \infty = \infty$ for every x, and $a < \infty$ for every real number a.

With this notation, we will write $W(u, u) = a$ if there is a loop at u of weight a, and write $W(u, u) = \infty$ otherwise. Then $W^*(u, u) < \infty$ means that there is a path [of length at least 1] from u to itself in G, and $W^*(u, u) = \infty$ means that there is no such path. The digraph G is acyclic if and only if $W^*(v, v) = \infty$ for every vertex v.

EXAMPLE 3   Using this notation, we would fill in all the blanks in the tables of Figures 3(a) and 3(b) with ∞'s. ∎

In Example 2 we simply announced the values of W^*, and the example is small enough that it is easy to check the values given. For more complicated digraphs the determination of W^* and of the min-paths can be nontrivial problems. In §§ 8.3 and 8.4 we will describe algorithms for finding both W^* and the corresponding min-paths, but until then we will stare at the picture until the answer is clear. For small digraphs this method is as good as any.

In many situations described by weighted graphs the min-weights and min-paths are the important concerns. However, one class of problems where weighted digraphs are useful, but where min-paths are irrelevant, is the scheduling of processes that involve a number of steps. Such scheduling problems are of considerable importance in business and manufacturing. The following example illustrates the sort of situation that arises.

EXAMPLE 4 Consider a cook preparing a simple meal of curry and rice. The curry recipe calls for the following steps.

(a) Cut up meat—about 10 minutes.
(b) Grate onion—about 2 minutes with a food processor.
(c) Peel and quarter potatoes—about 5 minutes.
(d) Marinate meat, onions and spices—about 30 minutes.
(e) Heat oil—4 minutes. Fry potatoes—15 minutes.
   Fry cumin seed—2 minutes.
(f) Fry marinated meat—4 minutes.
(g) Bake fried meat and potatoes—60 minutes.

In addition, there is

(h) Cook rice—20 minutes.

We have grouped three steps together in (e), since they must be done in sequence. Some of the other steps can be done simultaneously if enough help is available. We suppose that our cook has all the help needed.

Figure 4(a) gives a digraph that shows the sequence of steps and the possibilities for parallel processing. Cutting, grating, peeling and rice cooking can all go on at once. The dashed arrows after cutting and peeling indicate that frying and marinating cannot begin until cutting and peeling are completed. The other two dashed arrows have similar meanings. The picture has been redrawn in Figure 4(b) with weights on the edges to indicate time involved. [Ignore the numbers on the vertices for the moment.] The vertices denote stages of partial completion of the overall process, starting at the left and finishing at the right. In this case the min-path from left to right has weight 20, but there is much more total time required to prepare the meal than just the 20 minutes to cook the rice. The min-weight is no help. The important question here is: What is the smallest total time required to complete all steps in the process?

To answer the question we first examine vertices from left to right. Suppose that we start at s at time 0. What is the earliest time we can have

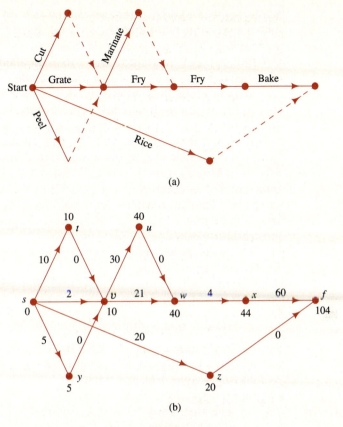

(a)

(b)

FIGURE 4

completed cutting, grating and peeling and arrive at vertex v? Clearly, 10 minutes, since we must wait for the cutting no matter how soon we start grating or peeling. In fact, 10 is the *largest* weight of a path from s to v. Now what is the earliest time we can arrive at vertex w? The shortest time to get from v to w is 30 minutes [the largest weight of a path from v to w], so the shortest time to get from s to w is $10 + 30 = 40$ minutes. Similarly, the earliest we can complete the whole process and arrive at f is $40 + 4 + 60 = 104$ minutes after we start.

In each instance, the smallest time to arrive at a given vertex is the largest weight of a path from s to that vertex. The numbers beside the vertices in Figure 4(b) give these smallest times. ∎

An acyclic digraph with nonnegative weights and with unique source and sink, such as the digraph in Figure 4, is called a **scheduling network**. For the rest of this section we suppose that we are dealing with a scheduling network G with source s [start] and sink f [finish]. For vertices u and v of G a **max-path** from u to v is a path of largest weight, and its weight is the **max-weight** from u to v, which we denote by $M(u, v)$.

Max-weights and max-paths can be analyzed in much the same way as min-weights and min-paths. In §8.3 we describe how to modify an algorithm for W^* to get one for M. For now we determine M by staring at the digraph. A max-path from s to f is called a **critical path**, and an edge belonging to such a path is a **critical edge**.

We introduce three functions on $V(G)$ that will help us understand scheduling networks. If we start from s at time 0, the earliest time we can arrive at a vertex v having completed all tasks preceding v is $M(s, v)$. We denote this earliest arrival time by $A(v)$. In particular, $A(f) = M(s, f)$, the time in which the whole process can be completed. Let $L(v) = M(s, f) - M(v, f) = A(f) - M(v, f)$. Since $M(v, f)$ represents the shortest time required to complete all steps from v to f, $L(v)$ is the latest time we can leave v and still complete all remaining steps by time $A(f)$. To calculate $L(v)$ we may work backward from f. Example 5(a) below gives an illustration.

The **slack time $S(v)$** of a vertex v is defined by $S(v) = L(v) - A(v)$. This is the maximum time that all tasks starting at v could be idle without delaying the entire process. So, of course, $S(v) \geq 0$ for all v. We can prove this formally by noting that

$$S(v) = L(v) - A(v) = M(s, f) - M(v, f) - M(s, v) \geq 0,$$

since $M(s, v) + M(v, f) \leq M(s, f)$. The last inequality holds because we can join max-paths from s to v and from v to f to obtain a path from s to f having weight $M(s, v) + M(v, f)$.

EXAMPLE 5 (a) The scheduling network in Figure 4(b) has only one critical path: $s\,t\,v\,u\,w\,x\,f$. Thus steps (a), (d), (f) and (g) are critical. The network is redrawn in Figure 5(a) with the critical path highlighted. Note that if the weights of the noncritical edges were decreased by improving efficiency, the entire process would still take 104 minutes. Even eliminating the noncritical tasks altogether would not speed up the process.

The functions A, L and S are given in the following table. Note that the values of A are written next to the vertices in Figure 4(b).

	s	t	u	v	w	x	y	z	f
A	0	10	40	10	40	44	5	20	104
L	0	10	40	10	40	44	10	104	104
S	0	0	0	0	0	0	5	84	0

Here are some sample calculations. Given that $A(u) = 40$, $W(u, v) = 0$, $A(v) = 10$ and $W(v, w) = 21$,

$$A(w) = \max\{40 + 0, 10 + 21\} = 40.$$

Given that $L(w) = 40$, $W(v, w) = 21$, $L(u) = 40$ and $W(v, u) = 30$,

$$L(v) = \min\{40 - 21, 40 - 30\} = 10.$$

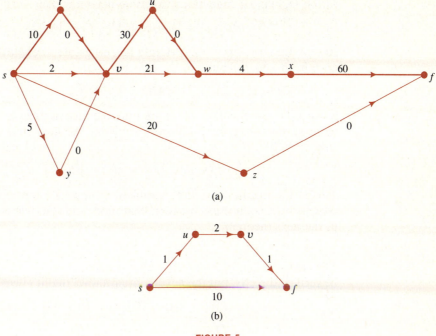

(a)

(b)

FIGURE 5

This calculation illustrates what we meant earlier when we spoke of working backward from f to compute $L(v)$. The entries in the $A(v)$ row are computed from left to right, and then once $A(f)$ is known the entries in the $L(v)$ row are computed from right to left. Exercise 21 asks for proofs that these methods for calculating $A(v)$ and $L(v)$ are correct in general.

Observe that the slack time is 0 at each vertex on the critical path. This always happens [Exercise 18]. Since $S(y) = 5$, the task of peeling potatoes can be delayed 5 minutes without delaying dinner. Since $S(z) = 84$, one could delay the rice 84 minutes; in fact, one would normally wait about 74 minutes before starting the rice to allow it to "rest" 10 minutes at the end.

A glance at edge (v, w) shows that we could delay this task 9 $[= 30 - 21]$ minutes, but we cannot deduce this information from the slack times. From $S(v) = 0$ we can only infer that we cannot delay *all* tasks starting at v. For this reason, we will introduce a more useful function on the set $E(G)$ of edges, called the float time.

(b) The slack time $S(v)$ is the amount of time all tasks at v can be delayed *provided that all other tasks are efficient*. That is, if one is "slack" at more than one vertex, the total time required might increase. To see this, consider the network in Figure 5(b). We have $S(u) = L(u) - A(u) = 7 - 1 = 6$ and $S(v) = L(v) - A(v) = 9 - 3 = 6$. A delay of 6 can occur at u before task (u, v) or at v after task (u, v), but not at both u and v. ∎

If (u, v) is an edge of a scheduling network, then

$$M(s, u) + W(u, v) + M(v, f) \le M(s, f)$$

since a path $s \cdots u v \cdots f$ certainly has no greater weight than a critical path from s to f. Thus

$$W(u, v) \le [M(s, f) - M(v, f)] - M(s, u) = L(v) - A(u).$$

Since $L(v) - A(u)$ is the maximum time the task at edge (u, v) could take without increasing the total time, we define the **float time $F(u, v)$** of the edge (u, v) by

$$F(u, v) = L(v) - A(u) - W(u, v).$$

This is the maximum delay possible for the task corresponding to the edge (u, v). A connection between float times and slack times will be given in the next theorem.

EXAMPLE 6   The critical edges in Figure 5(a) all have float time 0. This statement can be verified directly, but the next theorem shows that it always holds. Also,

$$F(s, v) = L(v) - A(s) - W(s, v) = 10 - 0 - 2 = 8;$$

$$F(v, w) = L(w) - A(v) - W(v, w) = 40 - 10 - 21 = 9;$$

$$F(s, y) = L(y) - A(s) - W(s, y) = 10 - 0 - 5 = 5;$$

$$F(y, v) = L(v) - A(y) - W(y, v) = 10 - 5 - 0 = 5;$$

etc. ∎

Theorem   Consider a scheduling network.

(a) The float time $F(u, v)$ is 0 if and only if (u, v) is a critical edge.
(b) $F(u, v) \ge \max\{S(u), S(v)\}$ for all edges of the network.

Proof. (a) Suppose that $F(u, v) = 0$; then

$$M(s, f) = M(s, u) + W(u, v) + M(v, f).$$

If we attach the edge (u, v) to max-paths from s to u and from v to f, we obtain a critical path for the network. Hence (u, v) is a critical edge.

If (u, v) is an edge of a critical path $s \cdots u v \cdots f$, then $s \cdots u$ must have weight $M(s, u)$, since otherwise $s \cdots u$ could be replaced by a heavier path. Similarly, $v \cdots f$ has weight $M(v, f)$, so

$$M(s, u) + W(u, v) + M(v, f) = M(s, f).$$

Therefore, we have $F(u, v) = 0$.

(b) The inequality $S(u) \leq F(u, v)$ is equivalent to each of the following:

$$L(u) - A(u) \leq L(v) - A(u) - W(u, v);$$

$$W(u, v) \leq L(v) - L(u);$$

$$W(u, v) \leq [M(s, f) - M(v, f)] - [M(s, f) - M(u, f)];$$

$$W(u, v) \leq M(u, f) - M(v, f);$$

$$W(u, v) + M(v, f) \leq M(u, f).$$

The last inequality is clear, since the edge (u, v) can be attached to a max-path from v to f to obtain a path from u to f of weight $W(u, v) + M(v, f)$. The inequality $S(v) \leq F(u, v)$ is slightly easier to verify [Exercise 18]. ∎

As we have seen, shortening the time required for a noncritical edge does not decrease the total time $M(s, f)$ required for the process. Identification of critical edges focuses attention on those steps in a process where improvement may make a difference and where delays will surely be costly. Since its introduction in the 1950s the method of critical path analysis, sometimes called PERT for Program Evaluation and Review Technique, has been a popular way of dealing with industrial management scheduling problems.

EXERCISES 8.2

Use the ∞ notation in all tables of W and W^*.

1. Give tables of W and W^* for the digraph of Figure 1 with the loop at B removed.

2. Give a table of W^* for the digraph of Figure 6(a).

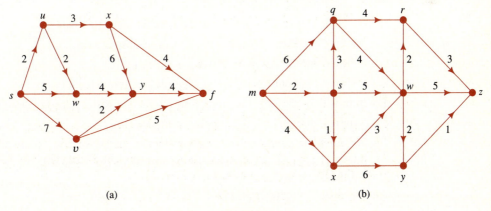

(a)   (b)

FIGURE 6

3. Give tables of W and W^* for the digraph of Figure 6(b).

4. The path $s\,w\,v\,y\,x\,z\,f$ is a min-path from s to f in the digraph of Figure 2. Find another min-path from s to f in that digraph.

5. Figure 7 shows a weighted digraph. The directions and weights have been left off the edges, but the number at each vertex v is $W^*(s, v)$.

   (a) Give three different weight functions W that yield these values of $W^*(s, v)$. [An answer could consist of three pictures with appropriate numbers on the edges.]

   (b) Do the different weight assigments yield different min-paths between points? Explain.

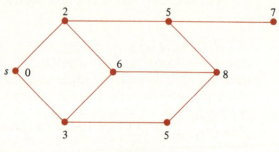

FIGURE 7

6. Suppose that u, v and w are vertices of a weighted digraph with min-weight function W^* and that $W^*(u, v) + W^*(v, w) = W^*(u, w)$. Explain why there is a min-path from u to w through v.

7. (a) Find a critical path for the digraph of Figure 2 with the edge from y to w removed.

   (b) Why does the critical path method apply only to acyclic digraphs?

8. Calculate the float times for the edges (s, z) and (z, f) in Figure 5(a). *Hint:* See Example 5(a) for some useful numbers.

9. (a) Give a table of A and L for the network in Figure 6(a).

   (b) Find the slack times for the vertices of this network.

   (c) Find the critical paths for this network.

   (d) Find the float times for the edges of this network.

10. Repeat Exercise 9 for the digraph in Figure 6(a) with each edge of weight 1.

11. Repeat Exercise 9 for the digraph in Figure 6(b).

12. (a) Calculate the float times for the edges in the network in Figure 5(b).

    (b) Can each task in a network be delayed by its float time without delaying the entire process? Explain.

13. Consider the network in Figure 8.

    (a) How many critical paths does this digraph have?

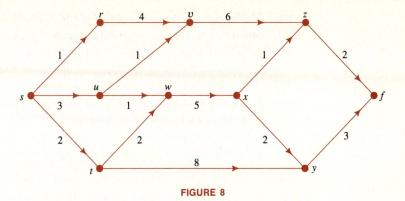

FIGURE 8

(b) What is the largest float time for an edge in this digraph?

(c) Which edges have the largest float time?

14. Find the slack times for the vertices of the network in Figure 8.

15. In Example 4 we used edges of weight 0 as a device to indicate that some steps could not start until others were finished.

   (a) Explain how to avoid such 0-edges if parallel edges are allowed.

   (b) Draw a digraph for the process of Example 4 to illustrate your answer.

16. If the cook in Example 4 has no helpers, then steps (a), (b) and (c) must be done one after the other, but otherwise the situation is as in the example.

   (a) Draw a scheduling network for the no-helper process.

   (b) Find a critical path for this process.

   (c) Which steps in the process are not critical?

17. (a) Give tables of W and W^* for the digraph of Figure 9.

   (b) Explain how to tell from the table for W^* whether or not this digraph is acyclic.

   (c) How would your answer to part (a) change if the edge of weight -2 had weight -6 instead?

   (d) Explain how to tell which are the sources and sinks of this digraph from the table for W.

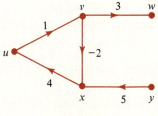

FIGURE 9

18. (a) Complete the proof of the theorem by showing $S(v) \leq F(u, v)$.

(b) Show that if (u, v) is a critical edge, then $S(u) = S(v) = 0$.

(c) If (u, v) is an edge with $S(u) = S(v) = 0$, must (u, v) be critical? Explain.

19. The float time $F(u, v)$ can be thought of as the amount of time we can delay starting along the edge (u, v) without delaying completion of the entire process. Define the **free-float time** $FF(u, v)$ to be the amount of time we can delay starting along (u, v) without increasing $A(v)$.

(a) Find an expression for $FF(u, v)$ in terms of the functions A and W.

(b) Find $FF(u, v)$ for all edges of the digraph in Figure 6(a).

(c) What is the difference between $F(u, v)$ and $FF(u, v)$?

20. (a) Show that increasing the weight of a critical edge in a scheduling network increases the max-weight from the source to the sink.

(b) Is there any circumstance in which reducing the amount of time for a critical step in a process does not reduce the total time required for the process? Explain.

21. (a) Show that $A(u) = \max\{A(w) + W(w, u) : (w, u) \in E(G)\}$.

(b) Show that $L(u) = \min\{L(v) - W(u, v) : (u, v) \in E(G)\}$.

§ 8.3   Digraph Algorithms

Our study of digraphs and graphs has led us to a number of concrete questions. Given a digraph, what is the length of a shortest path from one vertex to another? If the digraph is weighted, what's the minimum or maximum weight of such a path? Is there any path at all?

This section describes some algorithms for answering these questions and others, algorithms that can be implemented on computers as well as used to organize hand computations. The algorithms we have chosen are reasonably fast, and their workings are comparatively easy to follow. For a more complete discussion we refer the reader to books on the subject, such as *Data Structures and Algorithms* by Aho, Hopcroft and Ullman.

The min-weight problem is essentially a question about digraphs without loops or parallel edges, so we limit ourselves to that setting. Hence $E(G) \subseteq V(G) \times V(G)$, and we can describe the digraph with a table of the edge-weight function $W(u, v)$, as we did in §8.2. Our goal will be to obtain the min-weights $W^*(u, v)$.

The min-weight algorithms that we will consider all begin by looking at paths of length 1, i.e., single edges, and then systematically consider longer and longer paths between vertices. As they proceed, the algorithms find smaller and smaller path weights between vertices, and when they terminate the weights are the best possible.

Our first algorithm just finds min-weights from a selected vertex to the other vertices in the digraph G. To describe how it works, it will be convenient to suppose that $V(G) = \{1, \ldots, n\}$ and that 1 is the selected vertex. Starting with 1, the algorithm looks at additional vertices one at a time, chooses a new vertex w whose best known path weight from 1 is as small as possible, and then updates best known path weights from 1 to other vertices by considering paths through w. It puts the vertices that it has looked at in a set L, and it doesn't look at them again. At any given time, $D(j)$ is the smallest weight of a path from 1 to j whose vertices lie in L. Here is the recipe.

DIJKSTRA'S *Algorithm*

{input: digraph without loops or parallel edges, vertex set $\{1, \ldots, n\}$,
     nonnegative edge-weight function W}
{output: min-weights $W^*(1, j)$ for $j = 2, \ldots, n$}
{intermediate variables: sets L, V and function D}
Set $L := \varnothing$, $V := \{2, \ldots, n\}$.
For $i \in V$ do
   set $D(i) := W(1, i)$.
While $V \backslash L \neq \varnothing$ do
   choose k in $V \backslash L$ with $D(k)$ as small as possible
   put k in L
   for each j in $V \backslash L$ do
     if $D(j) > D(k) + W(k, j)$ then
       replace $D(j)$ by $D(k) + W(k, j)$
For $j \in V$ do
   $W^*(1, j) := D(j)$. ∎

We need to check that this algorithm stops, and that the final value of $D(j)$ is $W^*(1, j)$ for each j, i.e., the min-weight from 1 to j. We also want to get some estimate of the time it takes the algorithm to run. First, though, let's look at how it works.

EXAMPLE 1   Consider the weighted digraph G shown in Figure 1(a). Its edge-weight table is given in Figure 1(b).

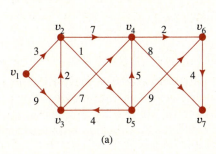

W	v_1	v_2	v_3	v_4	v_5	v_6	v_7
v_1	∞	3	9	∞	∞	∞	∞
v_2	∞	∞	∞	7	1	∞	∞
v_3	∞	2	∞	7	∞	∞	∞
v_4	∞	∞	∞	∞	∞	2	8
v_5	∞	∞	4	5	∞	9	∞
v_6	∞	∞	∞	∞	∞	∞	4
v_7	∞	∞	∞	∞	∞	∞	∞

(a)   (b)

FIGURE 1

L	$D(2)$	$D(3)$	$D(4)$	$D(5)$	$D(6)$	$D(7)$	Comment
$\varnothing$	3	9	∞	∞	∞	∞	Initial data
$\{2\}$	3	9	10	4	∞	∞	Found $v_1 v_2 v_4$ and $v_1 v_2 v_5$
$\{2,5\}$	3	8	9	4	13	∞	Found $v_1 v_2 v_5 v_3$, $v_1 v_2 v_5 v_4$ and $v_1 v_2 v_5 v_6$
$\{2,5,3\}$	3	8	9	4	13	∞	No improvement
$\{2,5,3,4\}$	3	8	9	4	11	17	Found $v_1 v_2 v_5 v_4 v_6$ and $v_1 v_2 v_5 v_4 v_7$
$\{2,5,3,4,6\}$	3	8	9	4	11	15	Found $v_1 v_2 v_5 v_4 v_6 v_7$

FIGURE 2

The table in Figure 2 shows how the values of L and of $D(2), \ldots, D(7)$ change as the algorithm progresses, starting with 1. Notice that the values in the $D(j)$ columns decrease with time and that once j gets into L the value of $D(j)$ doesn't change.   ∎

Theorem 1 | If the edge weights in G are nonnegative, then DIJKSTRA'S algorithm stops with $D(j) = W^*(1, j)$ for $j = 2, \ldots, n$.

Proof. Since each pass through the while loop in the algorithm adds one more vertex to L, the algorithm makes $n - 1$ trips through the loop and then stops. It is not at all obvious that $D(j) = W^*(1, j)$ at the end, because the algorithm makes greedy choices of new vertices whenever it gets a chance. Greed does not always pay—consider trying to get 40 cents out of a pile of dimes and quarters by picking a quarter first—but it works again this time.

We claim that the following two statements are loop invariants:

(a) $D(l) \leq D(j)$ whenever $l \in L$ and $j \in V \backslash L$;

(b) For every $j \in V$, $D(j)$ is the smallest weight of a path from 1 to j with intermediate vertices in L.

For short, we will call a path with intermediate vertices in L an **L-path** and such a path of minimum weight an **L-min-path**.

Assume that both (a) and (b) are true for L and D at the start of the pass in which k is chosen. We want to show that both are true for $L \cup \{k\}$ and the updated D at the end of the pass.

Suppose first that $l \in L \cup \{k\}$ and $j \in V \backslash (L \cup \{k\})$. Then $D(l) \leq D(k)$ by the assumption of (a) for L and D. Since $j \in V \backslash L$, $D(j) \geq D(k)$ at the start of the loop by the choice of k, and if $D(j)$ changes at all it changes to $D(k) + W(k, j)$, which is still at least $D(k)$ because $W(k, j) \geq 0$. Thus $D(l) \leq D(k) \leq D(j)$; so (a) holds for $L \cup \{k\}$ and the new D at the end of the loop.

To show that (b) is a loop invariant, consider j in V. At the start of the loop, $D(j)$ is the smallest weight of an L-path from 1 to j. For clarity, we write $D_L(j)$ for this starting value. We need to show that at the end of the loop $D(j)$ is equal to $D_{L \cup \{k\}}(j)$, the weight of an $L \cup \{k\}$-min-path from 1 to j.

First, suppose that $j = k$. Min-paths don't repeat vertices, so every $L \cup \{k\}$-min-path from 1 to k is an L-min-path from 1 to k. Thus $D_{L \cup \{k\}}(k) = D_L(k)$. Since $W(k, k) \geq 0$, we have $D(k) + W(k, k) \geq D(k)$. Thus the loop does not change the value of $D(k)$, so $D(k) = D_{L \cup \{k\}}(k)$ after the loop, as claimed.

Henceforth we consider j in V with $j \neq k$. If $D_{L \cup \{k\}}(j) = D_L(j)$, the weight of any $L \cup \{k\}$-path from 1 to j through k is at least $D_L(j) = D(j)$. In particular, $D(k) + W(k, j) \geq D(j)$ in this case, so the loop does not change the [correct] value of $D(j)$.

Finally, suppose that $D_{L \cup \{k\}}(j) < D_L(j)$. Let P be an $L \cup \{k\}$-min-path from 1 to j. Then P must go through k. Let l be the last vertex that the path P visits before reaching j. We will show that $l = k$. Otherwise, l is in L, and $D_L(l) \leq D_L(k)$ by (a) for L and D. The part of P from 1 to k must be an L-min-path from 1 to k, of weight $D_L(k)$, since otherwise we could replace it with an L-min-path and get a path in $L \cup \{k\}$ from 1 to j having weight less than that of P. It follows that

$$D_{L \cup \{k\}}(j) = \text{weight}(P) \geq D_L(k) + W(l, j) \geq D_L(l) + W(l, j).$$

Now an L-min-path from 1 to l followed by the edge from l to j is an L-path from 1 to j having weight $D_L(l) + W(l, j)$. Hence $D_L(l) + W(l, j) \geq D_L(j)$, so $D_{L \cup \{k\}}(j) \geq D_L(j)$, contrary to our assumption. Thus $l = k$, the path P has the form $1 \cdots kj$, and

$$D_{L \cup \{k\}}(j) = \text{weight}(P) = D(k) + W(k, j).$$

Since this value is less than $D(j) = D_L(j)$, the loop correctly updates $D(j)$ to be $D_{L \cup \{k\}}(j)$.

Both (a) and (b) are true on first entry into the loop. When the loop terminates, $L = V$ and (b) holds, so the final value of $D(j)$ is $W^*(1, j)$. ■

How long does DIJKSTRA'S algorithm take for a digraph with n vertices? The largest part of the time is spent going through the while loop, removing one of the original n vertices at each pass. Time to find the smallest $D(k)$ is $O(n)$ if we simply examine the vertices in $V(G)$ one by one, and in fact there are sorting algorithms that will locate k faster. For each chosen vertex k there are at most n comparisons and replacements, so the total time for one pass through the loop is $O(n)$. All told, the algorithm makes n passes, so it takes total time $O(n^2)$.

If the digraph is presented in terms of successor lists, the algorithm can be rewritten so that the replacement/update step only looks at successors of k. During the total operation, each edge is then considered just once in an update step. Such a modification speeds up the overall performance if $|E(G)|$ is much less than n^2.

DIJKSTRA'S algorithm finds the weights of min-paths from a given vertex. To find $W^*(v_i, v_j)$ for all pairs of vertices v_i and v_j we could just apply the algorithm n times, starting from each of the n vertices. There is another algorithm, originally due to Warshall and refined by Floyd, that produces all of the values $W^*(v_i, v_j)$ at the end, and that is easy to program. Like DIJKSTRA'S algorithm, it builds an expanding list of examined vertices and looks at paths through vertices on the list.

Suppose that $V(G) = \{v_1, \ldots, v_n\}$. WARSHALL'S algorithm works with an $n \times n$ matrix $\mathbf{W}$, which at the beginning is the edge-weight matrix $\mathbf{W}_0$ with $\mathbf{W}_0[i, j] = W(v_i, v_j)$ for all i and j, and at the end is the min-weight matrix $\mathbf{W}_n = \mathbf{W}^*$ with $\mathbf{W}^*[i, j] = W^*(v_i, v_j)$.

WARSHALL'S *Algorithm*

{input: nonnegative edge-weight matrix $\mathbf{W}_0$ of a weighted digraph
     without loops or parallel edges}
{output: min-weight matrix $\mathbf{W}^*$ for the digraph}
{intermediate variable: matrix $\mathbf{W}$}
$\mathbf{W} := \mathbf{W}_0$
For $k = 1$ to n do
    for $i = 1$ to n do
        for $j = 1$ to n do
            if $\mathbf{W}[i, j] > \mathbf{W}[i, k] + \mathbf{W}[k, j]$ then
               replace $\mathbf{W}[i, j]$ by $\mathbf{W}[i, k] + \mathbf{W}[k, j]$.
$\mathbf{W}^* := \mathbf{W}$. ∎

Theorem 2 | WARSHALL'S algorithm produces the min-weight matrix $\mathbf{W}^*$.

Proof. We have written the algorithm as a nest of for loops. We can also present it in the following form, which looks more like DIJKSTRA'S algorithm.

set $\mathbf{W} := \mathbf{W}_0$, $L := \varnothing$, $V := \{1, \ldots, n\}$
while $V \backslash L \neq \varnothing$ do
    choose $k \in V \backslash L$
    put k in L
    for each $i, j \in V$ do
        set $\mathbf{W}[i, j] := \min\{\mathbf{W}[i, j], \mathbf{W}[i, k] + \mathbf{W}[k, j]\}$
set $\mathbf{W}^* := \mathbf{W}$.

The assignment $\mathbf{W}[i, j] := \min\{\mathbf{W}[i, j], \mathbf{W}[i, k] + \mathbf{W}[k, j]\}$ here has the same effect as our previous "if $\cdots$ then replace" segment; the value of $\mathbf{W}[i, j]$ changes only if $\mathbf{W}[i, j] > \mathbf{W}[i, k] + \mathbf{W}[k, j]$.

We claim that for each i and j the statement "$\mathbf{W}[i, j]$ is the weight of an L-min-path from i to j" is an invariant of the while loop. Assume that this statement is true at the start of the pass in which k is chosen. Then the smallest weight of an $L \cup \{k\}$-path from i to j through k is $\mathbf{W}[i, k] + \mathbf{W}[k, j]$. If this number is less then $\mathbf{W}[i, j]$ it is used as the new value of $\mathbf{W}[i, j]$. Otherwise, we can do at least as well with an $L \cup \{k\}$-path that misses k, i.e., with an L-path, so the current value of $\mathbf{W}[i, j]$, which is not changed, is the correct weight of an $L \cup \{k\}$-min-path from i to j.

This argument proves that the statement is a loop invariant. It is true initially, since $\mathbf{W}_0[i, j]$ is the minimum weight of a path from i to j with no intermediate vertices at all. The loop terminates when all vertices are in L, so the final value of $\mathbf{W}[i, j]$ is the weight $\mathbf{W}^*[i, j]$ of a min-path from i to j. ∎

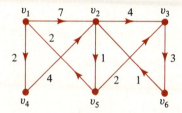

FIGURE 3

EXAMPLE 2   We apply WARSHALL'S algorithm to the digraph shown in Figure 3. Hand calculations with WARSHALL'S algorithm lead to n new matrices, one for each value of k. For this example, the matrices are the following.

$$\mathbf{W} = \mathbf{W}_0 = \begin{bmatrix} \infty & 7 & \infty & 2 & \infty & \infty \\ \infty & \infty & 4 & \infty & 1 & \infty \\ \infty & \infty & \infty & \infty & \infty & 3 \\ \infty & 4 & \infty & \infty & \infty & \infty \\ 2 & \infty & 2 & \infty & \infty & \infty \\ \infty & 1 & \infty & \infty & \infty & \infty \end{bmatrix} \qquad \mathbf{W}_1 = \begin{bmatrix} \infty & 7 & \infty & 2 & \infty & \infty \\ \infty & \infty & 4 & \infty & 1 & \infty \\ \infty & \infty & \infty & \infty & \infty & 3 \\ \infty & 4 & \infty & \infty & \infty & \infty \\ 2 & 9 & 2 & 4 & \infty & \infty \\ \infty & 1 & \infty & \infty & \infty & \infty \end{bmatrix}$$

$$\mathbf{W}_2 = \begin{bmatrix} \infty & 7 & 11 & 2 & 8 & \infty \\ \infty & \infty & 4 & \infty & 1 & \infty \\ \infty & \infty & \infty & \infty & \infty & 3 \\ \infty & 4 & 8 & \infty & 5 & \infty \\ 2 & 9 & 2 & 4 & 10 & \infty \\ \infty & 1 & 5 & \infty & 2 & \infty \end{bmatrix} \qquad \mathbf{W}_3 = \begin{bmatrix} \infty & 7 & 11 & 2 & 8 & 14 \\ \infty & \infty & 4 & \infty & 1 & 7 \\ \infty & \infty & \infty & \infty & \infty & 3 \\ \infty & 4 & 8 & \infty & 5 & 11 \\ 2 & 9 & 2 & 4 & 10 & 5 \\ \infty & 1 & 5 & \infty & 2 & 8 \end{bmatrix}$$

$$\mathbf{W}_4 = \begin{bmatrix} \infty & 6 & 10 & 2 & 7 & 13 \\ \infty & \infty & 4 & \infty & 1 & 7 \\ \infty & \infty & \infty & \infty & \infty & 3 \\ \infty & 4 & 8 & \infty & 5 & 11 \\ 2 & 8 & 2 & 4 & 9 & 5 \\ \infty & 1 & 5 & \infty & 2 & 8 \end{bmatrix} \qquad \mathbf{W}_5 = \begin{bmatrix} 9 & 6 & 9 & 2 & 7 & 12 \\ 3 & 9 & 3 & 5 & 1 & 6 \\ \infty & \infty & \infty & \infty & \infty & 3 \\ 7 & 4 & 7 & 9 & 5 & 10 \\ 2 & 8 & 2 & 4 & 9 & 5 \\ 4 & 1 & 4 & 6 & 2 & 7 \end{bmatrix}$$

$$\mathbf{W}^* = \mathbf{W}_6 = \begin{bmatrix} 9 & 6 & 9 & 2 & 7 & 12 \\ 3 & 7 & 3 & 5 & 1 & 6 \\ 7 & 4 & 7 & 9 & 5 & 3 \\ 7 & 4 & 7 & 9 & 5 & 10 \\ 2 & 6 & 2 & 4 & 7 & 5 \\ 4 & 1 & 4 & 6 & 2 & 7 \end{bmatrix}$$

We illustrate the computations by calculating $\mathbf{W}_4[5, 2]$. Here $k = 4$. The entry $\mathbf{W}_3[5, 2]$ is 9, corresponding to the shortest path $v_5 v_1 v_2$ with intermediate vertices from $\{v_1, v_2, v_3\}$. To find $\mathbf{W}_4[5, 2]$ we look at $\mathbf{W}_3[5, 4] + \mathbf{W}_3[4, 2]$, which is $4 + 4 = 8$, corresponding to the pair of paths $v_5 v_1 v_4$ and $v_4 v_2$. Since $9 > 8$, we replace $\mathbf{W}_3[5, 2]$ by $\mathbf{W}_3[5, 4] + \mathbf{W}_3[4, 2] = 8$, corresponding to $v_5 v_1 v_4 v_2$.

A given entry, such as $\mathbf{W}[5, 2]$, may change several times during the calculations, as k runs through all possible values. ■

It is easy to analyze how long WARSHALL'S algorithm takes. The comparison/replacement step inside the j-loop takes at most some fixed amount of time, say t. The step is done exactly n^3 times, once for each possible choice of the triple (k, i, j), so the total time to execute the algorithm is $n^3 t$, which is $O(n^3)$.

The comparison/replacement step for WARSHALL'S algorithm is the same as the one in DIJKSTRA'S algorithm, which includes other sorts of steps as well. Since DIJKSTRA'S algorithm can be done in $O(n^2)$ time, doing it once for each of the n vertices gives an $O(n^3)$ algorithm to find all min-weights. The time constants involved in this multiple DIJKSTRA'S algorithm are different from the ones for WARSHALL'S algorithm, though, so the choice of which algorithm to use may depend on the computer implementation available. If $|E(G)|$ is small compared with n^2, a successor list presentation of the digraph favors choosing DIJKSTRA'S algorithm.

This is probably the time to confess that we are deliberately ignoring one possible complication. Even if G has only a handful of vertices, the weights $W(i, j)$ could still be so large that it would take years to write them down, so both DIJKSTRA'S and WARSHALL'S algorithms might take a long time. In practical situations, though, the numbers that come up in the applications of these two algorithms are of manageable size. In any case, given the same set of weights, our comparison of the relative run times for the algorithms is still valid.

Nonnegativity of edge weights gets used in a subtle way in the proof of Theorem 2. If $\mathbf{W}[k, k] < 0$ for some k, then $\mathbf{W}[k, k] + \mathbf{W}[k, j] < \mathbf{W}[k, j]$, but there is no smallest $L \cup \{k\}$-path weight from k to j. We can go around some path of negative weight from k to itself as many times as we like before setting out for j. One way to rule out this possibility is to take nonnegative edge weights to begin with, as we have done, so that every replacement value is also nonnegative. One can permit negative input weights for WARSHALL'S algorithm by requiring the digraph to be acyclic, so that looping back through k cannot occur.

WARSHALL'S algorithm can be adapted to finding max-weights in an acyclic digraph. Replace all the ∞'s by $-\infty$'s, with $-\infty + x = -\infty = x + (-\infty)$ for all x, and $-\infty < a$ for all real a. Change the inequality in the replacement step to $\mathbf{W}[i, j] < \mathbf{W}[i, k] + \mathbf{W}[k, j]$. The resulting algorithm computes $\mathbf{W}_n[i, j] = M(v_i, v_j)$, where M is the max-weight function of §8.2.

If we just want max-weights from a single source, as we might for a scheduling network, we can simplify the algorithm somewhat, at the cost of first relabeling the digraph, using an algorithm such as NUMBERING VERTICES of §8.1 or LABEL of §7.3. It seems natural in this setting to use a **reverse sorted labeling**, in which the labels are arranged so that $i < j$ if there is a path from v_i to v_j. Such a labeling is easy to obtain from an ordinary sorted labeling; for $i = 1, \ldots, n$ simply relabel v_i as v_{n+1-i}.

To find max-weights from a given vertex v_s, for instance v_1, just fix $i = s$ in the max-modified WARSHALL'S algorithm. The reverse sorted labeling also means that $\mathbf{W}[k, j] = -\infty$ if $j \leq k$, so the j-loop does not need to go all the way from 1 to n. The resulting algorithm looks like this, with $i = s = 1$.

Algorithm MAX-WEIGHT

{input: edge-weight matrix $\mathbf{W}_0$ of an acyclic digraph with a reverse sorted labeling}
{output: max-weights $M(1, j)$ for $j = 2, \ldots, n$}
For $k = 2$ to $n - 1$ do
    for $j = k + 1$ to n do
        if $\mathbf{W}[1, j] < \mathbf{W}[1, k] + \mathbf{W}[k, j]$, then
            replace $\mathbf{W}[1, j]$ by $\mathbf{W}[1, k] + \mathbf{W}[k, j]$.
For $j = 2$ to n do
    set $M(1, j) := \mathbf{W}[1, j]$. ∎

We emphasize that this algorithm is only meant for acyclic digraphs with reverse sorted labelings. The proof that it works in time $O(n^2)$ [Exercise 12] is similar to the argument for WARSHALL'S algorithm. The algorithm can be speeded up somewhat if the digraph is given by successor lists and if we just consider j in SUCC(k). Then each edge gets examined exactly once to see if it enlarges max-weights. The algorithm then runs in a time of $O(\max\{|V(G)|, |E(G)|\})$, which is comparable to the time it takes to sort the vertices initially using algorithm LABEL.

DIJKSTRA'S algorithm does not work with negative weights [Exercise 10]. Moreover [Exercise 11], there seems to be no natural way to modify it to find max-weights.

EXAMPLE 3 We apply MAX-WEIGHT to the digraph of Figure 4(a). Figure 4(b) gives the initial matrix $\mathbf{W}_0$. For convenience, we use a row matrix $\mathbf{D}$ with $\mathbf{D}[j] = \mathbf{W}[1, j]$ for $j = 1, \ldots, 6$. The sequence of matrices is as follows.

$$\mathbf{D}_0 = \mathbf{D}_1 = [-\infty \quad 1 \quad 2 \quad -\infty \quad -\infty \quad -\infty]$$

$$\mathbf{D}_2 = [-\infty \quad 1 \quad 2 \quad 3 \quad 4 \quad -\infty]$$

$$\mathbf{D}_3 = [-\infty \quad 1 \quad 2 \quad 7 \quad 4 \quad -\infty]$$

$$\mathbf{D}_4 = \mathbf{D}_5 = [-\infty \quad 1 \quad 2 \quad 7 \quad 4 \quad 9]$$

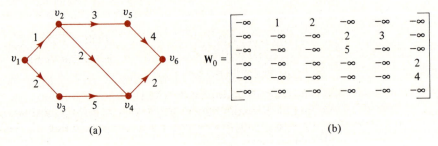

(a)   (b)

FIGURE 4

As an illustration, we compute $\mathbf{D}_4$, assuming that $\mathbf{D}_3$ gives the right values of $\mathbf{W}[1, j]$ for $k = 3$. The j-loop for $k = 4$ is

    for $j = 5$ to 6
        if $\mathbf{D}[j] < \mathbf{D}[4] + \mathbf{W}[4, j]$ then
            replace $\mathbf{D}[j]$ by $\mathbf{D}[4] + \mathbf{W}[4, j]$

Since $\mathbf{D}_3[5] = 4 > -\infty = 7 + (-\infty) = \mathbf{D}_3[4] + \mathbf{W}[4, 5]$, we make no replacement and get $\mathbf{D}_4[5] = \mathbf{D}_3[5] = 4$. Since $\mathbf{D}_3[6] = -\infty < 9 = 7 + 2 = \mathbf{D}_3[4] + \mathbf{W}[4, 6]$, we make a replacement and obtain $\mathbf{D}_4[6] = 9$. The values of $\mathbf{D}_3[1], \ldots, \mathbf{D}_3[4]$ are, of course, unchanged in $\mathbf{D}_4$. ∎

In the next section we consider how to modify the algorithms we have described here to answer some of our basic questions about digraphs, including how to find min-paths or max-paths corresponding to the min-weights or max-weights we have just been computing.

EXERCISES 8.3

1. (a) Give the min-weight matrix **W*** for the digraph shown in Figure 5(a). Any method is allowed, including staring at the picture.

   (b) Repeat part (a) for the digraph in Figure 5(b).

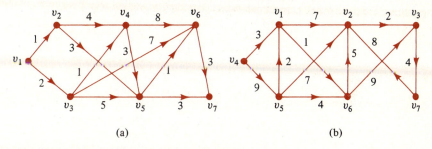

(a) (b)

FIGURE 5

2. Find the max-weight matrix for the digraph of Figure 5(a).

3. (a) Apply DIJKSTRA'S algorithm to the digraph of Figure 5(a). Start at v_1, and use the format of Figure 2.

   (b) Repeat part (a) with the digraph of Figure 5(b).

4. Apply DIJKSTRA'S algorithm to the digraph of Figure 3, starting at v_1. Compare your answer with the answer obtained by WARSHALL'S algorithm in Example 2.

5. (a) Use WARSHALL'S algorithm to find minimum path lengths in the digraph of Figure 6. Give the matrix for **W** at the start of each k-loop.

   (b) Use DIJKSTRA'S algorithm on this digraph to find minimum path lengths from v_1. Write your answer in the format of Figure 2.

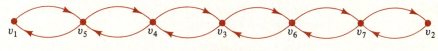

FIGURE 6

6. (a) Use WARSHALL'S algorithm to find **W*** for the digraph of Figure 4(a).

   (b) Find max-weights for the same digraph using the modified WARSHALL'S algorithm.

7. Give the final matrix **W** if WARSHALL'S algorithm is applied to the matrix

$$\mathbf{W_0} = \begin{bmatrix} \infty & 11 & 9 & \infty & 2 \\ \infty & \infty & \infty & \infty & \infty \\ \infty & 1 & \infty & \infty & \infty \\ \infty & \infty & 2 & \infty & \infty \\ \infty & \infty & 6 & 3 & \infty \end{bmatrix}.$$

8. Apply DIJKSTRA'S algorithm to find min-weights from v_3 in the digraph of Figure 5(a).

9. (a) Use algorithm MAX-WEIGHT to find max-weights from v_1 to the other vertices in the digraph of Figure 5(a).

   (b) Use MAX-WEIGHT to find max-weights from s to the other vertices in the digraph of Figure 7. [Start by sorting the digraph.]

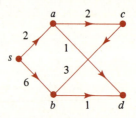

FIGURE 7

10. (a) Show that DIJKSTRA'S algorithm does not produce correct min-weights for the acyclic digraph shown in Figure 8. [So negative weights can cause trouble.]

    (b) Would DIJKSTRA'S algorithm give the correct min-weights for this digraph if the "for each j in $V \backslash L$ do" line of the algorithm were replaced by "for each j in V do"? Explain.

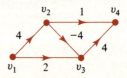

FIGURE 8

11. This exercise shows some of the difficulties in trying to modify DIJKSTRA'S algorithm to get max-weights. Change the replacement step in DIJKSTRA'S algorithm to the following:

    if $D(j) < D(k) + W(k, j)$
       replace $D(j)$ by $D(k) + W(k, j)$

    (a) Suppose that this modified algorithm chooses k in $V \backslash L$ with $D(k)$ as large as possible. Show that the new algorithm fails to give the right answer for the digraph in Figure 9(a). [Start at v_1.]

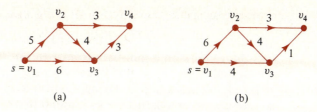

FIGURE 9

(b) Suppose that the modified algorithm instead chooses k in $V \backslash L$ with $D(k)$ nonnegative but as small as possible. Show that the new algorithm fails for the digraph in Figure 9(b).

(c) Would it help in either part (a) or (b) if "for each j in $V \backslash L$" were replaced by "for each j in V do"?

12. (a) Show that algorithm MAX-WEIGHT works; it finds max-weights from v_1 to other vertices.

(b) Show that MAX-WEIGHT operates in time $O(n^2)$.

§ 8.4 Modifications and Applications of the Algorithms

The versions of DIJKSTRA'S and WARSHALL'S algorithms that we saw in §8.3 produce min-weights and max-weights. We can easily modify the algorithms to produce the corresponding min-paths and max-paths as well. The key idea is to think of paths as linked lists.

To describe a path, we associate with each vertex a pointer to the next vertex along the path, until we come to the end. Think, perhaps, of a traveler in a strange city asking for directions to the railway station, going for a block and then asking again, and so on. For a path $x\,y\,z\,w$ we need pointers from x to y, from y to z and from z to w, which we can describe with a function p such that $p(x) = y$, $p(y) = z$ and $p(z) = w$. We can define $p(v)$ in some arbitrary way for other vertices v in $V(G)$ or just define p on $\{x, y, z\}$.

In the case of DIJKSTRA'S algorithm, which finds min-weights from some chosen vertex to each of the other vertices, it turns out to be technically easier to have the pointers point backward along the corresponding min-paths, rather than forward. Once the sequence of vertices backward along the path is known, it is easy to reverse the order and list the vertices in the forward direction.

To describe the modified DIJKSTRA'S algorithm we take $V(G)$ to be $\{1, 2, \ldots, n\}$ with 1 as the chosen vertex. We create a pointer function P defined on $\{2, \ldots, n\}$ such that at all times either $(P(j), j)$ is the final edge on a path of smallest known weight from 1 to j, or $P(j) = 0$ if no path from 1 to j has been discovered. When the algorithm stops, $P(j) = 0$

if there is no path from 1 to j in G. Otherwise, the sequence

$$j, P(j), P(P(j)), P(P(P(j))), \ldots$$

lists the vertices on a min-path from 1 to j in reverse order.

Initially, let $P(j) = 0$ if there is no edge from 1 to j; otherwise, let $P(j) = 1$. We add a line to the replacement loop in the original algorithm, so that whenever a path $1 \cdots kj$ is found that is better than the best path from 1 to j previously known, the pointer value $P(j)$ is set equal to k. Thus $P(j)$ is always the next-to-last stop on a best known path to j. Here is the revised algorithm.

DIJKSTRA'S *Algorithm with a Pointer*

{input: digraph without loops or parallel edges, vertex set $\{1, \ldots n\}$,
      nonnegative edge-weight function W}
{output: min-weights $W^*(1, j)$, pointers $P(j)$ for $j = 2, \ldots, n$}
Set $L := \varnothing$, $V := \{2, \ldots, n\}$.
For $i \in V$ do
   set $D(i) := W(1, i)$
   if $W(1, i) = \infty$ then set $P(i) := 0$
   else set $P(i) := 1$.
While $V \backslash L \neq \varnothing$ do
   choose k in $V \backslash L$ with $D(k)$ as small as possible
   put k in L
   for each j in $V \backslash L$ do
      if $D(j) > D(k) + W(k, j)$ then
         replace $D(j)$ by $D(k) + W(k, j)$
         replace $P(j)$ by k
For $j \in V$ do
   $W^*(1, j) := D(j)$. ∎

EXAMPLE 1   Consider the weighted digraph in Figure 1, which is the same one we looked at in Example 1 of § 8.3. The table in Figure 2 shows the successive values of D and P as each new vertex v_k is looked at and added to the list L. The reverse listing of a min-path from 1 to 7, derived from the last row of Figure 2, is

$$7, \quad P(7) = 6, \quad P(6) = 4, \quad P(4) = 5, \quad P(5) = 2, \quad P(2) = 1,$$

so the path is $v_1 v_2 v_5 v_4 v_6 v_7$. ∎

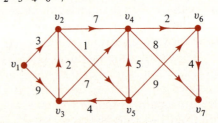

FIGURE 1

k	D(2)	D(3)	D(4)	D(5)	D(6)	D(7)	P(2)	P(3)	P(4)	P(5)	P(6)	P(7)
1	3	9	∞	∞	∞	∞	1	1	0	0	0	0
2	3	9	10	4	∞	∞	1	1	2	2	0	0
5	3	8	9	4	13	∞	1	5	5	2	5	0
3	3	8	9	4	13	∞	1	5	5	2	5	0
4	3	8	9	4	11	17	1	5	5	2	4	4
6	3	8	9	4	11	15	1	5	5	2	4	6

FIGURE 2

WARSHALL'S algorithm can also be given a pointer function. In this case, it is just as easy to have the pointer go in the forward direction, so that at all times $(v_i, v_{P(i, j)})$ is the first edge in a path of smallest known weight from v_i to v_j, if such a path has been found. When the modified algorithm stops, $P(i, j) = 0$ if there is no path from v_i to v_j. Otherwise, the sequence

$$i, \quad P(i, j), \quad P(P(i, j), j), \quad P(P(P(i, j), j), j), \quad \ldots$$

lists the indices of vertices on a min-path from v_i to v_j. Since each subscript k on the list is followed by $P(k, j)$, this sequence is easily recursively defined using the function P.

We can treat P as an $n \times n$ matrix with entries in $\{0, \ldots, n\}$. At the start, let $\mathbf{P}[i, j] = j$ if there is an edge from v_i to v_j and let $\mathbf{P}[i, j] = 0$ otherwise. Add a line to the replacement loop in WARSHALL'S algorithm so that whenever the algorithm discovers a path $v_i \cdots v_k v_j$ from v_i to v_j that is better than the best path known so far, the pointer value $\mathbf{P}[i, j]$ is set equal to $\mathbf{P}[i, k]$; the right way to head from v_i to v_j is to head for v_k first. The modified WARSHALL'S algorithm looks like this.

WARSHALL'S *Algorithm with a Pointer*

{input: nonnegative edge-weight matrix $\mathbf{W}_0$ of a weighted digraph without loops or parallel edges}
{output: min-weight matrix $\mathbf{W}^*$ and pointer matrix $\mathbf{P}^*$ for the digraph}
$\mathbf{W} := \mathbf{W}_0$
For $k = 1$ to n do
  for $i = 1$ to n do
    for $j = 1$ to n do
      if $\mathbf{W}[i, j] > \mathbf{W}[i, k] + \mathbf{W}[k, j]$ then
        replace $\mathbf{W}[i, j]$ by $\mathbf{W}[i, k] + \mathbf{W}[k, j]$
        replace $\mathbf{P}[i, j]$ by $\mathbf{P}[i, k]$.
$\mathbf{W}^* := \mathbf{W}, \mathbf{P}^* := \mathbf{P}$. ■

The same modifications will describe max-paths if WARSHALL'S algorithm has been revised to find max-weights.

Either DIJKSTRA'S or WARSHALL'S algorithm can be applied to undirected graphs, in effect by replacing each undirected edge $\{u, v\}$ for which $u \neq v$ by two directed edges (u, v) and (v, u). If the undirected edge has weight w, assign each directed edge the same weight w. If the graph is unweighted, assign weight 1 to all edges. Loops are really irrelevant in the applications of min-paths to undirected graphs, so it is convenient when using the algorithms to set $W(i, i) = 0$ for all i.

EXAMPLE 2   Consider the weighted graph shown in Figure 3. We use WARSHALL'S algorithm with a pointer to find min-weights and min-paths between vertices, allowing travel in either direction along the edges. Figure 4 gives the successive values of **W** and **P**. As a sample, we calculate $\mathbf{P}_4[2, 1]$.

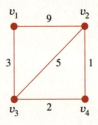

FIGURE 3

$$\mathbf{W}_0 = \mathbf{W}_1 = \begin{bmatrix} 0 & 9 & 3 & \infty \\ 9 & 0 & 5 & 1 \\ 3 & 5 & 0 & 2 \\ \infty & 1 & 2 & 0 \end{bmatrix} \qquad \mathbf{P}_0 = \mathbf{P}_1 = \begin{bmatrix} 0 & 2 & 3 & 0 \\ 1 & 0 & 3 & 4 \\ 1 & 2 & 0 & 4 \\ 0 & 2 & 3 & 0 \end{bmatrix}$$

$$\mathbf{W}_2 = \begin{bmatrix} 0 & 9 & 3 & 10 \\ 9 & 0 & 5 & 1 \\ 3 & 5 & 0 & 2 \\ 10 & 1 & 2 & 0 \end{bmatrix} \qquad \mathbf{P}_2 = \begin{bmatrix} 0 & 2 & 3 & 2 \\ 1 & 0 & 3 & 4 \\ 1 & 2 & 0 & 4 \\ 2 & 2 & 3 & 0 \end{bmatrix}$$

$$\mathbf{W}_3 = \begin{bmatrix} 0 & 8 & 3 & 5 \\ 8 & 0 & 5 & 1 \\ 3 & 5 & 0 & 2 \\ 5 & 1 & 2 & 0 \end{bmatrix} \qquad \mathbf{P}_3 = \begin{bmatrix} 0 & 3 & 3 & 3 \\ 3 & 0 & 3 & 4 \\ 1 & 2 & 0 & 4 \\ 3 & 2 & 3 & 0 \end{bmatrix}$$

$$\mathbf{W}^* = \mathbf{W}_4 = \begin{bmatrix} 0 & 6 & 3 & 5 \\ 6 & 0 & 3 & 1 \\ 3 & 3 & 0 & 2 \\ 5 & 1 & 2 & 0 \end{bmatrix} \qquad \mathbf{P}^* = \mathbf{P}_4 = \begin{bmatrix} 0 & 3 & 3 & 3 \\ 4 & 0 & 4 & 4 \\ 1 & 4 & 0 & 4 \\ 3 & 2 & 3 & 0 \end{bmatrix}$$

FIGURE 4

Since $\mathbf{W}_3[2, 4] + \mathbf{W}_3[4, 1] = 1 + 5 = 6 < 8 = \mathbf{W}_3[2, 1]$, $\mathbf{W}_4[2, 1] = 6$ and also $\mathbf{P}_4[2, 1] = \mathbf{P}_3[2, 4] = 4$. The min-path from v_2 to v_1 is described by the sequence 2, $\mathbf{P}^*[2, 1] = 4$, $\mathbf{P}^*[4, 1] = 3$, $\mathbf{P}^*[3, 1] = 1$.

The diagonal entries of $\mathbf{P}$ in this example are not especially meaningful. ∎

WARSHALL'S algorithm can be used to find reachable sets $R(v)$ for a digraph. Simply give all edges, including loops, weight 1. The final value $\mathbf{W}^*[i, j]$ is ∞ if there is no path from v_i to v_j and is a positive integer if a path exists. In particular, $\mathbf{W}^*[i, i] < \infty$ if and only if v_i is a vertex of a cycle in G. We can test whether a digraph G is acyclic by applying WARSHALL'S algorithm and looking at the diagonal entries of $\mathbf{W}^*$.

FLEURY'S algorithm in §6.2 involved checking an undirected graph G to see if removing a given edge e increased the number of components, i.e., to see if the endpoints of e were reachable from each other in the graph $G \backslash \{e\}$. WARSHALL'S algorithm can check reachability, but DIJKSTRA'S algorithm is faster in this instance; if e joins v_i and v_j we can apply DIJKSTRA'S algorithm to $G \backslash \{e\}$ with v_i as initial vertex. In fact, by Theorem 3 of §6.2, in doing this we are simply checking to see if the edge e belongs to a cycle of G, so by testing each edge in turn we can determine whether or not an undirected graph G is acyclic. In §6.6 we saw acyclicity checks based on FOREST and KRUSKAL'S algorithm that are even faster.

EXERCISES 8.4

1. (a) Give the initial and final min-path pointer matrices $\mathbf{P}$ for WARSHALL'S algorithm applied to the digraph in Figure 5(a). [Compare with Exercise 1 of §8.3.] You may use any method, including staring at the picture, to get your answer.

   (b) Repeat part (a) for max-paths instead of min-paths.

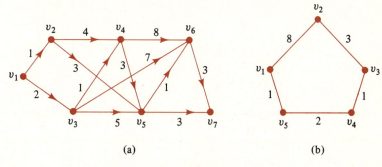

(a) (b)

FIGURE 5

2. Use WARSHALL'S algorithm with a pointer to find **W*** and **P*** for the graph of Figure 5(b).

3. (a) Apply DIJKSTRA'S algorithm with a pointer to the digraph shown in Figure 6(a), starting with vertex v_1. Display your answer in the format of Figure 2.

   (b) Repeat part (a) for the digraph of Figure 6(b).

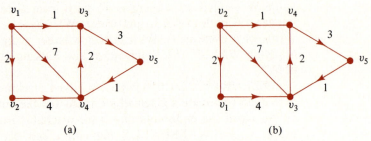

(a)   (b)

FIGURE 6

4. List the vertices of min-paths from v_1 to v_2, from v_1 to v_4 and from v_1 to v_6 for the digraph of Figure 1. *Hint:* See Figure 2.

5. (a) Use WARSHALL'S algorithm with a pointer to find **P*** for the digraph of Figure 7(a).

   (b) Repeat part (a) for max-paths instead of min-paths.

6. The **reachability matrix** $\mathbf{M}_R$ for a digraph is defined by $\mathbf{M}_R[i, j] = 1$ if $v_j \in R(v_i)$ and $\mathbf{M}_R[i, j] = 0$ otherwise.

   (a) Find the reachability matrix $\mathbf{M}_R$ for the digraph of Figure 7(b) by using WARSHALL'S algorithm.

   (b) Is this digraph acyclic?

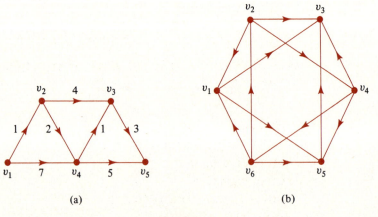

(a)   (b)

FIGURE 7

7. (a) Modify algorithm MAX-WEIGHT of §8.3 to get an algorithm that finds pointers along max-paths from a single vertex.

   (b) Apply your algorithm from part (a) to find max-path pointers from v_1 in the digraph of Figure 4(a) of §8.3.

8. (a) Draw a picture of the digraph with weight matrix

$$\mathbf{W}_0 = \begin{bmatrix} 0 & 0 & 0 & 0 & 0 & 0 \\ 1 & 0 & 0 & 0 & 0 & 0 \\ 0 & 1 & 0 & 0 & 0 & 0 \\ 0 & 0 & 1 & 0 & 0 & 0 \\ 0 & 0 & 0 & 1 & 0 & 0 \\ 0 & 0 & 0 & 0 & 1 & 0 \end{bmatrix}$$

   (b) Is this digraph acyclic?

   (c) Find the reachability matrix $\mathbf{M}_R$ [see Exercise 6] for this digraph.

9. Repeat Exercise 8 for the digraph with weight matrix

$$\mathbf{W}_0 = \begin{bmatrix} 0 & 0 & 1 & 0 & 0 & 0 \\ 0 & 0 & 0 & 1 & 0 & 0 \\ 1 & 0 & 0 & 0 & 1 & 0 \\ 0 & 1 & 0 & 0 & 0 & 1 \\ 0 & 0 & 1 & 0 & 0 & 0 \\ 0 & 0 & 0 & 1 & 0 & 0 \end{bmatrix}$$

10. Modify WARSHALL'S algorithm so that $\mathbf{W}[i, j] = 0$ if no path has been discovered from v_i to v_j and $\mathbf{W}[i, j] = 1$ if a path is known. *Hint:* Initialize suitably and use $\min\{\mathbf{W}[i, k], \mathbf{W}[k, j]\}$.

11. (a) Show that DIJKSTRA'S algorithm with a pointer takes $O(n^2)$ time for a digraph with n vertices.

    (b) Give the best estimate you can for the running time of WARSHALL'S algorithm with a pointer.

CHAPTER HIGHLIGHTS

As usual: What does it mean? Why is it here? How can I use it? Think of examples.

CONCEPTS AND NOTATION

sink, source

sorted labeling, reverse sorted labeling

indegree, outdegree

weighted digraph

   min-weight, min-path, $W^*(u, v)$

   max-weight, max-path, $M(u, v)$

scheduling network
  critical path, edge
  $A(v)$, $L(v)$
  slack time $S(v)$ of a vertex
  float time $F(u, v)$ of an edge
pointer

FACTS

A path in a digraph is acyclic if and only if its vertices are distinct.

Every finite acyclic digraph has a sorted labeling [also known from Chapter 7].

A digraph has an Euler circuit if and only if it is connected as a graph and every vertex has the same indegree as outdegree.

ALGORITHMS

SINK to find a sink in a finite acyclic digraph.

NUMBERING VERTICES to give a sorted labeling of an acyclic digraph in time $O(|V(G)|^2)$.

DIJKSTRA'S algorithm to compute min-weights from a selected vertex in time $O(|V(G)|^2)$ [or better], given a sorted labeling.

WARSHALL'S algorithm to compute min-weights or max-weights between all pairs of vertices, as well as determine reachability, in time, $O(|V(G)|^3)$.

MAX-WEIGHT to find max-weights from a selected vertex in time $O(|V(G)|^2)$ or $O(|V(G)| + |E(G)|)$, given a reverse sorted labeling.

DIJKSTRA'S and WARSHALL'S algorithms and MAX-WEIGHT with pointers to describe paths corresponding to min-weights or max-weights.

DIJKSTRA'S and WARSHALL'S algorithms to give min-weights and min-paths for undirected graphs [with each $W(i, i)$ initialized to 0].

PROBABILITY

9

This chapter expands on the brief introduction to probability in Chapter 5. The first section develops the ideas of independence and conditional probability of events. Starting with §9.2 the emphasis shifts from sample spaces to random variables and their distribution functions. Section 9.3 gives some of the basic properties of expectation and variance, as well as a discussion of independence of random variables. In the final section of the chapter we take a close look at binomial distributions and we point out, without proof, their connection with the normal distribution.

§ 9.1   Independence

As in §5.2, our setting for probability is a sample space Ω and a probability P: for events $E \subseteq \Omega$, $P(E)$ is a number in $[0, 1]$ and represents the probability of the event E.

The probabilities of events may change if more information is available, that is, if the outcome is known to be in some particular subset of Ω. For instance, if I happen to see that my opponent in poker has an ace, then I will think it's more likely that he has two aces than I would have thought before I got the information. On the other hand, if I learn that he has just won the lottery, that fact seems to be unrelated to the likelihood that he has two aces. Our goal in this section is to give mathematical meaning to statements such as "the probability of A given B," and "A is independent of B."

EXAMPLE 1   Figure 1 lists the set Ω of 36 equally likely outcomes when two fair dice are tossed, one black and one red. This situation is discussed in Example 4 of §5.2. Consider the events $B =$ "value on the black die is ≤ 3," $R =$ "value on the red die is ≥ 5," and $S =$ "the sum of the values is ≥ 8."

$$
\begin{array}{llll|ll}
(1,1) & (1,2) & (1,3) & (1,4) & (1,5) & (1,6) \\
(2,1) & (2,2) & (2,3) & (2,4) & (2,5) & (2,6) \\
(3,1) & (3,2) & (3,3) & (3,4) & (3,5) & (3,6) \\
(4,1) & (4,2) & (4,3) & (4,4) & (4,5) & (4,6) \\
(5,1) & (5,2) & (5,3) & (5,4) & (5,5) & (5,6) \\
(6,1) & (6,2) & (6,3) & (6,4) & (6,5) & (6,6)
\end{array}
$$

$\}\, B$

S is the set of outcomes
below the dashed line.

R

FIGURE 1

The outcomes in B are in the top three rows in Figure 1, those in R are in the last two columns, and the ones below the dotted line are in S.

(a) Observe that $P(R) = \frac{1}{3}$. The probability that the red die is 5 or 6 would be higher if we *knew* that the sum of the values were ≥ 8, i.e., if we knew that S occurred. The set S consists of the 15 possible outcomes that appear below the dotted line in Figure 1. The red die has value ≥ 5 in 9 of the 15 outcomes in S, so with this new knowledge the probability of R is $\frac{9}{15} = .6$, since all outcomes in S are equally likely.

(b) Note that $P(B) = \frac{1}{2}$ but the probability that the black die has value ≤ 3 drops to $\frac{3}{15} = .2$ if we know the sum is ≥ 8.

(c) How would $P(B) = \frac{1}{2}$ change if we knew that R occurred? That is, what's the probability that the black die is ≤ 3 if we know the red die is ≥ 5? Twelve outcomes in Figure 1 lie in R and six of them also lie in B, so the probability is still $\frac{1}{2}$. This is reasonable; we don't expect knowledge of the red die to tell us about the black die. The fact that we know the outcome is in R makes the set of possibilities smaller but does not change the likelihood that the outcome is in B.

The preceding paragraph is slightly misleading. It looks as if we've used probability to show that knowledge about the red die doesn't effect probabilities involving the black die. The truth is the other way around. We *believe* that the red die's outcome doesn't change the probabilities involving the black die. So, as we will explain in Example 7, we *arranged* for the probability P to reflect this fact. In other words, we view the 36 outcomes in Figure 1 as equally likely *because* this probability has the property that knowledge of the red die doesn't effect the probabilities of events involving only the black die, and vice versa. ∎

We formalize the idea in Example 1. Suppose that we know that an outcome ω is in some event $S \subseteq \Omega$. Then the outcome is in an event E if and only if it's in $E \cap S$, so the probability that ω is in E given that it's in S should depend only on the probability of $E \cap S$. If all outcomes in S are equally likely, then the probability of E given S, i.e., the probability of $E \cap S$ given S, is just the fraction of the outcomes in S that are in $E \cap S$, so it is $\dfrac{|E \cap S|}{|S|}$. More generally, the probability of $E \cap S$ given S should be the fraction of $P(S)$ associated with $E \cap S$. For $P(S) > 0$ we define the **conditional probability** $P(E|S)$, read "the probability of E given S," by

$$P(E|S) = \frac{P(E \cap S)}{P(S)} \qquad \text{for} \quad E \subseteq \Omega.$$

Then $P(\Omega|S) = \dfrac{P(\Omega \cap S)}{P(S)} = \dfrac{P(S)}{P(S)} = 1$, $P(E|S) = P(E \cap S|S)$ for $E \subseteq \Omega$, and it is not hard to see that the function $E \to P(E|S)$ satisfies the conditions that define a probability on Ω.

Since $P(S|S) = 1$, one could also think of S as the sample space for the conditional probability, instead of Ω, in which case we would only define $P(E|S)$ for $E \subseteq S$. Figure 2 suggests how we can view S as a cutdown version of Ω, with $S \cap E$ as the event in S corresponding to the event E in Ω. If $E \subseteq S$ and $P(S) < 1$, then $P(E|S) = \dfrac{P(E)}{P(S)} > P(E)$, as expected; the fact that S has occurred makes the event E more likely than it would have been without the extra information. For an event E not contained in S, $P(E|S)$ can be greater than, less than, or equal to $P(E)$, depending the circumstances.

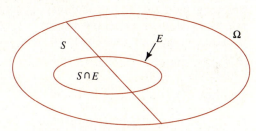

FIGURE 2

EXAMPLE 2 We return to the two-dice problem, so that Ω, B, R and S are as in Example 1 and Figure 1.

(a) With our new notation, the claim in Example 1(a) is simply $P(R|S) = \frac{9}{15}$. Indeed,

$$P(R|S) = \frac{P(R \cap S)}{P(S)} = \frac{9/36}{15/36} = \frac{9}{15}.$$

In Example 1(b) we observed that $P(B|S) = \frac{3}{15}$; this agrees with

$$P(B|S) = \frac{P(B \cap S)}{P(S)} = \frac{3/36}{15/36} = \frac{3}{15}.$$

It seems that our new definition of $P(E|S)$ has only complicated matters by introducing a bunch of 36's that get canceled anyway. But the new definition makes good sense even when the outcomes are not equally likely, in which case counting possible outcomes no longer helps.

(b) In Example 1(c) we saw that $P(B|R) = \frac{1}{2} = P(B)$. Similarly,

$$P(R|B) = \frac{P(R \cap B)}{P(B)} = \frac{6/36}{18/36} = \frac{1}{3} = P(R).$$

We repeat: The knowledge of R does not effect the probability of B, and vice versa. ■

Perhaps the most important question in probability and statistics is this: Does the occurrence of one event change the likelihood of the other? If not, the events are viewed as probabilistically independent. If so, the one event has a probabilistic effect on the other and, if the problem is important, one may look for causes for this effect.

EXAMPLE 3   Let Ω be the set of all adult American males. We assume that they are all equally likely to be selected in any study. If S is the event "he has smoked at least 10 years" and C = "he has lung cancer," then studies show that $P(C|S)$ is a lot larger than $P(C)$. There seem to be a cause and effect here. Certainly there are from the point of view of probability. Smoking increases the probability of getting lung cancer even though not every smoker gets lung cancer. Lung cancer doesn't "depend" on smoking in the sense that you must smoke to get lung cancer, but smoking does increase your chances. Lung cancer "depends" on smoking in the probabilistic sense.

Let M = "he studied college mathematics." As far as we know, $P(C|M) = P(C)$. In this case, getting lung cancer is probabilistically independent of studying mathematics. Briefly, the events C and M are [probabilistically] independent.

Suppose that it was discovered that $P(C|M) > P(C)$. Common sense tells us that mathematics doesn't directly cause cancer. So we would look for more reasonable explanations. Perhaps college-educated men lead more stressful lives and cancer thrives in such people. And so on. ■

For events A and B with $P(B) > 0$, we say that A and B are **independent** if $P(A|B) = P(A)$. This condition is equivalent to $\dfrac{P(A \cap B)}{P(B)} = P(A)$ and to $P(A \cap B) = P(A) \cdot P(B)$, and if $P(A) > 0$ to

$$P(B|A) = \frac{P(A \cap B)}{P(A)} = P(B).$$

Thus if A and B are independent, then B and A are independent; i.e., independence is a symmetric relation. The equivalent formulation

(I) $\qquad P(A \cap B) = P(A) \cdot P(B) \qquad$ for independent events A and B

is useful in computations. Also, it makes sense even if $P(A)$ or $P(B)$ is 0, so we will adopt (I) as the definition of **independence**. Following tradition, we will refer to pairs of independent events, but we will write $\{A, B\}$ to stress that order is immaterial.

EXAMPLE 4 Back to the two-dice problem in Examples 1 and 2. The events R and S are not independent since $P(R) = \frac{1}{3}$ while $P(R|S) = \frac{9}{15} = \frac{3}{5}$. Also,

$$P(S) = \tfrac{15}{36} = \tfrac{5}{12} \quad \text{while} \quad P(S|R) = \tfrac{9}{12} = \tfrac{3}{4}.$$

Similarly, B and S are not independent [check this].

However, B and R are independent:

$$P(B|R) = \tfrac{1}{2} = P(B) \quad \text{and} \quad P(R|B) = \tfrac{1}{3} = P(R). \quad \blacksquare$$

People sometimes refer to events as independent if they cannot both occur, i.e., if they are disjoint. This is **not** compatible with our usage and should be avoided. If $A \cap B = \varnothing$, then knowledge that B occurs *does* tell us something about A, namely that A does *not* occur: $P(A|B) = 0$. Disjoint events are not independent unless $P(A)$ or $P(B)$ is 0.

Before explaining what it means for a sequence $A_1, A_2, \ldots, A_n$ of events to be independent, let's consider what this should mean for three events A_1, A_2, A_3. It is not sufficient for each pair to be independent [called **pairwise independent**]. For example, to really have independence, the probability of A_1 shouldn't change if we know that both A_2 and A_3 have occurred [i.e., $P(A_1|A_2 \cap A_3)$ should equal $P(A_1)$].

EXAMPLE 5 We again consider the two-dice problem. Let $B_o =$ "black die's value is odd," $R_o =$ "red die's value is odd," and $E =$ "sum of the values is even." It is easy to check that the pairs $\{B_o, R_o\}$, $\{B_o, E\}$ and $\{R_o, E\}$ are all independent [Exercise 3]. On the other hand, if both B_o and R_o occur, then E is a certainty. In symbols,

$$P(E|B_o \cap R_o) = 1 \neq P(E) = \tfrac{1}{2}.$$

Thus we would not regard B_o, R_o, E as independent even though this sequence is pairwise independent. $\quad \blacksquare$

So for A_1, A_2, A_3 to be independent we require

$$P(A_1) = P(A_1|A_2 \cap A_3) = \frac{P(A_1 \cap A_2 \cap A_3)}{P(A_2 \cap A_3)},$$

and hence

$$P(A_1 \cap A_2 \cap A_3) = P(A_1) \cdot P(A_2 \cap A_3) = P(A_1) \cdot P(A_2) \cdot P(A_3),$$

in addition to pairwise independence. In short, we require

(I′)
$$P\left(\bigcap_{i \in J} A_i\right) = \prod_{i \in J} P(A_i)$$

for all nonempty subsets J of $\{1, 2, 3\}$. This is the right definition and works in general. Events $A_1, A_2, \ldots, A_n$ are **independent** if (I′) holds for all nonempty subsets J of $\{1, 2, \ldots, n\}$.

EXAMPLE 6   A fair coin is tossed n times, as in Example 5 of §5.2. For $k = 1, 2, \ldots, n$, let E_k be the event "kth toss is a head." We assume that these events are independent, since knowing what happened on the first and fifth tosses, say, shouldn't effect the probability of a head on the second or seventh tosses. Of course $P(E_k) = \frac{1}{2}$ for each k, since the coin is fair. From independence we obtain

$$P\left(\bigcap_{k=1}^{n} E_k\right) = \prod_{k=1}^{n} P(E_k) = \frac{1}{2^n}.$$

In other words, the probability of getting all heads is $1/2^n$. Similar reasoning shows that the probability of any particular sequence of n heads and tails is $1/2^n$. Thus the assumption that the outcomes of different tosses are independent implies that each n-tuple should have probability $1/2^n$, as in Example 5 of §5.2.   ■

EXAMPLE 7   We now show that the probability we've used for the two-dice problem is correct under the following assumptions: the probability of each value on the black die is $\frac{1}{6}$, the probability of each value on the red die is $\frac{1}{6}$, and the outcomes on the red and black dice are independent. Let's focus on the outcome $(4, 5)$. This is the unique outcome in $B_4 \cap R_5$ where $B_4 =$ "black die is 4" and $R_5 =$ "red die is 5." Then $P(B_4) = P(R_5) = \frac{1}{6}$ and by independence

$$P(4, 5) = P(B_4 \cap R_5) = P(B_4) \cdot P(R_5) = \frac{1}{36}.$$

This is why we used $P(k, l) = \frac{1}{36}$ for all (k, l) drawn in Figure 1.   ■

If A and B are not independent, the equation $P(A \cap B) = P(A) \cdot P(B)$ is *not* valid. If $P(A) \neq 0$ and $P(B) \neq 0$, though, we always have the equations

$$P(A \cap B) = P(A) \cdot P(B|A) \quad \text{and} \quad P(A \cap B) = P(B) \cdot P(A|B),$$

which are useful when one of $P(B|A)$ or $P(A|B)$ is easy to determine.

EXAMPLE 8 Two cards are drawn at random from a deck of cards.

(a) What is the probability that both cards are aces if the first card is put back into the deck [which is then shuffled] before the second card is drawn, i.e., drawn with replacement? The events $A_1 = $ "first card is an ace" and $A_2 = $ "second card is an ace" are independent, so

$$P(A_1 \cap A_2) = P(A_1) \cdot P(A_2) = \frac{1}{13} \cdot \frac{1}{13} = \frac{1}{169}.$$

The fraction $\frac{1}{13}$ comes from the fact that four of the 52 cards are aces.

(b) What is the probability that both cards are aces if the first card is not put back into the deck [drawing without replacement]? The events are no longer independent: if A_1 occurs, there is one less ace in the remaining deck. We find

$$P(A_1 \cap A_2) = P(A_1) \cdot P(A_2 \mid A_1) = \frac{4}{52} \cdot \frac{3}{51} = \frac{1}{221}.$$

W can also view this as the problem of selecting a random 2-element set from the 52-card deck. The probability of getting two aces is then

$$\frac{\binom{4}{2}}{\binom{52}{2}} = \frac{\frac{4 \cdot 3}{2 \cdot 1}}{\frac{52 \cdot 51}{2 \cdot 1}} = \frac{1}{221}. \quad \blacksquare$$

EXAMPLE 9 A company purchases a certain part from three firms and keeps a record of how many are defective. The facts are summarized in Figure 3. Thus 30 percent of the parts are purchased from firm C and 2 percent of them are defective. In terms of probability, if a purchased part is selected at random

$$P(A) = .50, \qquad P(B) = .20, \qquad P(C) = .30,$$

where $A = $ "part came from firm A," etc. If $D = $ "part is defective," we are also given

$$P(D \mid A) = .01, \qquad P(D \mid B) = .04, \qquad P(D \mid C) = .02.$$

Firm $\longrightarrow$	A	B	C
Fraction of parts purchased	.50	.20	.30
Fraction of defective parts	.01	.04	.02

FIGURE 3

(a) The probability that a part was purchased from firm A and was defective is $P(A \cap D) = P(A) \cdot P(D|A) = (.50) \cdot (.01) = .005$. Similarly,

$$P(B \cap D) = P(B) \cdot P(D|B) = (.20) \cdot (.04) = .008$$

and

$$P(C \cap D) = P(C) \cdot P(D|C) = (.30) \cdot (.02) = .006.$$

(b) What is the probability that a random part is defective? Since D is the disjoint union of $A \cap D$, $B \cap D$ and $C \cap D$, the answer is

$$P(D) = P(A) \cdot P(D|A) + P(B) \cdot P(D|B) + P(C) \cdot P(D|C)$$

$$= .005 + .008 + .006 = .019.$$

The tree in Figure 4 may be a helpful picture of what's going on.   ■

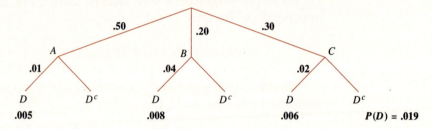

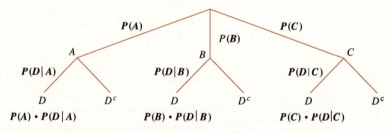

FIGURE 4

The calculation of $P(D)$ in Example 9(b) illustrates the next observation, with A, B, C being the partition.

Total Probability Formula

If events $A_1, A_2, \ldots, A_k$ partition the sample space Ω and $P(A_i) > 0$ for each i, then for each event B we have

$$P(B) = P(A_1) \cdot P(B|A_1) + P(A_2) \cdot P(B|A_2) + \cdots + P(A_k) \cdot P(B|A_k)$$

$$= \sum_{i=1}^{k} P(A_i) \cdot P(B|A_i).$$

Sometimes we know that event B has occurred and want to know the likelihood that each event A_j has occurred. That is, we want $P(A_1|B), \ldots,$

$P(A_k|B)$, numbers that are not obvious in the beginning. Since

$$P(A_j|B) = \frac{P(A_j \cap B)}{P(B)} = \frac{P(A_j) \cdot P(B|A_j)}{P(B)},$$

the Total Probability Formula implies the following.

Bayes'
Formula

> Suppose that the events $A_1, \ldots, A_k$ partition Ω, where $P(A_i) > 0$ for each i, and suppose that B is any event for which $P(B) > 0$. Then
>
> $$P(A_j|B) = \frac{P(A_j) \cdot P(B|A_j)}{P(B)}$$
>
> for each j, where
>
> $$P(B) = P(A_1) \cdot P(B|A_1) + P(A_2) \cdot P(B|A_2) + \cdots + P(A_k) \cdot P(B|A_k).$$

EXAMPLE 10   The randomly selected part in Example 9 is found to be defective. What is the probability that it came from firm A? We want $P(A|D)$, and Bayes' Formula gives

$$P(A|D) = \frac{P(A) \cdot P(D|A)}{P(D)} = \frac{(.50) \cdot (.01)}{.019} = \frac{.005}{.019} \approx .263.$$

Even though half the parts are purchased from firm A, only about a quarter of the defective parts come from this firm. Similarly,

$$P(B|D) = \frac{.008}{.019} \approx .421 \quad \text{and} \quad P(C|D) = \frac{.006}{.019} \approx .316.$$

Note how these fractions can be read from Figure 4. ■

The formula $P(A \cap B) = P(A) \cdot P(B|A)$ generalizes as follows:

$$P(A_1 \cap A_2 \cap \cdots \cap A_n)$$
$$= P(A_1) \cdot P(A_2|A_1) \cdot P(A_3|A_1 \cap A_2) \cdots P(A_n|A_1 \cap A_2 \cap \cdots \cap A_{n-1}).$$

Exercise 26 asks for a proof.

EXAMPLE 11   Four cards are drawn from a deck without replacement. What is the probability that the first two drawn will be aces and the second two will be kings? Using suggestive notation, we obtain

$$P(A_1 \cap A_2 \cap K_3 \cap K_4)$$
$$= P(A_1) \cdot P(A_2|A_1) \cdot P(K_3|A_1 \cap A_2) \cdot P(K_4|A_1 \cap A_2 \cap K_3)$$
$$= \frac{4}{52} \cdot \frac{3}{51} \cdot \frac{4}{50} \cdot \frac{3}{49} \approx .000022. \quad ■$$

EXERCISES 9.1

Wherever conditional probabilities like $P(A|B)$ occur, it is assumed that $P(B) > 0$.

1. A box has three red marbles and eight black marbles. Two marbles are drawn at random without replacement. Find the probability that
   (a) they are both red,   (b) they are both black,
   (c) one is red and one is black.

2. Repeat Exercise 1 if one marble is drawn and observed, then replaced before the second is drawn [i.e., drawn with replacement]. Compare the answers to those in Exercise 1.

3. (a) Show that the events B_o, R_o and E in Example 5 are pairwise independent.
   (b) Determine $P(B_o|E \cap R_o)$ and compare with $P(B_o)$.

4. Two dice, one black and one red, are tossed. Consider the events

   S = "the sum is ≥ 8,"

   L = "value on the black die is less than the value on the red die,"

   E = "values on the two dice are equal,"

   G = "value on the black die is greater than the value on the red die."

   Which of the following pairs of events are independent? $\{S, L\}, \{S, E\}, \{L, E\}, \{L, G\}$. Don't calculate unless necessary [but see Exercise 5].

5. By calculating suitable probabilities, determine which pairs of events in Exercise 4 are independent.

6. The two dice in Exercise 4 are tossed again.
   (a) Find the probability that the value on the red die is ≥ 5, given that the sum is 9.
   (b) Do the same, given that the sum is ≥ 9.

7. A fair coin is tossed four times. Consider the events A = "exactly one of the first two tosses is heads" and B = "exactly two of the four tosses are heads." Are the events A and B independent? Justify your answer.

8. A fair coin is tossed four times.
   (a) What is the probability of getting [at least two] consecutive heads?
   (b) What is the probability of getting consecutive heads, given that at least two of the tosses are heads?

9. Suppose that an experiment leads to events A, B, and C with $P(A) = .3$, $P(B) = .4$, $P(A \cap B) = .1$ and $P(C) = .8$.
   (a) Find $P(A|B)$.   (b) Find $P(A^c)$.
   (c) Are A and B independent? Explain.
   (d) Are A^c and B independent?

10. Given independent events A and B with $P(A) = .4$ and $P(B) = .6$, find

    (a) $P(A|B)$ (b) $P(A \cup B)$ (c) $P(A^c \cap B)$.

11. Three cards are drawn from an ordinary 52-card deck without replacement. Determine the probability that

    (a) three aces are drawn,

    (b) an ace, king and queen are drawn in that order,

    (c) at least one ace is drawn.

12. Recall that the probability of a poker hand being a flush is about .00197 [Example 1, §5.2]. What is the probability of a flush, given that all five of the cards are red?

13. Three urns contain marbles as indicated in Figure 5. An urn is selected at random and then a marble is selected at random from the urn.

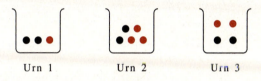

Urn 1   Urn 2   Urn 3

FIGURE 5

    (a) What is the probability that a black marble was selected?

    (b) Given that a black marble is selected, what is the probability that it was selected from urn 1? urn 2? urn 3?

    (c) What is the probability that a black marble was selected from urn 1?

14. An urn is selected at random from Figure 5 and then two marbles are selected at random [no replacement]. Find the probability that

    (a) both marbles are red,   (b) both marbles are black.

15. (a) Given that the two marbles selected in Exercise 14 are both black, find the probability that urn 1 was selected.

    (b) Repeat if both of the marbles are red.

16. One marble is selected at random from each urn in Figure 5. What is the probability that all three marbles are red? that all three are black?

17. Ellen, Frank and Gayle handle all the orders at Burger Queen. Figure 6 indicates the fraction of orders handled by each, and for each employee the fraction of complaints received.

Employee $\longrightarrow$	Ellen	Frank	Gayle
Fraction of orders	.25	.35	.40
Fraction of complaints	.04	.06	.03

FIGURE 6

(a) For a randomly selected order, what is the probability that there was a complaint?

(b) Given an order with a complaint, what is the probability that it was prepared by Ellen? Frank? Gayle?

18. A box contains two pennies, one fair and one with two heads. A coin is chosen at random and tossed.

(a) What is the probability that it comes up heads?

(b) Given that it comes up heads, what is the probability that the coin is the two-headed coin?

(c) If the selected coin is tossed twice, what is the probability that it comes up heads both times?

19. The following is observed for a test of a rare disease, where D = "subject has the disease," N = "subject tests negative," and N^c = "subject tests positive:" $P(N^c \cap D) = .004$, $P(N \cap D) = .0001$, $P(N^c) = .044$. Verify the following assertions.

(a) The test is over 97.5 percent accurate on diseased subjects. *Hint:* Find $P(N^c|D)$.

(b) The test is nearly 96 percent accurate on the general population. *Hint:* Find $P((N^c \cap D) \cup (N \cap D^c))$.

(c) Yet it is misleading over 90 percent of the time in cases where the test is positive! *Hint:* Find $P(D|N^c)$.

Moral: One must be very careful interpreting tests for rare diseases.

20. Show that if events A and B are independent so are the pairs $\{A, B^c\}$, $\{A^c, B\}$ and $\{A^c, B^c\}$.

21. An electronic device has n components. Each component has probability q of failure before the warranty is up. [Or think of light bulbs in your living unit.]

(a) What is the probability that some component will fail before the warranty is up? What assumptions are you making?

(b) What if $n = 100$ and $q = .01$?

(c) What if $n = 100$ and $q = .001$?

(d) What if $n = 100$ and $q = .1$?

22. Prove that if $P(A|B) > P(A)$, then $P(B|A) > P(B)$.

23. A box of 20 items is inspected by checking a sample of 5 items. If none of the items has a defect, the box is accepted.

(a) What is the probability that a box with two defective items will be accepted?

(b) Repeat part (a) if 10 items are checked instead of 5.

24. Prove that if $P(A) = P(B) = \frac{2}{3}$, then $P(A|B) \geq \frac{1}{2}$.

25. Suppose that A, B and C are independent events. Must $A \cap B$ and $A \cap C$ be independent?

26. Prove that

$$P(A_1 \cap A_2 \cap \cdots \cap A_n)$$
$$= P(A_1) \cdot P(A_2|A_1) \cdot P(A_3|A_1 \cap A_2) \cdots P(A_n|A_1 \cap A_2 \cap \cdots \cap A_{n-1}).$$

If the induction step is too complicated, verify the result for $n = 3$ and $n = 4$ instead.

27. Prove or disprove:

(a) If A and B are independent and if B and C are independent, then A and C are independent.

(b) Every event A is independent of itself.

(c) If A and B are disjoint events, then they are independent.

28. It is of course false that $(b \to a) \wedge a \Rightarrow b$. [Consider the case where a is true and b is false.] On the other hand, if $b \to a$ and a hold, then b is more likely than it was before, in the following sense.

(a) Show that if $P(A|B) = 1$ and $P(A) < 1$, then $P(B|A) > P(B)$.

(b) What if $P(A) = 1$?

29. Let S be an event in Ω with $P(S) > 0$.

(a) Show that if $P^*(E) = P(E|S)$ for $E \subseteq \Omega$, then P^* preserves ratios as follows:

$$\frac{P^*(E)}{P^*(F)} = \frac{P(E \cap S)}{P(F \cap S)} \quad \text{whenever } P(F \cap S) \neq 0.$$

(b) Show that if P^* is a probability on Ω that preserves ratios as in part (a), then $P^*(E) = P(E|S)$ for all $E \subseteq \Omega$.

§ 9.2 Random Variables

In this section we shift the point of view slightly. Rather than focus on events i.e., subsets of Ω, we look at numerical-valued functions defined on Ω. Since each event E corresponds to its characteristic function $\chi_E \colon \Omega \to \{0, 1\}$, we lose nothing by this change of perspective and we will gain a great deal.

A function from Ω into $\mathbb{R}$ is called a **random variable**. The terminology comes historically from the fact that the values of such a function vary as we go from one random sample ω in Ω to another. It is traditional to use capital letters near the end of the alphabet for random variables, with generic ones named X, Y or Z.

Given a random variable X on Ω and a condition C that some of its values may satisfy, we use the notation $\{X \text{ satisfies } C\}$ as an abbreviation for the event $\{\omega \in \Omega : X(\omega) \text{ satisfies } C\}$. For example,

$$\{5 < X \leq 7\} = \{\omega \in \Omega : 5 < X(\omega) \leq 7\},$$

$$\{X = a \text{ or } X^2 = b\} = \{\omega \in \Omega : X(\omega) = a \text{ or } X(\omega)^2 = b\},$$

$$\{|X - 4| < 3\} = \{\omega \in \Omega : |X(\omega) - 4| < 3\}, \text{ etc.}$$

For any $A \subseteq \mathbb{R}$, $\{X \in A\}$ represents the event $\{\omega \in \Omega : X(\omega) \in A\}$, a form that we'll often use. When writing probabilities of these events, we will usually drop the braces. So, for example,

$$P(X = 2) = P(\{X = 2\}) = P(\{\omega \in \Omega : X(\omega) = 2\}),$$

while

$$P(X \in A) = P(\{\omega \in \Omega : X(\omega) \in A\}) \qquad \text{for} \quad A \subseteq \mathbb{R}.$$

We will call the set of values that X takes on, i.e., the image set $X(\Omega) = \{X(\omega) : \omega \in \Omega\}$, the **value set** of X. If Ω is finite, the value set of X will be a finite subset of $\mathbb{R}$.

EXAMPLE 1   Let Ω consist of the 36 equally likely outcomes when two fair dice are tossed.

(a) The most interesting random variable on Ω gives the sum of the values on the two dice, namely

$$X_S(k, l) = k + l \qquad \text{for} \quad (k, l) \in \Omega.$$

The value set of X_S is $\{2, 3, 4, 5, 6, 7, 8, 9, 10, 11, 12\}$. All the numbers $P(X_S \in A)$ can be determined from Figure 1(a), which contains the same information as Figure 2(b) in §5.2. For example,

$$P(X_S \leq 5) = \frac{1}{36} + \frac{2}{36} + \frac{3}{36} + \frac{4}{36} = \frac{10}{36}$$

and

$$P(4 < X_S < 10) = \frac{4}{36} + \frac{5}{36} + \frac{6}{36} + \frac{5}{36} + \frac{4}{36} = \frac{2}{3}.$$

k	2	3	4	5	6	7	8	9	10	11	12
$P(X_S = k)$	$\frac{1}{36}$	$\frac{2}{36}$	$\frac{3}{36}$	$\frac{4}{36}$	$\frac{5}{36}$	$\frac{6}{36}$	$\frac{5}{36}$	$\frac{4}{36}$	$\frac{3}{36}$	$\frac{2}{36}$	$\frac{1}{36}$

(a)

k	1	2	3	4	5	6
$P(X_B = k) = P(X_R = k)$	$\frac{1}{6}$	$\frac{1}{6}$	$\frac{1}{6}$	$\frac{1}{6}$	$\frac{1}{6}$	$\frac{1}{6}$

(b)

FIGURE 1

If E is the set of even integers, then

$$P(X_S \text{ is even}) = P(X_S \in E)$$

$$= P(X_S = 2) + P(X_S = 4) + \cdots + P(X_S = 12)$$

$$= \frac{1}{36}(1 + 3 + 5 + 5 + 3 + 1) = \frac{1}{2}.$$

From the probabilistic point of view, everything we want to know about X_S can be determined from Figure 1(a).

(b) For each outcome (k, l) we define $X_B(k, l) = k$ and $X_R(k, l) = l$. Thus X_B is the value on the black die and X_R is the value on the red die. The random variables X_B and X_R each have value set $\{1, 2, 3, 4, 5, 6\}$. The possible values are equally likely; see Figure 1(b). These random variables are indistinguishable in the sense that they take the same values with the same probabilities.

(c) The random variable X_S is the sum of the random variables X_B and X_R. The events in Example 1 of §9.1 can be written in terms of these random variables:

$$B = \text{"value on the black die is} \le 3\text{"} = \{X_B \le 3\},$$

$$R = \text{"value on the red die is} \ge 5\text{"} = \{X_R \ge 5\},$$

$$S = \text{"sum is} \ge 8\text{"} = \{X_S \ge 8\}.$$

Calculations in Examples 1 and 2 of §9.1 can now be written

$$P(X_R \ge 5) = \frac{1}{3}, \qquad P(X_R \ge 5 \,|\, X_S \ge 8) = \frac{9}{15},$$

$$P(X_B \le 3) = \frac{1}{2}, \qquad P(X_B \le 3 \,|\, X_S \ge 8) = \frac{3}{15},$$

$$P(X_B \le 3 \,|\, X_R \ge 5) = \frac{1}{2} = P(X_B \le 3),$$

$$P(X_R \ge 5 \,|\, X_B \le 3) = \frac{1}{3} = P(X_R \ge 5).$$

The last two equations imply that the events $\{X_R \ge 5\}$ and $\{X_B \le 3\}$ are independent. This fact is not surprising and there is nothing special about the values 5 and 3 nor the senses of the inequalities $\ge$ and $\le$. We expect the random variables X_R and X_B to be independent, i.e., knowledge of X_R doesn't change probabilistic knowledge of X_B, and vice versa. We next formalize this notion. ■

We would like to say that the random variables X and Y on Ω are independent provided that $\{X \in A\}$ and $\{Y \in B\}$ are independent for all choices of subsets A and B of $\mathbb{R}$. Knowing that $X(\omega)$ belongs to some set A shouldn't affect the probability that $Y(\omega)$ belongs to some set B. Thus X and Y should be independent if

(I_1) $P(\{X \in A\} \cap \{Y \in B\}) = P(X \in A) \cdot P(Y \in B)$ for all $A, B \subseteq \mathbb{R}$.

We will usually abbreviate the left-hand term as $P(X \in A \text{ and } Y \in B)$. Unfortunately, in the most general situations, where Ω can be infinite, not all subsets A and B of $\mathbb{R}$ can be used in (I_1), although all the ones you can imagine, like intervals, can be. Since we do not want to pursue this technical issue, we will replace (I_1) with a requirement that is equivalent to (I_1) once the technical issue is clarified. Thus two random variables X and Y on Ω are **independent** if

(I_2) $P(X \in I \text{ and } Y \in J) = P(X \in I) \cdot P(Y \in J)$

for all intervals I and J in $\mathbb{R}$. If Ω is finite, (I_2) is equivalent to

(I_3) $P(X = x \text{ and } Y = y) = P(X = x) \cdot P(Y = y)$ for all $x, y \in \mathbb{R}$;

see Exercise 19. Of course, (I_3) only needs to be verified for x's in the value set of X and y's in the value set of Y.

EXAMPLE 2   (a) The random variables X_B and X_R that give the values on the black die and red die are independent. Indeed,

$$P(X_B = k \text{ and } X_R = l) = \frac{1}{36} = P(X_B = k) \cdot P(X_R = l) \qquad \text{for all}\quad (k, l).$$

(b) Let X_S be the sum of the values on the two dice. Then X_S and X_B are not independent; if X_B is big, X_S is more likely to be big. This intuitive explanation is *not* a proof and needs mathematical reinforcement. Rather than test (I_3) mindlessly, we'll use this comment as a guide. If X_B is big, like 6, the sum X_S surely can't be small, indeed it can't be less than 7. So, for example,

$$P(X_B = 6 \text{ and } X_S = 2) = 0 \neq P(X_B = 6) \cdot P(X_S = 2).$$

This is enough to *prove* that X_B and X_S are not independent.   ■

A sequence $X_1, X_2, \ldots, X_n$ of random variables on Ω is **independent** if

(I_2') $P(X_i \in J_i \text{ for } i = 1, 2, \ldots, n) = \prod_{i=1}^{n} P(X_i \in J_i)$

for all intervals $J_1, J_2, \ldots, J_n$ in $\mathbb{R}$. If Ω is finite, this condition is equivalent to

(I_3') $P(X_i = x_i \text{ for } i = 1, 2, \ldots, n) = \prod_{i=1}^{n} P(X_i = x_i)$

for $x_1, x_2, \ldots, x_n \in \mathbb{R}$.

EXAMPLE 3   (a) A fair coin is tossed n times, as in Example 6 of §9.1 [and Example 5 of §5.2]. For $i = 1, 2, \ldots, n$, let $X_i = 1$ if the ith toss is a head and $X_i = 0$ otherwise. Thus X_i is the characteristic function of the event $E_i =$ "ith toss is a head." In Example 6, §9.1, for each sequence $x_1, x_2, \ldots, x_n$ of 0's and 1's we assumed that the events $\{X_i = x_i\}$ were independent so that

$$P(X_i = x_i \text{ for } i = 1, 2, \ldots, n) = \prod_{i=1}^{n} P(X_i = x_i) = \frac{1}{2^n}.$$

In other words, we assumed that the random variables $X_1, X_2, \ldots, X_n$ were independent.

(b) The sum $S_n = X_1 + X_2 + \cdots + X_n$ is a very useful random variable. Since a member of the sample space is an n-tuple of H's and T's and since $X_i = 1$ if the ith entry is H and 0 otherwise, S_n counts the number of H's in the entire n-tuple. In other words, S_n counts the number of heads in the n tosses. The value set for S_n is $\{0, 1, 2, \ldots, n\}$. As we explained in Example 5(a) of §5.2,

$$P(S_n = k) = \frac{1}{2^n} \cdot \binom{n}{k} \qquad \text{for} \quad k \in \{0, 1, 2, \ldots, n\}. \quad \blacksquare$$

A random variable is called a **discrete random variable** if its value set is finite or an infinite sequence. All the random variables on a finite sample space Ω are discrete random variables. If X is a discrete random variable, all information about it can be obtained from knowing all the values of $P(X = x)$. The function f_X, defined on $\mathbb{R}$ by

$$f_X(x) = f(x) = P(X = x) \qquad \text{for} \quad x \in \mathbb{R},$$

is called the **probability density function** or **probability distribution** of X. It tells us how the values of X are distributed in $\mathbb{R}$, and it tells us the probabilities of the values.

EXAMPLE 4   (a) Let X_S, X_B and X_R be the random variables from our two fair dice. Their probability distributions can be read from Figure 1. The random variables X_B and X_R have the same probability distribution, namely

$$f(x) = \begin{cases} 1/6 & \text{if } x = 1, 2, 3, 4, 5 \text{ or } 6 \\ 0 & \text{for other } x. \end{cases}$$

The probability distribution f_S for X_S is harder to write down, but is clear from Figure 1(a):

$$f_S(x) = \begin{cases} 1/36 & \text{if } x = 2 \text{ or } 12 \\ 2/36 & \text{if } x = 3 \text{ or } 11 \\ \text{etc.} \\ 0 & \text{if } x \notin \{2, 3, \ldots, 12\}. \end{cases}$$

(b) Let $X_1, X_2, \ldots, X_n$ and S_n be the random variables that count heads as in Example 3. All the X_i's have the same probability distribution f where $f(0) = f(1) = \frac{1}{2}$ and $f(x) = 0$ for all other x in $\mathbb{R}$. The probability distribution for S_n is given by $f(k) = \frac{1}{2^n} \cdot \binom{n}{k}$ for $k = 0, 1, \ldots, n$ and $f(x) = 0$ for all other x. It is called the **binomial distribution** with parameter $\frac{1}{2}$, but it should be called "a" binomial distribution since there are infinitely many of them, one for each $n \geq 1$. A more general binomial distribution will appear in §9.4 where the coin will no longer have to be fair. ∎

Infinite sample spaces are introduced in and after Example 6 of §5.2. Random variables on infinite sample spaces might or might not be discrete.

EXAMPLE 5   (a) In Example 6(a), §5.2, a fair coin is tossed until a head is obtained. Let W be the random variable that counts the number of tosses needed to get a head; W is the "waiting time" for the first head. Then

$$P(W = k) = \frac{1}{2^k} \quad \text{for} \quad k = 1, 2, \ldots.$$

The value set for W is an infinite sequence, so W is a discrete random variable. Its probability distribution is the function f where $f(k) = 1/2^k$ for $k \in \mathbb{P}$ and $f(x) = 0$ for $x \in \mathbb{R} \backslash \mathbb{P}$.

(b) As in Example 6(b), §5.2, select a random number in $\Omega = [0, 1)$. Let U be the value of the number, i.e., let $U(x) = x$ for $x \in [0, 1)$. Such a U is sometimes called a **uniform random variable**; intervals of the same length are treated in a uniform fashion since $P(U \in [a, b)) = b - a$ for $[a, b) \subseteq [0, 1)$. The value set for U is the whole interval $[0, 1)$, which cannot be listed as a sequence [§13.3]. So U is not a discrete random variable. Moreover, if we try to define its "probability distribution" via

$$f(x) = P(U = x) = 0 \quad \text{for all} \quad x \in \mathbb{R},$$

we obtain a useless object. We certainly cannot recover information about U from this f. This example shows why we only define the probability distribution for discrete random variables. ∎

The next object we define is even more important than the probability distribution or distribution function. For any random variable X on any sample space, the **cumulative distribution function** or **cdf** of X is the function F_X, defined on $\mathbb{R}$ by

$$F_X(y) = F(y) = P(X \leq y) \quad \text{for} \quad y \in \mathbb{R}.$$

It accumulates or collects the values of the distribution function up to and including y. In fact, if X is a discrete random variable, it sums the values of the distribution function:

$$F_X(y) = \sum_{x \leq y} f_X(x).$$

EXAMPLE 6   (a) Consider random variables X_S, X_B and X_R with the probability distributions given in Example 4(a) and Figure 1. The cdf for X_B and X_R is F, where

$$F(y) = \begin{cases} 0 & \text{for } y < 1 \\ 1/6 & \text{for } 1 \le y < 2 \\ 2/6 & \text{for } 2 \le y < 3 \\ 3/6 & \text{for } 3 \le y < 4 \\ 4/6 & \text{for } 4 \le y < 5 \\ 5/6 & \text{for } 5 \le y < 6 \\ 1 & \text{for } y \ge 6. \end{cases}$$

The cdf F_S for X_S is given by

$$F_S(y) = \sum_{x \le y} f_S(x).$$

These cdf's are drawn in Figure 2 on page 458.

(b) Consider the random variables $X_1, X_2, \ldots, X_n$ and S_n that count heads when a fair coin is tossed. The cdf F for the X_i's is very simple:

$$F(y) = \begin{cases} 0 & y < 0 \\ 1/2 & 0 \le y < 1 \\ 1 & y \ge 1. \end{cases}$$

The cdf for S_n is given by

$$F(y) = \sum_{x \le y} \frac{1}{2^n} \binom{n}{x}.$$

For $n = 5$ this is drawn in Figure 2.

(c) The cdf F_W for the waiting time random variable W in Example 5(a) is defined by

$$F_W(y) = \sum_{k \le y} \frac{1}{2^k}.$$

It is sketched in Figure 2.

(d) The cdf F_U for the uniform random variable in Example 5(b) is called a **uniform distribution**. Since $U(x) \in [0, 1)$ for x, we find that

$$y < 0 \quad \text{implies} \quad F_U(y) = P(U \le y) = 0,$$

$$y \ge 1 \quad \text{implies} \quad F_U(y) = P(U \le y) = P(U \le 1) = 1,$$

$$0 \le y < 1 \quad \text{implies} \quad F_U(y) = P(U \le y)$$

$$= P(0 \le U \le y) = P([0, y]) = y.$$

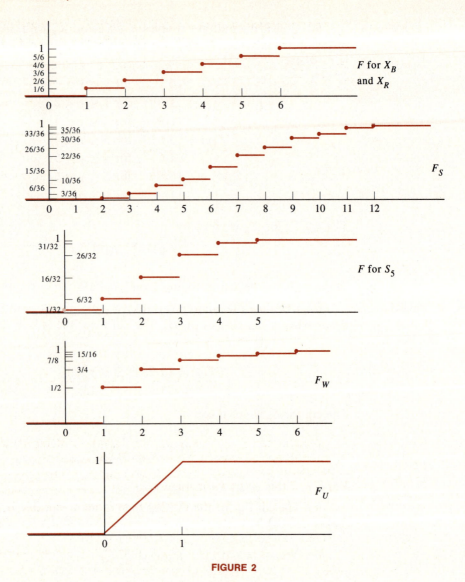

FIGURE 2

That is,

$$
F_U(y) = \begin{cases} 0 & y < 0 \\ y & 0 \le y < 1 \\ 1 & y \ge 1. \end{cases}
$$

See Figure 2. The cdf's for the discrete random variables have discrete jumps at points in the value set. In contrast, the cdf F_U is a nice continuous function with no jumps. ■

For a discrete random variable X, the cdf F_X can be defined in terms of the probability distribution f_X, since $F_X(y) = \sum_{x \leq y} f_X(x)$. We can also recover f_X from F_X. If the value set can be written $\{x_1, x_2, \ldots\}$ where $x_1 < x_2 < x_3 < \cdots$, as is the case in all of our examples, then we have

$$f_X(x_1) = F_X(x_1),$$
$$f_X(x_k) = F_X(x_k) - F_X(x_{k-1}) \quad \text{for} \quad k \geq 2,$$
$$f_X(x) = 0 \quad \text{for other } x.$$

Each $f_X(x)$ is the amount by which $F_X(x)$ jumps at the point x. See Figure 3.

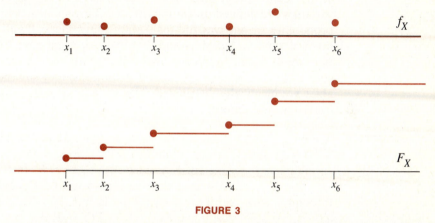

FIGURE 3

We summarize. A discrete random variable X determines a probability distribution f_X and a cdf F_X. Each of these contains all the essential probabilistic information about X, and either can be obtained from the other. Indeed, when probability and statistics texts provide tables for certain important distributions, like the binomial distribution, some give the probability distribution f_X and some give the cdf F_X. The reader is expected to be able to get from one to the other.

We believe that all the books should give the cdf and treat the cdf as the more fundamental object. At the computational level, it's easier to calculate differences like $F_X(x_k) - F_X(x_{k-1})$ than sums like $\sum_{x \leq y} f_X(x)$. That is, it's easier to get f_X from F_X than to get F_X from f_X. Second, the probabilities that are usually of interest have the form $P(a \leq X \leq b)$, $P(a < X < b)$, $P(X < b)$, etc. For instance, if a coin is tossed 1000 times and $X = S_{1000}$ counts the number of heads, numbers like $P(470 \leq X \leq 530)$ are much more interesting than numbers like $P(X = 501)$. The equations

$$P(470 \leq X \leq 530) = \sum_{x=470}^{530} f_X(x)$$

and

$$P(470 \leq X \leq 530) = F_X(530) - F_X(469)$$

show that the cdf is easier to use and more relevant to computations of this sort. The third reason we prefer the cdf is that probability distributions are only defined for discrete random variables whereas the cdf is defined [and useful!] for all random variables. The most important cdf of all is the normal distribution which, like the uniform distribution, is not the distribution of a discrete random variable. It will appear briefly in §9.4. Any reader who expects to study more probability is advised to adapt to the use of cdf's.

Did you notice how the sample spaces slowly disappeared from view as this section progressed? They are still there, of course, because random variables *are* defined on sample spaces. But one of the early triumphs in probability theory was the realization that practically all of probability involves random variables and that their cdf's contain all of the information about them. Indeed, probabilists regard two random variables as equivalent if they have the same distributions; they say the random variables are **identically distributed**. The underlying sample spaces are often irrelevant and ignored.

EXAMPLE 7   A coin is tossed once, so the sample space Ω is $\{H, T\}$. If we define $X(H) = 1$ and $X(T) = 0$, then $P(X = 0) = P(X = 1) = \frac{1}{2}$. This random variable has the same cdf F as each $X_1, X_2, \ldots, X_n$ in Example 6(b).

Two dice are tossed and the random variable Y is defined to be 1 if the sum is even and 0 otherwise. That is, $Y(k, l) = 1$ if $k + l$ is even and $Y(k, l) = 0$ otherwise. Then $P(Y = 0) = P(Y = 1) = \frac{1}{2}$ [see Example 1(a)] and the cdf for Y is the same as for X above, even though the sample spaces are quite different.

Let Ω be the infinite sample space $[0, 1)$. For a random x in $[0, 1)$ we define $W(x)$ to be 0 if x is in $[0, \frac{1}{2})$ and to be 1 if x is in $[\frac{1}{2}, 1)$. Again we have $P(W = 0) = P(W = 1) = \frac{1}{2}$, so the cdf for W is the same as for X and Y. ∎

Finally, we admit that the cdf's [and the probability distributions in the case of discrete random variables] do not tell us how different random variables interact on the same sample space. What we need are so-called joint distribution functions and joint probability distributions, topics that are dealt with in more complete treatments of the subject.

EXERCISES 9.2

1. (a) Three fair dice are tossed. What is the value set for the random variable that sums the values on the dice?

   (b) Repeat part (a) for *n* dice.

2. Two fair dice are tossed. Find

(a) the probability that the sum is less than or equal to 7,

(b) $P(5 \leq \text{sum} \leq 10)$,

(c) the probability that the sum is a multiple of 3.

3. Here are two random variables on the set Ω of 36 equally likely outcomes when two fair dice are tossed: $D(k, l) = |k - l|$ and $M(k, l) = \max\{k, l\}$.

(a) Give the value sets for D and M.

(b) Make a table of probability distributions for D and M as in Figure 1. *Hint:* Use Figure 1 in §9.1.

(c) Calculate $P(D \leq 1)$, $P(M \leq 3)$ and $P(D \leq 1 \text{ and } M \leq 3)$.

(d) Are the random variables D and M independent? Explain.

4. Give the cdf's for the random variables D and M in Exercise 3. Sketch them.

5. For Ω in Exercise 3, let T be the random variable defined by: $T(k, l) = k \cdot l$ [k times l].

(a) Give the value set for T.

(b) Calculate $P(T \leq 2)$.

(c) Calculate $P(T = 12)$.

6. Suppose that the independent random variables X and Y have the same probability distribution f where $f(0) = f(1) = \frac{1}{4}$, $f(2) = \frac{1}{2}$ and $f(x) = 0$ for other x. Calculate

(a) $P(X = 0 \text{ and } Y = 2)$

(b) $P(X = 0 \text{ or } Y = 2)$

(c) $P(X \leq 1 \text{ and } Y \geq 1)$

7. (a) Give the value set for the sum $X + Y$ of the random variables in Exercise 6.

(b) Calculate $P(X + Y = 2)$.

(c) Give the probability distribution for $X + Y$.

8. (a) For X as in Exercise 6, give the probability distribution for the random variable $3X + 2$.

(b) Do the same for $2 - X$.

9. An urn has 5 red marbles and 5 blue marbles.

(a) Four marbles are selected at random [without replacement]. Find the probability distribution for the random variable X that counts the number of red marbles selected.

(b) Repeat part (a) if seven marbles are selected at random.

10. A fair die is tossed and a player wins \$5 if a 5 appears and loses \$1 otherwise. Give the probability distribution for the random variable W that records the player's winnings. [A loss is a negative win.]

11. (a) Consider the function $f: \mathbb{R} \to \mathbb{R}$ where

$$f(x) = \begin{cases} 1/4 & \text{if } x = 1, 3, 5 \\ 0 & \text{otherwise.} \end{cases}$$

   Is f a probability distribution for some random variable? Explain.

   (b) Repeat part (a) for the function

$$f(x) = \begin{cases} 1/5 & \text{if } x = 1, 2, 3, 4, 5 \\ 0 & \text{otherwise.} \end{cases}$$

12. Draw the cdf for any random variable X satisfying $P(X = 0) = P(X = 1) = \frac{1}{2}$. This will be the cdf for the random variables $X_1, \ldots, X_n$ in Example 6(b) and for the random variables in Example 7.

13. Let F be the cdf for some random variable X. Verify:

   (a) $0 \le F(y) \le 1$ for all $y \in \mathbb{R}$.

   (b) F is nondecreasing, i.e., $y_1 < y_2$ implies that $F(y_1) \le F(y_2)$.

14. A fair coin is tossed n times. Let X be the random variable that counts the number of heads, and let Y be the one that counts tails.

   (a) Do X and Y have the same probability distributions? Explain.

   (b) What is the random variable $X + Y$?

15. A fair die is tossed. Let W be the random variable that counts the number of tosses until the first 6 appears. Give the probability distribution $f(x) = P(W = x)$ for W.

16. A fair die is tossed n times.

   (a) Give the probability that a 4, 5 or 6 appears at each toss.

   (b) Give the probability that a 5 or 6 appears at each toss.

   (c) Give the probability that a 6 appears at each toss.

17. A random number x is selected in $[0, 1)$. Find the probability that

   (a) $|x - \frac{1}{2}| \le \frac{1}{3}$ (b) $\min\{|x|, |x - 1|\} \le \frac{1}{3}$

18. Show that two events E and F are independent if and only if their characteristic functions χ_E and χ_F are independent random variables.

19. Prove that the independence conditions (I_2) and (I_3) prior to Example 2 are equivalent for random variables X and Y that have finite value sets.

§ 9.3   Expectation and Standard Deviation

Consider a random variable X on a finite sample space Ω, where all the outcomes are equally likely. Then the average value,

$$A = \frac{1}{|\Omega|} \sum_{\omega \in \Omega} X(\omega),$$

of X on Ω has a probabilistic interpretation: If elements ω of Ω are selected at random many times and the values $X(\omega)$ are recorded, then the average of the selected values will probably be close to A. This statement is actually a theorem that needs proof, but we hope you will accept it as reasonably intuitive.

EXAMPLE 1   A fair die is tossed and X is the value showing on the top side. The six outcomes 1, 2, 3, 4, 5 and 6 are equally likely:

$$P(X = 1) = P(X = 2) = \cdots = P(X = 6) = \frac{1}{6}.$$

The average outcome is

$$\frac{1}{6}(1 + 2 + 3 + 4 + 5 + 6) = \frac{7}{2} = 3.5.$$

If the die is tossed many times, we would expect the average of the outcomes to be close to 3.5. That is, the "expected average" is 3.5. ■

When the outcomes are not equally likely, we are interested in the probabilistic average, so we will weight the values of the random variable accordingly. Given a random variable X on a finite sample space Ω, its **expectation**, **expected value** or **mean** is defined by

$$E(X) = \mu = \sum_{\omega \in \Omega} X(\omega) \cdot P(\{\omega\}).$$

If all outcomes are equally likely, then $P(\{\omega\}) = 1/|\Omega|$ for all ω in Ω and so $E(X)$ is exactly the average A discussed prior to Example 1. Note that we use two notations and terms for the same concept: the expectation $E(X)$ of X and the mean μ [lowercase Greek mu], or μ_X if we need to specify the random variable.

EXAMPLE 2   An amateur dart player knows from experience that each time he throws a dart at the dart board in Figure 1, he will hit the regions with the probabilities indicated. Thus $P(A) = .02$, $P(B) = .06$, etc. A concessionaire owns the dart board and offers to pay \$10 whenever the dart hits region A, \$3 whenever it hits region B, and \$1 whenever it hits region C. What is the amateur's expected average income? It is certainly not the average $\frac{1}{4}(10 + 3 + 1 + 0) = \$3.50$. It is the probabilistic average or mean

$$\mu = 10(.02) + 3(.06) + 1(.10) + 0(.82) = .20 + .18 + .10 = \$0.48.$$

This is $E(X)$ for the random variable X defined on the sample space $\Omega = \{A, B, C, D\}$, where $X(A) = 10$, $X(B) = 3$, $X(C) = 1$ and $X(D) = 0$.

Normally, concessionaires charge for the privilege of playing games. A charge of 48 cents per try would be "fair" in the sense that the expected gain would then be 0. Since the concessionaire owns the dart board, it

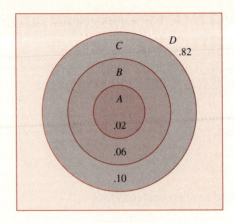

FIGURE 1

would be reasonable if the charge were somewhat more than 48 cents per try.

Let's clarify why the expected gain would be 0 if the charge were μ per try. The new random variable representing the player's net gain would be $X - \mu$. See Figure 2. One can check directly that $E(X - \mu) = 0$, but this equation always holds, as we will see after the next theorem. ∎

Region	X [no charge]	X − μ [charging μ]
A	10.00	9.52
B	3.00	2.52
C	1.00	0.52
D	0	−0.48

FIGURE 2

Theorem 1   Let X and Y be random variables on a finite sample space Ω. Then

(a) $E(X + Y) = E(X) + E(Y)$.
(b) $E(aX) = a \cdot E(X)$ for $a \in \mathbb{R}$.
(c) $E(a) = a$ for $a \in \mathbb{R}$.

In part (c), the first a in parentheses represents the constant function on Ω, which is a perfectly good random variable.

Proof. We have

$$E(X + Y) = \sum_{\omega \in \Omega} (X + Y)(\omega) \cdot P(\{\omega\}) = \sum_{\omega \in \Omega} \left[X(\omega) + Y(\omega)\right] \cdot P(\{\omega\}).$$

We now use the distributive law for real numbers to rearrange terms:

$$E(X + Y) = \sum_{\omega \in \Omega} X(\omega) \cdot P(\{\omega\}) + \sum_{\omega \in \Omega} Y(\omega) \cdot P(\{\omega\}) = E(X) + E(Y).$$

Thus (a) holds; the proof of (b) is even easier. Finally,

$$E(a) = \sum_{\omega \in \Omega} a \cdot P(\{\omega\}) = a \cdot \sum_{\omega \in \Omega} P(\{\omega\}) = a \cdot P(\Omega) = a. \quad \blacksquare$$

Corollary For any random variable X, we have $E(X - \mu) = 0$.

Proof. $E(X - \mu) = E(X) - E(\mu) = \mu - \mu = 0.$ $\blacksquare$

A simple induction shows that

$$E(X_1 + X_2 + \cdots + X_n) = E(X_1) + E(X_2) + \cdots + E(X_n)$$

for random variables $X_1, X_2, \ldots, X_n$. The next theorem gives the expectation in terms of the probability distribution. We are again suppressing the sample space.

Theorem 2 For a random variable X on a finite sample space, we have

$$E(X) = \sum_x x \cdot P(X = x) = \sum_x x \cdot f_X(x).$$

These are finite sums since the summands are 0 off of the value set of X.

Proof. Let A be the value set of X. Then the collection of sets $\{\{X = x\}: x \in A\}$ partitions Ω. [The set $\{X = x\}$ could also be written $X^{\leftarrow}(x)$.] Thus

$$E(X) = \sum_{\omega \in \Omega} X(\omega) \cdot P(\{\omega\}) = \sum_{x \in A} \left\{ \sum_{\omega \in \{X = x\}} X(\omega) \cdot P(\{\omega\}) \right\}.$$

Since $X(\omega) = x$ for ω in $\{X = x\}$ [by definition!], the inside sum equals

$$\sum_{\omega \in \{X = x\}} X(\omega) \cdot P(\{\omega\}) = x \cdot \sum_{\omega \in \{X = x\}} P(\{\omega\}) = x \cdot P(X = x).$$

Consequently, we have

$$E(X) = \sum_{x \in A} x \cdot P(X = x). \quad \blacksquare$$

The formula in Theorem 2 could have been taken as the definition of the expectation, and it is the definition that works for all discrete random variables. It cannot work in general, since the probability distribution f_X may be meaningless and worthless if X is not discrete. It turns out that for many random variables, the sums in Theorem 2 can be replaced by a generalized sum called an integral and denoted by an elongated S: $\int$. The probability distribution is replaced by a function f_X on $\mathbb{R}$ called the density. The expectation then looks like $E(X) = \int_{-\infty}^{\infty} x \cdot f_X(x)\, dx$, and the density determines the cdf F_X via $F_X(y) = \int_{-\infty}^{y} f_X(x)\, dx$. Note how these formulas look like generalizations of our earlier definitions. Even when these generalized formulas don't work, $E(X)$ can be defined in terms of the cdf F_X of X.

EXAMPLE 3 We illustrate the previous theorems by again considering the random variables X_S, X_B and X_R on the space Ω modeling the two-dice problem [Examples 1 and 2, §9.2].

(a) To calculate $E(X_S)$ from the definition we have to sum 36 numbers:

$$E(X_S) = \sum_{(k,l) \in \Omega} X_S(k, l) \cdot P(k, l) = \frac{1}{36}(2 + 3 + 3 + 4 + \cdots + 11 + 12).$$

We'd rather go to part (b).

(b) We calculate $E(X_S)$ using Theorem 2 and the values $P(X_S = k)$ in Figure 1(a) of §9.2:

$$E(X_S) = 2 \cdot \frac{1}{36} + 3 \cdot \frac{2}{36} + 4 \cdot \frac{3}{36} + 5 \cdot \frac{4}{36} + 6 \cdot \frac{5}{36} + 7 \cdot \frac{6}{36} + 8 \cdot \frac{5}{36}$$

$$+ 9 \cdot \frac{4}{36} + 10 \cdot \frac{3}{36} + 11 \cdot \frac{2}{36} + 12 \cdot \frac{1}{36}$$

$$= \frac{1}{36}[2 + 6 + 12 + 20 + 30 + 42 + 40 + 36 + 30 + 22 + 12]$$

$$= \frac{252}{36} = 7.$$

An even better solution appears in part (c).

(c) Recall that $X_S = X_B + X_R$. From Example 1, we have $E(X_B) = E(X_R) = 3.5$, and so $E(X_S) = E(X_B) + E(X_R) = 7$. In words, the expected value on the black die is 3.5 and the expected value on the red die is 3.5, so the expected sum is 7. ∎

EXAMPLE 4 A fair coin is tossed n times. What is the expected number of heads? If your intuition gives $n/2$, you are right. But please read on.

(a) The question is: What is $E(S_n)$ where S_n is the random variable that counts heads? Using Theorem 2, we find

$$E(S_n) = \sum_{k=0}^{n} k \cdot P(S_n = k) = \sum_{k=0}^{n} k \cdot \frac{1}{2^n} \binom{n}{k}.$$

This expression is too complicated, but we can learn from our experience in Example 3. The random variable S_n is the sum $X_1 + X_2 + \cdots + X_n$, where X_i is 1 if the ith toss is a head and 0 otherwise. Thus $P(X_i = 1) = P(X_i = 0) = \frac{1}{2}$. Since

$$E(X_i) = 1 \cdot P(X_i = 1) + 0 \cdot P(X_i = 0) = \tfrac{1}{2}$$

for each i, we see that

$$E(S_n) = \sum_{i=1}^{n} E(X_i) = \frac{n}{2},$$

as we expected.

(b) You may be puzzled by the first sum in part (a). It's correct, though, so what we have is an interesting proof of a "binomial identity:"

$$\sum_{k=0}^{n} k \cdot \binom{n}{k} = n \cdot 2^{n-1}.$$

Many similar intriguing identities are used and studied in probability and combinatorics. ∎

EXAMPLE 5   (a) Let W be the random variable that counts the number of tosses of a fair coin needed to get a head. What's the expected average wait? You might experiment to see. The answer is given by the infinite sum

$$\sum_{k=1}^{\infty} k \cdot P(W = k) = \sum_{k=1}^{\infty} k \cdot \frac{1}{2^k},$$

which turns out to be 2.

Is 2 the intuitively correct answer? We imagine that it is to some people, but not to most. Here's a nonsensical argument that isn't totally nonsense: You expect half a head in one toss, so you ought to get a whole head in two tosses!

(b) You will be happy to learn that if you select a number in $[0, 1)$ at random, the expected average value is $\frac{1}{2}$. That is, if you select several numbers in $[0, 1)$ at random, their average will be close to $\frac{1}{2}$. We will not pursue this topic, since we haven't even defined $E(X)$ for such random variables. ∎

Both formulas

$$E(X) = \sum_{\omega \in \Omega} X(\omega) \cdot P(\{\omega\}) \quad \text{and} \quad E(X) = \sum_{x} x \cdot P(X = x)$$

are special cases of the next lemma.

Lemma

Let Z be a random variable on a sample space Ω. Suppose that we are given a partition of Ω such that Z is constant on each set of the partition. Then $E(Z)$ is the sum of the products of the probability of each set times the [constant] value that Z takes on the set.

Proof. Let $E_1, \ldots, E_k$ be the sets of the partition. For each i, let $v(i)$ be the value of Z on E_i. Then $\{v(i): i = 1, \ldots, k\}$ is the value set A of Z; some values may be repeated. For x in A, $\{E_i: v(i) = x\}$ is a partition of $\{Z = x\}$, so

$$P(Z = x) = \sum_{\{i: v(i) = x\}} P(E_i).$$

Hence

$$x \cdot P(Z = x) = \sum_{\{i: v(i) = x\}} v(i) \cdot P(E_i),$$

so

$$E(Z) = \sum_{x \in A} x \cdot P(Z = x) = \sum_{x \in A} \sum_{\{i: v(i) = x\}} v(i) \cdot P(E_i).$$

Since $\{1, 2, \ldots, k\}$ is a disjoint union of the sets $\{i: v(i) = x\}$, $x \in A$, we conclude that

$$E(Z) = \sum_{i=1}^{k} v(i) \cdot P(E_i). \quad \blacksquare$$

It is useful to be able to calculate $E(X^2)$, $E(|X|)$, etc. in terms of X. Random variables like X^2 and $|X|$ have the form $\varphi \circ X$ where $\varphi: \mathbb{R} \to \mathbb{R}$. In fact, $X^2 = \varphi \circ X$ where $\varphi(x) = x^2$, and $|X| = \varphi \circ X$ where $\varphi(x) = |x|$.

Theorem 3

For a random variable X with a finite value set and a function $\varphi: \mathbb{R} \to \mathbb{R}$, we have

$$E(\varphi \circ X) = \sum_{x} \varphi(x) \cdot P(X = x).$$

The sum here is over all x in the value set of X.

Proof. Since $\varphi \circ X$ is constant on sets on which X is constant, we can apply the lemma to the random variable $Z = \varphi \circ X$ and the sets $\{X = x\}$, which partition Ω. Throughout the set $\{X = x\}$, the random variable $\varphi \circ X$ has the value $\varphi(x)$. $\quad \blacksquare$

Corollary For X as in Theorem 3,
$$E(X^2) = \sum_x x^2 \cdot P(X = x) \quad \text{and} \quad E((X - \mu)^2) = \sum_x (x - \mu)^2 \cdot P(X = x).$$

We also need a formula for $E(XY)$; it follows the next theorem. For random variables X and Y with finite value sets and a function $\psi: \mathbb{R} \times \mathbb{R} \to \mathbb{R}$, we write $\boldsymbol{\psi(X, Y)}$ for the random variable defined by $\psi(X, Y)(\omega) = \psi(X(\omega), Y(\omega))$ for $\omega \in \Omega$. For example, if $\psi(x, y) = xy$ for $x, y \in \mathbb{R}$, then $\psi(X, Y)(\omega) = X(\omega)Y(\omega)$ for $\omega \in \Omega$, i.e., $\psi(X, Y) = XY$.

Theorem 4 For X, Y and ψ as above, we have
$$E(\psi(X, Y)) = \sum_x \sum_y \psi(x, y) \cdot P(X = x \text{ and } Y = y).$$

The sums here are over the [finite] value sets of X and Y.

Proof. The sets $\{\omega \in \Omega : X(\omega) = x \text{ and } Y(\omega) = y\}$ partition Ω. Throughout such a set, the random variable $\psi(X, Y)$ has the value $\psi(x, y)$. The theorem now follows from the lemma prior to Theorem 3. ∎

Corollary
$$E(XY) = \sum_x \sum_y xy \cdot P(X = x \text{ and } Y = y).$$

Beware: The next result does not always hold; the random variables need to be independent.

Theorem 5 For independent random variables X and Y, we have
$$E(XY) = E(X) \cdot E(Y).$$

Proof. The statement is true in general, but our proof only works if X and Y have finite value sets. By the last corollary
$$E(XY) = \sum_x \sum_y xy \cdot P(X = x \text{ and } Y = y).$$

Since X and Y are independent,
$$E(XY) = \sum_x \sum_y x \cdot y \cdot P(X = x) \cdot P(Y = y).$$

Since these are finite sums, we can use the distributive law for real numbers to rewrite the equation as

$$E(XY) = \left\{\sum_x x \cdot P(X = x)\right\} \cdot \left\{\sum_y y \cdot P(Y = y)\right\} = E(X) \cdot E(Y). \quad \blacksquare$$

The expectation of a random variable X gives us its probabilistic average. However, it doesn't tell us how close to the average we expect to be. We need another measurement. A natural choice is the probabilistic average distance of X from its mean μ. This is the "mean deviation" $E(|X - \mu|)$, i.e., the mean of all the deviations $|X(\omega) - \mu|$, $\omega \in \Omega$. While this measure is sometimes used, it turns out that a similar measure, called the standard deviation, is much more manageable and useful. The **standard deviation** σ [or σ_X] is the square root of $E((X - \mu)^2)$. Like the "mean deviation," it lies between the smallest deviations $|X(\omega) - \mu|$ and the largest deviations, and serves as sort of an average of the deviations. The difficulty with $E(|X - \mu|)$ is that the absolute value function $|x|$ is troublesome, whereas x^2 is not. Students of calculus may recall that $|x|$ is difficult to handle primarily because of its abrupt behavior at 0, where its graph takes a right-angle turn.

For a random variable X, with mean μ, the square of the standard deviation σ is called the **variance** of X and written $V(X)$. Thus

$$V(X) = \sigma^2 = \sigma_X^2 = E((X - \mu)^2).$$

The corollary to Theorem 3 shows that

$$V(X) = \sum_x (x - \mu)^2 \cdot P(X = x)$$

if X has a finite value set. The variance and standard deviation measure how spread out the values of X are. The smaller $V(X)$ is, the more confident we can be that a randomly selected value of $X(\omega)$ is close to μ. In particular, $V(X) = 0$ if and only if X is the constant μ.

EXAMPLE 6   (a) Let X be the random variable that records the value of a tossed fair die, Since $\mu = 3.5$ from Example 1, we have

$$V(X) = \sum_{k=1}^{6} (k - \mu)^2 \cdot \frac{1}{6}$$

$$= \frac{1}{6}[(1 - 3.5)^2 + (2 - 3.5)^2 + (3 - 3.5)^2$$

$$+ (4 - 3.5)^2 + (5 - 3.5)^2 + (6 - 3.5)^2]$$

$$= \frac{1}{6}[17.5] = \frac{35}{12}.$$

The standard deviation is $\sigma = \sqrt{\dfrac{35}{12}} \approx 1.71.$

(b) Let X be the coin-tossing random variable so that $P(X = 0) = P(X = 1) = \frac{1}{2}$. Then

$$\mu = \frac{1}{2}, \quad V(X) = \left(0 - \frac{1}{2}\right)^2 \cdot \frac{1}{2} + \left(1 - \frac{1}{2}\right)^2 \cdot \frac{1}{2} = \frac{1}{4}, \quad \text{and} \quad \sigma = \sqrt{V(X)} = \frac{1}{2}.$$

∎

Another formula for $V(X)$ makes the computations a little easier because the mean μ only has to be handled once.

Theorem 6

For a random variable X, with mean μ, we have $V(X) = E(X^2) - \mu^2$.

Proof. Since $(X - \mu)^2 = X^2 - 2\mu X + \mu^2$, Theorem 1 implies that

$$V(X) = E(X^2) - 2\mu \cdot E(X) + \mu^2.$$

But $E(X) = \mu$, so

$$V(X) = E(X^2) - 2\mu^2 + \mu^2 = E(X^2) - \mu^2. \quad \blacksquare$$

EXAMPLE 6 revisited

(a) We have

$$E(X^2) = \sum_{k=1}^{6} k^2 \cdot \frac{1}{6} = \frac{1}{6}[1 + 4 + 9 + 16 + 25 + 36] = \frac{91}{6},$$

so

$$V(X) = E(X^2) - \mu^2 = \frac{91}{6} - \left(\frac{7}{2}\right)^2 = \frac{35}{12}.$$

(b) We have $E(X^2) = 1 \cdot \frac{1}{2} = \frac{1}{2}$, so

$$V(X) = E(X^2) - \mu^2 = \frac{1}{2} - \left(\frac{1}{2}\right)^2 = \frac{1}{4}. \quad \blacksquare$$

In general, the formula for $V(X + Y)$ is complicated but for independent random variables we have the following.

Theorem 7

For independent random variables $X_1, X_2, \ldots, X_n$ we have

$$V(X_1 + X_2 + \cdots + X_n) = V(X_1) + V(X_2) + \cdots + V(X_n).$$

Proof. This can be proved first for two random variables and then by an induction argument, but it's not entirely straightforward until it's known that $X_1 + X_2, X_3, \ldots, X_n$ are independent. See Exercise 21. The proof we give avoids induction.

Case 1: Suppose that $E(X_i) = 0$ for each i, so that $V(X_i) = E(X_i^2)$ for each i. Then

$$V\left(\sum_{i=1}^{n} X_i\right) = E\left(\left(\sum_{i=1}^{n} X_i\right)^2\right) = E\left(\sum_{i=1}^{n}\sum_{j=1}^{n} X_i X_j\right) = \sum_{i=1}^{n}\sum_{j=1}^{n} E(X_i X_j).$$

Since the X_i's are independent, Theorem 5 shows that $E(X_i X_j) = E(X_i) \cdot E(X_j) = 0 \cdot 0$ for $i \neq j$. Therefore in this case

$$V\left(\sum_{i=1}^{n} X_i\right) = \sum_{i=1}^{n} E(X_i^2) = \sum_{i=1}^{n} V(X_i).$$

Case 2: In the general case write $E(X_i) = \mu_i$ for each i. Since $E(X_i - \mu_i) = 0$ for each i, we can apply Case 1 to the random variables $X_i - \mu_i$ to obtain

$$V\left(\sum_{i=1}^{n} X_i - \sum_{i=1}^{n} \mu_i\right) = V\left(\sum_{i=1}^{n} (X_i - \mu_i)\right) = \sum_{i=1}^{n} V(X_i - \mu_i).$$

Since $V(Y + c) = V(Y)$ for any random variables Y and $c \in \mathbb{R}$ [Exercise 17], we can drop the constants. That is, $V\left(\sum_{i=1}^{n} X_i\right) = \sum_{i=1}^{n} V(X_i)$. ∎

EXAMPLE 7   (a) We seek the variance of the random variable X_S that adds the values on two fair dice. A direct assault using the definition or Theorem 6 would be hard work. But we know that $X_S = X_B + X_R$ where X_B and X_R are independent. We also know that $V(X_B) = V(X_R) = \frac{35}{12}$ from Example 6. Therefore, $V(X_S) = \frac{35}{6}$ by Theorem 7, so $\sigma_S = \sqrt{\frac{35}{6}} \approx 2.42$.

(b) To obtain the variance of the random variable S_n that counts the heads from n tosses of a fair coin, we recall that S_n is the sum of the independent random variables $X_1, \ldots, X_n$ where $P(X_i = 0) = P(X_i = 1) = \frac{1}{2}$. From Example 6, $V(X_i) = \frac{1}{4}$ for each i. Theorem 7 now implies that

$$V(S_n) = \sum_{i=1}^{n} V(X_i) = \sum_{i=1}^{n} \frac{1}{4} = \frac{n}{4}.$$

The standard deviation is $\frac{1}{2}\sqrt{n}$, which grows as n does, but much more slowly for large n. ∎

Theorems 1, 6 and 7 are true in general, but our proofs are only valid if the random variables are defined on a finite sample space, because of the way we proved Theorem 1.

EXERCISES 9.3

1. As in Exercise 9(a), §9.2, four marbles are selected at random [without replacement] from an urn containing 5 red marbles and 5 blue marbles.

(a) What is the expected number of red marbles selected?

(b) What is the standard deviation?

2. A newspaper article states that the average family has 2.1 children and 1.8 automobiles. How many average families are there? Discuss.

3. Calculate the mean deviations for the random variables in Example 6 and compare with the standard deviations.

4. Suppose that certain lottery tickets cost $1.00 each and that the only prize is $1,000,000. If each ticket has probability .0000005 of winning, what is the expected gain when one ticket is purchased?

5. Consider independent random variables X and Y with the same probability distribution f, where $f(0) = f(1) = \frac{1}{4}$ and $f(2) = \frac{1}{2}$. Find the mean and the standard deviation for X, Y and $X + Y$.

6. Find the mean and the standard deviation for the random variables in Exercise 3, §9.2, defined on the set Ω of 36 equally likely outcomes when two fair dice are tossed: $D(k, l) = |k - l|$ and $M(k, l) = \max\{k, l\}$.

7. Let X be a random variable with probability distribution f where $f(-2) = f(0) = f(1) = \frac{1}{5}$, $f(2) = \frac{2}{5}$ and $f(x) = 0$ for other x. Find the expectation of

   (a) X (b) $|X|$ (c) X^2 (d) $3X + 2$

8. Find the standard deviation for the random variables X and $|X|$ in Exercise 7.

9. Find the standard deviation for the random variable X^2 in Exercise 7.

10. Would you rather have a 50 percent chance of winning $1,000,000 or a 20 percent chance of winning $3,000,000? Discuss.

11. What is the expected number of aces in a five-card poker hand? *Hint:* The direct assault using Theorem 2 can be avoided.

12. A fair die is tossed until the first 5 appears; see Exercise 15, §9.2. Use the "nonsensical" argument in Example 5 to determine what the expected waiting time is for the first 5.

13. An urn has 4 red marbles and 1 blue marble.

   (a) Marbles are drawn from the urn without replacement until the blue marble is obtained. What is the expected waiting time for getting the blue marble?

   (b) Repeat part (a) if each marble is replaced after each drawing. *Hint:* See Exercise 12.

14. Show that $\sum_{k=1}^{n} k^2 \cdot \binom{n}{k} = n(n + 1) \cdot 2^{n-2}$ for $n \geq 0$. *Hint:* Let S_n be the random variable in Example 7, so that $V(S_n) = n/4$ and $\mu = E(S_n) = n/2$. Calculate $E(S_n^2)$ two ways: using Theorem 6 and using the corollary to Theorem 3.

15. Let X be the random variable that selects an integer at random from $\{1, 2, \ldots, n\}$.

   (a) Observe that $Y = n + 1 - X$ has the same probability distribution as X.

   (b) Use part (a) to show that $E(X) = \frac{1}{2}(n + 1)$.

   (c) Use Theorem 2 to derive the formula $1 + 2 + \cdots + n = \frac{1}{2}n(n + 1)$.

16. Show that $E(XY)$ does not always equal $E(X) \cdot E(Y)$.

17. Show that $V(X + c) = V(X)$ and $V(cX) = c^2 \cdot V(X)$ for c in $\mathbb{R}$ and a random variable X with finite value set.

18. Let X be a random variable with mean μ and standard deviation σ. Give the mean and the standard deviation for $-X$.

19. (a) Suppose that $X_1, X_2, \ldots, X_n$ are independent random variables, each with mean μ and standard deviation σ. Find the mean and the standard deviation of $S = X_1 + X_2 + \cdots + X_n$.

   (b) Do the same for the average $\dfrac{1}{n} S = \dfrac{1}{n}(X_1 + X_2 + \cdots + X_n)$.

20. (a) Show that if random variables X and Y on a finite sample space Ω satisfy $X(\omega) \leq Y(\omega)$ for all $\omega \in \Omega$, then $E(X) \leq E(Y)$.

   (b) Show that if $|X(\omega) - \mu| \leq |X(\omega_0) - \mu|$ for all $\omega \in \Omega$, then $|X(\omega_0) - \mu| \geq \sigma_X$. That is, the largest deviation of $|X - \mu|$ is greater than or equal to the standard deviation of X.

21. Let $X_1, X_2, \ldots, X_n$ be independent random variables with finite value sets.

   (a) Show that $X_1 + X_2, X_3, \ldots, X_n$ are independent.

   (b) Use Theorem 7 for $n = 2$ and part (a) to give an induction proof of Theorem 7.

§ 9.4   Binomial and Related Distributions

The motivating model for the general binomial distribution, which has numerous applications, is the following. We imagine an experiment with one possible outcome of interest, traditionally called **success**; the complementary event is called **failure**. We assume that $P(\text{success}) = p$ for some p, $0 < p < 1$. We set $q = P(\text{failure})$ so that $p + q = 1$. We further assume that the experiment is repeated several times, say n times, and that the outcomes of the different experiments are independent: a successful first experiment does not change the likelihood that the second experiment will be successful, and so on.

EXAMPLE 1   (a) The tossing of a fair coin n times can be viewed as an experiment where we regard heads as a success and tails as a failure. Then $p = P(\text{success}) = \frac{1}{2}$. We already know that

$$P(k \text{ heads in } n \text{ tosses}) = \frac{1}{2^n} \cdot \binom{n}{k}$$

for $k = 0, 1, \ldots, n$. In general,

$$P(k \text{ successes in } n \text{ experiments}) = \frac{1}{2^n} \cdot \binom{n}{k}$$

provided that $p = \frac{1}{2}$.

(b) A coin is said to be **unfair** or **biased** if the probability p of a head is different from $\frac{1}{2}$. Most coins are slightly biased; nickels are the worst. We can still ask for the probability of k heads in n tosses, but the formula in part (a) no longer applies. ∎

EXAMPLE 2   (a) An electronic device has n components, and the probability for each component to fail before the warranty is up is q. A component is successful, then, provided it is still working when the warranty is up. [If it fails the next day, we will still regard this a success!] With $p = 1 - q$ this situation fits our general scheme even though the n " experiments" will be running simultaneously. As in Exercise 21, §9.1, we are assuming that the survivals of the components are independent, which might or might not be a reasonable assumption. Unless $p = \frac{1}{2}$, we do not yet have a formula for

$$P(k \text{ successes in } n \text{ experiments}),$$

i.e., for

$$P(k \text{ components are working when the warranty is up}).$$

(b) If the probability of curing a certain disease is p, and if the cure is applied to n people with the disease, then the collection of "experiments" fits our general model. Again, we are assuming that the success rates for the various patients are independent. ∎

Given p and n, we now calculate the probability of k successes in n independent experiments where p is the probability of success for each experiment. Here $k = 0, 1, \ldots, n$. The sample space Ω consists of all n-tuples of S's and F's [for successes and failures]. There are 2^n such n-tuples but they are *not* equally likely. Compare the likelihood of all S's to that of all F's if $p = .001$. For a fixed k, there are $\binom{n}{k}$ n-tuples with exactly k S's and these turn out to be equally likely. To illustrate the idea, we consider a special case.

EXAMPLE 3   Given any p and $n = 5$, we find the probability of exactly 3 successes. First fix a particular 5-tuple with 3 successes, say (S, S, F, S, F). The probability of this particular outcome is

$$P(\text{first is S}) \cdot P(\text{second is S}) \cdot P(\text{third is F}) \cdot P(\text{fourth is S}) \cdot P(\text{fifth is F})$$

$$p \cdot p \cdot q \cdot p \cdot q = p^3 q^2.$$

For another such 5-tuple, the order of factors will be different but the result $p^3 q^2$ will be the same, since each of the 3 successes has probability p and each of the 2 failures has probability q. Thus we have $\binom{5}{3}$ outcomes, each with probability p^3q^2, and we conclude that

$$P(3 \text{ successes in 5 experiments}) = \binom{5}{3}p^3q^2. \quad \blacksquare$$

Exactly the same argument works in the general case; hence

$$P(k \text{ successes in } n \text{ experiments}) = \binom{n}{k}p^kq^{n-k}.$$

The general **binomial distribution** is the probability distribution given by

$$f(k) = \binom{n}{k}p^kq^{n-k}, \qquad k = 0, 1, \ldots, n.$$

[It is understood that $f(x) = 0$ for all other x.] Note that we have a binomial distribution for each p and n.

If $p = \frac{1}{2}$, as with the experiment where we tossed n fair coins, then

$$f(k) = \binom{n}{k}\left(\frac{1}{2}\right)^k\left(\frac{1}{2}\right)^{n-k} = \frac{1}{2^n} \cdot \binom{n}{k}.$$

This formula agrees with our earlier work.

The cdf F for f is defined by

$$F(y) = \sum_{k \le y} f(k) \qquad \text{for} \quad y \in \mathbb{R}.$$

This cdf is called the **cumulative binomial distribution**. Since F is constant between jumps, it is completely determined by its values at $k = 0, 1, \ldots, n$. Moreover, $f(0) = F(0)$ and $f(k) = F(k) - F(k-1)$ for $k = 1, 2, \ldots, n$, so f is easy to obtain from F.

To illustrate the ideas, we let $n = 10$ and consider $p = \frac{1}{2}$, $p = \frac{1}{3}$ and $p = \frac{1}{10}$. Table 1 gives the values of the cumulative binomial distribution for these choices of n and p. Much more complete tables can be found in probability and statistics books.

TABLE 1 Cumulative Binomial Distributions F for $n = 10$

p \ k	0	1	2	3	4	5	6	7	8	9	10
1/2	.001	.011	.055	.172	.377	.623	.828	.945	.989	.999	1.00
1/3	.017	.104	.299	.559	.787	.923	.980	.997	1.00	1.00	1.00
1/10	.349	.736	.930	.987	.998	1.00	1.00	1.00	1.00	1.00	1.00

While the cdf F is best for calculations, to help convey a feeling for the probabilities we have graphed the probability distributions in Figure 1 [vertical and horizontal axes are not on the same scale].

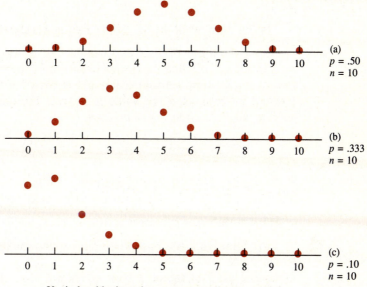

Vertical and horizontal axes are not on the same scale.

FIGURE 1

EXAMPLE 4   (a) A fair coin is tossed 10 times. Thus

$$P(k \text{ heads in 10 tosses}) = \frac{1}{2^{10}} \cdot \binom{10}{k} \quad \text{for} \quad k = 0, 1, \dots, 10.$$

We can calculate many probabilities quickly using Table 1. For example,

$P(\text{at most 5 heads}) = F(5) \approx .623,$

$P(7 \text{ heads}) = F(7) - F(6) \approx .945 - .828 = .117,$

$P(3 \le \text{number of heads} \le 7) = F(7) - F(2) \approx .945 - .055 = .890.$

(b) With a rather biased coin the probability of a head on each toss is $\frac{1}{3}$. Now

$$P(k \text{ heads in 10 tosses}) = \binom{10}{k} \cdot \left(\frac{1}{3}\right)^k \cdot \left(\frac{2}{3}\right)^{10-k}.$$

We can again use Table 1 to calculate some probabilities:

$P(\text{at least 5 heads}) = F(10) - F(4) \approx 1 - .787 = .213,$

$P(3 \le \text{number of heads} \le 7) = F(7) - F(2) \approx .997 - .299 = .698.$

The expected number of heads is $np = 3.33$, as we will see in Theorem 1. So we would expect the number of heads to be near 3. Indeed,

$$P(3 \text{ heads}) = F(3) - F(2) \approx .599 - .299 = .260,$$

while

$$P(2 \leq \text{number of heads} \leq 5) = F(5) - F(1) \approx .923 - .104 = .819.$$

(c) Even if our original experiment has several outcomes of interest, we can focus on some particular event and decree it to be a "success." For example, suppose that a fair die is tossed n times and we are interested in how often the value is 1 or 2. We decree this event to be success, so $P(\text{success}) = \frac{1}{3} = p$. Thus

$$P(k \text{ successes in } n \text{ experiments}) = \binom{n}{k} \cdot \left(\frac{1}{3}\right)^k \cdot \left(\frac{2}{3}\right)^{n-k}.$$

If the die is tossed 10 times we can use Table 1. For instance, the probability of getting a 1 or 2 at least 3 times in 10 tosses is

$$P(3 \leq \text{number of successes} \leq 10) = F(10) - F(2) \approx 1 - .299 = .701.$$

(d) Slim Hulk is a basketball player who makes $\frac{2}{3}$ of his free throws. If his successes are independent, what is the probability that he will make at least 7 of the next 10 free throws? Here $n = 10$ and $p = \frac{2}{3}$. We can't apply Table 1 directly, but we can if [to Slim's dismay] we call a miss a success and calculate $P(\text{number of misses} \leq 3)$. Now $P(\text{miss}) = P(\text{success}) = \frac{1}{3}$; so we use Table 1 with $p = \frac{1}{3}$ to obtain

$$P(\text{at least 7 free throws}) = P(\text{number of misses} \leq 3) = F(3) \approx .559.$$

Even though Slim only makes $\frac{2}{3}$ of his free throws, he has better than a 50 percent chance of making 7 or more in the next 10 tries! ■

EXAMPLE 5   We return to the n components, each of which has probability q of failure before the warranty is up. The probability that at least one component fails is

$$1 - P(n \text{ successes}) = 1 - \binom{n}{n} \cdot p^n = 1 - p^n = 1 - (1 - q)^n.$$

This result agrees with the answer to Exercise 21, §9.1.

The probability of at most 2 failures is

$$P(n - 2 \text{ successes}) + P(n - 1 \text{ successes}) + P(n \text{ successes})$$

$$= \binom{n}{n-2} \cdot p^{n-2}q^2 + \binom{n}{n-1} \cdot p^{n-1}q + \binom{n}{n} \cdot p^n$$

$$= \frac{n}{2}(n - 1) \cdot p^{n-2}q^2 + n \cdot p^{n-1}q + p^n.$$

If we interchange the roles of success and failure, we obtain

$P(2 \text{ failures}) + P(1 \text{ failure}) + P(0 \text{ failures})$

$$= \binom{n}{2} \cdot q^2 p^{n-2} + \binom{n}{1} \cdot qp^{n-1} + \binom{n}{n} \cdot p^n,$$

which is the same value, of course. ∎

The expectation calculated in the next theorem should come as no surprise.

Theorem 1

If S_n is a random variable with binomial distribution for some n and p, then

$$E(S_n) = np \quad \text{and} \quad V(S_n) = npq.$$

The standard deviation is $\sqrt{npq}$.

Proof. We may assume that S_n counts the successes of n independent experiments in which $P(\text{success}) = p$ for each experiment. Just as with $p = \frac{1}{2}$ in §9.3, a direct assault on $E(S_n)$ and $V(S_n)$ would be unwise.

For $i = 1, 2, \ldots, n$, define $X_i = 1$ if the ith experiment is a success, and define $X_i = 0$ otherwise. Then $P(X_i = 1) = p$, $P(X_i = 0) = q$ and

$$E(X_i) = 1 \cdot P(X_i = 1) + 0 \cdot P(X_i = 0) = p.$$

Also, $E(X_i^2) = p$, so

$$V(X_i) = E(X_i^2) - [E(X_i)]^2 = p - p^2 = p(1 - p) = pq.$$

Clearly, we have $S_n = X_1 + X_2 + \cdots + X_n$; hence

$$E(S_n) = \sum_{i=1}^{n} E(X_i) = \sum_{i=1}^{n} p = np.$$

Since the X_i's are independent, Theorem 7 of §9.3 implies that

$$V(S_n) = \sum_{i=1}^{n} V(X_i) = \sum_{i=1}^{n} pq = npq. \quad ∎$$

In §9.3 we discussed the "waiting time" for the first head when a fair coin is tossed. We can consider the same random variable even if the probability p of a head is different from $\frac{1}{2}$, $0 < p < 1$. [The probability distribution that we obtain will also be applicable to the problem of studying the first successful experiment in a sequence of independent experiments.] Let W be the number of tosses until a head is obtained. Then $P(W = 1) = p$. If $W = 2$, then the first toss is a tail and the second toss is a head; hence $P(W = 2) = qp$. If $W = 73$, the first 72 tosses are tails and the 73rd toss is a head; so $P(W = 73) = q^{72}p$. In general,

$$f(k) = P(W = k) = p \cdot q^{k-1} \qquad \text{for} \quad k = 1, 2, \ldots.$$

This distribution f is called the **geometric distribution**. The random variable W is a discrete random variable; its value set is the sequence $\{1, 2, \ldots\}$. Its expectation is the infinite series

$$\sum_{k=1}^{\infty} k \cdot p \cdot q^{k-1};$$

this sum turns out to be $1/p$. Thus $E(W) = 1/p$; compare Example 5(a), §9.3, where $p = \frac{1}{2}$.

EXAMPLE 6   (a) The probability that a certain unit [electronic component or light bulb, say] will burn out during an hour is known to be q. If q is small, it is a reasonable approximation to assume that the events "the unit burns out in the kth hour" are independent. Then the probability that the unit burns out during the kth hour is $p^{k-1}q$ where $p = 1 - q$. The expected waiting time for the burnout is $1/q$. Thus if $q = .01$, the expected lifetime of the unit is 100 hours. In other words, the average lifetime for a large number of units will be close to 100.

(b) We return to the electronic device with n components, each with probability q of failing before its warranty is up. Since all the components are operating simultaneously, the waiting time random variable and its geometric distribution [for q] are irrelevant! One could ask for the waiting time for the first failure of a component, but this is quite a different question and we haven't provided enough data or tools to answer it. ∎

Sometimes, even though we might wish otherwise, nice finite discrete objects lead to more complicated theoretical objects. Such is the case with binomial distributions. In applications the number n of experiments or samples is large and the computations of the binomial distribution get difficult. Clearly, it would be nice if, for large n, the cdf's F_n of the binomial distributions were close to a single cdf of some well-behaved random variable. [We will clarify what we mean here by "close" later.] As stated, this isn't true. The means and standard deviations [np and $\sqrt{npq}$, respectively] get large as n gets large, so the cdf's F_n drift off to the right on $\mathbb{R}$. For example, if $n = 1,000,000$ and $p = \frac{1}{2}$, the mean of the binomial distribution is 500,000 and its graph is centered at 500,000. The corresponding cdf doesn't reach height $\frac{1}{2}$ until $x = 500,000$.

Further thought suggests that if cdf's F_n are close together for large n, then their means and standard deviations are probably close too. This observation leads to the following idea. Given any random variable X, with mean μ and standard deviation $\sigma > 0$, the **normalization of** X is $\tilde{X} = \dfrac{X - \mu}{\sigma}$. This new random variable only differs from X by a constant multiple $1/\sigma$ and an additive constant, so knowledge of $\tilde{X}$ is easily transferred back to X. The nice properties of $\tilde{X}$ are stated in part (a) of the next theorem. The rest of the theorem shows the simple connection between the cdf's of X and $\tilde{X}$.

Theorem 2

Let X be a random variable with mean μ, standard deviation $\sigma > 0$, and cdf F. Let $\tilde{X}$ be the normalization of X and let $\tilde{F}$ be its cdf.

(a) $E(\tilde{X}) = 0$, $V(\tilde{X}) = 1$ and $\sigma_{\tilde{x}} = 1$.

(b) $F(y) = \tilde{F}\left(\dfrac{y - \mu}{\sigma}\right)$ for $y \in \mathbb{R}$.

(c) $\tilde{F}(y) = F(\sigma y + \mu)$ for $y \in \mathbb{R}$.

Proof. (a) $E(\tilde{X}) = \dfrac{1}{\sigma} E(X - \mu) = \dfrac{1}{\sigma}[E(X) - \mu] = \dfrac{1}{\sigma}[\mu - \mu] = 0$.

Exercise 17 of §9.3 shows that $V(X + c) = V(X)$ and $V(cX) = c^2 V(X)$ in general. Hence

$$V(\tilde{X}) = V\left(\frac{X}{\sigma}\right) = \frac{1}{\sigma^2} V(X),$$

and since $V(X) = \sigma^2$, we obtain $V(\tilde{X}) = 1$.

(b) By definition, $F(y) = P(X \leq y)$. Now

$$X(\omega) \leq y \Leftrightarrow X(\omega) - \mu \leq y - \mu \Leftrightarrow \frac{X(\omega) - \mu}{\sigma} \leq \frac{y - \mu}{\sigma}$$

so

$$\{X \leq y\} = \left\{\frac{X - \mu}{\sigma} \leq \frac{y - \mu}{\sigma}\right\} = \left\{\tilde{X} \leq \frac{y - \mu}{\sigma}\right\}.$$

It follows that

$$F(y) = P\left(\tilde{X} \leq \frac{y - \mu}{\sigma}\right) = \tilde{F}\left(\frac{y - \mu}{\sigma}\right).$$

(c) Replace y by $\sigma y + \mu$ in (b). ∎

EXAMPLE 7 Let S_n be a random variable with binomial distribution for n and some fixed p, $0 < p < 1$. The corresponding normalized random variable is

$$\tilde{S}_n = \frac{S_n - np}{\sqrt{npq}}. \quad \blacksquare$$

Since the random variables $\tilde{S}_n$ all have the same mean and standard deviation, it is reasonable to hope that their cdf's $\tilde{F}_n$ are close to some fixed cdf when n is large. A central result in probability theory is that there does exists a cdf Φ, called the **normal** or **Gaussian distribution**, such that

$$(*) \qquad \tilde{F}_n(y) \approx \Phi(y) \qquad \text{for large } n \text{ and for } y \in \mathbb{R}.$$

The distribution Φ does not depend on p; no matter what p is, the cdf's $\tilde{F}_n$ are close to Φ for large n. This extraordinary result and some powerful generalizations of it are known as "the central limit theorem." The proof is given more advanced texts.

Before explaining what Φ is, let's illustrate how this result can be used. But first note that assertion $(*)$ is very vague. How good an approximation do we get? How large must n be? These are sophisticated questions that we leave to the statisticians.

EXAMPLE 8 Suppose that an experiment is repeated $n = 10,000$ times with $P(\text{success}) = .1 = p$ each time. The expected number of successes is $\mu = np = 1000$, and we want to know

$$P(950 \le \text{number of successes} \le 1050).$$

This is $F_n(1050) - F_n(949)$. Since

$$\sigma = \sqrt{npq} = \sqrt{10{,}000 \cdot \frac{1}{10} \cdot \frac{9}{10}} = 30,$$

we have

$$F_n(1050) = \tilde{F}_n\left(\frac{1050 - 1000}{30}\right) = \tilde{F}_n\left(\frac{5}{3}\right) \approx \Phi(1.667) \approx .952$$

and

$$F_n(949) = \tilde{F}_n\left(\frac{949 - 1000}{30}\right) = \tilde{F}_n(-1.700) \approx \Phi(-1.700) \approx .048.$$

The approximations for the values of Φ were obtained from a table. We conclude that

$$P(950 \le \text{number of successes} \le 1050) \approx .904.$$

Note the power of the central limit theorem. For large n we only need to understand *one* cdf. At the computational level, this fact means that we only need one table of values or one function key on a calculator. ∎

Finally, what is Φ? Figure 2(a) is a picture of the so-called "bell curve." Its formula is

$$\varphi(x) = \frac{1}{\sqrt{2\pi}} e^{-x^2/2}$$

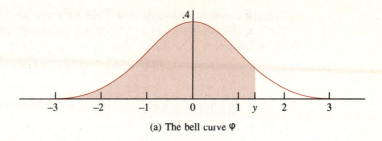

(a) The bell curve φ

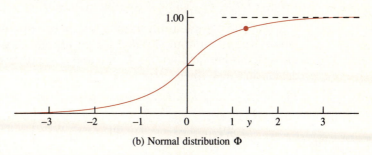

(b) Normal distribution Φ

FIGURE 2

where $e \approx 2.71828$. The total area under the bell curve is 1. For $y \in \mathbb{R}$,

$$\Phi(y) = \text{area under } \varphi \text{ to the left of } y;$$

see Figure 2(b). For the y indicated in Figure 2(a), this is the shaded area. Students of calculus would write

$$\Phi(y) = \frac{1}{\sqrt{2\pi}} \int_{-\infty}^{y} e^{-x^2/2} \, dx,$$

and might hope that there is a simple formula for this integral. Alas, there is no simple formula for $\Phi(y)$, but the values of Φ can be approximated as closely as desired and can be found in tables and on calculators. Finally, there is one value that is clear: $\Phi(0) = .500$ since half the area under φ lies to the left of 0.

The normal distribution arises in probability and statistics in many ways. It turns out that observed measured data, such as heights of all fourth graders or of all adult women, are approximately distributed like a normal distribution. This means that the normalized random variables have approximately the normal distribution.

EXAMPLE 9   (a) A random sample is taken of the heights of adult women. Thus we have a random variable X defined on the given set of women. It is observed that the mean μ of the heights is 66 inches and that the standard deviation σ is 2.5 inches. Assuming that this random variable is approximately normal, we estimate the probability that an adult woman has height between 61 and 71 inches. We will ignore those with height exactly 61 inches and estimate $P(61 < X \leq 71)$. This is $F(71) - F(61)$ where F is the cdf of X. If $\tilde{F}$ is the cdf for the normalized random variable $\tilde{X}$, then Theorem 2(b) shows that

$$F(71) - F(61) = \tilde{F}\left(\frac{71 - 66}{2.5}\right) - \tilde{F}\left(\frac{61 - 66}{2.5}\right) = \tilde{F}(2) - \tilde{F}(-2).$$

The cdf $\tilde{F}$ is approximately the normal distribution Φ by assumption, and so

$$P(61 < X \leq 71) \approx \Phi(2) - \Phi(-2).$$

From a table, this value is approximately $.9773 - .0227 = .9546$.

(b) The computation in part (a) applies in considerable generality. Let X be any random variable with a normal distribution, so that Φ is the distribution function for $\tilde{X}$. If μ and σ are the mean and the standard deviation for X, then

$$P(\mu - 2\sigma < X \leq \mu + 2\sigma) = F(\mu + 2\sigma) - F(\mu - 2\sigma)$$

$$= \Phi\left(\frac{\mu + 2\sigma - \mu}{\sigma}\right) - \Phi\left(\frac{\mu - 2\sigma - \mu}{\sigma}\right) = \Phi(2) - \Phi(-2) \approx .9546.$$

In part (a), $\mu = 66$ and $\sigma = 2.5$. Probabilists would say "the probability that a normally distributed random variable is within 2 standard deviations of its mean is approximately .95."

(c) The probability that a normally distributed random variable is within 1 standard deviation of its mean is approximately

$$\Phi(1) - \Phi(-1) \approx .8413 - .1587 = .6826.$$

Similarly, such a random variable is within 3 standard deviations of its mean with probability approximately equal to

$$\Phi(3) - \Phi(-3) \approx .9987 - .0013 = .9974. \quad \blacksquare$$

EXERCISES 9.4

1. Where is independence used in Example 3?

2. A baseball player has batting average .333. That is, he is successful in $\frac{1}{3}$ of his official attempts at hits. We assume that the attempts are independent.

   (a) What is the expected number of hits in 3 official attempts?

   (b) What is the probability that he will get at least one hit in his next three official attempts?

3. Suppose that the baseball player in Exercise 2 has 10 official attempts.

   (a) What is the expected number of hits?

   (b) What is the probability that he will get at most 3 hits? *Hint:* Use Table 1.

   (c) What is the probability that he will get at least 3 hits?

4. Half of Burger Queen customers order french fries. Assume that they order independently.

   (a) What is the probability that exactly five of the next ten customers will order french fries?

   (b) What is the probability that at least three of the next ten customers will order french fries?

5. An electronic device has 10 components. Each has probability .10 of burning out in the next year. Assume that the components burn out independently.

   (a) What is the probability that none of the components burn out next year?

   (b) What is the probability that at most two of the components burn out in the next year?

6. A random sample is taken of the heights of college men. The mean μ is 69 inches and the standard deviation σ is 3 inches.

   (a) Estimate the probability that a random college man will have height between 66 and 72 inches. *Hint:* Use estimates on the normal distribution in Example 9.

   (b) Estimate the probability that a random college man will have height between 63 and 75 inches.

   (c) Estimate the probability that a random college man will have height between 60 and 78 inches.

7. (a) What is the probability that a random college man in Exercise 6 will be at least 72 inches tall?

   (b) What is his probability of being less than 63 inches tall?

8. An experiment is repeated 30,000 times with probability of success $\frac{1}{4}$ each time. Thus the expected number of successes is 7500. Estimate, in terms of the normal distribution Φ, the probability that the number X of successes will be in the interval (7400, 7600].

9. An experiment is repeated 1800 times with $\frac{1}{3}$ probability of success for each experiment.

   (a) What is the expected number of successes?

   (b) What is the standard deviation?

   (c) Recall from Example 9 that $\Phi(2) - \Phi(-2) \approx .9546$. Give an interval in which we can be about 95 percent sure that the number of successes will lie.

10. For a large dinner party, 1000 invitations are sent out. The host estimates that each invitee has a 60 percent chance of attending the party and he believes that their decisions are independent. How many people should he plan for if he wants to be about 97 percent sure that he has enough place settings.

11. A fair coin is tossed 1000 times.

    (a) Estimate, in terms of the normal distribution, the probability that the fraction of heads is between 49 and 51 percent. *Hint:* Estimate the probability that the number X of heads is in the interval $(490, 510]$; see Example 9.

(b) Repeat part (a) if the coin is tossed 10,000 times.

(c) Repeat part (a) if the coin is tossed 1,000,000 times.

12. On a national test taken by 1,000,000 students, only 51 percent got the right answer on a particular true-false question. Did any of the students understand the question? Or was it just luck that 51 percent were successful?

13. A sequence of three independent experiments is performed. The probability of success on the first experiment is $\frac{1}{2}$, the probability of success on the second is $\frac{1}{3}$, and the probability of success on the third is only $\frac{1}{4}$.

(a) Find the expectation of the number X of successes.

(b) Find the standard deviation.

(c) Is X a binomial random variable? Explain.

14. Let X_1 and X_2 be random variables with means μ_1, μ_2 and standard deviations σ_1, σ_2, respectively. Suppose that X_1 and X_2 both have the same normalization $\tilde{X}$. Give a formula for the cdf F_2 of X_2 in terms of $F_1, \mu_1, \mu_2, \sigma_1$ and σ_2.

15. Let Φ be the normal distribution.

(a) Explain why $\Phi(y) + \Phi(-y) = 1$ for $y \in \mathbb{R}$.

(b) Show that $\Phi(y) - \Phi(-y) = 2 \cdot \Phi(y) - 1$ for $y \in \mathbb{R}$.

16. Let X be a normally distributed random variable with mean μ and standard deviation σ.

(a) Show that $P(\mu - \sigma < X \leq \mu + \sigma) \approx \Phi(1) - \Phi(-1) \approx .6826$.

(b) For any $c > 0$, give $P(\mu - c\sigma < X \leq \mu + c\sigma)$ in terms of Φ.

CHAPTER HIGHLIGHTS

As usual: What does it mean? Why is it here? How can I use it? Think of examples. Think of more examples.

CONCEPTS AND NOTATION

conditional probability, $P(E|S)$

independent, pairwise independent events

random variable

   value set

   discrete

   independent random variables

   distribution function = probability distribution, f_X

   cumulative distribution function = cdf, F_X

   uniform distribution

expectation = expected value = mean, $E(X) = \mu$

standard deviation, variance, $V(X) = \sigma^2$

binomial distribution, S_n

success, p, failure, q

geometric distribution

normalization, $\tilde{X}, \tilde{F}$

normal = Gaussian distribution, Φ

FACTS

Total Probability Formula for events that partition Ω.

Bayes' Formula

$$P(A_j|B) = \frac{P(A_j) \cdot P(B|A_j)}{P(B)}$$

with $P(B) = \sum_i P(A_i) \cdot P(B|A_i)$.

Either of f_X and F_X determines the other if X is discrete.

If all outcomes are equally likely, then $E(X)$ is the average value of X.

$E(X + Y) = E(X) + E(Y)$, $E(aX) = aE(X)$ and $E(a) = a$ for $a \in \mathbb{R}$.

If $X(\Omega)$ is finite and $\varphi \colon \mathbb{R} \to \mathbb{R}$, then

$$E(\varphi \circ X) = \sum_x \varphi(x) \cdot P(X = x).$$

Theorem 4 of §9.3 gives a similar formula for $\psi \colon \mathbb{R} \times \mathbb{R} \to \mathbb{R}$.

If X and Y are independent random variables, then $E(XY) = E(X) \cdot E(Y)$.

$V(X) = E(X^2) - \mu^2$.

If $X_1, \ldots, X_n$ are independent, then

$$V(X_1 + \cdots + X_n) = V(X_1) + \cdots + V(X_n).$$

If S_n has a binomial distribution, then $E(S_n) = np$ and $V(S_n) = npq$.

The normalization of X has mean 0 and variance 1; its distribution function $\tilde{F}$ is related to the distribution function F of X by

$$F(y) = \tilde{F}\left(\frac{y - \mu}{\sigma}\right) \quad \text{and} \quad \tilde{F}(y) = F(\sigma y + \mu) \qquad \text{for} \quad y \in \mathbb{R}.$$

The cdf of the normalization of a random variable with a binomial distribution is approximately the normal cdf Φ for large n.

METHODS

Use of tables of cdf's to calculate probabilities of events.

Use of a table of Φ to estimate $P(a \le S_n \le b)$ for binomially distributed random variables S_n.

BOOLEAN ALGEBRA 10

The term "Boolean algebra" has two distinct but related uses. On the one hand, it means the kind of symbolic arithmetic first developed by George Boole in the nineteenth century to manipulate logical truth values in an algebraic way, a kind of algebra well suited to describe two-valued computer logic. In this broad sense, Boolean algebra can be taken to include all sorts of mathematical methods that describe the operation of logical circuits.

"Boolean algebra" is also the name for a specific kind of algebraic structure whose operations satisfy an explicit set of rules. The rules and operations are chosen to provide concrete models for logical arithmetic.

This chapter begins with an account of Boolean algebras as structures, develops the link between the algebras and their logical interpretations, applies the resulting theory to logical networks, and concludes with an account of one method for simplifying complex logical expressions.

§ 10.1 Boolean Algebras

This section contains the definitions and basic properties of Boolean algebras, and introduces some important examples. The motivation comes from the symbolic logic that we saw in Chapter 2. One of the main achievements of the section is the theorem that every finite Boolean algebra is essentially the algebra of subsets of a finite set, the kind of algebra that we visualized with Venn diagrams back in § 1.2. This theorem establishes the theoretical basis for the link that we saw long ago between the set-theoretic operations $\cup$ and $\cap$ and their logical counterparts $\vee$ and $\wedge$. A good way to study this section would be to omit the proofs on the first reading and focus on the concepts and the statements of the theorems.

We will build the theory on its own foundation, but the analogies to the rules of set theory on the one hand and to truth-table arithmetic on the other will be clear. Before we proceed with the formal definition, here are some examples of Boolean algebras.

EXAMPLE 1   (a) The set $\mathscr{P}(S)$ of all subsets of the set S has the familiar operations $\cup$ and $\cap$. These combine two members A and B of $\mathscr{P}(S)$, i.e., two subsets of S, to give members $A \cup B$ and $A \cap B$ of $\mathscr{P}(S)$. The complementation operation ^{c} takes A to its complement $A^c = S \backslash A$, another member of $\mathscr{P}(S)$. The operations $\cup$, $\cap$ and ^{c} enjoy a variety of properties, some of which are listed in Table 1 of §1.2. The sets S and $\varnothing$ have special properties, such as $A \cap S = A$ and $A \cap \varnothing = \varnothing$ for all $A \in \mathscr{P}(S)$.

(b) The set $\mathbb{B} = \{0, 1\}$ has the usual logical operations $\vee$ and $\wedge$, together with the operation $'$ defined by $0' = 1$ and $1' = 0$. Here 0 and 1 have the interpretations "false" and "true," respectively, and $'$ corresponds to negation $\neg$. Figure 1 describes the operations $\vee$ and $\wedge$. In terms of ordinary integer arithmetic, $a \vee b = \max\{a, b\}$, $a \wedge b = \min\{a, b\} = a \cdot b$ and $a' = 1 - a$ for $a, b \in \mathbb{B}$.

$\vee$	0	1		$\wedge$	0	1
0	0	1	and	0	0	0
1	1	1		1	0	1

FIGURE 1

(c) The set $\mathbb{B}^n = \mathbb{B} \times \cdots \times \mathbb{B}$ [with n factors] of n-tuples of 0's and 1's has Boolean operations $\vee$, $\wedge$ and $'$ derived from the corresponding operations on $\mathbb{B}$. Thus

$$(a_1, \ldots, a_n) \vee (b_1, \ldots, b_n) = (a_1 \vee b_1, \ldots, a_n \vee b_n),$$

with similar coordinatewise definitions for $\wedge$ and $'$.

(d) We can view $\mathbb{B}^n$ as the set of functions f from $\{1, 2, \ldots, n\}$ to $\mathbb{B}$ by identifying f with the n-tuple $(f(1), f(2), \ldots, f(n))$. Using this identification, we have

$$(f \vee g)(k) = f(k) \vee g(k),$$

$$(f \wedge g)(k) = f(k) \wedge g(k) \quad \text{and}$$

$$f'(k) = (f(k))'$$

for $k = 1, 2, \ldots, n$ and $f, g \in \mathbb{B}^n$. More generally, for any set S we can define $\vee$, $\wedge$ and $'$ on the set FUN$(S, \mathbb{B})$ of all functions from S to $\mathbb{B}$ by $(f \vee g)(x) = f(x) \vee g(x)$, $(f \wedge g)(x) = f(x) \wedge g(x)$ and $f'(x) = (f(x))'$ for all functions f and g from S to $\mathbb{B}$. ∎

In general, we define a **Boolean algebra** to be a set with two binary operations $\vee$ and $\wedge$, a unary operation $'$ and distinct elements 0 and 1 satisfying the following laws.

$$
\left.\begin{array}{l}
\text{1Ba. } x \vee y = y \vee x \\
\text{b. } x \wedge y = y \wedge x
\end{array}\right\} \qquad \text{commutative laws}
$$

$$
\left.\begin{array}{l}
\text{2Ba. } (x \vee y) \vee z = x \vee (y \vee z) \\
\text{b. } (x \wedge y) \wedge z = x \wedge (y \wedge z)
\end{array}\right\} \qquad \text{associative laws}
$$

$$
\left.\begin{array}{l}
\text{3Ba. } x \vee (y \wedge z) = (x \vee y) \wedge (x \vee z) \\
\text{b. } x \wedge (y \vee z) = (x \wedge y) \vee (x \wedge z)
\end{array}\right\} \qquad \text{distributive laws}
$$

$$
\left.\begin{array}{l}
\text{4Ba. } x \vee 0 = x \\
\text{b. } x \wedge 1 = x
\end{array}\right\} \qquad \text{identity laws}
$$

$$
\left.\begin{array}{l}
\text{5Ba. } x \vee x' = 1 \\
\text{b. } x \wedge x' = 0
\end{array}\right\} \qquad \text{complementation laws}
$$

The operation $\vee$ is called **join**, $\wedge$ is called **meet**, and the unary operation $'$ is called **complementation**.

If we interchange $\vee$ and $\wedge$ in the laws defining Boolean algebras and at the same time switch 0 with 1, we get the same laws back again. In each case, rules *a* and *b* are interchanged. In particular, the properties of complementation are unchanged. The properties that hold in every Boolean algebra are the ones that are consequences of the defining laws, so we have the fundamental **duality principle**:

If $\wedge$ is interchanged with $\vee$ and 0 with 1 everywhere in a formula valid for all Boolean algebras, then the resulting formula is also valid for all Boolean algebras.

Here are some examples of consequences of the defining laws.

Theorem 1

The following properties hold in every Boolean algebra:

$$
\left.\begin{array}{l}
\text{6Ba. } x \vee x = x \\
\text{b. } x \wedge x = x
\end{array}\right\} \qquad \text{idempotent laws}
$$

$$
\left.\begin{array}{l}
\text{7Ba. } x \vee 1 = 1 \\
\text{b. } x \wedge 0 = 0
\end{array}\right\} \qquad \text{more identity laws}
$$

$$
\left.\begin{array}{l}
\text{8Ba. } (x \wedge y) \vee x = x \\
\text{b. } (x \vee y) \wedge x = x
\end{array}\right\} \qquad \text{absorption laws}
$$

Proof. Here is a derivation of 6Ba:

$$
\begin{array}{ll}
x \vee x = (x \vee x) \wedge 1 & \text{identity law 4Bb} \\[4pt]
\quad = (x \vee x) \wedge (x \vee x') & \text{complementation law 5Ba} \\[4pt]
\quad = x \vee (x \wedge x') & \text{distributive law 3Ba} \\[4pt]
\quad = x \vee 0 & \text{complementation law 5Bb} \\[4pt]
\quad = x & \text{identity law 4Ba.}
\end{array}
$$

For 7Ba, observe that

$$x \vee 1 = x \vee (x \vee x') \qquad \text{complementation law 5Ba}$$
$$= (x \vee x) \vee x' \qquad \text{associative law 2Ba}$$
$$= x \vee x' \qquad \text{idempotent law 6Ba just proved}$$
$$= 1 \qquad \text{complementation law 5Ba.}$$

And for 8Ba, we have

$$(x \wedge y) \vee x = (x \wedge y) \vee (x \wedge 1) \qquad \text{identity law 4Bb}$$
$$= x \wedge (y \vee 1) \qquad \text{distributive law 3Bb}$$
$$= x \wedge 1 \qquad \text{identity law 7Ba just proved}$$
$$= x \qquad \text{identity law 4Bb.}$$

Now 6Bb, 7Bb and 8Bb follow by the duality principle. ∎

It turns out that the associative laws are consequences of the other defining laws for a Boolean algebra, so they are redundant. In fact, Theorem 1 can be proved without using the associative laws. Proofs of these facts are tedious and not very informative, so we omit them.

Theorem 2 Every Boolean algebra satisfies the DeMorgan laws:

9Ba. $(x \vee y)' = x' \wedge y'$
  b. $(x \wedge y)' = x' \vee y'$.

Proof. We show first that if $w \vee z = 1$ and $w \wedge z = 0$, then $z = w'$; thus the two properties $w \vee w' = 1$ and $w \wedge w' = 0$ characterize w'. Indeed,

$$z = z \vee 0 \qquad \text{identity law 4Ba}$$
$$= z \vee (w \wedge w') \qquad \text{complementation law 5Bb}$$
$$= (z \vee w) \wedge (z \vee w') \qquad \text{distributive law 3Ba}$$
$$= (w \vee z) \wedge (w' \vee z) \qquad \text{commutative law 1Ba}$$
$$= 1 \wedge (w' \vee z) \qquad \text{hypothesis}$$
$$= (w \vee w') \wedge (w' \vee z) \qquad \text{complementation law 5Ba}$$
$$= (w' \vee w) \wedge (w' \vee z) \qquad \text{commutative law 1Ba}$$
$$= w' \vee (w \wedge z) \qquad \text{distributive law 3Ba}$$
$$= w' \vee 0 \qquad \text{hypothesis}$$
$$= w' \qquad \text{identity law 4Ba.}$$

This characterization of complements shows that to prove 9Ba it suffices to prove $(x \vee y) \vee (x' \wedge y') = 1$ and $(x \vee y) \wedge (x' \wedge y') = 0$. We have

$$(x \vee y) \vee (x' \wedge y') = [(x \vee y) \vee x'] \wedge [(x \vee y) \vee y'] \qquad \text{distributivity}$$

$$= [y \vee (x \vee x')] \wedge [x \vee (y \vee y')] \qquad \text{associativity and}$$
$$\qquad \qquad \qquad \qquad \qquad \qquad \qquad \qquad \text{commutativity}$$

$$= [y \vee 1] \wedge [x \vee 1] = 1 \wedge 1 = 1.$$

Similarly, $(x \vee y) \wedge (x' \wedge y') = 0$, so 9Ba holds. Formula 9Bb follows by duality. ■

EXAMPLE 2   It is not hard to verify that the sets and operations in Example 1 form Boolean algebras. By Theorem 2, they must also satisfy the DeMorgan laws. In the context of the algebra $\mathscr{P}(S)$ with operations $\cup$, $\cap$ and ^{c}, the DeMorgan laws are

$$(A \cup B)^c = A^c \cap B^c \quad \text{and} \quad (A \cap B)^c = A^c \cup B^c.$$

The laws for the two-element Boolean algebra $\mathbb{B}$ and for $\mathbb{B}^n$ correspond to the familiar logical DeMorgan laws

$$\neg(p \vee q) \Leftrightarrow (\neg p) \wedge (\neg q) \quad \text{and} \quad \neg(p \wedge q) \Leftrightarrow (\neg p) \vee (\neg q). \quad ■$$

The Boolean algebra $\mathscr{P}(S)$ has a relation $\subseteq$ that links some of its members. This relation can be expressed in terms of $\cup$, since $A \subseteq B$ if and only if $A \cup B = B$. This fact suggests a definition in the general case. Define the relation $\leq$ on a Boolean algebra by

$$x \leq y \quad \text{if and only if} \quad x \vee y = y.$$

It follows from the idempotent law $x \vee x = x$ that $x \leq x$ for every x. Define $x < y$ to mean $x \leq y$ but $x \neq y$. Let $x \geq y$ mean $y \leq x$ and let $x > y$ mean $y < x$.

EXAMPLE 3   (a) For the Boolean algebra $\mathscr{P}(S)$ the relations $\leq$, $<$, $\geq$ and $>$ are the containment relations $\subseteq$, $\subset$, $\supseteq$ and $\supset$, respectively.

(b) In the Boolean algebra $\mathbb{B}$ we have $0 \leq 0, 0 \leq 1, 1 \leq 1$ and $0 < 1$, so also $0 \geq 0, 1 \geq 0, 1 \geq 1$ and $1 > 0$. No surprises here!

(c) In $\mathbb{B}^n$ we have $(a_1, \ldots, a_n) \leq (b_1, \ldots, b_n)$ if and only if we have $(a_1 \vee b_1, \ldots, a_n \vee b_n) = (a_1, \ldots, a_n) \vee (b_1, \ldots, b_n) = (b_1, \ldots, b_n)$, i.e., if and only if $a_k \vee b_k = b_k$ for each k. Thus $(a_1, \ldots, a_n) \leq (b_1, \ldots, b_n)$ if and only if $a_k \leq b_k$ for each k. ■

Although we defined the relation $\leq$ in terms of the operation $\vee$, we could just as easily have used $\wedge$.

Lemma 1	In a Boolean algebra:
	$x \vee y = y$ if and only if $x \wedge y = x$.

Proof. If $x \vee y = y$, then

$$x = (x \vee y) \wedge x \qquad \text{absorption law 8Bb}$$

$$= y \wedge x \qquad \text{assumption}$$

$$= x \wedge y \qquad \text{commutative law 1Bb.}$$

The reverse implication follows from a dual argument. ∎

It may be a bit of a surprise to find that the relations $\leq$ and $<$ that we have defined abstractly in terms of $\vee$ or $\wedge$ satisfy some familiar properties of the relations $\leq$ and $<$ for numbers.

Lemma 2	In a Boolean algebra:
	(a) If $x \leq y$ and $y \leq z$, then $x \leq z$;
	(b) If $x \leq y$ and $y \leq x$, then $x = y$;
	(c) If $x < y$ and $y < z$, then $x < z$.

Properties (a) and (b) say, in the terminology of § 11.1, that the relation $\leq$ is a partial order relation. Property (a), **transitivity**, is familiar from our study of equivalence relations in § 3.5, but $\leq$ is quite different from an equivalence relation.

Proof of Lemma 2. (a) If $x \leq y$ and $y \leq z$, then

$$z = y \vee z \qquad \text{since } y \leq z$$

$$= (x \vee y) \vee z \qquad \text{since } x \leq y$$

$$= x \vee (y \vee z) \qquad \text{associative law 2Ba}$$

$$= x \vee z \qquad \text{since } y \leq z.$$

Thus $x \leq z$.

(b) If $x \leq y$ and $y \leq x$, then $x \vee y = y$ and $y \vee x = x$. By the commutative law 1Ba, $x = y$.

(c) If $x < y$ and $y < z$, then $x \leq z$ by (a). The case $x = z$ is impossible, by (b), since we would then have $x \leq y$, $y \leq z = x$, but $x \neq y$. ∎

The relation $\leq$ is also linked to the operations $\vee$ and $\wedge$ and to the special elements 0 and 1 in ways that parallel the structure of $\mathscr{P}(S)$.

Lemma 3

In a Boolean algebra:

(a) $x \wedge y \leq x \leq x \vee y$ for every x and y;

(b) $0 \leq x \leq 1$ for every x.

Proof. (a) Since $(x \wedge y) \vee x = x$ by rule 8Ba, $x \wedge y \leq x$. Similarly, rule 8Bb gives $(x \vee y) \wedge x = x$, so $x \leq x \vee y$ by Lemma 1.

(b) The statement follows from rules 4Ba, $x \vee 0 = x$, and 4Bb, $x \wedge 1 = x$. ∎

Finite sets can be built up as unions of 1-element subsets. Indecomposable building blocks play an important role in analyzing more general Boolean algebras as well. An **atom** of a Boolean algebra is a nonzero element a that cannot be written in the form $a = b \vee c$ with $a \neq b$ and $a \neq c$, i.e., that cannot be written as a join of two elements different from itself.

EXAMPLE 4

(a) The atoms of $\mathscr{P}(S)$ are the 1-element sets $\{s\}$; every A in $\mathscr{P}(S)$ with more than one member can be decomposed as $(A\backslash\{s\}) \cup \{s\}$ for $s \in A$.

(b) The only atom of $\mathbb{B}$ is 1.

(c) The atoms of $\mathbb{B}^n$ are the n-tuples with exactly one entry 1 and the rest of the entries 0. ∎

Atoms are like prime numbers in the sense that they have no nontrivial decompositions. We will see that we can also use them like primes to build up other elements of our algebras. First, we give an alternative characterization of atoms as minimal nonzero elements.

Proposition

A nonzero element x of a Boolean algebra is an atom if and only if there is no element y with $0 < y < x$.

Proof. Suppose that x is an atom and that $y < x$. Then $x = x \wedge 1 = (y \vee x) \wedge (y \vee y') = y \vee (x \wedge y')$. Since x is an atom, one of the terms y or $x \wedge y'$ must be x itself. Since $y \neq x$ by assumption, $x \wedge y' = x$. But then $y = x \wedge y = (x \wedge y') \wedge y = x \wedge (y' \wedge y) = x \wedge 0 = 0$.

On the other hand, if x is not an atom, then $x = y \vee z$ for some y and z not equal to x. Since $y \leq y \vee z = x$ by Lemma 3(a), we have

$y < x$. Also, $0 < y$ because otherwise $y = 0$, and hence $x = 0 \vee z = z \neq x$. ■

We are now ready to describe how the atoms form the basic building blocks of a finite Boolean algebra. First, let's look at some examples.

EXAMPLE 5   (a) Consider a finite set S. The atoms of $\mathscr{P}(S)$ are the 1-element sets $\{s\}$. If $T = \{t_1, \ldots, t_m\}$ is any m-element subset of S, then the atoms $\{s\}$ of $\mathscr{P}(S)$ that satisfy $\{s\} \subseteq T$ are the atoms $\{t_1\}, \ldots, \{t_m\}$, and T is their union $\{t_1\} \cup \cdots \cup \{t_m\}$.

(b) The atoms of $\mathbb{B}^n$ are the n-tuples in $\mathbb{B}^n$ with exactly one 1. Say a_i is the atom with its 1 in the ith position. From Example 3(c), if $x = (x_1, \ldots, x_n) \in \mathbb{B}^n$, then $a_i \leq x$ if and only if $x_i = 1$, and x is the join of the atoms a_i for which $x_i = 1$. For instance, if $x = (1, 1, 0, 1, 0)$ in $\mathbb{B}^5$, then $x = (1, 0, 0, 0, 0) \vee (0, 1, 0, 0, 0) \vee (0, 0, 0, 1, 0) = a_1 \vee a_2 \vee a_4$. ■

The next theorem shows that what occurred in these examples happens in general.

Theorem 3   Let B be a finite Boolean algebra with set of atoms $A = \{a_1, \ldots, a_n\}$. Each nonzero x in B can be written as a join of distinct atoms:

$$x = a_{i_1} \vee \cdots \vee a_{i_k}.$$

Moreover, such an expression is unique, except for the order of the atoms.

Proof. We first show that every nonzero element can be written in the form shown, with the atoms themselves being written as joins with only one term. Suppose not, and let S be the set of nonzero members of B that are not joins of atoms. If $x \in S$, then x is not itself an atom so, as in the proof of the Proposition, $x = y \vee z$ with $0 < y < x$ and $0 < z < x$. Moreover, at least one of y and z is also in S, since otherwise y and z would be joins of atoms, so that x, the join of y and z, would be too. Thus for each x in S there is some y (or z) in S with $x > y$. It follows that starting from any x in S there is a chain $x = x_0 > x_1 > x_2 > \cdots$ all of whose elements are in S. Since B is finite, the elements $x_0, x_1, x_2, \ldots$ cannot all be different; sooner or later $x_k = x_m$ with $k < m$. Then transitivity and $x_k > \cdots > x_m$ imply that $x_k > x_m$ by Lemma 2(c), contradicting $x_k = x_m$. [A little induction is hiding here.] Assuming that S is nonempty leads to a contradiction, so every nonzero x must be a join of atoms. [A recursive algorithm for finding an expression as a join of atoms can be based on this argument.]

We next show that each nonzero x in B is the join of all the atoms a for which $a \leq x$;

$$(*) \qquad\qquad x = \bigvee \{a \in A : a \leq x\}.$$

By what we just showed, the element 1 is the join of some set of atoms. It follows that

$$1 = \bigvee \{a \in A : a \leq 1\} = a_1 \vee \cdots \vee a_n,$$

since we could add more atoms to the join and still get 1. [Actually, in view of the theorem we are proving, there aren't any more atoms, but we don't know this yet.] Now

$$x = x \wedge 1 = x \wedge (a_1 \vee \cdots \vee a_n) = (x \wedge a_1) \vee \cdots \vee (x \wedge a_n).$$

Since $0 \leq x \wedge a_i \leq a_i$ and a_i is an atom, the Proposition says that $x \wedge a_i = a_i$ if $a_i \leq x$ and $x \wedge a_i = 0$ otherwise. This establishes the equality $(*)$.

To check uniqueness, suppose that $x = b_1 \vee \cdots \vee b_k$ is an expression for x as a join of atoms. Then $b_i \leq x$ for each i, so all the b_i's belong to $\{a \in A : a \leq x\}$. On the other hand, if $a \in A$ and $a \leq x$, then

$$0 \neq a = a \wedge x = a \wedge (b_1 \vee \cdots \vee b_k) = (a \wedge b_1) \vee \cdots \vee (a \wedge b_k).$$

Some $a \wedge b_i$ must be different from 0, so $a \wedge b_i = a = b_i$ since a and b_i are atoms. Thus a is one of the b_i's. Consequently, $\{b_1, \ldots, b_k\}$ is precisely the set of atoms $\leq x$. ∎

A one-to-one correspondence φ between Boolean algebras B_1 and B_2 that satisfies

$$(1) \qquad\qquad \varphi(x \vee y) = \varphi(x) \vee \varphi(y),$$

$$(2) \qquad\qquad \varphi(x \wedge y) = \varphi(x) \wedge \varphi(y)$$

and

$$(3) \qquad\qquad \varphi(x') = \varphi(x)'$$

for all $x, y \in B_1$ is called a **Boolean algebra isomorphism**. Two Boolean algebras are said to be **isomorphic** if there is an isomorphism between them. In that case their algebraic structures are essentially the same.

The next theorem tells us that a finite Boolean algebra is completely determined, up to isomorphism, by the number of atoms that it has.

Theorem 4 │ If B_1 is a finite Boolean algebra with set of atoms $A_1 = \{a_1, \ldots, a_n\}$ and if B_2 is another finite Boolean algebra with set of atoms $A_2 = \{b_1, \ldots, b_n\}$, then there is a Boolean algebra isomorphism φ of B_1 onto B_2 such that $\varphi(a_i) = b_i$ for each i.

Proof. By Theorem 3, every x in B_1 can be written uniquely in the form

$$x = a_{i_1} \vee \cdots \vee a_{i_k}.$$

We define $\varphi(a_i) = b_i$ for $i = 1, 2, \ldots, n$ and more generally

$$\varphi(a_{i_1} \vee \cdots \vee a_{i_k}) = b_{i_1} \vee \cdots \vee b_{i_k}.$$

By our definition and Theorem 3,

$$\varphi(a_{i_1} \vee \cdots \vee a_{i_k}) = \varphi(a_{i_1}) \vee \cdots \vee \varphi(a_{i_k})$$
$$= \bigvee \{\varphi(a) : a \in A_1 \text{ and } a \leq x\}.$$

But also

$$\varphi(x) = \bigvee \{b : b \in A_2 \text{ and } b \leq \varphi(x)\}.$$

Since the expression for $\varphi(x)$ is unique, we conclude that for $a \in A_1$,

$$a \leq x \quad \text{if and only if} \quad \varphi(a) \leq \varphi(x).$$

To verify (1) in the definition of isomorphism, consider x and y in B_1 and note that for $a \in A_1$:

$$\varphi(a) \leq \varphi(x \vee y) \Leftrightarrow a \leq x \vee y \qquad \text{by what we have just shown,}$$
with $x \vee y$ in place of x

$$\Leftrightarrow a \leq x \text{ or } a \leq y \qquad \text{[see Exercise 11(a)]}$$

$$\Leftrightarrow \varphi(a) \leq \varphi(x) \text{ or } \varphi(a) \leq \varphi(y).$$

That is, for $b \in A_2$:

$$b \leq \varphi(x \vee y) \Leftrightarrow b \leq \varphi(x) \text{ or } b \leq \varphi(y) \Leftrightarrow b \leq \varphi(x) \vee \varphi(y).$$

It follows from Theorem 3 applied to B_2 that $\varphi(x \vee y) = \varphi(x) \vee \varphi(y)$. Isomorphism condition (2) has a similar proof. Now

$$\varphi(x) \vee \varphi(x') = \varphi(x \vee x') = \varphi(1)$$

and

$$\varphi(x) \wedge \varphi(x') = \varphi(x \wedge x') = \varphi(0),$$

so $\varphi(x') = \varphi(x)'$ by the first argument in the proof of Theorem 2. ∎

If a set S has n elements, then the Boolean algebra $\mathscr{P}(S)$ [with operations $\cup$, $\cap$ and complementation] has exactly n atoms, namely the one-element subsets of S, so we have the following corollary.

Corollary | Every finite Boolean algebra with n atoms is isomorphic to the Boolean algebra $\mathscr{P}(S)$ of all subsets of an n-element set S, and hence has exactly 2^n elements. In particular, $\mathbb{B}^n$ is isomorphic to $\mathscr{P}(\{1, 2, \ldots, n\})$.

An **n-variable Boolean function** is a function

$$f: \mathbb{B}^n \to \mathbb{B}.$$

We write BOOL(n) for the set of all n-variable Boolean functions. If $f \in$ BOOL(n) and $(x_1, \ldots, x_n) \in \mathbb{B}^n$, we write $f(x_1, \ldots, x_n)$ for the value $f((x_1, \ldots, x_n))$.

EXAMPLE 6   A 3-variable Boolean function is an f such that $f(x, y, z) = 0$ or 1 for each of the 2^3 choices of x, y and z in $\mathbb{B} \times \mathbb{B} \times \mathbb{B}$. We could think of the input variables x, y and z as amounting to three switches, each in one of two positions. Then f behaves like a black box that produces an output of 0 or 1 depending on the settings of the switches and the internal structure of the box. Since there are 8 ways to set the switches and since each setting can lead to either of 2 outputs, depending on the function, there are $2^8 = 256$ 3-variable Boolean functions. That is, $|\text{BOOL}(3)| = 256$.

x	y	z	f
0	0	0	1
0	0	1	1
0	1	0	0
0	1	1	1
1	0	0	0
1	0	1	0
1	1	0	0
1	1	1	1

FIGURE 2

A 3-variable Boolean function f can also be viewed as a column in a truth table. For example, Figure 2 describes a unique function f, which is just one of the $2^8 = 256$ possible functions, since the column can contain any arrangement of eight 0's and 1's. Each of columns x, y and z itself describes a function in BOOL(3), i.e., a function from $\mathbb{B}^3$ to $\mathbb{B}$. For instance, column x describes the function g such that $g(a, b, c) = a$ for $(a, b, c) \in \mathbb{B}^3$. ∎

The counting argument in Example 6 works in general to give $|\text{BOOL}(n)| = 2^{(2^n)}$, a very big number unless n is very small. As in Example 1(d), BOOL(n) is a Boolean algebra with the Boolean operations defined coordinatewise.

EXAMPLE 7   Figure 3 illustrates the Boolean operations in BOOL(3) with a truth table for the indicated functions f and g. Note that, since $f \wedge g$ takes the value

x	y	z	f	g	$f \vee g$	$f \wedge g$	f'	$f' \wedge g$
0	0	0	1	0	1	0	0	0
0	0	1	1	1	1	1	0	0
0	1	0	0	0	0	0	1	0
0	1	1	1	0	1	0	0	0
1	0	0	0	1	1	0	1	1
1	0	1	0	0	0	0	1	0
1	1	0	0	1	1	0	1	1
1	1	1	1	0	1	0	0	0

FIGURE 3

1 at exactly one point in $\mathbb{B}^3$, it is an atom of BOOL(3). There are seven other atoms in BOOL(3). In the next section we will show how to write any member of a finite Boolean algebra, in particular any member of BOOL(n), as a join of atoms. ■

EXERCISES 10.1

1. (a) Verify that $\mathbb{B} = \{0, 1\}$ in Example 1(b) is a Boolean algebra by checking some of the laws 1Ba through 5Bb.

   (b) Do the same for FUN(S, $\mathbb{B}$) in Example 1(d).

2. (a) Let $S = \{a, b, c, d, e\}$ and write $\{a, c, d\}$ as a join of atoms in $\mathscr{P}(S)$.

   (b) Write $(1, 0, 1, 1, 0)$ as a join of atoms in $\mathbb{B}^5$.

   (c) Let f be the function in FUN(S, $\mathbb{B}$) that maps a, c and d to 1 and b and e to 0. Write f as a join of atoms in FUN(S, $\mathbb{B}$).

3. Find a set S so that $\mathscr{P}(S)$ and $\mathbb{B}^5$ are isomorphic Boolean algebras. Exhibit a Boolean algebra isomorphism from $\mathbb{B}^5$ to $\mathscr{P}(S)$.

4. Describe the atoms of FUN(S, $\mathbb{B}$) in Example 1(d). Is your description valid even if S is infinite?

5. (a) Give tables for the atoms of the Boolean algebra BOOL(2).

   (b) Write the function $g: \mathbb{B}^2 \to \mathbb{B}$ defined by $g(x, y) = x$ as a join of atoms in BOOL(2).

   (c) Write the function $h: \mathbb{B}^2 \to \mathbb{B}$ defined by $h(x, y) = x' \vee y$ as a join of atoms in BOOL(2).

6. (a) How many atoms are there in BOOL(4)?

   (b) Consider a function in BOOL(4) described by a column that has five 1's and the rest of its entries 0. How many atoms appear in its representation as a join of atoms?

   (c) How many different elements of BOOL(4) are joins of five atoms?

7. (a) Is there a Boolean algebra with 6 elements? Explain.

   (b) Is every finite Boolean algebra isomorphic to a Boolean algebra BOOL(n) of Boolean functions? Explain.

8. (a) Describe the atoms of the Boolean algebra $\mathscr{P}(\mathbb{N})$.

   (b) Is every nonempty member of $\mathscr{P}(\mathbb{N})$ a join of atoms? Discuss.

9. There is a natural way to draw pictures of finite Boolean algebras. For x and y in the Boolean algebra B we say that x **covers** y in case $x > y$ and there are no elements z with $x > z > y$. [Thus atoms are the elements that cover 0.] A **Hasse diagram** of B is a picture of the digraph whose vertices are the members of B and which has an edge from x to y if and only if x covers y.

   Draw Hasse diagrams of the following Boolean algebras. Draw the element 1 at the top, and direct the edges generally downward.

   (a) $\mathscr{P}(\{1, 2\})$ (b) $\mathbb{B}$ (c) $\mathbb{B}^2$ (d) $\mathbb{B}^3$

10. For an integer n greater than 1 let D_n be the set of divisors of n. Define $\vee$, $\wedge$ and $'$ on D_n by $a \vee b = \text{lcm}(a, b)$ [see the dictionary], $a \wedge b = \text{gcd}(a, b)$ and $a' = n/a$.

(a) The set $D_6 = \{1, 2, 3, 6\}$ with these operations $\vee$, $\wedge$ and $'$ is a Boolean algebra. What are its 0 and 1 elements?

(b) Find a set S so that D_6 and $\mathcal{P}(S)$ are isomorphic, and exhibit an isomorphism between them.

(c) Show that D_4 with these operations is not a Boolean algebra. [Try to think of an obvious reason.]

(d) Show that D_8 with these operations is not a Boolean algebra.

11. Let x and y be elements of a Boolean algebra, and let a be an atom.

(a) Show that $a \leq x \vee y$ if and only if $a \leq x$ or $a \leq y$.

(b) Show that $a \leq x \wedge y$ if and only if $a \leq x$ and $a \leq y$.

(c) Show that either $a \leq x$ or $a \leq x'$, but not both.

12. Let x and y be elements of a finite Boolean algebra, each written as a join of atoms:

$$x = a_1 \vee \cdots \vee a_n \quad \text{and} \quad y = b_1 \vee \cdots \vee b_m.$$

(a) Explain how to write $x \vee y$ and $x \wedge y$ as joins of distinct atoms. Illustrate with examples.

(b) How would you write x' as the join of distinct atoms?

13. Show that if φ is a Boolean algebra isomorphism between Boolean algebras B_1 and B_2, then $x \leq y$ if and only if $\varphi(x) \leq \varphi(y)$.

14. Let $S = [0, 1)$ and let $\mathscr{A}$ consist of the empty set $\varnothing$ and all subsets of S that can be written as finite unions of intervals of the form $[a, b)$.

(a) Show that each member of $\mathscr{A}$ can be written as a finite *disjoint* union of intervals of the form $[a, b)$.

(b) Show that $\mathscr{A}$ is a Boolean algebra with respect to the operations $\cup$, $\cap$ and complementation.

(c) Show that $\mathscr{A}$ has no atoms whatever.

§ 10.2   Boolean Expressions

The main purpose of this section is to introduce the mathematical terminology and ideas that are used in applying Boolean algebra methods to circuit design and logical analysis. The next section will discuss the applications themselves more fully.

A Boolean expression is a string of symbols involving the constants 0 and 1, some variables and the Boolean operations. To be more precise, we define **Boolean expressions in n variables** $x_1, x_2, \ldots, x_n$ recursively as follows:

(B)   The symbols 0, 1 and $x_1, x_2, \ldots, x_n$ are Boolean expressions in $x_1, \ldots, x_n$;

(R)   If E_1 and E_2 are Boolean expressions in $x_1, x_2, \ldots, x_n$, so are $(E_1 \vee E_2)$, $(E_1 \wedge E_2)$ and E_1'.

As usual, in practice we will normally omit the outside parentheses and will freely use the associative laws.

EXAMPLE 1   (a) Here are four Boolean expressions in the three variables x, y, z:

$$(x \vee y) \wedge (x' \vee z) \wedge 1; \quad (x' \wedge z) \vee (x' \wedge y) \vee z'; \quad x \vee y; \quad z.$$

The first two obviously involve all three variables. The last two don't. Whether we regard $x \vee y$ as an expression in two or three or more variables often doesn't matter. When it does matter and the context doesn't make the variables clear, we will be careful to say how we are viewing the expression.

The Boolean expressions 0 and 1 can be viewed as expressions in any number of variables, just as constant functions can be viewed as functions of one or of several variables.

(b) The expression

$$(x_1 \wedge x_2 \wedge \cdots \wedge x_n) \vee (x_1' \wedge x_2 \wedge \cdots \wedge x_n) \vee (x_1 \wedge x_2' \wedge x_3 \wedge \cdots \wedge x_n)$$

is an example of a Boolean expression in n variables. ∎

The usage of both symbols $\vee$ and $\wedge$ leads to bulky and awkward Boolean expressions, so we will usually replace the connective $\wedge$ by a dot or by no symbol at all.

EXAMPLE 2   (a) With this new convention for $\wedge$, the first two Boolean expressions in Example 1(a) can be written as

$$(x \vee y) \cdot (x' \vee z) \cdot 1 \quad \text{and} \quad (x'z) \vee (x'y) \vee z'$$

or, more simply, as

$$(x \vee y)(x' \vee z)1 \quad \text{and} \quad x'z \vee x'y \vee z';$$

just as in ordinary algebra, the "product" $\wedge$ or $\cdot$ takes precedence over the "sum" $\vee$.

(b) The Boolean expression in Example 1(b) is

$$x_1 x_2 \cdots x_n \vee x_1' x_2 \cdots x_n \vee x_1 x_2' x_3 \cdots x_n.$$

(c) The expression $xyz \lor xy'z \lor x'z$ is shorthand for

$$(x \land y \land z) \lor (x \land y' \land z) \lor (x' \land z). \quad \blacksquare$$

If we substitute 0 or 1 for each occurrence of each variable in a Boolean expression, we get an expression involving 0, 1, $\lor$, $\land$ and $'$ that has a meaning as a member of the Boolean algebra $\mathbb{B} = \{0, 1\}$. For example, replacing x by 0, y by 1 and z by 1 in the Boolean expression $x'z \lor x'y \lor z'$ gives

$$0'1 \lor 0'1 \lor 1' = (1 \land 1) \lor (1 \land 1) \lor 0 = 1 \lor 1 \lor 0 = 1.$$

In general, if E is a Boolean expression in the n variables $x_1, x_2, \ldots, x_n$, then E defines a **Boolean function** mapping $\mathbb{B}^n$ into $\mathbb{B}$ whose function value at $(a_1, a_2, \ldots, a_n)$ is the element of $\mathbb{B}$ obtained by replacing x_1 by a_1, x_2 by $a_2, \ldots$, and x_n by a_n in E.

EXAMPLE 3   The Boolean function mapping $\mathbb{B}^3$ into $\mathbb{B}$ that corresponds to $x'z \lor x'y \lor z'$ is given in the following table. Just as with truth tables for propositions, we first calculate the Boolean functions for some of the subexpressions. The fourth entry in the last column is the value we calculated a moment ago. Note that the Boolean expression z' corresponds to the function on $\mathbb{B}^3$ that maps each triple (a, b, c) to c' where $a, b, c \in \{0, 1\}$. Similarly, z corresponds to the function that maps each (a, b, c) to c.

x	y	z	$x'z$	$x'y$	z'	$x'z \lor x'y \lor z'$
0	0	0	0	0	1	1
0	0	1	1	0	0	1
0	1	0	0	1	1	1
0	1	1	1	1	0	1
1	0	0	0	0	1	1
1	0	1	0	0	0	0
1	1	0	0	0	1	1
1	1	1	0	0	0	0

$\blacksquare$

We will regard two Boolean expressions as **equivalent** provided that their corresponding Boolean functions are the same. For instance, $x(y \lor z)$ and $(xy) \lor (xz)$ are equivalent, since each corresponds to the function with value 1 at $(a, b, c) = (1, 1, 0), (1, 0, 1)$ or $(1, 1, 1)$ and 0 otherwise. We will write $x(y \lor z) = (xy) \lor (xz)$, and in general we will write $E = F$ if the two Boolean expressions E and F are equivalent. The usage of "$=$" to denote this equivalence relation is customary and seems to cause no confusion.

The use of notation in this way is familiar from our experience with algebraic expressions and algebraic functions on $\mathbb{R}$. Technically, the algebraic expressions $(x + 1)(x - 1)$ and $x^2 - 1$ are different [because they *look* different] but the functions f and g defined by

$$f(x) = (x + 1)(x - 1) \quad \text{and} \quad g(x) = x^2 - 1$$

are equal. We regard the two expressions as equivalent and commonly use either $(x + 1)(x - 1)$ or $x^2 - 1$ as a name for the function they define. Similarly, we will often use Boolean expressions as names for the Boolean functions they define.

EXAMPLE 4   The function in BOOL(3) named xy is defined by $xy(a, b, c) = ab$ for all (a, b, c) in $\mathbb{B}^3$, so

$$xy(a, b, c) = \begin{cases} 1 & \text{if } a = b = 1 \\ 0 & \text{otherwise.} \end{cases}$$

Similarly the functions named $x \vee z'$ and $xy'z$ satisfy

$$(x \vee z')(a, b, c) = a \vee c' = \begin{cases} 1 & \text{if } a = 1 \text{ or } c = 0 \\ 0 & \text{otherwise} \end{cases}$$

and

$$xy'z(a, b, c) = ab'c = \begin{cases} 1 & \text{if } a = 1, b = 0, c = 1 \\ 0 & \text{otherwise.} \end{cases}$$

Since $xy'z$ takes the value 1 at exactly one point in $\mathbb{B}^3$, it is an atom of the Boolean algebra BOOL(3). The other seven atoms in BOOL(3) are

$$xyz, \quad xyz', \quad xy'z', \quad x'yz, \quad x'yz', \quad x'y'z \quad \text{and} \quad x'y'z'. \quad \blacksquare$$

Suppose that E_1, E_2 and E_3 are Boolean expressions in n variables. Since BOOL(n) is a Boolean algebra, the Boolean expressions $E_1(E_2 \vee E_3)$ and $(E_1 E_2) \vee (E_1 E_3)$ define the same function. Thus the two expressions are equivalent, and we can write the distributive law

$$E_1(E_2 \vee E_3) = (E_1 E_2) \vee (E_1 E_3).$$

In the same way, Boolean expressions satisfy all the other laws of a Boolean algebra as well, as long as we are willing to write equivalences as if they were equations.

Boolean expressions consisting of a single variable or its complement, such as x or y', are called **literals**. The functions that correspond to them have the value 1 at half of the elements of $\mathbb{B}^n$. For example, the literal y' for $n = 3$ corresponds to the function with value 1 at all points $(a, 0, c)$ in $\mathbb{B}^3$ and value 0 at all points $(a, 1, c)$.

Just as in Example 7 of §10.1, the atoms of BOOL(n) are the functions that have the value 1 at exactly one member of $\mathbb{B}^n$. Each atom corresponds to a Boolean expression of a special form, called a minterm. A **minterm** in n variables is a meet [i.e., product] of exactly n literals, each involving a different variable.

EXAMPLE 5   (a) The expressions $xy'z'$ and $x'yz'$ are minterms in the three variables x, y, z. The corresponding functions in BOOL(3) have the value 1 only at $(1, 0, 0)$ and $(0, 1, 0)$, respectively.

(b) The expression xz' is a minterm in the two variables x, z. It is *not* a minterm in the three variables x, y, z; the corresponding function in BOOL(3) has value 1 at both $(1, 0, 0)$ and $(1, 1, 0)$.

(c) The expression $xyx'z$ is not a minterm since it involves the variable x in more than one literal. In fact, this expression is equivalent to 0. The expression $xy'zx$ is not a minterm either; it is equivalent to the minterm $xy'z$ in x, y, z, however.

(d) In the following table we list the eight elements of $\mathbb{B}^3$ and the corresponding minterms that take the value 1 at the indicated elements. Note that the literals corresponding to 0 entries are complemented, while the other literals are not.

(a, b, c)	Minterm with value 1 at (a, b, c)
$(0, 0, 0)$	$x'y'z'$
$(0, 0, 1)$	$x'y'z$
$(0, 1, 0)$	$x'yz'$
$(0, 1, 1)$	$x'yz$
$(1, 0, 0)$	$xy'z'$
$(1, 0, 1)$	$xy'z$
$(1, 1, 0)$	xyz'
$(1, 1, 1)$	xyz

According to Theorem 3 of §10.1, every member of BOOL(n) can be written as a join of atoms. Since atoms in BOOL(n) correspond to minterms, every Boolean expression in n variables is equivalent to a join of distinct minterms. Moreover, such a representation as a join is unique, apart from the order in which the minterms are written. We call the join of minterms equivalent to a given Boolean expression E the **minterm canonical form** of E. [Another popular term, which we will not use, is **disjunctive normal form**, or **DNF**.] Parts (b) and (c) of the next example illustrate two different procedures for finding minterm canonical forms.

EXAMPLE 6   (a) The Boolean expression

$$x'yz' \vee xy'z' \vee xy'z \vee xyz'$$

is a join of minterms in x, y, z as it stands, so this expression is its own minterm canonical form. The corresponding Boolean function has the values shown in the right-hand column of the table. The 1's in the column tell which atoms in BOOL(3) are involved, and hence determine the corresponding minterms. For instance, the 1 in the $(1, 1, 0)$ row corresponds to the minterm xyz'.

x	y	z	$x'yz'$	$xy'z'$	$xy'z$	xyz'	$x'yz' \vee xy'z' \vee xy'z \vee xyz'$
0	0	0	0	0	0	0	0
0	0	1	0	0	0	0	0
0	1	0	1	0	0	0	1
0	1	1	0	0	0	0	0
1	0	0	0	1	0	0	1
1	0	1	0	0	1	0	1
1	1	0	0	0	0	1	1
1	1	1	0	0	0	0	0

(b) The Boolean expression $(x \vee yz')(yz)'$ is not written as a join of minterms. To get its minterm canonical form we can calculate the values of the corresponding Boolean function. For instance, $x = 0$, $y = 0$, $z = 0$ gives the value

$$(0 \vee 01)(00)' = (0 \vee 0)0' = 01 = 0$$

and $x = 1$, $y = 0$, $z = 1$ gives

$$(1 \vee 00)(01)' = (1 \vee 0)0' = 11 = 1.$$

When we calculate all eight values of the function we get the right-hand column in the table in part (a). Thus $(x \vee yz')(yz)'$ is equivalent to the join of minterms in part (a), i.e., its minterm canonical form is $x'yz' \vee xy'z' \vee xy'z \vee xyz'$.

(c) We can attack $(x \vee yz')(yz)'$ directly and try to convert it into a join of minterms using Boolean algebra laws. Recall that we write $E = F$ in case the Boolean expressions E and F are equivalent. By the Boolean algebra laws,

$(x \vee yz')(yz)' = (x \vee yz')(y' \vee z')$ DeMorgan law

$ = (x(y' \vee z')) \vee ((yz')(y' \vee z'))$ distributive law

$ = (xy' \vee xz') \vee (yz'y' \vee yz'z')$ distributive law twice

$ = (xy' \vee xz') \vee (0 \vee yz')$ $yy' = 0$, $z'z' = z'$

$ = xy' \vee xz' \vee yz'$ associative law and property of 0.

We first applied the DeMorgan laws to get all complementation down to the level of the literals. Then we distributed $\vee$ across meets as far as possible.

Now we have an expression as a join of meets of literals, but not as a join of minterms in x, y, z. Consider the subexpression xy', which is missing the variable z. Since $z \vee z' = 1$ we have $xy' = xy'1 = xy'(z \vee z') = xy'z \vee xy'z'$, which is a join of minterms. We can do the same sort of thing to the other two terms and get

$$xy' \vee xz' \vee yz' = (xy'z \vee xy'z') \vee (xyz' \vee xy'z') \vee (xyz' \vee x'yz'),$$

which is a join of minterms. Deleting repetitions gives the minterm canonical form

$$xy'z \lor xy'z' \lor xyz' \lor x'yz'$$

for the expression $(x \lor yz')(yz)'$ we started with. This is of course the same as the answer obtained in part (b). ∎

The methods illustrated in this example work in general. Given a Boolean expression, we can calculate the values of the Boolean function it defines—in effect, find its truth table. Then each value of 1 corresponds to a minterm in the canonical form of the expression. This is the method of Example 6(b). From this point of view the minterm canonical form is just another way of looking at the Boolean function.

Alternatively, we can obtain the minterm canonical form as in Example 6(c). First use the DeMorgan laws to move all complementation to the literals. Then distribute $\lor$ over products wherever possible. Then replace xx by x and xx' by 0 as necessary and insert missing variables using $x \lor x' = 1$. Finally, eliminate duplicates.

It is not always clear which technique is preferable for a given Boolean expression. One would not want to do a lot of calculations by hand using either method. Fortunately, the minterm canonical form is primarily useful as a theoretical tool, and when calculations do arise in practice they can be performed by machine using simple algorithms.

From a theoretical point of view, the minterm canonical form of a Boolean expression is very valuable, since it gives the expression in terms of its basic parts, namely minterms or atoms. As we will illustrate in §10.3, Boolean expressions can be realized as electronic circuits, and equivalent Boolean expressions correspond to electronic circuits that perform identically, i.e., that give the same outputs for given inputs. Hence it is of interest to "simplify" Boolean expressions, to get corresponding "simplified" electronic circuits.

There are various ways to measure the simplification. It would be impossible to describe here all methods that have practical importance, but we can at least discuss one simple criterion. Let's say that a join of products [i.e., meets] of literals is **optimal** if there is no equivalent Boolean expression that is a join of fewer products and if, among all equivalent joins of the same number of products, there are none with fewer literals. Our task is to find an optimal join of products equivalent to a given Boolean expression. We can suppose that we have already found *one* equivalent join of products, namely the minterm canonical form.

EXAMPLE 7   (a) Consider the expression $(xy)'z$. The table shows the values of the Boolean function it defines. The minterm canonical form is thus $x'y'z \lor x'yz \lor xy'z$. This expression is not optimal. By the Boolean algebra laws,

$(xy)'z = (x' \vee y')z = x'z \vee y'z$, which is a join of only two terms with four literals. We will be able to show in Example 2(d) of §10.4 that $x'z \vee y'z$ is optimal [or see Exercise 13].

x	y	z	xy	$(xy)'$	$(xy)'z$
0	0	0	0	1	0
0	0	1	0	1	1
0	1	0	0	1	0
0	1	1	0	1	1
1	0	0	0	1	0
1	0	1	0	1	1
1	1	0	1	0	0
1	1	1	1	0	0

This example illustrates a problem that can arise in practice. It seems plausible that a circuit to produce $x'z \vee y'z$ might be simpler than one to produce $x'y'z \vee x'yz \vee xy'z$, but perhaps a circuit to produce the original expression $(xy)'z$ would be simplest of all. We return to this point in §10.3.

(b) Consider the join of products $E = x'z' \vee x'y \vee xy' \vee xz$. Is it optimal? We use Boolean algebra calculations, including the $x \vee x' = 1$ trick, to find its minterm canonical form:

$$E = x'yz' \vee x'y'z' \vee x'yz \vee x'yz' \vee xy'z \vee xy'z' \vee xyz \vee xy'z$$

$$= x'yz' \vee x'y'z' \vee x'yz \vee xy'z \vee xy'z' \vee xyz.$$

This has just made matters worse—more products and more literals. We want to repackage the expression in some clever way. Observe that we can group the six minterms together in pairs $x'yz'$ and $x'y'z'$, $x'yz$ and xyz, $xy'z$ and $xy'z'$, so that two minterms in the same pair differ in exactly one literal. Since

$$x'yz' \vee x'y'z' = x'(y \vee y')z' = x'z',$$

$$x'yz \vee xyz = yz \quad \text{and} \quad xy'z \vee xy'z' = xy',$$

we have $E = x'z' \vee yz \vee xy'$. A different grouping gives

$$x'yz' \vee x'yz = x'y, \quad x'y'z' \vee xy'z' = y'z' \quad \text{and} \quad xy'z \vee xyz = xz,$$

so that $E = x'y \vee y'z' \vee xz$. Each of these joins of products $x'z' \vee yz \vee xy'$ and $x'y \vee y'z' \vee xz$ will be shown to be optimal in Example 2(c) of §10.4. Thus no join of products that is equivalent to E has fewer than three products, and no join with three products has fewer than six literals. Whether or not we believe these claims now, the two expressions look simpler than the join of four products that we started with. ■

There is a method, called the **Quine–McCluskey procedure**, that builds optimal expressions by systematically grouping together products which differ in only one literal. The algorithm is tedious to use by hand but is readily programmed for computer calculation. Among other references, the textbooks *Applications-Oriented Algebra* by J. L. Fisher and *Modern Applied Algebra* by G. Birkhoff and T. C. Bartee contain readable accounts of the method.

Another procedure for finding optimal expressions, called the method of **Karnaugh maps**, has a resemblance to Venn diagrams. The method works pretty well for Boolean expressions in three or four variables, where the problems are fairly simple anyway, but is less useful for more than four variables. The textbook *Computer Hardware and Organization* by M. E. Sloan devotes several sections to Karnaugh maps and discusses their advantages and disadvantages in applications. We will illustrate the method in § 10.4, after we have described the elements of logical circuitry.

EXERCISES 10.2

1. Let $f: \mathbb{B}^3 \to \mathbb{B}$ be the Boolean function such that $f(0, 0, 0) = f(0, 0, 1) = f(1, 1, 0) = 1$ and $f(a, b, c) = 0$ for all other $(a, b, c) \in \mathbb{B}^3$. Write the corresponding Boolean expression in minterm canonical form.

2. Give the Boolean function corresponding to the Boolean expression in Example 7(b).

3. For each of the following Boolean expressions in x, y, z describe the corresponding Boolean function and write the minterm canonical form.

   (a) xy (b) z' (c) $xy \vee z'$ (d) 1

4. Consider the Boolean expression $x \vee yz$ in x, y, z.

   (a) Give a table for the corresponding Boolean function $f: \mathbb{B}^3 \to \mathbb{B}$.

   (b) Write the expression in minterm canonical form.

5. Find the minterm canonical form for the four-variable Boolean expressions

   (a) $(x_1 x_2 x_3') \vee (x_1' x_2 x_3 x_4')$ (b) $(x_1 \vee x_2) x_3' x_4$

6. Use the method of Example 6(c) to find the minterm canonical form of the 3-variable Boolean expression $((x \vee y)' \vee z)'$.

7. (a) Find a join of products involving a total of three literals that is equivalent to the expression

$$xz \vee [y' \vee y'z] \vee xy'z'.$$

   (b) Repeat part (a) for $[(xy \vee xyz) \vee xz] \vee z$.

8. The Boolean function $f: \mathbb{B}^3 \to \mathbb{B}$ is given by $f(a, b, c) = a +_2 b +_2 c$ for $(a, b, c) \in \mathbb{B}^3$. Recall that $+_2$ refers to addition modulo 2, which is defined in §3.6.

(a) Determine a Boolean expression corresponding to f.

(b) Write the expression in minterm canonical form with variables x, y, z.

9. Find an optimal expression equivalent to

$$(x \vee y)' \vee z \vee x(yz \vee y'z').$$

10. Group the three minterms in $xyz \vee xyz' \vee xy'z$ in two pairs to obtain an equivalent expression as a join of two products with two literals each.

11. There is a notion of maxterm dual to the notion of minterm. A **maxterm** in $x_1, \ldots, x_n$ is a join of n literals, each involving a different one of $x_1, \ldots, x_n$.

(a) Use the DeMorgan laws to show that every Boolean expression in variables $x_1, \ldots, x_n$ is equivalent to a product of maxterms.

(b) Write $xy' \vee x'y$ as a product of maxterms in x and y.

12. Consider a product E of k literals chosen from among $x_1, x_1', \ldots, x_n, x_n'$ and involving k different variables x_i. Show that E determines a function in BOOL(n) that takes the value 1 on a subset of $\mathbb{B}^n$ having 2^{n-k} elements. *Hint:* Get a minterm canonical expression for E by using the $x \vee x'$ trick of Example 6(c) with the variables $w_1, \ldots, w_{n-k}$ that are not involved in E.

13. (a) Show that $x'z \vee y'z$ is not equivalent to a product of literals. *Hint:* Use Exercise 12.

(b) Show that $x'z \vee y'z$ is not equivalent to a join of products of literals in which one "product" is a single literal. [Parts (a) and (b) together show that $x'z \vee y'z$ is optimal.]

14. Prove that if E_1 and E_2 are Boolean expressions in $x_1, \ldots, x_n$, then $E_1 \vee E_2$ and $E_2 \vee E_1$ are equivalent.

§ 10.3 Logic Networks

Computer science at the hardware level involves designing devices to produce appropriate outputs from given inputs. For inputs and outputs that are 0's and 1's, this becomes a problem of designing circuitry to transform input data according to the rules for Boolean functions. In this section we will look briefly at ways in which Boolean algebra methods can be applied to logical design. Some of the methods in this section and the next also have applications to software logic for parallel processors.

The basic building blocks of our logic networks are small units, called **gates**, that correspond to simple Boolean functions. Hardware versions of these units are available from manufacturers, packaged in a wide variety of configurations. Figure 1 shows the standard ANSI/IEEE symbols for the six most elementary gates. We use the convention that the lines entering the symbol from the left are input lines, and the line on the right is the output line. Placing a small circle on an input or output line complements the signal on that line. The table shows the Boolean function values associated with these six gates and gives the

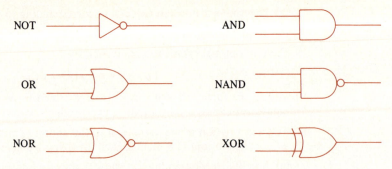

FIGURE 1

corresponding Boolean function names for inputs x and y. AND, OR, NAND and NOR gates are also available with more than two input lines.

x	y	x' NOT	$x \vee y$ OR	$(x \vee y)'$ NOR	xy AND	$(xy)'$ NAND	$x \oplus y$ XOR
0	0	1	0	1	0	1	0
0	1	1	1	0	0	1	1
1	0	0	1	0	0	1	1
1	1	0	1	0	1	0	0

EXAMPLE 1   (a) The gate shown in Figure 2(a) corresponds to the Boolean function $(x \vee y')'$, or equivalently $x'y$.

(b) The 3-input AND gate in Figure 2(b) goes with the function $x'yz$.

(c) The gate in Figure 2(c) gives $(x'y')'$ or $x \vee y$, so it acts like an OR gate. ∎

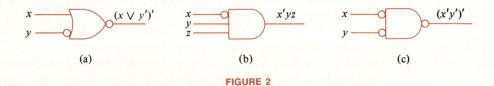

(a)   (b)   (c)

FIGURE 2

We consider the problem of designing a network of gates to produce a given complicated Boolean function of several variables. One major consideration is to keep the number of gates small. Another is to keep the length of the longest chain of gates small. Still other criteria arise in concrete practical applications.

EXAMPLE 2   Consider the foolishly designed network shown in Figure 3(a). [Dots indicate points where input lines divide.] There are four gates in the

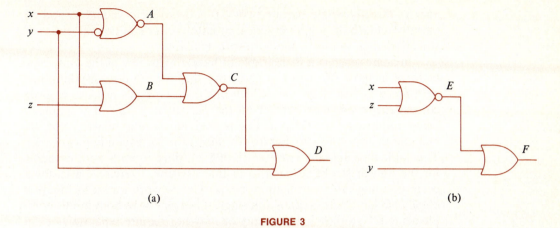

(a)   (b)

FIGURE 3

network. Reading from left to right there are two chains that are three gates long. We calculate the Boolean functions at A, B, C and D:

$$A = (x \vee y')'; \qquad B = x \vee z;$$

$$C = (A \vee B)' = ((x \vee y')' \vee (x \vee z))';$$

$$D = C \vee y = ((x \vee y')' \vee (x \vee z))' \vee y.$$

Boolean algebra laws give

$$D = ((x \vee y')(x \vee z)') \vee y = ((x \vee y')x'z') \vee y$$

$$= (xx'z' \vee y'x'z') \vee y = y'x'z' \vee y$$

$$= y'x'z' \vee yx'z' \vee y = (y' \vee y)x'z' \vee y = x'z' \vee y.$$

The network shown in Figure 3(b) produces the same output, since

$$E = (x \vee z)' = x'z' \quad \text{and} \quad F = E \vee y = x'z' \vee y. \quad \blacksquare$$

This simple example shows how it is sometimes possible to redesign a complicated network into one that uses fewer gates. One reason for trying to reduce the lengths of chains of gates is that in many situations, including programmed simulations of hard-wired circuits, the operation of each gate takes a fixed basic unit of time, and the gates in a chain must operate one after the other. Long chains mean slow operation.

The expression $x'z' \vee y$ that we obtained for the complicated expression D in the last example happens to be an optimal expression for D in the sense of §10.2. Optimal expressions do not always give the simplest networks. For example, one can show [Exercise 7(a) of §10.4] that $xz \vee yz$ is an optimal expression in x, y, z. Now $xz \vee yz = (x \vee y)z$, which can be implemented with an OR gate and an AND gate, whereas to implement

$xz \vee yz$ directly would require two AND gates to form xz and yz and an OR gate to finish the job. In practical situations our definition of "optimal" should change to match the hardware available.

In some settings it is desirable to have all gates be of the same type or be of at most two types. It turns out that we can do everything just with NAND or just with NOR. Which of these two types of gates is more convenient to use may depend on the particular transistor technology being employed. The table in Figure 4(a) shows how to write NOT, OR and AND in terms of NAND. Figure 4(b) shows the corresponding networks. This table also answers Exercise 18 of §2.4, since NAND is another name for the Sheffer stroke referred to in that exercise. Exercise 2 asks for a corresponding table and figure for NOR. The network for OR in Figure 4 could also have been written as a single NAND gate with both inputs complemented. Complementation may or may not require separate gates in a particular application, depending on the technology involved and the source of the inputs. In most of our discussion we proceed as if complementation can be done at no cost.

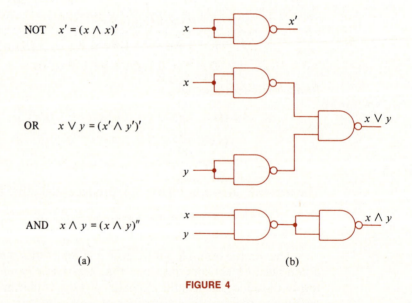

NOT $x' = (x \wedge x)'$

OR $x \vee y = (x' \wedge y')'$

AND $x \wedge y = (x \wedge y)''$

(a)   (b)

FIGURE 4

Combinations of AND and OR such as those that arise in joins of products can easily be done entirely with NAND's.

EXAMPLE 3   Figure 5 shows a simple illustration. Just replace all AND's and OR's by NAND's in an AND-OR 2-stage network to get an equivalent network. An OR-AND 2-stage network can be replaced by a NOR network in a similar way [Exercise 4]. ■

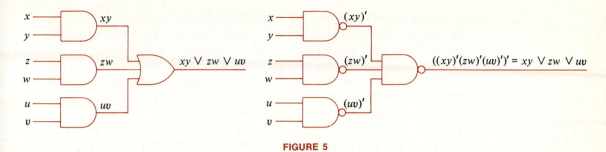

FIGURE 5

Logic networks can be viewed as acyclic digraphs with the sources labeled by variables $x_1, x_2, \ldots$, the other vertices labeled with $\vee$, $\wedge$ and $\oplus$, and some edges labeled $\neg$ for complementation. Each vertex then has an associated Boolean expression in the variables that label the sources.

EXAMPLE 4 The network of Figure 6(a) yields the digraph of Figure 6(b), with all edges directed from left to right. If we insert a 1-input $\wedge$-vertex in the middle of the edge from z to $(x \wedge y)' \vee z \vee (x \wedge z' \wedge w)$ we don't change the logic, and we get a digraph in which the vertices appear in columns—first a variable column, then an $\wedge$ column, then an $\vee$ column—and in which edges go only from one column to the next.

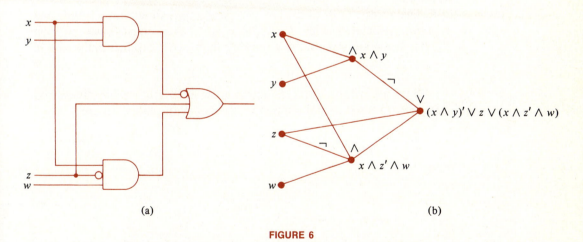

(a)   (b)

FIGURE 6

The labeled digraph in Example 4 describes a computation of the Boolean function $(x \wedge y)' \vee z \vee (x \wedge z' \wedge w)$. In a similar way, any such labeled digraph describes computations for the Boolean functions that are associated with its sinks. Since every Boolean expression can be written as a join of products of literals, every Boolean function has a computation that can be described by a digraph like the one in Example

4, with a variable column, an $\wedge$ column and an $\vee$ column [consisting of a single vertex]. Indeed, as we saw in Example 3, all corresponding gates can be made NAND gates, so the $\vee$ vertex can be made an $\wedge$ vertex.

In a digraph of this sort no path has length greater than 2. The interpretation is that the associated computation takes just 2 units of time. The price we pay may be an enormous number of gates.

EXAMPLE 5   Consider the Boolean expression $E = x_1 \oplus x_2 \oplus \cdots \oplus x_n$. The corresponding Boolean function on $\mathbb{B}^n$ takes the value 1 at $(a_1, a_2, \ldots, a_n)$ if and only if an odd number of the entries $a_1, a_2, \ldots, a_n$ are 1. The corresponding minterms are the ones with an odd number of uncomplemented literals. Hence the minterm canonical form for E uses half of all the possible minterms and is a join of 2^{n-1} terms.

We next show that the minterm canonical form for this E is optimal, i.e., whenever E is written as a join of products of literals the products that appear must be the minterms mentioned in the last paragraph. Otherwise, some term would be a product of fewer than n literals, say with x_k and x_k' both missing. Some choice of values of $a_1, a_2, \ldots, a_n$ makes this term have value 1, and an odd number of such values $a_1, a_2, \ldots, a_n$ must be 1. If we change a_k from 0 to 1 or from 1 to 0, the term will still have value 1 but an even number of the values $a_1, a_2, \ldots, a_n$ will be 1. No term of E can have this property, so each term for E must involve all n variables.

The observations of the last two paragraphs show that a length two digraph associated with $E = x_1 \oplus x_2 \oplus \cdots \oplus x_n$ must have at least $2^{n-1} + 1$ $\wedge$ and $\vee$ vertices, a number that grows exponentially with n. ∎

If we are willing to let the paths grow in length, we can divide and conquer to keep the total number of vertices manageable. Figure 7 shows

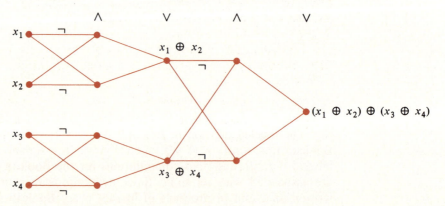

FIGURE 7

the idea for $x_1 \oplus x_2 \oplus x_3 \oplus x_4$. This digraph has 9 $\wedge$ and $\vee$ vertices. So does the digraph associated with the join-of-products computation of $x_1 \oplus x_2 \oplus x_3 \oplus x_4$, since $2^3 + 1 = 9$; we have made no improvement. But how about $x_1 \oplus x_2 \oplus \cdots \oplus x_8$? The join-of-products digraph has $2^7 + 1 = 129$ $\wedge$ and $\vee$ vertices, while the analog of the Figure 7 digraph only has $9 + 9 + 2 + 1 = 21$ $\wedge$ and $\vee$ vertices [Exercise 11].

For $x_1 \oplus x_2 \oplus \cdots \oplus x_n$ in general, the comparison is $2^{n-1} + 1$ gates for the 2-stage computation versus only $3(n-1)$ gates for the divide-and-conquer scheme, while the maximum path length increases from 2 to at most $2 \log_2 n$ [Exercise 13]. Thus doubling the number of inputs increases path length by at most 2.

EXAMPLE 6 Circuits to perform operations in binary arithmetic, for example to add two integers, are an important class of logic networks. We illustrate some of the methods that arise by adding the two integers 25 and 13, written in binary form as 11001 and 1101, respectively. This representation just means that

$$25 = 16 + 8 + 1 = 1 \cdot 2^4 + 1 \cdot 2^3 + 0 \cdot 2^2 + 0 \cdot 2 + 1 \cdot 1 \quad \text{and}$$

$$13 = 8 + 4 + 1 = 1 \cdot 2^3 + 1 \cdot 2^2 + 0 \cdot 2 + 1 \cdot 1,$$

so our problem looks like

$$\begin{array}{r} 11001 \\ + \ 1101 \\ \hline ? \end{array}$$

Binary addition is similar to ordinary decimal addition. Working from right to left, we can add the digits in each column. If the sum is 0 or 1 we write the sum in the answer line and carry a digit 0 one column to the left. If the sum is 2 or 3 [i.e., 10 or 11 in binary], we write 0 or 1, respectively, and go to the next column with a carry digit 1. Figure 8 gives the details for our illustration, with the top row inserted to show the carry digits. The answer represents $1 \cdot 2^5 + 0 \cdot 2^4 + 0 \cdot 2^3 + 1 \cdot 2^2 + 1 \cdot 2 + 0 \cdot 1 = 38$, as it should.

The rightmost column contains only two digits x and y [in our illustration $x = y = 1$]. The answer digit in this column is $(x + y)$ MOD 2,

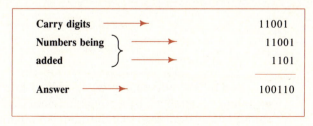

FIGURE 8

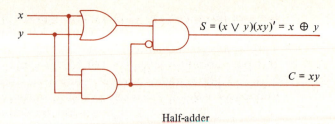

$$S = (x \vee y)(xy)' = x \oplus y$$

$$C = xy$$

Half-adder

FIGURE 9

i.e., $x \oplus y$, and the carry digit for the next column is $(x + y)\,\mathrm{DIV}\,2$, which is xy. The simple logic network shown in Figure 9, called a **half-adder**, produces the two outputs $S = x \oplus y$ and $C = xy$ from inputs x and y. Here S signifies "sum" and C signifies "carry."

For the more general case with a carry input, C_I, as well as a carry output, C_O, we can combine two half-adders and an OR gate to get the network of Figure 10, called a **full-adder**. Here C_O is 1 if and only if at least two of x, y and C_I are 1, i.e., if and only if either x and y are both 1 or exactly one of them is 1 and C_I is also 1.

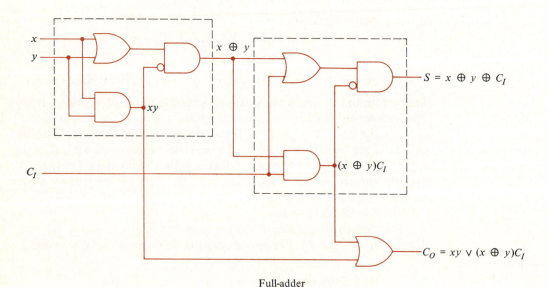

$$x \oplus y$$

$$xy$$

$$S = x \oplus y \oplus C_I$$

$$(x \oplus y)C_I$$

$$C_O = xy \vee (x \oplus y)C_I$$

Full-adder

FIGURE 10

Several full-adders can be combined into a network for adding n-digit binary numbers, or a single full-adder can be used repeatedly with suitable delay devices to feed the input data bits in sequentially. In practice, each of these two schemes is slower than necessary. Fancy networks have been designed to add more rapidly and to perform other arithmetic operations. ■

Our purpose in including Example 6 was to illustrate the use of logic networks in hardware design and also to suggest how partial results from parallel processes can be combined. The full-adder shows how networks to implement two or more Boolean functions can be blended together. The minterm canonical form of $S = x \oplus y \oplus C_I$ is

$$xyC_I \vee xy'C_I' \vee x'yC_I' \vee x'y'C_I$$

which turns out to be optimal [Example 5, or Exercise 7(c) of § 10.4]. It can be implemented with a logic network using four AND gates and one OR gate if we allow four input lines. The optimal join-of-products expression for C_O is $xy \vee xC_I \vee yC_I$, which can be produced with three AND gates and one OR gate. To produce S and C_O separately would appear to require $4 + 1 + 3 + 1 = 9$ gates, yet Figure 10 shows that we can get by with 7 gates if we want both S and C_O at once. Moreover, each gate in Figure 10 has only two input lines. As this discussion suggests, the design of economical logic networks is not an easy problem.

<h3 style="text-align:center">EXERCISES 10.3</h3>

Note. In these exercises, inputs may be complemented unless otherwise specified.

1. (a) Describe the Boolean function that corresponds to the logic network shown in Figure 11.

   (b) Sketch an equivalent network consisting of two 2-input gates.

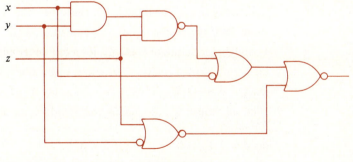

FIGURE 11

2. Write logical equations and sketch networks as in Figure 4 that show how to express NOT, OR and AND in terms of NOR without complementation of inputs.

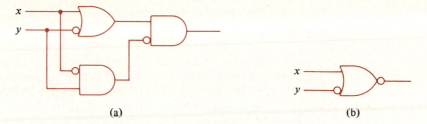

(a)   (b)

FIGURE 12

3. Sketch logic networks equivalent to those in Figure 12, but composed entirely of NAND gates.

4. Sketch logic networks equivalent to those in Figure 13, but composed entirely of NOR gates.

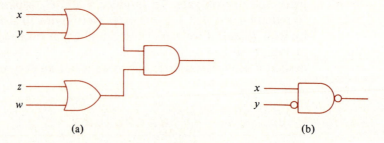

(a)   (b)

FIGURE 13

5. Sketch a logic network for the function XOR using

   (a) two AND gates and one OR gate.

   (b) two OR gates and one AND gate.

6. Sketch a logic network that has output 1 if and only if

   (a) exactly one of the inputs x, y, z has the value 1,

   (b) at least two of the inputs x, y, z, w have value 1.

7. Calculate the values of S and C_O for a full-adder with the given input values.

   (a) $x = 1, y = 0, C_I = 0$ (b) $x = 1, y = 1, C_I = 0$

   (c) $x = 0, y = 1, C_I = 1$ (d) $x = 1, y = 1, C_I = 1$

8. Find all values of x, y and C_I that produce the following outputs from a full-adder.

   (a) $S = 0, C_O = 0$ (b) $S = 0, C_O = 1$ (c) $S = 1, C_O = 1$

9. Consider the "triangle" and "circle" gates whose outputs are as shown in Figure 14. Show how to make a logic network from these two types of gates without complementation on input or output lines to produce the Boolean function.

   (a) x' (b) xy (c) $x \vee y$

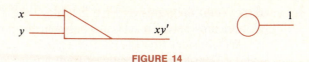

FIGURE 14

10. AND-OR-INVERT gates that produce the same effect as the logic network shown in Figure 15 are available commercially. What inputs should be used to make such a gate into an XOR gate?

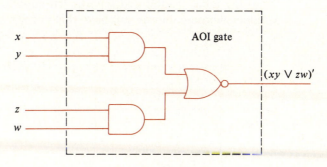

FIGURE 15

11. Draw a digraph like the one in Figure 7 for a divide-and-conquer computation of $x_1 \oplus x_2 \oplus \cdots \oplus x_8$.

12. (a) Draw the digraph for the 2-stage join-of-products computation of $x_1 \oplus x_2 \oplus x_3 \oplus x_4$.

    (b) How many $\wedge$ vertices are there in the digraph of the join-of-products computation of $x_1 \oplus x_2 \oplus \cdots \oplus x_8$?

    (c) Would you like to draw the digraph in part (b)?

13. Show by induction that for $n \geq 2$ there is a digraph for the computation of $x_1 \oplus x_2 \oplus \cdots \oplus x_n$ that has $3(n-1)$ $\wedge$ and $\vee$ vertices and is such that if $2^m \geq n$ then every path has length at most $2m$. *Suggestion:* Consider k with $2^{k-1} < n \leq 2^k$ and combine digraphs for 2^{k-1} and $n - 2^{k-1}$ variables.

14. Draw a digraph for the computation of $x_1 \oplus x_2 \oplus \cdots \oplus x_6$ with 15 $\wedge$ and $\vee$ vertices and all paths of length at most 6. *Suggestion:* See Exercise 13.

§ 10.4   Karnaugh Maps

Instead of trying to find the most economical or "best" logic network possible, we may decide to settle for a solution that just seems reasonably good. Optimal solutions in the sense of § 10.2 can be considered to be approximately best, so a technique for finding optimal solutions is worth having. The method of **Karnaugh maps**, which we now discuss briefly, is such a scheme. We can think of it as a sort of Boolean algebra mixture of the Venn diagrams and truth tables that we used earlier to visualize relationships between sets and between propositions.

We consider first the case of a three-variable Boolean function in x, y and z. The Karnaugh map of such a function is a 2×4 table, such as the ones in Figure 1. Each of the eight squares in the table corresponds to a minterm. The plus marks indicate which minterms are involved in the function described by the table. The columns of a Karnaugh map are arranged so that neighboring columns differ in just one literal. If we wrap the table around and sew the left edge to the right edge, we get a cylinder whose columns still have this property.

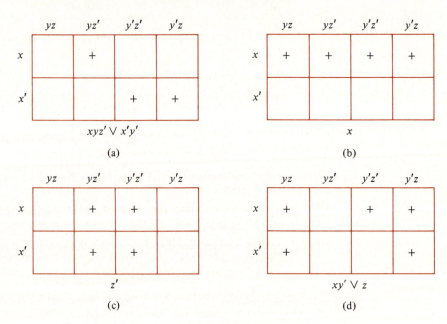

FIGURE 1

EXAMPLE 1   (a) In Figure 1(a) the minterm canonical form is $xyz' \vee x'y'z' \vee x'y'z$. Since $x'y'z' \vee x'y'z = x'y'(z' \vee z) = x'y'$, the function can also be written as $xyz' \vee x'y'$.

(b) The Karnaugh maps for literals are particularly simple. The map for x in Figure 1(b) has the whole first row marked; x' has the whole second row marked. The map for y has the left 2×2 block marked and the one for y' has the right 2×2 block marked. The map for z' in Figure 1(c) has just the entries in the middle 2×2 block marked. If we sew the left edge to the right edge, the columns involving z also form a 2×2 block.

(c) The map in Figure 1(d) has all entries involving z marked, and also the minterm $xy'z'$. Since both xy' boxes are marked, the function can be written as $xy' \vee z$.   ■

We now have a cylindrical map on which the literals x and x' correspond to 1×4 blocks, the literals y, z, y' and z' correspond to 2×2 blocks, products of two literals correspond to 1×2 or 2×1 blocks and products of three literals correspond to 1×1 blocks.

To find an optimal expression for a given Boolean function in x, y and z, we outline blocks corresponding to products by performing the following steps.

Step 1.   Mark the squares on the Karnaugh map corresponding to the function.

Step 2.   (a)   Outline each marked block with 8 squares. [If all 8 boxes are marked, the Boolean function is 1. Yawn.]
   (b)   Outline each marked block with 4 squares that is not contained in a larger outlined block.
   (c)   Outline each marked block with 2 squares that is not contained in a larger outlined block.
   (d)   Outline each marked square that is not contained in a larger outlined block.

Step 3.   Select a set of outlined blocks that
   (a)   has every marked square in at least one selected block,
   (b)   has as few blocks as possible, and
   (c)   among all sets satisfying (b) gives an expression with as few literals as possible.

We will say more about how to satisfy (b) and (c) in Step 3, but first we consider some examples.

EXAMPLE 2   (a)   Consider the Boolean function with Karnaugh map in Figure 2(a). The "rounded rectangles" outline three blocks, one with four squares, corresponding to y, and two with two squares, corresponding to xz' and $x'z$. The $x'z$ block is made from squares on the two sides of the seam where we sewed the left and right edges together. Since it takes all three outlined blocks to cover all marked squares, we must use all three blocks in Step 3. The resulting optimal expression is $y \vee xz' \vee x'z$.

(b)   The Boolean function $(x'y'z)'$ is mapped in Figure 2(b). Here the outlined blocks go with x, y and z'. Again, it takes all three to cover the marked squares, so the optimal expression is $x \vee y \vee z'$.

(c)   The Karnaugh map in Figure 2(c) has six outlined blocks, each with two squares. The marked squares can be covered with either of two

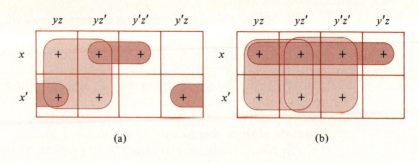

(a)   (b)

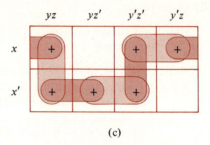

(c)

FIGURE 2

sets of three blocks, corresponding to

$$x'y \vee xz \vee y'z' \quad \text{and} \quad x'z' \vee yz \vee xy'.$$

Since no fewer than three of the blocks can cover six squares, both of these expressions are optimal. We saw this Boolean function in Example 7(b) of § 10.2.

(d) Draw the Karnaugh map for the Boolean function $x'z \vee y'z$ and outline blocks following Steps 1 to 3. Only the $x'z$ and $y'z$ blocks will be marked. Both will be needed to cover the marked squares, so $x'z \vee y'z$ is its own optimal Boolean expression. ∎

Example 2(c) shows a situation in which more than one choice is possible. To illustrate the problems that choices may cause in selecting the blocks in Step 3, we increase the number of variables to four, say w, x, y, z. Now the map is a 4×4 table, such as the ones in Figure 3, and we can think of sewing the top and bottom edges together to form a tube and then the left and right edges together to form a doughnut-shaped surface. The three-step procedure is the same as before, except that in Step 2 we start by looking for blocks with 16 squares.

EXAMPLE 3   (a) The map in Figure 3(a) has four outlined blocks, three with four squares corresponding to wy, yz' and $w'z'$, and one with two squares cor-

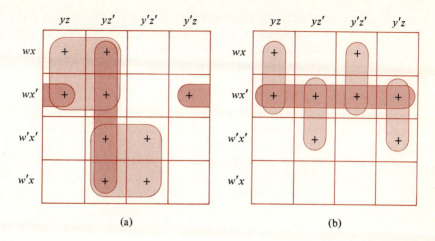

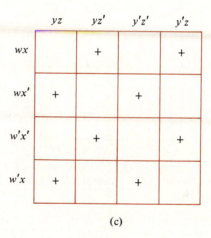

responding to $wx'z$. The two-square block is the only one containing the marked $wx'y'z$ square, and the blocks for wy and $w'z'$ are the only ones containing the squares for $wxyz$ and $w'x'y'z'$, respectively, so these three blocks must be used. Since they cover all the marked squares, they meet the conditions of Step 3. The optimal expression is $wx'z \vee wy \vee w'z'$.

(b) The map in Figure 3(b) has five blocks. Each two-square block is essential, since each is the only block containing one of the marked squares. The big four-square wx' block is superfluous, since its squares are already covered by the other blocks. The optimal expression is $wyz \vee x'yz' \vee wy'z' \vee x'y'z$. Greed does not pay here, since any algorithm that selected the biggest blocks first would pick up the superfluous four-square block.

(c) The checkerboard pattern in Figure 3(c) describes the Boolean function $w \oplus x \oplus y \oplus z$ in w, x, y, z. In this case all eight blocks are 1×1 and the optimal expression is just the minterm canonical form. A similar conclusion holds for $x_1 \oplus x_2 \oplus \cdots \oplus x_n$ in general, as noted in Example 5 of §10.3. ■

EXAMPLE 4   The maps in Example 3 offered no real choices; the essential blocks already covered all marked squares. The map of Figure 4(a) offers the opposite extreme. Every marked square is in at least two blocks. Clearly, we must use at least one two-square block to cover $wx'yz'$. Suppose that we choose the $wx'z'$ block. We can finish the job by choosing the four additional blocks shown in Figure 4(b). The resulting expression is

$$wx'z' \vee wy' \vee w'y \vee w'z \vee w'x.$$

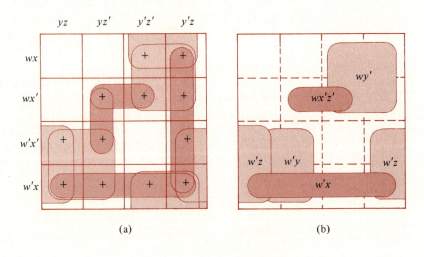

(a)   (b)

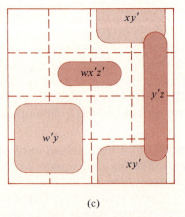

(c)

FIGURE 4

Figure 4(c) shows another choice of blocks, this time with only four blocks altogether. The corresponding expression is

$$wx'z' \vee w'y \vee xy' \vee y'z.$$

This expression is better, but is it optimal? The only possible improvement would be to reduce to one two-square and two four-square blocks. Since there are twelve squares to cover, no such improvement is possible, and so the expression is optimal. ■

The rules for deciding which blocks to choose in situations like this are fairly complicated. It is not enough simply to choose the essential blocks because we are forced to, and then cover the remaining squares with the largest blocks possible. Such a procedure could lead to the non-optimal solution of Figure 4(b).

For hand calculations, the tried-and-true method is to stare at the picture until the answer becomes clear. For machine calculations, which are necessary for more than five variables in any case, the Karnaugh map procedure is logically the same as the Quine–McCluskey method, for which software exists.

We close this discussion by emphasizing once again that the technical term "optimal" only refers to the complexity of a particular type of expression, as a join of products of literals. It is not synonymous with "best." An optimal expression for a Boolean function gives a way to construct a corresponding two-stage AND–OR network with as few gates as possible, but other kinds of networks may be cheaper to build.

EXERCISES 10.4

For each of the Karnaugh maps in Exercises 1 through 4, give the corresponding minterm canonical form and an optimal expression.

1.

	yz	yz'	$y'z'$	$y'z$
x	+	+	+	+
x'	+			+

2.

	yz	yz'	$y'z'$	$y'z$
x	+	+		+
x'	+		+	+

3.

	yz	yz'	$y'z'$	$y'z$
x	+	+		+
x'			+	+

4.

	yz	yz'	$y'z'$	$y'z$
x	+			+
x'			+	

5. Draw the Karnaugh maps and outline the blocks for the given Boolean functions of x, y and z using the three-step procedure.

   (a) $x \vee x'yz$

   (b) $(x \vee yz)'$

   (c) $y'z \vee xyz$

   (d) $y \vee z$

6. Suppose that the Boolean functions E and F each have Karnaugh maps consisting of a single block, and suppose that the block for E contains the block for F.

   (a) How are the optimal expressions for E and F related?

   (b) Give examples of E and F related in this way.

7. Draw the Karnaugh map of each of the following Boolean expressions in x, y, z and show that the expression is optimal.

   (a) $xz \vee yz$

   (b) $xy \vee xz \vee yz$

   (c) $xyz \vee xy'z' \vee x'yz' \vee x'y'z$

8. Repeat Exercise 7 for the following expressions in x, y, z, w.

   (a) $x' \vee yzw$

   (b) $x'z' \vee xy'z \vee w'xy$

   (c) $wxz \vee wx'z' \vee w'x'z \vee w'xz'$

9. Find optimal expressions for the Boolean functions with these Karnaugh maps.

(a)

	yz	yz'	y'z'	y'z
wx	+	+	+	
wx'		+	+	+
w'x'	+	+	+	+
w'x	+	+	+	

(b)

	yz	yz'	y'z'	y'z
wx	+			
wx'				
w'x'	+	+	+	+
w'x		+	+	+

(c)

	yz	yz'	y'z'	y'z
wx	+	+	+	
wx'	+		+	+
w'x'		+	+	
w'x			+	+

(d)

	yz	yz'	y'z'	y'z
wx	+			+
wx'	+			+
w'x'		+	+	
w'x	+			+

10. (a) Find an optimal expression for the Boolean function E of x, y, z, w that has the value 1 if and only if at least two of x, y, z, w have the value 1.

(b) Give a Boolean expression for the function E of part (a) that has eight $\vee$ and $\wedge$ operations. [Hence the optimal expression in part (a) does not minimize the number of gates in a logic network for E.]

CHAPTER HIGHLIGHTS

As usual: What does it mean? Why is it here? How can I use it? Think of examples.

CONCEPTS AND NOTATION

Boolean algebra
  join, meet, complement
  duality principle
  $\leq$ order
  atom
  isomorphism
$\mathbb{B}$, $\mathbb{B}^n$
Boolean function, BOOL(n)
Boolean expression
  equivalent expressions
  minterm, minterm canonical form
  optimal join of products of literals
logic network
    NOT, AND, OR, NAND, NOR, XOR gates
  equivalent networks
Karnaugh map, block

FACTS

Boolean algebra laws in Theorems 1 and 2 of § 10.1.

Properties of $\leq$ given in Lemmas 1, 2 and 3 of § 10.1.

Nonzero elements of finite Boolean algebras are uniquely expressible as joins of atoms.

Any two Boolean algebras with n atoms are isomorphic.

Every logic network is equivalent to one using just NAND gates or just NOR gates.

Boolean expressions and logic networks correspond to labeled acyclic digraphs [§ 10.3].

Optimal Boolean expressions may not correspond to simplest networks.

Choosing essential blocks first in a Karnaugh map and then greedily choosing the largest remaining blocks to cover may not give an optimal expression.

METHODS

Determination of minterm canonical form by calculating the corresponding Boolean function or by using Boolean algebra laws.

Use of a Karnaugh map to find all optimal expressions equivalent to a given Boolean expression.

MORE ON RELATIONS 11

This chapter continues the study of relations that we began long ago in Chapter 3. It may be wise to review that account quickly for terminology. The first two sections of this chapter discuss relations that order the elements of a set, beginning with general partial orderings and then turning to specific order relations on the sets $S_1 \times \cdots \times S_n$ and Σ^*. Section 11.3 discusses composition of relations in general and develops matrix analogs of statements about relations. The final section determines the smallest relations with various properties that contain a given relation R on a set S. It describes, in particular, the smallest equivalence relation containing R.

Sections 11.3 and 11.4 are independent of the first two sections of the chapter, and may be studied separately.

§ 11.1 Partially Ordered Sets

In this section we look at sets whose members can be compared with each other in some way. In typical instances we will think of one element as being smaller than another, or as coming before another in some sort of order.

EXAMPLE 1
(a) We are used to comparing real numbers. For example, 3 is less than 5, -1 is less than 4 and -1 is greater than -3. We compare two numbers by observing which is larger and which is smaller.

(b) If the members of a set S are listed using subscripts from $\mathbb{P}$ or $\mathbb{N}$ so that different elements have different subscripts, then we can compare two members of S by observing which of them has the smaller subscript. Different ways of subscripting members of S give different ways of ordering S. The element that has the lowest subscript of all for one scheme might have many members preceding it using another scheme. ∎

A set whose members can be compared in such a way is said to be **ordered**, and the specification of how its members compare with each other is called an **order relation** on the set. To say anything useful about ordered sets we need to make these definitions more precise. First note, however, that in many sets that arise naturally we know how to compare some elements with others but also have pairs that are not comparable.

EXAMPLE 2   (a) If we try to compare makes of automobiles we can perhaps agree that Make RR is better than Make H because it is better in every respect, but we may not be able to say that either Make F or Make C is better than the other, since each may be superior in some ways.

(b) We can agree to compare two numbers in $\{1, 2, 3, \ldots, 73\}$ if one is a factor of the other. Then 6 and 72 are comparable and so are 6 and 3. But 6 and 8 are not, since neither 6 nor 8 is a factor of the other.

(c) We can compare two subsets of a set S [i.e., members of $\mathscr{P}(S)$] if one is contained in the other. If S has more than one member, then it has some incomparable subsets. For example, if $s_1 \neq s_2$ and s_1 and s_2 belong to S, then the sets $\{s_1\}$ and $\{s_2\}$ are incomparable.

(d) We can compare functions pointwise. For instance, if f and g are defined on the set S and have values in $\{0, 1\}$, we could consider f to be less than or equal to g in case $f(s) \leq g(s)$ for every $s \in S$. This is essentially the order we gave $\mathbb{B}^n$ in Example 3 of §10.1. It is similar to the comparison of cars in (a). ∎

Sets with comparison relations that allow the possibility of incomparable elements, such as those in Example 2, are said to be partially ordered. They form an important class, which we now define precisely.
Recall that a relation R on a set S is a subset of $S \times S$. A **partial order** on a set S is a relation R that is reflexive, antisymmetric and transitive. These conditions mean that if we write $x \preceq y$ as an alternative notation for $(x, y) \in R$, then a partial order satisfies:

(R) $s \preceq s$ for every s in S;
(AS) $s \preceq t$ and $t \preceq s$ imply $s = t$;
(T) $s \preceq t$ and $t \preceq u$ imply $s \preceq u$.

These are exactly the properties of the usual order $\leq$ on $\mathbb{R}$ that we highlighted in Example 4, §3.1.
If $\preceq$ is a partial order on S, the pair $(S, \preceq)$ is called a **partially ordered set**, or **poset** for short. We use the notation "$\preceq$" as a general-purpose, generic name for a partial order. If there is already a notation, such as "$\leq$" or "$\subseteq$," for a particular partial order we will generally use it in preference to "$\preceq$."

In Example 2 the understood relations were "is not as good as," "is a factor of," "is a subset of" and "is never bigger than." We could just as well have considered the relations "is as good as," "is a multiple of," "contains" and "is always at least as big as," since these relations convey the same comparative information as the chosen ones. Each partial order on a set determines a **converse** relation, in which x and y are related if and only if y and x are related in the original way. The converse of a partial order $\leq$ is usually denoted by $\geq$. Thus $x \geq y$ means the same as $y \leq x$. The converse relation is also a partial order [Exercise 7(a)]. If we view $\leq$ on S as a subset R of $S \times S$, then $\geq$ corresponds to the converse relation $R^{\leftarrow}$ defined in §3.1.

Given a partial order $\leq$ on a set S we can define another relation $\prec$ on S by

$$x \prec y \quad \text{if and only if} \quad x \leq y \text{ and } x \neq y.$$

For example if $\leq$ is set inclusion $\subseteq$, then $A \prec B$ means that A is a proper subset of B, i.e., $A \subset B$. The relation $\prec$ is antireflexive and transitive:

(AR) $s \prec s$ is false for all s in S;
(T) $s \prec t$ and $t \prec u$ imply $s \prec u$.

We call an antireflexive transitive relation a **quasi-order**. Each partial order on S yields a quasi-order, and conversely if $\prec$ is a quasi-order on S, then the relation $\leq$ defined by

$$x \leq y \quad \text{if and only if} \quad x \prec y \text{ or } x = y$$

is a partial order on S [Exercise 7(b)]. Whether one chooses a partial order or its associated quasi-order to describe comparisons between members of a poset depends on the particular problem at hand. We will generally use the partial order, but switch back and forth as convenient.

It is possible, at least in principle, to draw a diagram that shows at a glance the order relation on a finite poset. Given a partial order $\leq$ on S, we say the element t **covers** the element s in case $s \prec t$ and there is no u in S with $s \prec u \prec t$. A **Hasse** [pronounced HAH-suh] **diagram** of the poset $(S, \leq)$ is a picture of the digraph whose vertices are the members of S and which has an edge from t to s if and only if t covers s. Hasse diagrams, like rooted trees, are generally drawn with their edges directed downward and with the arrows left off.

EXAMPLE 3   (a) Let $S = \{1, 2, 3, 4, 5, 6\}$. We write $m|n$ in case m divides n, i.e., in case n is an integer multiple of m. The diagram in Figure 1 is a Hasse diagram of the poset $(S, |)$. There is no edge between 1 and 6 because 6 does not cover 1. We can see from the diagram, though, that $1|6$ because the relation is transitive and there is a chain of edges corresponding to

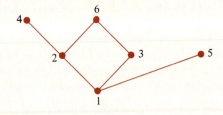

FIGURE 1

$1|2$ and $2|6$. Similarly, we can see that $1|4$ from the path $1|2|4$. Note that, in general, transitive relations can be run together without causing confusion: $x \preceq y \preceq z$ means $x \preceq y$, $y \preceq z$ and $x \preceq z$.

(b) Consider the power set $\mathscr{P}(\{a, b, c\})$ with $\subseteq$ as partial order. Figure 2 shows a Hasse diagram of $(\mathscr{P}(\{a, b, c\}), \subseteq)$. Note that the line joining $\{a, c\}$ to $\{a\}$ happens to cross the line joining $\{a, b\}$ to $\{b\}$, but this crossing is simply a feature of the drawing and has no significance as far as the partial order is concerned. In particular, the intersection of the two lines does *not* represent an element of the poset.

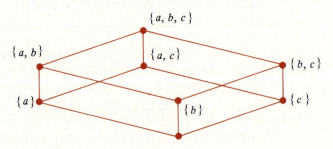

FIGURE 2

(c) The diagram in Figure 3 is not a Hasse diagram, because u cannot cover x if u also covers y and y covers x. If any of the three edges connecting u, x and y were removed, the figure would be a Hasse diagram.

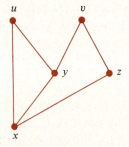

FIGURE 3

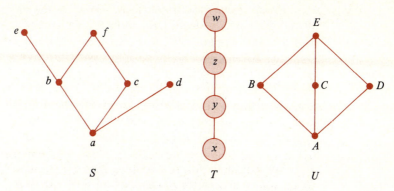

FIGURE 4

(d) The diagrams in Figure 4 are Hasse diagrams of posets, whose order relations can be read off directly from the diagrams. All elements are related to themselves. In addition:

For $S = \{a, b, c, d, e, f\}$ we have $a \preceq b$, $a \preceq c$, $a \preceq d$, $a \preceq e$, $a \preceq f$, $b \preceq e$, $b \preceq f$ and $c \preceq f$. We saw this picture before in part (a) of this example.

For $T = \{x, y, z, w\}$ we have $x \preceq y$, $x \preceq z$, $x \preceq w$, $y \preceq z$, $y \preceq w$ and $z \preceq w$. This is the picture we would get for divisors of 8 or of 27 or of 125 with order relation $|$.

For $U = \{A, B, C, D, E\}$ we have $A \preceq B$, $A \preceq C$, $A \preceq D$, $A \preceq E$, $B \preceq E$, $C \preceq E$ and $D \preceq E$. This picture is the Hasse diagram of the poset consisting of the sets $\{1\}$, $\{1, 2\}$, $\{1, 3\}$, $\{1, 4\}$, $\{1, 2, 3, 4\}$ with set inclusion as order relation.

(e) The relation $<$ that we defined on a Boolean algebra in §10.1 is a quasi-order. The atoms are the elements of the algebra that cover the element 0. The poset $(\mathscr{P}(\{a, b, c\}), \subseteq)$ in part (b) above is an example of a Boolean algebra viewed as a poset. ∎

In general, given a Hasse diagram for a poset, we see that $s \preceq t$ in case either $s = t$ or there is a [downward] path from t to s. The reflexive and transitive laws are understood, and the covering information tells us the rest.

The fact that every finite poset has a Hasse diagram may be intuitively obvious, but we will provide a proof anyway, using properties of acyclic digraphs.

Theorem | Every finite poset has a Hasse diagram.

Proof. Given the poset $(P, \preceq)$, let H be the digraph with vertex set P and with an edge from x to y whenever x covers y. A typical path in H has a vertex sequence $x_1 x_2 \cdots x_{n+1}$ in which x_1 covers x_2, x_2 covers x_3, etc., so $x_1 \succ x_2 \succ \cdots \succ x_{n+1}$. By transitivity and antisymmetry $x_1 \succ x_{n+1}$; in particular $x_1 \neq x_{n+1}$ and the path is not closed. Hence H is an acyclic digraph. We showed in §§ 7.3 and 8.1 that every finite ayclic digraph has a sorted labeling. By giving the digraph H such a labeling and drawing its picture so that the vertices with larger numbers are higher, we obtain a Hasse diagram for $(P, \preceq)$. ∎

Some infinite posets also have Hasse diagrams. A Hasse diagram of $\mathbb{Z}$ with the usual order $\leq$ is a vertical line with dots spaced along it. On the other hand, no real number covers any other in the usual $\leq$ order, so $(\mathbb{R}, \leq)$ has no Hasse diagram.

EXAMPLE 4   (a) Starting with an alphabet Σ, we can make the set Σ^* of all words using letters from Σ into an infinite poset as follows. For words w_1, w_2 in Σ^* define $w_1 \preceq w_2$ if w_1 is an **initial segment** of w_2, i.e., if there is a word w in Σ^* with $w_1 w = w_2$. For example, we have $ab \preceq abbaa$, since $w_1 w = w_2$ with $w_1 = ab$, $w = baa$ and $w_2 = abbaa$. Also, $\lambda \preceq w$ for *all* words because $w = \lambda w$. Note that $abbaa$ does not cover ab, since $u = abb$ and $u = abba$ both satisfy $ab \prec u \prec abbaa$. However, $abbaa$ covers $abba$, $abba$ covers abb, and abb covers ab. In general, if w_2 covers w_1, then $\text{length}(w_2) = 1 + \text{length}(w_1)$.

For $\Sigma = \{a, b\}$, part of the Hasse diagram for $(\Sigma^*, \preceq)$ is drawn in Figure 5. This Hasse diagram is a tree. In § 6.4 [Example 7] we viewed the diagram as a rooted tree, at which point tradition forced us to draw it upside down.

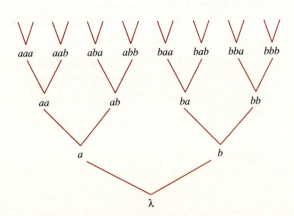

FIGURE 5

(b) A finite rooted tree T has a natural order in which the root r is the largest element. Define the relation $\preceq$ on the set V of vertices of T by saying that $v \preceq w$ in case $v = w$ or w is on the [unique] path from r to v. It is easy to check that $\preceq$ is a partial order on V. [See Exercise 8.] One Hasse diagram for $(V, \preceq)$ is the original tree T, drawn as usual with the root at the top and branches going downward. Pictures of rooted trees can be thought of as pictures of rather special posets.

(c) The examples in (a) and (b) showed trees with their roots at the bottom and at the top. There is a natural connection between the two sorts of presentation. A Hasse diagram determined by the converse relation $\succeq$ to the relation $\preceq$ on a poset S is simply a Hasse diagram for $(S, \preceq)$ turned top to bottom. The reason is that y covers x in the original relation if and only if x covers y in the converse relation, so the edges for a diagram of $(S, \succeq)$ are directed oppositely to the edges for a diagram of $(S, \preceq)$. ∎

The elements corresponding to points near the top or bottom of a Hasse diagram often turn out to be important. If $(P, \preceq)$ is a poset, we call an element x of P **maximal** in case there is no y in P with $x \prec y$, and call x **minimal** if there is no y in P with $y \prec x$. In the posets with Hasse diagrams shown in Figure 4 the elements d, e, f, w and E are maximal, while a, x and A are minimal. The infinite poset in Figure 5 has no maximal elements; the empty word λ is its only minimal element.

A subset S of a poset P inherits the partial order on P and is itself a poset, since the laws (R), (AS) and (T) apply to all members of P. We call S a **subposet** of P.

EXAMPLE 5   (a) The sets $\{2, 3, 4, 5, 6\}$ and $\{1, 2, 3, 6\}$ are subposets of the poset $\{1, 2, 3, 4, 5, 6\}$ given in Example 3(a), with Hasse diagrams shown in Figure 6. [Notice the placement of the primes in Figure 6(a).]

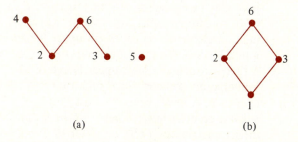

(a)   (b)

FIGURE 6

(b) The set of nonempty proper subsets of $\{a, b, c\}$ is a subposet of $\mathscr{P}(\{a, b, c\})$ with partial order $\subseteq$. Figure 7 shows a Hasse diagram for it. Compare with Figure 2. ■

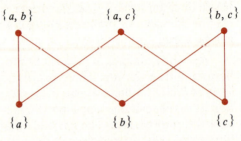

$\{a, b\}$ $\{a, c\}$ $\{b, c\}$

$\{a\}$ $\{b\}$ $\{c\}$

FIGURE 7

If S is a subposet of a poset $(P, \preceq)$, it may happen that S has a member M such that $s \preceq M$ for every s in S. In Figure 6(b), $s \preceq 6$ for every s, while no such element M exists in Figure 6(a) or 7. An element M with this property is called the **largest member** of S or the **maximum** of S and denoted **max(S)**. [There is at most one such M; why?] This notation is consistent with our usage of $\max\{m, n\}$ to denote the larger of the two numbers m and n. Similarly, if S has a member m such that $m \preceq s$ for every s in S, then m is called the **smallest member** of S or the **minimum** of S and is denoted **min(S)**.

EXAMPLE 6   (a) Consider again the poset $(\{1, 2, 3, 4, 5, 6\}, |)$ illustrated in Figure 1. This poset has no largest member or maximum even though 4, 6 and 5 are all maximal elements. The element 1 is a minimum of the poset and is the only minimal element. The subset $\{2, 3\}$ has no largest member; 3 is larger than 2 in the usual order, but not in the order under discussion.

(b) If a Boolean algebra is viewed as a poset, its largest element is 1 and its smallest element is 0. ■

Whether or not a subposet S of a poset $(P, \preceq)$ has a largest member, there may be elements x in the larger set P such that $s \preceq x$ for every s in S. [For example, both elements in the set $\{2, 3\}$ of Example 6 divide 6.] Such an element x is called an **upper bound** for S in P. If x is an upper bound for S in P and is such that $x \preceq y$ for every upper bound y for S in P, then x is called a **least upper bound** of S in P, and we write $x = $ **lub(S)**. Similarly, an element z in P such that $z \preceq s$ for all s in S is a **lower bound** for S in P. A lower bound z such that $w \preceq z$ for every lower bound w is called a **greatest lower bound** of S in P and is denoted by **glb(S)**. By the antisymmetric law (AS), a subset of P cannot have two different least upper bounds or two different greatest lower bounds.

EXAMPLE 7   (a) In the poset $(\{1, 2, 3, 4, 5, 6\}, |)$ the subset $\{2, 3\}$ has exactly one upper bound, namely 6, so lub$\{2, 3\} = 6$. Similarly, glb$\{2, 3\} = 1$. The subset $\{4, 6\}$ has no upper bounds in the poset; 2 and 1 are both lower bounds, so glb$\{4, 6\} = 2$. The subset $\{3, 6\}$ has 6 as an upper bound, and has 3 and 1 as lower bounds; hence lub$\{3, 6\} = 6$ and glb$\{3, 6\} = 3$. Thus least upper bounds and greatest lower bounds for a subset may or may not exist, and if they do exist they may or may not belong to the subset. For this particular poset, greatest lower bounds are greatest common divisors, and least upper bounds, when they exist in the poset, are least common multiples.

(b) In the poset P shown in Figure 8 the subset $\{b, c\}$ has d, e, g and h as upper bounds in P, and h is an upper bound for $\{d, f\}$. The set $\{b, c\}$ has no least upper bound in P [why?] but $h = \text{lub}\{d, f\}$. The elements a and c are lower bounds for $\{d, e, f\}$, which has no greatest lower bound because a and c are not comparable. Element a is the greatest lower bound of $\{b, d, e, f\}$. ■

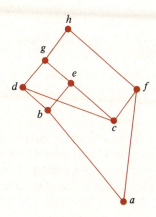

FIGURE 8

Many of the posets that come up in practice have the property that every 2-element subset has both a least upper bound and a greatest lower bound. A **lattice** is a poset in which lub$\{x, y\}$ and glb$\{x, y\}$ exist for every x and y. In a lattice $(P, \preceq)$ the equations

$$x \vee y = \text{lub}\{x, y\} \quad \text{and} \quad x \wedge y = \text{glb}\{x, y\}$$

define binary operations $\vee$ and $\wedge$ on P. As we will see in the next example, this usage is consistent with that introduced for Boolean algebras in §10.1. Note that glb$\{x, y\} = x \wedge y = x$ if and only if $x \preceq y$ if and only if lub$\{x, y\} = x \vee y = y$. In particular, we can recover the order relation $\preceq$ if we know either binary operation $\wedge$ or $\vee$; see Exercise 12. One can show by induction [Exercise 19(b)] that every finite subset of a lattice has both a least upper bound and a greatest lower bound.

EXAMPLE 8   (a) The poset $(\mathscr{P}(\{a, b, c\}), \subseteq)$ shown in Figure 2 is a lattice. For instance,

$$\text{lub}(\{a\}, \{c\}) = \{a\} \vee \{c\} = \{a, c\},$$

$$\text{lub}(\{a, b\}, \{a, c\}) = \{a, b\} \vee \{a, c\} = \{a, b, c\},$$

$$\text{glb}(\{a, b\}, \{c\}) = \{a, b\} \wedge \{c\} = \varnothing$$

and

$$\text{glb}(\{a, b\}, \{b, c\}) = \{a, b\} \wedge \{b, c\} = \{b\}.$$

In general, for any set S, $(\mathscr{P}(S), \subseteq)$ is a lattice with $\text{lub}\{A, B\} = A \cup B$ and $\text{glb}\{A, B\} = A \cap B$, so that

$$\text{lub}\{A, B, \ldots, Z\} = A \cup B \cup \cdots \cup Z$$

and

$$\text{glb}\{A, B, \ldots, Z\} = A \cap B \cap \cdots \cap Z.$$

The poset shown in Figure 7 is not a lattice; for example $\{a, b\}$ and $\{a, c\}$ have no least upper bound in this poset.

(b) Define the partial order $|$ on $\mathbb{P}$ by $m | n$ if and only if m divides n. The subposet $S = \{1, 2, 3, 4, 5, 6\}$ of $\mathbb{P}$, shown in Figure 1, is not a lattice, since $\{3, 4\}$ has no upper bound in S. The full poset $(\mathbb{P}, |)$ is a lattice, however. An upper bound for $\{m, n\}$ is an integer k in $\mathbb{P}$ such that m divides k and n divides k, i.e., a common multiple of m and n. The least upper bound $\text{lub}\{m, n\}$ is the **least common multiple** of m and n. Similarly the greatest lower bound $\text{glb}\{m, n\}$ is the **greatest common divisor** of m and n, the largest positive integer that divides both m and n. For instance, $\text{lub}\{12, 10\} = 60$ and $\text{glb}\{12, 10\} = 2$. The numbers $\text{lub}\{m, n\}$ and $\text{glb}\{m, n\}$ can be determined from the factorizations of m and n into products of primes. The primes themselves are the minimal members of the subposet $\mathbb{P} \backslash \{1\}$, i.e., they are the numbers that cover 1 in $\mathbb{P}$.

(c) Consider the set $\text{FUN}(\{a, b, c\}, \{0, 1\})$ of all functions from the 3-element set $\{a, b, c\}$ to $\{0, 1\}$. We obtain a partial order $\leq$ on this set by defining

$$f \leq g \quad \text{if and only if} \quad f(x) \leq g(x) \text{ for } x = a, b, c.$$

It is convenient to label the eight functions in this poset with subscripts, such as 101, that list the values the functions take at a, b and c, respectively. For example, f_{101} represents the function such that $f_{101}(a) = 1$, $f_{101}(b) = 0$ and $f_{101}(c) = 1$. The Hasse diagram for the poset $(\text{FUN}(\{a, b, c\}, \{0, 1\}), \leq)$ is given in Figure 9. This poset is a lattice. It is essentially the Boolean algebra $\text{BOOL}(3)$ of Boolean functions of three variables.

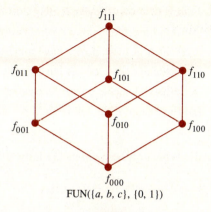

FUN($\{a, b, c\}$, $\{0, 1\}$)

FIGURE 9

(d) In §10.1 we started out with a Boolean algebra $(B, \vee, \wedge, ')$ and created the relation $\leq$ by defining $x \leq y$ if and only if $x \vee y = y$ if and only if $x \wedge y = x$. Lemma 2 of that section shows that $\leq$ is a partial order. We now show that, with respect to the order $\leq$, the element $a \wedge b$ is the greatest lower bound of $\{a, b\}$; similarly, $a \vee b$ is lub$\{a, b\}$. We will use here just the algebraic properties of $\wedge$ and $\vee$.

First, $a \wedge b \leq a$ because $(a \wedge b) \wedge a = a \wedge b$, and similarly, $a \wedge b \leq b$. Hence $a \wedge b$ is a lower bound for $\{a, b\}$. If $c \leq a$ and $c \leq b$, too, then $c \wedge a = c$ and $c \wedge b = c$. It follows that $c \wedge (a \wedge b) = (c \wedge a) \wedge b = c \wedge b = c$, and so $c \leq a \wedge b$. Thus $a \wedge b$ is the greatest lower bound of $\{a, b\}$. ■

Most of the posets that we have looked at in this section have had pairs of incomparable elements. In the next section we will consider those posets in which every element is related to every other element. We will also discuss how to use orders on relatively simple sets to produce orders for more complicated ones.

EXERCISES 11.1

1. Draw Hasse diagrams for the following posets.

   (a) $(\{1, 2, 3, 4, 6, 8, 12, 24\}, |)$ where $m|n$ means that m is a factor of [i.e., divides] n.

   (b) The set of subsets of $\{3, 7\}$ with $\subseteq$ as partial order.

2. (a) Give examples of two posets that come from everyday life or from other courses.

   (b) Do your examples have maximal or minimal elements? If so, what are they?

   (c) What are the converses of the partial orders in your examples?

3. Figure 10 shows the Hasse diagrams of three posets.

(a) What are the maximal members of these posets?

(b) Which of these posets have minimal elements?

(c) Which of these posets have smallest members?

(d) Which elements cover the element e?

(e) Find each of the following if it exists.

$$\text{lub}\{d, c\}, \qquad \text{lub}\{w, y, v\}, \qquad \text{lub}\{p, m\}, \qquad \text{glb}\{a, g\}.$$

(f) Which of these posets are lattices?

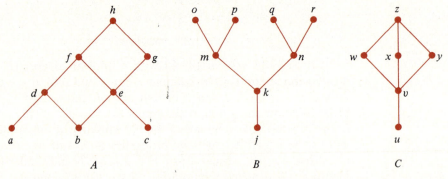

FIGURE 10

4. Find the maximal proper subsets of the 3-element set $\{a, b, c\}$. That is, find the maximal members of the subposet of $\mathscr{P}(\{a, b, c\})$ consisting of proper subsets of $\{a, b, c\}$.

5. Consider $\mathbb{R}$ with the usual order $\leq$.

(a) Is $\mathbb{R}$ a lattice? If it is, what are the meanings of $a \vee b$ and $a \wedge b$ in $\mathbb{R}$?

(b) Give an example of a nonempty subset of $\mathbb{R}$ that has no least upper bound.

(c) Find $\text{lub}\{x \in \mathbb{R} : x < 73\}$. \qquad (d) Find $\text{lub}\{x \in \mathbb{R} : x \leq 73\}$.

(e) Find $\text{lub}\{x \in \mathbb{R} : x^2 < 73\}$. \qquad (f) Find $\text{glb}\{x \in \mathbb{R} : x^2 < 73\}$.

6. Let S be a set of subroutines of a computer program. For A and B in S write $A \prec B$ if A must be completed before B can be completed. What sort of restriction must be placed on subroutine calls in the program to make $\prec$ a quasi-order on S?

7. (a) Show that if $\preceq$ is a partial order on a set S, then so is its converse relation $\succeq$.

(b) Show that if $\prec$ is a quasi-order on a set S, then the relation $\preceq$ defined by

$$x \preceq y \quad \text{if and only if} \quad x \prec y \text{ or } x = y$$

is a partial order on S.

8. Let G be an acyclic digraph. Show that the relation $\prec$, defined by $u \prec v$ if there is a path from u to v, is a quasi-order on $V(G)$. Thus the relation $\preceq$ in Example 4(b) is a partial order by Exercise 7(b).

9. Verify that the partial order $\preceq$ on Σ^* in Example 4(a) is reflexive and transitive.

10. Let Σ be an alphabet. For $w_1, w_2 \in \Sigma^*$ define $w_1 \preceq w_2$ if there are w and w' in Σ^* with $w_2 = w w_1 w'$. Is $\preceq$ a partial order on Σ^*? Explain.

11. Let Σ be an alphabet. For $w_1, w_2 \in \Sigma^*$, let $w_1 \preceq w_2$ mean length$(w_1) \le$ length(w_2). Is $\preceq$ a partial order on Σ^*? Explain.

12. The table in Figure 11 has been partially filled in. It gives the values of $x \vee y$ for x and y in a certain lattice $(L, \preceq)$. For example, $b \vee c = d$.

    (a) Fill in the rest of the table.

    (b) Which are the largest and smallest elements of L?

    (c) Show that $f \preceq c \preceq d \preceq e$.

    (d) Draw a Hasse diagram for L.

$\vee$	a	b	c	d	e	f
a			e	a	e	e	a
b				d	d	e	b
c					d	e	c
d						e	d
e							e
f							

FIGURE 11

13. Let $\mathscr{F}(\mathbb{N})$ be the collection of all *finite* subsets of $\mathbb{N}$. Then $(\mathscr{F}(\mathbb{N}), \subseteq)$ is a poset.

    (a) Does $\mathscr{F}(\mathbb{N})$ have a maximal element? If yes, give one. If no, explain.

    (b) Does $\mathscr{F}(\mathbb{N})$ have a minimal element? If yes, give one. If no, explain.

    (c) Given A, B in $\mathscr{F}(\mathbb{N})$, does $\{A, B\}$ have a least upper bound in $\mathscr{F}(\mathbb{N})$? If yes, specify it. If no, provide a specific counterexample.

    (d) Given A, B in $\mathscr{F}(\mathbb{N})$, does $\{A, B\}$ have a greatest lower bound in $\mathscr{F}(\mathbb{N})$? If yes, specify it. If no, provide a specific counterexample.

    (e) Is $\mathscr{F}(\mathbb{N})$ a lattice? Explain.

14. Repeat Exercise 13 for the collection $\mathscr{I}(\mathbb{N})$ of all *infinite* subsets of $\mathbb{N}$.

15. Define the relations $<$, $\le$ and $\preceq$ on the plane $\mathbb{R} \times \mathbb{R}$ by

    $$(x, y) < (z, w) \qquad \text{if} \quad x^2 + y^2 < z^2 + w^2,$$

    $$(x, y) \le (z, w) \qquad \text{if} \quad (x, y) < (z, w) \text{ or } (x, y) = (z, w),$$

    $$(x, y) \preceq (z, w) \qquad \text{if} \quad x^2 + y^2 \le z^2 + w^2.$$

    (a) Which of these relations are partial orders? Explain.

  (b) Which are quasi-orders? Explain.

  (c) Draw a sketch of $\{(x, y):(x, y) \le (3, 4)\}$.

  (d) Draw a sketch of $\{(x, y):(x, y) \preceq (3, 4)\}$.

16. Let $\mathscr{E}(\mathbb{N})$ be the set of all finite subsets of $\mathbb{N}$ that have an even number of elements, with partial order $\subseteq$.

  (a) Let $A = \{1, 2\}$ and $B = \{1, 3\}$. Find four upper bounds for $\{A, B\}$.

  (b) Does $\{A, B\}$ have a least upper bound in $\mathscr{E}(\mathbb{N})$? Explain.

  (c) Is $\mathscr{E}(\mathbb{N})$ a lattice?

17. Is every subposet of a lattice a lattice? Explain.

18. (a) Show that every nonempty finite poset has a minimal element. *Hint:* Use induction.

  (b) Give an example of a poset with a maximal element but with no minimal element.

19. (a) Consider elements x, y, z in a poset. Show that if $\text{lub}\{x, y\} = a$ and $\text{lub}\{a, z\} = b$, then $\text{lub}\{x, y, z\} = b$.

  (b) Show that every finite subset of a lattice has a least upper bound.

  (c) Show that if x, y and z are members of a lattice, then $(x \vee y) \vee z = x \vee (y \vee z)$.

20. Consider the poset C whose Hasse diagram is shown in Figure 10. Show that $w \vee (x \wedge y) \ne (w \vee x) \wedge (w \vee y)$ and $w \wedge (x \vee y) \ne (w \wedge x) \vee (w \wedge y)$. This example shows that lattices need not satisfy "distributive" laws for $\vee$ and $\wedge$.

§ 11.2   Special Orderings

Partially ordered sets arise in a variety of ways, and in many cases the fact that there are pairs of elements that cannot be compared is an essential feature. As we saw in the preceding section, rooted trees can be thought of as consisting of Hasse diagrams, such as those shown in Figure 1, in which $\text{lub}\{x, y\}$ exists for all x and y but $\text{glb}\{x, y\}$ exists only if $x \preceq y$ or $y \preceq x$. Trees are useful data structures, even though they have incomparable elements, because it is possible to start at the root and follow the ordering to get to any element fairly quickly.

  Lists, or sequences, are another common class of data structures. No matter which two elements are chosen in a list, one comes before the other. Such a structure is an example of a **chain**, which we define to be a poset in which every two elements are comparable. A partial order is called a **total order** or **linear order** if for each choice of s and t in S either $s \preceq t$ or $t \preceq s$. Thus a chain is a poset with a total order. The terms "totally ordered set" and "linearly ordered set" are sometimes used as synonyms for "chain."

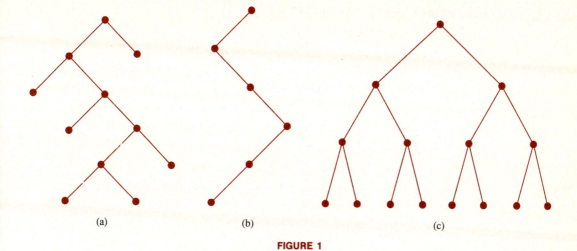

(a)   (b)   (c)

FIGURE 1

EXAMPLE 1   (a) The poset of Figure 1(b) is a chain, but the other posets in Figure 1 are not.

(b) The set $\mathbb{R}$ with the usual order $\leq$ is a chain.

(c) The lists of names in a phone book or words in a dictionary are chains if we define $w_1 \preceq w_2$ to mean that $w_1 = w_2$ or w_1 comes before w_2. ■

Every subposet of a chain is itself a chain. For example, the posets $(\mathbb{Z}, \leq)$ and $(\mathbb{Q}, \leq)$ are subposets of $(\mathbb{R}, \leq)$ and are linearly ordered by the orders they inherit from $\mathbb{R}$. The words in the dictionary between "start" and "stop" form a subchain of the chain of all words in Example 1(c).

Every poset, whether or not it is itself a chain, will have subposets that are chains. It is often useful to know something about these subposets.

EXAMPLE 2   (a) Let S be the set of all people at some family reunion, and write $m \prec n$ in case m is a descendant of n. Then $\prec$ is a quasi-order that gives a partial order $\preceq$ by defining $m \preceq n$ if and only if $m \prec n$ or $m = n$. A chain in the poset $(S, \preceq)$ is a set of the form $\{m, n, p, \ldots, r\}$ in which m is a descendant of n, n a descendant of p, and so on. It would be unusual for such a chain to have more than 5 members, though the set S itself might be quite large.

(b) The Hasse diagram shown in Figure 2 describes a poset with a number of subchains [49 if we count the 1-element chains but not the empty chain]. ■

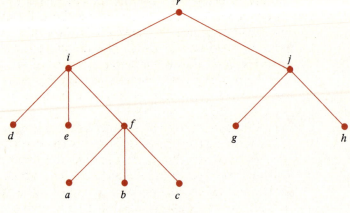

FIGURE 2

We are sometimes interested in chains in posets that cannot be made any larger. Observe that if $(S, \preceq)$ is a poset and if $\mathscr{C}(S)$ is the set of chains in S, then $(\mathscr{C}(S), \subseteq)$ is also a poset. A **maximal chain** in S is defined to be a maximal member of $\mathscr{C}(S)$, i.e., a chain that is not properly contained in another chain.

EXAMPLE 3   (a) In the poset of Figure 2 the maximal chains are $\{a, f, i, r\}$, $\{b, f, i, r\}$, $\{c, f, i, r\}$, $\{d, i, r\}$, $\{e, i, r\}$, $\{g, j, r\}$ and $\{h, j, r\}$. Notice that the maximal chains are not all of the same size.

(b) In the poset shown in Figure 3 the two maximal chains $\{a, c, d, e\}$ and $\{a, b, e\}$ containing a and e have different numbers of elements. This poset has four maximal chains in all.

(c) Paths from the root to the leaves in a rooted tree correspond to maximal chains with respect to the usual partial ordering of the tree. ∎

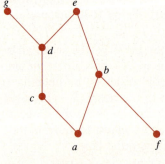

FIGURE 3

A finite chain must have a smallest member, and so must each of its nonempty subchains. Infinite chains, on the other hand, can exhibit a variety of behaviors. The infinite chains $(\mathbb{R}, \leq)$ and $(\mathbb{Z}, \leq)$ with their usual orders do not have smallest members. The chain $(\{x \in \mathbb{R}: 0 \leq x\}, \leq)$ has a smallest member, namely 0, but has subsets such as $\{x \in \mathbb{R}: 1 < x\}$ without smallest members. The infinite chain $(\mathbb{N}, \leq)$ has a smallest member, and the well-ordering property of $\mathbb{N}$ stated in §4.1 says that every nonempty subset does too. It was this property that we used when we observed that every decreasing chain in $\mathbb{N}$ is finite, so that the Division Algorithm terminates.

We say that a chain C is **well-ordered** in case each nonempty subset of C has a smallest member. If C is well-ordered, and if for each c in C we have a statement $p(c)$, we can hope to prove that all of the statements $p(c)$ are true by supposing that $\{c \in C: p(c)$ is false$\}$ is a nonempty subset of C, considering the smallest c for which $p(c)$ is false and deriving a contradiction. This was the idea behind our explanation of the principles of induction in §4.2 and in §4.5.

In the rest of this section we study how to build new partial orders from known ones. Suppose first that $(S, \preceq)$ is a given poset and that T is a nonempty set. We can define a partial order, which we also denote by $\preceq$, on the set of functions from T to S by defining

$$f \preceq g \quad \text{if} \quad f(t) \preceq g(t) \text{ for all } t \text{ in } T.$$

This is the partial order we used in Examples 2(d) and 8(c) of §11.1. The verification that this new relation is a partial order on $\text{FUN}(T, S)$ is straightforward [Exercise 7(a)]. If the order on S is $\leq$, we will write $\leq$ in place of $\preceq$ for the order relation on the set of functions.

EXAMPLE 4   (a) If $S = T = \mathbb{R}$ with the usual order, then $f \leq g$ means that the graph of f lies on or below the graph of g, as in Figure 4.

(b) Consider $S = \{0, 1\}$ with $0 < 1$. The functions in $\text{FUN}(T, \{0, 1\})$ are the characteristic functions of subsets of T. Each subset A of T has a corresponding function χ_A in $\text{FUN}(T, \{0, 1\})$, with $\chi_A(x) = 1$ if $x \in A$ and 0 if $x \notin A$. Then $\chi_A \leq \chi_B$ if and only if $x \in B$ whenever $x \in A$, i.e., if and only if $A \subseteq B$. Thus Hasse diagrams for $(\text{FUN}(T, \{0, 1\}), \leq)$ and $(\mathscr{P}(T), \subseteq)$ look alike. Figures 2 and 9 of §11.1 show the diagrams for $T = \{a, b, c\}$. These two posets are in fact isomorphic Boolean algebras. ■

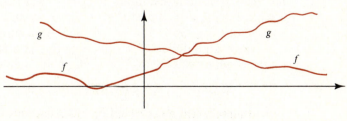

FIGURE 4

Example 4(b) shows that $(\text{FUN}(T, S), \preceq)$ need not be a chain even if S is a chain. The poset $(\text{FUN}(T, S), \preceq)$ does inherit some properties from S, however. If S has largest or smallest elements, so does $\text{FUN}(T, S)$, and if S is a lattice, so is $\text{FUN}(T, S)$. For more in this vein, see Exercise 7.

Another way to combine two sets into a new one is to form their product. Suppose that $(S, \preceq_1)$ and $(T, \preceq_2)$ are posets, where we use the subscripts to keep track of which partial order is which. There is more than one natural way to make $S \times T$ into a poset. Our preference will depend on the problem at hand.

The first partial order we describe for $S \times T$ is called the **product order**. For $s, s' \in S$ and $t, t' \in T$ define

$$(s, t) \preceq (s', t') \quad \text{if} \quad s \preceq_1 s' \text{ and } t \preceq_2 t'.$$

EXAMPLE 5   Let $S = T = \mathbb{N}$ with the usual order $\leq$ in each case. Then $(2, 5) \preceq (3, 7)$ since $2 \leq 3$ and $5 \leq 7$. Also, $(2, 5) \preceq (3, 5)$ since $2 \leq 3$ and $5 \leq 5$. But the pairs $(2, 7)$ and $(3, 5)$ are not comparable; $(2, 7) \preceq (3, 5)$ would mean $2 \leq 3$ and $7 \leq 5$, while $(3, 5) \preceq (2, 7)$ would mean $3 \leq 2$ and $5 \leq 7$. Figure 5 indicates the pairs (m, n) in $S \times T = \mathbb{N} \times \mathbb{N}$ with $(2, 1) \preceq (m, n) \preceq (3, 4)$.   ■

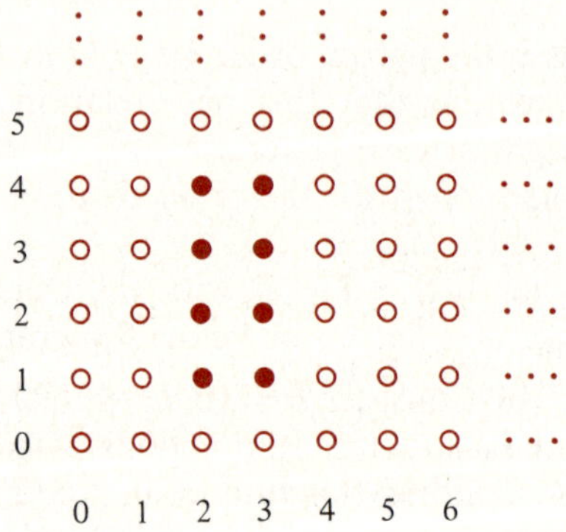

FIGURE 5

Consider, again, two posets $(S, \preceq_1)$ and $(T, \preceq_2)$. The fact that the product order $\preceq$ is a partial order on $S \times T$ is almost immediate from the definition. For example, if $(s, t) \preceq (s', t')$ and $(s', t') \preceq (s, t)$, then $s \preceq_1 s'$, $t \preceq_2 t'$, $s' \preceq_1 s$ and $t' \preceq_2 t$. Since $\preceq_1$ and $\preceq_2$ are antisymmetric, $s = s'$ and $t = t'$. So $(s, t) = (s', t')$. Thus $\preceq$ is antisymmetric.

There is no difficulty in extending this idea to define a partial order on the product $S_1 \times S_2 \times \cdots \times S_n$ of a finite number of posets. We define

$$(s_1, s_2, \ldots, s_n) \preceq (s'_1, s'_2, \ldots, s'_n) \quad \text{if} \quad s_i \leq s'_i \text{ for all } i.$$

In Example 5, $\mathbb{N}$ is a chain but $\mathbb{N} \times \mathbb{N}$ is not; for instance $(2, 4)$ and $(3, 1)$ are not related in the product order. In fact, the product order is almost never a total order [Exercise 12]. On the other hand, if $S_1, S_2, \ldots, S_n$ are chains, there is another natural order on $S_1 \times S_2 \times \cdots \times S_n$ that makes it a chain, which we now illustrate.

EXAMPLE 6 If $S = \{0, 1, 2, \ldots, 9\}$ with the usual order then the set $S \times S$ consists of pairs (m, n), which we can identify with the integers 00, 01, 02, ..., 98, 99 from 0 to 99. To make $S \times S$ a chain we can simply define $(m, n) \prec (m', n')$ if the corresponding integers are so related, i.e., if $m < m'$ or if $m = m'$ and $n < n'$. For example this order makes $(5, 7) \prec (6, 3)$ since $57 < 63$, and $(3, 5) \prec (3, 7)$ since $35 < 37$.

In a similar way we can identify the set $S \times S \times S$ with the set of integers from 0 to 999 and make $S \times S \times S$ into a chain by defining $(m, n, p) \prec (m', n', p')$ if $m < m'$ or if $m = m'$ and $n < n'$ or if $m = m'$, $n = n'$ and $p < p'$. ∎

The idea of this example works in general. If $(S_1, \preceq_1), \ldots, (S_n, \preceq_n)$ are posets we can define a relation $\prec$ on $S_1 \times \cdots \times S_n$ by

$$(s_1, s_2, \ldots, s_n) \prec (t_1, t_2, \ldots, t_n) \text{ if } s_1 \prec_1 t_1 \text{ or if there is an } r \text{ in} \\ \{2, \ldots, n\} \text{ such that } s_1 = t_1, \ldots, s_{r-1} = t_{r-1} \text{ and } s_r \prec_r t_r.$$

Then $\prec$ is a quasi-order [Exercise 19] that induces a partial order $\preceq$ on $S_1 \times S_2 \times \cdots \times S_n$ which we call the **filing order**.

EXAMPLE 7 If $S = \{a, b, c, \ldots, z\}$ and $T = \{0, 1, \ldots, 9\}$ with the usual orders, we can identify the members of $S \times T$ with 2-symbol strings such as $a5$ and $x3$. Imagine a device that has subassemblies, labeled by letters, each of which has at most 10 parts. It would be reasonable to label spare parts with letter–number combinations and to file them in bins arranged according to the filing order on $S \times T$. Then bin $a5$ would come before bin $x3$ because a precedes x, but $a5$ would come after $a3$ since $3 < 5$. ∎

The filing order is primarily useful if each S_i is a chain.

Theorem 1

> Let $(S_1, \preceq_1), \ldots, (S_n, \preceq_n)$ be chains. Then the filing order makes the product $S_1 \times \cdots \times S_n$ a chain.

Proof. We already know that $\preceq$ is a partial order on $S_1 \times \cdots \times S_n$. Let $s = (s_1, \ldots, s_n)$ and $t = (t_1, \ldots, t_n)$ be distinct elements in $S_1 \times \cdots \times S_n$. Since $s \neq t$, there is a first value of r for which $s_r \neq t_r$. Since $(S_r, \preceq_r)$ is a chain, either $s_r \prec_r t_r$ or $t_r \prec_r s_r$. In the first case $s \prec t$; in the second $t \prec s$. In either case, the two elements of $S_1 \times \cdots \times S_n$ are comparable. ∎

The special case in which all the posets $(S_i, \preceq_i)$ are the same is important enough to warrant a notation of its own. If $(S, \preceq)$ is a poset and $k \in \mathbb{P}$ we write $\preceq^k$ for the filing order on $S^k = S \times \cdots \times S$.

In the remainder of this section we are interested primarily in the case in which S is an alphabet. Accordingly, from now on we go Greek and write Σ instead of S. We also make the natural identification of k-tuples $(a_1, \ldots, a_k)$ in the product Σ^k with words $a_1 \cdots a_k$ of length k in Σ^*. We still have in mind some given partial order $\preceq$ on Σ. The set Σ^0 is $\{\lambda\}$, and we define $\preceq^0$ on Σ^0 in the only possible way: $\lambda \preceq^0 \lambda$.

EXAMPLE 8   If $\Sigma = \{a, b, c, \ldots, z\}$ with the usual linear order on the English alphabet, then Σ^k consists of all strings of letters of length k, and $\preceq^k$ is the usual alphabetical order on Σ^k. For example, if $k = 3$, then

$$fed \preceq^3 few \preceq^3 one \preceq^3 six \preceq^3 ten \preceq^3 two \preceq^3 won. \quad \blacksquare$$

Since $\Sigma^* = \Sigma^0 \cup \Sigma^1 \cup \Sigma^2 \cup \cdots$ we can piece the partial orders $\preceq^0, \preceq^1, \preceq^2, \ldots$ together to get an order $\preceq^*$ for Σ^* that we call the **standard order**, in which λ comes first, then all words of length 1, then all of length 2, and so on. More precisely,

$w_1 \preceq^* w_2$ if either $w_1 \in \Sigma^k$ and $w_2 \in \Sigma^r$ and $k < r$
or $w_1 \in \Sigma^k$ and $w_2 \in \Sigma^k$ for the same k and $w_1 \preceq^k w_2$.

If $(\Sigma, \preceq)$ is a chain, as it was in Example 8, then $(\Sigma^*, \preceq^*)$ is also a chain.

EXAMPLE 9   (a) If Σ is the English alphabet in the usual order, the first few terms of Σ^* in the standard order are

$$\lambda, a, b, \ldots, z, aa, ab, \ldots, az, ba, bb, \ldots, bz, ca, cb, \ldots, cz,$$
$$da, db, \ldots, dz, \ldots, za, zb, \ldots, zz, aaa, aab, aac, \ldots.$$

(b) Let $\Sigma = \{0, 1\}$ with $0 < 1$. The first few terms of Σ^* in the standard order are

$$\lambda, 0, 1, 00, 01, 10, 11, 000, 001, 010, 011,$$
$$100, 101, 110, 111, 0000, 0001, 0010, \ldots. \quad \blacksquare$$

Note that if a dictionary were constructed using the standard order in Example 9(a) all the short words would be at the beginning of the dictionary, and to find a word it would be essential to know its exact length. [In fact, some dictionaries designed for crossword puzzle solvers are arranged this way.] To find words in a dictionary with the ordinary alphabetical order, one scans words from left to right looking for differences and ignoring the lengths of the words. Thus "aardvark" is listed before "axe" and "break" precedes "breakfast." Alphabetical order is based on the usual alphabet, ordered in the usual way. The idea generalizes naturally to an arbitrary Σ with partial order $\preceq$.

The **lexicographic** or **dictionary order** $\preceq_L$ on Σ^* is defined as follows. For $a_1, \ldots, a_m$ and $b_1, \ldots, b_n$ in Σ let $k = \min\{m, n\}$. We define

$$a_1 \cdots a_m \prec_L b_1 \cdots b_n \quad \text{if} \quad a_1 \cdots a_k \prec^k b_1 \cdots b_k$$

$$\text{or if} \quad k = m < n \quad \text{and} \quad a_1 \cdots a_k = b_1 \cdots b_k.$$

Then $\prec_L$ is a quasi-order on Σ^* that defines the partial order $\preceq_L$. Thus *aardvark* $\prec_L$ *axe* because *aar* $\prec^3$ *axe*, while *break* $\prec_L$ *breakfast* since *break* has the form $a_1 \cdots a_5$, *breakfast* has the form $b_1 \cdots b_9$ and $a_1 \cdots a_5 = b_1 \cdots b_5$.

We can describe $\preceq_L$ in another way. For words w and z in Σ^*, $w \preceq_L z$ if and only if either

(a) w is an initial segment of z, i.e., $z = wu$ for some $u \in \Sigma^*$, or

(b) $w = xu$ and $z = xv$ for words u and v in Σ^* such that the first letter of u precedes the first letter of v in the ordering of Σ.

Note that in (b) x can be any word, possibly the empty word.

Lexicographic order, standard order and filing order all agree if we consider words in some fixed Σ^k, but lexicographic and standard orders differ in the treatment they give to words of different lengths.

EXAMPLE 10 Let $\Sigma = \{a, b\}$ with $a \prec b$. The first few terms of Σ^* in the lexicographic order are

$$\lambda, a, aa, aaa, aaaa, aaaaa, \ldots.$$

Any word using the letter b is preceded by an *infinite* number of words, including all the words using only the letter a. Moreover, Σ^* contains infinite decreasing sequences of words; for example,

$$b \succ_L ab \succ_L aab \succ_L aaab \succ_L \cdots.$$

And there are infinitely many words between members of this sequence; for example,

$$aaab, aaaba, aaabaa, aaabaaa, aaabaaaa, \ldots$$

all precede *aab*. Thus the lexicographic order on the infinite set Σ^* is very complicated and is difficult to visualize. Nevertheless, it defines a chain, as we show in the next theorem. ■

Theorem 2 If $(\Sigma, \preceq)$ is a chain, then $(\Sigma^*, \preceq^*)$ is a well-ordered chain and $(\Sigma^*, \preceq_L)$ is a chain.

Proof. We know from Theorem 1 that each $(\Sigma^k, \preceq^k)$ is a chain. The standard order $\preceq^*$ simply links these chains end to end for $k = 0, 1, 2, \ldots$, so $(\Sigma^*, \preceq^*)$ is a chain. To check that $\preceq^*$ well-orders Σ^*, consider a nonempty subset A of Σ^*. Let k be the shortest length of a word in A. Since $A \cap \Sigma^k$ is nonempty and finite, $A \cap \Sigma^k$ possesses a smallest element w_0 in $(\Sigma^k, \preceq^k)$. It follows that $w_0 \preceq^* w$ for all w in A so that w_0 is the smallest member of A.

For the lexicographic order $\preceq_L$, consider two elements $a_1 \cdots a_m$ and $b_1 \cdots b_n$ in Σ^* with $m \leq n$. If $a_1 \cdots a_m = b_1 \cdots b_m$, then $a_1 \cdots a_m \preceq_L b_1 \cdots b_n$, by definition. Otherwise, since $(\Sigma^m, \preceq^m)$ is a chain, one of $a_1 \cdots a_m$ and $b_1 \cdots b_m$ precedes the other in $(\Sigma^m, \preceq^m)$, so $a_1 \cdots a_m$ and $b_1 \cdots b_n$ are comparable in $(\Sigma^*, \preceq_L)$. Thus every two members of Σ^* are comparable under $\preceq_L$, and $\preceq_L$ is a total order. ∎

If Σ has more than one element, then $(\Sigma^*, \preceq_L)$ is *not* well-ordered. For example, the set $\{b, ab, aab, aaab, aaaab, \ldots\}$ in Example 10 has no smallest member. Of course, every *finite* subset of Σ^* has a smallest member, since it is itself a finite chain.

<h1 style="text-align:center">EXERCISES 11.2</h1>

1. Let P be the set of all subsets of $\{1, 2, 3, 4, 5\}$.

   (a) Give two examples of maximal chains in $(P, \subseteq)$.

   (b) How many maximal chains are there in $(P, \subseteq)$?

2. Let $A = \{1, 2, 3, 4\}$ with the usual order, and let $S = A \times A$ with the product order.

   (a) Find a chain in S with 7 members.

   (b) Can a chain in S have 8 members? Explain.

3. Let $(S, |)$ be the set $\{2, 3, 4, \ldots, 999, 1000\}$ with partial order "is a factor of."

   (a) There are exactly 500 maximal elements of $(S, |)$. What are they?

   (b) Give two examples of maximal chains in $(S, |)$.

   (c) Does every maximal chain contain a minimal element of S? Explain.

4. (a) Suppose that no chain in the poset $(S, \preceq)$ has more than 73 members. Must a chain in S with 73 members be a maximal chain? Explain.

   (b) Give an example of a poset that has two maximal chains with four members and four maximal chains with two members.

5. Is every chain a lattice? Explain.

6. Let $(C_1, \preceq_1), (C_2, \preceq_2), \ldots, (C_n, \preceq_n)$ be a set of disjoint chains. Describe a way to make $C_1 \cup C_2 \cup \cdots \cup C_n$ into a chain.

7. Let $(S, \preceq)$ be a poset and T a set. Define the relation $\preceq$ on FUN(T, S) by

$$f \preceq g \quad \text{if} \quad f(t) \preceq g(t) \text{ for all } t \text{ in } T.$$

(a) Show that $\preceq$ is a partial order.

(b) Show that if m is a maximal element in S, then the function f_m defined by $f_m(t) = m$ for all $t \in T$ is a maximal element of FUN(T, S).

(c) Show that if S is a lattice and if $f, g \in$ FUN(T, S), then the function h defined by $h(t) = f(t) \vee g(t)$ for $t \in T$ is the least upper bound of $\{f, g\}$.

(d) Describe the quasi-order $\prec$ associated with $\preceq$.

8. Let $\mathbb{N} \times \mathbb{N}$ have the product order. Draw a sketch like the one in Figure 5 that shows $\{(m, n):(m, n) \preceq (5, 2)\}$.

9. Let $S = \{0, 1, 2\}$ with the usual order, and let $T = \{a, b\}$ with $a < b$.

(a) Draw a Hasse diagram for the poset (FUN(T, S), $\preceq$) with order $\preceq$ described in Exercise 7. *Hint:* See Example 8(c) of §11.1.

(b) Draw a Hasse diagram for the poset ($S \times S$, $\preceq$) with the product order.

(c) Draw a Hasse diagram for $S \times T$ with the product order.

10. Suppose that (S, $\preceq_1$) and (T, $\preceq_2$) are posets and we define $\preceq$ on $S \times T$ by

$$(s, t) \preceq (s', t') \quad \text{if} \quad s \preceq_1 s' \quad \text{or} \quad t \preceq_2 t'.$$

Is $\preceq$ a partial order? Explain.

11. Let $S = \{0, 1, 2\}$ and $T = \{3, 4\}$ with both sets given the usual order. List the members of the following sets in increasing filing order.

(a) $S \times S$ (b) $S \times T$ (c) $T \times S$

12. Suppose that (S, $\preceq_1$) and (T, $\preceq_2$) are posets, each with more than one element. Show that $S \times T$ with the product order is not a chain.

13. Let $\mathbb{B} = \{0, 1\}$ with the usual order. List the element 101, 010, 11, 000, 10, 0010, 1000 of $\mathbb{B}^*$ in increasing order

(a) for the lexicographic order,

(b) for the standard order.

14. Let (Σ, $\preceq$) be a nonempty chain.

(a) Does (Σ^*, $\preceq^*$) have a maximal member? Explain.

(b) Does (Σ^*, $\preceq_L$) have a maximal member? Explain.

15. Let Σ be the English alphabet with the usual order.

(a) List the words of this sentence in increasing standard order.

(b) List the words of this sentence in increasing lexicographic order.

16. Under what conditions on Σ are the lexicographic order and standard order on Σ^* the same?

17. Show that in a finite poset (S, $\preceq$) every maximal chain contains a minimal element of S.

18. Let (S, $\preceq_1$) and (T, $\preceq_2$) be posets, and give $S \times T$ the filing order.

(a) Show that if m_1 is maximal in S and m_2 is maximal in T, then (m_1, m_2) is maximal in $S \times T$.

(b) Does $S \times T$ have other maximal elements besides the ones described in is part (a)? Explain.

(c) Suppose that $S \times T$ has a largest element. Must S or T have a largest element? Explain.

19. Let $(S_1, \preceq_1), \ldots, (S_n, \preceq_n)$ be posets and define $\prec$ on $S_1 \times \cdots \times S_n$ by

$$(s_1, \ldots, s_n) \prec (t_1, \ldots, t_n) \text{ if } s_1 \prec_1 t_1 \text{ or if there is an } r \text{ in}$$
$$\{2, \ldots, n\} \text{ such that } s_1 = t_1, \ldots, s_{r-1} = t_{r-1} \text{ and } s_r \prec_r t_r.$$

Show that $\prec$ is a quasi-order.

§ 11.3   Properties of General Relations

In Chapter 3 we studied equivalence relations, and in the preceding two sections we have discussed partial orders. Both of these types of relations are reflexive and transitive, but the difference between symmetry and antisymmetry gives the two subjects entirely different feelings. Partial orders have a sense of direction from small to large, whereas equivalence relations partition sets into unrelated blocks whose members are lumped together because they have something in common.

Each of these two types of binary relation provides us with some organized structure on a set S. In contrast, the theory of general binary relations, i.e., subsets of $S \times S$, is so abstract and nonspecific that there is really no structure to be developed. In this section we will be focusing on statements that are true for all relations. Not surprisingly, we will be limited to broad generalities. We begin by discussing how to compose two relations to get a third one. We then translate our account into matrix terms and revisit the links between relations, matrices and digraphs that we first discussed in Chapter 3.

Functions can be viewed as relations, as we saw in § 3.1. We identify the function $f: S \to T$ with the relation R_f from S to T defined by

$$R_f = \{(s, t) \in S \times T : f(s) = t\}.$$

If $g: T \to U$ is also a function, then the composite function $g \circ f: S \to U$ yields the relation

$$R_{g \circ f} = \{(s, u) \in S \times U : (g \circ f)(s) = g(f(s)) = u\},$$

which we can think of as the composite of R_f and R_g. The pair (s, u) is in $R_{g \circ f}$ if and only if $u = g(t)$, where $t = f(s) \in T$, so

$$R_{g \circ f} = \{(s, u) \in S \times T : (s, t) \in R_f \text{ and } (t, u) \in R_g \text{ for some } t \in T\}.$$

We generalize this fact about relations associated with functions to define the composition of two arbitrary relations. If R_1 is a relation from S to T and R_2 is a relation from T to U, then the **composite** of R_1 and R_2 is the relation $\boldsymbol{R_2 \circ R_1}$ from S to U defined by

$$\boldsymbol{R_2 \circ R_1} = \{(s, u) \in S \times U : \text{for some } t \in T, (s, t) \in R_1 \text{ and } (t, u) \in R_2\}.$$

For relations determined by functions this definition gives $R_g \circ R_f = R_{g \circ f}$.

Since we think of R_1 as the first relation and R_2 as the second, on aesthetic grounds alone this general definition seems backward. Moreover, it will turn out to be backward when we observe the connection between composition of relations and products of their matrices. Many people use $R_1 \circ R_2$ as a name for what we have called $R_2 \circ R_1$. Such usage is inconsistent with the notation for functional composition and is a source of confusion. To avoid misunderstanding, we will write $\boldsymbol{R_1 R_2}$ rather than $R_1 \circ R_2$ for $R_2 \circ R_1$. To summarize:

Given relations R_1 from S to T and R_2 from T to U, the **composite relation**

$$\{(s, u) \in S \times U : (s, t) \in R_1 \text{ and } (t, u) \in R_2 \text{ for some } t \in T\}$$

will be denoted by either $R_1 R_2$ or $R_2 \circ R_1$. Thus $(s, u) \in R_1 R_2$ precisely if there is a t in T such that $(s, t) \in R_1$ and $(t, u) \in R_2$.

EXAMPLE 1   Example 2 of §3.1 concerns university students, courses and departments. The relations are

$$R_1 = \{(s, c) \in S \times C : s \text{ is enrolled in } c\}$$

and

$$R_2 = \{(c, d) \in C \times D : c \text{ is required by } d\}.$$

Observe that

$$R_1 R_2 = \{(s, d) \in S \times D : (s, c) \in R_1 \text{ and } (c, d) \in R_2 \text{ for some } c \in C\}.$$

Therefore (s, d) belongs to $R_1 R_2$ provided that student s is taking some course that is required by department d.

Note that $R_2 R_1$ makes no sense, because the second entries of R_2 lie in the set D while the first entries of R_1 lie in S; it could not happen that $(c, t) \in R_2$ and $(t, c') \in R_1$. Of course, $R_2 \circ R_1$ makes sense; this is just another name for $R_1 R_2$. ■

EXAMPLE 2   Consider relations R_1 and R_2 from S to T and relations R_3 and R_4 from T to U.

(a) If $R_1 \subseteq R_2$ and $R_3 \subseteq R_4$, then $R_1 R_3 \subseteq R_2 R_4$. To see this, consider $(s, u) \in R_1 R_3$. Then for some $t \in T$ we have $(s, t) \in R_1$ and $(t, u) \in R_3$. Since $R_1 \subseteq R_2$ and $R_3 \subseteq R_4$ we also have $(s, t) \in R_2$ and $(t, u) \in R_4$. So $(s, u) \in R_2 R_4$. This argument shows that $R_1 R_3 \subseteq R_2 R_4$.

(b) The union of two relations from A to B, i.e., two subsets of $A \times B$, is a relation from A to B. We show that for R_1, R_2, R_3 and R_4 as above,

$$(R_1 \cup R_2) R_3 = R_1 R_3 \cup R_2 R_3.$$

Since $R_1 \subseteq R_1 \cup R_2$ we have $R_1 R_3 \subseteq (R_1 \cup R_2)R_3$ from part (a); similarly, $R_2 R_3 \subseteq (R_1 \cup R_2)R_3$, so

$$R_1 R_3 \cup R_2 R_3 \subseteq (R_1 \cup R_2)R_3.$$

To check the reverse inclusion, consider $(s, u) \in (R_1 \cup R_2)R_3$. For some $t \in T$ we have $(s, t) \in R_1 \cup R_2$ and $(t, u) \in R_3$. Then either $(s, t) \in R_1$ so that $(s, u) \in R_1 R_3$, or else $(s, t) \in R_2$ so that $(s, u) \in R_2 R_3$. Either way, $(s, u) \in R_1 R_3 \cup R_2 R_3$ and hence

$$(R_1 \cup R_2)R_3 \subseteq R_1 R_3 \cup R_2 R_3. \quad \blacksquare$$

In § 1.3 we observed that composition of functions is associative. So is compositions of relations.

Associative Law for Relations

If R_1 is a relation from S to T, R_2 is a relation from T to U, and R_3 is a relation from U to V, then

$$(R_1 R_2)R_3 = R_1(R_2 R_3).$$

Proof. We show that an ordered pair (s, v) in $S \times V$ belongs to $(R_1 R_2)R_3$ if and only if

(∗)   for some $t \in T$ and $u \in U$ we have $(s, t) \in R_1$, $(t, u) \in R_2$ and $(u, v) \in R_3$.

A similar argument shows that (s, v) belongs to $R_1(R_2 R_3)$ if and only if (∗) holds.

Consider (s, v) in $(R_1 R_2)R_3$. Since $R_1 R_2$ is a relation from S to U, this means that there exists $u \in U$ such that $(s, u) \in R_1 R_2$ and $(u, v) \in R_3$. Since $(s, u) \in R_1 R_2$ there exists $t \in T$ such that $(s, t) \in R_1$ and $(t, u) \in R_2$. Thus (∗) holds.

Now suppose that (∗) holds for an element (s, v) in $S \times V$. Then $(s, t) \in R_1$ and $(t, u) \in R_2$ so that $(s, u) \in R_1 R_2$. Since also $(u, v) \in R_3$, we conclude that $(s, v) \in (R_1 R_2)R_3$. $\quad \blacksquare$

In view of the associative law, we may write $R_1 R_2 R_3$ for either $(R_1 R_2)R_3$ or $R_1(R_2 R_3)$. As shown in the proof, (s, v) belongs to $R_1 R_2 R_3$ provided there exist $t \in T$ and $u \in U$ such that $(s, t) \in R_1$, $(t, u) \in R_2$ and $(u, v) \in R_3$.

The sets and relations that we have been considering so far could have been finite or infinite. We now restrict ourselves to finite sets S, T and U, and consider relations R_1 from S to T and R_2 from T to U. Just as we did in § 3.3, we can associate matrices of 0's and 1's with these relations. We list the members of each of the sets S, T and U in some order. Then the (s, t)-entry of the matrix $\mathbf{A}_1$ corresponding to the relation R_1 is 1 if $(s, t) \in R_1$ and is 0 otherwise. The matrix $\mathbf{A}_2$ for R_2 is defined similarly. Given $\mathbf{A}_1$ and $\mathbf{A}_2$ we want to find the matrix for the composite relation $R_1 R_2$.

EXAMPLE 3 Let $S = \{1, 2, 3, 4, 5\}$, $T = \{a, b, c\}$ and $U = \{e, f, g, h\}$. Consider the relations

$$R_1 = \{(1, a), (2, a), (2, c), (3, a), (3, b), (4, a), (4, b), (4, c), (5, b)\},$$

$$R_2 = \{(a, e), (a, g), (b, f), (b, g), (b, h), (c, e), (c, g), (c, h)\},$$

so that

$$R_1 R_2 = \{(1, e), (1, g), (2, e), (2, g), (2, h), (3, e), (3, f), (3, g), (3, h),$$

$$(4, e), (4, f), (4, g), (4, h), (5, f), (5, g), (5, h)\}.$$

The matrices $\mathbf{A}_1$, $\mathbf{A}_2$ and $\mathbf{A}$ for these relations are given in Figure 1. Compare the matrix $\mathbf{A}$ with the product $\mathbf{A}_1 \mathbf{A}_2$ in Figure 1. The 1's in the matrix $\mathbf{A}$ occur where the nonzero entries occur in $\mathbf{A}_1 \mathbf{A}_2$. This coincidence is not an accident, as we now explain.

$$
\mathbf{A}_1 =
\begin{array}{c c}
 & \begin{array}{c c c} a & b & c \end{array} \\
\begin{array}{c} 1 \\ 2 \\ 3 \\ 4 \\ 5 \end{array} &
\begin{bmatrix}
1 & 0 & 0 \\
1 & 0 & 1 \\
1 & 1 & 0 \\
1 & 1 & 1 \\
0 & 1 & 0
\end{bmatrix}
\end{array}
\qquad
\mathbf{A}_2 =
\begin{array}{c c}
 & \begin{array}{c c c c} e & f & g & h \end{array} \\
\begin{array}{c} a \\ b \\ c \end{array} &
\begin{bmatrix}
1 & 0 & 1 & 0 \\
0 & 1 & 1 & 1 \\
1 & 0 & 1 & 1
\end{bmatrix}
\end{array}
$$

$$
\mathbf{A} =
\begin{array}{c c}
 & \begin{array}{c c c c} e & f & g & h \end{array} \\
\begin{array}{c} 1 \\ 2 \\ 3 \\ 4 \\ 5 \end{array} &
\begin{bmatrix}
1 & 0 & 1 & 0 \\
1 & 0 & 1 & 1 \\
1 & 1 & 1 & 1 \\
1 & 1 & 1 & 1 \\
0 & 1 & 1 & 1
\end{bmatrix}
\end{array}
\qquad
\mathbf{A}_1 \mathbf{A}_2 =
\begin{array}{c c}
 & \begin{array}{c c c c} e & f & g & h \end{array} \\
\begin{array}{c} 1 \\ 2 \\ 3 \\ 4 \\ 5 \end{array} &
\begin{bmatrix}
1 & 0 & 1 & 0 \\
2 & 0 & 2 & 1 \\
1 & 1 & 2 & 1 \\
2 & 1 & 3 & 2 \\
0 & 1 & 1 & 1
\end{bmatrix}
\end{array}
$$

FIGURE 1

Consider $s \in \{1, 2, 3, 4, 5\}$ and $u \in \{e, f, g, h\}$. The (s, u)-entry of $\mathbf{A}_1 \mathbf{A}_2$ is

$$\sum_{t \in \{a, b, c\}} \mathbf{A}_1[s, t] \mathbf{A}_2[t, u].$$

The sum is positive if any product term is positive. A product $\mathbf{A}_1[s, t] \mathbf{A}_2[t, u]$ is 0 unless both $\mathbf{A}_1[s, t]$ and $\mathbf{A}_2[t, u]$ are 1, in which case we have $(s, t) \in R_1$ and $(t, u) \in R_2$. Thus the sum is 0 if there is no such t, i.e., if $(s, u) \notin R_1 R_2$, and is greater than 0 if $(s, u) \in R_1 R_2$. More precisely, the sum is exactly the number

$$|\{t \in \{a, b, c\} : (s, t) \in R_1 \text{ and } (t, u) \in R_2\}|.$$

For example, the $(2, e)$-entry is 2 because

$$\{t \in \{a, b, c\} : (2, t) \in R_1 \text{ and } (t, e) \in R_2\} = \{a, c\}.$$

The $(2, f)$-entry is 0 because

$$\{t \in \{a, b, c\} : (2, t) \in R_1 \text{ and } (t, f) \in R_2\} = \varnothing,$$

i.e., $(2, f) \notin R_1 R_2$. ■

As Example 3 illustrates, the (s, u)-entry of the product of the matrices for the relations R_1 and R_2 is the number of t's with $(s, t) \in R_1$ and $(t, u) \in R_2$. To find the matrix for $R_1 R_2$ we just need to know whether this number is nonzero; if it is, then the (s, u)-entry in the matrix for $R_1 R_2$ is 1, and otherwise the entry is 0.

To get the matrix for $R_1 R_2$ we could define a new product for matrices, letting $\mathbf{A}_1 * \mathbf{A}_2$ be the matrix obtained from $\mathbf{A}_1 \mathbf{A}_2$ by replacing each nonzero entry with 1. An equivalent but superior approach replaces integer arithmetic by more suitable operations. Matrices for relations have their entries in the set $\mathbb{B} = \{0, 1\}$, which has **Boolean operations** $\vee$ and $\wedge$ defined by Table 1. These operations are the same ones we have seen in our studies of truth tables in Chapter 2 and Boolean algebras in Chapter 10. Note that $m \vee n = \max\{m, n\}$ and $m \wedge n = \min\{m, n\}$.

TABLE 1

$\vee$	0	1		$\wedge$	0	1
0	0	1		0	0	0
1	1	1		1	0	1

Matrices with entries in $\mathbb{B}$ are called **Boolean matrices**. We define the **Boolean product** $\mathbf{A}_1 * \mathbf{A}_2$ of the $m \times n$ Boolean matrix $\mathbf{A}_1$ and the $n \times p$ Boolean matrix $\mathbf{A}_2$ by using the ordinary definition of matrix product, but with the operations addition and multiplication replaced by $\vee$ and $\wedge$, respectively. That is, the (i, k)-entry of $\mathbf{A}_1 * \mathbf{A}_2$ is

$$(\mathbf{A}_1 * \mathbf{A}_2)[i, k]$$
$$= (\mathbf{A}_1[i, 1] \wedge \mathbf{A}_2[1, k]) \vee (\mathbf{A}_1[i, 2] \wedge \mathbf{A}_2[2, k]) \vee \cdots \vee (\mathbf{A}_1[i, n] \wedge \mathbf{A}_2[n, k]),$$

which can be written more compactly as

$$\bigvee_{j=1}^{n} (\mathbf{A}_1[i, j] \wedge \mathbf{A}_2[j, k]).$$

EXAMPLE 4   The Boolean product $\mathbf{A}_1 * \mathbf{A}_2$ of the matrices in Figure 1 is the matrix $\mathbf{A}$ in Figure 1. For example, the $(3, g)$-entry of $\mathbf{A}_1 * \mathbf{A}_2$ is $(1 \wedge 1) \vee (1 \wedge 1) \vee (0 \wedge 1) = 1 \vee 1 \vee 0 = 1$. The $(5, e)$-entry is $(0 \wedge 1) \vee (1 \wedge 0) \vee (0 \wedge 1) = 0 \vee 0 \vee 0 = 0$. ■

By the definitions of $\vee$ and $\wedge$, the (i, k)-entry in $\mathbf{A}_1 * \mathbf{A}_2$ is 1 if and only if at least one of the terms $\mathbf{A}_1[i, j] \wedge \mathbf{A}_2[j, k]$ is 1, and such a term is 1 if and only if both $\mathbf{A}_1[i, j]$ and $\mathbf{A}_2[j, k]$ are 1. This fact and the discussion following Example 3 yield the following result.

Theorem 1 Consider relations R_1 from S to T and R_2 from T to U, where S, T and U are finite. If $\mathbf{A}_1$ and $\mathbf{A}_2$ are the corresponding Boolean matrices, then the Boolean product $\mathbf{A}_1 * \mathbf{A}_2$ is the matrix for the composite relation $R_1 R_2$.

The Boolean product operation $*$ on Boolean matrices is associative. This fact can be shown directly or by using Theorem 1 and the associativity of the corresponding relations, as suggested in Exercise 17.

For the remainder of this section we consider relations on a single set S. Since relations on S are simply subsets of $S \times S$, the collection of all relations on S is $\mathscr{P}(S \times S)$.

Theorem 2 The composition operation on $\mathscr{P}(S \times S)$ is associative, and $\mathscr{P}(S \times S)$ has an identity E, i.e., a member E such that $RE = ER = R$ for all $R \in \mathscr{P}(S \times S)$.

Proof. First note that if R_1 and R_2 are in $\mathscr{P}(S \times S)$, so is their composite, $R_1 R_2$. Associativity of composition has already been verified. The identity is the **equality relation**

$$E = \{(x, x) \in S \times S : x \in S\}. \quad \blacksquare$$

The usual notational conventions for associative operations apply. Thus if R is a relation on S, then $R^0 = E$ and, for $n \in \mathbb{P}$, R^n is the composition of R with itself n times. Note that if $n > 1$, then (x, z) belongs to R^n provided that there exist $y_1, y_2, \ldots, y_{n-1}$ in S such that (x, y_1), $(y_1, y_2), \ldots, (y_{n-1}, z)$ are all in R. In other words, $(x, z) \in R^n$ if x and z are R-related through a chain of length n.

EXAMPLE 5 (a) Consider a graph with no parallel edges. As in §3.2, define the adjacency relation R on the set V of vertices by $(u, v) \in R$ whenever $\{u, v\}$ is an edge. Figure 2 shows an example. We have $(u, w) \in R^n$ provided there exist vertices $v_1, \ldots, v_{n-1}$ such that $(u, v_1), (v_1, v_2), \ldots, (v_{n-1}, w)$ are all in R, i.e., provided that there is a path of length n connecting u and w. The Boolean matrix $\mathbf{A}$ for R is the adjacency matrix defined in §3.3. Its Boolean power

$$\mathbf{A} * \mathbf{A} * \cdots * \mathbf{A} \quad [n \text{ times}],$$

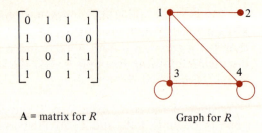

$$\begin{bmatrix} 0 & 1 & 1 & 1 \\ 1 & 0 & 0 & 0 \\ 1 & 0 & 1 & 1 \\ 1 & 0 & 1 & 1 \end{bmatrix}$$

A = matrix for R Graph for R

FIGURE 2

which is the Boolean matrix for R^n, tells us exactly which pairs of vertices are connected by paths of length n. If we also wanted the number of such paths, we would calculate the ordinary matrix power $\mathbf{A}^n$, as we did in §3.4 and will do next.

(b) If the relation R on $\{1, 2, 3, 4\}$ has the matrix and graph of Figure 2, then the relation R^2 has the matrix and graph of Figure 3. The (3, 4)-entry of $\mathbf{A} * \mathbf{A}$ tells us that there is at least one path in Figure 2 of length 2 from vertex 3 to vertex 4. That is why there is an edge from 3 to 4 in the graph of R^2. The (3, 4)-entry of $\mathbf{A}^2$ is 3, so there are exactly three paths of length 2 in Figure 2 from vertex 3 to vertex 4. What are they? ■

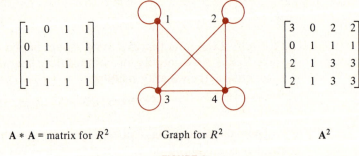

$$\begin{bmatrix} 1 & 0 & 1 & 1 \\ 0 & 1 & 1 & 1 \\ 1 & 1 & 1 & 1 \\ 1 & 1 & 1 & 1 \end{bmatrix} \qquad\qquad \begin{bmatrix} 3 & 0 & 2 & 2 \\ 0 & 1 & 1 & 1 \\ 2 & 1 & 3 & 3 \\ 2 & 1 & 3 & 3 \end{bmatrix}$$

$\mathbf{A} * \mathbf{A}$ = matrix for R^2 Graph for R^2 $\mathbf{A}^2$

FIGURE 3

Once we have agreed on an order in which to list the members of a finite set S, every relation R on S has its associated Boolean matrix $\mathbf{A}$. This matrix can be viewed as the adjacency matrix for a digraph whose vertices are the members of S and which has an edge from x to y if and only if $\mathbf{A}$ has a 1 as its (x, y)-entry. Hence the digraph has an edge from x to y if and only if $(x, y) \in R$, i.e., R is the adjacency relation for the digraph. We will call a picture of this digraph a **picture of** R. Note that although $\mathbf{A}$ depends on the order in which we list the members of S, the digraph does not.

If R is symmetric, then the pairs of oppositely directed edges in the associated digraph can be combined into edges of an undirected graph for which R is the adjacency relation. In Example 5 we started with such an undirected graph; the relation that we got from it was symmetric, of course.

The next theorem shows the connection between transitivity and the composition of relations on S.

Theorem 3 | If R is a relation on a set S, then R is transitive if and only if $R^2 \subseteq R$.

Proof. Suppose first that R is transitive, and consider $(x, z) \in R^2$. By the definition of R^2 there is a y in S such that $(x, y) \in R$ and $(y, z) \in R$. Since R is transitive, (x, z) is also in R. We have shown that every (x, z) in R^2 is in R, i.e., that $R^2 \subseteq R$.

For the converse, suppose that $R^2 \subseteq R$. Consider (x, y) and (y, z) in R. Then (x, z) is in R^2 and hence in R. This proves that R is transitive. ∎

For $m \times n$ Boolean matrices A_1 and A_2 we write $A_1 \leq A_2$ if every entry of A_1 is less than or equal to the corresponding entry of A_2, i.e., in case

$$A_1[i, j] \leq A_2[i, j] \qquad \text{for} \quad 1 \leq i \leq m \quad \text{and} \quad 1 \leq j \leq n.$$

If R_1 and R_2 are relations from S to T, with matrices A_1 and A_2, then

$$R_1 \subseteq R_2 \quad \text{if and only if} \quad A_1 \leq A_2;$$

think about where the 1's and 0's are in A_1 and A_2 [Exercise 16(a)]. It follows that $\leq$ is a partial order on the set of Boolean matrices. Moreover, a relation R on a set S satisfies $R^2 \subseteq R$ if and only if its matrix A satisfies $A * A \leq A$. Thus R is transitive if and only if $A * A \leq A$.

EXAMPLE 6   Consider the relation R on $\{1, 2, 3\}$ with matrix

$$A = \begin{bmatrix} 1 & 0 & 0 \\ 1 & 0 & 0 \\ 1 & 1 & 0 \end{bmatrix}.$$

Since

$$A * A = \begin{bmatrix} 1 & 0 & 0 \\ 1 & 0 & 0 \\ 1 & 1 & 0 \end{bmatrix} * \begin{bmatrix} 1 & 0 & 0 \\ 1 & 0 & 0 \\ 1 & 1 & 0 \end{bmatrix} = \begin{bmatrix} 1 & 0 & 0 \\ 1 & 0 & 0 \\ 1 & 0 & 0 \end{bmatrix} \leq \begin{bmatrix} 1 & 0 & 0 \\ 1 & 0 & 0 \\ 1 & 1 & 0 \end{bmatrix} = A,$$

R is transitive. The transitivity of R can also be seen from its corresponding digraph in Figure 4. Whenever a path of length 2 connects two vertices, a single edge also does. For example, 3 2 1 is a path and there is an edge from 3 to 1. ∎

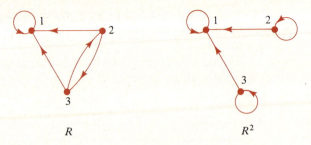

FIGURE 4

Since relations on finite sets correspond to Boolean matrices, properties of relations can be described in matrix terms. To list some of the more important matrix equivalents we introduce a bit more notation. For Boolean $m \times n$ matrices $\mathbf{A}_1$ and $\mathbf{A}_2$ we define $\mathbf{A}_1 \vee \mathbf{A}_2$ by

$$(\mathbf{A}_1 \vee \mathbf{A}_2)[i, j] = \mathbf{A}_1[i, j] \vee \mathbf{A}_2[i, j] \qquad \text{for} \quad 1 \le i \le m \quad \text{and} \quad 1 \le j \le n.$$

The matrix $\mathbf{A}_1 \wedge \mathbf{A}_2$ has a similar definition. For example, if

$$\mathbf{A}_1 = \begin{bmatrix} 1 & 0 & 1 & 1 \\ 0 & 1 & 1 & 0 \\ 1 & 1 & 0 & 1 \end{bmatrix} \quad \text{and} \quad \mathbf{A}_2 = \begin{bmatrix} 1 & 1 & 1 & 0 \\ 0 & 1 & 1 & 1 \\ 1 & 0 & 1 & 0 \end{bmatrix},$$

then

$$\mathbf{A}_1 \vee \mathbf{A}_2 = \begin{bmatrix} 1 & 1 & 1 & 1 \\ 0 & 1 & 1 & 1 \\ 1 & 1 & 1 & 1 \end{bmatrix} \quad \text{and} \quad \mathbf{A}_1 \wedge \mathbf{A}_2 = \begin{bmatrix} 1 & 0 & 1 & 0 \\ 0 & 1 & 1 & 0 \\ 1 & 0 & 0 & 0 \end{bmatrix}.$$

The summary below lists the matrix versions of some familiar properties that relations may have. Recall that $\mathbf{A}^T$ denotes the transpose of the matrix $\mathbf{A}$. The equivalences are straightforward to verify [see Exercises 15 and 16].

Summary. Let R be a relation on a finite set S with Boolean matrix $\mathbf{A}$. Then

(R) R is reflexive if and only if all the diagonal entries of $\mathbf{A}$ are 1;

(AR) R is antireflexive if and only if all the diagonal entries of $\mathbf{A}$ are 0;

(S) R is symmetric if and only if $\mathbf{A} = \mathbf{A}^T$;

(AS) R is antisymmetric if and only if $\mathbf{A} \wedge \mathbf{A}^T \le \mathbf{I}$, where $\mathbf{I}$ is the identity matrix;

(T) R is transitive if and only if $\mathbf{A} * \mathbf{A} \le \mathbf{A}$.

Let R_1 and R_2 be relations from a finite set S to a finite set T with Boolean matrices $\mathbf{A}_1$ and $\mathbf{A}_2$. Then

(a) $R_1 \subseteq R_2$ if and only if $\mathbf{A}_1 \le \mathbf{A}_2$;

(b) $R_1 \cup R_2$ has Boolean matrix $\mathbf{A}_1 \vee \mathbf{A}_2$;

(c) $R_1 \cap R_2$ has Boolean matrix $\mathbf{A}_1 \wedge \mathbf{A}_2$.

Finally, composition of relations corresponds to the Boolean product of their associated matrices, as explained in Theorem 1.

EXERCISES 11.3

1. For each of the following Boolean matrices, consider the corresponding relation R on $\{1, 2, 3\}$. Find the Boolean matrix for R^2 and determine whether R is transitive.

(a) $\begin{bmatrix} 1 & 1 & 0 \\ 0 & 1 & 1 \\ 1 & 0 & 1 \end{bmatrix}$
(b) $\begin{bmatrix} 1 & 0 & 1 \\ 0 & 1 & 0 \\ 1 & 0 & 1 \end{bmatrix}$
(c) $\begin{bmatrix} 0 & 0 & 1 \\ 0 & 1 & 0 \\ 1 & 0 & 0 \end{bmatrix}$

2. Draw pictures of the relations in Exercise 1.

3. Let $S = \{1, 2, 3\}$ and $R = \{(2, 1), (2, 3), (3, 2)\}$.

   (a) Find the matrices for R and R^2.

   (b) Draw pictures of the relations in part (a).

   (c) Is R transitive?

   (d) Is R^2 transitive?

   (e) Is $R \cup R^2$ transitive?

4. Let $S = \{1, 2, 3\}$ and $R = \{(1, 1), (1, 2), (1, 3), (3, 2)\}$.

   (a) Find the matrices for R and R^2.

   (b) Draw pictures of the relations in part (a).

   (c) Show that R is transitive, i.e., $R^2 \subseteq R$, but that $R^2 \neq R$.

   (d) Find R^n for all $n = 2, 3, \ldots$.

5. Let R be the relation on $\{1, 2, 3\}$ with Boolean matrix

$$\mathbf{A} = \begin{bmatrix} 1 & 0 & 0 \\ 0 & 1 & 1 \\ 1 & 0 & 1 \end{bmatrix}.$$

   (a) Find the Boolean matrix for R^n for $n \geq 0$.

   (b) Is R reflexive? symmetric? transitive?

6. Repeat Exercise 5 for

$$\mathbf{A} = \begin{bmatrix} 0 & 1 & 0 \\ 1 & 1 & 1 \\ 0 & 1 & 0 \end{bmatrix}.$$

7. Consider the functions f and g from $\{1, 2, 3, 4\}$ to itself defined by $f(m) = \max\{2, 4 - m\}$ and $g(m) = 5 - m$.

(a) Find Boolean matrices $\mathbf{A}_f$ and $\mathbf{A}_g$ for the relations R_f and E_g corresponding to f and g.

(b) Find Boolean matrices for $R_f R_g$ and $R_{f \circ g}$ and compare.

(c) Find Boolean matrices for the converse relations $R_f^{\leftarrow}$ and $R_g^{\leftarrow}$. Do these relations correspond to functions?

8. Give Boolean matrices for the following relations on $S = \{0, 1, 2, 3\}$.

(a) $(m, n) \in R_1$ if $m + n = 3$
(b) $(m, n) \in R_2$ if $m \equiv n \pmod 2$

(c) $(m, n) \in R_3$ if $m \leq n$
(d) $(m, n) \in R_4$ if $m + n \leq 4$

(e) $(m, n) \in R_5$ if $\max\{m, n\} = 3$

9. For each relation in Exercise 8, specify which of the properties (R), (AR), (S), (AS) and (T) the relation satisfies.

10. (a) Which of the relations in Exercise 8 are partial orders?

(b) Which of the relations in Exercise 8 are equivalence relations?

11. Let R_1 and R_2 be relations on a set S. Prove or disprove.

(a) If R_1 and R_2 are reflexive, so is $R_1 R_2$.

(b) If R_1 and R_2 are transitive, so is $R_1 R_2$.

(c) If R_1 and R_2 are symmetric, so is $R_1 R_2$.

12. What is the Boolean matrix for the equality relation E on a finite set S? Is this relation reflexive? symmetric? transitive?

13. Consider relations R_1 and R_2 from S to T and relations R_3 and R_4 from T to U.

(a) Show that $R_1(R_3 \cup R_4) = R_1 R_3 \cup R_1 R_4$.

(b) Show that $(R_1 \cap R_2)R_3 \subseteq R_1 R_3 \cap R_2 R_3$ and that equality need not hold.

(c) How are the relations $R_1(R_3 \cap R_4)$ and $R_1 R_3 \cap R_1 R_4$ related?

14. Let R_1 be a relation from S to T and R_2 be a relation from T to U. Show that the converse of $R_1 R_2$ is $R_2^{\leftarrow} R_1^{\leftarrow}$.

15. Verify statements (R), (AR), (S), (AS) and (T) in the summary at the end of this section.

16. Verify (a), (b) and (c) in the summary at the end of this section.

17. Use the associative law for relations to prove that the Boolean product is an associative operation.

18. Let R be a relation from a set S to a set T.

(a) Prove that $RR^{\leftarrow}$ is a symmetric relation on S. Don't use Boolean matrices, since S or T might be infinite.

(b) Use the fact proved in part (a) to quickly infer that $R^{\leftarrow}R$ is a symmetric relation on T.

(c) Under what conditions is $RR^{\leftarrow}$ reflexive?

19. Let R be an antisymmetric and transitive relation on a set S.

(a) Prove that $R \cup E$ is a partial order on S.

(b) Prove that $R \backslash E$ is a quasi-order on S.

§ 11.4   Closures of Relations

Sometimes we may want to form new relations out of those we already have. For example, we may have two equivalence relations, i.e., reflexive, symmetric and transitive relations, R_1 and R_2 on S, and want to find an equivalence relation containing them both. Since R_1 and R_2 are subsets of $S \times S$, the obvious candidate is $R_1 \cup R_2$. Unfortunately, $R_1 \cup R_2$ need not be an equivalence relation; the trouble is that $R_1 \cup R_2$ need not be transitive. Well then, what *is* the smallest transitive relation containing $R_1 \cup R_2$? This turns out to be a loaded question. How do we know there is such a relation? We will see in what follows that if R is a relation on a set S, then there is always a smallest transitive relation containing R, which we will denote by $t(R)$, and we will learn how to find it. There are also smallest relations containing R that are reflexive and symmetric; we'll denote them by $r(R)$ and $s(R)$.

EXAMPLE 1   Consider the relation R on $\{1, 2, 3, 4\}$ whose Boolean matrix is

$$\mathbf{A} = \begin{bmatrix} 0 & 0 & 1 & 1 \\ 0 & 1 & 0 & 0 \\ 0 & 0 & 1 & 0 \\ 1 & 0 & 0 & 0 \end{bmatrix}.$$

Figure 1 gives a picture of R.

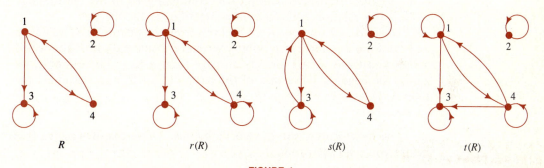

$$R \qquad\qquad r(R) \qquad\qquad s(R) \qquad\qquad t(R)$$

FIGURE 1

(a) The relation R is not reflexive, since neither 1 nor 4 is related to itself. To obtain the reflexive relation $r(R)$, we just need to add the two ordered pairs $(1, 1)$ and $(4, 4)$. The Boolean matrix $\mathbf{r}(\mathbf{A})$ of $r(R)$ is simply the matrix $\mathbf{A}$ with all the diagonal entries set equal to 1:

$$\mathbf{r}(\mathbf{A}) = \begin{bmatrix} 1 & 0 & 1 & 1 \\ 0 & 1 & 0 & 0 \\ 0 & 0 & 1 & 0 \\ 1 & 0 & 0 & 1 \end{bmatrix}.$$

To get the picture for $r(R)$ in Figure 1, we added the missing arrows from points to themselves.

(b) The relation R is not symmetric, since $(1, 3) \in R$ but $(3, 1) \notin R$. If we add the ordered pair $(3, 1)$ to R we get the symmetric relation $s(R)$. Its Boolean matrix is

$$s(\mathbf{A}) = \begin{bmatrix} 0 & 0 & 1 & 1 \\ 0 & 1 & 0 & 0 \\ 1 & 0 & 1 & 0 \\ 1 & 0 & 0 & 0 \end{bmatrix}.$$

To get the picture of $s(R)$ from the picture of R, we added the missing reverses of all the arrows.

(c) The relation R isn't transitive either. For example, we have $(4, 1) \in R$ and $(1, 3) \in R$ but $(4, 3) \notin R$. The scheme for finding $t(R)$ [or its Boolean matrix $\mathbf{t}(\mathbf{A})$] is not so simple as the methods for $r(R)$ and $s(R)$. Since $(4, 1) \in R$ and $(1, 3) \in R$, $t(R)$ will also contain $(4, 1)$ and $(1, 3)$. Since $t(R)$ must be transitive, $t(R)$ must contain $(4, 3)$, so we have to put $(4, 3)$ in $t(R)$. In general, if there is a path from x to y in the digraph of R, i.e., if there are points $x_1, x_2, \ldots, x_{m-1}$ so that $(x, x_1), (x_1, x_2), \ldots, (x_{m-1}, y)$ are all in R, then (x, y) must be in $t(R)$. If there is a path from x to y and also one from y to z, then there is one from x to z. So the set of all pairs (x, y) connected by paths in the digraph is a transitive relation and is the smallest transitive relation $t(R)$ containing R. In the terminology of §3.2, this relation is the reachable relation for the digraph.

To get the picture of $t(R)$ in Figure 1 from the picture of R, we added an edge from a point x to a point y whenever some path in R connected x to y and there wasn't already an edge from x to y. For example, we added the edge $(4, 3)$ because of the path $4\,1\,3$ in R, and we added the loop $(1, 1)$ because of the path $1\,4\,1$ in R. ∎

The next proposition is nearly obvious. Think about why it's true before you read the proof.

Proposition   Let R be a relation. Then $R = r(R)$ if and only if R is reflexive, $R = s(R)$ if and only if R is symmetric, and $R = t(R)$ if and only if R is transitive. Moreover,

$$r(r(R)) = r(R), \quad s(s(R)) = s(R), \quad \text{and} \quad t(t(R)) = t(R).$$

Proof. If R is reflexive, then R is clearly the smallest reflexive relation containing R, i.e., $R = r(R)$. Conversely, if $R = r(R)$, then R is reflexive, since $r(R)$ is. Since $r(R)$ is reflexive, $r(R) = r(r(R))$ by what we have just shown.

The proofs for $s(R)$ and $t(R)$ are similar. ∎

We can think of r, s and t as functions that map relations to relations; for instance s maps R to $s(R)$. Sometimes functions such as these are called "operators." The proposition shows that repeating any of these three operators gives nothing new; operators with this property are called **closure operators**. On the other hand, combining two or more of these operators can lead to new relations. The next theorem gives explicit descriptions of the relations $r(R)$, $s(R)$ and $t(R)$, which are called the **reflexive**, **symmetric** and **transitive closures** of R.

Theorem 1

If R is a relation on a set S and if $E = \{(x, x) : x \in S\}$, then

(r) $r(R) = R \cup E$;

(s) $s(R) = R \cup R^{\leftarrow}$;

(t) $t(R) = \displaystyle\bigcup_{k=1}^{\infty} R^k$.

Proof. (r) A relation on S is reflexive if and only if it contains E. Hence $R \cup E$ is reflexive, and every reflexive relation that contains R must contain $R \cup E$. So $R \cup E$ is the smallest reflexive relation containing R. This argument shows that $r(R) = R \cup E$.

(s) A relation is symmetric if and only if it is its own converse [Exercise 11 of §3.1]. If $(x, y) \in R \cup R^{\leftarrow}$, then $(y, x) \in R^{\leftarrow} \cup R = R \cup R^{\leftarrow}$; thus $R \cup R^{\leftarrow}$ is symmetric. Consider a symmetric relation R' that contains R. If $(x, y) \in R^{\leftarrow}$, then $(y, x) \in R \subseteq R'$ and, since R' is symmetric, $(x, y) \in R'$. This argument shows that $R^{\leftarrow} \subseteq R'$. Since $R \subseteq R'$, we conclude that $R \cup R^{\leftarrow} \subseteq R'$. Hence $R \cup R^{\leftarrow}$ is the smallest symmetric relation containing R, so $s(R) = R \cup R^{\leftarrow}$.

(t) First we show that the union $U = \displaystyle\bigcup_{k=1}^{\infty} R^k$ is transitive. Consider x, y, z in S such that $(x, y) \in U$ and $(y, z) \in U$. Then we must have $(x, y) \in R^k$ and $(y, z) \in R^j$ for some k and j in $\mathbb{P}$. Hence (x, z) belongs to $R^k R^j = R^{k+j}$, so that $(x, z) \in U$. Thus U is a transitive relation containing R.

Now consider any transitive relation R^* containing R. To show that $U \subseteq R^*$, we prove that $R^k \subseteq R^*$ for every $k \in \mathbb{P}$ by induction. The inclusion is given for $k = 1$. If the inclusion holds for some k, then

$$R^{k+1} = R^k R \subseteq R^* R \subseteq R^* R^* \subseteq R^*;$$

the last inclusion is valid because $R*$ is transitive [Theorem 3 of §11.3]. By induction, $R^k \subseteq R*$ for all $k \in \mathbb{P}$, so $U \subseteq R*$. Thus U is the smallest transitive relation containing R, and

$$t(R) = U = \bigcup_{k=1}^{\infty} R^k. \quad \blacksquare$$

If S is finite, then we can improve on the description of $t(R)$ in Theorem 1.

Theorem 2

If R is a relation on a set S with n elements, then

$$t(R) = \bigcup_{k=1}^{n} R^k.$$

Proof. Think of the digraph of R. The pair (x, y) is in $t(R)$ if and only if there is a path from x to y in the digraph. If there is such a path, there is one that doesn't visit the same vertex twice, unless $x = y$. It can't involve more than n vertices, so it can't have length more than n. Thus $(x, y) \in R^k$ for some $k \leq n$. [This argument is essentially the proof of Theorem 1 of §6.1.] $\blacksquare$

EXAMPLE 2   (a) Suppose that R is a relation on a set S with n elements and that **A** is a Boolean matrix for R. Theorem 2 and the matrix equivalents summarized in §11.3 show that the Boolean matrices of $t(R)$, $s(R)$ and $r(R)$ are

$$\mathbf{t(A)} = \mathbf{A} \vee \mathbf{A}^2 \vee \cdots \vee \mathbf{A}^n,$$

$$\mathbf{s(A)} = \mathbf{A} \vee \mathbf{A}^T$$

and

$$\mathbf{r(A)} = \mathbf{A} \vee \mathbf{I},$$

where I is the $n \times n$ identity matrix and the powers $\mathbf{A}^k$ are Boolean powers.

(b) For the relation R back in Example 1, it is easy to see that $\mathbf{s(A)} = \mathbf{A} \vee \mathbf{A}^T$ and $\mathbf{r(A)} = \mathbf{A} \vee \mathbf{I}$ where **I** is the 4×4 identity matrix. One can also verify that

$$\mathbf{t(A)} = \mathbf{A} \vee \mathbf{A}^2 = \mathbf{A} \vee \mathbf{A}^2 \vee \mathbf{A}^3 \vee \mathbf{A}^4 = \begin{bmatrix} 1 & 0 & 1 & 1 \\ 0 & 1 & 0 & 0 \\ 0 & 0 & 1 & 0 \\ 1 & 0 & 1 & 1 \end{bmatrix}.$$

Of course, for such a simple relation it's easier to find $t(R)$ by using the picture of R. $\blacksquare$

Given $\mathbf{A}$, it is easy to construct $\mathbf{s(A)}$ and $\mathbf{r(A)}$ using the formulas in Example 2(a). We can also use the formula in the example to compute $\mathbf{t(A)}$ by forming Boolean matrix powers $\mathbf{A}^k = \mathbf{A} * \cdots * \mathbf{A}$ and then combining the results. For large values of n such a computation involves a great number of products of large matrices, and is quite slow.

We can find $t(R)$ more quickly if S is large by thinking of the digraph for R. We observed in §3.2 and in Example 1 that $t(R)$ is the reachable relation for the digraph of R, whose adjacency relation is R. With each edge given weight 1, WARSHALL'S algorithm in §8.4 computes a matrix $\mathbf{W}^*$ whose (i, j)-entry is a positive integer if there is a path in the digraph from v_i to v_j and is ∞ otherwise. Replacing the ∞'s by 0's and the integers by 1's in the output produces the Boolean matrix $\mathbf{t(A)}$. In fact [Exercise 18], minor changes in the algorithm itself allow us to compute $\mathbf{t(A)}$ using Boolean entries and Boolean operations without the need for ∞'s or large integer calculations.

EXAMPLE 3   For the relation R in Example 1 we obtained

$$\mathbf{r(A)} = \begin{bmatrix} 1 & 0 & 1 & 1 \\ 0 & 1 & 0 & 0 \\ 0 & 0 & 1 & 0 \\ 1 & 0 & 0 & 1 \end{bmatrix} \quad \text{and} \quad \mathbf{s(A)} = \begin{bmatrix} 0 & 0 & 1 & 1 \\ 0 & 1 & 0 & 0 \\ 1 & 0 & 1 & 0 \\ 1 & 0 & 0 & 0 \end{bmatrix}.$$

The Boolean matrix for $sr(R) = s(r(R))$ is

$$\mathbf{sr(A)} = \begin{bmatrix} 1 & 0 & 1 & 1 \\ 0 & 1 & 0 & 0 \\ 1 & 0 & 1 & 0 \\ 1 & 0 & 0 & 1 \end{bmatrix}.$$

This is also the matrix $\mathbf{rs(A)}$ for $rs(R)$, so $rs(R) = sr(R)$. Equality here is not an accident [Exercise 11(b)]. The transitive closure of $sr(R) = rs(R)$ turns out to have the matrix

$$\mathbf{tsr(A)} = \mathbf{trs(A)} = \begin{bmatrix} 1 & 0 & 1 & 1 \\ 0 & 1 & 0 & 0 \\ 1 & 0 & 1 & 1 \\ 1 & 0 & 1 & 1 \end{bmatrix},$$

which is also the matrix of the equivalence relation on $\{1, 2, 3, 4\}$ whose equivalence classes are $\{2\}$ and $\{1, 3, 4\}$: Thus $tsr(R)$ is transitive, symmetric and reflexive. This fact may seem obvious from the notation tsr, but we must be careful here. It is conceivable, for example, that the transitive closure of a symmetric relation might not be symmetric. Our next example shows that the symmetric closure of a transitive relation need not be transitive, so we must be cautious about jumping to conclusions on the basis of notation. ∎

EXAMPLE 4   Let R be the relation on $\{1, 2, 3\}$ with Boolean matrix

$$\mathbf{A} = \begin{bmatrix} 1 & 1 & 1 \\ 0 & 1 & 0 \\ 0 & 0 & 1 \end{bmatrix}.$$

Then R is reflexive [since $\mathbf{I} \leq \mathbf{A}$] and transitive [since $\mathbf{A} * \mathbf{A} = \mathbf{A}$], so $\mathbf{A} = \mathbf{r}(\mathbf{A}) = \mathbf{t}(\mathbf{A}) = \mathbf{tr}(\mathbf{A}) = \mathbf{rt}(\mathbf{A})$. The relation $s(R)$ has matrix

$$s(\mathbf{A}) = \begin{bmatrix} 1 & 1 & 1 \\ 1 & 1 & 0 \\ 1 & 0 & 1 \end{bmatrix}.$$

Since $(2, 1)$ and $(1, 3)$ are in $s(R)$ but $(2, 3)$ is not, $s(R)$ is not transitive. We have

$$\mathbf{st}(\mathbf{A}) = \mathbf{str}(\mathbf{A}) = \mathbf{srt}(\mathbf{A}) = \mathbf{s}(\mathbf{A}) \neq \mathbf{ts}(\mathbf{A}) = \mathbf{tsr}(\mathbf{A}) = \mathbf{trs}(\mathbf{A}).$$

The order in which we form closures does matter.   ■

The next lemma shows that the trouble we encountered in Example 4 is the only kind there is. Except in the case of symmetric closures, which can destroy transitivity as we just saw, forming closures preserves the properties that we already have.

Lemma

> (a) If R is reflexive, so are $s(R)$ and $t(R)$.
> (b) If R is symmetric, so are $r(R)$ and $t(R)$.
> (c) If R is transitive, so is $r(R)$.

Proof. (a) This is obvious, because if $E \subseteq R$, then $E \subseteq s(R)$ and $E \subseteq t(R)$. Part (b) is left to Exercise 10.

(c) Suppose that R is transitive, and consider (x, y) and (y, z) in $r(R) = R \cup E$. If $(x, y) \in E$, then $x = y$, so $(x, z) = (y, z)$ is in $R \cup E$. If $(y, z) \in E$, then $y = z$, so $(x, z) = (x, y)$ is in $R \cup E$. If neither (x, y) nor (y, z) is in E, then both are in R, so $(x, z) \in R \subseteq R \cup E$ by the transitivity of R. Hence $(x, z) \in R \cup E$ in all cases.   ■

The next theorem gives an answer to the basic question with which we began this section.

Theorem 3

> For any relation R on a set S, $tsr(R)$ is the smallest equivalence relation containing R.

Proof. Since $r(R)$ is reflexive, two applications of (a) of the lemma show that $tsr(R)$ is reflexive. Since $sr(R)$ is automatically symmetric, one application of (b) of the lemma shows that $tsr(R)$ is symmetric. Finally, $tsr(R)$ is automatically transitive, so $tsr(R)$ is an equivalence relation.

Consider any equivalence relation R' such that $R \subseteq R'$. Then $r(R) \subseteq r(R') = R'$, hence $sr(R) \subseteq s(R') = R'$ and thus $tsr(R) \subseteq t(R') = R'$. Therefore, $tsr(R)$ is the smallest equivalence relation containing R. ∎

EXAMPLE 5   (a) In Example 3, $tsr(R)$ was shown to be the equivalence relation with equivalence classes $\{2\}$ and $\{1, 3, 4\}$.

(b) Let R be the relation on $\{1, 2, 3\}$ in Example 4. Then

$$\mathbf{r(A)} = \begin{bmatrix} 1 & 1 & 1 \\ 0 & 1 & 0 \\ 0 & 0 & 1 \end{bmatrix}, \qquad \mathbf{sr(A)} = \begin{bmatrix} 1 & 1 & 1 \\ 1 & 1 & 0 \\ 1 & 0 & 1 \end{bmatrix}, \qquad \mathbf{tsr(A)} = \begin{bmatrix} 1 & 1 & 1 \\ 1 & 1 & 1 \\ 1 & 1 & 1 \end{bmatrix}.$$

The smallest equivalence relation containing R is the universal relation $\{1, 2, 3\} \times \{1, 2, 3\}$. These computations can be double-checked by drawing pictures for the corresponding relations. ∎

Theorem 3 has a graph-theoretic interpretation. Starting with a digraph whose adjacency relation is R, we form the digraph for $r(R)$ by putting loops at all vertices. Then we form the digraph for $sr(R)$ by adding edges, if necessary, so that whenever there is an edge from x to y there is also one from y to x. The result is essentially an undirected graph with a loop at each vertex. Now the graph for $tsr(R)$ has edges joining pairs (x, y) that are joined by paths in the graph for $sr(R)$. Thus (x, y) is in the relation $tsr(R)$ if and only if x and y are in the same connected component of the graph for $sr(R)$. Any algorithm that computes connected components, such as FOREST in §6.6, can compute $tsr(R)$ as well.

EXERCISES 11.4

1. Consider the relation R on $\{1, 2, 3\}$ with Boolean matrix $\mathbf{A} = \begin{bmatrix} 0 & 1 & 0 \\ 0 & 0 & 0 \\ 0 & 0 & 1 \end{bmatrix}$.

   Find the Boolean matrices for

   (a) $r(R)$ (b) $s(R)$ (c) $rs(R)$ (d) $sr(R)$ (e) $tsr(R)$

2. Repeat Exercise 1 with $\mathbf{A} = \begin{bmatrix} 0 & 1 & 1 \\ 0 & 0 & 1 \\ 0 & 0 & 0 \end{bmatrix}$.

3. For R as in Exercise 1, list the equivalence classes of $tsr(R)$.

4. For R as in Exercise 2, list the equivalence classes of $tsr(R)$.

5. Repeat Exercise 1 for the relation R on $\{1, 2, 3, 4, 5\}$ with Boolean matrix

$$\mathbf{A} = \begin{bmatrix} 0 & 1 & 0 & 0 & 0 \\ 0 & 1 & 0 & 1 & 0 \\ 0 & 0 & 0 & 0 & 1 \\ 0 & 1 & 0 & 0 & 0 \\ 0 & 0 & 0 & 0 & 0 \end{bmatrix}.$$

6. List the equivalence classes of $tsr(R)$ for the relation R of Exercise 5.

7. Let R be the usual quasi-order relation on $\mathbb{P}$: $(m, n) \in R$ in case $m < n$. Find or describe

   (a) $r(R)$ (b) $sr(R)$ (c) $rs(R)$

   (d) $tsr(R)$ (e) $t(R)$ (f) $st(R)$

8. Repeat Exercise 7 where now $(m, n) \in R$ means that m divides n.

9. The Fraternal Order of Hostile Hermits is an interesting organization. Hermits know themselves. In addition, everyone knows the High Hermit, but neither he nor any of the other members knows any other member. Define the relation R on the F.O.H.H. by $(h_1, h_2) \in R$ if h_1 knows h_2. Determine $st(R)$ and $ts(R)$ and compare. [The High Hermit acts a little like a file server in a computer network.]

10. (a) Show that if (R_k) is a sequence of symmetric relations on a set S, then the union $\bigcup_{k=1}^{\infty} R_k$ is symmetric.

    (b) Let R be a symmetric relation on S. Show that R^n is symmetric for each $n \in \mathbb{P}$.

    (c) Show that if R is symmetric, so are $r(R)$ and $t(R)$.

11. Consider a relation R on a set S.

    (a) Show that $tr(R) = rt(R)$.   (b) Show that $sr(R) = rs(R)$.

12. Let R_1 and R_2 be binary relations on the set S.

    (a) Show that $r(R_1 \cup R_2) = r(R_1) \cup r(R_2)$.

    (b) Show that $s(R_1 \cup R_2) = s(R_1) \cup s(R_2)$.

    (c) Is $r(R_1 \cap R_2) = r(R_1) \cap r(R_2)$ always true? Explain.

    (d) Is $s(R_1 \cap R_2) = s(R_1) \cap s(R_2)$ always true? Explain.

13. Consider two equivalence relations R_1 and R_2 on a set S.

    (a) Show that $t(R_1 \cup R_2)$ is the smallest equivalence relation that contains both R_1 and R_2. *Hint:* Use Exercise 12(a) and (b).

    (b) Describe the largest equivalence relation contained in both R_1 and R_2.

14. Show that $st(R) \neq ts(R)$ for the relation R in Example 4.

15. Show that there does not exist a smallest antireflexive relation containing the relation R on $\{1, 2\}$ whose Boolean matrix is $\begin{bmatrix} 1 & 0 \\ 1 & 0 \end{bmatrix}$.

16. We say that a relation R on a set S is an **onto relation** if for every $y \in S$ there exists $x \in S$ such that $(x, y) \in R$. Show that there does not exist a smallest onto relation containing the relation R on $\{1, 2\}$ specified in Exercise 15.

17. Suppose that a property p of relations on a nonempty set S satisfies:

(i) the universal relation $S \times S$ has property p;

(ii) p is **closed under intersections**, i.e., if $\{R_i : i \in I\}$ is a nonempty family of relations on S possessing property p, then the intersection $\bigcap_{i \in I} R_i$ also possesses property p.

(a) Prove that for every relation R there is a smallest relation that contains R and has property p.

(b) Observe that the properties reflexivity, symmetry and transitivity satisfy both (i) and (ii).

(c) Which of (i) and (ii) does antireflexivity fail to satisfy?

(d) Which of (i) and (ii) does the property "onto relation" in Exercise 16 fail to satisfy?

18. Give a modified version of WARSHALL'S algorithm to compute $t(\mathbf{A})$ directly from $\mathbf{A}$ using Boolean operations.

CHAPTER HIGHLIGHTS

As usual: What does it mean? Why is it here? How can I use it? Think of examples.

CONCEPTS AND NOTATION

partial order, $\preceq$, poset, subposet
  converse, $\succeq$
  quasi-order, $\prec$
  Hasse diagram
  maximal, minimal, largest, smallest elements
  upper, lower bound
    least upper bound, $x \vee y = \text{lub}\{x, y\}$
    greatest lower bound, $x \wedge y = \text{glb}\{x, y\}$
  lattice
  chain = totally ordered set = linearly ordered set
    well-ordered set
  product order on $S_1 \times \cdots \times S_n$
  filing order on $S_1 \times \cdots \times S_n$, $\preceq^k$ on S^k
  standard order on Σ^*
  lexicographic = dictionary order $\preceq_L$ on Σ^*

composite relation $R_2 \circ R_1 = R_1 R_2$

equality relation E

Boolean matrix, Boolean product $*$

FACTS

Every finite poset has a Hasse diagram.

Filing order on $S_1 \times \cdots \times S_n$ is linear if each S_i is a chain.

If Σ is a chain, then standard order on Σ^* is a well-ordering and lexico-graphic order is linear but not a well-ordering.

Composition of relations is associative.

The matrix of the composite $R_1 R_2$ of relations is the Boolean product $\mathbf{A}_1 * \mathbf{A}_2$ of their matrices.

The relation R on S is transitive if and only if $R^2 \subseteq R$ if and only if $\mathbf{A} * \mathbf{A} \leq \mathbf{A}$.

Matrix analogs of some common relation statements are summarized at the end of §11.3.

The operators r, s and t given by $r(R) = R \cup E$, $s(R) = R \cup R^{\leftarrow}$ and $t(R) = \bigcup_{k=1}^{\infty} R^k$ are closure operators on the class of all relations.

$$t(R) = \bigcup_{k=1}^{n} R^k \text{ if } |S| = n.$$

The relation $st(R)$ may not be transitive.

The smallest equivalence relation containing R is $tsr(R)$.

METHODS

WARSHALL'S algorithm with all edge weights 1 can compute $t(R)$.

Algorithm FOREST can compute $tsr(R)$ by computing connected components of the graph of $sr(R)$.

12

ALGEBRAIC
STRUCTURES

The approach in this chapter is to go from the concrete to the abstract. First we study permutations and groups of permutations. We prove a couple of counting theorems and apply them to some nontrivial coloring problems. With this as a background, we then introduce general groups and other algebraic systems in §§ 12.5 to 12.8. We have written the later sections so that they can be read independently of §§ 12.1 to 12.4, by glossing over the examples that refer to groups of permutations. We find, though, that most readers learn the material better by studying concrete situations first.

§ 12.1 Permutations

A **permutation** is a one-to-one correspondence of some set onto itself. In this section we study the set S_n of all permutations of an n-element set. [S_n stands for the "Symmetric group on n letters."] The n-element set may be taken as $\{1, 2, \ldots, n\}$, although in applications it might be the set X of vertices of some graph, edges of some digraph, etc. In such a case we also write $\text{PERM}(X)$ for the set of all permutations of X.

We saw a long time ago that if f and g are permutations, so is $f \circ g$ and so are f^{-1} and g^{-1}. In fact, Exercises 9 and 10 of § 1.4 assert that the composition of one-to-one correspondences is a one-to-one correspondence and that the inverse of a one-to-one correspondence is also a one-to-one correspondence. In particular, $f \circ f^{-1} = f^{-1} \circ f = e$ for all $f \in S_n$, where e is the **identity permutation**: $e(k) = k$ for $k \in \{1, 2, \ldots, n\}$.

For $n \geq 3$, S_n does *not* satisfy the commutative law, i.e., $f \circ g$ need not equal $g \circ f$, as we will see in Example 2(c). For every n, the group S_n has $n!$ elements, a number that gets large fast. In our first examples we will see a few of the elements of S_6, which has 720 elements.

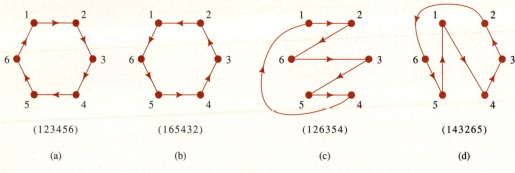

(123456)　　　(165432)　　　(126354)　　　(143265)

(a)　　　　　(b)　　　　　(c)　　　　　(d)

FIGURE 1

To work with permutations we need to establish some efficient notation. Consider, for example, the permutation in S_6 indicated in Figure 1(a). We will use the notation $(1\,2\,3\,4\,5\,6)$ for this permutation. Thus $(1\,2\,3\,4\,5\,6)$ is a name for the permutation that takes 1 to 2, 2 to 3, 3 to 4, 4 to 5, 5 to 6, and 6 to 1. Even without the picture, this notation gives us a device for remembering the meaning: Think of arrows from 1 to 2, from 2 to 3, etc. and one from 6 back to 1. In other words, visualize

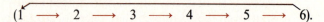

$$(1 \longrightarrow 2 \longrightarrow 3 \longrightarrow 4 \longrightarrow 5 \longrightarrow 6).$$

We can also write this permutation as $(2\,3\,4\,5\,6\,1)$, $(4\,5\,6\,1\,2\,3)$, etc. The permutation $(1\,6\,5\,4\,3\,2) = (6\,5\,4\,3\,2\,1)$ drawn in Figure 1(b) is the inverse of $(1\,2\,3\,4\,5\,6)$. Figures 1(c) and 1(d) illustrate the permutations $(1\,2\,6\,3\,5\,4)$ and $(1\,4\,3\,2\,6\,5)$, respectively.

It isn't necessary for a permutation to move all the members of the set $\{1, 2, 3, 4, 5, 6\}$. If a permutation p leaves some element s unmoved, i.e., if $p(s) = s$, then we say that p **fixes** s. We use the notation $(1\,6\,4)$ for the permutation shown in Figure 2(a) that takes 1 to 6, 6 to 4, and 4 to 1 but fixes 2, 3 and 5. Figure 2(b) shows the permutation $(2\,3)$, and Figure 2(c) shows the product $(1\,6\,4)(2\,3)$ of these two permutations. The

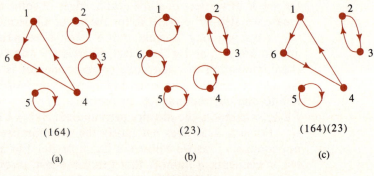

(164)　　　　　(23)　　　　　(164)(23)

(a)　　　　　(b)　　　　　(c)

FIGURE 2

product here is just the composition of the two permutations. We see that because {1, 6, 4} and {2, 3} are disjoint sets it doesn't matter whether we write (1 6 4)(2 3) or (2 3)(1 6 4).

A permutation that we can write in the form $(a_1 a_2 \cdots a_n)$ with $a_1, a_2, \ldots, a_n$ all different is called a **cycle**. Thus (1 6 4) and (2 3) are cycles. We will see that it is no accident that the pictures of these permutations are cycles in the graph-theoretic sense. The permutation (1 6 4)(2 3) is not a cycle; the elements it moves around belong to two different disjoint classes. We say that two cycles $(a_1 a_2 \cdots a_n)$ and $(b_1 b_2 \cdots b_m)$ are **disjoint** in case the sets $\{a_1, a_2, \ldots, a_n\}$ and $\{b_1, b_2, \ldots, b_m\}$ have no elements in common. The permutation (1 6 4)(2 3) is a product of two disjoint cycles.

EXAMPLE 1   (a) Consider the permutation in S_6 shown in Figure 3(a). To be orderly, we begin with 1 and follow the permutation until we return to 1. We get (1 5 3) since this permutation takes 1 to 5, 5 to 3, and 3 back to 1. Continuing to be orderly, we select the smallest value not yet visited, namely 2, and repeat the process. We obtain (2 4 6), and we conclude that Figure 3(a) represents the permutation (1 5 3)(2 4 6).

(b) In Figure 3(b) we first follow 1 to obtain (1 2). Since 3 is the smallest value not yet visited, we follow 3 to obtain (3 6). Finally, we get (4 5), so the permutation equals (1 2)(3 6)(4 5).

(c) In Figure 3(c) we follow 1 to obtain (1 3). Next we follow 2, but 2 is fixed, i.e., mapped to itself. To help our bookkeeping, let's write (2) to signify this. Next we follow 4 and obtain (4 6). Finally, 5 is fixed, so we write (5). The permutation is then (1 3)(2)(4 6)(5), but (2) and (5) are just the identity permutation in thin disguise. Hence we can write (1 3)(2)(4 6)(5) = (1 3)(4 6). ∎

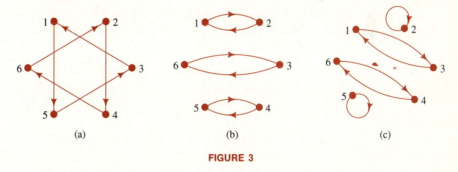

(a)   (b)   (c)

FIGURE 3

Theorem 1   Every permutation in S_n is a product of disjoint cycles.

Proof. Let g be in S_n and consider the digraph obtained using the vertex set $V = \{1, 2, \ldots, n\}$ and the edge set $\{(k, g(k)) : k \in V\}$. Figures 1,

2 and 3 are all digraphs of this sort and may serve as models for this proof. Each vertex k has outdegree 1 since the function value $g(k)$ is unique. Each vertex k has indegree 1 since $k = g(l)$ for *exactly one* choice of l; in fact, $l = g^{-1}(k)$.

Now momentarily ignore the arrows on the edges of this digraph and consider the connected components of the undirected graph so obtained. In our examples, each graph in Figure 1 is already connected and so there's only one component: the whole graph. The graphs in Figure 2 have 4, 5 and 3 components, respectively, while the graphs in Figure 3 have 2, 3 and 4 components, respectively. The digraph version of Euler's theorem [Theorem 4, §8.1] assures us that there is a closed directed path on each component. Since the in- and outdegrees are all 1, no vertex is repeated, so these closed directed paths are cycles in the graph-theoretic sense. These cycles correspond to cycles in the permutation sense and the original permutation is a product of all such cycles. ∎

To multiply permutations one must remember that we are really composing functions like $g_1 \circ g_2 \circ g_3$. To calculate $g_1 \circ g_2 \circ g_3(k)$, one calculates $g_3(k)$, then applies g_2 to this to get $g_2(g_3(k))$, and then applies g_1 to get $g_1(g_2(g_3(k)))$. One takes the permutations in order from right to left. Since the composite is still a permutation, to compute its cycle representation one follows individual elements as in Example 1 above.

EXAMPLE 2   (a) We illustrate in detail how to multiply [compose!] permutations in S_6. Let's calculate $(1\,6\,4)(2\,3)$ [from Figure 2(c)] times $(2\,5\,3\,4\,6)$. Thus we want

$$(1\,6\,4)(2\,3)(2\,5\,3\,4\,6).$$

The multiplication process is easy, with a little practice, and should require no paperwork, but the explanation is lengthy; read it carefully and practice by doing Exercises 5 and 6.

Step 1. We begin by finding the image of 1, taking the cycles in order from right to left. First, 1 is not moved by $(2\,5\,3\,4\,6)$ or by $(2\,3)$, and then it is mapped to 6 by $(1\,6\,4)$. We conclude that 1 goes to 6, and record "$(1\,6$".

Step 2. In order to pursue this cycle we next find the image of 6. First, $(2\,5\,3\,4\,6)$ takes 6 to 2, then $(2\,3)$ takes 2 to 3, and then $(1\,6\,4)$ does not move 3. We conclude that 6 goes to 3, and add this to our record: "$(1\,6\,3$".

Step 3. To find the image of 3, note that $(2\,5\,3\,4\,6)$ takes it to 4, then $(2\,3)$ leaves 4 alone, and then $(1\,6\,4)$ takes 4 to 1. Thus 3 goes to 1. This completes the cycle: "$(1\,6\,3)$". Note that we simply closed parentheses to signify that the cycle is complete.

Step 4. From Step 3 we have one cycle for the product. To find another, we choose an element not in the cycle and follow it. To be methodical we follow 2, the smallest number available. First, $(2\,5\,3\,4\,6)$ takes 2 to 5, then $(2\,3)$ and $(1\,6\,4)$ fix 5. So 2 goes to 5 and we record: "$(1\,6\,3)(2\,5$".

Step 5. To follow 5, note that $(2\,5\,3\,4\,6)$ takes 5 to 3, then $(2\,3)$ takes 3 to 2, and then $(1\,6\,4)$ leaves 2 alone. So 5 goes to 2, completing another cycle: "$(1\,6\,3)(2\,5)$".

Step 6. Since all numbers but 4 are accounted for, 4 must surely go to 4. However, it's wise to double-check: $(2\,5\,3\,4\,6)$ takes 4 to 6, $(2\,3)$ leaves 6 alone, and $(1\,6\,4)$ takes 6 back to 4. As we expected, 4 goes to 4. We add this to our record: "$(1\,6\,3)(2\,5)(4)$".

Step 7. All vertices are accounted for and we are done. Therefore

$$(1\,6\,4)(2\,3)(2\,5\,3\,4\,6) = (1\,6\,3)(2\,5)(4) = (1\,6\,3)(2\,5).$$

(b) A similar calculation shows that

$$(2\,5\,3\,4\,6)(1\,6\,4)(2\,3) = (1\,2\,4)(3\,5).$$

Comparison with the result in (a) shows that S_6 is not commutative.

(c) In fact, S_n is never commutative for $n \geq 3$. For example,

$$(1\,2)(1\,3) = (1\,3\,2) \quad \text{while} \quad (1\,3)(1\,2) = (1\,2\,3). \quad \blacksquare$$

For very small n, it is illuminating to write a multiplication table for S_n.

EXAMPLE 3 Figure 4 gives the multiplication tables for S_2 and S_3. We decline to give the table for S_4, which would be 24×24. Observe that each row and each column of the multiplication table for S_3 contains every element of S_3 exactly once. This property of the tables means that multiplication on the left or right by a member of S_3 is a one-to-one map of S_3 onto S_3.

	e	(123)	(132)	(23)	(12)	(13)
e	e	(123)	(132)	(23)	(12)	(13)
(123)	(123)	(132)	e	(12)	(13)	(23)
(132)	(132)	e	(123)	(13)	(23)	(12)
(23)	(23)	(13)	(12)	e	(132)	(123)
(12)	(12)	(23)	(13)	(123)	e	(132)
(13)	(13)	(12)	(23)	(132)	(123)	e

	e	(12)
e	e	(12)
(12)	(12)	e

S_2 $\qquad$ S_3

FIGURE 4

More generally, if h is an element of S_n, then the maps $g \rightarrow g \circ h$ and $g \rightarrow h \circ g$ are one-to-one correspondences of S_n onto S_n. This fact is a special case of Theorem 5 in § 12.5. ∎

The reason that S_n is often called the **symmetric group** on n letters is that it satisfies the definition we will give for the term "group" in § 12.5. In particular,

(i) $f, g \in S_n$ implies $f \circ g \in S_n$,
(ii) $f \in S_n$ implies $f^{-1} \in S_n$.

We will be interested in subsets of S_n that satisfy these properties. A non-empty subset G of S_n will be called a **subgroup** of S_n provided that

(i) $f, g \in G$ implies $f \circ g \in G$,
(ii) $f \in G$ implies $f^{-1} \in G$.

Actually, (ii) is redundant: Theorem 2 in § 12.5 shows that (i) implies (ii) for subsets G of S_n. Properties (i) and (ii) together imply that G contains the identity permutation e:

(iii) $e \in G$.

Indeed, given any member f of G, property (ii) implies that $f^{-1} \in G$, so $e = f \circ f^{-1} \in G$ by (i).

EXAMPLE 4   (a) Clearly, S_n and $\{e\}$ are subgroups of S_n, sometimes called the **improper subgroups** of S_n. The subgroup $\{e\}$ is also called the **trivial subgroup**.

(b) We can read off the proper subgroups of S_3 from Figure 4. They are $G_1 = \{e, (1\,2)\}$, $G_2 = \{e, (1\,3)\}$, $G_3 = \{e, (2\,3)\}$ and $G_4 = \{e, (1\,2\,3), (1\,3\,2)\}$. Subgroups also have multiplication tables. The multiplication tables of G_1, G_2 and G_3 look like the multiplication table for S_2. In fact, G_1 *is* S_2. To get G_2 we trade 2 for 3 in S_2, and to get G_3 we trade 1 for 3. The multiplication table for G_4 is given in Figure 5.

(c) Figure 5 also shows the multiplication table for the smallest subgroup G_5 of S_4 that contains $(1\,2\,3\,4)$. Observe that G_5 is commutative: $f \circ g = g \circ f$ for all $f, g \in G_5$. This is because the table is symmetric about the red diagonal.

(d) The multiplication table for the smallest subgroup G_6 of S_4 containing $(1\,2)$ and $(3\,4)$ also appears in Figure 5. Like G_5, this is a commutative subgroup with 4 elements, but the two subgroups are quite different. For example, $g \circ g = e$ for all $g \in G_6$ [see the diagonal of its

	e	(123)	(132)
e	e	(123)	(132)
(123)	(123)	(132)	e
(132)	(132)	e	(123)

$$G_4$$

	e	(1234)	(13)(24)	(1432)
e	e	(1234)	(13)(24)	(1432)
(1234)	(1234)	(13)(24)	(1432)	e
(13)(24)	(13)(24)	(1432)	e	(1234)
(1432)	(1432)	e	(1234)	(13)(24)

$$G_5$$

	e	(12)	(34)	(12)(34)
e	e	(12)	(34)	(12)(34)
(12)	(12)	e	(12)(34)	(34)
(34)	(34)	(12)(34)	e	(12)
(12)(34)	(12)(34)	(34)	(12)	e

$$G_6$$

FIGURE 5

table], but G_5 does not have this property. Another distinguishing feature is discussed in Example 7.

(e) Note that the tables of Figure 5 have the one-to-one properties mentioned in Example 3; each entry appears exactly once in each row and once in each column. ■

EXAMPLE 5 Figure 6 shows the multiplication table for the smallest subgroup J of S_6 that contains (1 6 4) and (2 3). Observe that the table is symmetric about the diagonal, so that J is commutative. This table is *not* like the table in Figure 4 for the noncommutative 6-element group S_3. ■

	e	(1 6 4)(2 3)	(1 4 6)	(2 3)	(1 6 4)	(1 4 6)(2 3)
e	e	(1 6 4)(2 3)	(1 4 6)	(2 3)	(1 6 4)	(1 4 6)(2 3)
(1 6 4)(2 3)	(1 6 4)(2 3)	(1 4 6)	(2 3)	(1 6 4)	(1 4 6)(2 3)	e
(1 4 6)	(1 4 6)	(2 3)	(1 6 4)	(1 4 6)(2 3)	e	(1 6 4)(2 3)
(2 3)	(2 3)	(1 6 4)	(1 4 6)(2 3)	e	(1 6 4)(2 3)	(1 4 6)
(1 6 4)	(1 6 4)	(1 4 6)(2 3)	e	(1 6 4)(2 3)	(1 4 6)	(2 3)
(1 4 6)(2 3)	(1 4 6)(2 3)	e	(1 6 4)(2 3)	(1 4 6)	(2 3)	(1 6 4)

$$J$$

FIGURE 6

As this chapter evolves we will be taking a more abstract approach to groups. The abstract approach can really make things clearer, as we now illustrate.

EXAMPLE 6   (a) Consider again the subgroup J in Figure 6. Let's write g for the permutation $(1\,6\,4)(2\,3)$, g^2 for $g \circ g = (1\,4\,6)$, g^3 for $g \circ g \circ g = (2\,3)$, etc. Figure 7(a) is Figure 6 rewritten using this notation. It is now evident that J consists of g and all its powers. The product entries are obviously correct too, once we observe that $g^6 = e$. In analogy with the convention $x^0 = 1$ for $x \neq 0$ from algebra, we agree that $g^0 = e$.

	e	g	g^2	g^3	g^4	g^5
e	e	g	g^2	g^3	g^4	g^5
g	g	g^2	g^3	g^4	g^5	e
g^2	g^2	g^3	g^4	g^5	e	g
g^3	g^3	g^4	g^5	e	g	g^2
g^4	g^4	g^5	e	g	g^2	g^3
g^5	g^5	e	g	g^2	g^3	g^4

(a) J

$+_6$	0	1	2	3	4	5
0	0	1	2	3	4	5
1	1	2	3	4	5	0
2	2	3	4	5	0	1
3	3	4	5	0	1	2
4	4	5	0	1	2	3
5	5	0	1	2	3	4

(b) $\mathbb{Z}(6)$

FIGURE 7

If we replace each g^k by k in the multiplication table [and $e = g^0$ by 0], we get the multiplication table in Figure 7(b). This is just the addition table for $\mathbb{Z}(6)$, which also appears in Figure 2 of § 3.6. Although $\mathbb{Z}(6)$ is not a set of permutations, it turns out that $\mathbb{Z}(6)$ is an additive group and that J and $\mathbb{Z}(6)$ are essentially the same groups; their tables for multiplication and addition, respectively, match up.

(b) The tables in Figure 5 can also be given in more abstract form. See Figure 8. ■

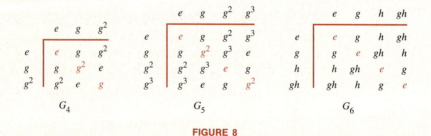

	e	g	g^2
e	e	g	g^2
g	g	g^2	e
g^2	g^2	e	g

G_4

	e	g	g^2	g^3
e	e	g	g^2	g^3
g	g	g^2	g^3	e
g^2	g^2	g^3	e	g
g^3	g^3	e	g	g^2

G_5

	e	g	h	gh
e	e	g	h	gh
g	g	e	gh	h
h	h	gh	e	g
gh	gh	h	g	e

G_6

FIGURE 8

The subgroup J is a special sort of subgroup because it consists of a single permutation and all its powers. For any permutation g in any S_n we write $\langle g \rangle$ for the subgroup $\{e, g, g^2, \ldots\}$, and call $\langle g \rangle$ the **group gener-**

ated by g. Thus $J = \{e, g, g^2, g^3, g^4, g^5\}$ is the group in S_6 generated by $g = (1\,6\,4)(2\,3)$. For any $g \in S_n$ there is a smallest positive integer m such that $g^m = e$ [the skeptical can look ahead to the proof of Theorem 2 in §12.5]. The integer m is called the **order** of g, and we have $\langle g \rangle = \{e, g, g^2, \ldots, g^{m-1}\}$. Note that the order of a cycle $(k_1\,k_2\,\cdots\,k_m)$ is just its length m. We call such a cycle an **m-cycle**.

EXAMPLE 7   (a) The subgroup G_4 in Figure 5 is generated by $g = (1\,2\,3)$, and the subgroup G_5 is generated by $g = (1\,2\,3\,4)$. See Figure 8.

(b) The subgroup G_6 is not of the form $\langle g \rangle$; that is, G_6 is not generated by any single element. This is because $g^2 = e$ for all $g \in G_6$ and so each $\langle g \rangle$ has two elements, or just one if $g = e$. However, G_6 is generated by the two permutations $g = (1\,2)$ and $h = (3\,4)$; see Figure 8. ∎

If a permutation g is written as a product of disjoint cycles, it is easy to determine its order and hence the size of $\langle g \rangle$. We've already observed that m-cycles have order m. If g is a product $c_1 c_2 \cdots c_k$ of disjoint cycles, then *using commutativity because of disjointness*, we see that

$$g^j = (c_1 c_2 \cdots c_k)^j = c_1^j c_2^j \cdots c_k^j \qquad \text{for} \quad j \geq 1.$$

If the exponent j is a multiple of the order m_i of each cycle c_i, then each $c_i^j = e$, so $g^j = e$. The smallest such j is the least common multiple of the various cycle orders. This argument makes the following theorem plausible.

Theorem 2
> If a permutation is written as a product of disjoint cycles, then its order is the least common multiple of the orders [i.e., the lengths] of the cycles.

The discussion before the theorem shows that $g^j = e$ for $j = \mathrm{lcm}(m_1, m_2, \ldots, m_k)$, so the order m of g satisfies $m \leq \mathrm{lcm}(m_1, m_2, \ldots, m_k)$. To complete the proof we would have to show that if $g^j = e$ and $j > 0$, then $j \geq \mathrm{lcm}(m_1, m_2, \ldots, m_k)$. The proof is outlined in Exercise 19.

EXERCISES 12.1

1. Write each permutation in Figure 9 as a cycle or a product of disjoint cycles.

2. Write the inverse of each permutation in Figure 9 as a cycle or a product of disjoint cycles.

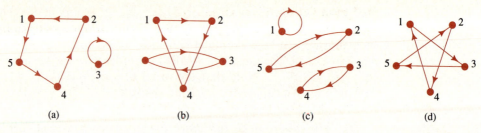

(a)   (b)   (c)   (d)

FIGURE 9

3. Draw pictures of the following permutations in S_4:
   (a) $(1\,2\,3\,4)$ (b) $(1\,4\,2)$
   (c) $(1\,2)(3\,4)$ (d) $(2\,3)$

4. Draw pictures of the following permutations in S_6:
   (a) $(1\,4)(3\,6\,5)$ (b) $(1\,5)(2\,6)(3\,4)$
   (c) $(1\,3)(2\,5\,4\,6)$ (d) $(1\,2\,5\,3\,4)$

5. Write each of the following as a cycle or a product of disjoint cycles, i.e.,
   multiply the permutations as in Example 2.
   (a) $(1\,2)(1\,4)$ (b) $(1\,2)(1\,4\,3)$
   (c) $(1\,3)(2\,3)(3\,4)$ (d) $(1\,3\,2)(1\,2\,4)$
   (e) $(1\,2\,3)(1\,2\,4)$ (f) $(1\,3)(2\,3\,4)(1\,4)(3\,4)$

6. Repeat Exercise 5 for the following:
   (a) $(1\,3\,6\,4)(1\,4\,2)$ (b) $(2\,5\,4)(1\,4)(1\,3)$
   (c) $(1\,5\,3\,4\,6)(2\,5)(1\,3\,5)$ (d) $(1\,2\,3\,4\,5)(1\,2\,3\,4\,5)$
   (e) $(1\,4\,2)(1\,4\,2)(1\,4\,2)$ (f) $(2\,6\,5\,4\,3)(4\,6)$

7. Give the inverses of the following cycles:
   (a) $(1\,2)$ (b) $(1\,4\,2\,6)$
   (c) $(1\,3\,6)$ (d) $(1\,6\,2\,5\,4\,3)$

8. Write the inverse of each permutation in Exercise 5 as a cycle or product of
   disjoint cycles. *Hint:* Use your answers to Exercise 5.

9. A cycle that interchanges two elements, like $(1\,3)$, is called a **transposition**.
   (a) Show that $(2\,3\,4\,6\,5) = (2\,5)(2\,6)(2\,4)(2\,3)$.
   (b) Write $(1\,7\,2\,5\,3\,4)$ as a product of transpositions.
   (c) State a generalization of parts (a) and (b) for any cycle $(k_1\,k_2\,\cdots\,k_m)$.
   (d) Explain why every permutation can be written as a product of trans-
       positions.

10. Prove the statement of part (c) in Exercise 9 by induction on m.

11. Consider the subgroup J in Figure 7. How many elements in J have order
    (a) 1?   (b) 2?   (c) 3?   (d) 6?

12. Repeat Exercise 11 for the group S_3.

13. List the members of the subgroup of S_6 that each of the following permutations generates.

   (a) $(2\,5)$

   (b) $(1\,5\,2\,4)$

   (c) $(1\,6)(2\,4\,3)$

   (d) $(1\,3\,5\,2)(4\,6)$

14. Find the orders of the following permutations:

   (a) $(1\,6)(3\,6)$

   (b) $(1\,3\,5)(2\,5\,1)$

   (c) $(1\,3)(2\,6\,3)$

15. (a) For each of the values $m = 1, 2, 3, 4, 5, 6$, give a permutation in S_6 having order m. Do not give a cycle when you can avoid doing so.

   (b) Show that every permutation in S_6 has order 1, 2, 3, 4, 5 or 6. *Hint:* Use Theorem 2. *Warning:* In general, S_n can have elements of order greater than n. See the next exercise.

16. (a) Give a permutation in S_5 of order 6.

   (b) Give permutations in S_7 of orders 10 and 12.

17. Is every subgroup of S_n of the form $\langle g \rangle$ commutative? Explain.

18. Let $g = (1\,2\,3)$ and $h = (2\,3)$ in S_3.

   (a) Show that $S_3 = \{e, g, g^2, h, gh, hg\}$.

   (b) Are ghg and hgh in S_3?

19. (a) Show that if g in S_n has order m, then $g^j = e$ if and only if j is a multiple of m. *Hint:* The Division Algorithm helps.

   (b) Show that if $c_1, c_2, \ldots, c_k$ are disjoint cycles and $c_1^j c_2^j \cdots c_k^j = e$, then $c_i^j = e$ for $i = 1, 2, \ldots, k$.

   (c) Use (a) and (b) to complete the proof of Theorem 2.

20. Give the multiplication tables for the subgroups in Exercises 13(a) and (b).

§ 12.2 Groups Acting on Sets

Consider a group G of permutations of some set X, i.e., a subgroup of PERM(X). The ideas of § 12.1 apply to G here, even though we allow X to be any finite set instead of $\{1, 2, \ldots, n\}$. We sometimes say that G **acts on** X, especially when we focus on what the permutations in G do to particular elements or subsets of X.

When we write a permutation f of X as a product of disjoint cycles we get a partition of X into disjoint subsets, one for each cycle, including the cycles of length 1. For example, the permutation $f = (1\,7\,4\,2)(6\,8) = (1\,7\,4\,2)(3)(5)(6\,8)$ in S_8 gives the partition of $\{1, 2, 3, 4, 5, 6, 7, 8\}$ into $\{\{1, 2, 4, 7\}, \{3\}, \{5\}, \{6, 8\}\}$. The blocks in the partition consist of members of X that f permutes among themselves in cycles, and a suitable power of f will take any member of a block to any other member. In our example, f takes 7 to 4, f^2 takes 7 to 2, f^3 takes 7 to 1 and $e = f^0$ takes 7 to itself.

In general, blocks in a partition can be viewed as equivalence classes. The permutation f gives an equivalence relation $\sim$ on X by defining

$$x \sim y \text{ in case } y = f^j(x) \text{ for some } f^j \text{ in } \langle f \rangle.$$

The idea can be extended from $\langle f \rangle$ to a general group G, whether or not G is generated by a single element.

Proposition 1

Let G be a group acting on a set X. For x, $y \in X$ define $x \sim y$ if $y = g(x)$ for some $g \in G$. Then $\sim$ is an equivalence relation on X. For $x \in X$ the equivalence class containing x is

$$\boldsymbol{Gx} = \{g(x) : g \in G\}.$$

For abstract equivalence relations we have usually written $[x]$ for equivalence classes, but Gx is a more natural notation here.

Proof. It suffices to verify that $\sim$ is reflexive, symmetric and transitive.

(R)   Let x be in X. Since $e \in G$ and $x = e(x)$ we have $x \sim x$.

(S)   Suppose that $x \sim y$ where x, $y \in X$. This means that $y = g(x)$ for some $g \in G$. Since $g^{-1} \in G$ and $x = g^{-1}(y)$ we have $y \sim x$.

(T)   Suppose that $x \sim y$ and $y \sim z$. Then $y = g(x)$ and $z = f(y)$ for some g and f in G. Since $f \circ g \in G$ and $z = f \circ g(x)$, we conclude that $x \sim z$.   ∎

These special equivalence classes Gx are called **orbits of G on X**, or **G-orbits**. Equivalence relations determine partitions into equivalence classes [Theorem 1 of §3.5], so we get the following.

Corollary

If G acts on X, then the G-orbits partition X.

EXAMPLE 1   Let $g = (1\,6\,4)(2\,3)$ in S_6. Then $\langle g \rangle = \{e, g, g^2, g^3, g^4, g^5\}$, so one orbit is

$$\langle g \rangle 1 = \{e(1), g(1), g^2(1), g^3(1), g^4(1), g^5(1)\} = \{1, 6, 4, 1, 6, 4\} = \{1, 4, 6\}.$$

Similarly, $\langle g \rangle 2 = \{2, 3\}$ and $\langle g \rangle 5 = \{5\}$ as expected. In short, the $\langle g \rangle$-orbits can be read off from the cycle decomposition $g = (1\,6\,4)(2\,3)$ without any computations. This sort of reasoning can be used to prove the next result.   ∎

Proposition 2

If g is a product of disjoint cycles, the sets of entries in the cycles, including all the 1-cycles, are the $\langle g \rangle$-orbits.

We will often be interested in G-orbits when G is not generated by a single permutation. In what follows, our groups will be finite groups acting on finite sets.

EXAMPLE 2   (a) If $G = \text{PERM}(X)$, then every element of X is equivalent to every other element of X, since given any x and y in X some permutation will map x to y. Hence there is only one big orbit, X itself. This G is too big to have interesting orbits.

(b) Let G_6 be the group in Figure 5 of §12.1. It is easy to see by inspection that the orbits are $\{1, 2\}$ and $\{3, 4\}$. Note that we cannot apply Proposition 2 because G_6 is not generated by a single permutation. ■

We will be most interested in G-orbits when G is a group of permutations that describes the symmetry of some object, such as a graph or a digraph. Consider a digraph D with no parallel edges. A permutation g of the set $V(D)$ of vertices is a **digraph automorphism** provided it preserves the edge structure, i.e., provided (x, y) is an edge if and only if $(g(x), g(y))$ is an edge. We can think of a digraph automorphism as moving the vertex labels around without essentially changing the structure of the digraph. We write $\text{AUT}(D)$ for the set of automorphisms of D. It is easy to verify that $\text{AUT}(D)$ is a group of permutations acting on $V(D)$ [Exercise 1]. Very roughly speaking, the larger $\text{AUT}(D)$ is, the more symmetry D has.

EXAMPLE 3   (a) The digraph D drawn in Figure 1(a) has two sources and two sinks, and digraph automorphisms must send sources to sources and sinks to sinks. One can verify that there are four automorphisms, namely e, $f = (q\,s)$, $g = (p\,r)$ and $fg = (p\,r)(q\,s)$. For convenience, their function values are listed in Figure 1(b). Notice that g, for example, switches the labels p and r, but we could have done the same thing by just flipping the digraph over, without changing its structure. The multiplication table is in Figure 1(c). This group is essentially the same as G_6 in Figure 5 of §12.1.

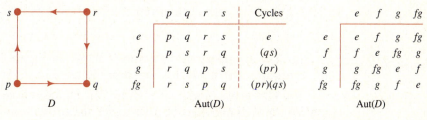

	p	q	r	s	Cycles
e	p	q	r	s	e
f	p	s	r	q	$(q\,s)$
g	r	q	p	s	$(p\,r)$
fg	r	s	p	q	$(p\,r)(q\,s)$

	e	f	g	fg
e	e	f	g	fg
f	f	e	fg	g
g	g	fg	e	f
fg	fg	g	f	e

D · Aut(D) · Aut(D)

FIGURE 1

(b) To find the orbits under AUT(D) in a systematic way we start with p and find

$$\text{AUT}(D)p = \{e(p), f(p), g(p), fg(p)\} = \{p, p, r, r\} = \{p, r\};$$

see the first column of Figure 1(b). Now we consider a vertex not yet found and repeat the process. We find that

$$\text{AUT}(D)q = \{q, s, q, s\} = \{q, s\}$$

from the second column of Figure 1(b). We conclude that there are two orbits, $\{p, r\}$ and $\{q, s\}$. ■

EXAMPLE 4   (a) The digraph E in Figure 2(a) has another kind of symmetry and a different automorphism group. Every automorphism of E must send x to itself, but the remaining vertices can be permuted in any way. So AUT(E) is essentially PERM($\{y, z, w\}$) or S_3. Figure 2(b) gives the function values of the automorphisms, where $g = (y\,z\,w)$ and $h = (z\,w)$.

(b) The orbits under AUT(E) can be read off from Figure 2(b) or from the picture. They are $\{x\}$ and $\{y, z, w\}$. ■

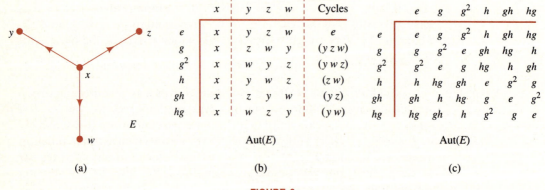

	x	y	z	w	Cycles
e	x	y	z	w	e
g	x	z	w	y	$(y\,z\,w)$
g^2	x	w	y	z	$(y\,w\,z)$
h	x	y	w	z	$(z\,w)$
gh	x	z	y	w	$(y\,z)$
hg	x	w	z	y	$(y\,w)$

Aut(E)

	e	g	g^2	h	gh	hg
e	e	g	g^2	h	gh	hg
g	g	g^2	e	gh	hg	h
g^2	g^2	e	g	hg	h	gh
h	h	hg	gh	e	g^2	g
gh	gh	h	hg	g	e	g^2
hg	hg	gh	h	g^2	g	e

Aut(E)

E

(a)   (b)   (c)

FIGURE 2

Automorphism groups of undirected graphs can be studied in the same way. For a graph H without parallel edges, a permutation g of $V(H)$ is a **graph automorphism** in case $\{x, y\}$ is an edge if and only if $\{g(x), g(y)\}$ is an edge. The group of all such automorphisms is denoted by AUT(H).

EXAMPLE 5   (a) The symmetry of the graph H shown in Figure 3(a) is described by its group AUT(H), which is listed in the table in Figure 3(b).

(b) The orbits under AUT(H) are $\{u, v\}$ and $\{w, x, y, z\}$. See Figure 3(b); one should also be able to see the orbits from staring at Figure 3(a) and visualizing the graph automorphisms. Notice that some of

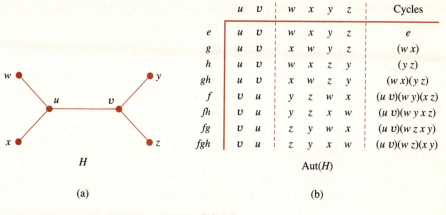

	u	v	w	x	y	z	Cycles
e	u	v	w	x	y	z	e
g	u	v	x	w	y	z	(w x)
h	u	v	w	x	z	y	(y z)
gh	u	v	x	w	z	y	(w x)(y z)
f	v	u	y	z	w	x	(u v)(w y)(x z)
fh	v	u	y	z	x	w	(u v)(w y x z)
fg	v	u	z	y	w	x	(u v)(w z x y)
fgh	v	u	z	y	x	w	(u v)(w z)(x y)

H

(a)

Aut(H)

(b)

FIGURE 3

the automorphisms are not just the result of rotating or flipping the graph as a whole.

(c) Table 1 gives the orbits under the subgroups of AUT(H) that are generated by single permutations. ■

TABLE 1

The subgroup	Its orbits	Number of orbits
$\langle e \rangle$	$\{u\}, \{v\}, \{w\}, \{x\}, \{y\}, \{z\}$	6
$\langle g \rangle$	$\{u\}, \{v\}, \{w, x\}, \{y\}, \{z\}$	5
$\langle h \rangle$	$\{u\}, \{v\}, \{w\}, \{x\}, \{y, z\}$	5
$\langle gh \rangle$	$\{u\}, \{v\}, \{w, x\}, \{y, z\}$	4
$\langle f \rangle$	$\{u, v\}, \{w, y\}, \{x, z\}$	3
$\langle fh \rangle$	$\{u, v\}, \{w, x, y, z\}$	2
$\langle fg \rangle$	$\{u, v\}, \{w, x, y, z\}$	2
$\langle fgh \rangle$	$\{u, v\}, \{w, z\}, \{x, y\}$	3

The G-orbit of an element x in X is closely related to a subgroup of G, namely the subgroup $\mathrm{FIX}_G(x)$ of permutations in G that leave x fixed. In symbols,

$$\mathrm{FIX}_G(x) = \{g \in G : g(x) = x\}.$$

Exercise 10 asks for a proof that $\mathrm{FIX}_G(x)$ is actually a subgroup of G. When the group G is understood we may write $\mathrm{FIX}(x)$ in place of $\mathrm{FIX}_G(x)$.

EXAMPLE 6 (a) For the group AUT(D) of automorphisms of the digraph D in Figure 1, the fixing subgroups are FIX(p) = $\{e, f\}$, FIX(q) = $\{e, g\}$, FIX(r) = $\{e, f\}$, and FIX(s) = $\{e, g\}$. These sets can be read from Figure 1(b) by finding the p's in the first column, the q's in the second column, etc.

(b) For the group AUT(E) in Figure 2, the fixing subgroups are FIX(x) = AUT(E), FIX(y) = $\{e, h\}$, FIX(z) = $\{e, hg\}$ and FIX(w) = $\{e, gh\}$.

(c) For the group AUT(H) in Figure 3, we get the fixing subgroups FIX(u) = FIX(v) = $\{e, g, h, gh\}$, FIX(w) = FIX(x) = $\{e, h\}$ and FIX(y) = FIX(z) = $\{e, g\}$. Note that FIX(u) has 4 elements and the orbit AUT(H)u = $\{u, v\}$ has 2 elements, whereas FIX(w) has 2 elements and the orbit AUT(H)w = $\{w, x, y, z\}$ has 4 elements. Note also that $2 \cdot 4 = 4 \cdot 2 = 8 = |\text{AUT}(H)|$. The next theorem shows that this is not an accident. ∎

Theorem Let G be a group acting on a set X, and let $x \in X$. The number of permutations in G is the product of the number of elements in the orbit Gx by the number of permutations in the fixing subgroup FIX(x). In symbols,

$$|G| = |Gx| \cdot |\text{FIX}_G(x)|.$$

In particular, $|Gx|$ divides $|G|$.

Proof. Each element in Gx has the form $g(x)$ for some $g \in G$. Hence there exist $g_1, \ldots, g_k$ in G so that $Gx = \{g_1(x), \ldots, g_k(x)\}$ with no elements duplicated. Here $k = |Gx|$. We claim that

(1) $G = \bigcup_{j=1}^{k} (g_j \circ \text{FIX}(x))$

where $g_j \circ \text{FIX}(x)$ signifies $\{g_j \circ g : g \in \text{FIX}(x)\}$. We also claim that

(2)   the sets $g_j \circ \text{FIX}(x)$ are disjoint

and

(3)   each set $g_j \circ \text{FIX}(x)$ is the same size as FIX(x).

With assertions (1) to (3), the proof is easy to finish:

$$|G| = \sum_{j=1}^{k} |g_j \circ \text{FIX}(x)| \qquad \text{by (1) and (2)}$$

$$= \sum_{j=1}^{k} |\text{FIX}(x)| \qquad \text{by (3)}$$

$$= k \cdot |\text{FIX}(x)| = |Gx| \cdot |\text{FIX}(x)|.$$

The proofs of (1) to (3) are straightforward exercises [Exercise 14]. Moreover, (3) is a general property of subgroups of groups, as we show in Theorem 5 of §12.5.   ∎

EXERCISES 12.2

1. Let D be a digraph.

   (a) Show that AUT(D) is a subgroup of PERM$(V(D))$.

   (b) Digraph automorphisms preserve digraph properties. As examples, show that automorphisms take sources to sources and sinks to sinks.

2. (a) Determine a *few* entries of the multiplication table for the group in Example 5 and Figure 3.

   (b) Is the group commutative?

3. Consider AUT(D) in Figure 1. The orbits are given in Example 3(b) and the fixing subgroups are in Example 6(a). Confirm the equality in the Theorem in these cases.

4. Repeat Exercise 3 for AUT(E) in Figure 2. The orbits are given in Example 4(b) and the fixing subgroups are in Example 6(b).

5. Consider the graph in Figure 4(a).

   (a) Convince yourself that the automorphism group for this graph has the eight permutations in Figure 4(b). Note that r signifies "rotation," h "horizontal flip," v "vertical flip," d "diagonal flip" and f "other diagonal flip."

   (b) For each of the eight subgroups generated by a single permutation, give its orbits.

   (c) Give the cycle notations for the permutations in this automorphism group.

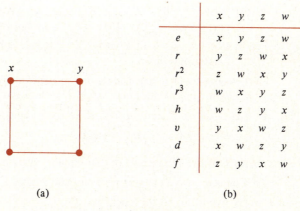

	x	y	z	w
e	x	y	z	w
r	y	z	w	x
r^2	z	w	x	y
r^3	w	x	y	z
h	w	z	y	x
v	y	x	w	z
d	x	w	z	y
f	z	y	x	w

(a)   (b)

FIGURE 4

6. What can you say about the sizes of the orbits of a group with 27 members?

7. Suppose that G acts on X. Show that if $|G|$ is 2^k for some k and $|X|$ is odd, then some member of X must be fixed by *all* members of G. *Hint:* Since the orbits of G partition X, we can choose one element from each orbit; let $x_1, \ldots, x_m$ be such a collection. Then $|X| = \sum\limits_{j=1}^{m} |Gx_j|$. Apply the Theorem to each x_j.

8. (a) How many automorphisms does the binary tree T_1 in Figure 5 have?

   (b) Repeat part (a) for the tree T_2 in Figure 5.

   (c) Use Exercise 7 to show that some vertex is sent to itself by every automorphism of T_2.

   (d) Find a vertex as described in part (c).

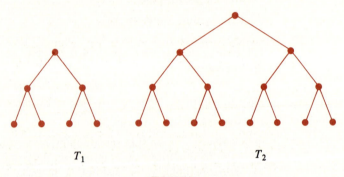

T_1 T_2

FIGURE 5

9. (a) Show how to construct, for each n in $\mathbb{P}$, a tree such that $|\text{AUT}(T)| = 2^n$. *Hint:* Keep attaching suitable graphs with two automorphisms.

   (b) Show how to construct, for each n in $\mathbb{P}$, a digraph such that $|\text{AUT}(D)| = n$.

10. Let G be a group acting on a set X.

    (a) Prove directly that each fixing subgroup $\text{FIX}(x)$ is a subgroup of G or do part (b) and observe that this is a special case.

    (b) For a subset Y of X let $\text{FIX}(Y) = \{g \in G : g(Y) = Y\}$. Show that $\text{FIX}(Y)$ is a subgroup of G.

11. Let $G = \text{AUT}(H)$ from Figure 3. For subsets Y of H, $\text{FIX}(Y)$ is as defined in Exercise 10(b). Determine $\text{FIX}(\{w, y\})$. Does $\text{FIX}(\{w, y\}) = \text{FIX}(w) \cap \text{FIX}(y)$?

12. Let G act on X and for $Y \subseteq X$ let $\text{FIX}(Y)$ be defined as in Exercise 10(b).

    (a) Show that $\bigcap\limits_{y \in Y} \text{FIX}(y) \subseteq \text{FIX}(Y)$.

    (b) Observe [Exercise 11] that equality need not hold in part (a).

13. Consider $\text{AUT}(H)$ in Example 6(c) and Figure 3.

    (a) According to the Theorem,

$$|\text{AUT}(H)| = |\text{AUT}(H)u| \cdot |\text{FIX}(u)|,$$

i.e., $8 = 2 \cdot 4$. Illustrate the proof of the Theorem by using suitable g_1 and g_2 in AUT(H) with AUT(H)$u = \{g_1(u), g_2(u)\}$.

(b) Is your choice of g_1 and g_2 unique?

(c) With u replaced by x, give an illustration similar to that in part (a).

14. Prove assertions (1) to (3) in the proof of the Theorem.

15. A group G acting on a set X is said to act **transitively** on X if $X = Gx$ for some x in X. Suppose that G acts transitively on X.

(a) Show that $X = Gx$ for every x in X.

(b) Show that if G is finite, so is X, and $|G| = |\text{FIX}(x)| \cdot |X|$ for each x in X.

16. Suppose that the group G acts on the set X and K is a subgroup of G. Then K also acts on X. Show that each orbit of G on X is a union of orbits of K.

17. Let G be a group acting on a set X. Define

$$R = \{(x, y) \in X \times X : g(x) = y \text{ for some } g \in G\}.$$

(a) Show that R is an equivalence relation on X.

(b) Describe the partition of X corresponding to R.

§ 12.3   Groups Acting on Sets, Part 2

We continue our preparation for the interesting counting applications in the next section, and begin by presenting a counting theorem in rather abstract terms. Later we will specialize the theorem so that it will be easy to apply. We will also need to extend our notion of a group acting on a set.

Consider again a group G acting on a set X. In §12.2 we looked at the group $\text{FIX}_G(x)$ of permutations which fixed a given element x of X. Now we switch perspective and for each g in G look at the set

$$\text{FIX}_X(g) = \{x \in X : g(x) = x\}$$

consisting of all members of X that g fixes. The subscript on FIX_X should help us remember that this is a subset of X; similarly, our old friends FIX_G are subsets of G. The sets $\text{FIX}_X(g)$ are connected with the orbits of G in a surprising way.

Theorem 1

Let G be a [finite] group acting on a set X. The number of orbits of G on X equals

$$\frac{1}{|G|} \left(\sum_{g \in G} |\text{FIX}_X(g)| \right).$$

The sum here has one term for each g in G. If you are moderately comfortable with the Theorem in §12.2, read the next proof and then go

592	Ch. 12	Algebraic Structures

on to see what this theorem says in particular cases. Otherwise, we advise you to skip the proof now and return to it after studying the examples.

Proof. We use an idea from the proof of the Generalized Pigeon-Hole Principle in § 5.5; count a set of pairs in two different ways. Finding a set to count can be tricky sometimes, but in our case the choice is rather natural:

$$S = \{(g, x) \in G \times X : g(x) = x\}.$$

First, for each g in G we count pairs (g, x) with $g(x) = x$—there are $|\text{FIX}_X(g)|$ of them—and then add the answers. We obtain

(1) $$|S| = \sum_{g \in G} |\text{FIX}_X(g)|.$$

We can also count the members of S by counting, for each x in X, the pairs (g, x) with $g(x) = x$; we get

(2) $$|S| = \sum_{x \in X} |\text{FIX}_G(x)|.$$

For each $x \in X$, $|\text{FIX}_G(x)| = \dfrac{|G|}{|Gx|}$ by the Theorem in § 12.2, so

$$|S| = \sum_{x \in X} \frac{|G|}{|Gx|} = |G| \sum_{x \in X} \frac{1}{|Gx|}.$$

Now we group together the terms in the sum that come from a given orbit Gx_0. Since $Gx = Gx_0$ for each x in Gx_0,

$$\sum_{x \in Gx_0} \frac{1}{|Gx|} = \sum_{x \in Gx_0} \frac{1}{|Gx_0|} = 1.$$

That is, the orbit contributes a total value of 1 to the sum $\sum_{x \in X} \dfrac{1}{|Gx|}$. Thus if there are m orbits, then

$$\sum_{x \in X} \frac{1}{|Gx|} = 1 + 1 + \cdots + 1 = m.$$

Hence $|S| = |G| \cdot m$, so by (1) we have

$$m = \frac{1}{|G|} \cdot |S| = \frac{1}{|G|} \left(\sum_{g \in G} |\text{FIX}_X(g)| \right),$$

as claimed in the theorem. ∎

In our first examples it will be easy to evaluate both sides of the equality in Theorem 1, so you might wonder about the value of the theorem. In nontrivial applications, as in § 12.4, the number of orbits can be difficult to calculate directly, whereas the numbers $|\text{FIX}_X(g)|$ will be relatively easy to determine.

EXAMPLE 1 We return to the group $G = \text{AUT}(H)$ of automorphisms of the graph H in Figure 1, which we saw in Example 5 of §12.2. This group acts on the set $V = \{u, v, w, x, y, z\}$ of vertices of H. To see which members of V each automorphism fixes, we look in the table of Figure 1. We find $\text{FIX}_V(e) = \{u, v, w, x, y, z\}$, $\text{FIX}_V(g) = \{u, v, y, z\}$, $\text{FIX}_V(h) = \{u, v, w, x\}$, $\text{FIX}_V(gh) = \{u, v\}$ and $\text{FIX}_V(f) = \text{FIX}_V(fh) = \text{FIX}_V(fg) = \text{FIX}_V(fgh) = \varnothing$. These sets have 6, 4, 4, 2, 0, 0, 0 and 0 vertices, respectively. Also, $|G| = 8$, so Theorem 1 asserts that G has

$$\frac{1}{8}(6 + 4 + 4 + 2 + 0 + 0 + 0 + 0) = 2$$

orbits. Indeed, from the picture of H or the table we see that there are exactly two orbits, namely $\{u, v\}$ and $\{w, x, y, z\}$ under G, so Theorem 1 confirms our observation. ■

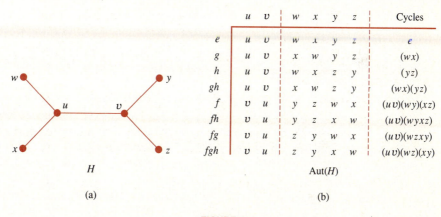

	u	v	w	x	y	z	Cycles
e	u	v	w	x	y	z	e
g	u	v	x	w	y	z	(wx)
h	u	v	w	x	z	y	(yz)
gh	u	v	x	w	z	y	$(wx)(yz)$
f	v	u	y	z	w	x	$(uv)(wy)(xz)$
fh	v	u	y	z	x	w	$(uv)(wyxz)$
fg	v	u	z	y	w	x	$(uv)(wzxy)$
fgh	v	u	z	y	x	w	$(uv)(wz)(xy)$

H Aut(H)

(a)   (b)

FIGURE 1

Consider any graph H with vertex set V, and assume that H has no parallel edges. So far, we have regarded $G = \text{AUT}(H)$ as acting on V, i.e., we have regarded G as a subset of $\text{PERM}(V)$. We can also view G as acting on the set E of edges of the graph as follows. For each g in $G \subseteq \text{PERM}(V)$ define g^* in $\text{PERM}(E)$ by setting $g^*(\{u, v\}) = \{g(u), g(v)\}$ for each edge $\{u, v\}$. Note that if $f, g \in G$, then

$(*)$ $$(f \circ g)^* = f^* \circ g^*;$$

the first composition is in $\text{PERM}(V)$ and the second is in $\text{PERM}(E)$. Property $(*)$ holds because

$$(f \circ g)^*(\{u, v\}) = \{f \circ g(u), f \circ g(v)\}$$

and

$$f^* \circ g^*(\{u, v\}) = f^*(\{g(u), g(v)\}) = \{f(g(u)), f(g(v))\}.$$

Usually, $g \to g^*$ is one-to-one and then (∗) shows that G and $G^* = \{g^*:g \in G\}$ are essentially the same. We will say that G acts on E even though technically it is G^* that acts on E. In fact, we will say that G acts on E even if $g \to g^*$ is not one-to-one, provided that (∗) still holds.

EXAMPLE 2   (a) In Figure 2(a) we redraw the graph H, but the edges are also labeled. The table in Figure 2(b) shows how G acts on E. To show how the table was created, let's check that f^* is correct: f is the automorphism that reflects the graph about a vertical line through e_3, so f^* interchanges e_1 and e_4, interchanges e_2 and e_5, and leaves e_3 fixed. We can check these claims formally by calculating

$$f^*(e_1) = f^*(\{u, w\}) = \{f(u), f(w)\} = \{v, y\} = e_4,$$

$$f^*(e_3) = f^*(\{u, v\}) = \{f(u), f(v)\} = \{v, u\} = e_3,$$

etc., but this shouldn't be necessary for graphs that we can draw.

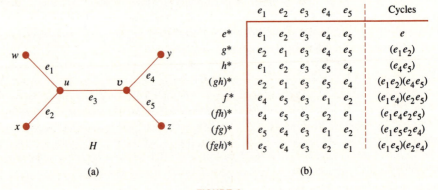

	e_1	e_2	e_3	e_4	e_5	Cycles
e^*	e_1	e_2	e_3	e_4	e_5	e
g^*	e_2	e_1	e_3	e_4	e_5	(e_1e_2)
h^*	e_1	e_2	e_3	e_5	e_4	(e_4e_5)
$(gh)^*$	e_2	e_1	e_3	e_5	e_4	$(e_1e_2)(e_4e_5)$
f^*	e_4	e_5	e_3	e_1	e_2	$(e_1e_4)(e_2e_5)$
$(fh)^*$	e_4	e_5	e_3	e_2	e_1	$(e_1e_4e_2e_5)$
$(fg)^*$	e_5	e_4	e_3	e_1	e_2	$(e_1e_5e_2e_4)$
$(fgh)^*$	e_5	e_4	e_3	e_2	e_1	$(e_1e_5)(e_2e_4)$

(a)   (b)

FIGURE 2

Now from Figure 2(b) we find $\text{FIX}_E(e^*) = \{e_1, e_2, e_3, e_4, e_5\}$, $\text{FIX}_E(g^*) = \{e_3, e_4, e_5\}$, $\text{FIX}_E(h^*) = \{e_1, e_2, e_3\}$ and $\text{FIX}_E(a^*) = \{e_3\}$ for the other automorphisms a. These sets have 5, 3, 3, 1, 1, 1, 1, 1 edges, respectively. According to Theorem 1, $G = \text{AUT}(H)$ has

$$\frac{1}{8}(5 + 3 + 3 + 1 + 1 + 1 + 1 + 1) = 2$$

orbits on E. Sure enough, the orbits are $\{e_3\}$ and $\{e_1, e_2, e_4, e_5\}$.

(b) The group G also acts on the set T of all two-element subsets of V using the same definition for g^*. Note that T contains the set E of edges and a lot more. In fact, T has $\binom{6}{2} = 15$ elements. This time the situation is more abstract and less intuitive than when we viewed G as acting on vertices or edges. We don't have a useful picture of T, and a table like that in Figure 2(b) would be very cumbersome. So we will just have to be mathematical. As always, $\text{FIX}_T(e^*)$ is the entire set on which G acts, T in this case. Now g maps each member of $\{u, v, y, z\}$ to itself, so g^* sends each of the 2-element subsets of $\{u, v, y, z\}$ back to itself. It also fixes $\{w, x\}$, since $g^*(\{w, x\}) = \{g(w), g(x)\} = \{x, w\}$. So $\text{FIX}_T(g^*)$ has $\binom{4}{2} + 1 = 7$ elements. Similarly, $|\text{FIX}_T(h^*)| = 7$. In addition, $\text{FIX}_T((gh)^*) = \{\{u, v\}, \{w, x\}, \{y, z\}\}$, $\text{FIX}_T(f^*) = \{\{u, v\}, \{w, y\}, \{x, z\}\}$, $\text{FIX}_T((fh)^*) = \text{FIX}_T((fg)^*) = \{\{u, v\}\}$, and $\text{FIX}_T((fgh)^*) = \{\{u, v\}, \{w, z\}, \{x, y\}\}$. So the eight sets $\text{FIX}_T(\quad)$ have 15, 7, 7, 3, 3, 1, 1, 3 elements.

Theorem 1 then says that T consists of

$$\frac{1}{8}(15 + 7 + 7 + 3 + 3 + 1 + 1 + 3) = 5$$

orbits under G. This was not so obvious to begin with. Now we can observe that $\{w, u\}, \{w, v\}, \{w, x\}, \{w, y\}$ and $\{u, v\}$ belong to five different orbits. [Look at Figure 1(a) to see that none of these subsets can be mapped to another one by automorphisms of the graph.] If we had been asked originally to find a representative of each orbit, we would know we were done when we had exhibited these five subsets. ■

Consider again a group G acting on a set X: $G \subseteq \text{PERM}(X)$. We sometimes want to restrict the action to a subset Y of X by just ignoring what the permutations in G do to elements outside of Y. For this to work, however, we need to be sure that the permutations in G map elements of Y into Y. So we consider a subset Y of X that is an orbit of G or a union of orbits. For each g in G we define $g^*: Y \to Y$ by $g^*(y) = g(y)$; g^* is called the **restriction** of g to Y. Compare the usage of this term in §1.4. Since Y is finite and g^* is a one-to-one map of Y into Y, g^* maps Y *onto* Y. Therefore, each g^* is a permutation of Y. Note also that

$$(*) \qquad\qquad (f \circ g)^* = f^* \circ g^*$$

holds for all $f, g \in G$. The correspondence $g \to g^*$ might be one-to-one and it might not, as we'll see in the next example.

EXAMPLE 3   (a) We return to the example in Figure 1 and restrict the action of $G = \text{AUT}(H)$ to the orbit $\{u, v\}$. To see the table of values of the restrictions to $\{u, v\}$, simply ignore the last four columns in Figure 1(b) and pretend that each permutation e, g, h, etc. has an asterisk $*$ on it. We have eight different names for the restricted permutations, but there are only two genuinely different ones: e^* and f^*. The correspondence $g \to g^*$ is not one-to-one; for example, $e^* = g^* = h^* = (gh)^*$. Even so, we say that G acts on the orbit $\{u, v\}$.

(b) We use the same group G but restrict to the other orbit $\{w, x, y, z\}$. The table of values is evident from Figure 1(b) by ignoring the u- and

v-columns. Note that the eight restricted permutations are all different, so $g \to g^*$ is one-to-one this time. ■

Let's return to Theorem 1. Since there are $|G|$ terms in the sum

$$\sum_{g \in G} |\text{FIX}_X(g)|,$$

when we divide the sum by $|G|$ we obtain the average value of $|\text{FIX}_X(g)|$ over all members g of G. If some values are larger than average, then some must be smaller. This observation leads to the following surprising corollaries.

Corollary 1

> If X is the only orbit of G on X, and if $|X| > 1$, then there exists a g in G such that $g(x) \neq x$ for all $x \in X$.

Proof. By Theorem 1, the average value of $|\text{FIX}_X(g)|$ is 1, since there is just one orbit. Moreover, $|\text{FIX}_X(e)| = |X| > 1$, so $|\text{FIX}_X(g)| < 1$ for at least one g. For such a g, the set $\text{FIX}_X(g) = \{x \in X : g(x) = x\}$ must be the empty set. ■

Exercise 15 in §12.2 concerns actions like the ones in Corollary 1.

Corollary 2

> If G acts on X, and if Y is an orbit of G on X with $|Y| > 1$, then there exists g in G so that $g(x) \neq x$ for all $x \in Y$.

Proof. The set Y is the only orbit under the group of restrictions g^* for $g \in G$. So by Corollary 1, there is a g^* such that $g^*(x) \neq x$ for all $x \in Y$. Thus $g(x) \neq x$ for all $x \in Y$. ■

EXAMPLE 4   We refer back to Example 3. According to Corollary 2, some automorphism must move every member of the orbit $\{u, v\}$. In fact, the automorphisms f, fh, fg and fgh all have this property.

The same corollary assures us that some automorphism moves every element in the orbit $\{w, x, y, z\}$. A glance at Figure 1(b) shows that gh, f, fh, fg and fgh all have this property. ■

Several times in this section we have viewed a group G as acting on a set even though the set was not involved in the original description of G. Here is a formal definition. If G is a group and $g \to g^*$ is a function from G into $\text{PERM}(X)$ satisfying

(*) $$(f \circ g)^* = f^* \circ g^*,$$

we say that G **acts on** X. Since the set $G^* = \{g^*:g \in G\}$ is a group of permutations on X, the results in this section and the preceding section remain valid using our extended meaning of G acting on X.

Sometimes groups act on sets of functions that are natural and useful. Suppose that G acts on a set X. Let FUN(X, C) be the set of all functions from X into some finite set, which we have called C because of our applications to colorings in the next section. For φ in FUN(X, C) we define $g^*(\varphi) = \varphi \circ g^{-1}$. Then for $f, g \in G$,

$$(f \circ g)^*(\varphi) = \varphi \circ (f \circ g)^{-1} = \varphi \circ g^{-1} \circ f^{-1}$$

while

$$f^* \circ g^*(\varphi) = f^*(\varphi \circ g^{-1}) = (\varphi \circ g^{-1}) \circ f^{-1}.$$

That is,

$$(*) \qquad\qquad (f \circ g)^* = f^* \circ g^* \qquad \text{for} \quad f, g \in G,$$

so G acts on FUN(X, C).

The G-orbit of a function φ in FUN(X, C) is $G\varphi = \{g^*(\varphi):g \in G\} = \{\varphi \circ g^{-1}:g \in G\}$. Since every h in G is an inverse of something, namely $h = (h^{-1})^{-1}$, this set is just $\{\varphi \circ h:h \in G\}$, which we could denote by $\varphi \circ G$.

EXAMPLE 5 In the next section we will use groups acting on sets FUN(X, C) to solve such problems as coloring the cube and counting essentially different logical circuits. To illustrate how we can interpret FUN(X, C) in a relevant way, we begin here with a simpler coloring problem:

In how many ways can we color the vertices of the graph in Figure 1 using red, black or both?

Each of the six vertices can be colored red or black, so there are $2^6 = 64$ possible ways to assign each vertex a color. This count ignores the graph structure, though. We want to regard two colorings as equivalent if a graph automorphism will move one to the other. Our task is to convert the preceding sentence into mathematics.

To see what we have in mind, consider the colored graphs shown in Figure 3. These correspond to four different ways of choosing the colors for the vertices, but we can see from the pictures that we can get from any one of these colored graphs to any other by applying a suitable graph

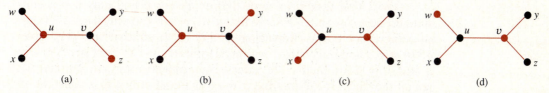

(a)   (b)   (c)   (d)

FIGURE 3

automorphism. For instance, the automorphism h in the table of Figure 1 switches the upper and lower edges on the right and converts (a) into (b), and rotating the picture a half turn by fgh converts (a) into (d) and (b) into (c).

Each coloring of a graph in Figure 3 corresponds to a function $\varphi: V \to C$ where V is the set of vertices and $C = \{B, R\}$ is the set of colors Black and Red; let φ_a, φ_b, φ_c, φ_d be the names for these four colorings. The colorings φ_a and φ_b are really equivalent, because $\varphi_b = \varphi_a \circ h$, i.e., we can get the (b) coloring by first performing h, to switch vertices, and then doing the (a) coloring. Similarly, $\varphi_a \circ f = \varphi_c$ and $\varphi_a \circ fgh = \varphi_d$.

When we say that **colorings** ψ and φ are **equivalent**, what we mean is that $\psi = \varphi \circ f$ for some f in the automorphism group G. In other words, ψ and φ are equivalent in case ψ and φ belong to the same orbit $\varphi \circ G$ of G acting on $\text{FUN}(V, C)$. ∎

The number of essentially different colorings in Example 5 is the number of G-orbits in $\text{FUN}(V, C)$. The answer, 21, is *not* obvious. To find this number we will need to apply Theorem 1 to $X = \text{FUN}(V, C)$, so we will need to be able to calculate numbers like $\left|\text{FIX}_{\text{FUN}(V,C)}(g^*)\right|$.

EXAMPLE 6 The automorphism fh of the graph in Example 5 has the cycle representation $(u\,v)(w\,y\,x\,z)$, with orbits $\{u, v\}$ and $\{w, x, y, z\}$. In order for $(fh)^*$ to fix a coloring, the vertices in each orbit must all be the same color; otherwise, applying fh would make a visible difference in the colors. There are four colorings that meet this condition: all black, all red, u and v black but w, x, y, z all red, and u, v red but w, x, y, z black. These four are the members of $\text{FIX}_{\text{FUN}(V,C)}(fh)^*$. In other words, $\text{FIX}_{\text{FUN}(V,C)}(fh)^*$ consists of the functions in $\text{FUN}(V, C)$ that are constant on the orbits of fh. ∎

The following theorem, which we will use in the next section, generalizes this example. Here $\langle g \rangle$ represents the group generated by g, as usual.

Theorem 2

> Suppose that G acts on X, and hence on $\text{FUN}(X, C)$. If g is in G, then $\text{FIX}_{\text{FUN}(X,C)}(g^*)$ consists of all the functions $\varphi: X \to C$ that are constant on the $\langle g \rangle$-orbits.

Proof. Our task is to show that $g^*(\varphi) = \varphi$ if and only if φ is constant on the $\langle g \rangle$-orbits. Observe that $g^*(\varphi) = \varphi$ if and only if $\varphi \circ g^{-1}(x) = \varphi(x)$ for every $x \in X$. Replacing $g^{-1}(x)$ by y, so that $x = g(y)$, gives $g^*(\varphi) = \varphi$ if and only if $\varphi(y) = \varphi(g(y))$ for every $y \in X$. That is, $g^*(\varphi) = \varphi$ if and only if for each y the values $\varphi(y)$, $\varphi(g(y))$, $\varphi(g^2(y))$, $\varphi(g^3(y))$, ... are all the same. It follows that $g^*(\varphi) = \varphi$ if and only if φ is constant on each orbit $\{g^n(y) : n \in \mathbb{P}\}$ under $\langle g \rangle$. ∎

EXERCISES 12.3

1. Consider the group $G = \text{AUT}(H)$ of Example 1 acting on the set V of vertices of H.

   (a) Find $|\text{FIX}_V(a)|$ for each automorphism a in G and add the results.

   (b) Find $|\text{FIX}_G(p)|$ for each vertex p of H and add the results.

   (c) Do the sums in (a) and (b) agree? Discuss.

2. Consider the square graph in Figure 4 of § 12.2. Confirm Theorem 1 for this example.

	w	x	y	z
e	w	x	y	z
g	y	x	w	z
h	w	z	y	x
gh	y	z	w	x

(a)   (b)   (c)

FIGURE 4

3. Consider the graph in Figure 4, and let G be the group of graph automorphisms acting on $\{w, x, y, z\}$.

   (a) Convince yourself that $G = \{e, g, h, gh\}$ where these automorphisms are given in Figure 4(b).

   (b) Confirm Theorem 1 for this example.

4. Let the group G in Exercise 3 act on the 5-element set E of edges. See Figure 4(c).

   (a) Give the table of values, as in Figure 2(b).

   (b) Confirm Theorem 1 for this example.

5. Show directly that Corollary 2 is true for each of the orbits of the group in Exercise 3.

6. Show directly that Corollary 2 is true for each of the orbits of the group in Exercise 4.

7. (a) Color the vertices in Figure 4(a) by coloring w and x black and y and z red. Give all the equivalent colorings.

   (b) Repeat where w is black and all the rest of the vertices are red.

8. (a) Color the edges in Figure 4(c) by coloring e_1, e_3, e_5 red and e_2, e_4 black. Give all equivalent colorings.

   (b) Repeat where e_1, e_2, e_3 are red and e_4, e_5 are black.

9. Consider the square graph in Figure 4 of § 12.2.

   (a) Color the vertices so that x, y are red and w, z are black. Give all equivalent colorings.

   (b) Do the same if x, z are red and w, y are black.

(c) Do the same if x is red and the other vertices are black.

(d) Can you guess how many essentially different colorings use red, black or both?

10. In Example 3(b), the group G^* of restrictions to the orbit $\{w, x, y, z\}$ has 8 elements while PERM($\{w, x, y, z\}$) has 24 elements. Give two permutations that are not in G^*.

11. Verify that $|\text{FIX}_T(h^*)| - 7$ in Example 2(b).

12. (a) Show that Exercise 7 in §12.2 can be applied to the group G acting on E in Example 2(a).

(b) Give an explicit illustration of the conclusion to part (a).

13. (a) Show that Exercise 7 in §12.2 applies to the group in Figure 4 acting on the edges.

(b) Give an explicit illustration of your conclusion to part (a).

14. Why didn't we define $g^*(\varphi) = \varphi \circ g$ when we showed how we can view $G \subseteq \text{PERM}(X)$ as acting on FUN(X, C)?

15. Let H be a connected graph with no parallel edges, V its set of vertices, and E its set of edges. Also let $G \subseteq \text{PERM}(V)$ be the group of graph automorphisms acting on V, and let $g \to g^*$ be the map of G into PERM(E) described prior to Example 2.

(a) Show that $g \to g^*$ is not one-to-one for the graph $u \bullet\!\!-\!\!\!-\!\!\bullet v$.

(b) Show that if H has more than one edge, then $g \to g^*$ is one-to-one. *Hint:* It suffices to show that $g^* = e^*$ implies $g = e$.

16. Consider a group G acting on an n-element set X. Show that if $|\text{FIX}_X(g)| \geq 1$ for each $g \in G$, then G has at least $1 + \dfrac{n-1}{|G|}$ orbits in X. For $n > 1$ this implies Corollary 1 to Theorem 1. *Hint:* Treat the element e of G separately from the others.

§ 12.4   Applications to Coloring Problems

Consider the problem of manufacturing logical circuits to compute all the different Boolean functions of four variables. On the one hand, since there are $2^4 = 16$ rows in a truth table for such a function, there are $2^{16} = 65{,}536$ such functions. On the other hand, just by switching the input connections around we can make one circuit compute various different functions, so we don't need as many circuits as there are functions. If we are willing to use external hardware to complement inputs and outputs, we need to manufacture still fewer circuits. How many can we get by with?

Or consider coloring the faces of a cube, with only three colors allowed. If one face is red and the rest are blue, it doesn't really matter which face is red, since we can rotate the cube to put the red face wherever we want. Given two ways of coloring the faces so that one is red, three are blue and two are green, it may or may not be possible to rotate the cube so that the two colorings are really the same. How many essentially different ways are there to color the cube? A natural group to consider for this question is the group G of all rotations of space that send the cube back onto itself, since this group describes the rotational symmetry of the cube. The group acts on the set X of faces of the cube, and it also acts on the set FUN(X, C) of **colorings** of the faces where $C = \{$red, blue, green$\}$. The equivalence classes of colorings are just the orbits of FUN(X, C) under G, and we can hope to count them using Theorem 1 of the preceding section. The method we will use applies quite widely, so instead of answering the cube-coloring question just now, we first prove a general theorem and then apply it in Example 3 to the special case of the cube.

Theorem 1

Consider a finite group G acting on a set X. If C is any set, then G also acts on FUN(X, C) by the definition $g*(\varphi) = \varphi \circ g^{-1}$ for $g \in G$ and $\varphi: X \to C$. For each g in G, let $m(g)$ be the number of orbits of the group $\langle g \rangle$ on X. Then the number of orbits of G on FUN(X, C) is

$$\frac{1}{|G|} \sum_{g \in G} |C|^{m(g)}.$$

Proof. According to Theorem 1 of §12.3, to show the formula is correct we just need to show that for each $g \in G$ we have

$$\left| \text{FIX}_{\text{FUN}(X,C)}(g*) \right| = |C|^{m(g)}.$$

But Theorem 2 of §12.3 shows that the set $\text{FIX}_{\text{FUN}(X,C)}(g*)$ consists of the functions $\varphi: X \to C$ that are constant on the $\langle g \rangle$-orbits. To describe such a φ we simply give its value on each $\langle g \rangle$-orbit. There are $|C|$ possible function values and $m(g)$ orbits under $\langle g \rangle$, so there are $|C|^{m(g)}$ functions φ that are constant on $\langle g \rangle$-orbits. The theorem follows. ∎

A coloring of a set X with colors from C is just a function from X to C, and we regard two colorings as equivalent under the action of G if they belong to the same G-orbit $\{\varphi \circ g^{-1} : g \in G\} = \varphi \circ G$ in FUN(X, C). The G-orbits are the G-equivalence classes of colorings. With this terminology, Theorem 1 gives the following information.

Theorem 2

Consider a finite group G acting on a set X and consider a set of k colors. Let $C(k)$ be the number of G-equivalence classes of colorings of X using some or all of the k colors. Then

$$C(k) = \frac{1}{|G|} \sum_{g \in G} k^{m(g)}$$

where $m(g)$ is the number of orbits of $\langle g \rangle$ on X.

EXAMPLE 1   We first color the vertices of the graph in Figure 3 of § 12.2 and Figure 1 of § 12.3. We have $|G| = 8$, and the numbers $m(g)$ are given in Table 1 of § 12.2. They are 6, 5, 5, 4, 3, 2, 2, 3, so by Theorem 2 we have

$$C(k) = \frac{1}{8}(k^6 + k^5 + k^5 + k^4 + k^3 + k^2 + k^2 + k^3)$$

$$= \frac{1}{8}(k^6 + 2k^5 + k^4 + 2k^3 + 2k^2).$$

The first few values of $C(k)$ are given in Figure 1. This problem for $k = 2$ was discussed in Example 5 of § 12.3, where it appeared moderately difficult to solve directly. The problem for $k \geq 3$ looks hopeless without the theory we have developed.

k	$C(k)$
1	1
2	21
3	171
4	820
5	2,850
6	8,001
7	19,306

FIGURE 1

EXAMPLE 2   Before we get back to the cube, we color the vertices of the square with k colors. We regard two colorings as the same if we can turn one into the other by a suitable rotation of the square or by flipping it over. Figure 2(a) shows the square, and Figure 2(b) lists the relevant group of permutations of its vertex set. The table in Figure 3 lists the orbits of the subgroups generated by one permutation; this solves Exercise 5, parts (b)

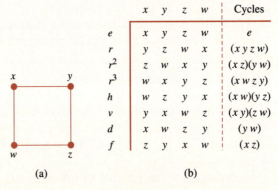

	x	y	z	w	Cycles
e	x	y	z	w	e
r	y	z	w	x	$(x\,y\,z\,w)$
r^2	z	w	x	y	$(x\,z)(y\,w)$
r^3	w	x	y	z	$(x\,w\,z\,y)$
h	w	z	y	x	$(x\,w)(y\,z)$
v	y	x	w	z	$(x\,y)(z\,w)$
d	x	w	z	y	$(y\,w)$
f	z	y	x	w	$(x\,z)$

(a)   (b)

FIGURE 2

$\langle e \rangle$	$\{x\}, \{y\}, \{z\}, \{w\}$	$m(e) = 4$
$\langle r \rangle$	$\{x, y, z, w\}$	$m(r) = 1$
$\langle r^2 \rangle$	$\{x, z\}, \{y, w\}$	$m(r^2) = 2$
$\langle r^3 \rangle$	$\{x, y, z, w\}$	$m(r^3) = 1$
$\langle h \rangle$	$\{x, w\}, \{y, z\}$	$m(h) = 2$
$\langle v \rangle$	$\{x, y\}, \{w, z\}$	$m(v) = 2$
$\langle d \rangle$	$\{x\}, \{z\}, \{y, w\}$	$m(d) = 3$
$\langle f \rangle$	$\{x, z\}, \{y\}, \{w\}$	$m(f) = 3$

FIGURE 3

and (c), of §12.2. One can check that $\langle e \rangle = \{e\}$, $\langle r \rangle = \langle r^3 \rangle = \{e, r, r^2, r^3\}$, $\langle r^2 \rangle = \{e, r^2\}$, $\langle h \rangle = \{e, h\}$, $\langle v \rangle = \{e, v\}$, $\langle d \rangle = \{e, d\}$ and $\langle f \rangle = \{e, f\}$. For example, since $\langle f \rangle = \{e, f\}$ the orbits of $\langle f \rangle$ are the sets $\{e(s), f(s)\}$ for s in $\{x, y, z, w\}$, so they are $\{e(x), f(x)\} = \{x, z\}$, $\{e(y), f(y)\} = \{y\}$, $\{e(z), f(z)\} = \{z, x\}$ and $\{e(w), f(w)\} = \{w\}$. There are just three different orbits; thus $m(f) = 3$. The orbits can also be read from the cycle descriptions of the permutations, not forgetting orbits of size 1.

According to Theorem 2 there are

$$C(k) = \frac{1}{8}(k^4 + k + k^2 + k + k^2 + k^2 + k^3 + k^3)$$

$$= \frac{1}{8}(k^4 + 2k^3 + 3k^2 + 2k)$$

different ways to color the vertices of the square with k colors. For $k = 1$, this number is of course 1. The table in Figure 4 gives the numbers of

k	Number of ways to color
1	1
2	6
3	21
4	55
5	120
6	231
7	406

(a)

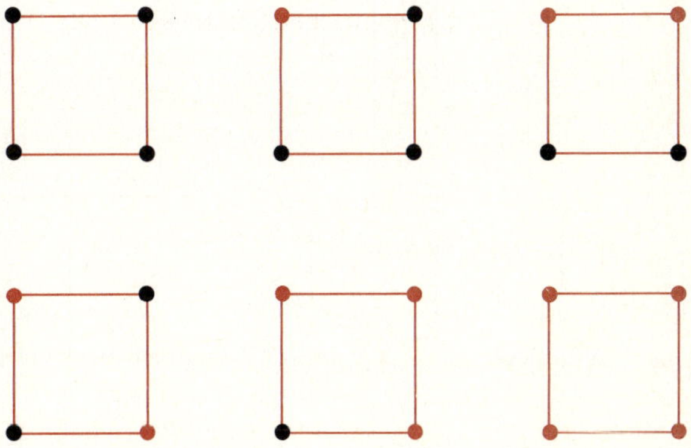

(b)

FIGURE 4

colorings possible for the first few values of k. Figure 4(b) indicates the six different possibilities for two colors, including both of the one-color colorings. ■

EXAMPLE 3   Now let us color the faces of the cube we started this section with. There are 24 rotations that send the cube back to itself. To list them all would take a fair amount of work, but in fact we only need to know their orbit sizes, and for that we can just count the rotations of the five types illustrated in Figure 5(a). The table in Figure 5(b) tells how many there are of each type. It also gives the number $m(g)$ of orbits of their cyclic groups acting on the set of faces of the cube. For instance, consider the 90° rotation of type b. Each such rotation has an axis through the centers of two opposite faces. There are 6 faces, so there are 3 opposite pairs and hence 3 such axes. Figure 5(b) lists 6 rotations of type b: each axis gives two 90° rotations, one in each direction. A rotation g of type b has 3 $\langle g \rangle$-orbits: the faces the axes go through form orbits of size 1 and the other four faces form an orbit of size 4. Thus $m(g) = 3$. The remaining entries in Figure 5(b) have been determined by similar reasoning. Theorem 2 gives the formula

$$C(k) = \frac{1}{24}\,(6k^3 + 6k^3 + 3k^4 + 8k^2 + k^6)$$

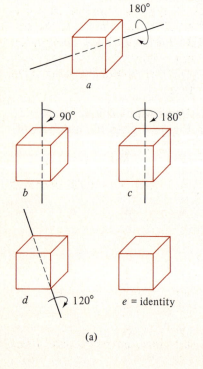

180°

90°   180°

b c

d 120° e = identity

(a)

Type	Number of that type	$m(g)$
a	6	3
b	6	3
c	3	4
d	8	2
e	1	6

(b)

k	$C(k)$
1	1
2	10
3	57
4	240
5	800
6	2226

(c)

FIGURE 5

for the number of colorings of the faces with k colors. Figure 5(c) lists the first few values of $C(k)$. ■

Theorem 2 yields the number $C(k)$ of colorings using at most k colors. In some applications we want to know the number of colorings using exactly k colors. For this the Inclusion–Exclusion Principle of § 5.3 is useful.

EXAMPLE 4   Let us calculate the number of [inequivalent] colorings of the vertices of the square in Example 2 using exactly four colors, say red, blue, green and yellow. Figure 2 gives us the number $C(k)$ of ways to color using at most k colors. For $i = 1, 2, 3, 4$ let A_i be the set of colorings that do not use the ith color. Then $A_1 \cup A_2 \cup A_3 \cup A_4$ is the set of colorings using three or fewer of the colors, and the answer we seek is $C(4) - |A_1 \cup A_2 \cup A_3 \cup A_4|$. Now by the Inclusion–Exclusion Principle,

$$|A_1 \cup A_2 \cup A_3 \cup A_4| = |A_1| + |A_2| + |A_3| + |A_4| - \{|A_1 \cap A_2|$$
$$+ |A_1 \cap A_3| + |A_1 \cap A_4| + |A_2 \cap A_3| + |A_2 \cap A_4| + |A_3 \cap A_4|\}$$
$$+ \{|A_1 \cap A_2 \cap A_3| + |A_1 \cap A_2 \cap A_4| + |A_1 \cap A_3 \cap A_4|$$
$$+ |A_2 \cap A_3 \cap A_4|\} - |A_1 \cap A_2 \cap A_3 \cap A_4|.$$

The set A_1 consists of all colorings using blue, green or yellow, so $|A_1|$ is $C(3)$. Similarly for A_2, A_3 and A_4. The set $A_1 \cap A_2$ consists of the colorings using green or yellow, so it has $C(2)$ elements; a similar observation applies to each intersection of two sets. Intersections like $A_1 \cap A_2 \cap A_3$ have only one coloring, i.e., they have $C(1)$ elements. Finally, $A_1 \cap A_2 \cap A_3 \cap A_4$ is the empty set. We conclude that

$$|A_1 \cup A_2 \cup A_3 \cup A_4| = 4C(3) - 6C(2) + 4C(1),$$

so the number of colorings using exactly four colors is

$$C(4) - 4C(3) + 6C(2) - 4C(1) = 55 - 4 \cdot 21 + 6 \cdot 6 - 4 \cdot 1 = 3.$$

Now that we know the answer, we can easily illustrate the different colorings in Figure 6. ■

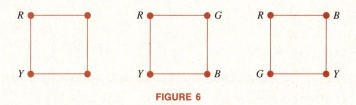

The next example uses exactly the same ideas as in Example 4, but the details are messier.

EXAMPLE 5 Consider now the problem of assigning different labels 1, 2, 3, 4, 5, 6 to the vertices of the graph last discussed in Example 1, as illustrated in Figure 7. Two assignments are regarded as the same if there is a graph automorphism that takes one to the other. Thus the labelings shown in Figure 7 are all considered the same.

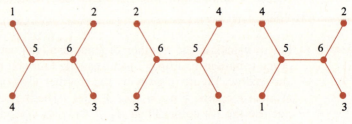

FIGURE 7

We can view this question as a coloring problem with exactly 6 colors. In Example 1 we already found the number $C(k)$ of ways to color using at most k colors. As in Example 4, for $i = 1, 2, \ldots, 6$ let A_i be the set of colorings that do *not* use the ith color. Then $A_1 \cup A_2 \cup \cdots \cup A_6$ is the set of colorings using 5 or fewer of the colors, and the answer to our question is $C(6) - |A_1 \cup A_2 \cup \cdots \cup A_6|$. The Inclusion–Exclusion Principle gives a formula for $|A_1 \cup A_2 \cup \cdots \cup A_6|$, namely

$$\sum_{i=1}^{6} |A_i| - \sum_{1 \le i < j \le 6} |A_i \cap A_j| + \sum_{1 \le i < j < k \le 6} |A_i \cap A_j \cap A_k| - \cdots,$$

where, for example, the third sum adds up the sizes of the intersections of three *distinct* sets from among $A_1, \ldots, A_6$. Now

$$|A_i| = C(5) \qquad \text{for each } i,$$
$$|A_i \cap A_j| = |\{\text{colorings not using } i\text{th and } j\text{th colors}\}|$$
$$= C(4) \qquad \text{for } i < j,$$
$$|A_i \cap A_j \cap A_k| = C(3) \qquad \text{for } i < j < k,$$

etc. Thus

$$|A_1 \cup A_2 \cup \cdots \cup A_6|$$
$$= \binom{6}{1} C(5) - \binom{6}{2} C(4) + \binom{6}{3} C(3) - \binom{6}{4} C(2) + \binom{6}{5} C(1),$$

so the number of colorings that use exactly 6 colors is

$$C(6) - \binom{6}{1} C(5) + \binom{6}{2} C(4) - \binom{6}{3} C(3) + \binom{6}{4} C(2) - \binom{6}{5} C(1)$$

$$= 8001 - 6 \cdot 2850 + 15 \cdot 820 - 20 \cdot 171 + 15 \cdot 21 - 6 \cdot 1 = 90.$$

Similarly, the number of colorings that use exactly two colors is just $C(2) - \binom{2}{1}C(1) = 21 - 2 = 19$ and the number using exactly seven colors is

$$19306 - 7 \cdot 8001 + 21 \cdot 2850 - 35 \cdot 820 + 35 \cdot 171 - 21 \cdot 21 + 7 \cdot 1 = 0. \quad \blacksquare$$

Now let us return to the problem of building logical circuits, at first for just two inputs. There are $2^4 = 16$ Boolean functions of two variables, namely the members of FUN$(\mathbb{B} \times \mathbb{B}, \mathbb{B})$, where $\mathbb{B} = \{0, 1\}$. We may think of a circuit as a black box with two input wires, one for x_1 and one for x_2, and one output wire. Each element (a_1, a_2) in $\mathbb{B} \times \mathbb{B}$ corresponds to a choice of values a_1 for x_1 and a_2 for x_2.

Interchanging the connections for x_1 and x_2 amounts to replacing (a_1, a_2) by (a_2, a_1), and corresponds to the permutation g of $\mathbb{B} \times \mathbb{B}$ that interchanges $(0, 1)$ and $(1, 0)$. We want to regard two black boxes as equivalent if one will produce the same results as the other, or will produce the same results if we interchange its input wires. That is, two boxes are equivalent if their Boolean functions f and f' are either the same or satisfy $f' = f \circ g$. Since $|\mathbb{B}| = 2$, the problem looks just like the 2-color question for a four-element set, $\mathbb{B} \times \mathbb{B}$, with two elements that are interchangeable. We apply Theorem 1 with $X = \mathbb{B} \times \mathbb{B}$, $C = \mathbb{B}$ and $G = \langle g \rangle = \{e, g\}$. The number of G-orbits is

$$\frac{1}{2}(2^4 + 2^3) = 12.$$

We can confirm this result using the table in Figure 8, which lists all sixteen Boolean functions from $\mathbb{B} \times \mathbb{B}$ to $\mathbb{B}$. Functions 2 and 4 can be performed with the same black box, as can functions 3 and 5, 10 and 12, and 11 and 13, so the number of orbits is $16 - 4 = 12$.

Function numbers

	0	1	2	3	4	5	6	7	8	9	10	11	12	13	14	15
$(0, 0)$	0	1	0	1	0	1	0	1	0	1	0	1	0	1	0	1
$(0, 1)$	0	0	1	1	0	0	1	1	0	0	1	1	0	0	1	1
$(1, 0)$	0	0	0	0	1	1	1	1	0	0	0	0	1	1	1	1
$(1, 1)$	0	0	0	0	0	0	0	0	1	1	1	1	1	1	1	1

FIGURE 8

Now suppose that we allow ourselves to complement inputs. Complementing the value on the first wire corresponds to interchanging $(0, 0)$ with $(1, 0)$, and $(0, 1)$ with $(1, 1)$. We denote this permutation of $\mathbb{B} \times \mathbb{B}$ by c_1 and the permutation that corresponds to complementing the second input by c_2. Altogether the permutations g, c_1 and c_2 generate the

	(0, 0)	(0, 1)	(1, 0)	(1, 1)		x	y	w	z
e	(0, 0)	(0, 1)	(1, 0)	(1, 1)	e	x	y	w	z
c_1	(1, 0)	(1, 1)	(0, 0)	(0, 1)	h	w	z	x	y
c_2	(0, 1)	(0, 0)	(1, 1)	(1, 0)	v	y	x	z	w
$c_1 \circ c_2$	(1, 1)	(1, 0)	(0, 1)	(0, 0)	r^2	z	w	y	x
g	(0, 0)	(1, 0)	(0, 1)	(1, 1)	d	x	w	y	z
$c_1 \circ g$	(1, 0)	(0, 0)	(1, 1)	(0, 1)	r^3	w	x	z	y
$c_2 \circ g$	(0, 1)	(1, 1)	(0, 0)	(1, 0)	r	y	z	x	w
$c_1 \circ c_2 \circ g$	(1, 1)	(0, 1)	(1, 0)	(0, 0)	f	z	y	w	x

(a)   (b)

FIGURE 9

group G of permutations of $\mathbb{B} \times \mathbb{B}$ described in Figure 9(a). [This fact is not expected to be obvious; take our word for it.] This group acts on the 4-element set $\mathbb{B} \times \mathbb{B}$ in the same way that the group in Example 2 acts on the vertices of the square, as we see by comparing Figure 9(a) with Figure 9(b), which is just Figure 2(b) rewritten with some rows and columns interchanged. The correspondence $(0, 0) \to x, (0, 1) \to y, (1, 0) \to w,$ $(1, 1) \to z$ converts one table into the other. From Figure 4 we know that there are $C(2) = 6$ ways to 2-color the square, so there are six orbits of Boolean functions under the action of the group G, i.e., six essentially different black boxes. Using the function numbers from Figure 8, the orbits in $\text{FUN}(\mathbb{B} \times \mathbb{B}, \mathbb{B})$ are

$$\{0\}, \quad \{1, 2, 4, 8\}, \quad \{3, 5, 10, 12\}, \quad \{6, 9\}, \quad \{7, 11, 13, 14\}, \quad \{15\}.$$

To build circuits, it would be enough to have a circuit to compute one function from each orbit, say the functions 0, 1, 3, 6, 7 and 15.

If we also allow ourselves to complement the output of a circuit, then a circuit that computes the function numbered n will also compute $15 - n$, and we need even fewer black boxes. A circuit for 0 will also compute 15. One for 1 will compute 14 and hence also 7, 11 or 13. A circuit for 3 will also compute 12, which we already knew, and similarly a circuit for 6 will compute 9. The classes of functions are now

$$\{0, 15\}, \quad \{1, 2, 4, 8, 7, 11, 13, 14\}, \quad \{3, 5, 10, 12\} \quad \text{and} \quad \{6, 9\}.$$

It still requires four different circuits to compute all 2-variable Boolean functions, allowing complementation on both input and output wires.

Our methods generalize, in theory, to count the number of black boxes needed for n-input Boolean functions. In practice, the detailed determination of orbits for all elements of G gets exceedingly complicated. For 4-input functions the answer is that there are 222 different circuits required, even if we allow free complementation on inputs and outputs.

This number is considerably smaller than $2^{16} = 65{,}536$. Knowing how many circuits there are does not help us find representative circuits, but it does tell us when we have found enough.

Our methods have not taken systematic advantage of the symmetry of the group G itself. By using such symmetry one can obtain a formula for the number of G-orbits in FUN($\mathbb{B}^n$, $\mathbb{B}$) whose members have exactly k of their values equal to 0 for $k = 1, 2, \ldots$. There is an extensive literature on the subject of using groups to count. For more in the spirit of this section, look in books on applied algebra, watching for the names Polyà and Burnside.

EXERCISES 12.4

1. The graph in Figure 10(a) has two automorphisms, which are described in Figure 10(b).

   (a) What is the average number of vertices fixed by the automorphisms of this graph?

   (b) Which of the automorphisms of this graph fix the average number of vertices?

   (c) Find the number of ways to color the vertices of this graph with k colors.

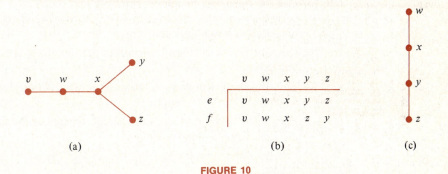

FIGURE 10

2. Verify Theorem 1 of §12.3 for

   (a) the group of automorphisms of the graph in Figure 10(a) acting on the set of vertices of the graph.

   (b) the group in part (a) acting on the set of edges of the graph in Figure 10(a).

   (c) the group of rotations in Example 3 acting on the set of faces of the cube.

3. The graph in Figure 10(c) has two automorphisms.

   (a) How many ways are there to color the vertices of this graph with k colors?

(b) How many ways are there to label the vertices of this graph with four different labels?

4. Compute $g \circ c_1 \circ g$ for members of the group described in Figure 9(a). Your answer implies that the table could be constructed using just c_1 and g.

5. The graph in Figure 11(a) has six automorphisms, which are described in Figure 11(b).

(a) Find the number of ways to color the vertices of this graph with k colors.

(b) Find the number of ways to color the edges of this graph with k colors.

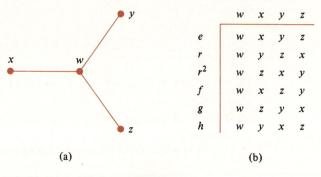

	w	x	y	z
e	w	x	y	z
r	w	y	z	x
r^2	w	z	x	y
f	w	x	z	y
g	w	z	y	x
h	w	y	x	z

(a) (b)

FIGURE 11

6. (a) Use the Inclusion–Exclusion Principle and the answer to Exercise 5(a) to find the number of ways to color the vertices of the graph in Figure 11(a) with exactly 3 colors.

(b) Describe all the different colorings in part (a), using the colors red, blue and green.

(c) Find the number of ways to color the vertices of this graph with exactly 4 colors.

7. (a) How many ways are there to color the vertices of the square in Example 2 with exactly four colors?

(b) List the different colorings that use all 4 of the colors red, blue, green, yellow.

8. How many different circular necklaces can be made from 5 beads of k different colors? Consider two necklaces to be the same if one looks just like the other when it is rotated or flipped over. *Hint:* The group here consists of e, four nontrivial rotations and five flips. See Example 2 for the 4-bead case.

9. Find the number of colorings of the vertices of the square in Example 2 using

(a) exactly 3 colors, (b) exactly 2 colors, (c) exactly 5 colors.

10. Draw representatives of each coloring of the types counted in Exercise 9.

11. Consider the graph of Exercise 3 of §12.3 and the group G acting on $V = \{w, x, y, z\}$.

(a) For each member g of G, find the number of $\langle g \rangle$-orbits in V.

(b) Find the number of ways to color the vertices of this graph with at most k colors.

(c) Use your answer to (b) to calculate the number of ways to color the vertices using red or black or both. Draw a picture of a representative of each coloring.

12. The group G in Exercise 11 also acts on the set $E = \{e_1, e_2, e_3, e_4, e_5\}$ of edges; see Exercise 4, §12.3.

(a) For each member g of G, find the number of $\langle g \rangle$-orbits in E.

(b) Find the number of ways to color the edges of this graph with at most k colors.

(c) How many ways are there to color the edges of this graph with exactly 2 colors?

13. (a) How many ways are there to color the vertices of a cube with k colors?

(b) How many ways are there to color the edges of a cube with k colors?
  Two colorings in parts (a) and (b) are considered the same if one can be turned into the other by a rotation of the cube. *Suggestion:* Use Figure 5(a) to create new tables like Figure 5(b) for the actions on vertices and edges.

14. Consider the problem of coloring the faces of a cube using crayons which are red, green and blue, as in Example 3.

(a) How many colorings use exactly two of the three colors?

(b) Use any method to find how many colorings have four red faces and two blue faces.

(c) How many colorings have exactly four red faces?

(d) Would you like to use the method of inspection to find all colorings with exactly two faces of each color?

15. (a) How many different 2-input logical circuits are there if we only regard two circuits as the same if they have the same or complementary outputs?

(b) How many are there if we also consider two circuits to be the same if they produce the same function when the input wires on one are interchanged?

§ 12.5 Groups

Sections 12.1 to 12.4 include an introduction to groups of permutations. Groups also arise in a variety of settings that have no obvious connection with permutations. Here is a general definition. A **group** $(G, \cdot)$ is a set G with a binary operation $\cdot$ [i.e., a map $(g, h) \to g \cdot h$ of $G \times G$ into G] with the following properties:

(i) $(g \cdot h) \cdot k = g \cdot (h \cdot k)$ for all $g, h, k \in G$;　　**[associative law]**

(ii) G has an element e, called the **identity**, such that $g \cdot e = e \cdot g = g$ for all $g \in G$;

(iii) to each g in G there corresponds an element g^{-1}, called its **inverse**, such that $g \cdot g^{-1} = g^{-1} \cdot g = e$.

For a generic group we will use $\cdot$, $\bullet$, or sometimes no symbol at all to denote the binary operation; i.e., we will write $g \cdot h$, $g \bullet h$ or just gh and call such an element the "product" of g and h. Inverses are unique in a strong sense. If $h \cdot g = e$, then

$$h = h \cdot e = h \cdot (g \cdot g^{-1}) = (h \cdot g) \cdot g^{-1} = e \cdot g^{-1} = g^{-1}.$$

Similarly, the equation $g \cdot h = e$ forces h to be g^{-1}. To check that h is g^{-1} then, it is enough to verify either of the conditions $h \cdot g = e$ or $g \cdot h = e$. The other will follow.

Nonempty subsets H of G are automatically groups in their own right provided that they are **closed** under the operation and under inverses: $g, h \in H$ implies $g \cdot h \in H$ and $g^{-1} \in H$. Such a subset is called a **subgroup** of G. Generic subgroups will usually be called H, J or K.

EXAMPLE 1　　(a) The symmetric group S_n on n letters studied in § 12.1 is a group where the binary operation is composition, $\circ$. In fact, the set PERM(X) of all permutations of a set X onto itself is a group under composition $\circ$. As we saw in the preceding sections, the subgroups of PERM(X) are a rich source of examples.

(b) The set GL(n) consisting of all $n \times n$ matrices that have inverses is a group under ordinary matrix multiplication; see § 3.4. Note that the product of invertible matrices is invertible; in fact, $(AB)^{-1} = B^{-1}A^{-1}$. At first glance, this group seems rather different from the groups with composition as operation, but there is a connection. In linear algebra $n \times n$ matrices are put into a one-to-one correspondence with linear transformations on $\mathbb{R}^n$, and matrix multiplication corresponds to the composition of linear transformations. ■

In a sense, Example 1(a) contains all the examples that there are. Two groups are **isomorphic** if there is a one-to-one mapping [called an **isomorphism**] of one onto the other that preserves the group operations. In symbols, if $(G_1, \cdot)$ and $(G_2, \bullet)$ are groups, a one-to-one correspondence φ of G_1 onto G_2 is an isomorphism provided that $\varphi(g \cdot h) = \varphi(g) \bullet \varphi(h)$ for all $g, h \in G$. Isomorphic groups are essentially the same groups; they have the same basic structure. A theorem, known as Cayley's theorem, states that every group is isomorphic to a group of permutations on some set; in fact, the set can be chosen to be G itself. [See Exercise 24.] Despite Cayley's theorem, we will study many groups in their natural setting where the permutation group is not in evidence.

A group G is **commutative** if $g \cdot h = h \cdot g$ for all $g, h \in G$. Commutative groups are often written additively, with operation $+$, because a number of important commutative groups traditionally use that notation. The **additive identity** is usually denoted **0**. Inverses with respect to $+$ are called **additive inverses** and written $-g$ instead of g^{-1}. Thus a commutative group $(G, +)$ has a binary operation $(g, h) \to g + h$ satisfying:

(i) $(g + h) + k = g + (h + k)$ for all $g, h, k \in G$;   [**associative law**]

(ii) G has an element **0**, called the **zero**, such that $g + 0 = g$ for all $g \in G$;

(iii) to each g in G there corresponds an additive inverse $-g$ such that $g + (-g) = 0$;

(iv) $g + h = h + g$ for all $g, h \in G$.   [**commutative law**]

EXAMPLE 2   (a) $(\mathbb{R}, +)$ is a commutative group. The additive identity is 0, of course, and the laws (i) to (iv) are all familiar.

(b) $(\mathbb{Z}, +)$ is also a commutative group; it's a subgroup of $(\mathbb{R}, +)$.

(c) The set $\mathfrak{M}_{m,n}$ of all $m \times n$ matrices is a commutative group under addition. This fact is spelled out in the theorem in §3.3.

(d) The set $\mathbb{R}$ under multiplication $\cdot$ is not a group even though it satisfies properties (i), (ii) and (iv). The trouble is that 0 has no inverse. Since the product of nonzero real numbers is nonzero, $\cdot$ is a binary operation on $\mathbb{R}\backslash\{0\}$, so $(\mathbb{R}\backslash\{0\}, \cdot)$ is a group. It is an example of a commutative group in which the operation is *not* $+$.

(e) The set $\mathbb{N}$ is not a group under addition because $n \in \mathbb{N}$ does not imply $-n \in \mathbb{N}$. In fact, if n and $-n$ are both in $\mathbb{N}$, then n must be 0.

(f) The set $\mathfrak{M}_{n,n}$ of square matrices is not a group under multiplication because many matrices, like the zero matrix, do not have inverses. A useful subset that is a group under multiplication was introduced in Example 1(b).   ■

Group theory is an amazingly rich subject when one realizes that all the results rely in the end on the three laws (i) to (iii). We will only scratch the surface. We begin with some simple consequences. Note that each equality of the proof is a consequence of one of the three laws.

Theorem 1   Let $(G, \cdot)$ be a group.

(a) If $g \cdot h = g \cdot k$ or $h \cdot g = k \cdot g$ for $g, h, k \in G$, then $h = k$. (These are the **cancellation laws**.)

(b) $(g \cdot h)^{-1} = h^{-1} \cdot g^{-1}$ for all $g, h \in G$.

Proof. (a) If $g \cdot h = g \cdot k$, then $h = e \cdot h = (g^{-1} \cdot g) \cdot h = g^{-1} \cdot (g \cdot h) = g^{-1} \cdot (g \cdot k) = (g^{-1} \cdot g) \cdot k = e \cdot k = k$. Similarly, $h \cdot g = k \cdot g$ implies $h = k$.

(b) Since the inverse of $g \cdot h$ is unique, it suffices to prove that $h^{-1} \cdot g^{-1}$ satisfies the equation $(g \cdot h) \cdot (h^{-1} \cdot g^{-1}) = e$. But $(g \cdot h) \cdot (h^{-1} \cdot g^{-1}) = g \cdot (h \cdot (h^{-1} \cdot g^{-1})) = g \cdot ((h \cdot h^{-1}) \cdot g^{-1}) = g \cdot (e \cdot g^{-1}) = g \cdot g^{-1} = e$. ■

As in the case of groups of permutations, for $g \in G$ we write $\langle g \rangle$ for the group $\{g^n : n \in \mathbb{Z}\}$ generated by g. As usual, $g^0 = e$. For positive n the symbol g^n represents the product of g by itself n times, and g^{-n} is $(g^{-1})^n$. With these agreements it can be shown [Exercise 14] that the familiar rule $g^{m+n} = g^m \cdot g^n$ holds for all $m, n \in \mathbb{Z}$. From this it is clear that $\langle g \rangle$ defined above is really a subgroup of G. Groups of the form $\langle g \rangle$ are often called **cyclic groups**. Since $g^m \cdot g^n = g^{m+n} = g^{n+m} = g^n \cdot g^m$, cyclic groups are commutative. A group element g is said to have [finite] **order** m if $\langle g \rangle$ has m elements and is said to have **infinite order** if it does not have finite order. We will see that if g has order m, then $g^m = e$ and m is the smallest positive integer with this property.

EXAMPLE 3   (a) Theorem 2 of §12.1 gives a recipe for finding the orders of permutations in S_n. Exercises 11 to 16 of that section involve calculating the orders of permutations.

(b) In §3.6 we defined the operations $+_p$ and $*_p$ on the set $\mathbb{Z}(p) = \{0, \ldots, p - 1\}$. Theorem 4 of §3.6 showed that both operations are associative and commutative. Moreover, 0 is the additive identity of $\mathbb{Z}(p)$ and 1 the multiplicative identity. Each m in $\mathbb{Z}(p)$ has an additive inverse, namely $(-m) \operatorname{MOD} p$, because by Theorem 3 of §3.6

$$m +_p (-m) \operatorname{MOD} p = m \operatorname{MOD} p +_p (-m) \operatorname{MOD} p$$

$$= (m + (-m)) \operatorname{MOD} p = 0 \operatorname{MOD} p = 0.$$

Thus $(\mathbb{Z}(p), +_p)$ is a commutative group. Because 1 is a generator, this group is cyclic. [In fact, k in $\mathbb{Z}(p)$ is a generator if and only if $\gcd(k, p) = 1$.]

Although $(\mathbb{Z}(p), *_p)$ has lots of nice properties, it is not a group; there is no m with $0 *_p m = 1$, so 0 has no multiplicative inverse.

(c) The infinite group $(\mathbb{Z}, +)$ is cyclic; both 1 and -1 are generators for this group. All the nonzero elements of $\mathbb{Z}$ have infinite order. In fact, if $n \in \mathbb{Z}$, the cyclic group generated by n is exactly $n\mathbb{Z} = \{nk : k \in \mathbb{Z}\}$ and this is infinite unless $n = 0$. ■

We will see in the next section that $(\mathbb{Z}, +)$ and the groups $(\mathbb{Z}(p), +_p)$ are the only cyclic groups there are, in the sense that every cyclic group is isomorphic to one of these. The isomorphisms are natural. If $\langle g \rangle$ is infinite, then $g^k \to k$ matches $\langle g \rangle$ with $\mathbb{Z}$, and of course $g^m \cdot g^n = g^{m+n} \to m + n$ shows that the correspondence is an isomorphism. If $\langle g \rangle$ has p

elements, i.e., if g has order p, the correspondence $g^k \to k \operatorname{MOD} p$ is an isomorphism of $\langle g \rangle$ onto $\mathbb{Z}(p)$. In this case the elements $e, g, \ldots, g^{p-1}$ are all different, $g^p = e$, and $g^{-1} = g^{p-1}$.

Normally, to check that a nonempty subset H of some known group G is a subgroup we would need to check closure of H under taking inverses, as well as under forming products. If H is finite, so in particular if G is finite, the following fact saves us the time of checking inverses.

Theorem 2

> If the nonempty finite subset H of the group G is closed under the group operation, i.e., if $g \cdot h \in H$ whenever g and h are in H, then H is a subgroup of G.

Proof. Given $g \in H$ we want to show that $g^{-1} \in H$. If you believed our last statement that $\langle g \rangle$ is $\{e, g, g^2, \ldots, g^{p-1}\}$ for some p, then $g^{-1} = g^{p-1} \in H$ because g^{p-1} is a product of elements of H.

Here is another argument for skeptics. The elements $e, g, g^2, \ldots$ are all in the finite set H, so they cannot all be different. Thus there are integers $k, n \in \mathbb{P}$ with $k < n$ and $g^k = g^n$. Hence

$$e = g^k \cdot (g^{-1})^k = g^n \cdot (g^{-1})^k = g^{n-k}.$$

If $n - k = 1$, then $g^{-1} = e = g \in H$. Otherwise, $n - k \geq 2$; hence $g^{-1} = e \cdot g^{-1} = g^{n-k} \cdot g^{-1} = g^{n-k-1} \in H$. ∎

Subgroups can be generated by subsets with more than one element. Let A be a nonempty subset of G. The **subgroup generated by** A is the set $\langle A \rangle$ of all products of elements from $A \cup A^{-1}$ where $A^{-1} = \{g^{-1} : g \in A\}$. It can be shown that $\langle A \rangle$ is a subgroup of G and that it is the smallest subgroup of G that contains A. The set $\langle A \rangle$ can also be defined recursively by:

(B) $A \subseteq \langle A \rangle$;
(R_1)   If $g, h \in \langle A \rangle$, then $g \cdot h \in \langle A \rangle$;
(R_2)   If $g \in \langle A \rangle$, then $g^{-1} \in \langle A \rangle$.

Conditions (R_1) and (R_2) show that $\langle A \rangle$ is closed under the operations $\cdot$ and inversion. Consider g in A. Then $g \in \langle A \rangle$ by (B), so $g^{-1} \in \langle A \rangle$ by (R_2) and thus the identity $e = g \cdot g^{-1}$ also belongs to $\langle A \rangle$ by (R_1). Hence $\langle A \rangle$ is a subgroup of G.

EXAMPLE 4   (a) Consider the subgroup $\langle \{4, -6\} \rangle$ of $(\mathbb{Z}, +)$. It contains $4 + (-6) = -2$ by (R_1), so it contains $-(-2) = 2$ by (R_2). Thus it contains all integer multiples of 2. This means that $\langle \{4, -6\} \rangle \supseteq \langle \{2\} \rangle = 2\mathbb{Z}$. Now every number we can form from 4 and -6 by adding two numbers or taking negatives is still going to be even, so $\langle\{4, -6\}\rangle$ consists only of even numbers. Thus $\langle\{4, -6\}\rangle \subseteq 2\mathbb{Z}$, so $\langle\{4, -6\}\rangle = 2\mathbb{Z}$.

(b) The subgroup $\langle\{3, 5\}\rangle$ of $(\mathbb{Z}, +)$ is $\mathbb{Z}$ itself, since we know that $\langle\{3, 5\}\rangle$ contains $3 + 3 - 5 = 1$, so $\langle\{3, 5\}\rangle \supseteq \langle1\rangle = \mathbb{Z}$. Notice that neither 3 nor 5 by itself generates $\mathbb{Z}$. ■

In fact, all subgroups of $(\mathbb{Z}, +)$ turn out to be cyclic, which is the simplest possible situation.

| *Theorem 3* | Every subgroup of $(\mathbb{Z}, +)$ is of the form $n\mathbb{Z}$ for some $n \in \mathbb{N}$. |

Proof. Consider a subgroup H of $(\mathbb{Z}, +)$. If $H = \{0\}$, then $H = 0\mathbb{Z}$, which is of the required form. Suppose that $H \neq \{0\}$. If $0 \neq m \in H$, then also $-m \in H$. Thus $H \cap \mathbb{P}$ is nonempty, so it has a smallest element, say n. We show that $H = n\mathbb{Z}$. Since $n \in H$ and H is a subgroup, we have $n\mathbb{Z} \subseteq H$. Consider an element m of H. By the Division Algorithm $m = qn + r$ with $0 \leq r < n$. Since $n\mathbb{Z} \subseteq H$, $qn \in H$ and thus $r = m - qn \in H$. Since $r < n$ and n is the smallest positive member of H, we must have $r = 0$. That is, $m = qn \in n\mathbb{Z}$. Since m was arbitrary in H, $H \subseteq n\mathbb{Z}$ as claimed. ■

In a general group we expect that a set A with more than one member will generate a subgroup that is not cyclic.

EXAMPLE 5   (a) For any group we have $\langle G \rangle = G$. For a generating set to be useful, however, it should be relatively small, certainly smaller than G.

(b) Section 12.2 contains several examples of groups with small specified sets of generators. The group in Figure 1 of that section is generated by $\{f, g\}$; it is commutative but not cyclic. The group in Figure 2 of that section is generated by $\{g, h\}$ and is not even commutative. The group in Figure 3 of that section is generated by $\{f, g, h\}$. It is actually generated by $\{f, g\}$ because $h = fgf$. To see this, show $gf = fh$ directly and then observe that $fgf = ffh = h$ since $f^2 = e$. This group is not commutative: $gf \neq fg$, for instance. ■

Consider a subgroup H of a group $(G, \cdot)$ with identity e. A **left coset** of H in G is a subset of the form

$$g \cdot H = gH = \{gh : h \in H\}$$

for some $g \in G$. The coset eH is H itself, and indeed $hH = H$ for every h in H [Exercise 21]. Since $e \in H$, the coset gH contains $ge = g$. Thus every

g in G belongs to at least one left coset. In fact, each g belongs to just one left coset.

Theorem 4 | The left cosets of a subgroup of a group form a partition of the group.

Proof. We just showed that G is the union of the various cosets gH, so we only need to show that overlapping left cosets are identical. First we show that if $k \in gH$, then $kH = gH$. Since $HH = H$ we have $kH \subseteq gHH = gH$. Also, $k = gh$ for some $h \in H$, so $g = kh^{-1} \in kH$. Hence $gH \subseteq kHH = kH$.

Now suppose that gH and $g'H$ overlap; say $k \in gH \cap g'H$. Then by what we have just shown, $gH = kH = g'H$. ■

We discuss an alternative proof of Theorem 4 at the end of this section.

Instead of the left cosets gH that we have been considering, we could just as easily have looked at **right cosets**, of the form $Hg = \{hg : h \in H\}$. If G is commutative, then $gH = Hg$ and we just refer to **cosets**. In general, the left coset gH and the right coset Hg are different subsets of G. The proof of Theorem 4 is still valid for right cosets, with the obvious left–right switches.

EXAMPLE 6 Consider the subgroup $3\mathbb{Z}$ of the group $(\mathbb{Z}, +)$. The cosets are the sets of the form $3\mathbb{Z} + r = \{3k + r : k \in \mathbb{Z}\}$. There are just three of them, namely the congruence classes $[0]_3 = 3\mathbb{Z}$, $[1]_3 = 3\mathbb{Z} + 1$ and $[2]_3 = 3\mathbb{Z} + 2$. Every n in $\mathbb{Z}$ can be written as $n = 3q + r$ with $r \in \{0, 1, 2\}$ and $q \in \mathbb{Z}$, so $\mathbb{Z}$ is the union of these three disjoint sets. Similarly, for every $p \in \mathbb{P}$ the cosets of $p\mathbb{Z}$ in $(\mathbb{Z}, +)$ are the sets $[r]_p = p\mathbb{Z} + r$ where $r = 0, 1, 2, \ldots, p - 1$. ■

EXAMPLE 7 Consider the permutation group $G = \text{PERM}(X)$ of all one-to-one functions from the nonempty set X onto itself, with composition as operation. For an element x_0 in X, the fixing subgroup $\text{FIX}_G(x_0) = \{f \in G : f(x_0) = x_0\}$ was introduced and used in §12.2. The left coset $e \circ \text{FIX}_G(x_0)$ is just $\text{FIX}_G(x_0)$ itself. Now consider any left coset $g \circ \text{FIX}_G(x_0)$. Any function in this coset has the form $g \circ f$ where $f \in \text{FIX}_G(x_0)$, so it satisfies $g \circ f(x_0) = g(x_0)$. Indeed, we claim that

$$g \circ \text{FIX}_G(x_0) = \{h \in G : h(x_0) = g(x_0)\}.$$

For suppose that $h(x_0) = g(x_0)$. Then $g^{-1} \circ h(x_0) = g^{-1} \circ g(x_0) = x_0$; hence $g^{-1} \circ h$ is in $\text{FIX}_G(x_0)$ and $h = g \circ (g^{-1} \circ h)$ belongs to $g \circ \text{FIX}_G(x_0)$.

If $g \notin \text{FIX}_G(x_0)$ and if X has at least three elements, then the right coset $\text{FIX}_G(x_0) \circ g$ is not the same as the left coset $g \circ \text{FIX}_G(x_0)$. To see this, note that $g(x_0) \neq x_0$, since $g \notin \text{FIX}_G(x_0)$. Choose h in G with $h(x_0) = x_0$ but $h(g(x_0)) \neq g(x_0)$. Then $h \in \text{FIX}_G(x_0)$, and so $h \circ g \in \text{FIX}_G(x_0) \circ g$, but $(h \circ g)(x_0) \neq g(x_0)$, so $h \circ g \notin g \circ \text{FIX}_G(x_0)$.

Exercise 17 deals with this example in detail for a three-element set X. ∎

Not only do the cosets of H in G partition G, they all have the same size, namely the size of H.

Theorem 5

> Let H be a subgroup of the group G and let $g \in G$. The function $h \to gh$ is a one-to-one correspondence of H onto the coset gH.

Proof. This function certainly maps H onto gH, and it is one-to-one thanks to the cancellation law: $gh = gh'$ implies $h = h'$. ∎

Our next result is one of the basic workhorses of finite group theory. To state it we need some notation. For H a subgroup of G, let G/H be the set of left cosets of H in G. As usual, $|H|$, $|G|$ and $|G/H|$ are the numbers of elements in H, G and G/H, respectively.

Lagrange's Theorem

> Let H be a subgroup of the finite group G. Then
>
> $$|G| = |G/H| \cdot |H|.$$
>
> In particular, $|H|$ and $|G/H|$ divide $|G|$.

Proof. There are $|G/H|$ left cosets of H, each of which has $|H|$ members, by Theorem 5. They partition G by Theorem 4. ∎

Theorem 5 and Lagrange's theorem have valid analogs for right cosets, so G has the same number of right cosets of H as left cosets. Exercise 23 asks for a direct proof of this fact.

EXAMPLE 8

(a) A group with 10 members can only have subgroups with 1, 2, 5 or 10 members.

(b) A group with 81 members can only have subgroups with 1, 3, 9, 27 or 81 members.

(c) Suppose that the set X has n elements. Then the group $G = \text{PERM}(X)$ of Example 7 has $n!$ elements, and the subgroup $\text{FIX}_G(x_0)$ has $(n-1)!$ elements. There are n left cosets $g \circ \text{FIX}_G(x_0)$, one for each possible value of $g(x_0)$. Lagrange's theorem in this case takes the form

$$n! = n \cdot (n-1)!.$$

(d) Let G be the group of automorphisms of the graph in Figure 1, with composition as operation. One can use Lagrange's theorem to help check that G has 10 members. We see at once the five "rotations," through angles of $0°$, $72°$, $144°$, $216°$ and $288°$, which form a subgroup, call it R. The correspondence g defined by $g(p) = p$, $g(q) = t$, $g(r) = s$, $g(s) = r$ and $g(t) = q$ is in G but not in R, so $|G| > |R| = 5$. By Lagrange's theorem $|G|$ is a multiple of 5, so it's at least 10. Now for an element f of G there are only five possible choices for $f(p)$, and then only two possible choices for $f(q)$ [one on each side of $f(p)$], after which the rest of the action of f is completely determined. So G has at most $5 \cdot 2 = 10$ members, and thus has exactly 10.

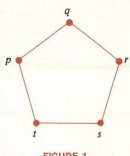

FIGURE 1

Half of the members of G are in R. The rest form the single coset $g \circ R$, where g is defined above. In this case $R \circ g = g \circ R$. [See also Exercise 22.]

The subgroup $\langle g \rangle$ consists just of e and g, since $g \circ g = e$, so $|\langle g \rangle| = 2$. Since $10 = |G| = |G/\langle g \rangle| \cdot |\langle g \rangle|$, there are 5 cosets of $\langle g \rangle$ in G. In fact, $\langle g \rangle$ is the subgroup of G fixing the point p. As in part (c) above, the cosets of $\langle g \rangle$ are simply the five sets of automorphisms in G taking p to each of its five possible images, namely $\{f \in G : f(p) = p\}$, $\{f \in G : f(p) = q\}$, $\{f \in G : f(p) = r\}$, etc. ∎

We end this section by indicating how some of the results can be viewed as special cases of results about groups of permutations. In the first place, G can be viewed as acting on itself if for each g_0 we define $g_0^*(g) = g_0 \cdot g$ for g in G. Each g_0^* is a permutation of G [check this]. Also,

$$(g_0 \cdot g_1)^*(g) = g_0 \cdot g_1 \cdot g = g_0 \cdot (g_1^*(g)) = g_0^*(g_1^*(g)) = g_0^* \circ g_1^*(g)$$

for all $g \in G$, so that $(g_0 \cdot g_1)^* = g_0^* \circ g_1^*$. In fact, $g \to g^*$ is an isomorphism of G onto a subgroup of PERM(G); this proves Cayley's theorem mentioned after Example 1.

A subgroup H of G also acts on G; for $h \in H$ we write $h^*(g) = h \cdot g$. Given g in G [*the set on which H acts*], its orbit under H is

$$\{h^*(g) : h \in H\} = \{h \cdot g : h \in H\} = H \cdot g = Hg.$$

In other words, the orbits under H are just the right cosets of H in G. The corollary to Proposition 1 of §12.2 tells us that the H-orbits partition G. So the right cosets of H partition G, which establishes the right-coset version of Theorem 4. Similarly, $h^*(g) = g \cdot h^{-1}$ defines another action of H on G for which the orbits are just the left cosets [Exercise 24]. So the left cosets of H also partition G; this provides another confirmation of Theorem 4.

EXERCISES 12.5

1. Describe each of the following subgroups of $(\mathbb{Z}, +)$.

    (a) $\langle 1 \rangle$ 　　　　　　　(b) $\langle 0 \rangle$ 　　　　　　　(c) $\langle \{-1, 2\} \rangle$

    (d) $\langle \mathbb{Z} \rangle$ 　　　　　　　(e) $\langle \{2, 3\} \rangle$ 　　　　　　(f) $\langle 6 \rangle \cap \langle 9 \rangle$

2. Which of the subgroups in Exercise 1 are cyclic groups? Justify your answers.

3. Which of the following subsets of $\mathbb{Z}$ are subgroups of $\mathbb{Z}$? Write the subgroups in the form $n\mathbb{Z}$; see Theorem 3.

    (a) $\{0, \pm 3, \pm 6, \pm 9, \ldots\}$ 　　　　　(b) $\{0, 5, 10, 15, 20, \ldots\}$

    (c) $\{0, \pm 2, \pm 4, \pm 8, \pm 16, \ldots\}$ 　　　(d) $\{k \in \mathbb{Z} : k \text{ is a multiple of } 4\}$

    (e) $\mathbb{N} \cup (-\mathbb{N})$

4. Which of the following subsets of $\mathbb{R}$ are subgroups of $\mathbb{R}$?

    (a) $\mathbb{Z}$ 　　　　　　　(b) $\mathbb{N}$ 　　　　　　　(c) $\mathbb{Q}$

    (d) $\{n\sqrt{2} : n \in \mathbb{Z}\}$ 　　　(e) $\{m + n\sqrt{2} : m, n \in \mathbb{Z}\}$

5. Complete the proof of Theorem 1(a). That is, elaborate on the sentence that begins "Similarly."

6. State and prove Theorem 1 for a commutative group using additive notation.

7. (a) Find all generators of the group $(\mathbb{Z}(5), +_5)$.

    (b) Find all generators of the group $(\mathbb{Z}(6), +_6)$.

8. (a) Find the intersection of all the subgroups $n\mathbb{Z}$ of $(\mathbb{Z}, +)$, where $n \in \mathbb{P}$.

    (b) Is the intersection in part (a) cyclic? Explain.

9. (a) Give an example of a one-to-one correspondence between $4\mathbb{Z}$ and the coset $4\mathbb{Z} + 3$ in $(\mathbb{Z}, +)$.

    (b) Give another example.

10. Consider the group $(\mathbb{Z}, +)$. Write $\mathbb{Z}$ as a disjoint union of five cosets of a subgroup.

11. The set S_4 of all permutations of $\{1, 2, 3, 4\}$ is a group under composition of functions. Which of the following are subgroups of $(S_4, \circ)$? Justify your answers.

    (a) $\{g \in S_4 : g(4) = 4\}$

    (b) $\{g \in S_4 : g(1) = 2\}$

    (c) $\{g \in S_4 : g(1) \in \{1, 2\}\}$

    (d) $\{g \in S_4 : g(1) \in \{1, 2\} \text{ and } g(2) \in \{1, 2\}\}$

12. (a) Explain why any 13-element subset of the group $(S_4, \circ)$ in Exercise 11 generates S_4. [There are actually some 2-element generating sets for S_4, but they are harder to find.]

    (b) Is $(S_4, \circ)$ a cyclic group? Justify your answer.

13. (a) Prove that $(g_1 \cdot g_2 \cdot g_3)^{-1} = g_3^{-1} \cdot g_2^{-1} \cdot g_1^{-1}$ for g_1, g_2, g_3 in a group G.

    (b) Prove a generalization for $(g_1 \cdot g_2 \cdots g_n)^{-1}$.

14. For an element g in a group G the powers g^n were defined for all $n \in \mathbb{Z}$ just prior to Example 3. Prove:

    (a) $(g^{-1})^{-k} = g^k$ for all $g \in G$ and $k \in \mathbb{Z}$.

    (b) $g^m \cdot g^1 = g^{m+1}$ for all $g \in G$ and $m \in \mathbb{Z}$. [The case $m \in \mathbb{N}$ is easy.]

    (c) $g^m \cdot g^n = g^{m+n}$ for all $g \in G$, $m \in \mathbb{Z}$ and $n \in \mathbb{N}$.

    (d) $g^m \cdot g^n = g^{m+n}$ for all $g \in G$, $m \in \mathbb{Z}$ and $n \in \mathbb{Z}$. [*Hint:* You can use part (c) with g^{-1} instead of g.]

15. (a) Show that the intersection of any family of subgroups of a group G is again a subgroup of G.

    (b) Give an example of a group G and subgroups H and K such that $H \cup K$ is not a subgroup of G.

16. (a) Show that the subgroup R in Example 8(d) is cyclic, and describe a generator for it.

    (b) Show that the group G in the example contains elements of orders 1, 2 and 5, but not 10.

17. Let $X = \{1, 2, 3\}$ and $x_0 = 1$ in Example 7.

    (a) Find $|\text{PERM}(X)|$.

    (b) List the permutations in $\text{FIX}_G(1)$.

    (c) The permutation $(1\,2\,3)$ is not in $\text{FIX}_G(1)$. List the members of the coset $\text{FIX}_G(1) \circ (1\,2\,3)$.

    (d) Show that $\text{FIX}_G(1) \circ (1\,2\,3) \neq (1\,2\,3) \circ \text{FIX}_G(1)$.

    (e) Show that in fact $\text{FIX}_G(1) \circ (1\,2\,3)$ is not a left coset at all.

    (f) How many cosets does $\text{FIX}_G(1)$ have in $\text{PERM}(X)$?

18. The table below describes a binary operation $\bullet$ on the set $G = \{a, b, c, d, e\}$ with e as identity element.

$\bullet$	e	a	b	c	d
e	e	a	b	c	d
a	a	e	c	d	b
b	b	d	a	e	c
c	c	b	d	a	e
d	d	c	e	b	a

    (a) Show that the set $\{e, a\}$ is a group under $\bullet$ as operation.

    (b) Without doing any calculations, use the result of part (a) and Lagrange's theorem to conclude that $(G, \bullet)$ is not a group.

19. The following table gives the binary operation for a group $(G, \bullet)$ with elements a, b, c, d, e, f.

$\bullet$	e	a	b	c	d	f
e	e	a	b	c	d	f
a	a	b	e	d	f	c
b	b	e	a	f	c	d
c	c	f	d	e	b	a
d	d	c	f	a	e	b
f	f	d	c	b	a	e

(a) List the members of the subgroup $\langle a \rangle$.

(b) Show that $\langle a \rangle \bullet c = c \bullet \langle a \rangle$.

(c) Find all the subgroups with two members.

(d) Find $|G/\langle d \rangle|$.

(e) Describe the right cosets of $\langle d \rangle$.

20. Repeat Exercise 19 for the group with the following table.

$\bullet$	e	a	b	c	d	f
e	e	a	b	c	d	f
a	a	b	e	d	f	c
b	b	e	a	f	c	d
c	c	d	f	a	b	e
d	d	f	c	b	e	a
f	f	c	d	e	a	b

21. Show that if H is a subgroup of a group $(G, \cdot)$, and if $g \in G$, then $g \cdot H = H$ if and only if $g \in H$. Try to be clever and use Theorem 4.

22. Consider a finite group $(G, \cdot)$ with a subgroup H such that $|G| = 2|H|$. Show that $g \cdot H = H \cdot g$ for every g in G. *Suggestion:* Consider the two cases $g \in H$ and $g \notin H$ separately.

23. Let H be a subgroup of the group $(G, \cdot)$.

(a) Show that for each g in G the right coset $H \cdot g^{-1}$ consists of the inverses of the elements in the left coset $g \cdot H$.

(b) Describe a one-to-one correspondence between the set of left cosets of H in G and the set of right cosets.

24. Let H be a subgroup of a group G. For $h \in H$ define $h^*: G \to G$ by $h^*(g) = g \cdot h^{-1}$ for $g \in G$.

(a) Prove that this defines an action of H on G.

(b) Show that $h \to h^*$ is one-to-one.

(c) Observe that, with $H = G$, part (b) completes the proof of Cayley's theorem, since this shows that G is isomorphic to a subgroup of PERM(G).

25. Let H be a subgroup of a group $(G, \cdot)$ and for $g_1, g_2 \in G$ define $g_1 \sim g_2$ if $g_2^{-1} \cdot g_1 \in H$.

   (a) Show that $\sim$ is an equivalence relation on G.

   (b) Show that the partition in Theorem 4 is precisely the partition of equivalence classes for $\sim$ described in Theorem 1 of §3.5.

26. Show that if H is a subgroup of $(G, \cdot)$, then $g \cdot H \cdot g^{-1}$ is also a subgroup for each g in G.

§ 12.6   The Fundamental Homomorphism Theorem

This section introduces homomorphisms, the functions that are most important in the study of groups. Recall that an isomorphism from one group $(G, \cdot)$ to another group $(G_0, \bullet)$ is a one-to-one correspondence $\varphi \colon G \to G_0$ satisfying $\varphi(g \cdot h) = \varphi(g) \bullet \varphi(h)$ for all $g, h \in G$. Isomorphisms preserve the group structure in a one-to-one way. Homomorphisms also preserve the group structure, but they don't need to be one-to-one or onto mappings.

A **homomorphism** from the group $(G, \cdot)$ to the group $(G_0, \bullet)$ is a function $\varphi \colon G \to G_0$ satisfying $\varphi(g \cdot h) = \varphi(g) \bullet \varphi(h)$ for all $g, h \in G$. Since the identity and inverses are part of the group structure, it appears that we should also require that φ preserve the identity and inverses. Happily, these are automatic consequences of the definition.

Proposition   Let $\varphi \colon G \to G_0$ be a homomorphism.

   (a) If e denotes the identity in G, then $\varphi(e)$ is the identity, e_0, of the group G_0.

   (b) $\varphi(g^{-1}) = \varphi(g)^{-1}$ for all $g \in G$.

   (c) $\varphi(G)$ is a subgroup of G_0.

Proof. (a) $\varphi(e) = \varphi(e) \bullet [\varphi(e) \bullet \varphi(e)^{-1}] = [\varphi(e) \bullet \varphi(e)] \bullet \varphi(e)^{-1} = \varphi(e \cdot e) \bullet \varphi(e)^{-1} = \varphi(e) \bullet \varphi(e)^{-1} = e_0$.

(b) Note that the first inverse is taken in the group G, while the second one is taken in G_0. Since $\varphi(g) \bullet \varphi(g^{-1}) = \varphi(g \cdot g^{-1}) = \varphi(e) = e_0$ and inverses are unique, $\varphi(g^{-1}) = \varphi(g)^{-1}$.

(c) Since φ takes inverses to inverses, $\varphi(G)$ contains the inverses of all of its elements. Since $\varphi(G)$ is also closed under products, it is a subgroup of G_0. ∎

EXAMPLE 1   (a) Let $(G, \cdot)$ and $(G_0, \bullet)$ both be $(\mathbb{Z}, +)$. The homomorphism condition is

$$\varphi(m + n) = \varphi(m) + \varphi(n) \qquad \text{for} \quad m, n \in \mathbb{Z}.$$

The function φ defined by $\varphi(n) = 5n$ for all n is an example of a homomorphism, since

$$\varphi(m + n) = 5 \cdot (m + n) = 5m + 5n = \varphi(m) + \varphi(n).$$

There is nothing special about 5; any other integer would define a homomorphism from $(\mathbb{Z}, +)$ to $(\mathbb{Z}, +)$ in the same way.

(b) Let $(G, \cdot)$ be $(\mathbb{Z}, +)$, let $(G_0, \bullet)$ be $(\mathbb{R} \backslash \{0\}, \cdot)$, and let $\varphi(m) = 2^m$ for m in $\mathbb{Z}$. Since

$$\varphi(m + n) = 2^{m+n} = 2^m \cdot 2^n = \varphi(m) \cdot \varphi(n),$$

φ is a homomorphism of $(\mathbb{Z}, +)$ into $(\mathbb{R} \backslash \{0\}, \cdot)$. Note that $\varphi(0) = 1$.

(c) Recall that $a = \log_2 b$ if and only if $b = 2^a$. The function φ from $(\{x \in \mathbb{R} : x > 0\}, \cdot)$ to $(\mathbb{R}, +)$ given by $\varphi(x) = \log_2 x$ is a homomorphism since $\log_2 (xy) = \log_2 (x) + \log_2 (y)$.

(d) We saw the equation

$$(f \circ g)^* = f^* \circ g^*$$

several times when we were looking at graph automorphisms in § 12.3. In each case, there was a homomorphism in the background. The first time, $f \to f^*$ was a mapping of AUT(H) into the group of permutations of the edge set of the graph H, determined by $f^*(\{u, v\}) = \{f(u), f(v)\}$. Restriction of the members of AUT(H) to some union of orbits for AUT(H) gave another homomorphism $f \to f^*$ in Example 3 of § 12.3.

To say that a group G "acts on" a set X in the sense defined in § 12.3 is simply to say that we have a homomorphism from G into PERM(X), i.e., a structure-preserving way of associating permutations with members of G. ∎

EXAMPLE 2   (a) Consider a positive integer $p \geq 2$. As in § 3.6, let n MOD p be the remainder when n is divided by p. Recall that MOD p is a function from $\mathbb{Z}$ into $\mathbb{Z}(p)$. In fact, MOD p is a homomorphism of $(\mathbb{Z}, +)$ onto $(\mathbb{Z}(p), +_p)$, by Theorem 3 in § 3.6. That is,

$$(m + n) \text{ MOD } p = (m \text{ MOD } p) +_p (n \text{ MOD } p) \qquad \text{for} \quad m, n \in \mathbb{Z}.$$

(b) The theorem also states that

$$(m \cdot n) \text{ MOD } p = (m \text{ MOD } p) *_p (n \text{ MOD } p).$$

It is tempting to conclude that MOD p is also a homomorphism of $(\mathbb{Z}, \cdot)$ onto $(\mathbb{Z}(p), *_p)$. But $(\mathbb{Z}(p), *_p)$ is *not* a group! Its 0 does not have a multiplicative inverse. This example shows how important it is in the Proposition that G_0 be a *group*. ∎

EXAMPLE 3   Let $(G, \cdot)$ be a group and let $g \in G$. Exercise 14 of § 12.5 asserts that $g^{m+n} = g^m \cdot g^n$ for all $m, n \in \mathbb{Z}$. That is, the function φ from $(\mathbb{Z}, +)$ to $(G, \cdot)$ given by $\varphi(n) = g^n$ is a homomorphism. Its image $\varphi(\mathbb{Z})$ is the cyclic subgroup $\langle g \rangle$ of G. ■

Each homomorphism φ defined on G has associated with it a special subgroup of G, called its **kernel**, which is defined to be $\{g \in G : \varphi(g) = \varphi(e)\}$. In a sense, the kernel tells us what we are ignoring when we pass from G to $\varphi(G)$. Conclusion (c) of the next theorem can be viewed as saying that two members of G have the same image under φ if and only if they differ by a member of the kernel of φ.

Theorem 1

Let $\varphi : G \to G_0$ be a homomorphism, where G has operation $\cdot$ and G_0 has operation $\bullet$. Let e and e_0 be the identities of G and G_0, respectively, and let K be the kernel of φ. Then

(a) K is a subgroup of G.

(b) $g \cdot K = K \cdot g$ for each g in G.

(c) $g \cdot K = \{h \in G : \varphi(h) = \varphi(g)\}$ for each g in G.

Proof. (a) If $g, h \in K$, then $\varphi(g \cdot h) = \varphi(g) \bullet \varphi(h) = \varphi(e) \bullet \varphi(e) = e_0 \bullet e_0 = e_0$, so $g \cdot h$ is in K. Moreover, $\varphi(g^{-1}) = \varphi(g)^{-1} = e_0^{-1} = e_0$, so g^{-1} is in K. Thus K is closed under products and inverses; so it is a subgroup of G.

(b) We show first that $K \cdot g \subseteq g \cdot K$. It is enough to consider $k \in K$ and show that $k \cdot g \in g \cdot K$. Since $k \cdot g = g \cdot (g^{-1} \cdot k \cdot g)$, it suffices to show that $g^{-1} \cdot k \cdot g$ is in K. But $\varphi(g^{-1} \cdot k \cdot g) = \varphi(g^{-1}) \bullet \varphi(k) \bullet \varphi(g) = \varphi(g)^{-1} \bullet \varphi(e) \bullet \varphi(g) = \varphi(g^{-1} \cdot e \cdot g) = \varphi(e)$. A similar argument shows that $g \cdot K \subseteq K \cdot g$.

(c) For $k \in K$, $\varphi(g \cdot k) = \varphi(g) \bullet \varphi(k) = \varphi(g) \bullet e_0 = \varphi(g)$, so $g \cdot K \subseteq \{h \in G : \varphi(h) = \varphi(g)\}$. In the other direction, if $\varphi(h) = \varphi(g)$, then $\varphi(g^{-1} \cdot h) = \varphi(g)^{-1} \bullet \varphi(h) = e_0$, which means that $g^{-1} \cdot h \in K$, so $h = g \cdot (g^{-1} \cdot h) \in g \cdot K$. ■

Corollary 1

In the setting of Theorem 1, the cosets $g \cdot K$ are the equivalence classes for the equivalence relation that φ defines by $g \sim h$ if and only if $\varphi(g) = \varphi(h)$.

Proof. This is just a restatement of (c). ■

We saw in Theorem 5 of §12.5 that all cosets $g \cdot K$ have the same number of elements, namely $|K|$, and this fact gives us a useful test for one-to-oneness of homomorphisms.

Corollary 2

> A homomorphism is one-to-one if and only if its kernel is just the identity element.

Proof. Since K is a subgroup, it contains the identity e for sure. Moreover, by Theorem 1(c), φ is one-to-one if and only if all cosets $g \cdot K$ have exactly one element, which is true if and only if K itself has just one member. ∎

EXAMPLE 4   (a) The homomorphism φ from $(\mathbb{Z}, +)$ to $(\mathbb{Z}, +)$ defined by $\varphi(n) = 5n$ is one-to-one and its kernel is $\{n \in \mathbb{Z} : 5n = 0\} = \{0\}$. The homomorphism in part (b) of Example 1 is also one-to-one and its kernel is $\{n \in \mathbb{Z} : 2^n = 1\} = \{0\}$. The homomorphism $\varphi(x) = \log_2 x$ in Example 1(d) is one-to-one and its kernel is $\{x \in \mathbb{R} : \log_2 x = 0\} = \{1\}$.

(b) The homomorphism MOD p [Example 2(a)] from $(\mathbb{Z}, +)$ to $(\mathbb{Z}(p), +_p)$ is not one-to-one. Its kernel is $\{n \in \mathbb{Z} : n \text{ MOD } p = 0\} = \{n \in \mathbb{Z} : n \text{ is a multiple of } p\} = p\mathbb{Z}$. Two integers are in the same coset of the kernel if and only if their difference is a multiple of p, and the cosets are just the congruence classes $[r]_p = p\mathbb{Z} + r$. ∎

EXAMPLE 5   The homomorphism $n \to g^n$ from $(\mathbb{Z}, +)$ into $(\langle g \rangle, \cdot)$ that we saw in Example 3 is the key to understanding cyclic groups. Its kernel is $\{n \in \mathbb{Z} : g^n = e\}$. This subgroup of $\mathbb{Z}$ is just $\{0\}$ if $\langle g \rangle$ is infinite, but is $p\mathbb{Z}$ for some nonzero p otherwise. In the first case, the homomorphism is one-to-one, but in the second case it is not. We will exploit these facts in Example 7, after we have built up a little more machinery. ∎

A subgroup K of a group $(G, \cdot)$ with the property that $g \cdot K = K \cdot g$ for every g in G is called a **normal subgroup** of G. Theorem 1(b) shows that kernels of homomorphisms are normal. If G is commutative, every subgroup is normal, but in the general case there will be nonnormal subgroups. When K is normal, its left and right cosets coincide, so we simply refer to its **cosets**.

When K is a normal subgroup of the group G the set G/K of left cosets $g \cdot K$ has a natural and useful group operation $*$ of its own. The members of G/K that we multiply are sets themselves [indeed, cosets], and the product of two cosets will be another such coset. This is not quite a new idea. We are used to making new sets out of two old ones, for instance by taking A and B and forming $A \cap B$ and $A \cup B$, and we have added and multiplied congruence classes in Example 6(a) of §3.6. Our product of cosets generalizes that example.

The natural way to try to multiply two cosets $g \cdot K$ and $h \cdot K$ is to define

$$(g \cdot K) * (h \cdot K) = g \cdot K \cdot h \cdot K = \{g \cdot k_1 \cdot h \cdot k_2 : k_1, k_2 \in K\}.$$

For an arbitrary subgroup K this product may not be a coset [see Exercise 19], but if K is normal, then it must be one.

Theorem 2

Let K be a normal subgroup of the group $(G, \cdot)$. Then

(a) $g \cdot K \cdot h \cdot K = (g \cdot h) \cdot K$ for all $g, h \in G$.

(b) The set G/K of cosets of K in G is a group under the operation $*$ defined by

$$(g \cdot K) * (h \cdot K) = g \cdot h \cdot K.$$

(c) The function $v\colon G \to G/K$ defined by $v(g) = g \cdot K$ is a homomorphism with kernel K. [v is the lowercase Greek nu.]

Proof. (a) Since K is a subgroup of G, $K = K \cdot e \subseteq K \cdot K \subseteq K$, so $K = K \cdot K$. Since K is normal, we have $h \cdot K = K \cdot h$, so

$$g \cdot K \cdot h \cdot K = g \cdot h \cdot K \cdot K = g \cdot h \cdot K \in G/K.$$

(b) According to (a), $(g \cdot K) * (h \cdot K)$ is the set $(g \cdot K) \cdot (h \cdot K)$ and thus $*$ is a well-defined binary operation on G/K. It is easy to check that it is associative, that K is the identity and that $(g \cdot K)^{-1} = g^{-1} \cdot K$.

(c) We have $v(g \cdot h) = g \cdot h \cdot K = (g \cdot K) * (h \cdot K) = v(g) * v(h)$ by definition of $*$, so v is a homomorphism. If $v(g) = v(e)$, then $g \in g \cdot K = e \cdot K = K$, and if $g \in K$, then $v(g) = g \cdot K = K = v(e)$ [Exercise 21, §12.5]. Thus K is the kernel of v. ∎

The mapping v is called the **natural homomorphism** of G onto G/K. Naming it with the Greek letter nu is supposed to help us remember that it's natural. Theorems 1 and 2 together tell us that kernels of homomorphisms are normal subgroups and, conversely, every normal subgroup is the kernel of some homomorphism. If K is the kernel of φ then the natural mapping v in Theorem 2 is the same one, $s \to [s]$, that we saw in §3.5, since $g \cdot K$ can be viewed as the equivalence class $[g]$ of g, by Corollary 1 of Theorem 1.

EXAMPLE 6   Let $(G, \cdot) = (\mathbb{Z}, +)$ and let $K = 6\mathbb{Z}$. Theorem 2(b) tells us that $\mathbb{Z}/6\mathbb{Z}$ is a group under the operation

$$(k + 6\mathbb{Z}) * (m + 6\mathbb{Z}) = k + m + 6\mathbb{Z}.$$

The identity of $\mathbb{Z}/6\mathbb{Z}$ is $6\mathbb{Z}$.

Theorem 2(c) tells us that if $v(k) = k + 6\mathbb{Z}$, then v maps $\mathbb{Z}$ onto $\mathbb{Z}/6\mathbb{Z}$ and the kernel of v is $6\mathbb{Z}$, i.e.,

$$\{k \in \mathbb{Z} : v(k) = 6\mathbb{Z}\} = \{k \in \mathbb{Z} : k + 6\mathbb{Z} = 6\mathbb{Z}\} = 6\mathbb{Z}.$$

In Example 4(b) we observed that $6\mathbb{Z}$ is also the kernel of $\varphi = \text{MOD } 6$, which maps $(\mathbb{Z}, +)$ onto $(\mathbb{Z}(6), +_6)$. If we define

$$\varphi^*(k + 6\mathbb{Z}) = k \qquad \text{for} \quad k \in \{0, 1, 2, 3, 4, 5\} = \mathbb{Z}(6),$$

we obtain a one-to-one correspondence of $\mathbb{Z}/6\mathbb{Z}$ onto $\mathbb{Z}(6)$. In the notation of §3.6, this correspondence is $[m]_6 \leftrightarrow m \text{ MOD } 6$. The mapping φ^* is an isomorphism because

$$\varphi^*((k + 6\mathbb{Z}) * (m + 6\mathbb{Z})) = \varphi^*(k + m + 6\mathbb{Z}) = \varphi^*(k +_6 m + 6\mathbb{Z})$$
$$= k +_6 m = \varphi^*(k + 6\mathbb{Z}) +_6 \varphi^*(m + 6\mathbb{Z}).$$

Thus the groups $\mathbb{Z}/6\mathbb{Z}$ and $\mathbb{Z}(6)$ are isomorphic. Using our G and K notation, this says that G/K and $\varphi(G)$ are isomorphic, and illustrates the next theorem. ∎

The Fundamental Homomorphism Theorem

Let φ be a homomorphism from the group $(G, \cdot)$ to the group $(G_0, \bullet)$, with kernel K. Then G/K is isomorphic to $\varphi(G)$ under the isomorphism φ^* defined by

$$\varphi^*(g \cdot K) = \varphi(g).$$

Proof. Theorem 1(c) says that $g \cdot K$ is the set of all h in G for which $\varphi(h) = \varphi(g)$. Thus $\varphi^*(g \cdot K)$ is the common value that φ has on all members of $g \cdot K$. The function φ has different values on different cosets, so φ^* is one-to-one. Its image is clearly $\varphi(G)$, so all we need to do is to observe that φ^* is a homomorphism:

$$\varphi^*((g \cdot K) * (h \cdot K))$$

$$= \varphi^*(g \cdot h \cdot K) \qquad \text{definition of } *$$

$$= \varphi(g \cdot h) \qquad \text{definition of } \varphi^*$$

$$= \varphi(g) \bullet \varphi(h) \qquad \varphi \text{ is a homomorphism}$$

$$= \varphi^*(g \cdot K) \bullet \varphi^*(h \cdot K) \qquad \text{definition of } \varphi^* \text{ again.} \quad ∎$$

The diagrams in Figure 1 show the Fundamental Homomorphism Theorem schematically. Figure 1(a) shows the groups and mappings. The theorem says that the homomorphism $\varphi: G \to G_0$ induces an isomorphism $\varphi^*: G/K \to \varphi(G)$ such that $\varphi = \varphi^* \circ \nu$. Here ν is the natural homomorphism of G onto G/K in Theorem 2. Figure 1(b) shows the images of the elements.

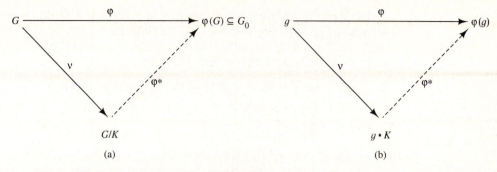

FIGURE 1

EXAMPLE 7   We show that every cyclic group is either isomorphic to a group $(\mathbb{Z}(p), +_p)$
for some $p \in \mathbb{P}$ or to the group $(\mathbb{Z}, +)$.

Given the cyclic group $\langle g \rangle = \{g^n : n \in \mathbb{Z}\}$ define $\varphi : \mathbb{Z} \to \langle g \rangle$ by
$\varphi(n) = g^n$. Since $\varphi(m + n) = g^{m+n} = g^m \cdot g^n$ [Exercise 14 of §12.5], φ is a
homomorphism of $(\mathbb{Z}, +)$ onto $(\langle g \rangle, \cdot)$. So φ has a kernel, K, and by the
Fundamental Homomorphism Theorem the image group $\langle g \rangle$ is iso-
morphic to $\mathbb{Z}/K$ under $\varphi^*(n + K) = g^n$.

Since K is a subgroup of $\mathbb{Z}$, by Theorem 3 of §12.5 we have $K = p\mathbb{Z}$
for some $p \in \mathbb{N}$. Thus $\langle g \rangle$ is isomorphic to $\mathbb{Z}/p\mathbb{Z}$.

If $p > 0$, then $|\mathbb{Z}/p\mathbb{Z}| = |\{p\mathbb{Z}, 1 + p\mathbb{Z}, \ldots, p - 1 + p\mathbb{Z}\}| = p$, so
$|\langle g \rangle| = p$. The mapping MOD p: $\mathbb{Z} \to \mathbb{Z}(p)$ is also a homomorphism with
kernel $p\mathbb{Z}$, so $\mathbb{Z}/p\mathbb{Z}$ is also isomorphic to $\mathbb{Z}(p)$. Hence the composite mapping $k \to k + p\mathbb{Z} \to g^k$ is an isomorphism of $(\mathbb{Z}(p), +)$ onto $(\langle g \rangle, \cdot)$. [See
Example 6 for the case $p = 6$.]

If $p = 0$, then the kernel of φ is $\{0\}$, so φ is one-to-one [Corollary 2
of Theorem 1] and $\langle g \rangle$ is isomorphic to $\mathbb{Z}$ under $\varphi(n) = g^n$. ∎

If G is finite, then Lagrange's theorem in §12.5 says that $|G/K| = |G|/|K|$. Since $|\varphi(G)| = |G/K|$ by the Fundamental Homomorphism Theo-
rem, we have the following.

Corollary   Let φ be a homomorphism defined on a finite group G, with kernel
K. Then $|\varphi(G)| = |G|/|K|$. In particular, $|\varphi(G)|$ divides $|G|$.

EXAMPLE 8   (a) The mapping φ from $\mathbb{Z}(30)$ to $\mathbb{Z}(5)$ defined by $\varphi(n) = n$ MOD 5
turns out to be a well-defined homomorphism. The reason is that throwing
away 30's doesn't hurt (mod 5), so $(m +_{30} n)$ MOD $5 = (m + n)$ MOD $5 = (m$ MOD $5) +_5 (n$ MOD $5)$. The kernel K of φ is

$$\{n \in \mathbb{Z}(30) : n \text{ MOD } 5 = 0\} = \{0, 5, 10, 15, 20, 25\}.$$

Sure enough, $|\mathbb{Z}(5)| = 5 = 30/6 = |\mathbb{Z}(30)|/|K|$.

(b) If $(G, \cdot)$ and $(H, \bullet)$ are groups, we can make $G \times H$ into a group by defining $(g_1, h_1) \,\square\, (g_2, h_2) = (g_1 \cdot g_2, h_1 \bullet h_2)$. [See Exercise 11.] Define $\varphi \colon \mathbb{Z}(6) \to \mathbb{Z}(3) \times \mathbb{Z}(2)$ by $\varphi(n) = (n \bmod 3, n \bmod 2)$. As in part (a), one checks that φ is a homomorphism. We have

$$\varphi(0) = (0, 0) = \text{the identity}, \qquad \varphi(3) = (0, 1),$$

$$\varphi(1) = (1, 1), \qquad\qquad\qquad \varphi(4) = (1, 0),$$

$$\varphi(2) = (2, 0), \qquad\qquad\qquad \varphi(5) = (2, 1).$$

The kernel of φ is just $\{0\}$, φ is one-to-one and φ maps onto $\mathbb{Z}(3) \times \mathbb{Z}(2)$, so φ is an isomorphism. Now $\mathbb{Z}(6)$ is cyclic, generated by 1. Since $\varphi(\mathbb{Z}(6)) = \varphi(\langle 1 \rangle) = \langle \varphi(1) \rangle$, $\mathbb{Z}(3) \times \mathbb{Z}(2)$ is also cyclic. Surprise? Not really. We saw this example in another form in Examples 5 and 6 of § 12.1.

(c) The mapping $\theta \colon \mathbb{Z}(6) \to \mathbb{Z}(3) \times \mathbb{Z}(2)$ given by $\theta(n) = (n \bmod 3, 0)$ is also a homomorphism. Its kernel is $\{0, 3\}$, and $\theta(\mathbb{Z}(6)) = \mathbb{Z}(3) \times \{0\} = \{(m, 0) \colon m \in \mathbb{Z}(3)\}$. As expected, $|\mathbb{Z}(3) \times \{0\}| = 3 = 6/2 = |\mathbb{Z}(6)|/|\{0, 3\}|$. The group $\mathbb{Z}(6)/\{0, 3\}$ consists of the three cosets $\{0, 3\}$, $\{1, 4\}$ and $\{2, 5\}$ and is isomorphic to $\mathbb{Z}(3)$ in a natural way.   ■

The message of the Fundamental Homomorphism Theorem is that to study homomorphic images of G, it is enough to look at the various groups G/K that can be made from cosets of certain subgroups K of G. The message of Theorem 2 is that we can identify the interesting subgroups; they are the ones satisfying the equations $g \cdot K = K \cdot g$ for all g in G. If you fully understand the group and its normal subgroups, then you fully understand all its homomorphic images.

Instead of comparing left and right cosets, we can test a subgroup for normality by looking at its **conjugates**, sets of the form $g \cdot K \cdot g^{-1}$. If K is a subgroup of G, so is $g \cdot K \cdot g^{-1}$ for each g in G [Exercise 26 of § 12.5]. Moreover, $g \cdot K = K \cdot g$ if and only if $g \cdot K \cdot g^{-1} = K \cdot g \cdot g^{-1} = K$, so K is normal if and only if $K = g \cdot K \cdot g^{-1}$ for every g, i.e., if and only if all conjugates of K are equal to K. In fact, to show K normal it is enough to show $g \cdot K \cdot g^{-1} \subseteq K$ for all g in G, for in that case we have

$$K = g^{-1} \cdot (g \cdot K \cdot g^{-1}) \cdot g \subseteq g^{-1} \cdot K \cdot g \subseteq K$$

since g^{-1} is also in G.

EXERCISES 12.6

1. Which of the following functions φ from $(\mathbb{Z}, +)$ to $(\mathbb{Z}, +)$ are homomorphisms?

(a) $\varphi(n) = 6n$

(b) $\varphi(n) = n + 1$

(c) $\varphi(n) = -n$

(d) $\varphi(n) = n^2$

(e) $\varphi(n) = (6n^2 + 3n)/(2n + 1)$

2. Which of the following functions φ are homomorphisms from $(\mathbb{Z}, +)$ to $(\mathbb{R}\backslash\{0\}, \cdot)$?

   (a) $\varphi(n) = 6^n$ (b) $\varphi(n) = n$

   (c) $\varphi(n) = (-6)^n$ (d) $\varphi(n) = n^2$

   (e) $\varphi(n) = 2^{n+1}$

3. Which of the homomorphisms in Exercise 1 are isomorphisms? Explain briefly.

4. Which of the homomorphisms in (a), (b) and (c) of Example 1 are isomorphisms? Explain briefly.

5. Let $F = \text{FUN}(\mathbb{R}, \mathbb{R})$, and define $+$ on F by

$$(f + g)(x) = f(x) + g(x) \qquad \text{for} \quad x \in \mathbb{R}.$$

   (a) Show that $(F, +)$ is a group.

   (b) Is F commutative?

   (c) Define φ from F to $\mathbb{R}$ by $\varphi(f) = f(73)$. Show that φ is a homomorphism of $(F, +)$ onto $(\mathbb{R}, +)$.

6. Complete the proof of Theorem 2(b).

7. Find the kernel of each of the following group homomorphisms φ.

   (a) From $(\mathbb{Z}, +)$ to $(\mathbb{Z}, +)$, defined by $\varphi(n) = 73n$.

   (b) From $(\mathbb{Z}, +)$ to $(\mathbb{Z}, +)$, defined by $\varphi(n) = 0$ for all n.

   (c) From $(\mathbb{Z}, +)$ to $(\mathbb{Z}(5), +_5)$, defined by $\varphi(n) = n \text{ MOD } 5$.

   (d) From $(\mathbb{Z}, +)$ to $(\mathbb{Z}, +)$, defined by $\varphi(n) = n$.

8. For each homomorphism in Exercise 7 describe the coset of the kernel that contains 73.

9. Suppose that φ is a homomorphism defined on a group G and that $|G| = 12$ and $|\varphi(G)| = 3$.

   (a) Find $|K|$, where K is the kernel of φ.

   (b) How many members of G does φ map onto each member of $\varphi(G)$?

   (c) What is $|G/K|$?

10. (a) Check that the mapping φ from $\mathbb{Z}(p \cdot q)$ to $\mathbb{Z}(p)$ defined by $\varphi(n) = n \text{ MOD } p$ is a homomorphism.

    (b) What is the image of φ?

    (c) What is the kernel of φ?

11. The **direct product** of two groups $(G, \cdot)$ and $(H, \bullet)$ is the group $(G \times H, \square)$ with operation $\square$ defined by

$$(g_1, h_1) \square (g_2, h_2) = (g_1 \cdot g_2, h_1 \bullet h_2) \qquad \forall g_1, g_2 \in G, \forall h_1, h_2 \in H.$$

    (a) Describe the identity element of $G \times H$ and the inverse of an element (g, h) in $G \times H$.

    (b) Verify that the mapping $\pi \colon G \times H \to G$ defined by $\pi(g, h) = g$ is a homomorphism. [Here π is for "projection."]

(c) Find the kernel of the homomorphism π in part (b).

(d) Find a normal subgroup of $G \times H$ that is isomorphic to H.

12. Define the mapping φ from $(\mathbb{Z}, +)$ to $\mathbb{Z}(2) \times \mathbb{Z}(2)$ [see Exercise 11] by $\varphi(k) = (k \text{ MOD } 2, k \text{ MOD } 2)$.

(a) Verify that φ is a homomorphism.

(b) Verify that $\varphi(\mathbb{Z})$ is a subgroup of $\mathbb{Z}(2) \times \mathbb{Z}(2)$.

(c) Find the kernel of φ.

(d) Is $\mathbb{Z}(2) \times \mathbb{Z}(2)$ isomorphic to $\mathbb{Z}(4)$? Explain.

13. Let H be the set of 2×2 matrices of the form $\begin{bmatrix} 1 & x \\ 0 & 1 \end{bmatrix}$, with matrix multiplication as operation.

(a) Verify that H is a group, with

$$\begin{bmatrix} 1 & x \\ 0 & 1 \end{bmatrix}^{-1} = \begin{bmatrix} 1 & -x \\ 0 & 1 \end{bmatrix}.$$

(b) Verify that

$$\begin{bmatrix} 0 & 1 \\ 1 & 0 \end{bmatrix} \cdot H \neq H \cdot \begin{bmatrix} 0 & 1 \\ 1 & 0 \end{bmatrix}.$$

This shows that H is not a normal subgroup of the multiplicative group G of all 2×2 matrices that have inverses.

(c) Show that H is a normal subgroup of the group T of all 2×2 matrices of the form

$$\begin{bmatrix} y & z \\ 0 & 1/y \end{bmatrix}, \qquad y \neq 0,$$

with multiplication as operation.

(d) The mapping φ from T to $(\mathbb{R}\backslash\{0\}, \cdot)$ defined by

$$\varphi\left(\begin{bmatrix} y & z \\ 0 & 1/y \end{bmatrix}\right) = y$$

is a group homomorphism. Find its kernel.

(e) Show that T/H is isomorphic to the group of nonzero real numbers under multiplication.

14. An **antihomomorphism** from $(G, \cdot)$ to $(G_0, \bullet)$ is a function ψ such that

$$\psi(g \cdot h) = \psi(h) \bullet \psi(g) \quad \text{for all} \quad g, h \in G.$$

(a) Show that the mapping $g \to g^{-1}$ is always an antihomomorphism of a group onto itself. [This ψ could be called an **anti-isomorphism** since it is one-to-one and onto.]

(b) When is the anti-isomorphism in part (a) also an isomorphism?

(c) Show that if ψ_1 and ψ_2 are antihomomorphisms for which the composition $\psi_1 \circ \psi_2$ is defined, then $\psi_1 \circ \psi_2$ is a homomorphism.

15. Show that if φ is a homomorphism defined on the group G and if $\varphi(g)$ has just one pre-image under φ for some g in G, then φ is one-to-one.

16. Show that if J and K are normal subgroups of a group $(G, \cdot)$, then $J \cap K$ is a normal subgroup. *Suggestion:* Consider $g \cdot (J \cap K) \cdot g^{-1}$.

17. Let H be a subgroup of $(G, \cdot)$.

(a) Show that $\{g \in G : g \cdot H \cdot g^{-1} = H\}$ is a subgroup of G.

(b) Conclude that if G is generated by a subset A and if $g \cdot H \cdot g^{-1} = H$ for all $g \in A$, then H is a normal subgroup.

18. Consider a finite group G with a subgroup H such that $|G| = 2|H|$. Show that H must be a normal subgroup of G. *Suggestion:* See Exercise 22 of § 12.5.

19. Let G be the symmetric group S_3 and let $K = S_2 = \langle (1\,2) \rangle$.

(a) Show that $e \cdot K \cdot (1\,3) \cdot K$ is not a coset of K in G.

(b) Is K a normal subgroup of G? Explain.

§ 12.7 Semigroups

One frequently encounters structures with some but not all of the properties of groups. For instance, the important associative law may hold even though elements don't have inverses. There might not even be an identity. To discuss such structures we will use a little square □ as a symbol for a generic binary operation. This operation *might* be addition or multiplication or some other familiar operation like union or intersection, but it doesn't have to be. A set S with a binary associative operation □ is called a **semigroup**. Thus $(S, \square)$ is a semigroup in case $s_1 \square s_2$ is in S for all $s_1, s_2 \in S$ and

$$s_1 \square (s_2 \square s_3) = (s_1 \square s_2) \square s_3 \qquad \text{for all} \quad s_1, s_2, s_3 \in S.$$

In view of associativity, expressions like $s_1 \square s_2 \square s_3$ are unambiguous in a semigroup. Not much can be said without the associative law, and wherever we have encountered it, there has always been a semigroup nearby. All that's new now is the emphasis on the overall structure.

As with groups, if the semigroup is **commutative** [i.e., if $s_1 \square s_2 = s_2 \square s_1$ for all $s_1, s_2 \in S$], we sometimes use additive notation $+$ instead of □. An element e in S is an **identity for** S provided that

$$s \square e = e \square s = s \qquad \text{for all} \quad s \in S.$$

A semigroup with an identity is called a **monoid**. A **group**, then, is a monoid in which every element has an inverse. In this section we focus primarily on semigroups that are not groups. A **subsemigroup** of a semigroup is a subset that is closed under the operation.

EXAMPLE 1   (a) $(\mathbb{N}, +)$ is a semigroup, since the sum of two numbers in $\mathbb{N}$ is in $\mathbb{N}$ and since addition is associative. It has an identity, namely 0, since $n + 0 = 0 + n = n$ for all $n \in \mathbb{N}$, so it is a monoid. None of the positive numbers in $\mathbb{N}$ have additive inverses [i.e., negatives] in $\mathbb{N}$, so $(\mathbb{N}, +)$ is not a group. The monoid $(\mathbb{N}, +)$ is commutative.

(b) $(\mathbb{P}, +)$ is also a commutative semigroup, indeed, a subsemigroup of $(\mathbb{N}, +)$, but it has no identity. It makes no sense to inquire further about inverses. ∎

EXAMPLE 2   (a) Multiplication $\cdot$ in $\mathbb{R}$ is associative and commutative, so any subset of $\mathbb{R}$ that is closed under multiplication is a semigroup. Thus $(\mathbb{R}, \cdot)$, $(\mathbb{Q}, \cdot)$, $(\mathbb{Z}, \cdot)$ and $(\mathbb{P}, \cdot)$ are all commutative semigroups. They all include the multiplicative identity 1, so all are monoids. None of these semigroups is a group: $\mathbb{R}$, $\mathbb{Q}$ and $\mathbb{Z}$ all contain 0, and 0 has no multiplicative inverse [i.e., reciprocal]. The semigroup $\mathbb{P}$ contains 2 but not its inverse $\frac{1}{2}$.

(b) Since the product of nonzero real numbers is nonzero, $\cdot$ is also a binary operation on $\mathbb{R}\backslash\{0\}$. Moreover, $(\mathbb{R}\backslash\{0\}, \cdot)$ is a bona fide commutative group.

(c) The set $\mathbb{Z}(p)$ is a commutative semigroup under $*_p$. This fact is proved in Theorem 4 of §3.6. We already observed in Example 3(b) of §12.5 that $\mathbb{Z}(p)$ is a commutative group under $+_p$. ∎

EXAMPLE 3   Let Σ be an alphabet. The set Σ^* of all words in the letters of Σ is informally defined in §1.1 and is recursively defined in Example 2 of §7.1. Two words w_1 and w_2 in Σ^* are multiplied by **concatenation**; i.e., $w_1 w_2$ is the word obtained by placing the string w_2 right after the string w_1. Thus if $w_1 = a_1 a_2 \cdots a_m$ and $w_2 = b_1 b_2 \cdots b_n$ with the a_j's and b_k's in Σ, then $w_1 w_2 = a_1 a_2 \cdots a_m b_1 b_2 \cdots b_n$. For example, if $w_1 = cat$ and $w_2 = nip$, then $w_1 w_2 = catnip$ and $w_2 w_1 = nipcat$. Multiplication by the empty word λ leaves a word unchanged:

$$w\lambda = \lambda w = w \qquad \text{for all} \quad w \in \Sigma^*.$$

It is evident from the definition that concatenation is an associative binary operation. Since the empty word λ serves as an identity for Σ^*, Σ^* is a monoid. It is certainly not a group; only the empty word has an inverse. ∎

EXAMPLE 4   (a) Let $\mathfrak{M}_{m,n}$ be the set of all $m \times n$ matrices. Matrix addition $+$ is a binary operation on $\mathfrak{M}_{m,n}$ since

$$\mathbf{A} + \mathbf{B} \text{ is in } \mathfrak{M}_{m,n} \qquad \text{for all} \quad \mathbf{A}, \mathbf{B} \in \mathfrak{M}_{m,n}.$$

The set $\mathfrak{M}_{m,n}$ of all $m \times n$ matrices is a commutative group under addition. This fact is spelled out in the theorem in §3.3.

(b) Matrix multiplication is also a binary operation on the set $\mathfrak{M}_{n,n}$ of $n \times n$ matrices. Under multiplication $\mathfrak{M}_{n,n}$ is a monoid, with identity $\mathbf{I}_n$. The associative law is discussed at the end of § 3.4. Except for the trivial case $n = 1$, this monoid is not commutative: $\mathbf{AB}$ does not necessarily equal $\mathbf{BA}$. Some, but not all, matrices in $\mathfrak{M}_{n,n}$ have inverses. See Exercises 14 and 15 of § 3.4. The subsemigroup of invertible matrices is a group in its own right [Example 1(b) of § 12.5]. ■

EXAMPLE 5   (a) Let $\mathscr{P}(U)$ be the set of all subsets of some set U. With the operation $\cup$, $\mathscr{P}(U)$ is a commutative semigroup with identity $\varnothing$; see laws 1a, 2a and 5a in Table 1 of § 1.2. Only the empty set itself has an inverse, since $A \cup B \neq \varnothing$ whenever $A \neq \varnothing$.

(b) $\mathscr{P}(U)$ is a commutative semigroup under the operation $\cap$ with identity U; see laws 1b, 2b and 5d in Table 1 of § 1.2.

(c) $\mathscr{P}(U)$ is also a semigroup using symmetric difference $\oplus$; see Exercise 2. ■

EXAMPLE 6   Consider a set $T = \{a, b, \ldots\}$ with at least two members. The set FUN(T, T) of all functions mapping T into T is a semigroup under composition; the associative law is discussed in § 1.3. The identity function 1_T on T is the identity for this semigroup, since

$$(1_T \circ f)(t) = 1_T(f(t)) = f(t) \qquad \text{for all} \quad t \in T,$$

so that $1_T \circ f = f$ and similarly, $f \circ 1_T = f$. This monoid is not commutative. For example, let f and g be the constant functions defined by $f(t) = a$ and $g(t) = b$ for all t in T. Then $(f \circ g)(t) = f(g(t)) = a$ for all t, and $(g \circ f)(t) = b$ for all t. That is, $f \circ g = f \neq g = g \circ f$.

The group PERM(T) of all permutations of T is an important subsemigroup of FUN(T, T). ■

EXAMPLE 7   (a) Let T be a nonempty set. The set FUN$(T, \mathbb{N})$ of functions mapping T into $\mathbb{N}$ is a semigroup under the operation $+$ defined by

$$(f + g)(t) = f(t) + g(t) \qquad \text{for all} \quad t \in T.$$

[Observe, for instance, that $(f + g) + h$ is defined by $((f + g) + h)(t) = (f + g)(t) + h(t) = (f(t) + g(t)) + h(t)$ for all $t \in T$.] The identity element of FUN$(T, \mathbb{N})$ is the constant function z defined by $z(t) = 0$ for all $t \in T$. An inverse for f would be a function g with $f(t) + g(t) = 0$ for all t, i.e., with $g(t) = -f(t)$ for all t. Since f and g have values in $\mathbb{N}$, no such g exists if $f(t) > 0$ for some t.

(b) Replace $\mathbb{N}$ by $\mathbb{Z}$ in (a). Now inverses exist; the inverse of f is the function $-f$ defined by $(-f)(t) = -f(t)$ for all $t \in T$. Thus FUN$(T, \mathbb{Z})$ is a group under $+$.

(c) We can also make FUN(T, $\mathbb{N}$) into a semigroup under an operation $*$ defined by $(f * g)(t) = f(t) \cdot g(t)$. Associativity of $*$ comes from associativity of $\cdot$ on $\mathbb{N}$. The identity element is the constant function e with $e(t) = 1$ for all t.

(d) The usual definitions for adding and multiplying functions, namely

$$(f + g)(x) = f(x) + g(x) \quad \text{for all } x, \quad \text{and}$$

$$(f \cdot g)(x) = f(x) \cdot g(x) \quad \text{for all } x,$$

make FUN($\mathbb{R}$, $\mathbb{R}$) into a semigroup in two different ways. ∎

Consider once again an arbitrary semigroup $(S, \square)$. For $s \in S$ and $n \in \mathbb{P}$, we continue a familiar convention and write s^n for the $\square$-product of s with itself n times. If S has an identity we also write s^0 for e. The next theorem is obvious, although a formal induction proof can be given based on the recursive definition in which $s^{n+1} = s^n \square s$.

Theorem 1

Let $(S, \square)$ be a semigroup. For $s \in S$ and $m, n \in \mathbb{P}$, we have

(a) $s^m \square s^n = s^{m+n}$

(b) $(s^m)^n = s^{mn}$.

If $(S, \square)$ is a monoid, these formulas hold for $m, n \in \mathbb{N}$.

We have treated the generic operation $\square$ the same way we would treat multiplication. Let's see how the foregoing would look when $\square$ is replaced by $+$. The general product s^n is replaced by the general sum $ns = s + s + \cdots + s$ and the theorem now says

(a) $ms + ns = (m + n)s$

(b) $n(ms) = (mn)s$

for $s \in S$ and $m, n \in \mathbb{P}$.

Starting with a nonempty subset A of S, we define the set A^+ **generated by** A recursively as follows.

(B) $A \subseteq A^+$;

(R) If $s, t \in A^+$, then $s \square t \in A^+$.

Theorem 3 will tell us that A^+ consists of all products of elements from A. We say that A **generates** S if $A^+ = S$. Whether or not A generates S, A^+ is a subsemigroup of S by the following fact.

Theorem 2

> Let A be a nonempty subset of the semigroup $(S, \square)$. Then A^+ is the unique smallest subsemigroup of S that contains A.

Proof. By (B), A^+ contains A. By (R), A^+ is closed under the operation $\square$, so by definition A^+ is a subsemigroup of S.

Now consider an arbitrary subsemigroup T of S that contains A. We'll show $A^+ \subseteq T$, from which it will follow that A^+ is the unique smallest subsemigroup containing A. We want to show that $s \in T$ for every s in A^+. Since A^+ is defined recursively, it's natural to use the Generalized Principle of Mathematical Induction introduced in § 7.1. Let $p(s)$ be the proposition "$s \in T$." To show that $p(s)$ is true for all s in A^+ we must establish that:

(B′) $p(s)$ is true for the members of A^+ specified in the basis.
(I)   If a member u of A^+ is specified by (R) in terms of previously defined members, say $u = s \square t$, then $p(s) \wedge p(t) \Rightarrow p(u)$.

Now (B′) is true, since $A \subseteq T$ by the choice of T. Condition (I) just says "if $u = s \square t$ and if $s \in T$ and $t \in T$, then $u \in T$," which is true because T is a subsemigroup. The conditions for the Generalized Principle are met, so $s \in T$ for all s in A^+. ■

Theorem 2 helps us describe the members of A^+ without recursion, as follows.

Theorem 3

> Let A be a nonempty subset of the semigroup $(S, \square)$. Then A^+ consists of all elements of S of the form $a_1 \square \cdots \square a_n$ for $n \in \mathbb{P}$ and $a_1, \ldots, a_n \in A$.

Proof. Let X be the set of all products $a_1 \square \cdots \square a_n$. We want to show that $A^+ = X$. If $a_1, \ldots, a_n, b_1, \ldots, b_m$ are in A, then $a_1 \square \cdots \square a_n \square b_1 \square \cdots \square b_m$ is a product of this same form. Thus X is closed under $\square$, i.e., it's a subsemigroup of S. The case $n = 1$ gives $A \subseteq X$, so $A^+ \subseteq X$ by Theorem 2.

To show that $X \subseteq A^+$, we use ordinary induction on n. Since $A \subseteq A^+$, $a_1 \in A^+$ whenever $a_1 \in A$. Assume inductively that $a_1 \square \cdots \square a_n \in A^+$ for some n and some $a_1, \ldots, a_n$ in A and let $a_{n+1} \in A$. Since $a_{n+1} \in A^+$, we have $a_1 \square \cdots \square a_n \circ a_{n+1} \in A^+$ by (R) in the recursive definition of A^+. By induction, every member of X is in A^+. ■

EXAMPLE 8

(a) The subsemigroup $\{2\}^+$ of the semigroup $(\mathbb{Z}, +)$ consists of all "products" of members of $\{2\}$. Since the operation here is $+$, $\{2\}^+$ consists of all sums $2 + 2 + \cdots + 2$, i.e., of all positive even integers. Thus $\{2\}^+ = 2\mathbb{P} = \{2n : n \in \mathbb{P}\}$.

(b) More generally, the subsemigroup of $(S, \square)$ generated by a single element s is the set

$$\{s\}^+ = \{s^n : n \in \mathbb{P}\}$$

[or $\{s\}^+ = \{ns : n \in \mathbb{P}\}$ if $\square$ is $+$]. Such a single-generator semigroup is called a **cyclic semigroup**.

(c) The subsemigroup $\{2\}^+$ of $(\mathbb{Z}, \cdot)$ is $\{2^n : n \in \mathbb{P}\} = \{2, 4, 8, 16, \ldots\}$. Notice that our notation is deficient; we can't tell from the expression $\{2\}^+$ alone whether we mean this subsemigroup or the one in part (a).

(d) The subsemigroup $\{2, 7\}^+$ of $(\mathbb{Z}, \cdot)$ generated by $\{2, 7\}$ is the set $\{2^m 7^n : m, n \in \mathbb{N} \text{ and } m + n \geq 1\}$.

(e) We show that the cyclic subsemigroup of $(\mathfrak{M}_{2,2}, \cdot)$ generated by the matrix

$$\mathbf{M} = \begin{bmatrix} 1 & 1 \\ 0 & 1 \end{bmatrix}$$

is

$$\{\mathbf{M}\}^+ = \left\{ \begin{bmatrix} 1 & n \\ 0 & 1 \end{bmatrix} : n \in \mathbb{P} \right\}.$$

By part (b) it will be enough to show that $\mathbf{M}^n = \begin{bmatrix} 1 & n \\ 0 & 1 \end{bmatrix}$ for $n \in \mathbb{P}$. This is clear for $n = 1$. Assume inductively that $\mathbf{M}^n = \begin{bmatrix} 1 & n \\ 0 & 1 \end{bmatrix}$ for some $n \in \mathbb{P}$. Then matrix multiplication gives

$$\mathbf{M}^{n+1} = \mathbf{M}^n \cdot \mathbf{M} = \begin{bmatrix} 1 & n \\ 0 & 1 \end{bmatrix} \begin{bmatrix} 1 & 1 \\ 0 & 1 \end{bmatrix} = \begin{bmatrix} 1 & 1+n \\ 0 & 1 \end{bmatrix}.$$

The result now follows by mathematical induction. ∎

The intersection of any collection of subsemigroups of a semigroup S is either empty or is itself a subsemigroup of S. Indeed, if s and t belong to the intersection, then they both belong to each subsemigroup in the collection and so their product $s \square t$ does too.

EXAMPLE 9   (a) Both $2\mathbb{Z}$ and $3\mathbb{Z}$ are subsemigroups of $(\mathbb{Z}, \cdot)$, so their intersection $6\mathbb{Z}$ is too. Similarly, the intersection $\mathbb{P} \cap 2\mathbb{Z} = \{2k : k \in \mathbb{P}\}$ is a subsemigroup of $(\mathbb{Z}, \cdot)$.

(b) Both $\mathbb{P}$ and $\{-k : k \in \mathbb{P}\}$ are subsemigroups of $(\mathbb{Z}, +)$. Their intersection is empty.

(c) Consider a nonempty subset A of a semigroup S. By Theorem 2 the intersection of all of the subsemigroups that contain A [including S itself, naturally] is a subsemigroup that contains A. It must be the smallest subsemigroup containing A, namely A^+. This observation gives a way to define A^+ without describing its elements in terms of A. ∎

EXAMPLE 10   Lagrange's theorem can fail spectacularly for semigroups. Consider an arbitrary nonempty set S, and define □ on S by $s \, \square \, t = t$ for all $s, t \in S$; i.e., the product of two elements is always just the second element. Then $(s \, \square \, t) \, \square \, u = u = s \, \square \, u = s \, \square \, (t \, \square \, u)$, so □ is associative and $(S, \square)$ is a semigroup. *Every* nonempty subset of S is a subsemigroup, since it's closed under □. ∎

As with groups, the important functions between semigroups are the ones that preserve the semigroup structure. Let $(S, \bullet)$ and $(T, \square)$ be semigroups. A function $\varphi \colon S \to T$ is a **semigroup homomorphism** if

$$\varphi(s_1 \bullet s_2) = \varphi(s_1) \, \square \, \varphi(s_2) \qquad \text{for all} \quad s_1, s_2 \in S.$$

If φ is also a one-to-one correspondence, then φ is called a **semigroup isomorphism**. If there is a semigroup isomorphism of S onto T, we say that S and T are **isomorphic** and write $S \simeq T$ [or $(S, \bullet) \simeq (T, \square)$ if the operations need to be mentioned].

EXAMPLE 11   (a) Let $(S, \bullet)$ be $(\mathbb{P}, +)$, let $(T, \square)$ be $(\mathbb{P}, \cdot)$, and let $\varphi(m) = 2^m$ for m in $\mathbb{P}$. Since

$$\varphi(m + n) = 2^{m+n} = 2^m \cdot 2^n = \varphi(m) \cdot \varphi(n),$$

φ is a homomorphism of $(\mathbb{P}, +)$ into $(\mathbb{P}, \cdot)$.

(b) In Example 2 of § 12.6 we observed that MOD p is a group homomorphism of $(\mathbb{Z}, +)$ onto $(\mathbb{Z}(p), +_p)$, but that MOD p is not a group homomorphism of $(\mathbb{Z}, \cdot)$ onto $(\mathbb{Z}(p), *_p)$ for the very good reason that $(\mathbb{Z}(p), *_p)$ is not a group. But $(\mathbb{Z}(p), *_p)$ *is* a semigroup and, as noted in Example 2(b) in § 12.6, MOD p is a semigroup homomorphism. ∎

EXAMPLE 12   Let U be any set. We define the "complementation function" φ from $\mathscr{P}(U)$ into $\mathscr{P}(U)$ by $\varphi(A) = A^c$ for $A \in \mathscr{P}(U)$. By a DeMorgan law in Table 1 of § 1.2,

$$\varphi(A \cup B) = (A \cup B)^c = A^c \cap B^c = \varphi(A) \cap \varphi(B),$$

so φ is a homomorphism from the semigroup $(\mathscr{P}(U), \cup)$ into the semigroup $(\mathscr{P}(U), \cap)$. The function φ is a one-to-one correspondence of $\mathscr{P}(U)$ onto $\mathscr{P}(U)$, so it is an isomorphism of $(\mathscr{P}(U), \cup)$ onto $(\mathscr{P}(U), \cap)$. In fact, φ is its own inverse [why?], so φ also gives an isomorphism of $(\mathscr{P}(U), \cap)$ onto $(\mathscr{P}(U), \cup)$. Note that φ maps the identity $\varnothing$ of $(\mathscr{P}(U), \cup)$ onto the identity U of $(\mathscr{P}(U), \cap)$. ∎

Isomorphisms get used in two different ways, as we saw when we looked at graph isomorphisms. Sometimes we want to call attention to the fact that two apparently different semigroups are actually identical in structure. At other times the identical structure is obvious, but we want to examine the various isomorphisms that are possible, for instance from S back onto itself, to see how much symmetry is present.

EXERCISES 12.7

1. $(\mathbb{N}, \cdot)$ is a semigroup.

   (a) Is this semigroup commutative?

   (b) Is there an identity for this semigroup?

   (c) If yes, do inverses exist? If yes, specify them.

   (d) Is this a monoid? A group?

2. $\mathscr{P}(U)$ is a semigroup with respect to symmetric difference $\oplus$. Repeat Exercise 1 for this semigroup.

3. Repeat Exercise 1 for the semigroup $\text{FUN}(\mathbb{R}, \mathbb{R})$ of real-valued functions on $\mathbb{R}$ under addition.

4. Repeat Exercise 1 for the semigroup $\text{FUN}(\mathbb{R}, \mathbb{R})$ under multiplication.

5. The set $\mathfrak{M}_{2,2}$ of 2×2 matrices is a semigroup with respect to matrix multiplication.

   (a) Is this semigroup commutative?

   (b) Is there an identity for this semigroup?

   (c) If yes, do inverses exist?

   (d) Is this semigroup a monoid? A group?

6. Let $\Sigma = \{a, b, c, d\}$ and consider the words: $w_1 = bad$, $w_2 = cab$ and $w_3 = abcd$.

   (a) Determine $w_1 w_2$, $w_2 w_1$, $w_2 w_3 w_2 w_1$ and $w_3 w_2 w_3$.

   (b) Determine w_1^2, w_2^3 and λ^4.

7. Let Σ be the usual English alphabet and consider the words: $w_1 = break$, $w_2 = fast$, $w_3 = lunch$ and $w_4 = food$.

   (a) Determine λw_1, $w_2 \lambda$, $w_2 w_4$, $w_3 w_1$ and $w_4 \lambda w_4$.

   (b) Compare $w_1 w_2$ and $w_2 w_1$.

   (c) Determine w_2^2, w_4^2, $w_2^2 w_4 w_1^2$ and λ^{73}.

8. Show that $\mathbb{R}^+ = \{x \in \mathbb{R} : x > 0\}$ is not a semigroup with the binary operation $(x, y) \rightarrow x/y$.

9. (a) Convince yourself that $\mathbb{N}$ is a semigroup under the binary operation $(m, n) \rightarrow \min\{m, n\}$ and also under $(m, n) \rightarrow \max\{m, n\}$.

   (b) Are the semigroups in part (a) monoids?

10. (a) Show that $\mathbb{P}$ is a semigroup with respect to $(m, n) \rightarrow \gcd(m, n)$ where $\gcd(m, n)$ represents the greatest common divisor of m and n.

(b) Show that $\mathbb{P}$ is a semigroup with respect to $(m, n) \to \text{lcm}(m, n)$ where lcm(m, n) represents the least common multiple of m and n.

(c) Are the semigroups in parts (a) and (b) monoids?

11. Describe each of the following subsemigroups of $(\mathbb{Z}, +)$.

(a) $\{1\}^+$ (b) $\{0\}^+$ (c) $\{-1, 2\}^+$

(d) $\mathbb{P}^+$ (e) $\mathbb{Z}^+$ (f) $\{2, 3\}^+$

(g) $\{6\}^+ \cap \{9\}^+$

12. Describe each of the following subsemigroups of $(\mathbb{Z}, \cdot)$.

(a) $\{1\}^+$ (b) $\{0\}^+$ (c) $\{-1, 2\}^+$

(d) $\mathbb{P}^+$ (e) $\mathbb{Z}^+$ (f) $\{2, 3\}^+$

13. Which of the semigroups in Exercise 11 are cyclic semigroups? Justify your answers.

14. Which of the semigroups in Exercise 12 are cyclic semigroups? Justify your answers.

15. If A is a subset of a monoid $(M, \square)$ with identity e, the **submonoid generated by** A is defined to be $A^+ \cup \{e\}$. Find each of the following.

(a) The submonoid of $(\mathbb{Z}, +)$ generated by $\{2\}$.

(b) The submonoid of $(\mathbb{Z}, +)$ generated by $\{1, -1\}$.

(c) The submonoid of $(\mathbb{Z}, +)$ generated by $\{0\}$.

(d) The submonoid of $(\mathbb{Z}, \cdot)$ generated by $\{1\}$.

(e) The submonoid of Σ^* generated by Σ, using concatenation on Σ^*.

16. (a) Give an example of a cyclic semigroup and a subsemigroup of it that is not cyclic.

(b) Give an example of a cyclic group that is not a cyclic semigroup.

(c) Give an example of a cyclic group that *is* a cyclic semigroup.

17. List the members of each of the following finite subsemigroups of $(\mathfrak{M}_{3,3}, \cdot)$.

(a) $\left\{ \begin{bmatrix} 0 & 1 & 0 \\ 1 & 0 & 0 \\ 0 & 0 & 1 \end{bmatrix} \right\}^+$

(b) $\left\{ \begin{bmatrix} 0 & 1 & 0 \\ 1 & 0 & 0 \\ 0 & 0 & 1 \end{bmatrix}, \begin{bmatrix} 0 & 0 & 1 \\ 1 & 0 & 0 \\ 0 & 1 & 0 \end{bmatrix} \right\}^+$

(c) $\left\{ \begin{bmatrix} 0 & 2 & 3 \\ 0 & 0 & 4 \\ 0 & 0 & 0 \end{bmatrix} \right\}^+$

18. (a) Which of the subsemigroups in Exercise 17 are groups?

(b) Which are commutative?

(c) Which are cyclic?

19. (a) Find the intersection of the subsemigroups $2\mathbb{P}$ and $3\mathbb{P}$ of the semigroup $(\mathbb{P}, +)$.

(b) Is the intersection in part (a) cyclic? Explain.

(c) Repeat part (b) with the semigroup $(\mathbb{P}, \cdot)$.

20. (a) Find the intersection of the subsemigroups of $(\mathbb{P}, \cdot)$ generated by 2 and 3, respectively.

    (b) Is the intersection in part (a) cyclic? Explain.

21. (a) Find the intersection of the three subsemigroups $4\mathbb{P}$, $6\mathbb{P}$, $10\mathbb{P}$ of the semigroup $(\mathbb{P}, \cdot)$.

    (b) Is the intersection in part (a) cyclic? Explain.

    (c) Repeat part (b) with the semigroup $(\mathbb{P}, +)$.

22. Which of the following functions φ are homomorphisms from $(\mathbb{P}, +)$ to $(\mathbb{P}, \cdot)$?

    (a) $\varphi(n) = 2^n$ (b) $\varphi(n) = n$

    (c) $\varphi(n) = (-1)^n$ (d) $\varphi(n) = 2n$

    (e) $\varphi(n) = 2^{n+1}$

23. Which of the homomorphisms in Exercise 22 are isomorphisms? Explain briefly.

24. Let Σ be the English alphabet. Define φ on Σ^* by $\varphi(w) = $ length of w. Explain why φ is a homomorphism from Σ^* with its usual operation to $(\mathbb{N}, +)$.

25. Show that if S is a semigroup, if A generates S and if φ is a homomorphism defined on S, then $\varphi(A)$ generates $\varphi(S)$.

26. (a) Show that if φ is a homomorphism from a semigroup $(S, \bullet)$ to a semigroup $(T, \square)$, and if ψ is a homomorphism from $(T, \square)$ to a semigroup $(U, \triangle)$, then $\psi \circ \varphi$ is a homomorphism.

    (b) Show that if φ is an isomorphism of $(S, \bullet)$ onto $(T, \square)$, then φ^{-1} is an isomorphism of $(T, \square)$ onto $(S, \bullet)$.

    (c) Use the results of parts (a) and (b) to show that the relation $\simeq$ is reflexive, symmetric and transitive.

27. An element z of a semigroup $(S, \bullet)$ is called a **zero element** or **zero** of S in case

$$z \bullet s = s \bullet z = z \qquad \text{for all } s \text{ in } S.$$

    (a) Show that a semigroup cannot have more than one zero element.

    (b) Give an example of an infinite semigroup that has a zero element.

    (c) Give an example of a finite semigroup that has at least two members and has a zero element.

28. Let z be a zero element of a semigroup $(S, \bullet)$ and let φ be a homomorphism from $(S, \bullet)$ to a semigroup $(T, \square)$.

    (a) Show that $\varphi(z)$ is a zero element of $\varphi(S)$.

    (b) Must $\varphi(z)$ be a zero element of $(T, \square)$? Justify your answer.

29. Suppose that $(S, \bullet)$ is a monoid with identity e and that φ is a semigroup homomorphism from $(S, \bullet)$ to $(T, \square)$.

    (a) Show that $(\varphi(S), \square)$ is a monoid with identity $\varphi(e)$.

    (b) Must $\varphi(e)$ be an identity of $(T, \square)$? Justify your answer.

§ 12.8   Other Algebraic Systems

So far in this chapter we have been looking at sets with just one binary operation on them. A number of important and familiar algebraic structures have two binary operations, usually written + and ·. The "additive" operation + is typically very well-behaved, while the "multiplicative" operation · is generally less so, and there are usually distributive laws relating the two operations to each other. In this section we briefly introduce rings and fields, which are the two main kinds of algebraic structures with two operations, and we give some examples and discuss the basic facts about homomorphisms of such systems.

EXAMPLE 1   (a) The sets $\mathbb{Z}$, $\mathbb{Q}$ and $\mathbb{R}$ are each closed under ordinary addition and multiplication. Both + and · are commutative and associative for each of these sets. Moreover, these operations satisfy the distributive laws:

$$a \cdot (b + c) = (a \cdot b) + (a \cdot c) \quad \text{and} \quad (a + b) \cdot c = (a \cdot c) + (b \cdot c).$$

(b) The set $\mathfrak{M}_{n,n}$ of $n \times n$ matrices with real entries is closed under matrix addition and also under matrix multiplication. Both operations are associative, and addition is commutative, but matrix multiplication is not commutative. For instance,

$$\begin{bmatrix} 2 & 2 \\ 1 & 1 \end{bmatrix} \begin{bmatrix} 3 & -1 \\ -3 & 1 \end{bmatrix} = \begin{bmatrix} 0 & 0 \\ 0 & 0 \end{bmatrix} \neq \begin{bmatrix} 5 & 5 \\ -5 & -5 \end{bmatrix} = \begin{bmatrix} 3 & -1 \\ -3 & 1 \end{bmatrix} \begin{bmatrix} 2 & 2 \\ 1 & 1 \end{bmatrix}.$$

The distributive laws

$$\mathbf{A(B + C)} = \mathbf{(AB)} + \mathbf{(AC)} \quad \text{and} \quad \mathbf{(A + B)C} = \mathbf{(AC)} + \mathbf{(BC)}$$

are valid.

(c) For $p \geq 2$ the operations $+_p$ and $*_p$ on $\mathbb{Z}(p)$ are commutative and associative, and the distributive law holds by Theorem 4 of § 3.6. ■

Structures like the ones in Example 1 come up frequently enough to deserve a name. A **ring $(R, +, \cdot)$** is a set R closed under two binary operations, generally denoted + and ·, such that

(a) $(R, +)$ is a commutative group,

(b) $(R, \cdot)$ is a semigroup,

(c) $a \cdot (b + c) = (a \cdot b) + (a \cdot c)$ and $(a + b) \cdot c = (a \cdot c) + (b \cdot c)$ for all $a, b, c \in R$.

If $(R, \cdot)$ is also commutative, we say the ring $(R, +, \cdot)$ is **commutative**. The ring $(\mathfrak{M}_{n,n}, +, \cdot)$ in Example 1(b) is not commutative if $n > 1$. The other rings in Example 1 are commutative. A ring always has an additive identity element, denoted 0. If it has a multiplicative identity that is different from 0, we usually call the multiplicative identity 1 and we say the ring is a **ring with identity**. Each of the rings in Example 1 is a ring with identity. The ring $(2\mathbb{Z}, +, \cdot)$ of even integers, with the usual sum and product, is an example of a commutative ring without identity.

EXAMPLE 2   The set POLY($\mathbb{R}$) of all polynomials in x with real coefficients is closed under $+$ and $\cdot$. Addition is easy. For example,

$$(2x^2 + 3x + 1) + (-5x^4 + 4x - 2) = -5x^4 + 2x^2 + 7x - 1.$$

Given polynomials $p(x)$ and $q(x)$, we just add the coefficients of x^k in $p(x)$ and $q(x)$ to get the coefficient of x^k in $p(x) + q(x)$. Addition is clearly associative and commutative; indeed, (POLY($\mathbb{R}$), $+$) is a commutative group.

Multiplication is a little more complicated. For instance,

$$(2x^2 - 3x + 1) \cdot (4x^3 - 2x^2)$$
$$= 2x^2 \cdot (4x^3 - 2x^2) - 3x \cdot (4x^3 - 2x^2) + 1 \cdot (4x^3 - 2x^2)$$
$$= 8x^5 - 4x^4 - 12x^4 + 6x^3 + 4x^3 - 2x^2$$
$$= 8x^5 - 16x^4 + 10x^3 - 2x^2.$$

We use the distributive laws several times and then collect terms with the same power of x.

The coefficient of x^k in the product

$$(a_m x^m + \cdots + a_1 x + a_0) \cdot (b_n x^n + \cdots + b_1 x + b_0)$$

turns out to be

$$(*) \qquad a_0 \cdot b_k + a_1 \cdot b_{k-1} + \cdots + a_i \cdot b_{k-i} + \cdots + a_k \cdot b_0,$$

using the notational convention that $a_i = 0$ for $i > m$ and $b_j = 0$ for $j > n$. One can check that this multiplication is associative and commutative, and that multiplication distributes over addition.

Distributivity here may seem to be a sure thing. Didn't we use it to *define* the product? No, we just used distributivity in motivating the definition of product. In fact, if we just start out defining $\cdot$ by $(*)$ a tedious proof shows that all the other good properties follow.

With these operations, (POLY($\mathbb{R}$), $+$, $\cdot$) is a commutative ring. So are the structures (POLY($\mathbb{Z}$), $+$, $\cdot$), (POLY($\mathbb{Q}$), $+$, $\cdot$) and (POLY($\mathbb{Z}(p)$), $+$, $\cdot$) that we get by taking polynomials with coefficients in $\mathbb{Z}$, $\mathbb{Q}$ or $\mathbb{Z}(p)$, respectively. [Of course, for $\mathbb{Z}(p)$ we use the operations $+_p$ and $*_p$ on the coefficients.] The identities in these rings are the constant polynomials $e(x) = 1$. ∎

The distributive laws make calculations in a ring behave very much like the arithmetic we are used to in $\mathbb{Z}$, allowing for the obvious fact that we need to watch out for noncommuting elements. For example, we have

$$(a + b)^2 = (a + b) \cdot (a + b) = [a \cdot (a + b)] + [b \cdot (a + b)]$$
$$= (a \cdot a) + (a \cdot b) + (b \cdot a) + (b \cdot b) = a^2 + a \cdot b + b \cdot a + b^2,$$

but this is not $a^2 + 2(a \cdot b) + b^2$ unless $a \cdot b = b \cdot a$. As another example, we get

$$(0 \cdot a) + (0 \cdot a) = (0 + 0) \cdot a = 0 \cdot a,$$

so

$$0 \cdot a = (0 \cdot a) + (0 \cdot a) - (0 \cdot a) = (0 \cdot a) - (0 \cdot a) = 0.$$

Similarly, $a \cdot 0 = 0$ for all a in a ring. One can also show that $(-a) \cdot b = -(a \cdot b) = a \cdot (-b)$ [Exercise 8].

We can never hope to divide by 0, but in the rings $(\mathbb{Q}, +, \cdot)$ and $(\mathbb{R}, +, \cdot)$ we can divide by every non-0 element. In $(\mathbb{Z}, +, \cdot)$, on the other hand, we can divide 6 by 3 successfully but cannot divide 6 by 5 to get an answer that is still in $\mathbb{Z}$. A **field** is a commutative ring $(R, +, \cdot)$ in which the non-0 elements form a group under multiplication. In a field the inverse of a non-0 element a is usually written a^{-1} or $1/a$, and it has the property that $a^{-1} \cdot a = a \cdot a^{-1} = 1$. We also often write b/a for $b \cdot a^{-1}$, a notation that we justify by the fact that $(b/a) \cdot a = b \cdot a^{-1} \cdot a = b$.

Since the set of non-0 elements in a field is closed under multiplication, fields have the property

(ID)   if $a \cdot b = 0$,   then $a = 0$ or $b = 0$.

This property might hold even if inverses do not exist—for example, in $(\mathbb{Z}, +, \cdot)$. A commutative ring with identity in which property (ID) holds is called an **integral domain**. These rings form an important class intermediate between fields and more general commutative rings with identity. They are the ones that satisfy the **cancellation law**

if $a \cdot c = a \cdot d$ and $a \neq 0$,   then $c = d$,

since $a \cdot (c - d) = 0$ and $a \neq 0$ imply that $c - d = 0$.

One can show that every finite integral domain is a field [Exercise 14(c)].

EXAMPLE 3   (a) Groups are always nonempty, so the multiplicative identity in a field is always non-0. The smallest possible field is $(\mathbb{Z}(2), +_2, *_2)$ which has just two elements 0 and 1 and operations as shown in the tables.

$+_2$	0	1		$*_2$	0	1
0	0	1		0	0	0
1	1	0		1	0	1

(b) The set FUN$(\mathbb{R}, \mathbb{R})$ is a ring with $f + g$ and $f \cdot g$ defined by

$$(f + g)(x) = f(x) + g(x) \quad \text{and} \quad (f \cdot g)(x) = f(x) \cdot g(x)$$

for all $x \in \mathbb{R}$. The zero element is the constant function **0** defined by $\mathbf{0}(x) = 0$ for all x. [See Exercise 5 of §12.6 for the additive structure.]

This ring is commutative but is not an integral domain, even though it gets its multiplication from the field $\mathbb{R}$. For example, let $f(x) = 0$ for $x \leq 0$ and $f(x) = 1$ for $x > 0$, and let $g(x) = 1$ for $x \leq 0$ and $g(x) = 0$ for $x > 0$. Then $(f \cdot g)(x) = 0$ for every x, so $f \cdot g = 0$ but $f \neq 0$ and $g \neq 0$.

(c) The commutative ring $(\mathbb{Z}(4), +_4, *_4)$ is not an integral domain; $2 *_4 2 = 0$, but $2 \neq 0$.

(d) The polynomial rings POLY($\mathbb{R}$), POLY($\mathbb{Q}$) and POLY($\mathbb{Z}$) are integral domains: If $a(x) \neq 0$ and $b(x) \neq 0$, then $a(x) \cdot b(x) \neq 0$. [A polynomial is 0 only if all of its coefficients are 0.] It is not hard to see that

$$(a_m x^m + \cdots + a_1 x + a_0) \cdot (b_n x^n + \cdots + b_1 x + b_0)$$

$$= (a_m \cdot b_n) x^{m+n} + \text{terms with lower powers of } x.$$

If $a_m \neq 0$ and $b_n \neq 0$, then $a_m \cdot b_n \neq 0$; i.e., if $a(x) \neq 0$ and $b(x) \neq 0$, then $a(x) \cdot b(x) \neq 0$.

(e) The polynomial ring POLY($\mathbb{Z}(4)$) is not an integral domain. Since $(\mathbb{Z}(4), +_4, *_4)$ itself is not an integral domain, there are non-0 constant polynomials whose product is the zero polynomial.

(f) It turns out that if p is a prime then $\mathbb{Z}(p)$ is a field [see Exercise 16]. The argument in (d) shows that POLY($\mathbb{Z}(p)$) is an integral domain in this case. ∎

A **subring** of a ring R is simply a subset of R that is itself a ring under the two operations of R. A **subfield** of a field F is a subring of F that is itself a field; in particular, it contains the multiplicative identity 1 of F and is closed under taking inverses.

EXAMPLE 4   (a) The ring $(\mathbb{Z}, +, \cdot)$ has subrings $2\mathbb{Z}$, $73\mathbb{Z}$ and $\{0\}$, among others. In fact, the subrings of $\mathbb{Z}$ are precisely the rings $n\mathbb{Z}$ for n an integer, in view of Theorem 3 of §12.5. The ring $(\mathbb{Z}(p), +_p, *_p)$ is *not* a subring of $\mathbb{Z}$; in fact, the two rings have quite different structures. For example, for each a in $\mathbb{Z}(p)$ we have $a +_p a +_p \cdots +_p a = 0$ if there are p terms in the sum, whereas $a + a + \cdots + a = pa$ in $\mathbb{Z}$.

(b) Given a field, every subring with identity is clearly an integral domain. In particular, the subring $(\mathbb{Z}, +, \cdot)$ of the field $(\mathbb{R}, +, \cdot)$ is an integral domain. It is not a field, since only 1 and -1 have multiplicative inverses in $\mathbb{Z}$. The field $(\mathbb{Q}, +, \cdot)$ is a subfield of $(\mathbb{R}, +, \cdot)$.

(c) The polynomial ring (POLY($\mathbb{Z}$), $+, \cdot$) is a subring of (POLY($\mathbb{Q}$), $+, \cdot$), which is itself a subring of (POLY($\mathbb{R}$), $+, \cdot$). ∎

The appropriate mappings to use in studying rings are the ones that are compatible with both the additive and multiplicative structures. A **ring homomorphism** from a ring $(R, +, \cdot)$ to a ring $(S, +, \cdot)$ is a function

$\varphi: R \to S$ such that

$$\varphi(a + b) = \varphi(a) + \varphi(b) \quad \text{and} \quad \varphi(a \cdot b) = \varphi(a) \cdot \varphi(b)$$

for all $a, b \in R$. The operations on the left sides of these equations are the operations in R; those on the right are in S. Thus a ring homomorphism is just a function that is both a group homomorphism from $(R, +)$ to $(S, +)$ and a semigroup homomorphism from $(R, \cdot)$ to $(S, \cdot)$.

EXAMPLE 5   (a) The function φ from $\mathbb{Z}$ to $\mathbb{Z}(p)$, defined by $\varphi(m) = m \text{ MOD } p$, is a ring homomorphism, as observed in the corollary to Theorem 2 of § 3.6.

(b) The function φ from $\mathbb{Z}$ to $\mathbb{Z}$ defined by $\varphi(m) = 3m$ is an additive group homomorphism but is not a ring homomorphism because $\varphi(m \cdot n) = 3mn$, while $\varphi(m) \cdot \varphi(n) = 3m \cdot 3n = 9mn$.

(c) The mapping φ of FUN($\mathbb{R}$, $\mathbb{R}$) into $\mathbb{R}$ given by $\varphi(f) = f(10)$ is a ring homomorphism, since

$$\varphi(f + g) = (f + g)(10) = f(10) + g(10) = \varphi(f) + \varphi(g)$$

and

$$\varphi(f \cdot g) = (f \cdot g)(10) = f(10) \cdot g(10) = \varphi(f) \cdot \varphi(g).$$

More generally, for any set S and ring R we can make FUN(S, R) into a ring just as we made FUN($\mathbb{R}$, $\mathbb{R}$) into one. Each s in S gives rise to an **evaluation homomorphism** φ from FUN(S, R) to R defined by $\varphi(f) = f(s)$ for all f in FUN(S, R).

(d) We can think of POLY($\mathbb{R}$) as a subring of FUN($\mathbb{R}$, $\mathbb{R}$), since each polynomial defines a unique function and since one can show that two different polynomials must give different functions [i.e., have different graphs]. An evaluation homomorphism such as $f \to f(73)$ from FUN($\mathbb{R}$, $\mathbb{R}$) to $\mathbb{R}$ yields a homomorphism from POLY($\mathbb{R}$) to $\mathbb{R}$, defined in this instance by $p \to p(73)$. The evaluation homomorphism $p \to p(0)$ assigns to each polynomial p its constant coefficient a_0.

Evaluation is important in the design of algorithms for fast multiplication of very large integers. To see the idea, observe that evaluation at 10 gives the correspondences

$$3x^4 + 2x^3 + x + 1 \quad \longrightarrow \quad 32011$$

and

$$2x^2 + 7x + 8 \quad \longrightarrow \quad 278.$$

To multiply 32011 by 278 we could multiply the two polynomials to get $(3x^4 + 2x^3 + x + 1) \cdot (2x^2 + 7x + 8) = p(x)$ and then evaluate $p(10)$. [See Exercise 4 for another example.] This looks like the hard way to multiply. The key fact, though, is that the coefficients are small, and the Fast

Fourier Transform gives a very fast way to multiply polynomials of high degrees. ∎

Since ring homomorphisms are additive group homomorphisms, they have kernels. Suppose that φ is a homomorphism from $(R, +, \cdot)$ to $(S, +, \cdot)$. The kernel of φ is the set $K = \{a \in R : \varphi(a) = \varphi(0)\}$, and for $a, b \in R$ we have

$$\varphi(a) = \varphi(b) \quad \Leftrightarrow \quad a - b \in K \quad \Leftrightarrow \quad a + K = b + K.$$

Everything is in additive dress here, so $a + K$ is the coset $\{a + k : k \in K\}$ of the subgroup K of $(R, +)$. As before, φ is one-to-one if and only if $K = \{0\}$. The kernel of φ is a special kind of subring of R. It is an additive subgroup of course, but it is also closed under multiplication not only by its own elements but even under multiplication by other elements in R:

$$a \in K \quad \text{and} \quad r \in R \quad \text{imply} \quad a \cdot r, r \cdot a \in K.$$

The reason is that $\varphi(a) = \varphi(0)$, and if $r \in R$, then

$$\varphi(a \cdot r) = \varphi(a) \cdot \varphi(r) = \varphi(0) \cdot \varphi(r) = \varphi(0 \cdot r) = \varphi(0)$$

and likewise $\varphi(r \cdot a) = \varphi(0)$.

An additive subgroup I of a ring $(R, +, \cdot)$ is called an **ideal** of R if $r \cdot a \in I$ and $a \cdot r \in I$ for all $a \in I$ and $r \in R$. Kernels of ring homomorphisms are ideals, as we noted in the preceding paragraph, and one can show [Exercise 10] that every ideal is the kernel of a homomorphism.

EXAMPLE 6   (a) If the ring R is commutative and if $a \in R$, then the set $R \cdot a = \{r \cdot a : r \in R\}$ is an ideal of R. To check this, we observe that $r \cdot a + s \cdot a = (r + s) \cdot a \in R \cdot a$ and $-(r \cdot a) = (-r) \cdot a \in R \cdot a$ for every $r, s \in R$, so $R \cdot a$ is an additive subgroup of R. Since $s \cdot (r \cdot a) = (s \cdot r) \cdot a \in R \cdot a$ for $r, s \in R$, $R \cdot a$ is closed under multiplication by elements of R. An ideal of the form $R \cdot a$ is called a **principal ideal**.

All of the subgroups of $(\mathbb{Z}, +)$ are of the form $n\mathbb{Z}$ by Theorem 3 of §12.5, so every ideal of $(\mathbb{Z}, +, \cdot)$ is principal. So are the ideals of POLY($\mathbb{R}$), as it turns out, but such a situation is very special. For example, in the commutative ring POLY($\mathbb{Z}$) consisting of polynomials with integer coefficients, the set of all polynomials $a_n x^n + \cdots + a_1 x + a_0$ in which a_0 is even is an ideal that is not principal [Exercise 17].

(b) Ideals of fields are boring. Suppose that I is a non-0 ideal of a field F. Let $0 \neq a \in I$. For every $b \in F$ we have $b = (b \cdot a^{-1}) \cdot a \in F \cdot a \subseteq I$, so $I = F$. That is, F has only the obvious ideals $\{0\}$ and F. ∎

If R is a ring with ideal I, then the group R/I consisting of additive cosets $r + I = \{r + i : i \in I\}$ can be made into a ring in a natural way. We define

$$(r + I) + (s + I) = (r + s) + I$$

and

$$(r + I) \cdot (s + I) = r \cdot s + I.$$

We have already seen in Theorem 2(b) of § 12.6 that the addition on R/I is well-defined; we check multiplication. If $r + I = r' + I$ and $s + I = s' + I$, then $r - r'$ and $s - s'$ are in I and hence

$$r \cdot s - r' \cdot s' = r \cdot s - r \cdot s' + r \cdot s' - r' \cdot s'$$

$$= r \cdot (s - s') + (r - r') \cdot s' \in r \cdot I + I \cdot s' \subseteq I.$$

Thus $r \cdot s + I = r' \cdot s' + I$ and our definition of product is independent of the choice of representatives we take in the cosets $r + I$ and $s + I$. The rest of the properties of a ring are easy to check.

The fundamental homomorphism theorem for groups leads to a corresponding result for rings. Consider a ring homomorphism φ from R to S with kernel I. Then φ is an additive group homomorphism of $(R, +)$ into $(S, +)$, so we already know from the Fundamental Homomorphism Theorem in § 12.6 that the mapping φ^* from R/I to $\varphi(R)$ defined by $\varphi^*(r + I) = \varphi(r)$ for $r \in R$ is a group isomorphism. Since

$$\varphi^*((r + I) \cdot (r' + I)) = \varphi^*((r \cdot r') + I) = \varphi(r \cdot r')$$

$$= \varphi(r) \cdot \varphi(r') = \varphi^*(r + I) \cdot \varphi^*(r' + I),$$

φ^* is, in fact, a ring homomorphism. Therefore, φ^* is a **ring isomorphism** between R/I and $\varphi(R)$, i.e., a ring homomorphism that is one-to-one and onto. We have shown the following.

Theorem 1   Let φ be a ring homomorphism with kernel I from the ring R to the ring S. Then the mapping $r + I \rightarrow \varphi(r)$ is an isomorphism of the ring R/I onto $\varphi(R)$.

EXAMPLE 7   (a) The ring homomorphism $n \rightarrow n \bmod 2$ of $\mathbb{Z}$ onto $\mathbb{Z}(2)$ has kernel $2\mathbb{Z} = $ EVEN. The ring $\mathbb{Z}/2\mathbb{Z}$ has just two members, EVEN and $2\mathbb{Z} + 1 = $ ODD, with operations as shown in Figure 1. The ring isomorphism from $\mathbb{Z}/2\mathbb{Z}$ to $\mathbb{Z}(2)$ is EVEN $\rightarrow 0$, ODD $\rightarrow 1$.

+	EVEN	ODD		$\cdot$	EVEN	ODD
EVEN	EVEN	ODD		EVEN	EVEN	EVEN
ODD	ODD	EVEN		ODD	EVEN	ODD

FIGURE 1

(b) The evaluation homomorphism $p(x) \rightarrow p(0)$ maps POLY($\mathbb{R}$) onto $\mathbb{R}$. Its kernel is the set of polynomials with constant coefficient 0, i.e., the principal ideal $x \cdot$ POLY($\mathbb{R}$) of all multiples of the polynomial x. The cosets are of the form $r + x \cdot$ POLY($\mathbb{R}$) for constants r in $\mathbb{R}$. The isomorphism from POLY($\mathbb{R}$)/($x \cdot$ POLY($\mathbb{R}$)) to $\mathbb{R}$ is simply

$$r + x \cdot \text{POLY}(\mathbb{R}) \rightarrow r.$$

(c) More generally, the kernel of the evaluation map $p(x) \rightarrow p(a)$ is the principal ideal $(x - a) \cdot$ POLY($\mathbb{R}$) of multiples of $x - a$ [Exercise 5]. Its cosets are of the form $r + (x - a) \cdot$ POLY($\mathbb{R}$) and the isomorphism from POLY($\mathbb{R}$)/(($x - a) \cdot$ POLY($\mathbb{R}$)) onto $\mathbb{R}$ is

$$r + (x - a) \cdot \text{POLY}(\mathbb{R}) \rightarrow r. \quad \blacksquare$$

EXAMPLE 8   We can add another operation to Example 8(b) of §12.6 and make $\mathbb{Z}(2) \times \mathbb{Z}(3)$ into a ring by defining

$$(m, n) + (j, k) = (m +_2 j, n +_3 k)$$

and

$$(m, n) \cdot (j, k) = (m *_2 j, n *_3 k).$$

The mapping $\varphi: m \rightarrow (m \text{ MOD } 2, m \text{ MOD } 3)$ is a ring homomorphism from $\mathbb{Z}$ onto $\mathbb{Z}(2) \times \mathbb{Z}(3)$. Its kernel is

$$\{m \in \mathbb{Z} : m \equiv 0 \ (\text{mod } 2) \text{ and } m \equiv 0 \ (\text{mod } 3)\}$$

$$= \{m \in \mathbb{Z} : m \equiv 0 \ (\text{mod } 6)\} = 6\mathbb{Z}.$$

Thus by Theorem 1, the ring $\varphi(\mathbb{Z})$ is isomorphic to $\mathbb{Z}/6\mathbb{Z}$, i.e., is isomorphic to $\mathbb{Z}(6)$. That is,

$$\mathbb{Z}(2) \times \mathbb{Z}(3) \text{ is ring-isomorphic to } \mathbb{Z}(6). \quad \blacksquare$$

The ideas in Example 8 generalize. If $R_1, \ldots, R_n$ is a list of rings, not necessarily distinct, we can make the product $R_1 \times \cdots \times R_n$ into a ring by defining

$$(r_1, \ldots, r_n) + (s_1, \ldots, s_n) = (r_1 + s_1, \ldots, r_n + s_n)$$

and

$$(r_1, \ldots, r_n) \cdot (s_1, \ldots, s_n) = (r_1 \cdot s_1, \ldots, r_n \cdot s_n),$$

where the operations in the kth coordinate are the operations defined on the corresponding ring R_k. If $\varphi_1, \ldots, \varphi_n$ are homomorphisms from some ring R to $R_1, \ldots, R_n$, respectively, then one can check [Exercise 12] that the mapping φ from R to $R_1 \times \cdots \times R_n$ defined by $\varphi(r) = (\varphi_1(r), \ldots, \varphi_n(r))$ is a homomorphism. Its kernel is

$$\{r \in R : \varphi_k(r) = 0 \text{ for all } k = 1, \ldots, n\},$$

i.e., it is the intersection of the kernels of $\varphi_1, \ldots, \varphi_n$.

Suppose now that $I_1, \ldots, I_n$ are ideals of R, and that for $k = 1, \ldots, n$ each φ_k is the natural homomorphism from R onto R/I_k given by $\varphi_k(r) = r + I_k$. Then the homomorphism φ described in the preceding paragraph is defined by $\varphi(r) = (r + I_1, \ldots, r + I_n)$ for $r \in R$. Since I_k is the kernel of φ_k, we obtain the following.

Theorem 2

> Let R be a ring with ideals $I_1, \ldots, I_n$. Then $I_1 \cap \cdots \cap I_n$ is an ideal of R, and $R/(I_1 \cap \cdots \cap I_n)$ is isomorphic to a subring of $(R/I_1) \times \cdots \times (R/I_n)$.

In Example 8, with $I_1 = 2\mathbb{Z}$ and $I_2 = 3\mathbb{Z}$, the ring $\mathbb{Z}/(2\mathbb{Z} \cap 3\mathbb{Z})$ was isomorphic to the whole ring $(\mathbb{Z}/2\mathbb{Z}) \times (\mathbb{Z}/3\mathbb{Z})$, but in general $\varphi(R)$ is only a subring of $R_1 \times \cdots \times R_n$. For example, in $\mathbb{Z}$ we have $6\mathbb{Z} \cap 10\mathbb{Z} \cap 15\mathbb{Z} = 30\mathbb{Z}$ [check this]. Hence $\mathbb{Z}/(6\mathbb{Z} \cap 10\mathbb{Z} \cap 15\mathbb{Z}) = \mathbb{Z}/30\mathbb{Z}$ has 30 members, while $(\mathbb{Z}/6\mathbb{Z}) \times (\mathbb{Z}/10\mathbb{Z}) \times (\mathbb{Z}/15\mathbb{Z})$ has $6 \cdot 10 \cdot 15 = 900$ elements. Exercise 9 gives another example.

Corollary

> If $p_1, \ldots, p_n$ are distinct primes, then $\mathbb{Z}(p_1 \cdots p_n)$ is isomorphic to $\mathbb{Z}(p_1) \times \cdots \times \mathbb{Z}(p_n)$.

Proof. Consider the ideals $I_1 = p_1\mathbb{Z}, \ldots, I_n = p_n\mathbb{Z}$. The intersection $I_1 \cap \cdots \cap I_n$ is $p_1 \cdots p_n\mathbb{Z}$, because every number divisible by each p_k has to be divisible by their product. Now $\mathbb{Z}/I_k = \mathbb{Z}/p_k\mathbb{Z} \simeq \mathbb{Z}(p_k)$, so $\mathbb{Z}/I_1 \times \cdots \times \mathbb{Z}/I_n$ is isomorphic to $\mathbb{Z}(p_1) \times \cdots \times \mathbb{Z}(p_n)$, which has $p_1 \cdots p_n$ elements, just as $\mathbb{Z}/(p_1 \cdots p_n\mathbb{Z})$ does. By Theorem 2, $\mathbb{Z}/(p_1 \cdots p_n\mathbb{Z})$ is isomorphic to a subring of $\mathbb{Z}(p_1) \times \cdots \times \mathbb{Z}(p_n)$. Since the two rings have the same number of elements, and since $\mathbb{Z}(p_1 \cdots p_n) \simeq \mathbb{Z}/(p_1 \cdots p_n\mathbb{Z})$, the Corollary follows. ∎

EXAMPLE 9

The Corollary is a slightly special case of the **Chinese Remainder Theorem**, which may have been used to count an army in ancient China. Consider a small army, one known to have fewer than 1000 members. Observe that $7 \cdot 11 \cdot 13 = 1001$ and that the Corollary says that the numbers $0, 1, \ldots, 1000$ in $\mathbb{Z}(1001)$ correspond one-to-one with the triples in $\mathbb{Z}(7) \times \mathbb{Z}(11) \times \mathbb{Z}(13)$. The correspondence is

$$m \quad \longleftrightarrow \quad (m \text{ MOD } 7, m \text{ MOD } 11, m \text{ MOD } 13),$$

so if N is the army size and if we know $N \text{ MOD } 7$, $N \text{ MOD } 11$ and $N \text{ MOD } 13$, then in principle we have determined the value of N.

Ask the soldiers to group together in bunches of 7 and tell you how many soldiers are left over. Say it's 1. Then group them by 11's and find that there are, say, 2 left over. Finally, group them by 13's and find 8 left. This information is enough to determine N. It turns out to be 827.

One needs an algorithm to find N, or a staff of clerks. These days we would use a fast method based on the Euclidean Algorithm and the ideas in §4.6. We could also use primes a lot larger than 7, 11 and 13. A 9-decimal-digit prime p fits into a 32-bit computer word, which means that computation mod p can be done quickly. The Chinese Remainder Theorem lets us use three such large primes p, q, r to describe numbers up to $p \cdot q \cdot r - 1$, i.e., up to about 10^{30}. Ring theory lets us perform integer computations mod p, mod q and mod r, and then fit the results together at the end. The idea is rather like taking snapshots of an object from three directions to get a sense of the object as a whole. ∎

EXAMPLE 10   Polynomial interpolation can be viewed in the context of Theorem 2. Here is a simple illustration. Suppose that we want to find $p(x)$ in POLY($\mathbb{R}$) such that $p(1) = 5$, $p(4) = 8$ and $p(6) = 7$, and suppose we also ask that $p(x)$ have degree 2 or less, so that $p(x) = ax^2 + bx + c$ for some unknown coefficients a, b and c.

Evaluation at 1 is a homomorphism, call it φ_1, from POLY($\mathbb{R}$) onto $\mathbb{R}$. We are given $\varphi_1(p(x)) = p(1) = 5$. Evaluation at 4 and at 6 give two more homomorphisms, φ_4 and φ_6, with $\varphi_4(p(x)) = 8$ and $\varphi_6(p(x)) = 7$. As we noted in Example 7(c), the kernel of φ_1 is the ideal $I_1 = (x - 1) \cdot$ POLY($\mathbb{R}$), while φ_4 and φ_6 have kernels $I_4 = (x - 4) \cdot$ POLY($\mathbb{R}$) and $I_6 = (x - 6) \cdot$ POLY($\mathbb{R}$).

By Theorem 2 the mapping

$$p(x) + I_1 \cap I_4 \cap I_6 \quad \longrightarrow \quad (p(x) + I_1, p(x) + I_4, p(x) + I_6)$$

is an isomorphism of POLY($\mathbb{R}$)/$(I_1 \cap I_4 \cap I_6)$ into the product ring POLY($\mathbb{R}$)/$I_1 \times$ POLY($\mathbb{R}$)/$I_4 \times$ POLY($\mathbb{R}$)/I_6, which is itself isomorphic to $\mathbb{R} \times \mathbb{R} \times \mathbb{R}$. The composite mapping from POLY($\mathbb{R}$)/$(I_1 \cap I_4 \cap I_6)$ to $\mathbb{R} \times \mathbb{R} \times \mathbb{R}$ is

$$p(x) + I_1 \cap I_4 \cap I_6 \quad \longrightarrow \quad (p(1), p(4), p(6)).$$

The interpolation problem is to find $p(x)$ to give the right image $(p(1), p(4), p(6))$. In our example, $(5, 8, 7)$ corresponds to $p(x) = -0.3x^2 + 2.5x + 2.8$.

Fast algorithms to solve the Chinese remainder problem can be adapted to solve the polynomial interpolation problem. Interpolation, in turn, gives the following method for multiplying polynomials. To find $a(x) \cdot b(x)$, we can evaluate $a(r)$ and $b(r)$ at lots of r's, compute $a(r) \cdot b(r)$ in each case, and then interpolate to find the $p(x)$ so that $p(r) = a(r) \cdot b(r)$ for each r. Techniques such as this are at the heart of fast algorithms for computing with "infinite-precision" large integers. ∎

A great deal more can be said about rings and fields. We have only introduced the most basic ideas and a few examples, but we hope to have given some feeling for the kinds of questions it might be reasonable to ask about these systems and the kinds of answers one might get. The study of groups, rings and fields makes up a large part of the area of mathematics called abstract algebra. At this point you are in a good position to read an introductory book in this area.

EXERCISES 12.8

In these exercises, the words "homomorphism" and "isomorphism" mean "ring homomorphism" and "ring isomorphism."

1. Which of the following sets are subrings of $(\mathbb{R}, +, \cdot)$?

(a) $2\mathbb{Z}$ (b) $2\mathbb{R}$

(c) $\mathbb{N}$ (d) $\{m + n\sqrt{2} : m, n \in \mathbb{Z}\}$

(e) $\{m/2 : m \in \mathbb{Z}\}$ (f) $\{m/2^a : m \in \mathbb{Z}, a \in \mathbb{P}\}$

2. (a) For each subset in Exercise 1 that is a subring of $\mathbb{R}$ verify closure under addition and multiplication.

(b) For each subset in Exercise 1 that is not a subring of $\mathbb{R}$ give a property of subrings that the subset does not satisfy.

3. Which of the following functions are homomorphisms? Justify your answer in each case.

(a) φ: FUN$(\mathbb{R}, \mathbb{R}) \to \mathbb{R}$ defined by $\varphi(f) = f(0)$.

(b) φ: $\mathbb{R} \to \mathbb{R}$ defined by $\varphi(r) = r^2$.

(c) φ: $\mathbb{R} \to$ FUN$(\mathbb{R}, \mathbb{R})$ defined by $(\varphi(r))(x) = r$. That is, $\varphi(r)$ is the constant function on $\mathbb{R}$ having value r at every x.

(d) φ: $\mathbb{Z} \to \mathbb{R}$ defined by $\varphi(n) = n$.

(e) φ: $\mathbb{Z}/3\mathbb{Z} \to \mathbb{Z}/6\mathbb{Z}$ defined by $\varphi(n + 3\mathbb{Z}) = 2n + 6\mathbb{Z}$.

4. Let $p(x) = 8x^4 + 3x^3 + x + 7$ and $q(x) = x^4 + x^2 + x + 1$.

(a) Evaluate $p(10)$ and $q(10)$.

(b) Compute the product 83017×10111 the hard way, by finding $p(x) \cdot q(x) = r(x)$ and evaluating $r(10)$. Notice the problem with "carries."

5. (a) Show that if $p(x) = q(x) \cdot (x - a) + b$ in POLY$(\mathbb{R})$ with a and b in $\mathbb{R}$, then $b = p(a)$.

(b) Show that if

$$q_k(x) = \sum_{i=1}^{k} a^{i-1} x^{k-i} = x^{k-1} + ax^{k-2} + \cdots + a^{k-2}x + a^{k-1},$$

then $x^k - a^k = q_k(x) \cdot (x - a)$.

(c) Use part (b) to show that if

$$p(x) = \sum_{k=0}^{n} c_k x^k \in \text{POLY}(\mathbb{R}) \text{ and } a \in \mathbb{R},$$

then $p(x) = q(x) \cdot (x - a) + p(a)$ for some $q(x) \in$ POLY($\mathbb{R}$). That is, prove the Remainder Theorem from algebra.

(d) Show that the kernel of the evaluation mapping $p(x) \to p(a)$ from POLY($\mathbb{R}$) to $\mathbb{R}$ is the ideal $(x - a) \cdot$ POLY($\mathbb{R}$) consisting of all multiples of $x - a$.

6. Find the kernel of the homomorphism φ from FUN($\mathbb{R}, \mathbb{R}$) to $\mathbb{R}$ in Example 5(c).

7. Consider the ring $\mathbb{Z}$. Write each of the following in the form $n\mathbb{Z}$ with $n \in \mathbb{N}$.

(a) $6\mathbb{Z} \cap 8\mathbb{Z}$ (b) $6\mathbb{Z} + 8\mathbb{Z}$

(c) $3\mathbb{Z} + 2\mathbb{Z}$ (d) $6\mathbb{Z} + 10\mathbb{Z} + 15\mathbb{Z}$

8. Show that in a ring $(-a) \cdot b = -(a \cdot b) = a \cdot (-b)$ for every a and b.

9. (a) Verify that the mapping φ from $\mathbb{Z}(12)$ to $\mathbb{Z}(4) \times \mathbb{Z}(6)$ given by $\varphi(m) = (m \text{ MOD } 4, m \text{ MOD } 6)$ is a well-defined homomorphism.

(b) Find the kernel of φ.

(c) Find an element of $\mathbb{Z}(4) \times \mathbb{Z}(6)$ that is not in the image of φ.

(d) Which elements in $\mathbb{Z}(12)$ are mapped to $(1, 3)$ by φ?

10. Let I be an ideal of a ring R.

(a) Show that the mapping $r \to r + I$ is a homomorphism of R onto R/I.

(b) Find the kernel of this homomorphism.

11. (a) Show that $(\mathbb{Z}(6), +_6, *_6)$ is not a field.

(b) Show that $(\mathbb{Z}(5), +_5, *_5)$ is a field.

(c) Show that if F and K are fields, then $F \times K$ is *not* a field.

12. If $R_1, \ldots, R_n$ are rings and if $\varphi_1, \ldots, \varphi_n$ are homomorphisms from a ring R into $R_1, \ldots, R_n$, respectively, then the mapping φ defined by $\varphi(r) = (\varphi_1(r), \ldots, \varphi_n(r))$ is a homomorphism of R into $R_1 \times \cdots \times R_n$. Verify this fact for $n = 2$.

13. (a) Show that if φ is a homomorphism from a field F to a ring R, then either $\varphi(a) = 0$ for all $a \in F$ or φ is one-to-one.

(b) Show that if I is an ideal of the ring R and if θ is a homomorphism from R onto a ring S, then $\theta(I)$ is an ideal of S.

(c) Show that if S is a field in part (b), then either $\theta(I) = S$ or $\theta(I) = \{0\}$.

14. Let R be a commutative ring with identity.

(a) Show that R is an integral domain if and only if the mapping $r \to a \cdot r$ from R to R is one-to-one for each non-0 a in R.

(b) Show that R is a field if and only if this mapping is a one-to-one correspondence of R onto itself for each non-0 a in R.

(c) Show that every finite integral domain is a field.

15. (a) Find an ideal I of $\mathbb{Z}$ for which $\mathbb{Z}/I$ is isomorphic to $\mathbb{Z}(3) \times \mathbb{Z}(5)$.

(b) Describe an isomorphism between $\mathbb{Z}(12)$ and $\mathbb{Z}(3) \times \mathbb{Z}(4)$.

16. Let p be a prime. This exercise shows that the commutative ring $\mathbb{Z}(p)$ is a field.

    (a) Show that if $0 < k < p$, then there are integers s and t with $k \cdot s + p \cdot t = 1$.

    (b) With notation as in part (a), show that s MOD p is the multiplicative inverse of k in $\mathbb{Z}(p)$. [We have a fast algorithm to compute s and t, so we can compute effectively in $\mathbb{Z}(p)$, even if p is very large.]

17. Consider the ring $R = $ POLY$(\mathbb{Z})$ of polynomials in x with integer coefficients.

    (a) Describe the members of the ideals $R \cdot 2$, $R \cdot x$ and $R \cdot 2 + R \cdot x$.

    (b) Show that there is no polynomial p in R for which $R \cdot p = R \cdot 2 + R \cdot x$.

18. The set $\mathbb{B} \times \mathbb{B}$ can be made into a ring in another way besides the one described in the text. Define $+$ and $\cdot$ by the tables:

$+$	(0, 0)	(1, 0)	(0, 1)	(1, 1)
(0, 0)	(0, 0)	(1, 0)	(0, 1)	(1, 1)
(1, 0)	(1, 0)	(0, 0)	(1, 1)	(0, 1)
(0, 1)	(0, 1)	(1, 1)	(0, 0)	(1, 0)
(1, 1)	(1, 1)	(0, 1)	(1, 0)	(0, 0)

$\cdot$	(0, 0)	(1, 0)	(0, 1)	(1, 1)
(0, 0)	(0, 0)	(0, 0)	(0, 0)	(0, 0)
(1, 0)	(0, 0)	(1, 0)	(0, 1)	(1, 1)
(0, 1)	(0, 0)	(0, 1)	(1, 1)	(1, 0)
(1, 1)	(0, 0)	(1, 1)	(1, 0)	(0, 1)

Verify that $\mathbb{B} \times \mathbb{B}$ is an additive group and that the non-0 elements form a group under multiplication. *Suggestion:* Save work by exhibiting known groups isomorphic to your alleged groups. [Finite fields such as this are important in algebraic coding to minimize the effects of noise on transmission channels.]

CHAPTER HIGHLIGHTS

As usual: What does it mean? Why is it here? How can I use it? Think of examples. This chapter covers a lot of ideas. The main themes are subsystems, homomorphisms and actions. As you review, try to see how individual topics connect with these themes.

CONCEPTS AND NOTATION

permutation, permutation group, symmetric group S_n

   cycle, disjoint cycles, m-cycle

   group acting on a set X, on FUN(X, C)

   orbit

   fix, FIX$_G(x)$, FIX$_X(g)$

   restriction

applications of permutation groups

   digraph or graph automorphism, AUT(D), AUT(H)

   coloring, equivalent colorings

algebraic system
  semigroup, monoid, group
    identity, inverse, g^{-1}, $-g$
  ring, integral domain, field
    zero, identity
  subgroup [subring, etc.], proper, trivial
    subgroup $\langle A \rangle$ generated by A
    subsemigroup A^{+} generated by A
  cyclic group, order of an element
homomorphism, isomorphism, $\simeq$
  of groups, semigroups, rings
  kernel
    normal subgroup, ideal
    principal ideal
coset, gH, Hg, $g + H$
natural operation on G/K, natural operations on R/I
natural homomorphism $\nu: G \to G/K$ or $R \to R/I$

FACTS ABOUT ACTIONS

Every permutation in S_n is the product of disjoint cycles; its order is the lcm of the cycle lengths.

The G-orbits partition a set on which G acts.

The entries in the disjoint cycles of a permutation g are the $\langle g \rangle$-orbits.

$|G| = |Gx| \cdot |\text{FIX}_G(x)|$ for every x in the set X on which the group G acts.

The number of G-orbits in X is the average number of elements of X fixed by members of G.

The number of orbits of G on $\text{FUN}(X, C)$ is

$$\frac{1}{|G|} \sum_{g \in G} |C|^{m(g)},$$

where $m(g)$ is the number of orbits of $\langle g \rangle$ on X.

The number of G-classes of colorings of X with k colors is

$$C(k) = \frac{1}{|G|} \sum_{g \in G} k^{m(g)}.$$

GENERAL ALGEBRAIC FACTS

Cancellation is legal in a group. Cancellation of non-0 factors is legal in an integral domain.

A finite subset of a group that is nonempty and closed under the group operation is a subgroup.

Intersections of subgroups [subsemigroups, subrings, normal subgroups, etc.] are subgroups [subsemigroups, etc.].

The subsemigroup [subgroup] generated by A consists of all products of members of A [and their inverses].

The subgroups of $(\mathbb{Z}, +)$ are the cyclic groups $n\mathbb{Z}$. They are also the ideals of $(\mathbb{Z}, +, \cdot)$.

The cosets of a subgroup partition a group into sets of equal size [only one of which is a subgroup].

Lagrange's Theorem: $|G| = |G/H| \cdot |H|$.

The order of an element of a finite group G divides $|G|$.

Normal subgroups are the kernels of group homomorphisms; ideals are the kernels of ring homomorphisms.

Homomorphisms take identities to identities and inverses to inverses.

Fundamental Theorem: If K is the kernel of a homomorphism φ on a group G, then $G/K \simeq \varphi(G)$. A similar statement is true for rings.

A group or ring homomorphism is one-to-one if and only if its kernel consists of just one element.

Every cyclic group is isomorphic to $(\mathbb{Z}, +)$ or to some $(\mathbb{Z}(p), +_p)$.

If $I_1, \ldots, I_n$ are ideals of R, then $R/(I_1 \cap \cdots \cap I_n)$ is isomorphic to a subring of $(R/I_1) \times \cdots \times (R/I_n)$.

If $p_1, \ldots, p_n$ are distinct primes, then

$$\mathbb{Z}(p_1 \cdots p_n) \simeq \mathbb{Z}(p_1) \times \cdots \times \mathbb{Z}(p_n)$$

[a special case of the Chinese Remainder Theorem].

The theory of rings applies to polynomial interpolation.

PREDICATE CALCULUS AND INFINITE SETS

13

This chapter contains topics that might well have been included in Chapters 2 and 5, but which we deferred until now in order not to break the flow of ideas in the core of the book. The informal introduction to ∀ and ∃ that we did give in Chapter 2 was adequate for everyday use, and even a moderately complete introduction to the predicate calculus would have been substantially more sophisticated than the level of the discussion at that stage. As it is, the account in the next two sections only begins to suggest the questions that arise in a formal development of the subject.

The reason for leaving infinite sets out of the chapter on counting is simply that the ideas and methods they require are completely different from those appropriate to finite sets. Finite intuition and infinite intuition should be kept quite separate.

§ 13.1 Quantifiers

The propositional calculus developed in Chapter 2 is a nice, complete, self-contained theory of logic, but it is totally inadequate for most of mathematics. The problem is that the propositional calculus does not allow the use of an infinite number of propositions, and the notation is even awkward for handling a large finite set of propositions. For example, we frequently encounter an infinite sequence of propositions $p(n)$ for n in $\mathbb{N}$. The informal statement "$p(n)$ is true for all n" means "$p(0)$ is true, $p(1)$ is true, $p(2)$ is true, etc." The only symbolism available from the propositional calculus would be something like $p(0) \wedge p(1) \wedge p(2) \wedge \cdots$, which is not acceptable. Similarly, the informal statement "$p(n)$ is true for some n" would correspond to the unacceptable $p(0) \vee p(1) \vee p(2) \vee \cdots$. To get around this sort of problem we will use the **quantifiers** ∀ and ∃ that were introduced in §2.1. We need to develop the rules for using the new symbols and for combining them with the old ones. The augmented system of symbols and rules is called the **predicate calculus**.

Recall that the quantifiers are applied to families $\{p(x) : x \in U\}$ of propositions; the nonempty set U is called the **universe of discourse** or **domain of discourse**. The compound proposition $\forall x \, p(x)$ is assigned truth values as follows:

> $\forall x \, p(x)$ is true if $p(x)$ is true for every x in U;
> otherwise, $\forall x \, p(x)$ is false.

The compound proposition $\exists x \, p(x)$ has these truth values:

> $\exists x \, p(x)$ is true if $p(x)$ is true for at least one x in U;
> $\exists x \, p(x)$ is false if $p(x)$ is false for every x in U.

EXAMPLE 1   Compound propositions are not completely defined unless the universe of discourse is specified. Thus $\forall x [x^2 \geq x]$ is ambiguous until we specify its universe of discourse, which might be $\mathbb{R}$, $[0, \infty)$ or some other subset of $\mathbb{R}$. In general, the truth value depends on the universe of discourse. For example, this proposition is true for the universe of discourse $[1, \infty)$ but false for $[0, \infty)$ and for $\mathbb{R}$. ∎

EXAMPLE 2   Occasionally, we are confronted with a constant function in algebra, such as $f(x) = 2$ for $x \in \mathbb{R}$, so that the variable x does not appear in the right side of the definition of f. Although the value of $f(x)$ does not depend on the choice of x, we nevertheless regard f as a function of x. Similarly, in logic we occasionally encounter propositions $p(x)$ whose truth values do not depend on the choice of x in U. As artificial examples, consider $p(n) = $ "2 is prime" and $q(n) = $ "16 is prime," with universe of discourse $\mathbb{N}$. Since all propositions $p(n)$ are true, $\exists n \, p(n)$ and $\forall n \, p(n)$ are both true. Since all propositions $q(n)$ are false, $\exists n \, q(n)$ and $\forall n \, q(n)$ are both false. Those propositions $p(x)$ whose truth values do not depend on x are essentially the ones we studied earlier in the propositional calculus. In a sense, the propositional calculus fills the same place in the predicate calculus that the constant functions fill in the study of all functions. ∎

Let's analyze the proposition $\forall x \, p(x)$ more closely. The expression $p(x)$ is called a **predicate**. In ordinary grammatical usage a predicate is the part of a sentence that says something about the subject of the sentence. For example, "_____ went to the moon" and "_____ is bigger than a bread box" are predicates. To make a sentence, we supply the subject. For instance, the predicate "_____ is bigger than a bread box" becomes the sentence "This book is bigger than a bread box" if we supply the subject "This book." If we call the predicate p, the sentence could be denoted p(This book). Each subject x yields a sentence $p(x)$.

A **predicate** in our symbolic logic setting is a function that produces a proposition whenever we feed it a member of the universe, that is, a proposition-valued function with domain U. We follow our usual practice and denote such a function by $p(x)$. The variable x in the expression $p(x)$ is called a **free variable** of the predicate. As x varies over U the truth value of $p(x)$ may vary. In contrast, the proposition $\forall x\, p(x)$ has a well-defined truth value that does not depend on x. The variable x in $\forall x\, p(x)$ is called a **bound variable**; it is bound by the quantifier $\forall$. Since $\forall x\, p(x)$ has a fixed meaning and truth value, it would be pointless and unnatural to quantify it again. That is, it would be pointless to introduce $\forall x[\forall x\, p(x)]$ and $\exists x[\forall x\, p(x)]$ since their truth values are the same as that of $\forall x\, p(x)$.

We can also consider predicates that are functions of more than one variable, perhaps from more than one universe of discourse, and in such cases multiple use of quantifiers is natural.

EXAMPLE 3   (a) Let $\mathbb{N}$ be the universe of discourse, and for each m and n in $\mathbb{N}$ let $p(m, n)$ be the proposition "$m < n$." We could think of these propositions as being labeled by $\mathbb{N} \times \mathbb{N}$ and then think of $\mathbb{N} \times \mathbb{N}$ as the universe of discourse, but for the present we prefer to treat the variables m and n separately. Both variables m and n are free, in the sense that the meanings and truth values of $p(m, n)$ vary with both m and n. In the expression $\exists m\, p(m, n)$, the variable m is bound but the variable n is free. The proposition $\exists m\, p(m, n)$ reads "there is an m in $\mathbb{N}$ with $m < n$," so $\exists m\, p(m, 0)$ is false, $\exists m\, p(m, 1)$ is true, $\exists m\, p(m, 2)$ is true, etc. For each choice of n the proposition $\exists m\, p(m, n)$ is either true or false; its truth value does not depend on m but depends on n alone. It is meaningful to quantify $\exists m\, p(m, n)$ with respect to the free variable n to obtain $\forall n[\exists m\, p(m, n)]$ and $\exists n[\exists m\, p(m, n)]$. The proposition $\forall n[\exists m\, p(m, n)]$ is false because $\exists m\, p(m, 0)$ is false and the proposition $\exists n[\exists m\, p(m, n)]$ is true because, for example, $\exists m\, p(m, 1)$ is true. Henceforth we will usually omit the brackets [   ] and write $\forall n\, \exists m\, p(m, n)$ and $\exists n\, \exists m\, p(m, n)$.

There are eight ways to apply the two quantifiers to the two variables: $\forall m\, \forall n$, $\forall n\, \forall m$, $\exists m\, \exists n$, $\exists n\, \exists m$, $\forall m\, \exists n$, $\exists n\, \forall m$, $\forall n\, \exists m$, $\exists m\, \forall n$. The first two turn out to be logically equivalent, in a precise sense that we will define in §13.2, since they have the same meaning as $\forall (m, n)\, p(m, n)$ where (m, n) varies over the new universe of discourse $\mathbb{N} \times \mathbb{N}$. Similarly, $\exists m\, \exists n\, p(m, n)$ and $\exists n\, \exists m\, p(m, n)$ are logically equivalent. The remaining four must be approached carefully. In the case of our last example, we already observed that $\forall n\, \exists m\, p(m, n)$ is false. No matter what m is, $p(m, 0)$ is false, so $\forall n\, p(m, n)$ is false, and therefore $\exists m\, \forall n\, p(m, n)$ is also false. To analyze $\forall m\, \exists n\, p(m, n)$, note that for each m, $\exists n\, p(m, n)$ is true because $p(m, m + 1)$ is true. Therefore $\forall m\, \exists n\, p(m, n)$ is also true. To analyze $\exists n\, \forall m\, p(m, n)$, note that for each n $\forall m\, p(m, n)$ is false because, for example, $p(n, n)$ is false. Therefore, $\exists n\, \forall m\, p(m, n)$ is false. We repeat:

for this example, $\forall m\, \exists n\, p(m, n)$ is true, while $\exists n\, \forall m\, p(m, n)$ is false.

The left proposition asserts, correctly, that for every m, there is a bigger n. The right proposition asserts, incorrectly, that there is an n bigger than all m.

(b) Here is a less mathematical example illustrating the importance of order when using both quantifiers $\forall$ and $\exists$. Let the universe of discourse consist of all people and consider

$$p(x, y) = \text{``}y \text{ is a mother of } x.\text{''}$$

Then $\forall x \exists y \, p(x, y)$ asserts that everyone has a mother, which is true. On the other hand, $\exists y \forall x \, p(x, y)$ asserts that some one person is the mother of everyone, which is false.

The proposition $\forall y \exists x \, p(x, y)$ asserts that everyone is a mother, and $\exists x \forall y \, p(x, y)$ asserts that someone has everyone for his or her mother. Clearly, these are both false statements. ∎

With these examples in mind we turn now to a more formal account. Let $U_1, U_2, \ldots, U_n$ be nonempty sets. An **n-place predicate** over $U_1 \times U_2 \times \cdots \times U_n$ is a function $p(x_1, x_2, \ldots, x_n)$ with domain $U_1 \times U_2 \times \cdots \times U_n$ that has propositions as its function values. The variables $x_1, x_2, \ldots, x_n$ for $p(x_1, x_2, \ldots, x_n)$ are all **free variables** for the predicate, and each x_j varies over the corresponding universe of discourse U_j. The term "free" is short for "free for substitution," meaning that the variable x_j is available in case we wish to substitute a particular value from U_j for all occurrences of x_j.

If we substitute a value for x_j, say for definiteness we substitute a for x_1 in $p(x_1, x_2, \ldots, x_n)$, we get the predicate $p(a, x_2, \ldots, x_n)$ that is free on the $n - 1$ remaining variables $x_2, \ldots, x_n$ but no longer free on x_1. An application of a quantifier $\forall x_j$ or $\exists x_j$ to a predicate $p(x_1, x_2, \ldots, x_n)$ gives a predicate $\forall x_j \, p(x_1, x_2, \ldots, x_n)$ or $\exists x_j \, p(x_1, x_2, \ldots, x_n)$ whose value depends only on the values of the remaining $n - 1$ variables other than x_j. We say the quantifier **binds** the variable x_j, making x_j a **bound variable** for the predicate. Application of n quantifiers, one for each variable, makes all variables bound and yields a proposition whose truth value can be determined by applying to the universes $U_1, U_2, \ldots, U_n$ the rules for $\forall x$ and $\exists x$ specified prior to Example 1.

EXAMPLE 4 (a) Consider the proposition

(1) $$\forall m \, \exists n [n > 2^m];$$

here $p(m, n) = \text{``}n > 2^m\text{''}$ is a 2-place predicate over $\mathbb{N} \times \mathbb{N}$. That is, m and n are both allowed to vary over $\mathbb{N}$. Recall our bracket-dropping convention that (1) represents

$$\forall m [\exists n [n > 2^m]].$$

Both variables m and n are bound. To decide the truth value of (1), we consider the inside expression $\exists n [n > 2^m]$ in which n is a bound variable and m is a free variable. We mentally fix the free variable m and note that the proposition "$n > 2^m$" is true for some choices of n in $\mathbb{N}$, for example $n = 2^m + 1$. It follows that $\exists n[n > 2^m]$ is true. This thought process is valid for each m in $\mathbb{N}$, so we conclude that $\exists n[n > 2^m]$ is true for *all* m. That is, (1) is true.

If we reverse the quantifiers in (1), we obtain

(2) $\exists n \, \forall m[n > 2^m]$.

This is false because $\forall m[n > 2^m]$ is false for each n, since "$n > 2^m$" is false for $m = n$.

(b) Consider the propositions

(3) $\forall x \, \exists y[x + y = 0]$,

(4) $\exists y \, \forall x[x + y = 0]$,

(5) $\forall x \, \exists y[xy = 0]$,

(6) $\exists y \, \forall x[xy = 0]$,

where each universe of discourse is $\mathbb{R}$.

To analyze (3) we consider a fixed x. Then $\exists y[x + y = 0]$ is true, because the choice $y = -x$ makes "$x + y = 0$" true. That is, $\exists y[x + y = 0]$ is true for all x, so (3) is true.

To analyze (4) we consider a fixed y. Then $\forall x[x + y = 0]$ is not true, because the choice $x = 1 - y$ makes "$x + y = 0$" false. That is, for each y, $\forall x[x + y = 0]$ is false, so (4) is false.

Proposition (5) is true, because $\exists y[xy = 0]$ is true for all x. In fact, the choice $y = 0$ makes "$xy = 0$" true.

To deal with (6) we analyze $\forall x[xy = 0]$. If $y = 0$, this proposition is clearly true. Since $\forall x[xy = 0]$ is true for some y, namely $y = 0$, the proposition (6) is true.

In the next section we will see that the truth of (6) implies the truth of (5) on purely logical grounds; that is,

$$\exists y \, \forall x \, p(x, y) \rightarrow \forall x \, \exists y \, p(x, y)$$

is always true. ∎

We have already noted that an n-place predicate becomes an $(n - 1)$-place predicate when we bind one of the variables with a quantifier. Its truth value depends on the truth values of the remaining $n - 1$ free variables, and in particular doesn't depend on what name we choose to call the bound variable. Thus if $p(x)$ is a 1-place predicate with universe of discourse U, then $\forall x \, p(x)$, $\forall y \, p(y)$ and $\forall t \, p(t)$ all have the same truth value, namely true if $p(u)$ is true for every u in U and false otherwise. Similarly, if $q(x, y)$ is a 2-place predicate with universes U and V, then $\exists y \, q(x, y)$,

$\exists t \, q(x, t)$ and $\exists s \, q(x, s)$ all describe the same 1-place predicate, namely the predicate that has truth value true for a given x in U if and only if $q(x, v)$ is true for some v in the universe V in which the second variable lies. On the other hand, the predicate $\exists x \, q(x, x)$ is *not* the same as these last three. The difference is that the quantifier in this instance binds both of the free variables.

EXAMPLE 5   Let U and V be $\mathbb{N}$ and let $q(x, y) = $ "$x > y$." Then $\exists x \, q(x, y)$ is the 1-place predicate "some member of $\mathbb{N}$ is greater than y," and so is $\exists t \, q(t, y)$. The predicate $\exists y \, q(x, y)$ is the 1-place predicate "there is a member of $\mathbb{N}$ less than x," which is the same predicate as $\exists s \, q(x, s)$ and has the value true for $x > 0$ and false for $x = 0$. But $\exists x \, q(x, x)$ is the proposition "$x > x$ for some x," which is quite different and has the value false.   ■

EXERCISES 13.1

As in Chapter 2, the truth values "true" and "false" may be written as 1 and 0, respectively.

1. Determine the truth values of the following, where the universe of discourse is $\mathbb{N}$.

   (a) $\forall m \, \exists n [2n = m]$ (b) $\exists n \, \forall m [2m = n]$

   (c) $\forall m \, \exists n [2m = n]$ (d) $\exists n \, \forall m [2n = m]$

   (e) $\forall m \, \forall n [\neg \{2n = m\}]$

2. Determine the truth values of the following, where the universe of discourse is $\mathbb{R}$.

   (a) $\forall x \, \exists y [xy = 1]$ (b) $\exists y \, \forall x [xy = 1]$

   (c) $\exists x \, \exists y [xy = 1]$ (d) $\forall x \, \forall y [(x + y)^2 = x^2 + y^2]$

   (e) $\forall x \, \exists y [(x + y)^2 = x^2 + y^2]$ (f) $\exists y \, \forall x [(x + y)^2 = x^2 + y^2]$

   (g) $\exists x \, \exists y [(x + 2y = 4) \wedge (2x - y = 2)]$

   (h) $\exists x \, \exists y [x^2 + y^2 + 1 = 2xy]$

3. Write the following sentences in logical notation. Be sure to bind all variables. When using quantifiers, specify universes; use $\mathbb{R}$ if no universe is indicated.

   (a) If $x < y$ and $y < z$, then $x < z$.

   (b) For every $x > 0$, there exists an n in $\mathbb{N}$ such that $n > x$ and $x > 1/n$.

   (c) For every $m, n \in \mathbb{N}$, there exists $p \in \mathbb{N}$ such that $m < p$ and $p < n$.

   (d) There exists $u \in \mathbb{N}$ so that $un = n$ for all $n \in \mathbb{N}$.

   (e) For each $n \in \mathbb{N}$, there exists $m \in \mathbb{N}$ such that $m < n$.

   (f) For every $n \in \mathbb{N}$, there exists $m \in \mathbb{N}$ such that $2^m \le n$ and $n < 2^{m+1}$.

4. Determine the truth values of the propositions in Exercise 3.

5. Write the following sentences in logical notation; the universe of discourse is the set Σ^* of words using letters from a finite alphabet Σ.

   (a) If $w_1 w_2 = w_1 w_3$, then $w_2 = w_3$.

   (b) If $\text{length}(w) = 1$, then $w \in \Sigma$.

   (c) $w_1 w_2 = w_2 w_1$ for all $w_1, w_2 \in \Sigma^*$.

6. Determine the truth values of the propositions in Exercise 5.

7. Specify the free and bound variables in the following expressions.

   (a) $\forall x \, \exists z [\sin(x + y) = \cos(z - y)]$

   (b) $\exists x [xy = xz \to y = z]$

   (c) $\exists x \, \exists z [x^2 + z^2 = y]$

8. Consider the expression $x + y = y + x$.

   (a) Specify the free and bound variables in the expression.

   (b) Apply universal quantifiers over the universe $\mathbb{R}$ to get a proposition. Is the proposition true?

   (c) Apply existential quantifiers over the universe $\mathbb{R}$ to get a proposition. Is the proposition true?

9. Repeat Exercise 8 for the expression $(x - y)^2 = x^2 - y^2$.

10. Consider the proposition $\forall m \, \exists n [m + n = 7]$.

    (a) Is the proposition true for the universes of discourse $\mathbb{N}$?

    (b) Is the proposition true for the universes of discourse $\mathbb{Z}$?

11. Repeat Exercise 10 for $\forall n \, \exists m [m + 1 = n]$.

12. Consider the proposition $\forall x \, \exists y [(x^2 + 1)y = 1]$.

    (a) Is the proposition true for the universes of discourse $\mathbb{N}$?

    (b) Is the proposition true for the universes of discourse $\mathbb{Q}$?

    (c) Is the proposition true for the universes of discourse $\mathbb{R}$?

13. Another useful quantifier is $\exists!$ where $\exists! x \, p(x)$ is read "there exists a unique x such that $p(x)$." This compound proposition is assigned truth value true if $p(x)$ is true for exactly one value of x in the universe of discourse; otherwise, it is false. Write the following sentences in logical notation.

    (a) There is a unique x in $\mathbb{R}$ such that $x + y = y$ for all $y \in \mathbb{R}$.

    (b) The equation $x^2 = x$ has a unique solution.

    (c) Exactly one set is a subset of all sets in $\mathscr{P}(\mathbb{N})$.

    (d) If $f: A \to B$, then for each $a \in A$ there is exactly one $b \in B$ such that $f(a) = b$.

    (e) If $f: A \to B$ is a one-to-one function, then for each $b \in B$ there is exactly one $a \in A$ such that $f(a) = b$.

14. Determine the truth values of the propositions in Exercise 13.

15. In this problem $A = \{0, 2, 4, 6, 8, 10\}$ and the universe of discourse is $\mathbb{N}$. True or False.

    (a) A is the set of even integers in $\mathbb{N}$ less than 12.

    (b) $A = \{0, 2, 4, 6, \ldots\}$

    (c) $A = \{n \in \mathbb{N} : 2n < 24\}$

    (d) $A = \{n \in \mathbb{N} : \forall m [2m = n \rightarrow m < 6]\}$

    (e) $A = \{n \in \mathbb{N} : \forall m [2m = n \wedge m < 6]\}$

    (f) $A = \{n \in \mathbb{N} : \exists m [2m = n \rightarrow m < 6]\}$

    (g) $A = \{n \in \mathbb{N} : \exists m [2m = n \wedge m < 6]\}$

    (h) $A = \{n \in \mathbb{N} : \exists ! m [2m = n \wedge m < 6]\}$

    (i) $A = \{n \in \mathbb{N} : n \text{ is even and } n^2 \le 100\}$

    (j) $\forall n [n \in A \rightarrow n \le 10]$

    (k) $3 \in A \rightarrow 3 < 10$

    (l) $12 \in A \rightarrow 12 < 10$

    (m) $8 \in A \rightarrow 8 < 10$

16. With universe of discourse $\mathbb{N}$, let $p(n) = $ "n is prime" and $e(n) = $ "n is even." Write the following in ordinary English.

    (a) $\exists m \forall n [e(n) \wedge p(m + n)]$

    (b) $\forall n \exists m [\neg e(n) \rightarrow e(m + n)]$

    Translate the following into logical notation using p and e.

    (c) There are two prime integers whose sum is even.

    (d) If the sum of two primes is even, then neither of them equals 2.

    (e) The sum of two prime integers is odd.

17. Determine the truth values of the propositions in Exercise 16.

§ 13.2   Elementary Predicate Calculus

One of our aims in Chapter 2 was to learn how to tell when two compound propositions were logically equivalent, or when one proposition logically implied another. Questions like that become even more important, and also somewhat more difficult, when we allow ourselves to use quantifiers $\forall$ and $\exists$ to build up compound expressions. When we just had $\neg, \wedge, \vee, \rightarrow$ and $\leftrightarrow$ to deal with, we used truth tables to break our analysis into small steps. To get a similar breakdown for predicate logic we need to determine first exactly what kinds of expressions we have to handle and then learn the rules for how $\forall$ and $\exists$ interact with each other and with $\neg, \wedge, \vee, \rightarrow$ and $\leftrightarrow$.

The ideas of "proof" and "rule of inference" that we discussed in § 2.4 for the propositional calculus can also be extended to the predicate calculus setting. Not surprisingly, more possible expressions means more complications. A moderately thorough account of the subject would form a substantial part of another book. In this section we will just discuss some of the most basic and useful connections between quantifiers and logical operators.

EXAMPLE 1   When we hear the television advertiser say, "All cars are not created equal," we know that what he *really* means is "Not all cars are created equal." The distinction would be even more obvious if we were comparing "All cars are not yellow" with "Not all cars are yellow." The first of these two statements is false—there *are* some yellow cars—but the second is true. The statements have different truth values. Given a universe of cars, if $y(c)$ is the predicate "c is yellow," then the first statement is $\forall c(\neg y(c))$ while the second is $\neg(\forall c\, y(c))$.

These are simple examples of predicates, in this case propositions, built using the quantifier $\forall$ and the logical operator $\neg$. More complicated predicates come up frequently in debugging computer programs that have branching instructions, to see how the programs will behave with different kinds of input data.

Our simple predicates do not have the same truth value for the universe of cars and the particular interpretation "c is yellow" for $y(c)$. Could they be logically equivalent if we had a different universe or a different meaning for $y(c)$? This is one kind of general question that we plan to answer. See Example 4(b) below for a continuation of this discussion. ∎

To begin, we need to describe the logical expressions that we will be dealing with. In Chapter 2 we used the term "compound proposition" in an informal way to describe propositions built up out of simpler ones. In Exercise 15 of §7.2 we gave a recursive definition of wff's for the propositional calculus; these are the compound propositions of Chapter 2. In the same way, we will use a recursive definition in order to precisely define "compound propositions" for the predicate calculus. We will also look more closely at what we did in the propositional calculus.

In Chapter 2 we used logical operators to build up propositions from the symbols p, q, r, etc., which we considered to be names for other propositions. The symbols could be considered to be variables, and a compound proposition such as $p \wedge (\neg q)$ could be considered a function of the two variables p and q; its value is true or false, depending on the truth values of p and q and the form of the compound proposition. The truth table for the compound proposition is simply a list of the function values as the variables range independently over the set {true, false}. For the propositional calculus such variables are all that we need, but for the predicate calculus we must also consider variables associated with infinite universes of discourse, such as $\mathbb{R}$, $\mathbb{Z}$ and $\mathfrak{M}_{m,n}$.

Suppose that we have available a collection of nonempty universes of discourse with which the free variables of all predicates we consider are associated. We define the class of **compound predicates** as follows:

(B_1)   Logical variables are compound predicates.

(B_2) n-place predicates are compound predicates for $n \geq 1$.

(R_1)   If $p(x)$ and $q(x)$ are compound predicates with free variable x, then so are

$$\neg p(x), \quad (p(x) \vee q(x)), \quad (p(x) \wedge q(x)),$$

$$(p(x) \to q(x)) \quad \text{and} \quad (p(x) \leftrightarrow q(x)).$$

(R_2)   If $p(x)$ is a compound predicate with free variable x, then

$$(\forall x\, p(x)) \quad \text{and} \quad (\exists x\, p(x))$$

are compound predicates for which x is not a free variable.

When we write "$p(x)$" here we mean to indicate that the given compound predicate has x as one of its free variables. We admit the possibility that the truth value of $p(x)$ is actually independent of the choice of x. In particular, we can view a proposition as trivially having a free variable, so that $p(x)$ and $q(x)$ in (R_1) might just be propositions p and q. If we delete (B_2) and (R_2) and the reference to free variables in (R_1) we obtain the recursive description of wff's for the propositional calculus in Exercise 15 of §7.2. In our present setting there are compound predicates, besides those made from (B_1) and (R_1), that are propositions. If all of the variables in a compound predicate are bound, then the predicate is a proposition. We extend our definition and call a compound predicate with no free variables a **compound proposition**. For example,

$$((\exists x(\exists z\, p(x, z))) \to (\forall y(\neg r(y))))$$

is a compound proposition with no free variables. In contrast,

$$(p(x) \vee (\neg \forall y\, q(x, y)))$$

and

$$((\exists z\, p(x, z)) \to (\forall y(\neg r(y))))$$

are compound predicates with free variable x.

The number and the placement of parentheses in a compound predicate are explicitly prescribed by our recursive definition. In practice, for clarity, we may add or suppress some parentheses. For example, we may write $((\forall x\, p(x)) \to (\exists x\, p(x)))$ as $\forall x\, p(x) \to \exists x\, p(x)$ and we may write $(\exists x \neg p(x))$ as $\exists x(\neg p(x))$. We sometimes also use brackets or braces instead of parentheses.

The truth value of a compound proposition ordinarily depends on the choices of the universes of discourse that the bound variables are quantified over, but there are important instances in which the truth value not only does not depend on the universe choices but is in fact independent of the values of the logical variables as well. A compound proposition that has the value true for all universes of discourse and all values of its logical variables is called a **tautology**. This definition extends the usage in Chapter 2, where there were no universes to worry about.

EXAMPLE 2   (a) An important class of tautologies consists of the generalized DeMorgan laws; compare rules 8a to 8d in Table 1 of §2.2. These are

(1) $$\neg \forall x\, p(x) \leftrightarrow \exists x[\neg p(x)],$$

(2) $$\neg \exists x\, p(x) \leftrightarrow \forall x[\neg p(x)],$$

(3) $$\forall x\, p(x) \leftrightarrow \neg \exists x[\neg p(x)],$$

(4) $$\exists x\, p(x) \leftrightarrow \neg \forall x[\neg p(x)].$$

To see that (1) is a tautology, note that $\neg \forall x\, p(x)$ has truth value true exactly when $\forall x\, p(x)$ has truth value false, and this occurs whenever there exists an x in the universe of discourse such that $p(x)$ is false, i.e., such that $\neg p(x)$ is true. Thus $\neg \forall x\, p(x)$ is true precisely when $\exists x[\neg p(x)]$ is true. This argument does not rely on the choice of universe, so (1) is a tautology.

The DeMorgan law (2) can be analyzed in a similar way. Alternatively, we can derive (2) from (1) by substituting the 1-place predicate $\neg p(x)$ in place of $p(x)$ to obtain

$$\neg \forall x[\neg p(x)] \leftrightarrow \exists x[\neg \neg p(x)].$$

The substitution rules in §2.4 are still valid, so we may substitute $p(x)$ for $\neg \neg p(x)$ to obtain the equivalent expression

$$\neg \forall x[\neg p(x)] \leftrightarrow \exists x[p(x)].$$

This is DeMorgan's law (4) and, if we negate both sides, we obtain (2). An application of (2) to $\neg p(x)$ yields (3).

(b) Consider again the predicate $y(c) =$ "c is yellow," where c ranges over the universe of cars. DeMorgan law (1) tells us that

$$\neg (\forall c\, y(c)) \leftrightarrow \exists c(\neg y(c))$$

is a tautology. We conclude that $\neg (\forall c\, y(c))$ and $\exists c(\neg y(c))$ must have the same truth value on purely logical grounds; we do not need to consider the context of cars. In Example 1 we decided that $\neg (\forall c\, y(c))$, i.e., "Not all cars are yellow," is true, so $\exists c(\neg y(c))$ must also be true. Sure enough, it's true that "There exists a car that is not yellow." ∎

EXAMPLE 3   (a) The following compound predicate is true for every 2-place predicate $p(x, y)$:

(∗) $$\exists x\, \forall y\, p(x, y) \rightarrow \forall y\, \exists x\, p(x, y).$$

In other words, (∗) is a tautology. To see this, suppose that the left side, $\exists x \, \forall y \, p(x, y)$, of (∗) has truth value true. Then there exists an x_0 in the universe of discourse such that $\forall y \, p(x_0, y)$ is true, so $p(x_0, y)$ is true for all y. Thus for each y, $\exists x \, p(x, y)$ is true; in fact, the same x_0 works for each y. Since $\exists x \, p(x, y)$ is true for all y, the right side of (∗) has truth value true. Since the right side is true whenever the left side is, (∗) is a tautology.

(b) The converse of (∗), namely

$$\forall y \, \exists x \, p(x, y) \rightarrow \exists x \, \forall y \, p(x, y),$$

is not a tautology, as we noted in Example 3 of § 13.1. Here is another very simple example. Let $p(x, y)$ be the 2-place predicate "$x = y$" on the two-element universe $U = \{a, b\}$. Observe that $\exists x \, p(x, a)$ is true since $p(x, a)$ is true for $x = a$. Similarly, $\exists x \, p(x, b)$ is true, so $\forall y \, \exists x \, p(x, y)$ is true.

On the other hand, as noted in the proof of (∗), $\exists x \, \forall y \, p(x, y)$ is true only if $\forall y \, p(x_0, y)$ is true for some x_0. Since x_0 must be a or b, either $\forall y \, p(a, y)$ or $\forall y \, p(b, y)$ would be true. But $\forall y \, p(a, y)$ is false since $p(a, y)$ is false for $y = b$, and similarly, $\forall y \, p(b, y)$ is false. Thus, in this setting, the proposition $\forall y \, \exists x \, p(x, y) \rightarrow \exists x \, \forall y \, p(x, y)$ is false. Hence this compound proposition is not a tautology. ■

As in the propositional calculus, we say that two compound propositions P and Q are **logically equivalent**, and we write $P \Leftrightarrow Q$, in case $P \leftrightarrow Q$ is a tautology. Also, P **logically implies** Q provided that $P \rightarrow Q$ is a tautology, in which case we write $P \Rightarrow Q$. Table 1 lists some useful logical equivalences and implications. We begin numbering the rules with 35, since Chapter 2 contains rules 1 through 34.

In Examples 2 and 3 we discussed the tautologies corresponding to rules 37 and 36. The remaining rules are easy to verify.

**TABLE 1. Logical Relationships
in the Predicate Calculus**

35a.	$\forall x \, \forall y \, p(x, y) \Leftrightarrow \forall y \, \forall x \, p(x, y)$
b.	$\exists x \, \exists y \, p(x, y) \Leftrightarrow \exists y \, \exists x \, p(x, y)$
36.	$\exists x \, \forall y \, p(x, y) \Rightarrow \forall y \, \exists x \, p(x, y)$
37a.	$\neg \forall x \, p(x) \Leftrightarrow \exists x [\neg p(x)]$
b.	$\neg \exists x \, p(x) \Leftrightarrow \forall x [\neg p(x)]$
c.	$\forall x \, p(x) \Leftrightarrow \neg \exists x [\neg p(x)]$
d.	$\exists x \, p(x) \Leftrightarrow \neg \forall x [\neg p(x)]$
38.	$\forall x \, p(x) \Rightarrow \exists x \, p(x)$

Rules 37a–d are the DeMorgan Laws.

EXAMPLE 4   (a) To verify rule 35b, that is, to verify that

$$\exists x \, \exists y \, p(x, y) \leftrightarrow \exists y \, \exists x \, p(x, y)$$

is a tautology, we must check that this proposition has the value true for all possible universes of discourse. By the definition of $\leftrightarrow$, we need only check that $\exists x \, \exists y \, p(x, y)$ has the value true for a given universe if and only if $\exists y \, \exists x \, p(x, y)$ has the value true for that universe. Suppose that $\exists x \, \exists y \, p(x, y)$ is true. Then $\exists y \, p(x_0, y)$ is true for some x_0 in the universe, so $p(x_0, y_0)$ is true for some y_0 in the universe. Hence $\exists x \, p(x, y_0)$ is true and thus $\exists y \, \exists x \, p(x, y)$ is true. The implication in the other direction follows similarly. Moreover, both $\exists x \, \exists y \, p(x, y)$ and $\exists y \, \exists x \, p(x, y)$ are logically equivalent to the proposition $\exists (x, y) \, p(x, y)$ where (x, y) varies over $U_1 \times U_2$, with U_1 and U_2 the universes of discourse for the variables x and y.

(b) Rule 38, applied to $\neg p(x)$ in place of $p(x)$, gives

$$\forall x \, \neg p(x) \Rightarrow \exists x \, \neg p(x).$$

Then the DeMorgan Law 37a applied to $\exists x \, \neg p(x)$ gives

$$\forall x \, \neg p(x) \Rightarrow \neg \forall x \, p(x).$$

The reverse implication is, of course, false, as we saw in Example 1. ■

The DeMorgan laws 37a to 37d can be used repeatedly to negate any quantified proposition. For example,

$$\neg \exists w \, \forall x \, \exists y \, \exists z \, p(w, x, y, z)$$

is successively logically equivalent to

$$\forall w [\neg \forall x \, \exists y \, \exists z \, p(w, x, y, z)]$$

$$\forall w \, \exists x [\neg \exists y \, \exists z \, p(w, x, y, z)]$$

$$\forall w \, \exists x \, \forall y [\neg \exists z \, p(w, x, y, z)]$$

$$\forall w \, \exists x \, \forall y \, \forall z [\neg p(w, x, y, z)].$$

This example illustrates the general rule: The negation of a quantified predicate is logically equivalent to the proposition obtained by replacing each $\forall$ by $\exists$, replacing each $\exists$ by $\forall$, and by replacing the predicate itself by its negation.

EXAMPLE 5   (a) The proposition

(1) $$\forall x \, \forall y \, \exists z [x < z < y]$$

states that for every x and y there is a z between them. Its truth value, which we really don't care about here, depends on the choices of the universes for x, y and z [see Exercise 14]. The negation of (1) is

$$\exists x \, \exists y \, \forall z \{\neg [x < z < y]\}.$$

Since "$x < z < y$" means "$(x < z) \wedge (z < y)$," we can apply an elementary DeMorgan Law from Chapter 2 to get

$$\neg[x < z < y] \Leftrightarrow \neg(x < z) \vee \neg(z < y) \Leftrightarrow (x \geq z) \vee (z \geq y).$$

Hence the negation of (1) is logically equivalent to

$$\exists x \, \exists y \, \forall z [(z \leq x) \vee (z \geq y)].$$

This states that for some choice of x_0 and y_0, every z is either less than or equal to x_0 or else bigger than or equal to y_0.

(b) Consider universes U_1, U_2 and U_3 made up of companies, components and devices, respectively. Let $p(x, y)$ be the predicate "x produces y" and let $q(y, z)$ be "y is a component part of z." The predicate

$$p(x, y) \wedge q(y, z)$$

has the meaning "x produces y, which is a component of z." The proposition

$$\forall x \, \forall z \, \exists y [p(x, y) \wedge q(y, z)]$$

means that each company produces some component of each device. Its negation is

$$\neg \forall x \, \forall z \, \exists y [p(x, y) \wedge q(y, z)]$$

which is logically equivalent, by the DeMorgan laws, to

$$\exists x \, \exists z \, \forall y [\neg p(x, y) \vee \neg q(y, z)].$$

This negation has the interpretation that there exist a company x_0 and a device z_0 so that for each choice of a component either x_0 does not produce it or it's not a component of z_0. An equivalent form of the negation is

$$\exists x \, \exists z \, \neg \exists y [p(x, y) \wedge q(y, z)],$$

with the interpretation that there exist a company x_0 and a device z_0 so that no component part is both produced by x_0 and a component of z_0.

Compare this example with part (a). In the case of part (a), the negation is also equivalent to

$$\exists x \, \exists y \, \neg \exists z [x < z < y],$$

i.e., there exist x_0 and y_0 with no z strictly between them.

(c) The negation of

(2) $$\forall x \, \forall y [x < y \rightarrow x^2 < y^2]$$

is

$$\exists x \, \exists y \{\neg [x < y \rightarrow x^2 < y^2]\}.$$

By rule 10a in Table 1 of §2.2 and DeMorgan's law, $\neg(p \rightarrow q) \Leftrightarrow \neg(\neg p \vee q) \Leftrightarrow p \wedge \neg q$. So $\neg[x < y \rightarrow x^2 < y^2] \Leftrightarrow (x < y) \wedge (x^2 \geq y^2)$.

Therefore, the negation of (2) is logically equivalent to

$$\exists x\, \exists y[(x < y) \wedge (x^2 \geq y^2)].$$

This proposition has the value true if the universe for x and y is $\mathbb{Z}$ and false if the universe is $\mathbb{N}$, just the opposite of (2), of course. ■

EXAMPLE 6   Let the universe of discourse U consist of two members, a and b. The DeMorgan law 37a then becomes

$$\neg[p(a) \wedge p(b)] \Leftrightarrow [\neg p(a)] \vee [\neg p(b)].$$

Except for the names $p(a)$ and $p(b)$, in place of p and q, this is the De-Morgan law 8b in Table 1 of §2.2. ■

A general proposition often has the form $\forall x\, p(x)$ where x ranges over some universe of discourse. This proposition is false if and only if $\exists x[\neg p(x)]$ is true, by DeMorgan's law 37a. Thus $\forall x\, p(x)$ is false if some x_0 can be exhibited for which $p(x_0)$ is false. As we pointed out in §2.1 [after Example 11], such an x_0 is called a **counterexample** to the proposition $\forall x\, p(x)$. Some illustrations are given in Example 12 of that section. Here are a couple more.

EXAMPLE 7   (a) The matrices

$$\begin{bmatrix} 1 & 0 \\ 0 & 0 \end{bmatrix}, \quad \begin{bmatrix} 0 & 0 \\ 1 & 1 \end{bmatrix}$$

provide a counterexample to the [false] assertion "If 2×2 matrices $\mathbf{A}$ and $\mathbf{B}$ satisfy $\mathbf{AB} = \mathbf{0}$, then $\mathbf{A} = \mathbf{0}$ or $\mathbf{B} = \mathbf{0}$." This general assertion could have been written

$$\forall \mathbf{A}\, \forall \mathbf{B}[\mathbf{AB} = \mathbf{0} \rightarrow (\mathbf{A} = \mathbf{0} \vee \mathbf{B} = \mathbf{0})].$$

(b) The proposition "Every connected graph has an Euler circuit" is false. In view of Euler's theorem in §6.2, any connected graph having a vertex of odd degree will serve as a counterexample to this assertion. The simplest counterexample has two vertices with one edge connecting them. ■

It is worth stressing that implication 38 in Table 1 and its equivalent versions in Example 4(b) are valid because we have restricted our attention to nonempty universes of discourse. If we were to allow the empty universe of discourse, then implication 38 would be false. In that case $\forall x\, p(x)$ would have the value true vacuously, whereas $\exists x\, p(x)$ would be false. It is true that everyone with three heads is rich. [You disagree? Give a counterexample.] But it is not true that there is a rich person with three heads. Here the universe consists of all three-headed people and $p(x)$ denotes "x is rich."

The second implication in Example 4(b) also fails for the empty universe, but it is a bit slippery to analyze. It is true that every three-headed person is not rich, but false that not every three-headed person is rich. Three-headed people have amazing properties.

EXERCISES 13.2

1. Consider a universe U_1 consisting of members of a club, and a universe U_2 of airlines. Let $p(x, y)$ be the predicate "x has been a passenger on y" or equivalently "y has had x as a passenger." Write out the meanings of the following.

   (a) rule 35a   (b) rule 35b   (c) rule 36

2. Consider the universe U of all university professors. Let $p(x)$ be the predicate "x likes thrash metal."

   (a) Express the proposition "not all university professors like thrash metal" in predicate calculus symbols.

   (b) Do the same for "every university professor does not like thrash metal."

   (c) Does either of the propositions in part (a) or (b) imply the other? Explain.

   (d) Write out the meaning of rule 37b for this U and $p(x)$.

   (e) Do the same for rule 37d.

3. Interpret DeMorgan Laws 37b, 37c and 37d for the predicate $y(c) =$ "c is yellow," where c ranges over the universe of cars.

4. Let $p(x, y)$ be the predicate "$x \neq y$" on the domain of discourse $\mathbb{N}$. Give the truth value of each of the following.

   (a) $\forall y \, \exists x \, p(x, y) \rightarrow \exists x \, \forall y \, p(x, y)$

   (b) $\neg \forall x \, \exists y \, p(x, y) \rightarrow \neg \exists y \, \forall x \, p(x, y)$

   (c) $\forall y \, p(x, y)$

   (d) $\forall x [\neg \forall y \, p(x, y) \vee \exists y \, p(x, y)]$

5. Show that the following rules in Table 1 collapse to rules from Table 1 of §2.2 when the universe of discourse U has two elements, a and b.

   (a) rule 37d   (b) rule 37b

6. (a) Show that the logical implication

   $$[\exists x \, p(x)] \wedge [\exists x \, q(x)] \Rightarrow \exists x [p(x) \wedge q(x)]$$

   is false. You may do this by defining predicates $p(x)$ and $q(x)$ where this implication fails.

   (b) Do the same for the logical implication

   $$\exists x \, \forall y \, p(x, y) \Rightarrow \forall x \, \exists y \, p(x, y).$$

   [Compare this with the true implication of rule 36.]

7. For the universe of discourse $\mathbb{N}$ write the negation of $\forall n [p(n) \rightarrow p(n + 1)]$ without using the quantifier $\forall$.

8. Write the negation of $\exists x \, \forall y \, \exists z [z > y \rightarrow z < x^2]$ without using the connective $\neg$.

9. (a) Write the negation of

   $$P = \forall x \, \forall y [x < y \rightarrow \exists z \{x < z < y\}]$$

   without using the connective $\neg$.

   (b) Determine the truth value of P when the universe of discourse for x, y and z is $\mathbb{R}$ or $\mathbb{Q}$.

   (c) Determine the truth value of P when the universe of discourse is $\mathbb{N}$ or $\mathbb{Z}$.

10. Give a counterexample for each of the following assertions.

    (a) Every even integer is the product of two even integers.

    (b) $|S \cup T| = |S| + |T|$ for every two finite sets S and T.

    (c) Every positive integer of the form $6k - 1$ is a prime.

    (d) Every graph has an even number of edges.

    (e) All mathematics courses are fun.

11. Our definition of compound predicate does not permit expressions such as $\exists x \, p(x, x)$ with $p(x, y)$ a 2-place predicate. Describe a predicate $q(x, y)$ such that

    $$\exists x \, \exists y [p(x, y) \wedge q(x, y)]$$

    is true if and only if $p(x, x)$ is true for some x.

12. In the case that the universe of discourse is empty, $\forall x \, p(x)$ vacuously has the value true regardless of $p(x)$, and $\exists x \, p(x)$ is false. Describe the situation for a universe with exactly one member.

13. The statement "There are arbitrarily large integers n such that $p(n)$ is true" translates into the proposition

    $$\forall N \, \exists n [(n \geq N) \wedge p(n)]$$

    with universe of discourse $\mathbb{P}$. Write the negation of this proposition using the connective $\rightarrow$ but without using the connective $\neg$. Your answer should translate into a statement which implies that $p(n)$ is true for only a finite set of n's.

14. (a) Choose universes of discourse for x, y and z so that proposition (1) in Example 5 is true.

    (b) Choose universes of discourse so that proposition (1) in Example 5 is false.

§ 13.3   Infinite Sets

Chapter 5 focused on counting finite sets. Counting infinite sets is another matter. Perhaps you think they are all the same size, that there can only be one infinity. Read on to find out what mathematicians think. And leave your intuition behind; it won't help much here. The subject is fascinating, and we will just scratch the surface in this section.

Mathematicians have a way of classifying infinite sets according to their "size." First they generalize the concept "two sets are of the same size." The clue to the commonly accepted correct approach is the following elementary observation: Two finite sets are of the same size if and only if there exists a one-to-one correspondence between them. Following this guide, we define two sets S and T, finite or infinite, to be of the **same size** if there is a one-to-one correspondence between them. In this book we will not study the classification scheme for sets in detail, but we will distinguish between two kind of infinite sets.

Any set that is the same size as the set $\mathbb{P}$ of positive integers will be called **countably infinite**. Thus a set S is countably infinite if and only if there exists a one-to-one correspondence between $\mathbb{P}$ and S. A set is **countable** if it is finite or is countably infinite. One is able to count or list such a nonempty set by matching it with $\{1, 2, \ldots, n\}$ for some $n \in \mathbb{P}$, or with the whole set $\mathbb{P}$. In the infinite case, the list will never end. As one would expect, a set is **uncountable** if it is not countable.

EXAMPLE 1   (a) The set $\mathbb{N}$ is countably infinite because $f(n) = n - 1$ defines a one-to-one function f mapping $\mathbb{P}$ onto $\mathbb{N}$. Its inverse f^{-1} is a one-to-one mapping of $\mathbb{N}$ onto $\mathbb{P}$; note that $f^{-1}(n) = n + 1$ for $n \in \mathbb{N}$. Even though $\mathbb{P}$ is a proper subset of $\mathbb{N}$, by our definition $\mathbb{P}$ is the same size as $\mathbb{N}$. This may be surprising, since a similar situation does not occur for finite sets. Oh well, $\mathbb{N}$ has only one element that is not in $\mathbb{P}$.

(b) The set $\mathbb{Z}$ of *all* integers is also countably infinite. A one-to-one function f from $\mathbb{Z}$ onto $\mathbb{P}$ is indicated in Figure 1, where we have found

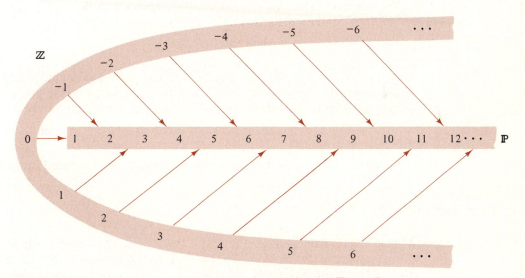

A one-to-one correspondence of $\mathbb{Z}$ onto $\mathbb{P}$

FIGURE 1

it convenient to bend the picture of $\mathbb{Z}$. This function can be given by a formula, if desired:

$$f(n) = \begin{cases} 2n + 1 & \text{for } n \geq 0 \\ -2n & \text{for } n < 0. \end{cases}$$

Even though $\mathbb{Z}$ looks about twice as big as $\mathbb{P}$, these sets are of the same size. Beware! For infinite sets, your intuition may be unreliable. Or, to take a more positive approach, you may need to refine your intuition when dealing with infinite sets.

(c) Even the set $\mathbb{Q}$ of all rational numbers is countably infinite. This fact is striking, because the set of rational numbers is distributed evenly throughout $\mathbb{R}$. To give a one-to-one correspondence between $\mathbb{P}$ and $\mathbb{Q}$, a picture is worth a thousand formulas. See Figure 2. The function f is obtained by following the arrows and skipping over repetitions. Thus $f(1) = 0, f(2) = 1, f(3) = \frac{1}{2}, f(4) = -\frac{1}{2}, f(5) = -1, f(6) = -2, f(7) = -\frac{2}{3}$, etc. ■

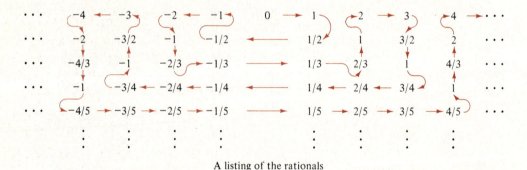

A listing of the rationals

FIGURE 2

EXAMPLE 2   Almost all of our examples of graphs and trees have had finitely many vertices and edges. However, there is no such restriction in the general definitions. Figure 3 contains partial pictures of some infinite graphs.

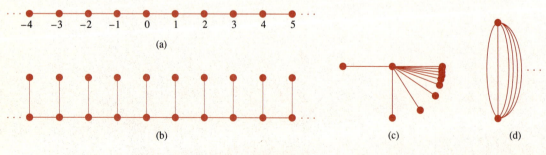

FIGURE 3

The set of vertices in Figure 3(a) is $\mathbb{Z}$, and only consecutive integers are connected by an edge. Note that this infinite tree has *no* leaves, whereas every finite tree with more than one vertex has leaves. The set of vertices in Figure 3(b) is $\mathbb{Z} \times \{0, 1\}$; this tree has infinitely many leaves. The central vertex in the tree in Figure 3(c) has infinite degree; all the other vertices are leaves. There are only two vertices in the graph of Figure 3(d), but they are connected by infinitely many edges.

In all of these examples, the sets of vertices and edges are countable. Graphs don't have to be countable, but it is hard to draw or visualize uncountable ones. ■

In the next example we illustrate a technique that goes back to Georg Cantor, the father of set theory, and is called "Cantor's diagonal procedure." You may find the result in part (b) more interesting, but the details in part (a) are easier to follow.

EXAMPLE 3   (a) We claim that the set FUN($\mathbb{P}$, $\{0, 1\}$) of all functions from $\mathbb{P}$ into $\{0, 1\}$ is uncountable. Equivalently, the set of all infinite strings of 0's and 1's is uncountable. Obviously, FUN($\mathbb{P}$, $\{0, 1\}$) is infinite, so if it were countable there would exist an infinite listing $\{f_1, f_2, \ldots\}$ of *all* the functions in this set. We define a function f^* on $\mathbb{P}$ as follows:

$$f^*(n) = \begin{cases} 0 & \text{if } f_n(n) = 1 \\ 1 & \text{if } f_n(n) = 0. \end{cases}$$

For each n in $\mathbb{P}$, $f^*(n) \neq f_n(n)$ by construction, so the function f^* must be different from every f_n. Thus $\{f_1, f_2, \ldots\}$ is not a listing of *all* functions in FUN($\mathbb{P}$, $\{0, 1\}$). This contradiction shows that FUN($\mathbb{P}$, $\{0, 1\}$) is uncountable.

(b) The interval $[0, 1)$ is uncountable. If it were countable, there would exist a one-to-one function f mapping $\mathbb{P}$ onto $[0, 1)$. We show that this is impossible. Each number in $[0, 1)$ has a decimal expansion $.d_1 d_2 d_3 \cdots$ where each d_j is a digit in $\{0, 1, 2, 3, 4, 5, 6, 7, 8, 9\}$. In particular, each number $f(k)$ has the form $.d_{1k} d_{2k} d_{3k} \cdots$; here d_{nk} represents the nth digit in $f(k)$. Consider Figure 4 and focus on the indicated "diagonal

$$
\begin{aligned}
f(1) &= .\, d_{11}\ d_{21}\ d_{31}\ d_{41} \cdots \\
f(2) &= .\, d_{12}\ d_{22}\ d_{32}\ d_{42} \cdots \\
f(3) &= .\, d_{13}\ d_{23}\ d_{33}\ d_{43} \cdots \\
f(4) &= .\, d_{14}\ d_{24}\ d_{34}\ d_{44} \cdots \\
&\ \vdots
\end{aligned}
$$

FIGURE 4

digits." We define the sequence $d*$ whose nth term d_n^* is constructed as follows: If $d_{nn} \neq 1$, let $d_n^* = 1$, and if $d_{nn} = 1$, let $d_n^* = 2$. The point is that $d_n^* \neq d_{nn}$ for all $n \in \mathbb{P}$. Now $.d_1^* d_2^* d_3^* \cdots$ represents a number x in $[0, 1)$ that is different from $f(n)$ in the nth digit for each $n \in \mathbb{P}$. Thus x cannot be one of the numbers $f(n)$; i.e., $x \notin \text{Im}(f)$, so f does not map $\mathbb{P}$ onto $[0, 1)$. Thus $[0, 1)$ is uncountable.

Note that we arranged for all of the digits of x to be 1's and 2's. This choice was quite arbitrary, except that we deliberately avoided 0's and 9's since there are some numbers, whose expansions involve strings of 0's and 9's, that have two decimal expansions. For example, $.250000 \cdots$ and $.249999 \cdots$ represent the same number in $[0, 1)$. ∎

The proof in Example 3(b) can be modified to prove that $\mathbb{R}$ and $(0, 1)$ are uncountable; in fact, all intervals $[a, b]$, $[a, b)$, $(a, b]$ and (a, b) are uncountable for $a < b$. In view of Exercise 9, another way to show that these sets are uncountable is to show that they are in one-to-one correspondence with each other. In fact, they are also in one-to-one correspondence with unbounded intervals. Showing the existence of such one-to-one correspondences can be challenging. We provide a couple of the trickier arguments in the next example, and ask for some easier ones in Exercise 3.

EXAMPLE 4   (a) We show that $\mathbb{R}$ and $(0, 1)$ are in one-to-one correspondence, and hence are of the same size. Though trigonometric functions are not

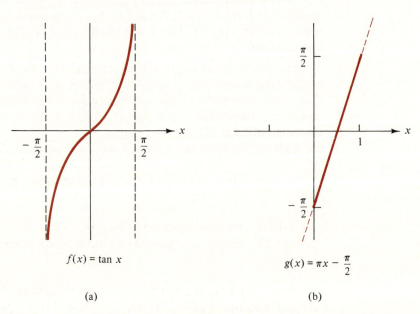

$f(x) = \tan x$

(a)

$g(x) = \pi x - \dfrac{\pi}{2}$

(b)

FIGURE 5

necessary here [Exercise 5], we will use a ready-made function from trigonometry, namely the tangent; see Figure 5(a). The function f given by $f(x) = \tan x$ is one-to-one on $(-\pi/2, \pi/2)$ and maps this interval onto $\mathbb{R}$. It is easy to find a linear function g mapping $(0, 1)$ onto $(-\pi/2, \pi/2)$, namely $g(x) = \pi x - \pi/2$; see Figure 5(b). The composite function $f \circ g$ is a one-to-one correspondence between $(0, 1)$ and $\mathbb{R}$.

(b) We show that $[0, 1)$ and $(0, 1)$ have the same size. No simple formula provides us with a one-to-one mapping between these sets. The trick is to isolate some infinite sequence in $(0, 1)$, say $\frac{1}{2}, \frac{1}{3}, \frac{1}{4}, \dots$, and then map this sequence onto $0, \frac{1}{2}, \frac{1}{3}, \frac{1}{4}, \dots$, while leaving the complement fixed. That is, let

$$C = (0, 1) \setminus \left\{ \frac{1}{n} : n = 2, 3, 4, \dots \right\}$$

and define

$$f(x) = \begin{cases} 0 & \text{if } x = 1/2 \\ \dfrac{1}{n-1} & \text{if } x = 1/n \text{ for some integer } n \geq 3 \\ x & \text{if } x \in C. \end{cases}$$

See Figure 6. ∎

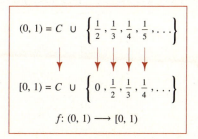

FIGURE 6

We next prove two basic facts about countable sets.

Theorem

(a) Subsets of countable sets are countable.

(b) The countable union of countable sets is countable.

Proof. (a) It is enough to show that subsets of $\mathbb{P}$ are countable. Consider a subset A of $\mathbb{P}$. Clearly, A is countable if A is finite. Suppose that A is infinite. We will use the Well-Ordering Principle. Define $f(1)$

to be the least element in A. Then define $f(2)$ to be the least element in $A \setminus \{f(1)\}$, $f(3)$ to be the least element in $A \setminus \{f(1), f(2)\}$, etc. Continue this process, so that $f(n+1)$ is the least element in the nonempty set $A \setminus \{f(k): 1 \leq k \leq n\}$ for each $n \in \mathbb{P}$. It is easy to verify that this recursive definition provides a one-to-one function f mapping $\mathbb{P}$ onto A [Exercise 10], so A is countable.

(b) The statement in part (b) means that if I is a countable set and if $\{A_i : i \in I\}$ is a family of countable sets, then the union $\bigcup_{i \in I} A_i$ is countable. We may assume that each A_i is nonempty and that $\bigcup_{i \in I} A_i$ is infinite, and we may assume that $I = \mathbb{P}$ or that I has the form $\{1, 2, \ldots, n\}$. If $I = \{1, 2, \ldots, n\}$ we can define $A_i = A_n$ for $i > n$ and obtain a family $\{A_i : i \in \mathbb{P}\}$ with the same union. Thus we may assume that $I = \mathbb{P}$. Each set A_i is finite or countably infinite. By repeating elements if A_i is finite, we can list each A_i as follows:

$$A_i = \{a_{1i}, a_{2i}, a_{3i}, a_{4i}, \ldots\}.$$

The elements in $\bigcup_{i \in I} A_i$ can be listed in an array as in Figure 7. The arrows in the figure suggest a single listing for $\bigcup_{i \in I} A_i$:

$(*)$ $\qquad a_{11}, a_{12}, a_{21}, a_{31}, a_{22}, a_{13}, a_{14}, a_{23}, a_{32}, a_{41}, \ldots$.

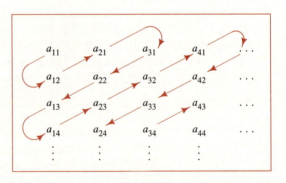

FIGURE 7

Some elements may be repeated, but the list includes infinitely many distinct elements since $\bigcup_{i \in I} A_i$ is infinite. Now a one-to-one mapping f of $\mathbb{P}$ onto $\bigcup_{i \in I} A_i$ is obtained as follows: $f(1) = a_{11}$, $f(2)$ is the next element listed in $(*)$ different from $f(1)$, $f(3)$ is the next element listed in $(*)$ different from $f(1)$ and $f(2)$, etc. ∎

EXAMPLE 5 The argument in Example 1(c) which shows that $\mathbb{Q}$ is countable is similar
to the proof of part (b) of the theorem. In fact, we can use the theorem
to give another proof that $\mathbb{Q}$ is countable. For each n in $\mathbb{P}$, let

$$A_n = \left\{\frac{m}{n} : m \in \mathbb{Z}\right\}.$$

Thus A_n consists of all integer multiples of $1/n$. Each A_n is clearly in one-
to-one correspondence with $\mathbb{Z}$ [map m to m/n], so each A_n is countable.
By part (b) of the theorem, the union

$$\bigcup_{n \in \mathbb{P}} A_n = \mathbb{Q}$$

is also countable. ∎

EXAMPLE 6 (a) If Σ is a finite alphabet, the set Σ^* of all words using letters from
Σ is countably infinite. Note that Σ is nonempty by definition. We already
know that Σ^* is infinite. Recall that

$$\Sigma^* = \bigcup_{k=0}^{\infty} \Sigma^k$$

where each Σ^k is finite. Thus Σ^* is a countable union of countable sets,
and hence Σ^* itself is countable by part (b) of the theorem.

(b) Imagine, if you can, a countably infinite alphabet Σ, and let Σ^*
consist of all words using letters of Σ, i.e., all finite strings of letters from
Σ. For each $k \in \mathbb{P}$, the set Σ^k of all words of length k is in one-to-one
correspondence with the product set $\Sigma^k = \Sigma \times \Sigma \times \cdots \times \Sigma$ (k times).
In fact, the correspondence maps each word $a_1 a_2 \cdots a_k$ to the k-tuple
$(a_1, a_2, \ldots, a_k)$. So each set Σ^k is countable by Exercise 15. The 1-element
set $\Sigma^0 = \{\lambda\}$ is countable too. Hence $\Sigma^* = \bigcup_{k=0}^{\infty} \Sigma^k$ is countable by part
(b) of the theorem. ∎

EXAMPLE 7 Consider a graph with sets of vertices and edges V and E. Even if V or
E is infinite, each path has finite length by definition. Let $\mathscr{P}$ be the set
of all paths of the graph.

(a) If E is nonempty, then $\mathscr{P}$ is infinite. For if e is any edge, then
e, ee, eee, etc. all describe paths in the graph.

(b) If E is finite, then $\mathscr{P}$ is countable. For purposes of counting, let's
view E as an alphabet. Since each path is a sequence of edges, it corresponds to a word in E^*. Of course, not all words in E^* correspond to
paths, since endpoints of adjacent edges must match up. But the set $\mathscr{P}$
is in one-to-one correspondence with some *subset* of E^*. The set E^* is

countably infinite by Example 6(a), so $\mathscr{P}$ is countable by part (a) of the theorem.

(c) If E is countably infinite, then $\mathscr{P}$ is still countable. Simply apply Example 6(b) instead of Example 6(a) in the discussion of part (b). ■

EXERCISES 13.3

1. Let A and B be finite sets with $|A| < |B|$. True or False.

   (a) There is a one-to-one map of A into B.

   (b) There is a one-to-one map of A onto B.

   (c) There is a one-to-one map of B into A.

   (d) There is a function mapping A onto B.

   (e) There is a function mapping B onto A.

2. True or False.

   (a) The set of positive rational numbers is countably infinite.

   (b) The set of all rational numbers is countably infinite.

   (c) The set of positive real numbers is countably infinite.

   (d) The intersection of two countably infinite sets is countably infinite.

   (e) There is a one-to-one correspondence between the set of all even integers and the set $\mathbb{N}$ of natural numbers.

3. Give one-to-one correspondences between the following pairs of sets.

   (a) $(0, 1)$ and $(-1, 1)$ (b) $[0, 1)$ and $(0, 1]$

   (c) $[0, 1]$ and $[-5, 8]$ (d) $(0, 1)$ and $(1, \infty)$

   (e) $(0, 1)$ and $(0, \infty)$ (f) $\mathbb{R}$ and $(0, \infty)$

4. Let $E = \{n \in \mathbb{N} : n \text{ is even}\}$. Show that E and $\mathbb{N}/E$ are countable by exhibiting one-to-one correspondences $f: \mathbb{P} \to E$ and $g: \mathbb{P} \to \mathbb{N}/E$.

5. Here is another one-to-one function f mapping $(0, 1)$ onto $\mathbb{R}$:

$$f(x) = \frac{2x - 1}{x(1 - x)}.$$

   (a) Sketch the graph of f.

   (b) If you know some calculus, prove that f is one-to-one by showing that its derivative is positive on $(0, 1)$.

6. Which of the following sets are countable? countably infinite?

   (a) $\{0, 1, 2, 3, 4\}$ (b) $\{n \in \mathbb{N} : n \le 73\}$

   (c) $\{n \in \mathbb{Z} : n \le 73\}$ (d) $\{n \in \mathbb{Z} : |n| \le 73\}$

   (e) $\{5, 10, 15, 20, 25, \ldots\}$ (f) $\mathbb{N} \times \mathbb{N}$

   (g) $\left[\frac{1}{4}, \frac{1}{3}\right]$

7. Let Σ be the alphabet $\{a, b, c\}$. Which of the following sets are countably infinite?

   (a) Σ^{73} (b) $\Sigma*$

   (c) $\bigcup_{k=0}^{\infty} \Sigma^{2k} = \{w \in \Sigma* : \text{length}(w) \text{ is even}\}$

   (d) $\bigcup_{k=0}^{3} \Sigma^k$ (e) $\bigcup_{k=1}^{3} \Sigma^{2k}$

8. A set A has m elements and a set B has n elements. How many functions from A into B are one-to-one? *Hint:* Consider the cases $m \le n$ and $m > n$ separately.

9. (a) Show that if there is a one-to-one correspondence of a set S onto some countable set, then S itself is countable.

   (b) Show that if there is a one-to-one correspondence of a set S onto some uncountable set, then S is uncountable.

10. Complete the proof of part (a) of the theorem by showing that f is one-to-one and that f maps $\mathbb{P}$ onto A.

11. (a) Prove that if S and T are countable, then $S \times T$ is countable.

    (b) Prove that if f maps S onto T and S is countable, then T is countable.

    (c) Use parts (a) and (b) to give another proof that $\mathbb{Q}$ is countable. *Suggestion:* For (m, n) in $\mathbb{Z} \times \mathbb{P}$, define $f(m, n) = m/n$.

12. Show that if S and T have the same size, then so do $\mathscr{P}(S)$ and $\mathscr{P}(T)$.

13. (a) Show that $\text{FUN}(\mathbb{P}, \{0, 1\})$ is in one-to-one correspondence with the set $\mathscr{P}(\mathbb{P})$ of all subsets of $\mathbb{P}$.

    (b) Show that $\mathscr{P}(\mathbb{P})$ is uncountable.

14. Show that any disjoint family of nonempty subsets of a countable set is countable.

15. Show that if S is countable, then $S^n = S \times S \times \cdots \times S$ [n times] is countable for each n. *Hint:* Use Exercise 11(a) and induction.

16. Here's an elegant explicit one-to-one correspondence of the set $\mathbb{Q}^+$ of positive rationals onto $\mathbb{P}$. Given m/n in $\mathbb{Q}^+$ where m and n are relatively prime, write $m = p_1^{m_1} \cdots p_k^{m_k}$ and $n = q_1^{n_1} \cdots q_l^{n_l}$ as products of primes. Define

$$f\left(\frac{m}{n}\right) = p_1^{2m_1} \cdots p_k^{2m_k} \cdot q_1^{2n_1 - 1} \cdots q_l^{2n_l - 1}.$$

   In particular, $f(m) = p_1^{2m_1} \cdots p_k^{2m_k} = m^2$ for positive integers m.

   (a) Calculate $f(\frac{1}{8})$, $f(\frac{1}{9})$, $f(\frac{1}{10})$, $f(\frac{1}{100})$ and $f(\frac{21}{20})$.

   (b) What fraction is mapped to 23? to 24? to 25? to 26? to 27? to 28?

   (c) Explain why f is one-to-one and why f maps $\mathbb{Q}^+$ onto $\mathbb{P}$.

   This example was given by Yoram Sagher in the November 1989 issue of *The American Mathematical Monthly*, page 823.

CHAPTER HIGHLIGHTS

As usual: What does it mean? Why is it here? How can I use it? Think of examples.

CONCEPTS AND NOTATION

quantifiers, $\forall$, $\exists$
universe [= domain] of discourse
predicate = proposition-valued function
n-place predicate
compound predicate
   free, bound variable
   compound proposition
tautology
logical equivalence, implication, $\Leftrightarrow$, $\Rightarrow$
counterexample
infinite sets
   same size
   countable
   countably infinite
   uncountable

FACTS

$\forall$ and $\exists$ do not commute with each other:

   $\exists x \, \forall y \, p(x, y) \rightarrow \forall y \, \exists x \, p(x, y)$ is a tautology, but

   $\forall x \, \exists y \, p(x, y) \rightarrow \exists y \, \forall x \, p(x, y)$ is not.

$\mathbb{N}$, $\mathbb{Z}$ and $\mathbb{Q}$ are countable.

$[0, 1)$ and $\text{FUN}(\mathbb{P}, \{0, 1\})$ are uncountable.

$\mathbb{R}$, $[0, 1)$ and $(0, 1)$ are the same size.

Subsets and countable unions of countable sets are countable.

METHODS

Use of generalized DeMorgan laws to negate quantified predicates.
Cantor's diagonal procedure for showing certain sets are uncountable.

DICTIONARY

The words listed below are of three general sorts: English words with which the reader may not be completely familiar, common English words whose usage in mathematical writing is specialized, and technical mathematical terms which are assumed background for this book. For technical terms introduced in this book, see the index.

absurd. Clearly impossible, being contrary to some evident truth.

all. See *every*.

ambiguous. Capable of more than one interpretation or meaning.

anomaly. Something that is, or appears to be, inconsistent.

any. See *every*.

assume. Assume and suppose mean the same thing and ask that we imagine a situation for the moment.

axiom. An assertion that is accepted and used without a proof. Obvious or self-evident axioms are preferred.

bona fide. Genuine or legitimate.

calls. If algorithm A contains the instruction to use algorithm B, we say that A calls B.

cf. Compare.

chicanery. Trickery.

class. See *set*.

collapse. To fall or come together.

collection. See *set*.

common factor. The integer m is a common factor or common divisor of two integers if it divides them both. See *divisible by*.

comparable. Capable of being compared.

conjecture. A guess or opinion, preferably based on some experience or other source of wisdom.

corollary. See *theorem*.

define. Often this looks like an instruction [as in "Define $f(x) = x^2$"] when it is merely a [bad] mathematical way of saying "We define" or "Let."

disprove. The instruction "prove or disprove" means that the assertion should either be proved true or proved false. Which you do depends, of course, on whether the assertion is actually true or not.

distinct. Different.

distinguishable. A collection of objects is regarded as distinguishable if there is some property or characteristic that makes it possible to tell different objects apart. In contrast, we would regard ten red marbles of the same size as indistinguishable.

divisible by. Consider integers m and n. We say that "n is divisible by m," that "m divides n," that "n is a multiple of m," or that "m is a factor of n" if $n = mk$ for some integer k. We write $m|n$ to signify any of these statements. For example, $3|6$, $4|20$ and $8|8$. Also $m|0$ for all m since $0 = mk$ for $k = 0$.

e.g. For example.

entries. The individual numbers or objects in ordered pairs, in matrices or in sequences.

even number. An integer that is exactly divisible by 2, i.e., any integer that can be written as $2k$ where $k \in \mathbb{Z}$. Note that 0 is an even number.

every. The expressions "for every," "for any" and "for all" mean the same thing. They all mean "for all choices of the variable in question," and so they correspond to the quantifier $\forall$ in §2.1. The expression "for some" means "for at least one" and corresponds to the quantifier $\exists$ in §2.1. It is generally good practice to avoid the use of "any," which is sometimes misunderstood. For example, "If $p(n)$ is true for any n" usually means "If $p(n)$ is true for some n," not "If $p(n)$ is true for every n."

factor. See *divisible by*.

family. See *set*.

fictitious. Imaginary, not actual.

gcd. If m and n are positive integers, then $\gcd(m, n)$ is the greatest common divisor of m and n, that is, the largest integer that m and n are both divisible by. $\mathrm{lcm}(m, n)$ is their least common multiple, i.e., the smallest positive integer that is a multiple of both m and n. For example $\gcd(10, 25) = 5$ and $\mathrm{lcm}(10, 25) = 50$. gcd and lcm can be calculated by writing m and n as products of primes. For example, for $m = 168$ and $n = 450$ we have $m = 2^3 \cdot 3 \cdot 7$ and $n = 2 \cdot 3^2 \cdot 5^2$, so $\gcd(m, n) = 2 \cdot 3 = 6$ and $\mathrm{lcm}(m, n) = 2^3 \cdot 3^2 \cdot 5^2 \cdot 7 = 12{,}600$.

greatest common divisor. See *gcd*.

inclusion. We sometimes refer to the relation $A \subseteq B$ as an "inclusion" just as we may refer to $A = B$ as an "equality."

initialize. Set the starting conditions.

inspection. Something can be seen "by inspection" if it can be seen directly, without calculation or modification.

invertible. Having an inverse.

irrational. An irrational number is a real number that is not rational, i.e., that cannot be written as m/n for $m, n \in \mathbb{Z}$, $n \neq 0$. Examples include $\sqrt{2}$, $\sqrt{3}$, $\sqrt[3]{2}$, π and e.

lcm. See *gcd*.

least common multiple. See *gcd*.

lemma. See *theorem*.

loop. A computer program or algorithm that is repeated under specified conditions.

matrices. Plural of matrix.

max. For two real numbers a and b, we write $\max\{a, b\}$ for the larger of the two. If $a = b$, then $\max\{a, b\} = a = b$.

min. For two real numbers a and b, we write $\min\{a, b\}$ for the smaller of the two. If $a = b$, then $\min\{a, b\} = a = b$.

multiple of. See *divisible by*.

necessary. We say a property p is necessary for a property q if p must hold whenever q holds, i.e., if $q \Rightarrow p$. The property p is sufficient for q if p is enough to guarantee q, i.e., if $p \Rightarrow q$. So p is necessary and sufficient for q provided that $p \Leftrightarrow q$.

odd number. An integer that is not an even integer. An odd number can be written as $2k + 1$ for some $k \in \mathbb{Z}$.

permute. To change the order of a sequence of elements.

prime number. An integer ≥ 2 that cannot be written as the product of two integers that are both ≥ 2. The first primes are 2, 3, 5, 7, 11, 13, 17, 19.

proposition. See *theorem*.

redundant. Unnecessary or excessive.

relatively prime. Two positive integers m and n are relatively prime if they have no common factors, i.e., if $\gcd(m, n) = 1$. See *gcd*.

sequence. A list of things following one another. A formal definition is given in §1.5.

set. The terms "set," "collection," "class" and "family" are used interchangeably. We tend to refer to families of sets, for example, to avoid the expression "sets of sets."

some. See *every*.

sufficient. See *necessary*.

suppose. See *assume*.

synonym. A word having the same meaning as another word. Two words that have the same meaning are *synonymous*.

theorem. A theorem, proposition, lemma or corollary is some assertion that has been or can be proved. The term "proposition" also has a special use in logic; see §2.1. Theorems are usually the most important facts. Lemmas are usually not of primary interest and are used to prove later theorems or propositions. Corollaries are usually easy consequences of theorems or propositions just presented.

truncate. To shorten by cutting.

unambiguous. Not ambiguous. See ***ambiguous***.

underlying set. The basic set on which the objects [like functions or operations] are defined.

vertices. Plural of vertex.

ANSWERS AND HINTS

Wise students will only look at these answers and hints after seriously working on the problems. Often only hints are given; you should write out the details to check understanding as well as to get practice in communicating mathematical ideas.

Section 1.1

1. (a) 0, 5, 10, 15, 20, say.
   (c) $\varnothing$, {1}, {2, 3}, {3, 4}, {5}, say.
   (e) 1, 1/2, 1/3, 1/4, 1/73, say.
   (g) 1, 2, 4, 16, 18, say.
3. (a) λ, a, ab, cab, ba, say.
   (c) $aaaa$, $aaab$, $aabb$, etc.
5. (a) 0.
   (c) 138.
   (e) 73.
   (g) 0.
   (i) ∞.
   (k) ∞.
   (m) ∞.
   (o) ∞.
7. $A \subseteq A$, $B \subseteq B$, C is a subset of A, and C, D are subsets of A, B and D.
9. (a) aba is in all three and has length 3 in each.
   (c) cba is in Σ_1^* and length(cba) = 3.
   (e) $caab$ is in Σ_1^* with length 4 and is in Σ_2^* with length 3.
11. (a) Yes.
   (c) Delete first letters from the string until no longer possible. If λ is reached, the original string is in Σ^*. Otherwise, it isn't.

Section 1.2

1. (a) {1, 2, 3, 5, 7, 9, 11}.
   (c) {1, 5, 7, 9, 11}.
   (e) {3, 6, 12}.
   (g) 16.
3. (a) [2, 3].
   (c) [0, 2).
   (e) $(-\infty, 0) \cup (3, \infty)$.
5. (a) $\varnothing$.
   (c) $\varnothing$.
   (e) {λ, ab, ba}.
   (g) $B^c \cap C^c$ and $(B \cup C)^c$ are equal by a DeMorgan law [or by calculation] as are $(B \cap C)^c$ and $B^c \cup C^c$.
7. $A \oplus A = \varnothing$ and $A \oplus \varnothing = A$.
9. $(A \cap B \cap C)^c = (A \cap B)^c \cup C^c$ and $(A \cap B)^c = A^c \cup B^c$ by DeMorgan law 9b. Now substitute.

11. (a) (a, a), (a, b), (a, c), (b, a), etc. There are nine altogether.

(c) (a, a), (b, b).

13. (a) $(0, 0)$, $(1, 1)$, $(2, 2)$, . . . , $(6, 6)$, say.   (c) $(6, 1)$, $(6, 2)$, $(6, 3)$, . . . , $(6, 7)$, say.

(e) The set has 5 elements. List them.

15. (a) False. Try $A = \varnothing$.

(c) True. Show $x \in B$ implies $x \in C$ by considering two cases: $x \in A$ and $x \notin A$. Similarly, $x \in C$ implies $x \in B$.

(e) The hint to part (c) also applies here.

17. (a) Any example with $A \neq B$ shows the failure.

(c) The Venn diagrams are

$$(A \backslash B) \backslash C \qquad\qquad A \backslash (B \backslash C)$$

Section 1.3

1. (a) $f(3) = 27$, $f(1/3) = 1/3$, $f(-1/3) = 1/27$, $f(-3) = 27$.

(c) $\mathrm{Im}(f) = [0, \infty)$.

3. (a) No; S is bigger than T.

(c) Yes. For example, let $f(1) = a$, $f(2) = b$, $f(3) = c$, $f(4) = f(5) = d$.

(e) No. This follows from either part (a) or part (d).

5. (a) $f(2, 1) = 2^2 3^1 = 12$, $f(1, 2) = 2^1 3^2 = 18$, etc.

(c) Consider 5, for instance.

7. (a) Pick b_0 in B. For every $a \in A$, $\mathrm{PROJ}(a, b_0) = a$ so every a in A is in the image of PROJ.

9. $\{n \in \mathbb{Z} : n \text{ is even}\}$.

11. (a) $f \circ g \circ h(x) = (x^8 + 1)^{-3} - 4(x^8 + 1)^{-1}$.

(c) $h \circ g \circ f(x) = [(x^3 - 4x)^2 + 1]^{-4}$.

(e) $g \circ g(x) = (x^2 + 1)^2 / [1 + (x^2 + 1)^2]$.

(g) $g \circ h(x) = (x^8 + 1)^{-1}$.

13. Since $g \circ f : S \to U$ and $h : U \to V$, the composition $h \circ (g \circ f)$ is defined and maps S to V. A similar remark applies to $(h \circ g) \circ f$. Show that the functions' values agree at each $x \in S$.

15. (a) $1, 0, -1$ and 0.

(c) $g \circ f$ is the characteristic function of $\mathbb{Z} \backslash E$. $f \circ f(n) = n - 2$ for all $n \in \mathbb{Z}$.

Section 1.4

1. (a) $f^{-1}(y) = (y - 3)/2$.   (c) $h^{-1}(y) = 2 + \sqrt[3]{y}$.

3. (a) All of them; verify this.

(c) $\text{SUM}^{\leftarrow}(4)$ has 5 elements, $\text{PROD}^{\leftarrow}(4)$ has 3 elements, $\text{MAX}^{\leftarrow}(4)$ has 9 elements, and $\text{MIN}^{\leftarrow}(4)$ is infinite.

5. (a) $f(0) = 1$, $f(1) = 2$, $f(2) = 3$, $f(3) = 4$, $f(4) = 5$, $f(73) = 74$.

(c) For one-to-oneness, observe that if $f(n) = f(n')$, then $n = f(n) - 1 = f(n') - 1 = n'$. f does not map onto $\mathbb{N}$ because $0 \notin \text{Im}(f)$.

(e) $g(f(n)) = \max\{0, (n + 1) - 1\} = n$, but $f(g(0)) = f(0) = 1$.

7. (a) $(f \circ f)(x) = 1/(1/x) = x$.   (b)–(d) are similar verifications.

9. Since f and g are invertible, the functions $f^{-1} \colon T \to S$, $g^{-1} \colon U \to T$ and $f^{-1} \circ g^{-1} \colon U \to S$ exist. So it suffices to show that $(g \circ f) \circ (f^{-1} \circ g^{-1}) = 1_U$ and $(f^{-1} \circ g^{-1}) \circ (g \circ f) = 1_S$.

11. (a) Prove: if $s_1, s_2 \in S$ and $f(s_1) = f(s_2)$, then $s_1 = s_2$. The proof will be very short.

13. (a) Suppose that $t \in f(f^{\leftarrow}(B))$. Then $t = f(s)$ for some $s \in f^{\leftarrow}(B)$. $s \in f^{\leftarrow}(B)$ means that $f(s) \in B$. So $t \in B$. This works for any t, so $f(f^{\leftarrow}(B)) \subseteq B$.

(c) First, $f(f^{\leftarrow}(B_1) \cap f^{\leftarrow}(B_2)) \subseteq f(f^{\leftarrow}(B_1)) \cap f(f^{\leftarrow}(B_2)) = B_1 \cap B_2$, so $f^{\leftarrow}(B_1) \cap f^{\leftarrow}(B_2) \subseteq f^{\leftarrow}(B_1 \cap B_2)$. In the other direction, $f(f^{\leftarrow}(B_1 \cap B_2)) = B_1 \cap B_2 \subseteq B_1$, so $f^{\leftarrow}(B_1 \cap B_2) \subseteq f^{\leftarrow}(B_1)$, and similarly, $f^{\leftarrow}(B_1 \cap B_2) \subseteq f^{\leftarrow}(B_2)$. Thus $f^{\leftarrow}(B_1 \cap B_2) \subseteq f^{\leftarrow}(B_1) \cap f^{\leftarrow}(B_2)$.

15. (a) The first sentence shows that the f in the example cannot be one-to-one. At the other extreme, if f is constant, $f \circ g = f \circ h$ for all g and h. Provide a specific example.

(c) If g and h have the same domain and if f maps onto it, then $g \circ f = h \circ f$ implies $g = h$.

Section 1.5

1. (a) 42.   (c) 1.   (e) 154.

3. (a) 3, 12, 39 and 120.   (c) 3, 9 and 45.

5. (a) -2, 2, 0, 0 and 0.

7. (a) 0, 1/3, 1/2, 3/5, 2/3, 5/7.

(c) Note that $a_{n+1} = \dfrac{(n + 1) - 1}{(n + 1) + 1} = \dfrac{n}{n + 2}$ for $n \in \mathbb{P}$.

9. (a) 0, 0, 2, 6, 12, 20, 30.

(c) Just substitute the values into both sides.

11. (a) 0, 0, 1, 1, 2, 2, 3, 3, ...

(c) (0, 0), (0, 1) (1, 1), (1, 2) (2, 2), (2, 3) (3, 3), ...

13. (a)

n	n^4	4^n	n^{20}	20^n	$n!$
5	625	1024	$9.54 \cdot 10^{13}$	$3.2 \cdot 10^6$	120
10	10^4	$1.05 \cdot 10^6$	10^{20}	$1.02 \cdot 10^{13}$	$3.63 \cdot 10^6$
25	$3.91 \cdot 10^5$	$1.13 \cdot 10^{15}$	$9.09 \cdot 10^{27}$	$3.36 \cdot 10^{32}$	$1.55 \cdot 10^{25}$
50	$6.25 \cdot 10^6$	$1.27 \cdot 10^{30}$	$9.54 \cdot 10^{33}$	$1.13 \cdot 10^{65}$	$3.04 \cdot 10^{64}$

15. $2^n = 10^{n \cdot \log_{10} 2}$. Why? The table values look a little different because the exponents $n \cdot \log_{10} 2$ are not integers.

Section 1.6

1. (a) $k = 2$.   (c) $k = 12$.

3. (a) $n!$ Note that $3^n \neq O(2^n)$ but $3^n = O(n!)$; see Example 3(b) or Exercise 8.
   (c) $\log_2 n$.

5. (a) True. Take $C \geq 2$.
   (c) False. $2^{2n} \leq C \cdot 2^n$ only for $2^n \leq C$, i.e., only for $n \leq \log_2 C$.

7. (a) False. If $40^n \leq C \cdot 2^n$ for all large n, then $20^n \leq C$ for all large n.
   (c) False. If $(2n)! \leq C \cdot n!$ for all large n, then $(n + 1)! \leq C \cdot n!$ for large n, so $n + 1 \leq C$ for large n, which is impossible.

9. (a) The inequality $\dfrac{1}{k^2} \leq \left(\dfrac{1}{k-1} - \dfrac{1}{k} \right)$ is equivalent to $(k - 1)k \leq k^2$.

   To check the equality of the hint, write out several terms of the sum.

11. (a) We are given $f(n) = 3n^4 + a(n)$ and $g(n) = 2n^3 + b(n)$ where $a(n) = O(n)$ and $b(n) = O(n)$. Now $f(n) + g(n) = 3n^4 + [a(n) + 2n^3 + b(n)]$ and $a(n) + 2n^3 + b(n) = O(n^3)$ since $a(n)$, $2n^3$ and $b(n)$ are all $O(n^3)$ sequences. Theorem 2(b) with $g(n) = n^3$ is being used here twice.

13. (a) Assume that $c \neq 0$ since the result is obvious for $c = 0$. There is a $C > 0$ so that $|f(n)| \leq C \cdot |g(n)|$ for large n. Then $|c \cdot f(n)| \leq C \cdot |c| \cdot |g(n)|$ for large n.

15. (a) $a(n)/b(n) = n^4$ is not $O(n^3)$, by Example 8(e).

17. (a) Let $\text{DIGIT}(n) = m$. Then $10^{\text{DIGIT}(n)}$ is written as a 1 followed by m 0's, so it's larger than any m-digit number, such as n. And 10^{m-1} is a 1 followed by $m - 1$ 0's, so it's the smallest m-digit number.
   (c) Use part (a). Show $\text{DIGIT}(n) \leq 1 + \log_{10} n = O(\log_{10} n)$.

Section 2.1

1. (a) $p \wedge q$.   (c) $\neg p \to (\neg q \wedge r)$.   (e) $\neg r \to q$.

3. (a) Parts (b) and (c) are true. The other three are false.

5. The proposition is true for all $x, y \in [0, \infty)$.

7. (a) $\neg r \to \neg q$.   (c) If it is false that $x = 0$ or $x = 1$, then $x^2 \neq x$.

9. (a) $3^3 < 3^3$ is false.

11. (a) $(-1 + 1)^2 = 0 < 1 = (-1)^2$.
    (c) No. If $x \geq 0$, then $(x + 1)^2 = x^2 + 2x + 1 > x^2$.

13. (a) $(0, -1)$.

15. (a) $p \to q$.   (c) $\neg r \to p$.   (e) $r \to q$.

17. (a) The probable intent is "If you touch those cookies, then I will spank you." It is easier to imagine p of $p \to q$ being true for this meaning than it is for "If you want a spanking, then touch those cookies."
    (c) If you do not leave, then I will set the dog on you.
    (e) If you do not stop that, then I will go.

Section 2.2

1. (a) Converse: $(q \wedge r) \to p$. Contrapositive: $\neg(q \wedge r) \to \neg p$.

   (c) Converse: If $3 + 3 = 8$, then $2 + 2 = 4$.
   Contrapositive: If $3 + 3 \neq 8$, then $2 + 2 \neq 4$.

3. (a) $q \to p$.   (c) $p \to q$, $\neg q \to \neg p$, $\neg p \vee q$.

5. (a) 0.   (c) 1.

   Note: For truth tables, only the final columns are given.

7. (a)

p	q	$\neg(p \wedge q)$
0	0	1
0	1	1
1	0	1
1	1	0

(c)

p	q	$\neg p \wedge \neg q$
0	0	1
0	1	0
1	0	0
1	1	0

9.

p	q	r	final column
0	0	0	1
0	0	1	1
0	1	0	1
0	1	1	0
1	0	0	1
1	0	1	1
1	1	0	1
1	1	1	0

11.

p	q	r	part (a)	part (b)
0	0	0	0	0
0	0	1	1	0
0	1	0	1	1
0	1	1	1	0
1	0	0	1	1
1	0	1	1	0
1	1	0	1	1
1	1	1	1	0

13. (b)

p	$p \oplus p$
0	0
1	0

(d)

p	$(p \oplus p) \oplus p$
0	0
1	1

15. (a) No fishing is allowed and no hunting is allowed. The school will not be open in July and the school will not be open in August.

19. (a) One need only consider rows in which $[(p \wedge r) \to (q \wedge r)]$ is false, i.e., $(p \wedge r)$ is true and $(q \wedge r)$ is false. This leaves one row to consider:

p	q	r
1	0	1

(c) One need only consider rows in which $[(p \wedge r) \to (q \wedge s)]$ is false, i.e., $(p \wedge r)$ is true and $(q \wedge s)$ is false. This leaves three rows to consider:

p	q	r	s
1	0	1	0
1	0	1	1
1	1	1	0

21. Let $p =$ "He finished dinner" and $q =$ "He was sent to bed." Then p is true and q is true, so the logician's statement $\neg p \to \neg q$ has truth value True. She was logically correct, but not very nice.

23. (a) Consider the truth tables. B has truth value 1 on every row that A does, and C has truth value 1 on every row that B does, so C has truth value 1 on every row that A does.

(c) We are given that $P \Rightarrow Q$. Since $Q \Rightarrow R$ and $R \Rightarrow P$, by (a) $Q \Rightarrow P$. Thus $P \Leftrightarrow Q$.

Section 2.3

1. Give a direct proof using the following fact. If m and n are even integers, then there exist j and k in $\mathbb{Z}$ so that $m = 2j$ and $n = 2k$.

3. This can be done using four cases: see Example 5.

5. This can be done using three cases: (i) $n = 3k$ for some $k \in \mathbb{N}$; (ii) $n = 3k + 1$ for some $k \in \mathbb{N}$; (iii) $n = 3k + 2$ for some $k \in \mathbb{N}$.

7. (a) Give a direct proof, as in Exercise 1.

(c) False. $2 + 3$ or $2 + 5$ or $2 + 11$, for instance.

(e) False.

9. (a) Trivially true.   (c) Vacuously true.

11. Example 2 shows that the set of primes is infinite. An argument like the one in Example 10 shows that some two primes give the same last six digits. This proof is nonconstructive.

13. (a) None of the numbers in the set

$$\{k \in \mathbb{N} : (n + 1)! + 2 \leq k \leq (n + 1)! + (n + 1)\}$$

is prime, since if $2 \leq m \leq n + 1$, then m divides $(n + 1)!$, so m also divides $(n + 1)! + m$.

(b) Yes. Since $7! = 5040$, the proof shows that all the numbers from 5042 to 5047 are nonprime.

(c) Simply adjoin 5048 to the list obtained in part (b). Another sequence of seven non-primes starts with 90.

15. (a) $14 = 2 \cdot 7$ and 7 is odd. So $14 = 2^1 \cdot 7$.

(c) $96 = 2 \cdot 48 = 2 \cdot 2 \cdot 24 = 2 \cdot 2 \cdot 2 \cdot 12 = 2 \cdot 2 \cdot 2 \cdot 2 \cdot 6 = 2 \cdot 2 \cdot 2 \cdot 2 \cdot 2 \cdot 3$, so $96 = 2^5 \cdot 3$.

Section 2.4

1. (a) $\neg(p \to q) \to ((p \to q) \to p)$ (c) $p \vee \neg p$

3. (a) Rule 14 with $q \wedge r$ replacing q.

(c) Rule 8a with $\neg p \wedge r$ replacing p and $q \to r$ replacing q.

5. (a) Rule 2a and Substitution Rule (b).

(c) Rules 10a and 1 and Rule (b).

7. (a) Rule 10a [with s for q, using Rule (a)] and Rule 11a [with s for p and t for q, using Rule (a)] and Rule (b).

(c) Rule 3a [with $\neg p$ for p, s for q, s for r, using Rule (a)] and Rule (b).

(e) Rule 3a again, using Rule (a).

9. We obtain successive equivalences using the indicated rules and suitable substitutions.

(a) $[(p \vee r) \wedge (q \to r)]$
 $\quad [(p \vee r) \wedge (\neg q \vee r)]$ Rule 10a
 $\quad [(r \vee p) \wedge (r \vee \neg q)]$ Rule 2a twice
 $\quad r \vee (p \wedge \neg q)$ Rule 4a
 $\quad (p \wedge \neg q) \vee r$ Rule 2a

(c) $p \vee (\neg p \wedge \neg q)$
 $\quad (p \vee \neg p) \wedge (p \vee \neg q)$ Rule 4a
 $\quad t \wedge (p \vee \neg q)$ Rule 7a
 $\quad p \vee \neg q$ Rules 2b and 6d

11. Consider the cases

p	q	r	or	p	q	r	and	p	q	r	or	p	q	r
1	0	1		1	1	0		0	0	1		0	1	0

.

13. 1, 2, 3: hypothesis
 4: rule 16, a tautology
 5: 4, 1 and hypothetical syllogism (rule 33)
 6: 5, 3 and rule 33
 7: 6, 2 and modus tollens (rule 31)

15. (a) Take the original proof and change the reason for A from "hypothesis" to "tautology;" i.e., the proof itself needs no change.

(b) Let A be a tautology in Example 8, and use (a).

17. (a) See rules 11a and 11b.

(c) No. Any proposition involving only p, q, $\wedge$ and $\vee$ will have truth value 0 whenever p and q both have truth values 0. See Exercise 17 of §7.2.

19. Use truth tables.

Section 2.5

1. The argument is not valid, since the hypotheses are true if C is true and A is false. The error is in treating $A \to C$ and $\neg A \to \neg C$ as if they were equivalent.

3. (a) and (b). See (c).

(c) The case is no stronger. If C and all A_i's are false, then every hypothesis $A_i \to C$ is true, whether or not C is true.

5. With suggestive notation, the hypotheses are $\neg b \to \neg s$, $s \to p$ and $\neg p$. We can infer $\neg s$ using the contrapositive. We cannot infer either b or $\neg b$. Of course, we can infer more complex propositions, like $\neg p \vee s$ or $(s \wedge b) \to p$.

(c) The hypotheses are $(m \vee f) \to c$, $n \to c$ and $\neg n$. No interesting conclusions, such as m or $\neg c$, can be inferred.

7. (a) True. $A \to B$ is a hypothesis. We showed that $\neg A \to \neg B$ follows from the hypotheses.

(c) True. Since $(B \vee \neg Y) \to A$ is given, $\neg Y \to A$ follows, or equivalently $\neg\neg Y \vee A$.

9. (a) Let $c =$ "my computations are correct," $b =$ "I pay the electric bill," $r =$ "I run out of money," and $p =$ "the power stays on." Then the theorem is: If $(c \wedge b) \to r$ and $\neg b \to \neg p$, then $(\neg r \wedge p) \to \neg c$.
   Here's one proof; you supply the missing reasons.

   1. $(c \wedge b) \to r$
   2. $\neg b \to \neg p$
   3. $\neg r \to \neg(c \wedge b)$ 1; contrapositive rule 9
   4. $\neg r \to (\neg c \vee \neg b)$
   5. $p \to b$
   6. $(\neg r \wedge p) \to [(\neg c \vee \neg b) \wedge b]$ 4, 5; rule of inference corresponding to rule 26b
   7. $(\neg r \wedge p) \to [b \wedge (\neg c \vee \neg b)]$
   8. $(\neg r \wedge p) \to [(b \wedge \neg c) \vee (b \wedge \neg b)]$
   9. $(\neg r \wedge p) \to [(b \wedge \neg c) \vee \text{contradiction}]$
   10. $(\neg r \wedge p) \to (b \wedge \neg c)$ 9; identity law 6a
   11. $(\neg r \wedge p) \to (\neg c \wedge b)$
   12. $(\neg c \wedge b) \to \neg c$ simplification (rule 17)
   13. $(\neg r \wedge p) \to \neg c$

   (c) Let $j =$ "I get the job," $w =$ "I work hard," $p =$ "I get promoted," and $h =$ "I will be happy." Then the theorem is: If $(j \wedge w) \to p$, $p \to h$ and $\neg h$, then $\neg j \vee \neg w$. Here's one proof; you supply the reasons.

   1. $(j \wedge w) \to p$ 4. $\neg p$
   2. $p \to h$ 5. $\neg(j \wedge w)$
   3. $\neg h$ 6. $\neg j \vee \neg w$

11. Establish s quickly. Then the hypothetical deductions from Example 3 can be replaced by modus ponens inferences.

13. (a) Here is one proof; you supply the reasons.

    1. $A \wedge \neg B$ 4. P
    2. A 5. $\neg B$
    3. $A \to P$ 6. $P \wedge \neg B$

Section 3.1

1. (a) R_1 satisfies (AR) and (S).
   (c) R_3 satisfies (R), (AS) and (T).
   (e) R_5 satisfies only (S).

3. The relations in (a) and (c) are reflexive. The relations in (c), (d), (f), (g) and (h) are symmetric.

5. R_1 satisfies (AR) and (S). R_2 and R_3 satisfy only (S).

7. (a) The empty relation satisfies (AR), (S), (AS) and (T). The last three properties hold vacuously.

9. (a) If $E \subseteq R_1$ and $E \subseteq R_2$, then $E \subseteq R_1 \cap R_2$.
   (c) Suppose that R_1 and R_2 are transitive. If $(x, y), (y, z) \in R_1 \cap R_2$, then $(x, y), (y, z) \in R_1$, so $(x, z) \in R_1$. Similarly, $(x, z) \in R_2$.

11. (a) Suppose that R is symmetric. If $(x, y) \in R$, then $(y, x) \in R$ by symmetry, so $(x, y) \in R^{\leftarrow}$. Similarly, $(x, y) \in R^{\leftarrow}$ implies $(x, y) \in R$ [check], so that $R = R^{\leftarrow}$. For the converse, suppose that $R = R^{\leftarrow}$ and show that R is symmetric.

13. (a)  **(c)** See Figure 1(a). **(e)**

Section 3.2

1. (a)

e	a	b	c	d	e	f
$\gamma(e)$	(x, v)	(v, x)	(v, w)	(w, y)	(w, y)	(y, x)

(c)

e	a	b	c	d
$\gamma(e)$	(x, w)	(w, x)	(y, z)	(z, y)

3. (a) Yes.   **(c)** No. There is no edge from t to x.   **(e)** Yes.

5. (a) $x\,w\,y$ or $x\,w\,v\,z\,y$.

(c) $v\,x\,w$ or $v\,z\,w$ or $v\,z\,x\,w$ or $v\,z\,y\,x\,w$.

(e) $z\,y\,x\,w\,v$ or $z\,w\,v$ or $z\,x\,w\,v$ or $z\,y\,v$.

7. Here is one:

$$x \quad \overset{\longrightarrow}{\underset{\longleftarrow}{}} \quad y \quad \overset{\longrightarrow}{\underset{\longleftarrow}{}} \quad z$$

9. (a) (v, w), (v, y), (v, z). Note that (v, z) is in the reachable relation but not in the adjacency relation.

(c) All of them. The reachable relation is the universal relation.

11. (a) 2.   **(c)** 3.

(e) Not the vertex sequence for a path.   **(g)** 3.

13. (a) Edges e, f and g are parallel.

15. (a) $A = \{(w, w), (w, x), (x, w), (y, y), (w, y), (y, w), (x, z), (z, x)\}$, while R consists of all sixteen ordered pairs of vertices.

17. (a) $c\,a\,d$ or $c\,b\,d$. Both have vertex sequence $y\,v\,w\,w$.

(c) There are 4 such paths, each with vertex sequence $v\,w\,w\,v\,y$.

Section 3.3

1. (a) 1.   **(c)** 2.

3. (a) $\begin{bmatrix} -1 & 1 & 4 \\ 0 & 3 & 2 \\ 2 & -2 & 3 \end{bmatrix}$.   **(c)** $\begin{bmatrix} 5 & 8 & 7 \\ 5 & 1 & 5 \\ 7 & 3 & 5 \end{bmatrix}$.   **(e)** $\begin{bmatrix} 5 & 5 & 7 \\ 8 & 1 & 3 \\ 7 & 5 & 5 \end{bmatrix}$.

(g) $\begin{bmatrix} 12 & 12 & 8 \\ 12 & -4 & 8 \\ 8 & 8 & 4 \end{bmatrix}$.   **(i)** $\begin{bmatrix} 4 & 8 & 9 \\ 6 & 4 & 3 \\ 11 & 5 & 8 \end{bmatrix}$.

5. (a) $\begin{bmatrix} 1 & -1 & 1 & -1 \\ -1 & 1 & -1 & 1 \\ 1 & -1 & 1 & -1 \end{bmatrix}$. (c) Not defined.

(e) $\begin{bmatrix} 3 & 2 & 5 & 4 \\ 2 & 5 & 4 & 7 \\ 5 & 4 & 7 & 6 \end{bmatrix}$.

7. (a) $\begin{bmatrix} 1 & 0 & 0 \\ 0 & 1 & 0 \\ 0 & 0 & 1 \end{bmatrix}$, $\begin{bmatrix} 1 & 0 & 0 \\ 0 & 0 & 1 \\ 0 & 1 & 0 \end{bmatrix}$, $\begin{bmatrix} 0 & 1 & 0 \\ 1 & 0 & 0 \\ 0 & 0 & 1 \end{bmatrix}$, $\begin{bmatrix} 0 & 1 & 0 \\ 0 & 0 & 1 \\ 1 & 0 & 0 \end{bmatrix}$, $\begin{bmatrix} 0 & 0 & 1 \\ 1 & 0 & 0 \\ 0 & 1 & 0 \end{bmatrix}$,

$\begin{bmatrix} 0 & 0 & 1 \\ 0 & 1 & 0 \\ 1 & 0 & 0 \end{bmatrix}$.

9. (a) $\begin{bmatrix} 1 & 0 \\ n & 1 \end{bmatrix}$. (c) $\{n \in \mathbb{N} : n \text{ is odd}\}$.

11. (a) The (i, j) entry of $a\mathbf{A}$ is $a\mathbf{A}[i, j]$. Similarly for $b\mathbf{B}$, so the (i, j) entry of $a\mathbf{A} + b\mathbf{B}$ is $a\mathbf{A}[i, j] + b\mathbf{B}[i, j]$. So the (i, j) entry of $c(a\mathbf{A} + b\mathbf{B})$ is $ca\mathbf{A}[i, j] + cb\mathbf{B}[i, j]$. A similar discussion shows that this is the (i, j) entry of $(ca)\mathbf{A} + (cb)\mathbf{B}$. Since their entries are equal, the matrices $c(a\mathbf{A} + b\mathbf{B})$ and $(ca)\mathbf{A} + (cb)\mathbf{B}$ are equal.

(c) Compare the (j, i) entries of $(a\mathbf{A})^T$ and $a\mathbf{A}^T$.

13. (a) $\begin{bmatrix} 0 & 0 & 1 & 0 \\ 0 & 0 & 1 & 0 \\ 0 & 0 & 0 & 0 \\ 0 & 0 & 1 & 0 \end{bmatrix}$. (c) $\begin{bmatrix} 0 & 0 & 0 & 0 \\ 0 & 0 & 0 & 2 \\ 0 & 0 & 1 & 0 \\ 0 & 1 & 0 & 0 \end{bmatrix}$.

15. (a) 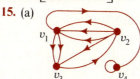 (c)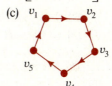

17. (a) $\begin{bmatrix} 0 & 0 & 0 & 1 \\ 0 & 0 & 1 & 0 \\ 0 & 1 & 0 & 0 \\ 1 & 0 & 0 & 0 \end{bmatrix}$. (c) See Example 5(a). (e) $\begin{bmatrix} 0 & 0 & 0 & 1 \\ 0 & 0 & 0 & 1 \\ 0 & 0 & 0 & 1 \\ 1 & 1 & 1 & 1 \end{bmatrix}$.

19. (a) $\begin{bmatrix} 1 & 1 & 1 \\ 0 & 1 & 1 \\ 0 & 0 & 1 \end{bmatrix}$. (c) $\begin{bmatrix} 1 & 0 & 0 \\ 0 & 1 & 0 \\ 0 & 0 & 1 \end{bmatrix}$. (e) $\begin{bmatrix} 1 & 1 & 1 \\ 0 & 1 & 0 \\ 0 & 1 & 0 \end{bmatrix}$.

(g) $\begin{bmatrix} 0 & 0 & 0 \\ 0 & 1 & 0 \\ 0 & 0 & 0 \end{bmatrix}$. (i) $\begin{bmatrix} 0 & 0 & 0 \\ 1 & 1 & 0 \\ 0 & 0 & 1 \end{bmatrix}$.

Section 3.4

1. (a) $\begin{bmatrix} -8 & 13 \\ 2 & 9 \end{bmatrix}$.   (c) $\begin{bmatrix} 31 & -16 & -6 \\ 29 & 4 & 26 \end{bmatrix}$.   (e) $\begin{bmatrix} -1 & 7 \\ 8 & 0 \\ 16 & 0 \end{bmatrix}$.

3. The products written in parts (a), (c) and (e) do not exist.

5. (a) $\begin{bmatrix} 1 & 10 \\ 11 & 19 \end{bmatrix}$.

7. (a) $\begin{bmatrix} 7 & 14 \\ 8 & 11 \\ 2 & -6 \end{bmatrix}$.

9. (a) 2.   (c) 2.

11. (a) $\mathbf{M}^3 = \begin{bmatrix} 3 & 20 & 2 & 3 \\ 0 & 8 & 0 & 0 \\ 2 & 9 & 1 & 2 \\ 0 & 4 & 0 & 0 \end{bmatrix}$.

(c) $fab, fac, fbd, fbe, fcd, fce, fhj, kjd, kje$.

13. (a) Simply remove the arrows from Figure 1.
   (c) $dd, ee, de, ed, bb, cc, bc, cb, jj$.

15. (a) $\mathbf{I}^{-1} = \mathbf{I}$.   (c) Not invertible.   (e) $\mathbf{D}^{-1} = \mathbf{D}$.

17. (b) Correct guess: $\mathbf{A}^n = \begin{bmatrix} 1 & 0 \\ n & 1 \end{bmatrix}$.

19. For $1 \le k \le p$ and $1 \le i \le m$,

$$(\mathbf{B}^T\mathbf{A}^T)[k, i] = \sum_{j=1}^{n} \mathbf{B}^T[k, j]\mathbf{A}^T[j, i] = \sum_{j=1}^{n} \mathbf{B}[j, k]\mathbf{A}[i, j].$$

Compare with the (k, i) entry of $(\mathbf{AB})^T$.

21. (a) In fact, $\mathbf{AB} = \mathbf{BA} = a\mathbf{B}$ for all $\mathbf{B}$ in $\mathfrak{M}_{2,2}$.

(b) $\mathbf{AB} = \mathbf{BA}$ with $\mathbf{B} = \begin{bmatrix} 1 & 0 \\ 0 & 0 \end{bmatrix}$ forces $\begin{bmatrix} a & 0 \\ c & 0 \end{bmatrix} = \begin{bmatrix} a & b \\ 0 & 0 \end{bmatrix}$, so $b = c = 0$.

So $\mathbf{A} = \begin{bmatrix} a & 0 \\ 0 & d \end{bmatrix}$. Now try $\mathbf{B} = \begin{bmatrix} 0 & 1 \\ 0 & 0 \end{bmatrix}$.

23. (a) Consider $1 \le i \le m$ and $1 \le k \le p$ and compare the (i, k) entries of $(\mathbf{A} + \mathbf{B})\mathbf{C}$ and $\mathbf{AC} + \mathbf{BC}$.

Section 3.5

1. (a) is an equivalence relation.

(c) is not reflexive because some Americans don't live in any state.
(e) is not an equivalence relation because $\approx$ is not transitive.

3. Very much so.

5. (a) The possibilities are

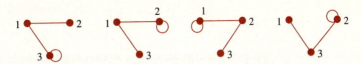

7. (a) Verify directly or apply Theorem 2(a) with $f(m) = m^2$ for $m \in \mathbb{Z}$.

9. (a) There are infinitely many classes: $\{0\}$ and the classes $\{n, -n\}$ for $n \in \mathbb{P}$.

11. Apply Theorem 2, using the length function. The equivalence classes are the sets Σ^k, $k \in \mathbb{N}$.

13. (a) Use brute force or Theorem 2(a) with part (b).

15. (a) Not well-defined: depends on the representative. For example, $[3] = [-3]$ and $-3 \le 2$. If the definition made sense, we would have $[3] = [-3] \le [2]$ and hence $3 \le 2$.

(c) Nothing wrong. If $[m] = [n]$, then $m^4 + m^2 + 1 = n^4 + n^2 + 1$.

17. (a) Since $\sim$ is reflexive and symmetric, so is $\approx$. Given chains $f = f_1 \sim f_2 \sim \cdots \sim f_n = g$ and $g = g_1 \sim g_2 \sim \cdots \sim g_m = h$, we see that $f = f_1 \sim f_2 \sim \cdots \sim f_n \sim g_2 \sim \cdots \sim g_m = h$ is a chain, so $f \approx h$. Thus $\approx$ is transitive.

Section 3.6

1. (a) $q = 6$, $r = 2$.   (c) $q = -7$, $r = 1$.   (e) $q = 5711$, $r = 31$.

3. (a) $-4, 0, 4$.   (c) $-2, 2, 6$.   (e) $-4, 0, 4$.

5. (a) 1.   (c) 1.   (e) 0.

7. (a) 3 and 2.   (c) $m *_{10} k$ is the last [decimal] digit of $m * k$.

9.

$+_4$	0	1	2	3
0	0	1	2	3
1	1	2	3	0
2	2	3	0	1
3	3	0	1	2

$*_4$	0	1	2	3
0	0	0	0	0
1	0	1	2	3
2	0	2	0	2
3	0	3	2	1

11. Solutions are 1, 3, 2 and 4, respectively.

13. (a) $m \equiv n \pmod{1}$ for all $m, n \in \mathbb{Z}$. There is only one equivalence class.

(c) $0 = 0 +_1 0$ and $0 = 0 *_1 0$.

15. (a) $n = 1000a + 100b + 10c + d = a + b + c + d + 9 \cdot (111a + 11b + c) \equiv a + b + c + d \pmod{9}$, or use Theorem 2 together with $1000 \equiv 100 \equiv 10 \equiv 1 \pmod{9}$.

17. Like Exercise 15. Note that $1000 = 91 \cdot 11 - 1$, $100 = 9 \cdot 11 + 1$, $10 = 1 \cdot 11 - 1$, so $1000a + 100b + 10c + d \equiv -a + b - c + d \equiv 0 \pmod{11}$ if and only if $a - b + c - d \equiv 0 \pmod{11}$.

19. We have $q \cdot p - q' \cdot p = r' - r$, so $r' - r$ is a multiple of p. But $-p < -r \le r' - r \le r' < p$, and 0 is the only multiple of p between $-p$ and p. Thus $r' = r$, so $0 = (q - q') \cdot p$ and $q = q'$.

21. (a) By Theorem 3(a)

$$(m \operatorname{MOD} p) +_p (n \operatorname{MOD} p) = (m + n) \operatorname{MOD} p = (n + m) \operatorname{MOD} p$$

$$= (n \operatorname{MOD} p) +_p (m \operatorname{MOD} p).$$

Since $m, n \in \mathbb{Z}(p)$, $m \operatorname{MOD} p = m$ and $n \operatorname{MOD} p = n$.

Section 4.1

1. (a) 0, 3, 9, 21, 45.   (c) 1, 1, 1, 1, 1.

3. (a)

	m	n
initially	0	0
after first pass	1	1
after second pass	4	2
after third pass	9	3
after fourth pass	16	4

5. (a) 4, 16, 36, 64.

7. (a) If $m + n$ is even, so is $(m + 1) + (n + 1) = (m + n) + 2$. Of course, we didn't need the guard $1 \leq m$ to see this.

9. (a) Yes. If $i < j^2$ and $j \geq 1$, then $i + 2 < j^2 + 2 < j^2 + 2j + 1 = (j + 1)^2$.

(c) No. Consider the case $i = j = 0$.

11. (a) $b \geq 2$.   (c) $b \in \mathbb{N}$.

13. No. The sequence "$k := k^2$, print k" changes the value of k. Algorithm A prints 1, 4 and stops. Algorithm B prints 1, 4, 9, 16, because "for" resets k each time.

15. (a) Yes; new $r < 73$ by definition of MOD.

(c) This is an invariant vacuously, because $r \leq 0$ and $r > 0$ cannot both hold at the start of the loop.

17. The sets in (a), (c) and (f) have smallest elements; the others don't.

19. (a) This is an invariant: If $r > 0$ and a, b and r are multiples of 5, then the new values b, r and $b \operatorname{MOD} r$ are multiples of 5. Note that $b \operatorname{MOD} r = b - (b \operatorname{DIV} r) \cdot r$.

(c) This is an invariant.

21. (a) If m is even $2^{(\text{new } k)} \cdot (\text{new } m) = 2^{k+1} \cdot m/2 = 2^k \cdot m = n$.

23. (a)

$a = 2, n = 11$	p	q	i
initially	1	2	11
after first pass	2	4	5
after second pass	8	16	2
after third pass	8	256	1
after fourth pass	$256 \cdot 8$	256^2	0

(b) Show that if $q^i p = a^n$, then $(\text{new } q)^{(\text{new } i)}(\text{new } p) = a^n$; consider the cases when i is odd and when i is even. Since i runs through a decreasing sequence of nonnegative integers, eventually $i = 0$ and the algorithm exits the loop. At this point $p = a^n$.

Section 4.2

Induction proofs should be written carefully and completely. These answers will serve only as guides, *not* as models.

1. Check the basis. For the inductive step, assume that the equality holds for k. Then

$$\sum_{i=1}^{k+1} i^2 = \sum_{i=1}^{k} i^2 + (k+1)^2 = \frac{k(k+1)(2k+1)}{6} + (k+1)^2.$$

Some algebra shows that the right-hand side equals

$$\frac{(k+1)(k+2)(2k+3)}{6},$$

so the equality holds for $k+1$ whenever it holds for k.

3. (a) Take $n = 37^{20}$ in Example 1.

(c) By (a), (b) and Example 1,

$$(37^{500} - 37^{100}) + (37^{100} - 37^{20}) + (37^{20} - 37^4)$$

is a multiple of 10.

(e) By (c) and (d), as in (c).

5. The basis is "$s_0 = 2^0 a + (2^0 - 1)b$." Assume inductively that $s_k = 2^k a + (2^k - 1)b$ for some $k \in \mathbb{N}$. The algebra in the inductive step is

$$2 \cdot [2^k a + (2^k - 1)b] + b = 2^{k+1} a + 2^{k+1} b - 2b + b.$$

7. Show that $11^{k+1} - 4^{k+1} = 11 \cdot (11^k - 4^k) + 7 \cdot 4^k$. Imitate Example 2(d).

9. (a) Suppose that $\sum_{i=0}^{k} 2^i = 2^{k+1} - 1$ and $0 \le k$. Then

$$\sum_{i=0}^{k+1} 2^i = \left(\sum_{i=0}^{k} 2^i\right) + 2^{k+1} = 2^{k+1} - 1 + 2^{k+1} = 2^{k+2} - 1,$$

so the equation still holds for the new value of k.

(c) Yes. $\sum_{i=0}^{0} 2^i = 1 = 2^1 - 1$ initially, so the loop never exits and the invariant is true for every value of k in $\mathbb{N}$.

11. (a) $1 + 3 + \cdots + (2n - 1) = n^2$.

(b) For the inductive step

$$k^2 + [2(k+1) - 1] = k^2 + 2k + 1 = (k+1)^2.$$

13. (a) Assume that $p(k)$ is true. Then $(k+1)^2 + 5(k+1) + 1 = (k^2 + 5k + 1) + (2k + 6)$. Since $k^2 + 5k + 1$ is even by assumption and $2k + 6$ is clearly even, $p(k+1)$ is true.

(b) All propositions $p(n)$ are false. *Moral:* The basis of induction is crucial for mathematical induction. Also, see Exercises 8 and 9.

15. *Hint:* $5^{k+1} - 4(k+1) - 1 = 5(5^k - 4k - 1) + 16k$.

17. *Hints:*

$$\frac{1}{n+2} + \cdots + \frac{1}{2n+2} = \left(\frac{1}{n+1} + \cdots + \frac{1}{2n}\right) + \left(\frac{1}{2n+1} + \frac{1}{2n+2} - \frac{1}{n+1}\right)$$

and

$$\frac{1}{2n+1} + \frac{1}{2n+2} - \frac{1}{n+1} = \frac{1}{2n+1} - \frac{1}{2n+2}.$$

Alternatively, to avoid induction, let $f(n) = \sum_{i=1}^{n} \frac{1}{i}$ and show that both sides are equal to $f(2n) - f(n)$.

19. *Hints:* $5^{k+2} + 2 \cdot 3^{k+1} + 1 = 5(5^{k+1} + 2 \cdot 3^k + 1) - 4(3^k + 1)$. Show that $3^n + 1$ is always even.

21. Here $p(n)$ is the proposition "$|\sin nx| \leq n|\sin x|$ for all $x \in \mathbb{R}$."
Clearly, $p(1)$ holds. By algebra and trigonometry,

$$|\sin(k+1)x| = |\sin(kx + x)| = |\sin kx \cos x + \cos kx \sin x|$$

$$\leq |\sin kx| \cdot |\cos x| + |\cos kx| \cdot |\sin x| \leq |\sin kx| + |\sin x|.$$

Now assume $p(k)$ is true and show that $p(k+1)$ is true.

Section 4.3

1. (a) $1, 2, 1, 2, 1, 2, 1, 2, \ldots$.

3. (a) $\text{SEQ}(n) = 3^n$.

5. No. It's okay up to $\text{SEQ}(100)$, but $\text{SEQ}(101)$ is not defined, since we cannot divide by zero. If, in (R), we restricted n to be ≤ 100, we would obtain a recursively defined *finite* sequence.

7. (a) $1, 3, 8$.   (c) $s_3 = 22$, $s_4 = 60$.

9. (a) $1, 1, 2, 4$.   (c) Ours doesn't, since $t_0 \neq 1/2$.

11. (a) $a_6 = a_5 + 2a_4 = a_4 + 2a_3 + 2a_4 = 3(a_3 + 2a_2) + 2a_3 = 5(a_2 + 2a_1) + 6a_2 = 11(a_1 + 2a_0) + 10a_1 = 11 \cdot 3 + 10 = 43$. This calculation uses only two intermediate value addresses at any given time. Other recursive calculations are possible that use more.

(b) Use induction.

13. Apply the Well-Ordering Principle to S and use the recursive definition of FIB in Example 3(a).

15. $\text{SEQ}(n) = 2^{n-1}$ for $n \geq 1$.

17. (a) $A(1) = 1$. $A(n) = n \cdot A(n-1)$.   (c) Yes.

19. (a) $\{1, 110, 1200\}$.

21. (a) (B) $\text{UNION}(1) = A_1$;

      (R) $\text{UNION}(n) = A_n \cup \text{UNION}(n-1)$ for $n \geq 2$.

(b) The "empty union" is $\varnothing$.

Section 4.4

1. $s_n = 3 \cdot (-2)^n$ for $n \in \mathbb{N}$.

3. We prove this by induction. $s_n = a^n \cdot s_0$ holds for $n = 0$ because $a^0 = 1$. If it holds for some n, then $s_{n+1} = as_n = a(a^n \cdot s_0) = a^{n+1} \cdot s_0$, and so the result holds for $n + 1$.

5. $s_0 = 3^0 - 2 \cdot 0 \cdot 3^0 = 1$, $s_1 = 3^1 - 2 \cdot 1 \cdot 3^1 = -3$. For $n \geq 2$,
$$6s_{n-1} - 9s_{n-2} = 6[3^{n-1} - 2(n-1) \cdot 3^{n-1}] - 9[3^{n-2} - 2(n-2) \cdot 3^{n-2}]$$
$$= 3^n[1 - 2n] = s_n.$$

7. This time $c_1 = 3$ and $c_2 = 0$, so $s_n = 3 \cdot 2^n$ for $n \in \mathbb{N}$.

9. Solve $1 = c_1 + c_2$ and $2 = c_1 r_1 + c_2 r_2$ for c_1 and c_2 to obtain $c_1 = (1 + r_1)/\sqrt{5}$ and $c_2 = -(1 + r_2)/\sqrt{5}$. Hence
$$s_n = \frac{1}{\sqrt{5}}(r_1^n + r_1^{n+1} - r_2^n - r_2^{n+1}) \qquad \text{for all } n,$$
where r_1, r_2 are as in Example 3.

11. (a) $r_1 = -3$, $r_2 = 2$, $c_1 = c_2 = 1$; hence $s_n = (-3)^n + 2^n$ for $n \in \mathbb{N}$.
    (c) Here the characteristic equation has one solution, $r = 2$; $c_1 = 1$ and $c_2 = 3$, so $s_n = 2^n + 3n \cdot 2^n$ for $n \in \mathbb{N}$.
    (e) $s_{2n} = 1$, $s_{2n+1} = 4$ for all $n \in \mathbb{N}$.
    (g) $s_n = (-3)^n$ for $n \in \mathbb{N}$.

13. (a) $s_{2m} = 2^m + 3 \cdot (2^m - 1) = 2^{m+2} - 3$.
    (c) $s_{2m} = \frac{5}{2} \cdot 2^m \cdot m$.   (e) $s_{2m} = 7 - 6 \cdot 2^m$.   (g) $s_{2m} = (6 - m)2^{m-1}$.

15. $s_{2m} = 2^m[s_1 + \frac{1}{2}(2^m - 1)]$. Verify that this formula satisfies $s_{20} = s_1$ and $s_{2m+1} = 2s_{2m} + (2^m)^2$.

17. (a) $t_{2m} = b^m t_1 + b^{m-1} \cdot \sum_{i=0}^{m-1} \frac{f(2^i)}{b^i}$.

Section 4.5

1. The First Principle is adequate for this. For the inductive step, use the identity
$$4n^2 - n + 8(n + 1) - 5 = 4n^2 + 7n + 3 = 4(n + 1)^2 - (n + 1).$$

3. Show that $n^5 - n$ is always even. Then use the identity
$$(n + 1)^5 = n^5 + 5n^4 + 10n^3 + 10n^2 + 5n + 1$$
[from the binomial theorem].

5. Yes. The oddness of a_n depends only on the oddness of a_{n-1}, since $2a_{n-2}$ is even whether a_{n-2} is odd or not.

7. (b) $a_n = 1$ for all $n \in \mathbb{N}$.
   (c) The basis needs to be checked for $n = 0$ and $n = 1$. For the inductive step, consider $n \geq 2$ and assume $a_k = 1$ for $0 \leq k < n$. Then
$$a_n = \frac{a_{n-1}^2 + a_{n-2}}{a_{n-1} + a_{n-2}} = \frac{1^2 + 1}{1 + 1} = 1.$$

This completes the inductive step, so $a_n = 1$ for all $n \in \mathbb{N}$ by the Second Principle of Induction.

9. (b) $a_n = n^2$ for all $n \in \mathbb{N}$.

(c) The basis needs to be checked for $n = 0$ and $n = 1$. For the inductive step, consider $n \geq 2$ and assume that $a_k = k^2$ for $0 \leq k < n$. To complete the inductive step, show that $a_n = n^2$.

11. (b) The basis needs to be checked for $n = 0$, 1 and 2. For the inductive step, consider $n \geq 3$ and assume that a_k is odd for $0 \leq k < n$. Show that a_n is also odd.

(c) Since the inequality is claimed for $n \geq 1$ and since you will want to use the identity $a_n = a_{n-1} + a_{n-2} + a_{n-3}$ in the inductive step, you will need $n - 3 \geq 1$ in the inductive step. So check the basis for $n = 1$, 2 and 3. For the inductive step, consider $n \geq 4$ and assume that $a_k \leq 2^{k-1}$ for $1 \leq k < n$. To complete the inductive step, show that $a_n < 2^{n-1}$.

13. (a) 2, 3, 4, 6.

(b) The inequality must be checked for $n = 3$, 4 and 5 before applying the Second Principle of Mathematical Induction to $b_n = b_{n-1} + b_{n-3}$.

(c) The inequality must be checked for $n = 2$, 3 and 4. Then use the Second Principle of Mathematical Induction and part (b). For the inductive step, consider $n \geq 5$ and assume that $b_k \geq (\sqrt{2})^{k-2}$ for $2 \leq k < n$. Then

$$b_n = b_{n-1} + b_{n-3} \geq 2b_{n-3} + b_{n-3} = 3b_{n-3}$$

$$\geq 3(\sqrt{2})^{n-5} > (\sqrt{2})^3(\sqrt{2})^{n-5} = (\sqrt{2})^{n-2}.$$

This can also be proved without using part (b).

15. Check for $n = 0$ and 1 before applying induction. It may be simpler to prove "SEQ$(n) \leq 1$ for all n" separately from "SEQ$(n) \geq 0$ for all n." For example, assume that $n \geq 2$ and that SEQ$(k) \leq 1$ for $0 \leq k < n$. Then

$$\text{SEQ}(n) = (1/n) * \text{SEQ}(n-1) + ((n-1)/n) * \text{SEQ}(n-2)$$

$$\leq (1/n) + ((n-1)/n) = 1.$$

The proof that SEQ$(n) \geq 0$ for $n \geq 0$ is almost the same.

17. The First Principle is enough. Use (R) to check for $n = 2$. For the inductive step from n to $n + 1$,

$$\text{FIB}(n+1) = \text{FIB}(n) + \text{FIB}(n-1)$$

$$= 1 + \sum_{k=0}^{n-2} \text{FIB}(k) + \text{FIB}(n-1) = 1 + \sum_{k=0}^{n-1} \text{FIB}(k).$$

19. For $n > 0$, let $L(n)$ be the largest integer 2^k with $2^k \leq n$. Show that $L(n) = T(n)$ for all n by showing first that $L(\lfloor n/2 \rfloor) = L(n/2)$ for $n \geq 2$ and then using the Second Principle of Induction.

21. Show that $S(n) \leq n$ for every n by the Second Principle of Induction.

Section 4.6

1. (a) 20.   (c) 1.   (e) 4.   (g) 6.

3. (a) (20, 14), (14, 6), (6, 2), (2, 0); gcd = 2.

   (c) (20, 30), (30, 20), (20, 10), (10, 0); gcd = 10.

5. (a) gcd(20, 14) = 2, $s = -2$, $t = 3$.   (c) gcd(20, 30) = 10, $s = -1$, $t = 1$.

a	q	s	t
20		1	0
14	1	0	1
6	2	1	−1
2	3	−2	3
0			

a	q	s	t
20		1	0
30	0	0	1
20	1	1	0
10	2	−1	1
0			

7. (a) $x = 21$ $[\equiv -5 \,(\text{mod } 26)]$.

   (c) No solution exists, because 4 and 26 are not relatively prime.

   (e) $x = 23$ $[\equiv -3 \,(\text{mod } 26)]$.

9. (a) $x \equiv 5 \,(\text{mod } 13)$.   (c) Same as (a), since $99 \equiv 8 \,(\text{mod } 13)$.

11. (a) $x = 99$. Details: $x = 99y$, so $8y \equiv 99y \equiv 8 \,(\text{mod } 13)$. Cancel a factor of 8 to get $y \equiv 1 \,(\text{mod } 13)$. Taking $y = 1$ gives $x = 99$.

   (b) $x = 65$.

   (c) $x = 164$. It is not an accident that the answers for (a) and (b) sum to the answer for (c). If x solves (a) and x' solves (b), then $x + x'$ solves (c).

13. Assume that $a = s \cdot m + t \cdot n$ and $a' = s' \cdot m + t' \cdot n$ at the start of the loop. The equation $a' = s' \cdot m + t' \cdot n$ becomes $a_{\text{next}} = s_{\text{next}} \cdot m + t_{\text{next}} \cdot n$ at the end, and

$$a'_{\text{next}} = a' - q \cdot a = s' \cdot m + t' \cdot n - q \cdot s \cdot m - q \cdot t \cdot n$$
$$= (s' - q \cdot s) \cdot m + (t' - q \cdot t) \cdot n = s'_{\text{next}} \cdot m + t'_{\text{next}} \cdot n$$

at the end.

15. (a) $1 = s \cdot (m/d) + t \cdot (n/d)$ for some integers s and t. Apply Exercise 14 to m/d and n/d in place of m and n. A longer proof can be based on prime factorization.

   (c) Let $x = s \cdot a/d$.

17. (a) Check for $l = 1$. For the inductive step it suffices to show that if $a = m = \text{FIB}(l + 2)$ and $b = n = \text{FIB}(l + 1)$, $l \geq 1$, then after the first pass of the while loop, $a = \text{FIB}(l + 1)$ and $b = \text{FIB}(l)$. For then, by the inductive hypothesis, exactly l more passes would be needed before terminating the algorithm. By the definition of (a, b) in the while loop, it suffices to show that

$$\text{FIB}(l + 2) \,\text{MOD}\, \text{FIB}(l + 1) = \text{FIB}(l).$$

   (c) By part (a), GCD makes l passes through the loop. By part (b), $\log_2 (m + n) = \log_2 \text{FIB}(l + 2) \leq l$ if $l \geq 3$.

Section 5.1

1. (a) 56.   (c) 56.   (e) 1.

3. (a) Draw a Venn diagram and work from the inside out. The 10 is given as $|S \cap B \cap J|$; the 17 is calculated from $27 = |J \cap S|$; etc. Answer $= 15$.

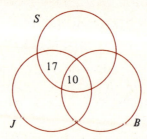

5. (a) 126.   (b) 105.

7. (a) 0.   (b) 840.   (c) 2401.

9. (a) This is the same as the number of ways of drawing ten cards so that the *first* one is not a repetition. Hence $52 \cdot (51)^9$.

    (b) $52^{10} - 52(51)^9$.

11. (a) $13 \cdot \binom{4}{4} \cdot \binom{48}{1} = 624$.   (b) 5108.

    (c) $13 \cdot \binom{4}{3} \cdot \binom{12}{2} \cdot 4 \cdot 4 = 54{,}912$.   (b) 1,098,240.

13. (a) It is the $n \times n$ matrix with 0's on the diagonal and 1's elsewhere.

15. (a) $n \cdot (n-1)^3$.   (b) $n(n-1)(n-2)(n-3)$.

    (c) $n(n-1)(n-2)(n-2)$.

Section 5.2

1. (a) $8/25 = .32$.   (b) .20.   (c) $9/25 = .36$.

3. (a) $\dfrac{5 \cdot 4 \cdot 3 \cdot 2}{5^4} = .192$.   (b) $\dfrac{3^4}{5^4} = .1296$.   (c) $2/5 = .40$.

5. Note that $\binom{7}{3} = 35$ is the number of ways to select three balls from the urn.

    (a) 1/35.   (b) 4/35.   (c) 18/35.   (d) 12/35.

7. (a) .2.   (b) .9.   (c) .6.

9. Here $N = 2{,}598{,}960$.

    (a) $624/N \approx .000240$.   (b) $54{,}912/N \approx .0211$.

    (c) $10{,}200/N \approx .00392$.   (d) $123{,}552/N \approx .0475$.

    (e) $1{,}098{,}240/N \approx .423$.

11. (a) 1/2.   (b) 15/36.   (c) 1/12.

13. We have $P(E_1 \cup E_2 \cup E_3) = P(E_1 \cup E_2) + P(E_3) - P((E_1 \cup E_2) \cap E_3)$.

Now

$$P(E_1 \cup E_2) = P(E_1) + P(E_2) - P(E_1 \cap E_2) \quad \text{and}$$

$$P((E_1 \cup E_2) \cap E_3) = P((E_1 \cap E_3) \cup (E_2 \cap E_3))$$
$$= P(E_1 \cap E_3) + P(E_2 \cap E_3) - P(E_1 \cap E_3 \cap E_2 \cap E_3).$$

Substitute. Alternatively, use

$$E_1 \cup E_2 \cup E_3 = (E_1 \backslash E_2) \cup (E_2 \backslash E_3) \cup (E_3 \backslash E_1) \cup (E_1 \cap E_2 \cap E_3).$$

15. (a) 1/64.   (b) 6/64.   (c) 15/64.   (d) 20/64.   (e) 22/64.

17. Let Ω_n be all n-tuples of H's and T's and E_n all n-tuples in Ω_n with an even number of H's. It suffices to show $|E_n| = 2^{n-1}$, which can be done by induction.

19. (a) If Ω is the set of all 4-element subsets of S, the outcomes are equally likely. If E_2 is the event "exactly 2 are even," then

$$|E_2| = \binom{4}{2} \cdot \binom{4}{2} = 36.$$

Since

$$|\Omega| = \binom{8}{4} = 70, \qquad P(E_2) = \frac{36}{70} \approx .514.$$

(b) 1/70.   (c) 16/70.   (d) 16/70.   (e) 1/70.

21. (a) The sample space Ω is all triples (k, l, m) where $k, l, m \in \{1, 2, 3\}$, so $|\Omega| = 3^3$. The triples where $k, l, m \in \{2, 3\}$ correspond to no selection of 1. So $P(\text{1 not selected}) = 2^3/3^3$ and answer $= 1 - \frac{8}{27} \approx .704$.

(b) $1 - \dfrac{3^4}{4^4} \approx .684.$ (c) $1 - \left(\dfrac{n-1}{n}\right)^n.$

(d) $1 - (.999999)^{1,000,000} \approx .632120.$

Section 5.3

1. 125.

3. There are 466 such numbers in the set, so the probability is .466. Remember that $D_4 \cap D_6 = D_{12}$, not D_{24}. Hence

$$|D_4| + |D_5| + |D_6| - |D_4 \cap D_5| - |D_4 \cap D_6| - |D_5 \cap D_6| + |D_4 \cap D_5 \cap D_6|$$
$$= 250 + 200 + 166 - 50 - 83 - 33 + 16 = 466.$$

5. (a) .142.   (b) .09.   (c) .78.   (d) .208.

7. (a) $\dbinom{12 + 4 - 1}{4 - 1} = 455.$ (b) $\dbinom{4 + 4 - 1}{4 - 1} = 35.$

9. (a) $x^4 + 8x^3 y + 24x^2 y^2 + 32xy^3 + 16y^4.$

(c) $81x^4 + 108x^3 + 54x^2 + 12x + 1.$

11. (a) $\dbinom{n}{r} = \dfrac{n!}{(n-r)!\,r!} = \dfrac{n!}{r!\,(n-r)!} = \dfrac{n!}{(n-(n-r))!\,(n-r)!} = \dbinom{n}{n-r}.$

13. (b) There are $\binom{n}{r}$ subsets of size r for each r, so there are $\sum_{r=0}^{n}\binom{n}{r}$ subsets in all.

(c) If true for n, then

$$\sum_{r=0}^{n+1}\binom{n+1}{r} = 1 + \sum_{r=1}^{n}\binom{n+1}{r} + 1 = 1 + \sum_{r=1}^{n}\binom{n}{r-1} + \sum_{r=1}^{n}\binom{n}{r} + 1$$

$$= \sum_{r=1}^{n+1}\binom{n}{r-1} + \sum_{r=0}^{n}\binom{n}{r} = 2\sum_{r=0}^{n}\binom{n}{r} = 2 \cdot 2^n = 2^{n+1}.$$

15. (a) Both sides equal 15.

(b) Let $p(n)$ be " $\sum_{k=m}^{n}\binom{k}{m} = \binom{n+1}{m+1}$ " for $n \geq m$. Check that $p(m)$ is true. Assume that $p(n)$ is true for some $n \geq m$. Then

$$\sum_{k=m}^{n+1}\binom{k}{m} = \left[\sum_{k=m}^{n}\binom{k}{m}\right] + \binom{n+1}{m}$$

$$= \binom{n+1}{m+1} + \binom{n+1}{m} \qquad \text{[by the inductive assumption]}$$

$$= \binom{n+2}{m+1} \qquad \text{[give reason]}.$$

(c) The set $\mathscr{A}$ of $(m+1)$-element subsets of $\{1, 2, \ldots, n+1\}$ is the disjoint union $\bigcup_{k=m}^{n}\mathscr{A}_k$, where $\mathscr{A}_k$ is the collection of $(m+1)$-element subsets whose largest element is $k+1$. A set in $\mathscr{A}_k$ is an m-element subset of $\{1, 2, \ldots, k\}$ with $k+1$ added to it, so $|\mathscr{A}| = \sum_{k=m}^{n}\binom{k}{m}$.

17. (a) Put 8 in one of the 3 boxes, then distribute the remaining 6 in three boxes. Answer is $3 \cdot \binom{8}{2} = 84$ ways.

(b) 36.

(c) Since $1 + 9 + 9 < 20$, each digit is at least 2. Then

$$(d_1 - 2) + (d_2 - 2) + (d_3 - 2) = 20 - 6 = 14$$

With $2 \leq d_i \leq 9$ for $i = 1, 2, 3$. By part (b) there are 36 numbers. Another way to get this number is to hunt for a pattern:

$$299; \quad 398, 389; \quad 497, 488, 479; \quad 596, 587, 578, 569; \quad \ldots$$

Looks like $1 + 2 + 3 + \cdots + 8 = \binom{9}{2} = 36.$

Section 5.4

1. (a) $\dfrac{15!}{3!\,4!\,5!\,3!}$.   (b) $\binom{15}{3}\binom{15}{4}\binom{15}{5}$.

3. (a) As in Example 4, count ordered partitions (A, B, C, D) where $|A| = 5$, $|B| = 3$, $|C| = 2$ and $|D| = 3$. Answer $= \dfrac{13!}{5! \cdot 3! \cdot 2! \cdot 3!} = 720{,}720$.

(b) $\dfrac{1}{2} \cdot \dfrac{13!}{4! \cdot 3! \cdot 3! \cdot 3!} = 600{,}600$.

(c) Count ordered partitions where $|A| = |B| = |C| = 3$ and $|D| = 4$, but note that permutations of A, B and C give equivalent sets of committees.
Answer $= \dfrac{1}{6} \cdot \dfrac{13!}{3! \cdot 3! \cdot 3! \cdot 4!} = 200{,}200$.

5. (a) $3^{10} = 59{,}049$.   (b) 252.   (c) $\dbinom{10}{3} = 120$.

(d) 15,360.   (e) $\dfrac{10!}{3! \cdot 4! \cdot 3!} = 4200$.

(f) 55,980, using the Inclusion–Exclusion Principle on the sets of sequences with no 0's, no 1's and no 2's.

7. (a) 625.   (b) 505.   (c) 250.   (d) 303.

9. There are $\dfrac{1}{2}\dbinom{2n}{n}$ unordered such partitions and $\dbinom{2n}{n}$ ordered partitions.

11. (a) $10 \cdot \dbinom{8}{2} \cdot \dbinom{5}{2} \cdot \dbinom{2}{2} = 2800$. Just choose a third member of their team and then choose 3 teams from the remaining 9 contestants.

(b) $2/11 \approx .18$.

13. (a) Think of putting 9 objects in 4 boxes, starting with 1 object in the first box. $\dbinom{8 + 4 - 1}{4 - 1} = 165$.   (b) 56.

15. Fifteen. Just count. We know no clever trick, other than breaking them up into types 4, 3-1, 2-2, 2-1-1 and 1-1-1-1. There are 1, 4, 3, 6 and 1 partitions of these respective types. This solves the problem because there is a one-to-one correspondence between equivalence relations and partitions by Theorem 1 of § 3.5.

Section 5.5

1. (a) Apply the Pigeon-Hole Principle to the partition $\{A_0, A_1, A_2\}$ of the set S of four integers where $A_i = \{n \in S : n \equiv i \,(\text{mod } 3)\}$. Alternatively, apply the second version of the Pigeon-Hole Principle to the function MOD 3: $S \to \mathbb{Z}(3)$.

(b) Apply the Pigeon-Hole Principle to the function
$f : \{1, 2, \dots, p + 1\} \to \mathbb{Z}(p)$ defined by $f(m) = a_m \,\text{MOD}\, p$.

3. Here $|S| = 73$ and $73/8 > 9$, so some box has more than 9 marbles.

5. For each 4-element subset B of A, let $f(B)$ be the sum of the numbers in B. Explain why the function f maps the set of 4-element subsets of A into $\{10, 11, 12, \dots, 194\}$. Note that A has $\dbinom{10}{4} = 210$ 4-element subsets. Apply the second version of the Pigeon-Hole Principle to f.

7. For each 2-element subset T of A, let $f(T)$ be the sum of the two elements. Then f maps the 300 2-element subsets of A into $\{3, 4, 5, \ldots, 299\}$.

9. The repeating blocks are various permutations of 142857.

11. (a) Look at the six blocks

$$(n_1, n_2, n_3, n_4), \quad (n_5, n_6, n_7, n_8), \quad \ldots, \quad (n_{21}, n_{22}, n_{23}, n_{24}).$$

(b) Use Example 2(b) of §4.2.

(c) Look at the 8 blocks

$$(n_1, n_2, n_3), \quad (n_4, n_5, n_6), \quad \ldots, \quad (n_{22}, n_{23}, n_{24}).$$

(d) Look at the five blocks

$$(n_1, \ldots, n_5), \quad (n_6, \ldots, n_{10}), \quad (n_{11}, \ldots, n_{15}), \quad (n_{16}, \ldots, n_{20}), \quad (n_{21}, \ldots, n_{24}).$$

13. If $0 \in \operatorname{Im}(f)$, then $n_1, n_1 + n_2$ or $n_1 + n_2 + n_3$ is divisible by 3. Otherwise, f is not one-to-one and there are three cases:

$$n_1 \equiv n_1 + n_2 \text{ (mod 3)}; \qquad n_1 \equiv n_1 + n_2 + n_3 \text{ (mod 3)};$$

$$n_1 + n_2 \equiv n_1 + n_2 + n_3 \text{ (mod 3)}.$$

15. (a) 262,144.   (b) 73,502.   (c) 20,160.   (d) 60.

17. (a) Show that S must contain both members of some pair $(2k - 1, 2k)$.

(b) For each $m \in S$, write $m = 2^k \cdot j$ where j is odd and let $f(m) = j$. See Exercise 21, §4.1. Then $f{:}S \to \{1, 3, 5, \ldots, 2n - 1\}$. Apply the second version of the Pigeon-Hole Principle.

(c) Let $S = \{2, 4, 6, \ldots, 2n\}$.

19. (a) By the remark at the end of the proof of the Generalized Pigeon-Hole Principle, the average size is $2 \cdot 21/7 = 6$.

Section 6.1

1. (a) $s\,t\,v$ or $s\,u\,v$; length 2.   (c) $u\,v\,w\,y$; length 3.

3. (a) true.

5. (a) true.

7. See Figure 1(c).

9. (a)

e	a	b	c	d	e	f	g	h	k
$\gamma(e)$	$\{w, x\}$	$\{x, u\}$	$\{t, u\}$	$\{t, v\}$	$\{u, v\}$	$\{u, y\}$	$\{v, z\}$	$\{x, y\}$	$\{y, z\}$

11. (a) $e\,b\,h\,k\,g$ and its reversal $g\,k\,h\,b\,e$.   (c) $e\,c\,d$, $b\,h\,f$ and their reversals.

13. (a)

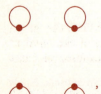

(c) There are none by Theorem 3, since $5 \cdot 3$ is not even.

15. (a), (c) and (d) are regular, but (b) is not. (a) and (c) have cycles of length 3, but (d) does not. Or count edges. (a) and (c) are isomorphic; the labels show an isomorphism between (a) and (c).

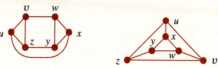

17. (a) $\binom{8}{5} = 56.$ (b) 37.   (c) $8 \cdot 7 + 8 \cdot 7 \cdot 6 + 8 \cdot 7 \cdot 6 \cdot 6 = 2{,}408.$

19. (a)   (e)

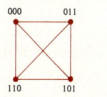

(c) Use Theorem 3 [consider Exercise 6(b)].   (g) K_4.

21. Assume no loops or parallel edges. Consider a longest path $v_1 \cdots v_m$ with distinct vertices. There is another edge at v_m. Adjoin it to the path to get a closed path and use Proposition 1.

23. Use $|V(G)| = D_0(G) + D_1(G) + D_2(G) + \cdots$ and Theorem 3.

Section 6.2

1. Only Figure 7(b) has an Euler circuit. To find one, do Exercise 2.

3. It won't break down until the second visit to the other vertex of degree 3, namely t.

5. (a) $v_3 v_1 v_2 v_3 v_6 v_2 v_4 v_6 v_5 v_1 v_4 v_5 v_3 v_4$ is one.

7. No. The edges and corners form a graph with eight vertices, each of degree 3. It has no Euler path. See Figure 5(c) of §6.5.

9. $\{0, 1\}^3$ consists of 3-tuples of 0's and 1's, which we may view as binary strings of length 3. The graph is then

(a) 2.   (b) All eight vertices have degree 3.   (c) No.

11. (a) Join the odd-degree vertices in pairs, with k new edges. The new graph has an Euler circuit, by Theorem 2. The new edges do not appear next to each other in the circuit and they partition the circuit into k simple paths of G.

(c) Imitate the proof. That is, add edges $\{v_2, v_3\}$ and $\{v_5, v_6\}$, say, create an Euler circuit, and then remove the two new edges.

13. (a) No such walk is possible. Create a graph as follows. Put a vertex in each room and one vertex outside the house. For each door draw an edge joining the vertices for the regions on its two sides. The resulting graph has two vertices of degree 4, three of degree 5 and one of degree 9. Apply the corollary to Theorem 1.

Section 6.3

1.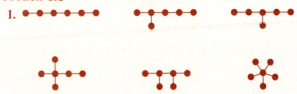

3. (a) 4.   (c) $4 + 2 \cdot 2 = 8$. Draw their pictures.

   (e) $8 \cdot 8 = 64$. Any spanning tree can be viewed as a pair of spanning trees, one for the upper half and one for the lower half of the graph.

5. (a) Since $2n - 2$ is the sum of the degrees of the vertices, we must have $2n - 2 = 4 + 4 + 3 + 2 + 1 \cdot (n - 4)$. Solve for n.

7. (a)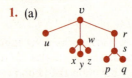

   (b) Prove by induction on n. The cases $n = 1, 2, 3$ are easy. Assume that the result is true for some $n \geq 4$. Suppose that

   $$d_1 + \cdots + d_{n+1} = 2(n + 1) - 2.$$

   At least one d_k is 1, say $d_{n+1} = 1$. At least one d_k exceeds 1, say d_1. Define $d_1^* = d_1 - 1$ and $d_k^* = d_k$ for $2 \leq k \leq n$. Then

   $$d_1^* + \cdots + d_n^* = 2n - 2$$

   and by the inductive hypothesis there is a tree with n vertices whose vertices have degrees $d_1^*, \ldots, d_n^*$. Attach a leaf to the vertex of degree d_1^* to obtain a tree with $n + 1$ vertices. The new vertex has degree $1 = d_{n+1}$ and the vertex of degree d_1^* now has degree $d_1^* + 1 = d_1$.

9. (a) Suppose that the components have $n_1, n_2, \ldots, n_m$ vertices, so that altogether $n_1 + n_2 + \cdots + n_m = n$. Apply Theorem 3 to each component. How many edges are there altogether?

11. By Lemma 1 to Theorem 3 it must be infinite.

Section 6.4

1. (a)

   v, *u*, *w*, *r*, *s*, *x*, *y*, *z*, *p*, *q*

   (c) 3.

3. (a) Rooted trees in Figures 5(b) and 5(c) have height 2; the one in 5(d) has height 3.

5. (a) There are seven of them.

(c) There is exactly one full binary tree.

(d) 21.

7. (a)

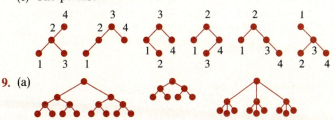

(c) The possibilities are

9. (a)

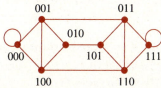

11. From Example 5 we have $(m - 1)p = m^h - 1$ and $m^h = t$.

13. There are 2^k words of length k.

15. (a) Move either Lyon's or Ross's records to the Rose node and delete the empty leaf created.

Section 6.5

1. (a) In the vertex sequence for a Hamilton circuit [if one existed], the vertex w would have to both precede and follow the vertex v. That is, the vertex w would have to be visited twice.

3. (a) Yes. Try $v_1 v_2 v_6 v_5 v_4 v_3 v_1$, for example.   (c) No.

5. (a) $2(n!)^2$. Note that the initial vertex can be in either V_1 or V_2.

(c) m and n even, or m odd and $n = 2$, or $m = n = 1$.

7. Here is the graph:

One possible Hamilton circuit has vertex sequence

$$000, 001, 011, 111, 110, 101, 010, 100, 000$$

corresponding to the circular arrangement 00011101. Although there are four essentially different Hamilton circuits in $\{0, 1\}^3$, there are only two different circular arrangements, 00011101 and 00010111, which are just the reverses of each other.

9. There is no Hamilton path because the graph is not connected. The graph is drawn in the answer to Exercise 9, §6.2.

11. (a) K_n^+ has n vertices and just one more edge than K_{n-1} has, so it has exactly $\frac{1}{2}(n-1)(n-2) + 1$ edges.

13. (a)

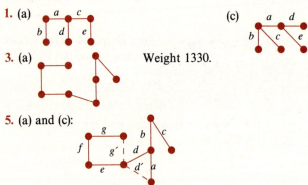

(c) Choose two vertices u and v in G. If they are *not* joined by an edge in G, they are joined by an edge in the complement. If they *are* joined by an edge in G, they are in the same component of G. Choose w in some other component. Then uwv is a path in the complement. In either case, u and v are joined by a path in the complement.

(e) No. Consider

15. Given G_{n+1}, consider the subgraph H_0, where $V(H_0)$ consists of all binary $(n+1)$-tuples with 0 in the $(n+1)$st digit and $E(H_0)$ is the set of all edges of G_{n+1} connecting vertices in $V(H_0)$. Define H_1 similarly. Show that H_0 and H_1 are isomorphic to G_n, and so have Hamilton circuits. Use these to construct a Hamilton circuit for G_{n+1}. For $n = 2$, see how this works in Figure 5.

Section 6.6

1. (a)

(c)

3. (a)   Weight 1330.

5. (a) and (c):

Either d or d' can be chosen, and either g or g', so there are four possible answers to (a).

7. (a) $e_1, e_2, e_3, e_5, e_6, e_7, e_9$.   (b) $e_7, e_5, e_2, e_1, e_3, e_6, e_9$.

9. (a)

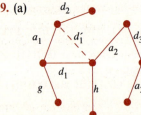

Edges a_1, a_2, a_3 can be chosen in any order. So can d_1, d_2, d_3. Edge d_1' can be chosen instead of d_1. The weight is 16.

11. 1687 miles.

13. Here is one possible algorithm:

> Set $E := \emptyset$.
> Choose w in $V(G)$ and set $V := \{w\}$.
> While $V \neq V(G)$
> > if there is an edge $\{u, v\}$ in $E(G)$ with $u \in V$ and $v \in V(G)\backslash V$
> > > choose such an edge of smallest weight
> > > > put $\{u, v\}$ in E and put v in V
> > otherwise choose $v \in V(G)\backslash V$ and put v in V.

15. *Outline:* Follow the hint. Look at minimum spanning trees S and T with $e \in T\backslash S$. Then $S \cup \{e\}$ has a cycle that contains an edge f in $S\backslash T$. $(S \cup \{e\})\backslash\{f\}$ is a spanning tree. $W(e) > W(f)$, contrary to the choice of e. Supply reasons.

Section 7.1

1. (a) Use induction on m. Clearly, 2^0 is in S by (B). If 2^m is in S, then $2 \cdot 2^m = 2^{m+1}$ is in S by (R).

(b) Use the Generalized Principle of Induction where $p(n) = $ "n has the form 2^m for some $m \in \mathbb{N}$."

3. (a) $S = \{(m, n) : m \leq n\}$.

(c) Yes. A pair $(0, n)$ is in S only by (B) or because $(n - 1, n - 1) \in S$, and a pair (m, n) with $0 < m \leq n$ can only come from $(m - 1, n)$.

5. (a) (B) λ is in S;

(R) If $w \in S$, then $aw \in S$ and $wb \in S$.

(c) $\lambda \in S$ by (B), so a, aa and aab are in S by (R).

(d) Our definition is not uniquely determined.

7. (a) Two possible constructions are

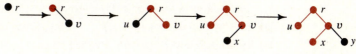

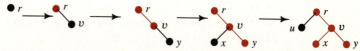

9. Use the Generalized Principle of Induction on $p(n) = $ "$n \equiv 0 \pmod 3$ or $n \equiv 1 \pmod 3$." Certainly, $p(1)$ is true. It suffices to show that $p(n) \Rightarrow p(3n)$ and $p(2n + 1) \Rightarrow p(n)$. The first implication is trivial because $3n \equiv 0 \pmod 3$ for all n. For the second implication, show the contrapositive, i.e., $n \equiv 2 \pmod 3$ implies $(2n + 1) \equiv 2 \pmod 3$.

11. (a) Use (B) and (R) to show that the sequence 1, 2, 4, 8, 16, 5, 10, 3, 6 lies in S.

13. Follow the hint. Let $p(w) = $ "$l'(w)$ is the number of letters in w." By (1) of Example 10(b), $p(w)$ is true if $w \in X = \{\lambda\} \cup \Sigma$. Suppose that $w = uv$ with $p(u)$ and $p(v)$ true and show that $p(w)$ is then true. Thus $p(w)$ is true for every w in Σ^* by the Generalized Principle of Induction.

15. (a) $(2, 3)$, $(4, 6)$, etc.

(b) **(B)** is clear since 5 divides $0 + 0$. For **(R)** you need to check

"if 5 divides $m + n$, then 5 divides $(m + 2) + (n + 3)$."

Alternatively, prove that every member of S is of the form $(2k, 3k)$ for $k \in \mathbb{N}$.

(c) No. Why?

17. (a) Obviously, $A \subseteq \mathbb{N} \times \mathbb{N}$. To show $\mathbb{N} \times \mathbb{N} \subseteq A$, apply the ordinary First Principle of Mathematical Induction to the propositions

$$p(k) = \text{"if } (m, n) \in \mathbb{N} \times \mathbb{N} \text{ and } m + n = k, \text{ then } (m, n) \in A.\text{"}$$

(b) Let $p(m, n)$ be a proposition-valued function defined on $\mathbb{N} \times \mathbb{N}$. To show that $p(m, n)$ is true for all (m, n) in $\mathbb{N} \times \mathbb{N}$, it is enough to show:

    (B) $p(0, 0)$ is true, and

    (I)   if $p(m, n)$ is true, then $p(m + 1, n)$ and $p(m, n + 1)$ are true.

19. (a) For w in Σ^*, let

$$p(w) = \text{"length}(\bar{w}) = \text{length}(w).\text{"}$$

Apply the Generalized Principle of Induction. Since $\tilde{\lambda} = \lambda$, $p(\lambda)$ is clearly true. You need to show that if $p(w)$ is true, then so is $p(wx)$:

$$\text{length}(\bar{w}) = \text{length}(w) \quad \text{implies} \quad \text{length}(\overleftarrow{wx}) = \text{length}(wx).$$

Section 7.2

1. (a) TEST(20;)

    TEST(10;)

       TEST(5;)

          $b := \text{false}; m := -\infty$

       $b := \text{false}; m := -\infty + 1 = -\infty$

    $b := \text{false}; m := -\infty + 1 = -\infty$

3. (a) $((x + y) + z)$ or $(x + (y + z))$.   (c) $((xy)z)$ or $(x(yz))$.

5. (a) By **(B)**, x, y and 2 are wff's. By the (f^g) part of **(R)**, we conclude that (x^2) and (y^2) are wff's. So by the $(f + g)$ part of **(R)**, $((x^2) + (y^2))$ is a wff.

(c) By **(B)**, X and Y are wff's. By the $(f + g)$ part of **(R)**, $(X + Y)$ is a wff. By the $(f - g)$ part of **(R)**, $(X - Y)$ is a wff. Finally, by the $(f * g)$ part of **(R)**, $((X + Y) * (X - Y))$ is a wff.

7. Blank entries below signify that the algorithm doesn't terminate with the indicated input n.

n	FOO	GOO	BOO	MOO	TOO	ZOO
8	8	40,320		4	4	8
9	9	362,880			4	8

9. Let $p(k)$ be the statement "ZOO produces 2^k whenever $2^k \leq n < 2^{k+1}$." For $k = 0$, this asserts that "ZOO produces 1 whenever $n = 1$," which is clear. Assume that $p(k)$ is true and consider n where $2^{k+1} \leq n < 2^{k+2}$. Then $2^k \leq n \,\text{DIV}\, 2 < 2^{k+1}$, so ZOO($n \,\text{DIV}\, 2$;) has output $s = 2^k$. The else branch of

ZOO(n;) sets $r = 2 * s = 2 * 2^k = 2^{k+1}$. That is, ZOO produces 2^{k+1} whenever $2^{k+1} \le n < 2^{k+2}$, so that $p(k+1)$ is true. Hence all statements $p(k)$ are true by induction.

11. EUCLID$^+$(108, 30;)
   EUCLID$^+$(30, 18;)
     EUCLID$^+$(18, 12;)
       EUCLID$^+$(12, 6;)
         EUCLID$^+$(6, 0;)

$$d := 6; \; s := 1; \; t := 0$$
$$d := 6; \; s := 0; \; t := 1 - 0 \cdot (12 \text{ DIV } 6) = 1$$
$$d := 6; \; s := 1; \; t := 0 - 1 \cdot (18 \text{ DIV } 12) = -1$$
$$d := 6; \; s := -1; \; t := 1 - (-1) \cdot (30 \text{ DIV } 18) = 2$$
$$d := 6; \; s := 2; \; t = -1 - 2 \cdot (108 \text{ DIV } 30) = -7.$$

Sure enough, $108 \cdot 2 + 30 \cdot (-7) = 6$.

13. (a) Since $\gcd(m, n) = m$ when $n = 0$, we may assume that $n \ne 0$. We need to check that the algorithm terminates and that if the output d' from EUCLID(n, $m \text{ MOD } n$;) is correct, then so is the output $d = d'$ from EUCLID(m, n;). The latter assertion follows from the equality $\gcd(m, n) = \gcd(n, m \text{ MOD } n)$, which was verified in §4.6.

To check that the algorithm terminates, we need to be sure that the second variable n in EUCLID(, n;) is eventually 0. This is clear, since at each step the new $n' = m \text{ MOD } n$ is less than n, so the values must eventually decrease to 0.

15. (a) p and q are wff's by (B). $p \lor q$ is a wff by (R). $\neg(p \lor q)$ is a wff by (R).

(c) p, q and r are wff's by (B). $(p \leftrightarrow q)$ and $(r \to p)$ are wff's by (R). $((r \to p) \lor q)$ is a wff by (R), so $((p \leftrightarrow q) \to ((r \to p) \lor q))$ is a wff by (R).

17. (a) $p, q \in \mathscr{F}$ by (B). $(p \lor q) \in \mathscr{F}$ by (R) with $P = p, Q = q$. Hence $(p \land (p \lor q)) \in \mathscr{F}$ by (R) with $P = p, Q = (p \lor q)$.

(b) There are too many p's and P's around, so use, say, $r(P)$ for the proposition-valued function on $\mathscr{F}$; then apply the general principle of induction. Thus you need to prove all $r(P)$ are true where

$$r(P) = \text{``if } p, q \text{ are false, then } P \text{ is false.''}$$

To prove $r(P)$ for all $P \in \mathscr{F}$ it is enough to show:

(B) $r(p)$ and $r(q)$ are true;
(I)   if $r(P)$ and $r(Q)$ are true, then so are $r((P \land Q))$ and $r((P \lor Q))$.

(c) If p and q are false, then $(p \to q)$ is true, so $r((p \to q))$ is false. Thus $(p \to q)$ cannot be logically equivalent to a proposition in $\mathscr{F}$, by part (b).

This exercise provides a negative answer to the question in Exercise 17(c) of §2.4: $p \to q$ cannot be written in some way using just $p, q, \land$ and $\lor$.

Section 7.3

1. Preorder: $r\,x\,w\,v\,y\,z\,s\,u\,t\,p\,q$.   Postorder: $v\,y\,w\,z\,x\,t\,p\,u\,q\,s\,r$.

3. Preorder: $r\,t\,x\,v\,y\,z\,w\,u\,p\,q\,s$.   Postorder: $v\,y\,z\,x\,w\,t\,p\,q\,s\,u\,r$.

5. The order of traversing the tree is

$$r, w, v, w, x, y, x, z, x, w, r, u, t, u, s, p, s, q, s, u, r.$$

(a) Preorder: $r\,w\,v\,x\,y\,z\,u\,t\,s\,p\,q.$ $L(w) = w\,v\,x\,y\,z.$ $L(u) = u\,t\,s\,p\,q.$

7. The order of traversing the tree is

$$u, x, w, v, w, y, w, x, r, z, r, t, r, x, u, s, p, s, q, s, u.$$

(a) Postorder: $v\,y\,w\,z\,t\,r\,x\,p\,q\,s\,u.$ (c) Inorder: $v\,w\,y\,x\,z\,r\,t\,u\,p\,s\,q.$

9. (a)

(c)
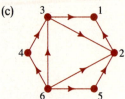

(b) Start with a, choose its successor b, choose its successor c, and label c with 1.

Return to b, choose its successor d and label d with 2.

Return to b and label b with 3.

Return to a and label a with 4.

11. (a)
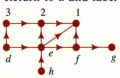
(c)
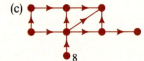

13. (a)
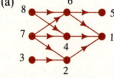

15. (a) The base cases are the ones in which the tree consists just of the root r.

(c) The number of descendants of v is a good measure. Each child w of v has fewer descendants than v has. The measure of a base case is 0.

17. Since the digraph is acyclic there can be at most one edge joining each of the $n(n-1)/2$ pairs of distinct vertices.

Section 7.4

1. Reverse Polish: $x\,4\,2\,{}^\wedge - y * 2\,3\,/ +.$ Polish: $+ * - x\,{}^\wedge 4\,2\,y\,/\,2\,3.$

3. (a) Polish: $- * + a\,b - a\,b - {}^\wedge a\,2\,{}^\wedge b\,2;$

   Infix: $(a+b)*(a-b)-((a{}^\wedge 2)-(b{}^\wedge 2)).$

5. (a) 20.

7. (a) $3\,x * 4 - 2\,{}^\wedge.$

(c) The answer depends on how the terms are associated. For the choice $(x - x^2) + (x^3 - x^4)$, the answer is $x\,x\,2\,{}^\wedge - x\,3\,{}^\wedge x\,4\,{}^\wedge - +.$

9. (a) $a\,b\,c * *$ and $a\,b * c *.$

(c) The associative law is $abc** = ab*c*$. The distributive law is $abc+* = ab*ac*+$.

11. (a) $p \to (q \lor (\neg p))$.

13. (a) Infix: $(p \land (p \to q)) \to q$.

15. (a) Both give $a/b + c$.

17. (a) **(B)** Numerical constants and variables are wff's.
     (R) If f and g are wff's, so are $+fg$, $-fg$, $*fg$, $/fg$ and $\hat{\ }fg$.

19. (a) **(B)** Variables, such as p, q, r, are wff's.
     (R) If P and Q are wff's, so are $PQ\lor$, $PQ\land$, $PQ\to$, $PQ\leftrightarrow$ and $P\neg$.

   (b) Argue, in turn, that $q\neg$, $pq\neg\land$ and $pq\neg\land\neg$ are wff's.
Similarly, $pq\neg\to$ is a wff. Thus $pq\neg\land\neg pq\neg\to\lor$ is a wff.

   (c) **(B)** Variables, such as p, q, r, are wff's.
     (R) If P and Q are wff's, so are $\lor PQ$, $\land PQ$, $\to PQ$, $\leftrightarrow PQ$ and $\neg P$.

Section 7.5

1. (a) 35, 56, 70, 82.

3. We recommend the procedure in Examples 4 and 5.
   (a) Weight $= 84$.   (c) Weight $= 244$.

5. All but (b) are prefix codes. In (b), 01 consists of the first two digits of 0111.

7. (b) 269.

9. (a) The vertex labeled 0 has only one child, 00.
   (b) Consider any string beginning with 01.

11. (a) 484.   (c) 373.   (e) It will involve at most 354 comparisons.

13. (b) 221.

15. For example, in Exercise 1(a), the first tree has weight 35. The leaf of weight $21 = 12 + 9$ was replaced by the subtree with weights 12 and 9 and the weight of the whole tree increased to 56, i.e., to $35 + 21$.

Section 8.1

1. Sinks are t and z. Only source is u.

3. (a) $R(s) = \{s, t, u, w, x, y, z\} = R(t)$, $R(u) = \{w, x, y, z\}$, $R(w) = \{z\}$, $R(x) = \varnothing$, $R(y) = \{x, z\}$, $R(z) = \varnothing$.
   (c) sts is a cycle, so G is not acyclic.

5. (a)

Supply labels.

   (c) s, u, x, z.

7. Use NUMBERING VERTICES. One example labels $t = 1$, $z = 2$, $y = 3$, $w = 4$, $v = 5$, $x = 6$, $u = 7$.

9. (a) One example is $r s w t v w v u x y z v r u y v s r$.

11. (a) One such digraph is drawn in Figure 3(b), where $w = 00$, $x = 01$, $z = 11$ and $y = 10$.

(b) One possible sequence is $1\,1\,1\,0\,1\,0\,0\,0$ placed in a circle.

13. Show that $\hat{G}$ is also acyclic. Apply Theorem 2 to $\hat{G}$. A sink for $\hat{G}$ is a source for G.

15. (a) See the second proof of Theorem 2.

(b) If a finite acyclic digraph has just one source, then there is a path to each vertex from the source.

17. (a) In the proof of Theorem 1 [given in §6.1], choose a shortest path consisting of edges of the given path.

(b) For $u \neq v$, apply Theorem 1 and Corollary 2. For $u = v$, apply Corollary 1.

Section 8.2

1.

W^*	A	B	C	D
A	∞	1.0	1.4	1.2
B	0.4	∞	0.4	0.2
C	0.7	0.3	∞	0.5
D	0.8	0.5	0.2	∞

3.

W	m	q	r	s	w	x	y	z
m	∞	6	∞	2	∞	4	∞	∞
q	∞	∞	4	∞	4	∞	∞	∞
r	∞	∞	∞	∞	∞	∞	∞	3
s	∞	3	∞	∞	5	1	∞	∞
w	∞	∞	2	∞	∞	∞	2	5
x	∞	∞	∞	∞	3	∞	6	∞
y	∞	∞	∞	∞	∞	∞	∞	1
z	∞	∞	∞	∞	∞	∞	∞	∞

W^*	m	q	r	s	w	x	y	z
m	∞	5	8	2	6	3	8	9
q	∞	∞	4	∞	4	∞	6	7
r	∞	∞	∞	∞	∞	∞	∞	3
s	∞	3	6	∞	4	1	6	7
w	∞	∞	2	∞	∞	∞	2	3
x	∞	∞	5	∞	3	∞	5	6
y	∞	∞	∞	∞	∞	∞	∞	1
z	∞	∞	∞	∞	∞	∞	∞	∞

5. (a) If the digraph has all arrows going to the right, the weights must be as shown:

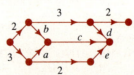

with $\min\{a + 1, b\} = 4$ and $\min\{c + 1, d, e\} = 3$.

7. (a) $s\,v\,x\,f$ is one. There is another.

9. (a)

	s	u	v	w	x	y	f
A	0	2	7	5	5	11	15
L	0	2	9	7	5	11	15

(b) $S(v) = S(w) = 2$. $S(t) = 0$ for all other vertices t.

(c) $s\,u\,x\,y\,f$ is the only critical path.

(d) Edges on the critical path have float time 0. Also $F(x, f) = 6$, $F(u, w) = F(v, f) = 3$ and $F(s, w) = F(w, y) = F(s, v) = F(v, y) = 2$.

11. (a)

	m	s	q	x	w	r	y	z
A	0	2	6	4	10	12	12	15
L	0	3	6	7	10	12	14	15

(c) There are two critical paths. Find them.

13. (a) The two critical paths are $s\,u\,w\,x\,y\,f$ and $s\,t\,w\,x\,y\,f$.

(b) 2.   (c) There are two edges with float time 2.

15. (a) Shrink the 0-edges to make their two endpoints the same.

17. (a)

W	u	v	w	x	y
u	∞	1	∞	∞	∞
v	∞	∞	3	-2	∞
w	∞	∞	∞	∞	∞
x	4	∞	∞	∞	∞
y	∞	∞	∞	5	∞

W^*	u	v	w	x	y
u	3	1	4	-1	∞
v	2	3	3	-2	∞
w	∞	∞	∞	∞	∞
x	4	5	8	3	∞
y	9	10	13	5	∞

(c) There would be no min-weights at all for paths involving u, v or x, because going around the cycle $u\,v\,x\,u$ repeatedly would keep reducing the weight by 1.

19. (a) $FF(u, v) = A(v) - A(u) - W(u, v)$.   (c) The slack time at v.

21. (a) $A(u) = M(s, u) =$ weight of a max-path from s to u. If there is an edge (w, u), a max-path from s to w followed by that edge has total weight at most $M(s, u)$. That is, $A(w) + W(w, u) \le A(u)$. If (w, u) is an edge in a max-path from s to u, then $A(w) + W(w, u) = A(u)$.

Section 8.3

1. (a) $\mathbf{W}^* = \begin{bmatrix} \infty & 1 & 2 & 3 & 4 & 5 & 7 \\ \infty & \infty & \infty & 4 & 3 & 4 & 6 \\ \infty & \infty & \infty & 1 & 4 & 5 & 7 \\ \infty & \infty & \infty & \infty & 3 & 4 & 6 \\ \infty & \infty & \infty & \infty & \infty & 1 & 3 \\ \infty & \infty & \infty & \infty & \infty & \infty & 3 \\ \infty & \infty & \infty & \infty & \infty & \infty & \infty \end{bmatrix}$.

3. (a)

L	$D(2)$	$D(3)$	$D(4)$	$D(5)$	$D(6)$	$D(7)$
$\emptyset$	1	2	∞	∞	∞	∞
$\{2\}$	1	2	5	4	∞	∞
$\{2, 3\}$	1	2	3	4	9	∞
$\{2, 3, 4\}$	1	2	3	4	9	∞
$\{2, 3, 4, 5\}$	1	2	3	4	5	7
$\{2, 3, 4, 5, 6\}$	1	2	3	4	5	7

No change now

5. (a)
$$\mathbf{W}_2 = \begin{bmatrix} \infty & \infty & \infty & \infty & 1 & \infty & \infty \\ \infty & \infty & \infty & \infty & \infty & \infty & 1 \\ \infty & \infty & \infty & 1 & \infty & 1 & \infty \\ \infty & \infty & 1 & \infty & 1 & \infty & \infty \\ 1 & \infty & \infty & 1 & 2 & \infty & \infty \\ \infty & \infty & 1 & \infty & \infty & \infty & 1 \\ \infty & 1 & \infty & \infty & \infty & 1 & 2 \end{bmatrix},$$

$$\mathbf{W}_4 = \begin{bmatrix} \infty & \infty & \infty & \infty & 1 & \infty & \infty \\ \infty & \infty & \infty & \infty & \infty & \infty & 1 \\ \infty & \infty & 2 & 1 & 2 & 1 & \infty \\ \infty & \infty & 1 & 2 & 1 & 2 & \infty \\ 1 & \infty & 2 & 1 & 2 & 3 & \infty \\ \infty & \infty & 1 & 2 & 3 & 2 & 1 \\ \infty & 1 & \infty & \infty & \infty & 1 & 2 \end{bmatrix}, \mathbf{W}_7 = \begin{bmatrix} 2 & 6 & 3 & 2 & 1 & 4 & 5 \\ 6 & 2 & 3 & 4 & 5 & 2 & 1 \\ 3 & 3 & 2 & 1 & 2 & 1 & 2 \\ 2 & 4 & 1 & 2 & 1 & 2 & 3 \\ 1 & 5 & 2 & 1 & 2 & 3 & 4 \\ 4 & 2 & 1 & 2 & 3 & 2 & 1 \\ 5 & 1 & 2 & 3 & 4 & 1 & 2 \end{bmatrix}.$$

7.
$$\mathbf{W}^* = \begin{bmatrix} \infty & 8 & 7 & 5 & 2 \\ \infty & \infty & \infty & \infty & \infty \\ \infty & 1 & \infty & \infty & \infty \\ \infty & 3 & 2 & \infty & \infty \\ \infty & 6 & 5 & 3 & \infty \end{bmatrix}.$$

9. (a) $\mathbf{D}_0 = \mathbf{D}_1 = [-\infty \quad 1 \quad 2 \quad -\infty \quad -\infty \quad -\infty \quad -\infty]$

$\mathbf{D}_2 = [-\infty \quad 1 \quad 2 \quad 5 \quad 4 \quad -\infty \quad -\infty]$

$\mathbf{D}_3 = [-\infty \quad 1 \quad 2 \quad 5 \quad 7 \quad 9 \quad -\infty]$

$\mathbf{D}_4 = [-\infty \quad 1 \quad 2 \quad 5 \quad 8 \quad 13 \quad -\infty]$

$\mathbf{D}_5 = [-\infty \quad 1 \quad 2 \quad 5 \quad 8 \quad 13 \quad 11]$

$\mathbf{D}_6 = [-\infty \quad 1 \quad 2 \quad 5 \quad 8 \quad 13 \quad 16].$

11. (a) The algorithm would give

L	$D(2)$	$D(3)$	$D(4)$
$\emptyset$	5	6	$-\infty$
$\{3\}$	5	6	9

No change

whereas $M(1, 3) = 9$ and $M(1, 4) = 12$.

(c) Both algorithms would still fail to give correct values of $M(1, 4)$.

Section 8.4

1. (a) $\mathbf{P}_0 = \begin{bmatrix} 0 & 2 & 3 & 0 & 0 & 0 & 0 \\ 0 & 0 & 0 & 4 & 5 & 0 & 0 \\ 0 & 0 & 0 & 4 & 5 & 6 & 0 \\ 0 & 0 & 0 & 0 & 5 & 6 & 0 \\ 0 & 0 & 0 & 0 & 0 & 6 & 7 \\ 0 & 0 & 0 & 0 & 0 & 0 & 7 \\ 0 & 0 & 0 & 0 & 0 & 0 & 0 \end{bmatrix}$, $\mathbf{P}_{\text{final}} = \begin{bmatrix} 0 & 2 & 3 & 3 & 2 & 2 & 2 \\ 0 & 0 & 0 & 4 & 5 & 5 & 5 \\ 0 & 0 & 0 & 4 & 4 & 4 & 4 \\ 0 & 0 & 0 & 0 & 5 & 5 & 5 \\ 0 & 0 & 0 & 0 & 0 & 6 & 7 \\ 0 & 0 & 0 & 0 & 0 & 0 & 7 \\ 0 & 0 & 0 & 0 & 0 & 0 & 0 \end{bmatrix}$.

3. (a)

k	$D(2)$	$D(3)$	$D(4)$	$D(5)$	$P(2)$	$P(3)$	$P(4)$	$P(5)$
1	2	1*	7	∞	1	1	1	0
3	2*	1	7	4	1	1	1	3
2	2	1	6	4*	1	1	2	3
5	2	1	5*	4	1	1	5	3

* in column $D(k)$ marks the time that k is chosen for L and $D(k)$ is frozen.

5. (a) $\mathbf{W}_0 = \mathbf{W}_1 = \begin{bmatrix} \infty & 1 & \infty & 7 & \infty \\ \infty & \infty & 4 & 2 & \infty \\ \infty & \infty & \infty & \infty & 3 \\ \infty & \infty & 1 & \infty & 5 \\ \infty & \infty & \infty & \infty & \infty \end{bmatrix}$, $\mathbf{P}_0 = \mathbf{P}_1 = \begin{bmatrix} 0 & 2 & 0 & 4 & 0 \\ 0 & 0 & 3 & 4 & 0 \\ 0 & 0 & 0 & 0 & 5 \\ 0 & 0 & 3 & 0 & 5 \\ 0 & 0 & 0 & 0 & 0 \end{bmatrix}$,

$\mathbf{W}_2 = \begin{bmatrix} \infty & 1 & 5 & 3 & \infty \\ \infty & \infty & 4 & 2 & \infty \\ \infty & \infty & \infty & \infty & 3 \\ \infty & \infty & 1 & \infty & 5 \\ \infty & \infty & \infty & \infty & \infty \end{bmatrix}$, $\mathbf{P}_2 = \begin{bmatrix} 0 & 2 & 2 & 2 & 0 \\ 0 & 0 & 3 & 4 & 0 \\ 0 & 0 & 0 & 0 & 5 \\ 0 & 0 & 3 & 0 & 5 \\ 0 & 0 & 0 & 0 & 0 \end{bmatrix}$,

$\mathbf{W}_3 = \begin{bmatrix} \infty & 1 & 5 & 3 & 8 \\ \infty & \infty & 4 & 2 & 7 \\ \infty & \infty & \infty & \infty & 3 \\ \infty & \infty & 1 & \infty & 4 \\ \infty & \infty & \infty & \infty & \infty \end{bmatrix}$, $\mathbf{P}_3 = \begin{bmatrix} 0 & 2 & 2 & 2 & 2 \\ 0 & 0 & 3 & 4 & 3 \\ 0 & 0 & 0 & 0 & 5 \\ 0 & 0 & 3 & 0 & 3 \\ 0 & 0 & 0 & 0 & 0 \end{bmatrix}$,

$\mathbf{W}_4 = \begin{bmatrix} \infty & 1 & 4 & 3 & 7 \\ \infty & \infty & 3 & 2 & 6 \\ \infty & \infty & \infty & \infty & 3 \\ \infty & \infty & 1 & \infty & 4 \\ \infty & \infty & \infty & \infty & \infty \end{bmatrix}$, $\mathbf{P}_4 = \begin{bmatrix} 0 & 2 & 2 & 2 & 2 \\ 0 & 0 & 4 & 4 & 4 \\ 0 & 0 & 0 & 0 & 5 \\ 0 & 0 & 3 & 0 & 3 \\ 0 & 0 & 0 & 0 & 0 \end{bmatrix}$.

Also, $\mathbf{W}_5 = \mathbf{W}^* = \mathbf{W}_4$ and $\mathbf{P}_5 = \mathbf{P}^* = \mathbf{P}_4$.

7. (a) Create a row matrix $\mathbf{P}$, with $\mathbf{P}[j] = 1$ initially if there is an edge from v_1 to v_j and $\mathbf{P}[j] = 0$ otherwise. Add the line

   replace $\mathbf{P}[j]$ by k.

   (b) Part of this exercise is solved in Example 3 of §8.3.

9. (a)

(c) $\mathbf{M}_R = \begin{bmatrix} 1 & 0 & 1 & 0 & 1 & 0 \\ 0 & 1 & 0 & 1 & 0 & 1 \\ 1 & 0 & 1 & 0 & 1 & 0 \\ 0 & 1 & 0 & 1 & 0 & 1 \\ 1 & 0 & 1 & 0 & 1 & 0 \\ 0 & 1 & 0 & 1 & 0 & 1 \end{bmatrix}.$

11. (a) Initialization of **P** takes time $O(n)$, and the replacement step still only takes constant time during each pass through the loop.

(b) $O(n^3)$. Why?

Section 9.1

1. (a) $\dfrac{\dbinom{3}{2}}{\dbinom{11}{2}} = \dfrac{3}{55}.$ (b) $\dfrac{28}{55}.$ (c) $\dfrac{3 \cdot 8}{\dbinom{11}{2}} = \dfrac{24}{55}.$

3. (a) $P(B_o) = P(R_o) = P(E) = \dfrac{1}{2}$ and $P(B_o \cap R_o) = P(B_o \cap E) = P(R_o \cap E) = \dfrac{1}{4}.$

5. $\{S, L\}$ are dependent since $P(S \cap L) = \dfrac{6}{36}$ while $P(S) \cdot P(L) = \dfrac{15}{36} \cdot \dfrac{15}{36}.$

$\{S, E\}$ are dependent since $P(S \cap E) = \dfrac{3}{36}$ while $P(S) \cdot P(E) = \dfrac{15}{36} \cdot \dfrac{6}{36}.$

$\{L, E\}$ are dependent since $P(L \mid E) = 0 \neq P(L)$. Similarly for $\{L, G\}$.

7. No. $P(B \mid A) = \dfrac{1}{2}$ while $P(B) = \dfrac{\dbinom{4}{2}}{2^4} = \dfrac{3}{8}.$

9. (a) .25.   (b) .7.

(c) No, $P(A \mid B) = .25 \neq P(A)$.

11. (a) $\dfrac{4}{52} \cdot \dfrac{3}{51} \cdot \dfrac{2}{50} \approx .00018.$ (b) $\approx .00048.$

(c) $1 - \dfrac{48}{52} \cdot \dfrac{47}{51} \cdot \dfrac{46}{50} \approx .217.$

13. (a) $P(B) = \dfrac{1}{3} \cdot \dfrac{2}{3} + \dfrac{1}{3} \cdot \dfrac{2}{5} + \dfrac{1}{3} \cdot \dfrac{1}{2} = \dfrac{47}{90}.$

(b) $P(U_1 \mid B) = \dfrac{20}{47}, \quad P(U_2 \mid B) = \dfrac{12}{47}, \quad P(U_3 \mid B) = \dfrac{15}{47}.$

(c) $P(B \cap U_1) = \dfrac{1}{3} \cdot \dfrac{2}{3} = \dfrac{2}{9}.$

15. (a) $\dfrac{5}{9}$.

17. (a) $P(C) = P(E) \cdot P(C|E) + P(F) \cdot P(C|F) + P(G) \cdot P(C|G) = .043$.

(b) $P(E|C) = \dfrac{10}{43} \approx .23$, etc.

19. (a) $P(D) = P(N^c \cap D) + P(N \cap D) = .0041$, so

$$P(N^c|D) = \frac{P(N^c \cap D)}{P(D)} = \frac{.004}{.0041} \approx .9756.$$

(b) $P((N^c \cap D) \cup (N \cap D^c)) = .9599$. This is the probability that the test confirms the subject's condition.

(c) $P(D|N^c) = \dfrac{P(D \cap N^c)}{P(N^c)} = \dfrac{.004}{.044} \approx .091$. Thus the probability of having the

disease, given a positive test is less than .10. The following tabulation may help clarify the situation.

	D [diseased]	D^c [not diseased]
N^c [tests positive]	.004	.04
N [tests negative]	.0001	.9559

21. (a) $1 - (1 - q)^n$. We are assuming that the failures of the components are independent.

(b) $1 - (.99)^{100} \approx .634$.   (c) $1 - (.999)^{100} \approx .0952$.

23. (a) $\dfrac{\dbinom{18}{5}}{\dbinom{20}{5}} = \dfrac{15 \cdot 14}{20 \cdot 19} = \dfrac{21}{38} \approx .55$.   (b) $\dfrac{9}{38}$.

25. No. For example, toss a fair coin three times and let $A_k = $ "kth toss is a head." Then A_1, A_2, A_3 are independent, but $A_1 \cap A_2$ and $A_1 \cap A_3$ are not. To see this, show that $P(A_1 \cap A_3 | A_1 \cap A_2) \neq P(A_1 \cap A_3)$.

27. (a) No. If true, then A and B independent and B and A independent would imply that A and A are independent, which generally fails by part (b).

(b) No.   (c) Absolutely not, unless $P(A) = 0$ or $P(B) = 0$.

29. (a) $\dfrac{P^*(E)}{P^*(F)} = \dfrac{P(E|S)}{P(F|S)} = \dfrac{\dfrac{P(E \cap S)}{P(S)}}{\dfrac{P(F \cap S)}{P(S)}} = \dfrac{P(E \cap S)}{P(F \cap S)}$.

Section 9.2

1. (a) $\{3, 4, 5, 6, \ldots, 18\}$.

3. (a) $\{0, 1, 2, 3, 4, 5\}$ and $\{1, 2, 3, 4, 5, 6\}$.

(c) $P(D \leq 1) = 4/9$, $P(M \leq 3) = 1/4$, $P(D \leq 1$ and $M \leq 3) = 7/36$.

5. (a) $\{1, 2, 3, 4, 5, 6, 8, 9, 10, 12, 15, 16, 18, 20, 24, 25, 30, 36\}$.

(c) $1/9$.

7. (a) $\{0, 1, 2, 3, 4\}$.   (b) 5/16.

(c)

k	0	1	2	3	4
$P(X + Y = k)$	1/16	1/8	5/16	1/4	1/4

9. (a) Sample calculation: $P(X = 2) = \dfrac{\binom{5}{2} \cdot \binom{5}{2}}{\binom{10}{4}} = \dfrac{10 \cdot 10}{210} = \dfrac{20}{42}.$

k	0	1	2	3	4
$P(X = k)$	1/42	10/42	20/42	10/42	1/42

(b) Sample calculation: $P(X = 2) = \dfrac{\binom{5}{2} \cdot \binom{5}{5}}{\binom{10}{7}} = \dfrac{10}{120} = \dfrac{1}{12}.$

11. (a) No. The values $f(x)$ must sum to 1.

13. (a) $0 \le P(X \le y) \le 1$, since $0 \le P(E) \le 1$ for all events E.

15. $f(k) = P(W = k) = \left(\dfrac{5}{6}\right)^{k-1} \cdot \dfrac{1}{6}$ for $k = 1, 2, \ldots$ and $f(x) = 0$ for all other x.

17. (a) $P\left(\left[\dfrac{1}{6}, \dfrac{5}{6}\right]\right) = \dfrac{2}{3}.$ (b) $P\left(\left[0, \dfrac{1}{3}\right] \cup \left[\dfrac{2}{3}, 1\right)\right) = \dfrac{2}{3}.$

19. Let $x_1, x_2, \ldots, x_m$ and $y_1, y_2, \ldots, y_n$ be the value sets for X and Y.
$(I_2) \Rightarrow (I_3)$. To show (I_3), we may assume that $x = x_i$ and $y = y_j$ for some i and j. Let I be an interval containing x_i and no other number in X's value set, and let J be an interval containing y_j and no other number in Y's value set. Then

$$P(X = x_i \text{ and } Y = y_j) = P(X \in I \text{ and } Y \in J)$$
$$= P(X \in I) \cdot P(Y \in J) = P(X = x_i) \cdot P(Y = y_j).$$

$(I_3) \Rightarrow (I_2)$. Given intervals I and J, let $A = \{i : x_i \in I\}$ and $B = \{j : y_j \in J\}$. Then $\{X \in I \text{ and } Y \in J\}$ is the disjoint union

$$\bigcup_{i \in A} \bigcup_{j \in B} \{X = x_i \text{ and } Y = y_j\};$$

now show that $P(X \in I \text{ and } Y \in J) = P(X \in I) \cdot P(Y \in J)$.

Section 9.3

1. (a) The answer is rather obviously 2. To confirm this, use the probability distribution in the answer to Exercise 9, §9.2, and calculate $\displaystyle\sum_{k=0}^{4} k \cdot P(X = k) = 2$.

(b) $\sigma = \sqrt{\dfrac{2}{3}} \approx .82.$

3. (a) Mean deviation $= \sum\limits_{k=1}^{6} |k - 3.5| \cdot \dfrac{1}{6} = 1.5 < 1.71 \approx \sigma$.

(b) Mean deviation $= 1/2 = \sigma$.

5. $\mu_X = \mu_Y = 5/4$ and $\mu_{X+Y} = 5/2$. Show $E(X^2) = 9/4$; use this to show $V(X) = 11/16$ and hence $\sigma_X = \frac{1}{4}\sqrt{11} \approx .83$. Same for Y. Finally, by Theorem 7, $V(X + Y) = V(X) + V(Y) = 11/8$, so $\sigma_{X+Y} = \sqrt{\frac{11}{8}} \approx 1.17$.

7. (a) 3/5. (b) 7/5. (c) 13/5. (d) 19/5.

9. Show that $E(X^4) = 49/5$; use this to show that $V(X^2) = 76/25$. Hence the standard deviation of X^2 is $\frac{1}{5}\sqrt{76} \approx 1.74$.

11. $E(X) = 5/13$. Note that the random variable X is the sum $X_1 + X_2 + X_3 + X_4 + X_5$, where $X_i = 1$ if the ith card is an ace and $X_i = 0$ otherwise. What is $E(X_i)$?

13. (a) Let W be the waiting time random variable. If we imagine that all five marbles are drawn from the urn, it is clear that the blue marble is as likely to be the first marble as it is the second marble, etc. That is, $P(W = k) = 1/5$ for $k = 1, 2, 3, 4, 5$. [These equalities can also be easily verified directly. For example, $P(W = 3) = \dfrac{4}{5} \cdot \dfrac{3}{4} \cdot \dfrac{1}{3} = \dfrac{1}{5}$.] Thus

$$E(W) = \frac{1}{5}(1 + 2 + 3 + 4 + 5) = 3.$$

(b) The answer is 5 using the "nonsensical" argument that 1/5 of a blue marble is expected on each draw.

15. (a) Y also takes each value $1, 2, \ldots, n$ with probability $1/n$.

(c) By Theorem 2,

$$E(X) = \frac{1}{n} + \frac{2}{n} + \cdots + \frac{n}{n}.$$

So $1 + 2 + \cdots + n = n \cdot E(X) = n \cdot \frac{1}{2}(n + 1)$.

17. If $\mu = E(X)$, then $E(X + c) = \mu + c$. Hence

$$V(X + c) = \sum_{x} (x - \mu - c)^2 \cdot P(X + c = x) = \sum_{x} (x - c - \mu)^2 \cdot P(X = x - c).$$

Replace each $x - c$ by y to get $\sum\limits_{y} (y - \mu)^2 \cdot P(X = y) = V(X)$. The result just shown is intuitively obvious: $X + c$ shifts all values of X by c, but does not modify the spread of its values.

Since $V(cX) = c^2 \cdot V(X)$ is obvious for $c = 0$, assume that $c \neq 0$. Since $E(cX) = c \cdot \mu$,

$$V(cX) = \sum_{x} (x - c \cdot \mu)^2 \cdot P(cX = x) = c^2 \sum_{x} \left(\frac{x}{c} - \mu\right)^2 \cdot P\left(X = \frac{x}{c}\right)$$

$$= c^2 \sum_{y} (y - \mu)^2 \cdot P(X = y) = c^2 \cdot V(X).$$

19. (a) $E(S) = n \cdot \mu$ and $\sigma_S = \sqrt{n} \cdot \sigma$.

   (b) $E\left(\dfrac{1}{n} S\right) = \mu$ and the standard deviation of $\dfrac{1}{n} S$ is $\dfrac{1}{\sqrt{n}} \cdot \sigma$.

21. (a) Since all these random variables have finite value sets, it suffices to show that

   (1) $\quad P(X_1 + X_2 = x$ and $X_i = x_i$ for $i = 3, \ldots, n)$

   $$= P(X_1 + X_2 = x) \cdot \prod_{i=3}^{n} P(X_i = x_i)$$

   for real numbers $x, x_3, \ldots, x_n$. Let A be the set of all pairs (u, v) of real numbers so that u is in the value set of X_1, v is in the value set of X_2, and $u + v = x$. Then $\{X_1 + X_2 = x\}$ is the disjoint union $\bigcup\limits_{(u,v) \in A} \{X_1 = u$ and $X_2 = v\}$; hence

   (2) $\quad P(X_1 + X_2 = x) = \sum\limits_{(u,v) \in A} P(X_1 = u$ and $X_2 = v).$

   Similarly,

   $P(X_1 + X_2 = x$ and $X_i = x_i$ for $i = 3, \ldots, n)$

   $$= \sum\limits_{(u,v) \in A} P(X_1 = u, X_2 = v \text{ and } X_i = x_i \text{ for } i = 3, \ldots, n).$$

   Use independence of $X_1, X_2, \ldots, X_n$ and then (2) to show that (1) holds.

   (b) Use induction on n and part (a).

Section 9.4

1. If the outcomes were not independent, we would have to use conditional probabilities to conclude that the probability of (S, S, F, S, F) is

   $P(\text{first is S}) \cdot P(\text{second is S} | \text{first is S}) \cdot P(\text{third is F} | \text{first and second are S}) \cdot$ etc.

3. (a) Expected number is $np = 10/3$, since $n = 10$ and $p = 1/3$.

   (c) $1 - P(\text{at most 2 hits}) = 1 - F(2) \approx 1 - .299 = .701$ from Table 1.

5. (a) $(.9)^{10} \approx .349$ or $F(0) \approx .349$ from Table 1.

7. (a) $1 - \Phi(1) \approx 1 - .8413 = .1587.$ (b) $\approx .0227.$

9. (a) $\mu = 600.$ (b) $\sigma = 20.$

   (c) As in Example 9, one such interval is

   $$(\mu - 2\sigma, \mu + 2\sigma] = (560{,}640].$$

11. (a) $\mu = 500$ and $\sigma \approx 15.81$. Thus $10 \approx .632 \cdot \sigma$, so

   $$P(490 < X \le 510) \approx \Phi(.63) - \Phi(-.63) \approx .47.$$

   (b) $\approx .95.$

   (c) $\approx \Phi(20) - \Phi(-20)$. This is very, very close to 1: $.9999 \ldots$, where the first 88 digits are 9. For all practical purposes, the event

   $$\{490{,}000 < X \le 510{,}000\}$$

   is a certainty.

13. (a) 13/12. $X = X_1 + X_2 + X_3$ where $X_i = 1$ if the ith experiment is a success and $X_i = 0$ otherwise. So

$$E(X) = E(X_1) + E(X_2) + E(X_3).$$

(b) By Theorem 7 of §9.3, $V(X) = V(X_1) + V(X_2) + V(X_3) = 95/144$. Hence $\sigma_X = \frac{1}{12}\sqrt{95} \approx .81$.

(c) No. Why?

15. (a) Since y or $-y$ is nonnegative, we may assume that one of them, say y, is nonnegative. Now

$$1 = \text{area under the bell curve } \varphi$$

$$= \text{area under } \varphi \text{ to left of } y \quad + \quad \text{area under } \varphi \text{ to right of } y$$

$$= \Phi(y) + \text{area under } \varphi \text{ to the right of } y.$$

By symmetry of the graph of φ,

the area under φ to the right of $y =$ the area under φ to the left of $-y = \Phi(-y)$.

Section 10.1

1. (a) Since the operations $\vee$ and $\wedge$ treat 0 and 1 just as if they represent truth values, checking the laws 1Ba through 5Bb for all cases amounts to checking corresponding truth tables. Do enough until the situation is clear to you.

3. One solution is to set $S = \{1, 2, 3, 4, 5\}$ and define

$$\varphi(x_1, x_2, x_3, x_4, x_5) = \{i \in S : x_i = 1\}.$$

5. (a) The atoms are given by the four columns on the right in the table.

x	y	a	b	c	d
0	0	1	0	0	0
0	1	0	1	0	0
1	0	0	0	1	0
1	1	0	0	0	1

(c) In the notation of the answer to (a), $h = a \vee b \vee d$.

7. (a) No. A finite Boolean algebra has 2^n elements for some n.

9. (a) {1, 2}   (c) (1, 1)

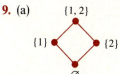

{1}   {2}

∅

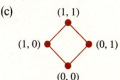

(1, 0)   (0, 1)

(0, 0)

11. (a) If $a \le x$ or $a \le y$, then surely $a \le x \vee y$ by Lemmas 3(a) and 2(a). Suppose that $a \le x \vee y$. Then $a = a \wedge (x \vee y) = (a \wedge x) \vee (a \wedge y)$. One of $a \wedge x$ and $a \wedge y$, say $a \wedge x$, must be different from 0. But $0 < a \wedge x \le a$, so $a \wedge x = a$ and $a \le x$.

(c) $a \le 1 = x \vee x'$, so $a \le x$ or $a \le x'$ by part (a). Both $a \le x$ and $a \le x'$ would imply $a \le x \wedge x' = 0$ by part (b), a contradiction.

13. $x \le y \Leftrightarrow x \vee y = y \Leftrightarrow \varphi(x \vee y) = \varphi(y) \Leftrightarrow \varphi(x) \vee \varphi(y) = \varphi(y) \Leftrightarrow \varphi(x) \le \varphi(y)$. Supply reasons.

Section 10.2

1. $x'y'z' \vee x'y'z \vee xyz'$.

3.

x	y	z	xy	xy ∨ z'
0	0	0	0	1
0	0	1	0	0
0	1	0	0	1
0	1	1	0	0
1	0	0	0	1
1	0	1	0	0
1	1	0	1	1
1	1	1	1	1

(a) $xyz' \vee xyz$.

(c) $x'y'z' \vee x'yz' \vee xy'z' \vee xyz' \vee xyz$.

5. (a) $x_1x_2x_3'x_4 \vee x_1x_2x_3'x_4' \vee x_1'x_2x_3x_4'$.

7. (a) $xz \vee y'$.

9. $y' \vee z$.

11. (a) Find the minterm canonical form for E'. Then find $E = (E')'$ using DeMorgan laws, first on joins and then on products.

   (b) $(x' \vee y')(x \vee y)$.

13. (a) The Boolean function for $x'z \vee y'z$ takes the value 1 at three elements in $\mathbb{B}^3$. The Boolean functions for products of literals take the value 1 at 1, 2 or 4 elements in $\mathbb{B}^3$ by Exercise 12.

Section 10.3

1. (a) xyz.

3. (a)

   (b) x —

5. (a)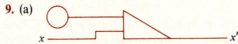

   $x \oplus y$

7. (a) $S = 1, C_O = 0$.   (c) $S = 0, C_O = 1$.

9. (a)

   x ———— x'

(c)

$$x \lor y$$

11.

Graph like Figure 7

Graph like Figure 7

13. It is convenient to view the result as valid for $n = 1$. Apply the Second Principle of Mathematical Induction. A glance at a piece of Figure 7 shows that the result is valid for $n = 2$. Assume that the result is true for all j with $1 \le j < n$. Consider k with $2^{k-1} < n \le 2^k$, and let $n' = 2^{k-1}$, $n'' = n - 2^{k-1}$. By the inductive assumption, there are digraphs D' and D'' for computing $x_1 \oplus x_2 \oplus \cdots \oplus x_{n'}$ and $x_{n'+1} \oplus \cdots \oplus x_n$. D' has $3(n' - 1)$ $\land$ and $\lor$ vertices and D'' has $3(n'' - 1)$ $\land$ and $\lor$ vertices. Also, every path in D' and in D'' has length at most $2(k - 1)$. Now create D as shown.

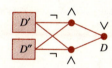

Section 10.4

1. $xyz \lor xyz' \lor xy'z' \lor xy'z \lor x'yz \lor x'y'z = x \lor z$.

3. $xyz \lor xyz' \lor xy'z \lor x'y'z' \lor x'y'z = xz \lor xy \lor x'y' = xy \lor y'z \lor x'y'$.

5. (a)

	yz	yz'	$y'z'$	$y'z$
x	+	+	+	+
x'	+			

(c)

	yz	yz'	$y'z'$	$y'z$
x	+			+
x'				+

7. (a)

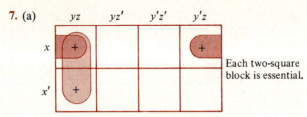

Each two-square block is essential.

(c)

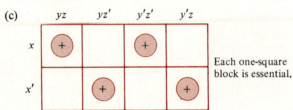

Each one-square block is essential.

9. (a) $z' \vee xy \vee x'y' \vee w'y$ or $z' \vee xy \vee x'y' \vee w'x'$.

   (c) $w'x'z' \vee w'xy' \vee wxy \vee wx'z \vee y'z'$.

Section 11.1

1. (a)

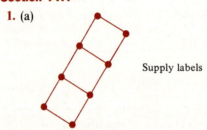

Supply labels

3. (a) h, o, p, q, r, z. (c) B and C.

   (e) f, z, p, does not exist.

5. (a) $a \vee b = \mathrm{lub}(a, b) = \max\{a, b\}$, $a \wedge b = \mathrm{glb}(a, b) = \min\{a, b\}$.

   (c) 73. (e) $\sqrt{73}$.

7. (a) Suppose that $\preceq$ is a partial order on S and that $\succeq$ is defined by $x \succeq y$ if and only if $y \preceq x$. Then $x \preceq x$, so $x \succeq x$. If $x \succeq y$ and $y \succeq x$, then $y \preceq x$ and $x \preceq y$, so $x = y$. If $x \succeq y$ and $y \succeq z$, then $y \preceq x$ and $z \preceq y$, so $z \preceq x$ and thus $x \succeq z$. Thus $\succeq$ satisfies (R), (AS) and (T).

   (b) Clearly, $x \preceq x$, so (R) holds for $\preceq$. If $x \preceq y$ and $y \preceq z$, there are four possible cases:

   $$x = y = z, \quad x = y \prec z, \quad x \prec y = z \quad \text{and} \quad x \prec y \prec z.$$

   Show that $x \preceq z$ in each case, so that (T) holds. If $x \preceq y$ and $y \preceq x$, the cases are:

   $$x = y = x, \quad x = y \prec x, \quad x \prec y = x \quad \text{and} \quad x \prec y \prec x.$$

   Only the first is possible. The other three cases violate (AR) for $\prec$. Thus (AS) holds for $\preceq$.

9. Since $w = w\lambda$, $w \preceq w$ and (R) holds. If $w_1 \preceq w_2$ and $w_2 \preceq w_3$, there are words u and v with $w_2 = w_1 u$ and $w_3 = w_2 v$. Then $w_3 = w_1 uv$ with $uv \in \Sigma^*$, and so $w_1 \preceq w_3$. Thus (T) holds.

11. Not if Σ has more than one element. Show that antisymmetry fails.

13. (a) No. Every finite subset of $\mathbb{N}$ is a subset of a larger finite subset of $\mathbb{N}$.

(c) $\text{lub}\{A, B\} = A \cup B$. Note that $A \cup B \in \mathcal{F}(\mathbb{N})$ for all $A, B \in \mathcal{F}(\mathbb{N})$.

(e) Yes.

15. (a) Only $\leq$. $<$ is not reflexive and $\leq$ is not antisymmetric.

(c)

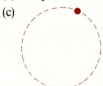

17. See Figures 2 and 7 or Exercise 16 for two different sorts of failure.

19. (a) Show that b satisfies the definition of $\text{lub}\{x, y, z\}$, i.e., $x \leq b$, $y \leq b$, $z \leq b$, and if $x \leq c$, $y \leq c$, $z \leq c$, then $b \leq c$.

(b) Show by induction on n that every n-element subset of a lattice has a least upper bound.

(c) Use part (a) and commutativity of $\vee$.

Section 11.2

1. (a) $\{\varnothing, \{1\}, \{1, 4\}, \{1, 4, 3\}, \{1, 4, 3, 5\}, \{1, 4, 3, 5, 2\}\}$ is one.

3. (a) $501, 502, \dots, 1000.$ (c) Yes. Think of primes or see Exercise 17.

5. Yes. If $a \leq b$, then $\text{lub}(a, b) = b$ and $\text{glb}(a, b) = a$.

7. (a) Transitivity, for example. If $f \preceq g$ and $g \preceq h$, then $f(t) \preceq g(t)$ and $g(t) \preceq h(t)$ in S, for all t in T. Since $\preceq$ is transitive on S, $f(t) \preceq h(t)$ for all t, so $f \preceq h$ in $\text{FUN}(T, S)$.

(c) $f(t) \preceq f(t) \vee g(t) = h(t)$ for all t, so $f \preceq h$. Similarly, $g \preceq h$, so h is an upper bound for $\{f, g\}$. Show that if $f \preceq k$ and $g \preceq k$, then $h \preceq k$, so that h is the least upper bound for $\{f, g\}$.

9. Supply labels. (a)   (c)

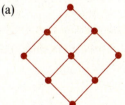

11. (a) $(0, 0), (0, 1), (0, 2), (1, 0), (1, 1), (1, 2), (2, 0), (2, 1), (2, 2).$

(c) $(3, 0), (3, 1), (3, 2), (4, 0), (4, 1), (4, 2).$

13. (a) $000, 0010, 010, 10, 1000, 101, 11.$

15. (a) in of the list this order words sentence standard increasing.

17. Consider a maximal chain $a_1 \prec a_2 \prec \cdots \prec a_n$ in S. There is no chain $b \prec a_1 \prec a_2 \prec \cdots \prec a_n$ in S, so there is no b with $b \prec a_1$. That is, a_1 is minimal. [Finiteness is essential. The chain $(\mathbb{Z}, \leq)$ is a maximal chain in itself.]

19. Antisymmetry is immediate. For transitivity consider cases. Suppose that

$$(s_1, \ldots, s_n) \prec (t_1, \ldots, t_n) \quad \text{and} \quad (t_1, \ldots, t_n) \prec (u_1, \ldots, u_n).$$

If $s_1 \prec_1 t_1$, then $s_1 \prec_1 t_1 \leq_1 u_1$, so $(s_1, \ldots, s_n) \prec (u_1, \ldots, u_n)$. If $s_1 = t_1, \ldots, s_{r-1} = t_{r-1}, s_r \prec_r t_r$, and $t_1 = u_1, \ldots, t_{p-1} = u_{p-1}, t_p \prec_p u_p$ and $r < p$, then $s_1 = u_1, \ldots, s_{r-1} = u_{r-1}$ and $s_r \prec_r t_r = u_r$, and again $(s_1, \ldots, s_n) \prec (u_1, \ldots, u_n)$. The remaining cases are similar.

Section 11.3

1. (a) $\mathbf{A} * \mathbf{A} = \begin{bmatrix} 1 & 1 & 1 \\ 1 & 1 & 1 \\ 1 & 1 & 1 \end{bmatrix}$. Since $\mathbf{A} * \mathbf{A} \leq \mathbf{A}$ is not true, R is not transitive.

(c) Not transitive. Note that $\mathbf{A} * \mathbf{A} = \begin{bmatrix} 1 & 0 & 0 \\ 0 & 1 & 0 \\ 0 & 0 & 1 \end{bmatrix}$.

3. (a) The matrices for R and R^2 are

$$\mathbf{A} = \begin{bmatrix} 0 & 0 & 0 \\ 1 & 0 & 1 \\ 0 & 1 & 0 \end{bmatrix} \quad \text{and} \quad \mathbf{A} * \mathbf{A} = \begin{bmatrix} 0 & 0 & 0 \\ 0 & 1 & 0 \\ 1 & 0 & 1 \end{bmatrix}.$$

(c) No; compare $\mathbf{A}$ and $\mathbf{A} * \mathbf{A}$ and note that $\mathbf{A} * \mathbf{A} \leq \mathbf{A}$ fails.

(e) Yes.

5. (a) Matrix for R^0 is the identity matrix. Matrix for R^1 is $\mathbf{A}$, of course. Matrix for R^n is $\mathbf{A} * \mathbf{A}$ for $n \geq 2$, as should be checked by induction.

(b) R is reflexive, but not symmetric or transitive.

7. (a) $\mathbf{A}_f = \begin{bmatrix} 0 & 0 & 1 & 0 \\ 0 & 1 & 0 & 0 \\ 0 & 1 & 0 & 0 \\ 0 & 1 & 0 & 0 \end{bmatrix}$ and $\mathbf{A}_g = \begin{bmatrix} 0 & 0 & 0 & 1 \\ 0 & 0 & 1 & 0 \\ 0 & 1 & 0 & 0 \\ 1 & 0 & 0 & 0 \end{bmatrix}$.

(b) They will be different, since the Boolean matrix for $R_f R_g$ is

$$\mathbf{A}_f * \mathbf{A}_g = \begin{bmatrix} 0 & 1 & 0 & 0 \\ 0 & 0 & 1 & 0 \\ 0 & 0 & 1 & 0 \\ 0 & 0 & 1 & 0 \end{bmatrix};$$

this is the Boolean matrix for $R_{g \circ f}$ but not for $R_{f \circ g}$.

(c) One does and one doesn't.

9. (a) R_1 satisfies (AR) and (S).

(c) R_3 satisfies (R), (AS) and (T).

(e) R_5 satisfies only (S).

11. (a) True. For each x, $(x, x) \in R_1 \cap R_2$, so $(x, x) \in R_1 R_2$.

(c) False. Consider the equivalence relations R_1 and R_2 on $\{1, 2, 3\}$ with Boolean matrices

$$\mathbf{A}_1 = \begin{bmatrix} 1 & 1 & 0 \\ 1 & 1 & 0 \\ 0 & 0 & 1 \end{bmatrix} \quad \text{and} \quad \mathbf{A}_2 = \begin{bmatrix} 1 & 0 & 0 \\ 0 & 1 & 1 \\ 0 & 1 & 1 \end{bmatrix}.$$

13. Don't use Boolean matrices; the sets S, T, U might be infinite.

(a) $R_1 R_3 \cup R_1 R_4 \subseteq R_1(R_3 \cup R_4)$ by Example 2(a). For the reverse inclusion, consider (s, u) in $R_1(R_3 \cup R_4)$ and show that (s, u) is in $R_1 R_3$ or $R_1 R_4$.

(c) Show that $R_1(R_3 \cap R_4) \subseteq R_1 R_3 \cap R_1 R_4$. Equality need not hold. For example, consider R_1, R_3, R_4 with Boolean matrices

$$\mathbf{A}_1 = \begin{bmatrix} 1 & 1 \\ 0 & 0 \end{bmatrix}, \quad \mathbf{A}_3 = \begin{bmatrix} 0 & 0 \\ 0 & 1 \end{bmatrix}, \quad \mathbf{A}_4 = \begin{bmatrix} 0 & 1 \\ 0 & 0 \end{bmatrix}.$$

15. (R) R is reflexive if and only if $(x, x) \in R$ for every x, if and only if $\mathbf{A}[x, x] = 1$ for every x.

(AR) Similar to the argument for (R), with $(x, x) \notin R$ and $\mathbf{A}[x, x] = 0$.

(S) Follows from $\mathbf{A}^T[x, y] = \mathbf{A}[y, x]$ for every x, y.

(AS) R is antisymmetric if and only if $x = y$ whenever $(x, y) \in R$ and $(y, x) \in R$, i.e., whenever $\mathbf{A}[x, y] = \mathbf{A}^T[x, y] = 1$. Thus R is antisymmetric if and only if all of the off-diagonal entries of $\mathbf{A} \wedge \mathbf{A}^T$ are 0.

(T) This follows from Theorem 3 and (a) of the summary.

17. Given $m \times n$, $n \times p$ and $p \times q$ Boolean matrices $\mathbf{A}_1, \mathbf{A}_2, \mathbf{A}_3$, they correspond to relations R_1, R_2, R_3 where R_1 is a relation from $\{1, 2, \ldots, m\}$ to $\{1, 2, \ldots, n\}$, etc. The matrices for $(R_1 R_2)R_3$ and $R_1(R_2 R_3)$ are $(\mathbf{A}_1 * \mathbf{A}_2) * \mathbf{A}_3$ and $\mathbf{A}_1 * (\mathbf{A}_2 * \mathbf{A}_3)$ by four applications of Theorem 1.

19. (a) To show that $R \cup E$ is a partial order, show

(R) $(x, x) \in R \cup E$ for all $x \in S$,

(AS) $(x, y) \in R \cup E$ and $(y, x) \in R \cup E$ imply $x = y$,

(T) $(x, y) \in R \cup E$ and $(y, z) \in R \cup E$ imply $(x, z) \in R \cup E$.

To verify (T), consider cases. The four cases for (T) are:

$$(x, y) \in R \quad \text{and} \quad (y, z) \in R,$$

$$(x, y) \in R \quad \text{and} \quad (y, z) \in E \quad [\text{so } y = z],$$

$$(x, y) \in E \quad \text{and} \quad (y, z) \in R,$$

$$(x, y) \in E \quad \text{and} \quad (y, z) \in E.$$

The last two can be grouped together, since if $(x, y) \in E$ and $(y, z) \in R \cup E$, then $(x, z) = (y, z) \in R \cup E$.

Section 11.4

1. (a) $\begin{bmatrix} 1 & 1 & 0 \\ 0 & 1 & 0 \\ 0 & 0 & 1 \end{bmatrix}$. (c) $\begin{bmatrix} 1 & 1 & 0 \\ 1 & 1 & 0 \\ 0 & 0 & 1 \end{bmatrix}$. (e) $\begin{bmatrix} 1 & 1 & 0 \\ 1 & 1 & 0 \\ 0 & 0 & 1 \end{bmatrix}$.

3. $\{1, 2\}$, $\{3\}$.

5. (a) $\begin{bmatrix} 1 & 1 & 0 & 0 & 0 \\ 0 & 1 & 0 & 1 & 0 \\ 0 & 0 & 1 & 0 & 1 \\ 0 & 1 & 0 & 1 & 0 \\ 0 & 0 & 0 & 0 & 1 \end{bmatrix}$. (c) $\begin{bmatrix} 1 & 1 & 0 & 0 & 0 \\ 1 & 1 & 0 & 1 & 0 \\ 0 & 0 & 1 & 0 & 1 \\ 0 & 1 & 0 & 1 & 0 \\ 0 & 0 & 1 & 0 & 1 \end{bmatrix}$. (e) $\begin{bmatrix} 1 & 1 & 0 & 1 & 0 \\ 1 & 1 & 0 & 1 & 0 \\ 0 & 0 & 1 & 0 & 1 \\ 1 & 1 & 0 & 1 & 0 \\ 0 & 0 & 1 & 0 & 1 \end{bmatrix}$.

7. (a) $r(R)$ is the usual order $\leq$.

(c) $rs(R)$ is the universal relation on $\mathbb{P}$.

(e) R is already transitive.

9. $(h_1, h_2) \in st(R)$ if $h_1 = h_2$ or if one of h_1, h_2 is the High Hermit. On the other hand, $ts(R)$ is the universal relation on F.O.H.H.

11. (a) Show that $t(R) \cup E \subseteq tr(R)$. For the reverse inclusion $tr(R) \subseteq rt(R)$, it is enough to show that $r(R) \subseteq rt(R)$ and that $rt(R)$ is transitive, since then $rt(R)$ contains the transitive closure of $r(R)$.

(b) Compare $(R \cup E) \cup (R \cup E)^\leftarrow$ and $(R \cup R^\leftarrow) \cup E$; see Exercise 12 of §3.1.

13. (a) By Exercise 12(a) and (b) $sr(R_1 \cup R_2) = sr(R_1) \cup sr(R_2) = R_1 \cup R_2$. Thus $tsr(R_1 \cup R_2) = t(R_1 \cup R_2)$. Apply Theorem 3.

(b) See Exercise 9 in §3.1.

15. Any relation that contains R will include the pair $(1, 1)$, so will not be antireflexive.

17. (a) The intersection of all relations that contain R and have property p is the smallest such relation.

(c) (i) fails. $S \times S$ is not antireflexive.

(d) (ii) fails.

Section 12.1

1. (a) $(1\,5\,4\,2)$. (c) $(2\,5)(3\,4)$.

3. (a)   (c)

5. (a) $(1\,4\,2)$. (c) $(1\,3\,4\,2)$. (e) $(1\,3)(2\,4)$.

7. (a) $(1\,2)$ itself. (c) $(1\,6\,3)$.

9. (b) $(1\,4)(1\,3)(1\,5)(1\,2)(1\,7)$.

(c) $(k_1\,k_2\cdots k_m) = (k_1\,k_m)(k_1\,k_{m-1})(k_1\,k_{m-2})\cdots(k_1\,k_3)(k_1\,k_2)$.

11. Answers are 1, 1, 2 and 2, respectively. Specify which elements have which orders.

13. (a) $\{e, (2\ 5)\}$.

(c) $\{e, (1\ 6)(2\ 4\ 3), (2\ 3\ 4), (1\ 6), (2\ 4\ 3), (1\ 6)(2\ 3\ 4)\}$.

15. (a) e, $(1\ 2)$, $(1\ 2\ 3)$, $(1\ 2\ 3\ 4)$, $(1\ 2\ 3\ 4\ 5)$ and $(1\ 2)(3\ 4\ 5)$, say. $(1\ 2\ 3\ 4\ 5\ 6)$ also has order 6, but it's a cycle.

(b) Clearly, cycles have orders 1, 2, 3, 4, 5 or 6. So consider a permutation written as a product of two or more disjoint nontrivial cycles; here we re-gard the 1-cycles as trivial. In view of Theorem 2 the problem reduces to analyzing the possible ways to write 6 as a sum with at least two summands greater than 1.

17. Yes. $g^i \circ g^j = g^{i+j} = g^j \circ g^i$ for all powers of g.

19. (a) If j is a multiple of m, say $j = qm$, then $g^j = (g^m)^q = e^q = e$. Conversely, suppose that $g^j = e$. Use the Division Algorithm to write $j = qm + r$ where $0 \le r < m$. Then as above $g^{qm} = e$, so $g^r = g^{qm+r} = g^j = e$. Since m is the least positive exponent for g^m to equal to e, r must be 0. So j is a multiple of m.

(c) Use the notation established prior to Theorem 2. Suppose that $g^j = e$ and $j > 0$. Then $c_1^j c_2^j \cdots c_k^j = e$. By part (b) each $c_i^j = e$. Apply part (a) k times to conclude that j is a multiple of each m_i. So j is a common multiple of $m_1, m_2, \ldots, m_k$ and hence $j > \mathrm{lcm}(m_1, m_2, \ldots, m_k)$.

Section 12.2

1. (a) Let g, f be in $\mathrm{AUT}(D)$. Then (x, y) is an edge if and only if $(g(x), g(y))$ is an edge. Similarly, $(g(x), g(y))$ is an edge if and only if $(f(g(x)), f(g(y)))$ is an edge. Hence (x, y) is an edge if and only if $(f(g(x)), f(g(y)))$ is. So $f \circ g$ is in $\mathrm{AUT}(D)$. As we noted in §12.1, we don't need to check that $g \in \mathrm{AUT}(D)$ implies $g^{-1} \in \mathrm{AUT}(D)$, although this is easy to do directly.

3. $|\mathrm{AUT}(D)| = 4$, $|\mathrm{AUT}(D)p| = |\{p, r\}| = 2$, $|\mathrm{FIX}(p)| = |\{e, f\}| = 2$ and $2 \cdot 2 = 4$; the cases for q, r and s are similar.

5. (b) and (c) See Example 2 in §12.4.

7. The Theorem shows that each $|Gx_j|$ must be a divisor of $|G| = 2^k$. So each $|Gx_j|$ must be 1, 2, 4, ..., or 2^k. Since $|X|$ is odd, for some j we must have $|Gx_j| = 1$. By the Theorem again, $|G| = |\mathrm{FIX}(x_j)|$, so $G = \mathrm{FIX}(x_j)$. Hence $g(x_j) = x_j$ for all $g \in G$.

9. (a) Here is one possible sequence.

Etc.

11. $\mathrm{FIX}(\{w, y\}) = \{e, f\}$ since both $f(w)$ and $f(y)$ belong to $\{w, y\}$. However, $\mathrm{FIX}(w) \cap \mathrm{FIX}(y) = \{e\}$.

13. (a) $\mathrm{AUT}(H)u = \{u, v\}$. Since $u = e(u)$ and $v = f(u)$, the choice $g_1 = e$, $g_2 = f$ will work.

(c) AUT$(H)x = \{w, x, y, z\}$. Since $w = g(x)$, $x = e(x)$, $y = f(x)$ and $z = fg(x)$, the choice $g_1 = e$, $g_2 = f$, $g_3 = g$, $g_4 = fg$ works.

15. (a) Say $X = Gx_0$. For any x in X, obviously $Gx \subseteq X$. Consider $y \in X$. Then $y = g_1(x_0)$ for some $g_1 \in G$. Also, $x = g_2(x_0)$ for some $g_2 \in G$. So $y = g_1(x_0) = g_1 \circ g_2^{-1}(x) \in Gx$.

(b) Since G is finite, so is $Gx = X$. Apply the Theorem and part (a).

17. (a) Since $e \in G$ and $e(x) = x$, R contains all (x, x) and is reflexive. Since $g(x) = y \Rightarrow y = g^{-1}(x)$, we have $(x, y) \in R \Rightarrow (y, x) \in R$, so R is symmetric. For transitivity, note that $g(x) = y$ and $g'(y) = z$ imply that $g' \circ g(x) = z$.

Section 12.3

1. (a) From Example 1, the numbers are 6, 4, 4, 2, 0, 0, 0, 0, with sum 16.

(b) See Example 6(c), §12.2.

(c) The sums should agree; they represent the two calculations of $|S|$ in the proof of Theorem 1, using formulas (1) and (2).

3. Graph automorphisms can interchange w and y and they can interchange x and z.

(b) FIX$(e) = \{w, x, y, z\}$, FIX$(g) = \{x, z\}$, FIX$(h) = \{w, y\}$ and FIX$(gh) = \varnothing$. So by Theorem 1, there are $\frac{1}{4}(4 + 2 + 2 + 0)$ orbits under G. Give them.

5. For the orbit $\{w, y\}$, the automorphisms g and gh both move each element. For the orbit $\{x, z\}$, the automorphisms h and gh both move each element.

7. (a)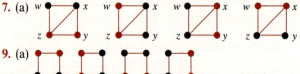

9. (a)

(c)

(d) There are three essentially different colorings in parts (a), (b) and (c). The situation with 3 vertices red is similar to part (c). This gives us $4 + 2 + 4 + 4 = 14$ distinct colorings. There are $2^4 = 16$ altogether. The remaining two colorings use all red or all black. So there are 6 essentially different colorings: (a), (b), (c), (c) with 3 red, all red, and all black. In the next section we will have a *much* better method for solving this and much harder problems. See Example 2, §12.4.

11. h takes each element of $\{w, x, u, v\}$ to itself, so h^* takes each 2-element subset of this set to itself. There are $\binom{4}{2} = 6$ such sets. The only other 2-element set mapped to itself is $\{y, z\}$, so FIX$_T(h^*)$ has 7 elements.

13. (a) $|G| = 2^2$ and $|E| = 5$ is odd, so Exercise 7, §12.2, shows that some edge must be fixed by all members of G.

15. (a) $G = \{e, g\}$ where e is the identity permutation of $V = \{u, v\}$ and $g(u) = v$, $g(v) = u$. Both e^* and g^* are the identity on the one-element edge set E.

(b) Since $g^* = h^*$ implies $(h^{-1} \circ g)^* = e^*$, it suffices to show that $g^* = e^*$ implies $g = e$. [For then $g^* = h^*$ implies $(h^{-1} \circ g)^* = e^*$ implies $h^{-1} \circ g = e$ implies $g = h$.] So assume that $g^* = e^*$. Since $g^*(\{u, v\}) = \{g(u), g(v)\} = \{u, v\}$, all that g^* can do to an edge is leave it alone or turn it end for end. If $g \neq e$, then g^* must turn some edge $\{u, v\}$ end for end. Then it has to take any *other* edge $\{u, w\}$ attached to u and attach it to v, while at the same time leaving it alone or switching it end for end. Think about this. We would have $\{u, w\} \to \{v, t\}$ for some t, but also $\{u, w\} \to \{u, w\}$, so $v = w$, $t = u$ and $\{u, w\} = \{u, v\}$. Thus $\{u, w\}$ could not really be an edge different from $\{u, v\}$. So there *are* no other edges attached to u or, by symmetry, to v. But then, since H is connected, $\{u, v\}$ is the only edge there is, and H is the graph of part (a).

Section 12.4

1. (a) 4.

(c) Apply Theorem 2, noting that $m(e) = 5$, $m(f) = 4$, $|G| = 2$. Answer is $(k^5 + k^4)/2$.

3. (a)

	w	x	y	z
e	w	x	y	z
a	z	y	x	w

Here $m(e) = 4$, $m(a) = 2$, $|G| = 2$. Theorem 2 gives $C(k) = (k^4 + k^2)/2$.

(b) As in Example 4, the answer is

$$C(4) - 4 \cdot C(3) + 6 \cdot C(2) - 4 \cdot C(1) = 136 - 4 \cdot 45 + 6 \cdot 10 - 4 \cdot 1 = 12.$$

Alternatively, consider the 24 permutations of the labels and note that two permutations give equivalent labels only if they are reverses of each other. For example, the labels $w\,x\,y\,z$ and $z\,y\,x\,w$ are reverses of each other.

5. (a) Apply Theorem 2, noting that $m(e) = 4$, $m(r) = m(r^2) = 2$, $m(f) = m(g) = m(h) = 3$. Answer is $C(k) = (k^4 + 2k^2 + 3k^3)/6$.

(b) Apply Theorem 2 to get $C(k) = (k^3 + 2k + 3k^2)/6$.

7. (a) Using Figure 4(a),

$$C(4) - 4 \cdot C(3) + 6 \cdot C(2) - 4 \cdot C(1) = 55 - 4 \cdot 21 + 6 \cdot 6 - 4 \cdot 1 = 3.$$

9. (a) $C(3) - 3 \cdot C(2) + 3 \cdot C(1) = 21 - 3 \cdot 6 + 3 \cdot 1 = 6$.

(b) 4.

(c) $C(5) - 5 \cdot C(4) + 10 \cdot C(3) - 10 \cdot C(2) + 5 \cdot C(1)$
    $= 120 - 5 \cdot 55 + 10 \cdot 21 - 10 \cdot 6 + 5 = 0$, of course.

11. (a) From Figure 4(b), §12.3, we find: orbits under $\langle e \rangle$ are $\{w\}, \{x\}, \{y\}, \{z\}$; orbits under $\langle g \rangle$ are $\{w, y\}, \{x\}, \{z\}$; orbits under $\langle h \rangle$ are $\{w\}, \{x, z\}, \{y\}$; and orbits under $\langle gh \rangle$ are $\{w, y\}, \{x, z\}$. So $m(e) = 4$, $m(g) = m(h) = 3$ and $m(gh) = 2$.

(b) $C(k) = \frac{1}{4}(k^4 + 2k^3 + k^2)$ using Theorem 2 and part (a).

(c) $C(2) = \frac{1}{4}(16 + 16 + 4) = 9$.

13. Following the suggestion:

Type	Number of that type	$m(g)$ when group acts on vertices	$m(g)$ when group acts on edges
a	6	4	7
b	6	2	3
c	3	4	6
d	8	4	4
e	1	8	12

(a) $C(k) = (k^8 + 17k^4 + 6k^2)/24$. For example, there are $6 + 3 + 8 = 17$ rotations with $m(g) = 4$.

(b) $C(k) = (k^{12} + 6k^7 + 3k^6 + 8k^4 + 6k^3)/24$.

15. (a) In this case, functions numbered n and $15 - n$ in Figure 8 are regarded as equivalent. So there are 8 equivalence classes, which correspond to 8 essentially different 2-input logical circuits.

(b) Now, as noted in the discussion of Figure 8, 2 and 4 are equivalent, as are 3 and 5. There are now 6 distinct classes. List them.

Section 12.5

1. (a) $\mathbb{Z}$.   (c) $\mathbb{Z}$.   (e) $\mathbb{Z}$.

3. (a) $3\mathbb{Z}$.

(c) Not a subgroup; 2 and 4 are in the subset, but $2 + 4$ is not.

(e) $\mathbb{N} \cup (-\mathbb{N}) = \mathbb{Z} = 1 \cdot \mathbb{Z}$.

5. $h \cdot g = k \cdot g$ implies $h = h \cdot e = h \cdot (g \cdot g^{-1}) = (h \cdot g) \cdot g^{-1} = (k \cdot g) \cdot g^{-1} = k \cdot (g \cdot g^{-1}) = k \cdot e = k$.

7. (a) We know that 1 is a generator. Show that 2, 3 and 4 are also generators.

(b) In this case 1 and 5 are the only generators. Show that none of $\langle 2 \rangle$, $\langle 3 \rangle$ and $\langle 4 \rangle$ equals $\mathbb{Z}(6)$.

9. (a) $\varphi(n) = n + 3$ defines one.

11. Show that the sets in (b) and (c) both contain the permutation $(1\,2\,3)$, but neither contains $(1\,2\,3)(1\,2\,3)$. The subsets in (a) and (d) are the subgroups $\text{FIX}_G(4)$ and $\text{FIX}_G(\{1, 2\})$; see Exercise 10 of § 12.2.

13. (a) Apply Theorem 1(b) twice.

(b) An easy induction shows that $(g_1 \cdot g_2 \cdots g_n)^{-1} = g_n^{-1} \cdots g_2^{-1} \cdot g_1^{-1}$. The inductive step uses $(g_1 \cdot g_2 \cdots g_n \cdot g_{n+1})^{-1} = g_{n+1}^{-1} \cdot (g_1 \cdot g_2 \cdots g_n)^{-1}$.

15. (a) If g and h belong to the intersection, they both belong to each subgroup in the collection, so their product $g \cdot h$ does too. Since each subgroup contains the identity, so does the intersection. Every member of the intersection belongs to each subgroup, so its inverse does too. Thus its inverse also belongs to the intersection. Hence the intersection is closed under products and inverses and contains the identity.

(b) Any example in which neither H nor K contains the other will work. For example, if $G = (\mathbb{Z}, +)$, let $H = 2\mathbb{Z}$ and $K = 3\mathbb{Z}$.

17. (a) $6 = 3!$.

(b) $\text{FIX}_G(1) = \{e, (2\,3)\}$.

(c) $\text{FIX}_G(1) \circ (1\,2\,3) = \{(1\,2\,3), (1\,3)\}$.

(d) $(1\,2\,3) \circ \text{FIX}_G(1) = \{(1\,2\,3), (1\,2)\} \neq \{(1\,2\,3), (1\,3)\}$.

(e) It contains $(1\,2\,3)$, so the only left coset it *could* be is $(1\,2\,3) \circ \text{FIX}_G(1)$. Apply part (d).

(f) 3.

19. (a) $\langle a \rangle = \{e, a, b\}$.

(b) $\langle a \rangle \bullet c = \{c, d, f\}$ while $c \bullet \langle a \rangle = \{c, f, d\}$.

(c) $\langle c \rangle, \langle d \rangle, \langle f \rangle$.

(d) $|G|/|\langle d \rangle| = 6/2 = 3$. Also, see part (e).

(e) $\{e, d\}, \{a, c\}, \{b, f\}$.

21. $g \cdot H$ contains $g \cdot e = g$. So if $g \cdot H = H$, then $g \in H$. If $g \in H$, then H and $g \cdot H$ both contain g, so are not disjoint. By Theorem 4, $g \cdot H = H$ in this case.

23. (a) For $h \in H$, $(g \cdot h)^{-1} = h^{-1} \cdot g^{-1} \in H \cdot g^{-1}$, so

$$\{f^{-1} : f \in g \cdot H\} \subseteq H \cdot g^{-1}.$$

Moreover, $h \cdot g^{-1} = (g \cdot h^{-1})^{-1}$ is in $\{f^{-1} : f \subset g \cdot H\}$ for $h \in H$, so

$$H \cdot g^{-1} \subseteq \{f^{-1} : f \in g \cdot H\}.$$

(b) $g \cdot H \to H \cdot g^{-1}$. Show that this is one-to-one.

25. (a) (R)   Note that $g^{-1} \cdot g = e \in H$.

(S)   Note that $g_1^{-1} \cdot g_2 = (g_2^{-1} \cdot g_1)^{-1}$.

(T)   Note that $g_3^{-1} \cdot g_1 = (g_3^{-1} \cdot g_2) \cdot (g_2^{-1} \cdot g_1)$.

(b) Show that $g_1 \cdot H = \{g \in G : g \sim g_1\}$.

Section 12.6

1. (a), (c), (e).

3. (a) Not an isomorphism, since it doesn't map $\mathbb{Z}$ *onto* $\mathbb{Z}$.

(c) Is an isomorphism, being a one-to-one and onto homomorphism.

(e) Not an isomorphism, since $\varphi(n) = 3n$ does not map $\mathbb{Z}$ *onto* $\mathbb{Z}$.

5. (a) Show $[f + (g + h)](x) = [(f + g) + h](x)$ for all $x \in \mathbb{R}$; this will show that F is associative. The zero function $\mathbf{0}$, where $\mathbf{0}(x) = 0$ for all $x \in \mathbb{R}$, is the additive identity for F. The additive inverses are just the negatives of the functions.

(b) Yes. Explain.

(c) $\varphi(f + g) = (f + g)(73) = f(73) + g(73) = \varphi(f) + \varphi(g)$.

7. (a) $\{0\}$.   (c) $5\mathbb{Z} = \{5n : n \in \mathbb{Z}\}$.

9. (a) 4.   (c) 3.

11. (a) The identity is (e_G, e_H), where e_G and e_H are the respective identities of G and H, and $(g, h)^{-1} = (g^{-1}, h^{-1})$.

(c) $\{(e_G, h) : h \in H\}$.

(d) $\{(e_G, h) : h \in H\}$. Part (c) shows that this subgroup is normal.

13. (b) $\begin{bmatrix} 0 & 1 \\ 1 & 0 \end{bmatrix} \cdot H = \left\{ \begin{bmatrix} 0 & 1 \\ 1 & x \end{bmatrix} : x \in \mathbb{R} \right\}$ while $H \cdot \begin{bmatrix} 0 & 1 \\ 1 & 0 \end{bmatrix} = \left\{ \begin{bmatrix} x & 1 \\ 1 & 0 \end{bmatrix} : x \in \mathbb{R} \right\}$.

(c) $\begin{bmatrix} y & z \\ 0 & 1/y \end{bmatrix} \begin{bmatrix} 1 & x \\ 0 & 1 \end{bmatrix} \begin{bmatrix} 1/y & -z \\ 0 & y \end{bmatrix} = \begin{bmatrix} 1 & xy^2 \\ 0 & 1 \end{bmatrix}$ is in H.

(d) Kernel of φ is H.

(e) Use the result of part (d) and the Fundamental Homomorphism Theorem.

15. The pre-image of $\varphi(g)$ is $g \cdot K$ where K is the kernel of φ. By assumption, $|g \cdot K| = 1$. So $|K| = 1$ and φ is one-to-one as noted in the corollary to Theorem 1.

17. (a) Use the identity $(g \cdot h) \cdot H \cdot (g \cdot h)^{-1} = g \cdot (h \cdot H \cdot h^{-1}) \cdot g^{-1}$. Also, if $H = g \cdot H \cdot g^{-1}$, then $g^{-1} \cdot H \cdot g = g^{-1} \cdot (g \cdot H \cdot g^{-1}) \cdot g = H$.

(b) $\{g \in G : g \cdot H \cdot g^{-1} = H\}$ is a subgroup [part (a)] of G containing A, and G is the smallest subgroup containing A. So $\{g \in G : g \cdot H \cdot g^{-1} = H\} = G$.

19. (a) $e \cdot K \cdot (1\,3) \cdot K = e \cdot \{e, (1\,2)\} \cdot (1\,3) \cdot \{e, (1\,2)\} = \{(1\,3), (1\,3\,2), (1\,2\,3), (2\,3)\}$, but cosets only have 2 members.

(b) No. Explain.

Section 12.7

1. (a) Yes.

(c) No. Only 1 itself has an inverse.

3. (a) Yes.   **(c)** Yes; compare Example 7(b).

5. (a) No. Illustrate.

(c) No, the zero matrix has no inverse, for example.

7. (a) *break, fast, fastfood, lunchbreak, foodfood.*

(c) *fastfast, foodfood, fastfastfoodbreakbreak, λ.*

9. (b) $(\mathbb{N}, \max)$ is a monoid because 0 is an identity. $(\mathbb{N}, \min)$ has no identity, so it is not a monoid. Check these claims.

11. (a) $\mathbb{P}$.   **(c)** $\mathbb{Z}$.   **(e)** $\mathbb{Z}$.

(g) $18\mathbb{P} = \{18k : k \in \mathbb{P}\}$.

13. $\mathbb{P} = \{1\}^+$, $\{0\} = \{0\}^+$, $18\mathbb{P} = \{18\}^+$.

15. (a) $2\mathbb{N} = \{2k : k \in \mathbb{N}\}$.   **(c)** $\{0\}$.   **(e)** Σ^*.

17. (a) $\begin{bmatrix} 0 & 1 & 0 \\ 1 & 0 & 0 \\ 0 & 0 & 1 \end{bmatrix}$ and $\begin{bmatrix} 1 & 0 & 0 \\ 0 & 1 & 0 \\ 0 & 0 & 1 \end{bmatrix}$.

(b) It consists of the six "permutation matrices," namely

$$\begin{bmatrix} 1 & 0 & 0 \\ 0 & 1 & 0 \\ 0 & 0 & 1 \end{bmatrix}, \begin{bmatrix} 1 & 0 & 0 \\ 0 & 0 & 1 \\ 0 & 1 & 0 \end{bmatrix}, \begin{bmatrix} 0 & 1 & 0 \\ 1 & 0 & 0 \\ 0 & 0 & 1 \end{bmatrix}, \begin{bmatrix} 0 & 1 & 0 \\ 0 & 0 & 1 \\ 1 & 0 & 0 \end{bmatrix}, \begin{bmatrix} 0 & 0 & 1 \\ 1 & 0 & 0 \\ 0 & 1 & 0 \end{bmatrix}, \begin{bmatrix} 0 & 0 & 1 \\ 0 & 1 & 0 \\ 1 & 0 & 0 \end{bmatrix}.$$

This semigroup is isomorphic to the group S_3 of all permutations of a 3-element set.

(c) It has three matrices; give them.

19. (a) $6\mathbb{P} = \{6k : k \in \mathbb{P}\}$.

(c) No. For example, 6 and 12 are not both powers of the same member of $6\mathbb{P}$, so cannot both lie in the same cyclic subgroup.

21. (a) $60\mathbb{P}$.

(c) 60 generates the *additive* semigroup $60\mathbb{P}$.

23. Only the function φ in (a) is a homomorphism. Although it is one-to-one, it does not map $\mathbb{P}$ *onto* $\mathbb{P}$, so it is not an isomorphism.

25. More generally, if $A \subseteq S$, then

$$\varphi(A^+) = \varphi(\{s : s \text{ is a product } a_1 \cdots a_n \text{ of members of } A\})$$

$$= \{\varphi(s) : s = a_1 \cdots a_n \text{ and } a_1, \ldots, a_n \in A\}$$

$$= \{t : t = \varphi(a_1) \cdots \varphi(a_n) \text{ and } a_1, \ldots, a_n \in A\} = \varphi(A)^+.$$

27. (a) If z' is also a zero, then $z' = z \bullet z' = z$.

(c) $(\{0, 1\}, \cdot)$. Can you find another example?

29. (a) $\varphi(S)$ is closed under products because

$$\varphi(s) \mathbin{\square} \varphi(s') = \varphi(s \bullet s') \in \varphi(S).$$

And $\varphi(e)$ is an identity for $\varphi(S)$ because

$$\varphi(s) \mathbin{\square} \varphi(e) = \varphi(s \bullet e) = \varphi(s) = \varphi(e \bullet s) = \varphi(e) \mathbin{\square} \varphi(s)$$

for all $\varphi(s) \in \varphi(S)$.

(b) No. Consider, for example, any monoid $(S, \bullet)$ and define $\varphi : (S, \bullet) \to (\mathbb{Z}, \cdot)$ by $\varphi(s) = 0$ for all $s \in S$.

Section 12.8

1. (a), (b), (d), (f).

3. All but (b) and (e). (b) is not an additive homomorphism and (e) is not multiplicative.

5. (a) Evaluate both sides of the equation at a.

(c) By (b),

$$p(x) = \sum_{k=0}^{n} c_k x^k = \sum_{k=0}^{n} c_k \{q_k(x) \cdot (x - a) + a^k\}$$

$$= \left(\sum_{k=0}^{n} c_k q_k(x) \right) \cdot (x - a) + \sum_{k=0}^{n} c_k a^k.$$

Let $q(x) = \sum_{k=0}^{n} c_k q_k(x)$.

(d) $p(x)$ is in the kernel if and only if $p(a) = 0$. Use part (c).

7. (a) $24\mathbb{Z}$.   (b) $2\mathbb{Z}$.

(c) $3\mathbb{Z} + 2\mathbb{Z} = 1\mathbb{Z} = \mathbb{Z}$, since $1 = 3 \cdot 1 + 2 \cdot (-1) \in 3\mathbb{Z} + 2\mathbb{Z}$.

(d) $\mathbb{Z}$.

9. (a) Verify well-definedness directly, or apply Theorem 1 to the homomorphism $m \to (m \, \text{MOD} \, 4, m \, \text{MOD} \, 6)$ from $\mathbb{Z}$ to $\mathbb{Z}(4) \times \mathbb{Z}(6)$, as in Example 8. As noted in Exercise 12, it suffices to check that $m \to m \, \text{MOD} \, 4$ and $m \to m \, \text{MOD} \, 6$ are homomorphisms on $\mathbb{Z}(12)$.

(c) $(1, 4)$ is one of the twelve; find another one.

(d) Just 9.

11. (a) Since $2 *_6 3 = 0$, 2 has no inverse.

(b) Exhibit an inverse for each non-0 element. The inverses for non-0 elements in $\mathbb{Z}(5)$ can be read off of Figure 3 in § 3.6. [See Exercise 16 for the general argument.]

(c) $F \times K$ isn't even an integral domain since $(1, 0) \cdot (0, 1) = (0, 0)$.

13. (a) The kernel of φ is either F or $\{0\}$ by Example 6(b).

(c) Use part (b) and Example 6(b).

15. (a) $I = 15\mathbb{Z}$.

(b) See Example 8. Let $\varphi(m) = (m \, \text{MOD} \, 3, m \, \text{MOD} \, 4)$ for $m \in \mathbb{Z}(12)$.

17. (a) $R \cdot 2 = \{a_0 + a_1 x + \cdots + a_n x^n \in R : \text{every } a_i \text{ is even}\}$,
$R \cdot x = \{a_0 + a_1 x + \cdots + a_n x^n \in R : a_0 = 0\}$,
$R \cdot 2 + R \cdot x = \{a_0 + a_1 x + \cdots + a_n x^n \in R : a_0 \text{ is even}\}$.

(b) Suppose that $R \cdot p = R \cdot 2 + R \cdot x$ for some $p \in R$. Since $2 \in R \cdot p$, p must be constant, and since $x \in R \cdot p$, p is 1 or -1. But then $R \cdot p = R$, a contradiction.

Section 13.1

1. (a) 0.   (b) 0.   (c) 1.   (d) 0.   (e) 0.

3. (a) $\forall x \, \forall y \, \forall z [((x < y) \wedge (y < z)) \to (x < z)]$; universes $\mathbb{R}$.

(c) $\forall m \, \forall n \, \exists p [(m < p) \wedge (p < n)]$; universes $\mathbb{N}$.

(e) $\forall n \, \exists m [m < n]$; universes $\mathbb{N}$.

5. (a) $\forall w_1 \, \forall w_2 \, \forall w_3 [(w_1 w_2 = w_1 w_3) \to (w_2 = w_3)]$.

(c) $\forall w_1 \, \forall w_2 [w_1 w_2 = w_2 w_1]$.

7. (a) x, z are bound; y is free.

(c) Same answers as part (a).

9. (a) x, y are free; there are no bound variables.

(b) $\forall x \, \forall y [(x - y)^2 = x^2 - y^2]$ is false.

(c) $\exists x \, \exists y [(x - y)^2 = x^2 - y^2]$ is true.

11. (a) No. $\exists m [m + 1 = n]$ is false for $n = 0$.

(b) Yes.

13. (a) $\exists ! x \, \forall y [x + y = y]$.

(c) $\exists ! A \, \forall B [A \subseteq B]$. Here A, B vary over the universe of discourse $\mathscr{P}(\mathbb{N})$. Note that $\forall B [A \subseteq B]$ is true if and only if $A = \varnothing$.

(e) "$f : A \to B$ is a one-to-one function" $\to \forall b \, \exists ! a [f(a) = b]$. Here a ranges over A and b ranges over B. One way to make this clear is to write $\forall b \in B \, \exists ! a \in A [f(a) = b]$.

15. (a) True.

(c) False; e.g., 3 is in the right-hand set.

(e) False; the right-hand set is empty.

(g) True.   (i) True.   (k) True.   (m) True.

17. (a) 0.   (c) 1.   (e) 0.

Section 13.2

1. (a) Every club member has been a passenger on every airline if and only if every airline has had every club member as a passenger.

(c) If there is a club member who has been a passenger on every airline, then every airline has had a club member as a passenger.

3. Rule 37b says that "There does not exist a yellow car" is logically equivalent to "Every car is not yellow." In fact, both are false. Rule 37d says that "There exists a yellow car" is logically eqivalent to "Not every car is nonyellow." Both are true.

5. (a) See rule 8c.

7. $\exists n[\neg\{p(n) \to p(n+1)\}]$ or $\exists n[p(n) \wedge \neg p(n+1)]$.

9. (a) $\exists x\, \exists y[(x < y) \wedge \forall z\{(z \le x) \vee (y \le z)\}]$.

(c) 0; for example, $[x < y \; \to \; \exists z\{x < z < y\}]$ is false for $x = 3$ and $y = 4$.

11. One can let $q(x, y)$ be the predicate "$x = y$." Another way to handle $\exists x\, p(x, x)$ is to let $r(x)$ be the 1-place predicate $p(x, x)$. Then $\exists x\, r(x)$ is a compound predicate.

13. $\exists N\, \forall n[p(n) \to (n < N)]$.

Section 13.3

1. (a) True.   (c) False.

(e) True. Compare Exercise 3 of § 1.3.

3. (a) A function of the form $f(x) = ax + b$ will work if you choose a and b so that $f(0) = -1$ and $f(1) = 1$. Sketch your answer to see that it works.

(b) Use g where $g(x) = 1 - x$.

(c) Modify suggestion for part (a).

(d) Use $x \to 1/x$.

(e) Map $(1, \infty)$ onto $(0, \infty)$ using $h(x) = x - 1$ and compose with your answer in part (d).

(f) $f(x) = 2^x$, say. Sketch f to see that it works.

5. (a) Use the data:

x	.1	.2	.3	.4	.5	.6	.7	.8	.9
$f(x)$	-8.89	-3.75	-1.90	$-.83$	0	.83	1.90	3.75	8.89

7. Only the sets in (b) and (c) are countably infinite.

9. (a) We may assume that S is infinite. Let $f\colon S \to T$ be a one-to-one correspondence where T is a countable set. There is a one-to-one correspondence $g\colon T \to \mathbb{P}$ [why?]. Then $g \circ f$ is a one-to-one correspondence of S onto $\mathbb{P}$.

11. (a) Apply part (b) of the theorem to $S \times T = \bigcup_{t \in T} (S \times \{t\})$. Why is each $S \times \{t\}$ countable?

(b) For each $t \in T$, let $g(t)$ be an element in S such that $f(g(t)) = t$. Show that g is one-to-one and apply part (a) of the theorem.

13. (a) For each f in FUN($\mathbb{P}$, $\{0, 1\}$), define $\varphi(f)$ to be the set $\{n \in \mathbb{P} : f(n) = 1\}$. Show φ is a one-to-one function from FUN($\mathbb{P}$, $\{0, 1\}$) onto $\mathscr{P}(\mathbb{P})$.

(b) Use Example 2(a) and Exercise 9.

15. For the inductive step, use the identity $S^n = S^{n-1} \times S$.

INDEX

GREEK ALPHABET

A	α	alpha	I	ι	iota	P	ρ	rho
B	β	beta	K	κ	kappa	Σ	σ	sigma
Γ	γ	gamma	Λ	λ	lambda	T	τ	tau
Δ	δ	delta	M	μ	mu	Υ	υ	upsilon
E	ϵ	epsilon	N	ν	nu	Φ	ϕ	phi
Z	ζ	zeta	Ξ	ξ	xi	X	χ	chi
H	η	eta	O	o	omicron	Ψ	ψ	psi
Θ	θ	theta	Π	π	pi	Ω	ω	omega